Chasles and the Projective Geometry

Paolo Bussotti

Chasles and the Projective Geometry

The Birth of a Global Foundational Programme for Mathematics, Mechanics and Philosophy

Birkhäuser

Paolo Bussotti
DIUM
University of Udine
Udine, Italy

ISBN 978-3-031-54265-7 ISBN 978-3-031-54266-4 (eBook)
https://doi.org/10.1007/978-3-031-54266-4

This book is published under the imprint Birkhäuser, www.birkhauser-science.com by the registered company Springer Nature Switzerland AG
The registered company address is: Gewerbestrasse 11, 6330 Cham, Switzerland

Paper in this product is recyclable.

Foreword

Paolo Bussotti's book should change everyone's appreciation of Michel Chasles, an important mathematician of the first half of the nineteenth century whose work has not had the reception among historians of mathematics that it deserves. His achievements often come a poor third to the innovative contributions of his predecessors Gaspard Monge and Jean-Victor Poncelet. His prize-winning book, the *Aperçu Historique sur l'origine et le Développement des Méthodes en Géométrie*, is raided for its historical insights, his later work on Euclid's porisms is disparaged for being unconvincing. His original contributions to mathematics have been largely neglected, but as Bussotti shows, they go beyond pure geometry and form a coherent, evolving programme to show how geometrical thinking is fundamental across a range of issues in mechanics and gravitational attraction.

Chasles rightly saw his work as improving the foundations of projective geometry. To that end, he used both synthetic and analytic methods, discussed the role of geometrical transformations, the invariance of cross-ratio (a concept he introduced) under homographies, and emphasised more and more the principle of duality. He also explained how metrical geometry fits into the projective framework and showed how to extend these ideas to give simple solutions to complicated problems in physics.

In particular, Bussotti's focus on the systematic nature of Chasles's work is, in my view, the most important single feature of this book. Chasles saw the principle of duality as the common root of projective geometry, the motion of a rigid body, and the study of systems of forces; as Bussotti shows, Chasles developed a whole philosophy of duality, and Bussotti emphasises the foundational character of this work, in which geometrical ideas were at the heart of mathematical concepts and physical phenomena. In this connection, Chasles moved away from the study of figures to the study of transformations, and this is perhaps the most influential part of his contribution. Geometers of subsequent generations followed him in this regard, whereas Chasles' approaches to physical problems did little to deflect the force of numerous analytical studies.

Bussotti also offers us a number of detailed comparisons between Chasles' work and those of other mathematicians, for example Steiner. Steiner was drawn independently to the synthetic position in geometry, which he arguably developed, as Bussotti shows, in a somewhat purer way. Both men saw projective geometry as lying at the basis of all geometry, although the two differed over the relationship of the new geometry to metrical (Euclidean) geometry. In fact, the extended comparisons Bussotti makes between Chasles and Steiner bring out advantageously the subtle differences between the two and the specific qualities of each one. There are also instructive comparisons between Chasles' work and Poinsot's on systems of forces, and to that of Rodrigues and de Jonquières on the displacement of rigid bodies.

In short, Bussotti restores to us a fuller picture of Chasles both as an ambitious mathematician with a vision of geometry that, while full of valuable discoveries, was also a programme or a way of thinking about much of the mathematics of his time, and as a mathematician in a historical context.

Warwick, UK Jeremy Gray

Preface

This book is a contribution to the internalist history of mathematics. Contextual indications will be given, although these are conceived as a frame within which the main scene takes place, where "the main scene" is intended as the development of Chasles' foundational programme for mathematics, mechanics and philosophy. A complete reading of the book requires a general knowledge of the mathematical techniques which were used in geometry and in mechanics in the nineteenth century, while no knowledge of Chasles' work is necessary since all his discoveries will be explained in detail. To favour the reader's understanding of Chasles' mathematics, I have enriched the book with figures, commentaries and explanations. It follows that the scholars who deal with the technical history of mathematics are the ideal readers of this work. On the other hand, a partial reading of the book is also possible if the aim of the reader is to achieve a general idea of the thesis I support and of the argumentative technique I use, but she/he will possibly not be interested in all the mathematical details. In this perspective, I suggest reading the following sections: 1) the Introduction; 2) the initial part of every section and subsection where I present the general picture of a topic and my thesis before dealing with the mathematical details; 3) the conclusive considerations of every chapter; 4) the entire sixth chapter which concerns the philosophy of duality and presents few mathematical details; 5) the Conclusions. With this selection, the book can be of interest not only to the historians of mathematics, but also to the historians of science interested in understanding the way in which the concept of foundation was conceived by a great mathematician who developed his thought on the subject mainly in the first half of the nineteenth century. The philosophers of mathematics who study the foundations of this discipline between the end of the nineteenth century and the beginning of the twentieth can also be interested in this book in order to compare the meaning of the locution "foundation of mathematics" in Chasles and in the period they study. Finally, the philosophers and the historians of philosophy can draw from this book some ideas regarding the philosophy of duality.

Udine, Italy Paolo Bussotti

Acknowledgement

I wish to express my gratitude to Jeremy Gray for his Foreword and for the numerous and precious suggestions he gave me while writing this book. I am grateful to Philippe Nabonnand for several tips he gave me. I am indebted to Brunello Lotti for his suggestions concerning the Introduction and to Danilo Capecchi for his advice on the section dedicated to the principle of virtual velocities. I express my gratitude to Martina Zamparo for her precious corrections of the English form of my text and to Alessandro Chimenti for his valuable contribution in drawing numerous figures that appear in the book. I am grateful to Frida Trotter for her continuous editorial support during the composition of this text.

Contents

Chapter 1
Introduction

Abstract In this introduction I present the contextual elements that may be useful to the reader in framing the work of Chasles that is expounded in the following chapters. Therefore, after an explanation of the purposes of this work, I will offer the state of the art with respect to the literature on Chasles. Since he was profoundly influenced by the milieu of the École Polytechnique, where he was student, in the second section the main aspects concerning the history of this institution until 1816 will be dealt with. Chasles contributed to the development of geometry, and in his years as a student and as a young researcher geometry was developing quickly thanks to the introduction of entire new branches and to deep changes in those existing. Therefore, in the fourth section a summary of the most important investigations in the period 1800–1825 is proposed. The Introduction is closed by a hint to Chasles' life.

1.1 Aim and Structure of This Book

From an initial phase of his scientific career, Michel Chasles (1793–1880) envisioned and developed a foundational programme for geometry. In this programme, projective geometry is considered the most general and fundamental discipline, to which metric geometry should be reduced. The problem for this reduction to be realized is that the concepts of measure of a segment and of an angle are not graphical concepts. Chasles tried to find the metric concepts and transformations on a projective basis and his work can be interpreted as the initial phase of a train of thought which reached its conclusions with the results obtained by Arthur Cayley and Felix Klein on projective metrics. In connection with such foundational research—though not completely identifiable with it—Chasles also addressed the issue concerning the relations between synthetic and analytical methods in geometry.[1] He was one of the most important members of the great

[1] I refer here to synthetic geometry as that branch of geometry which starts from a set of given axioms, does not resort to system of coordinates and studies the geometrical properties through the methods of figures' intersection, of transformations and of constructions. Analytic geometry is,

P. Bussotti, *Chasles and the Projective Geometry*,
https://doi.org/10.1007/978-3-031-54266-4_1

French school of projective geometry—whose founders were Carnot and Poncelet. Their aim was to devise the pure or synthetic methods general enough to compete with the analytical ones. Nonetheless, Chasles believed that the foundational concepts were even more important than the methods through which they are obtained. Hence, in his studies on the relations between metric and graphical properties, he also resorted to analytical reasonings. Thus, Chasles highlighted the foundational role of projective geometry within geometry, but his studies were not restricted to this issue: he believed that geometry also had to be the basis for entire sections of physics. This is the case for statics, kinematics and some aspects of the theory of gravitation. He reduced many of the basic concepts of these three branches of physics to projective notions and constructions. As to other concepts—namely, those in which metric properties are involved—he introduced some transformations and mathematical objects which, though not directly graphical, can be also referred to projective geometry, as I will clarify. Since Chasles considered projective geometry as the basis of the entire geometry and geometry itself as the basis of such important branches of physics, the coherent conclusion is that he believed projective geometry to be the basis of scientific knowledge as a whole.

In this context, the question of method is a crucial one since Chasles developed most of his arguments using the synthetic method. While dealing with the foundation of the basic projective concepts, he also sometimes resorted to analytical methods because these allowed him to simplify his treatment, though the same results can be obtained by means of pure geometry, as Chasles himself pointed out in several circumstances. In contrast, while dealing with the foundation of physics, he systematically tried to reduce its notions to geometrical concepts and constructions. One can even go as far as to say that, in this respect, he was indeed a disciple of Newton!

Chasles' ideas reached a higher plane when he claimed that one of the leading principles of projective geometry, the principle of duality, was not only a mathematical and gnoseological law: it was also an ontological law concerning the universe and its structure. In regard to its fundamental laws, the universe is neither a holistic entity nor a pluralistic one; it is dualistic. Every law of the universe has to have a dual counterpart.

Therefore, the scope of this work is to prove the following thesis:

> Chasles' foundational programme concerned the reduction of the basic metric concepts within a projective context. Furthermore, it also regarded important branches of physics and was connected to an ontological view. *Projective geometry is at the core of such a general research project.*

As to physics, basically the subjects dealt with by Chasles were three:

1) With regard to kinematics, he addressed the problem of the infinitesimal movement and displacement of a rigid body and that of the finite displacements.

---

instead, based on the use of coordinate systems and on the transcription of geometrical properties in equations or systems of equations. In the course of the book, we will see that, in the period I analyse, this distinction was not completely plane, although I will try to prove that it made perfectly sense.

2) As to statics, he studied the translational and rotational equilibrium of different equivalent systems of forces and the principle of virtual velocities.
3) As to gravity, he gave important contributions to the problems of the attraction exerted by an ellipsoid on any point of space.

The problem related to caustics should also be added; however, the three issues mentioned above are by far the most important ones in order to grasp Chasles' conception. Some of the results obtained by Chasles are new; but in this case, the methods are possibly even more important than the results. For as we will see, several scholars dealt with items 1)–3) giving decisive contributions, but Chasles' methodology was absolutely original. A restricted number of mathematicians—essentially in France and Italy—worked relying upon Chasles' ideas, developing his synthetic methods and trying to create a School of "synthetic mechanics". However, since these methods are difficult, complicated, and, in fact, less general and malleable than the analytic ones, they were used only by a minority of physicists and, starting from around the 60s–70s of the nineteenth century, they were abandoned in favour of a completely analytical approach.

The period in which Chasles wrote his contributions concerning the three issues mentioned above is rather long: it begins with 1829 and ends with 1871. The number of papers specifically dedicated to these subjects is about 20. Important considerations are also present in the *Aperçu historique* (Chasles, 1837a) and in the *Rapport* (Chasles, 1870).

Given the overall picture, this book will present six chapters: Chap. 2 concerns Chasles' foundational programme for geometry; Chap. 3 regards his contributions to the analysis of the rigid body's infinitesimal and finite displacements; Chap. 4 is dedicated to the equilibrium of equivalent systems of forces; Chap. 5 focuses on the principle of virtual velocities; Chap. 6 discusses Chasles' philosophy of duality; Chap. 7 concerns the attraction of an ellipsoid. Finally, the Conclusions will follow, where an epistemological and historical evaluation of Chasles' foundational work will be provided.

I have decided to adopt a comparative method because I am convinced that the work of an author can be understood and appreciated in all its nuances only if it is compared to those of the authors who developed similar ideas. Therefore, in each chapter the reader will find not only Chasles' conceptions, but also the ones of those mathematicians and scientists whose train of thought was closer to Chasles'. Only in this way will it be possible to appreciate the personal contributions given by each scholar to a certain problem or order of ideas.

In the rest of this Introduction, I will offer a concise view concerning those works on Chasles that address foundational issues in order to clarify the place of this book within the literature. Furthermore, I will also briefly present the socio-cultural and mathematical context in which Chasles was educated as well as some aspects of his scientific biography that will be useful for understanding his conception of mathematics.

1.2 Chasles and the Foundations of Mathematics in the Secondary Literature

The literature on Chasles is relatively abundant, as the discussions of various aspects of his work developed in this book will make clear. Here, I would like to briefly focus on the contributions which ascribe a foundational value to Chasles' work. They highlight that his ideas are not only original from a mathematical point of view, but that many of them also have the aim to enucleate a section of mathematics whose concepts, methods and principles are at the basis of other sections of this discipline. In this perspective, the relevant literature is scarce.

All scholars agree that Chasles, almost contemporarily with Steiner and Möbius, introduced the anharmonic ratio as the fundamental invariant element of projective transformations. However, some authors go further and highlight an approach to projective geometry that is typical of Chasles. Karine Chemla writes:

> [...] in his *Traité des propriétés projectives des figures*, Poncelet had, for instance, introduced the "principle of continuity". Chasles, however, does more than just introduce principles. He provides personal reflections on their nature, their properties and their justification. These principles are intimately related to his approach to the issue of generality. [...] As we saw above, when discussing various methods of transformation of some figures into others, Chasles does not consider these methods independently from each other, but seeks to understand their relationships and to organize them as a set. His analysis of methods of transformation between figures of the "same genre" leads him to state that they "all derive from a single *fundamental principle*, with respect to which they are only particular applications (Chasles, 1837, p. 219 and pp. 223–4. Chemla's emphasis). Chasles refers here precisely to the "principle of *homographic deformation*, or simply *principle of homography*" (Chasles, 1837, p. 262. Chasles' emphasis), which is one of the two main topics of his *Mémoire*" (Chemla, 2016, p. 70).

Therefore, Chemla identifies a foundational aspect of Chasles' thought—even if she does not use the term "foundational"—that is the principle of homographic transformation, which, jointly with that of duality, is, in Chasles' view, at the basis of projective geometry. Thus, homography, duality and, more in general, the idea of transformation represent, Chemla argues, the fundamental concepts through which Chasles was going to found pure geometry. Ergo, although ascribing to Chemla the idea that Chasles was developing a complete foundational programme would mean to overread her contribution, she nonetheless identifies some significant foundational aspects in his mathematics.

Chemla's picture of Chasles' thought captures the situation when the *Aperçu* (Chasles, 1837a) was published. Philippe Nabonnand focuses, instead, on a later work by Chasles, the *Traité de géométrie supérieure* (first edition 1852). The author identifies two principles which Chasles posed at the base of his geometry: the principle of signs jointly with the property of the anharmonic ratio (Nabonnand, 2011a, p. 20). These two principles, Nabonnand writes, "allow Chasles to claim for his geometry the same level of generality as that connoting analytic geometry" (Ibid., p. 23). Even more: they make projective geometry more general than algebra because the theorems obtained through the two principles can be generalized by duality (ibid., p. 23). Thus, they offer a precise and general foundation to projective geometry without falling into the logical problems connected to Carnot's principle of correlation and to Poncelet's principle of continuity. This means that, after having ascribed a sign to the anharmonic ratio, Chasles considered this relation and the signs principle as foundational with respect to the theory of transformations itself (ibid., pp. 20–21).

In Bussotti (2019), I proposed that Chasles developed a foundational programme for the entire field of geometry in the period 1828–1837. The programme had five steps: 1) the foundation of the graphical properties connoting plane projective geometry (period 1828–1829); 2) the foundation of spatial projective geometry with regard to graphical properties; 3) the dependence of plane projective geometry on spatial projective geometry; 4) the reduction of metric properties to graphical constructions, through the theory of parabolic transformations; 5) the derivation of the projective properties from the anharmonic ratio (1837). Thus I argued that Chasles was trying to reduce the metric properties within a graphical context. In this respect his work was the first step of a history which, passing through the fundamental results by Von Staudt, reached Cayley and Klein with their projective metrics.

In his PhD dissertation Nicolas Michel offers many insights with regard to the possible existence of a foundational programme in Chasles, although the author focuses on enumerative geometry, hence, on a late phase of his scientific career, but many interesting considerations also regard the previous results obtained by Chasles. After having explained the various meaning of the notions of simplicity and generality in Chasles' *Aperçu*, Michel writes:

> In other words, mathematical life for Chasles is not aimed toward the production of truths, but rather toward the refining of acquired truths, by means of the search of the most natural theoretical setting within which it can be inserted. (Michel, 2020a, p. 43).

This is a clear assertion of the foundational character of a significant part of Chasles' production. After an analysis of the *Aperçu*, the author focuses on the novelties introduced in the *Traité de géométrie supérieure* and concludes his treatment with the specification of a foundational idea already expressed by Nabonnand:

> Chasles' *Higher Geometry* can be viewed first and foremost as the construction of a new language for geometrical discourse, which captures the level of abstraction and generality of algebra whilst retaining the advantage of intrinsicality. (Ibid., p. 81).

In my perspective, this assertion by Michel means that Chasles' foundational programme was so wide and profound as to change the usual language of geometry, creating a new language whose fundamental expression is the signed cross ratio. Michel speaks explicitly of "Chasles' programme for the development of pure geometry" (ibid., p. 111). He refers basically to enumerative geometry, but this assertion holds for the whole of Chasles' production. After having analysed his theory of characteristics, Michel confirms that "Chasles has systematized the very language of geometry, both for its propositions and its proofs" (ibid., p. 184). Thus, the foundation of a new language for geometry was, according to Michel, one Chasles' main tenets.

My approach is less concentrated on epistemological aspects than Chemla's and Michel's and is more concerned with the fine-tuning of the development of Chasles' mathematics. My thesis is that Chasles conceived a foundational programme which, starting from the analysis of the basis of projective geometry, ends up envisioning the metric properties as expressible in graphical terms when the nature of some projective objects and transformations is specified. Even more: Chasles' programme is extended to mechanics and to philosophy. Therefore, though this book can be regarded as a part of what can be called "the foundational literature on Chasles", it presents new elements which are lacking in this literature. The reader can have a complete picture of the foundational part of Chasles' production by reading the works I have mentioned and this book, because they focus on different aspects of this topic.

1.3 The École Polytechnique from Its Foundation to 1816

As Jeremy Gray claims:

> Projective geometry, by any standard the most remarkable success story of nineteenth-century geometry, in both its synthetic and analytic modes deployed points, lines, and ultimately hyperplanes at infinity in a manner unintelligible to classical geometry (Gray, 2008, p. 19).

Projective geometry was the discipline on which Chasles thought to found the whole of geometry and a large part of physics. Many of the mathematicians who, between the 1810s and the 1830s, gave crucial contributions to this discipline were *élèves de l'École Polytechnique*. This was not a coincidence. Gaspard Monge (1746–1818), with his original conceptions and his work, most contributed to the birth of this school.[2] His ideas dominated the programmes of the École in the revolutionary and Napoleonic eras. He was the inventor of descriptive geometry.[3] The teaching of this discipline was an element strongly connoting the École Polytechnique. In several regards, descriptive geometry can be considered as the "mother" of projective geometry. Thus, it is understandable that the mathematicians who studied at the École were among the inventors of modern projective geometry. Chasles was an *élèves de l'École* from 1812 to 1815. This means that he was profoundly influenced by the cultural ambitions of this institution, by the different subjects that were taught (among which descriptive geometry had a prominent role),

[2]For this summary concerning the École Polytechnique, the other engineering schools and the relation between the development of mathematics and French revolution, I used: Belhoste (1989, 2003), Bottazzini (1990, pp. 61–81), Charbonneau (1993), Dhombres (2018), Fourcy (1828), Glas (1986), Grattan-Guinness (1990), Langins (1990), Smith (1935).

[3]The locution "géométrie descriptive" was coined by Monge himself in September 1793 (Barbin, 2019, p. 4).

and by the way in which the alumni recognized themselves as belonging to a cultural-scientific elite.

Therefore, to understand Chasles' mathematics, it is helpful to retrace the steps that led to the foundation of the École as well as its organization until 1816. Hence, in this section, I will deal with such topic. In the next section a synthetic overview of the development of geometry between the revolutionary period and the late 1820s will be given.

The École Polytechnique was founded to train French engineers through high-level science education based on mathematics. Before the French revolution, the training of the engineers took place in specialized schools. The most prestigious was the *École royale du génie de Mézières,* which was a military engineering school. It was founded in 1748. The duration of studies was 2 years. The candidates were subject to an oral examination by a member of the Royal Academy of Science. There was also an examination for the promotion from the first to the second year as well as a final examination to obtain the diploma. The candidates had to belong to the *second ordre* (noblesse). Therefore, there were both meritocratic criteria and criteria based on the candidates' birth. In the first year four subjects were taught: 1) course of mathematics, statics and hydraulic taught for many years by Charles Bossut (1730–1814); 2) drawing course concerning the pictorial technique of lavis used by Vauban to draw his system of fortifications; 3) drawing course concerning architecture according to Claude Perrault's ideas; 4) course of stereotomy and wood cutting. The second year was dedicated to more practical activities. Important French scientists were *élèves*; for example, Borda (1733–1799) was *élève* in the 2 years 1758–1759 and Coulomb (1736–1806) was *élève* in 1760–1761;

In our context the *École de Mézières* is significant because Gaspard Monge became professor (*répétituer*) of stereotomy. As Arago recalls:

> The branch of applied mathematics, today known as *descriptive geometry* dates back to this period in which Monge became *répétiteur* at the *École de Mézières.*[4]

Monge's influence in the 1770s became predominant: the Instruction-Regulation (Règlement-Instructions) approved in 1777 was largely inspired by Monge's ideas. Geometrical drawing—as a matter of fact, descriptive geometry—and mechanics became the most important subjects in the final examination. The course of Monge consisted in the techniques of descriptive geometry, of stones cutting and carpentry, in the teaching of perspective and shadows theory. Among the students who had Monge as teacher there was also Lazare Carnot (1753–1823). He was a mathematician who contributed to the birth of projective geometry and who, like Monge, held important institutional positions during the revolutionary and Napoleonic period. Though less known and important than Carnot, Meusnier de la Place (1754–1793) was *élève* at the *École* in 1774–1775 and had Monge as professor. He studied aerostatic, descriptive geometry and differential geometry. He fought in the

[4] Arago (1853) (read at the Academy of Science on 11 May 1846), p. IX: "C'est de l'époque où Monge entra en fonction comme répétiteur à l'école de Mézières, que date réellement la branche des mathématiques appliquées, connue aujourd'hui sous le nom de *Géométrie descriptive*".

revolutionary war and died in Mainz while fighting. Thus, Monge's personality and teaching were influential already before the French Revolution.

The civil engineers were, instead, trained at the *École des Ponts et Chaussées* (School of Bridges and Streets). The *Corp des Ponts et Chaussées* was officially founded in 1716, but became important at the beginning of the 1740s, when the Reign organized a general policy of the streets and the *Bureau des dessinateurs* was founded. From this period onwards, the work of Jean-Rodolphe Perronet (1708–1794) was crucial. As a consequence of his efforts, the *École des Ponts et Chaussées* was founded in 1747. Differently from the *École de Mézières*, noblesse was not necessary to enter this *École* and the admission was not based on an exam, but on recommendation. The scientific level of the *École des Ponts et Chaussées* was not comparable to that of the *École de Mézières* (Belhoste, 2003, p. 23).[5] All these schools had few students. For example, the *École de Mézières* admitted 20 students each year (Fourcy, 1828, p. 4). As I will clarify below, this represents a significant difference with respect to the *École Polytechnique*. Mathematics had a prominent role at the *École de Mézières*. However, the most recent mathematical discoveries and techniques were not taught there. Significantly, infinitesimal calculus was not taught. Mathematics was essentially regarded as an instrument for engineering.

The Revolution had a great impact on all areas of public life. Education was certainly no exception. To understand the reasons which led to the foundation of the *École Polytechnique* (1795), it is necessary to take into account the following elements:

1) *The role ascribed by many Illuminists to mathematics*: D'Alembert (1717–1783), e.g., thought that mathematics had a high and general value. It was not only a technical discipline, but a humanistic one as well; it represented the fundamental element of human progress:

> Key to d'Alembert's vision was a connection between human thought and mathematical reasoning tight enough to ground mathematics and give it meaning. In d'Alembert's view of mathematics, rigor and insight worked together as part of a larger quest for comprehensive understanding. When wholly understood, valid mathematical arguments were the highest achievements of the human mind and spirit, quintessential triumphs of l'*esprit humain* (Richards, 2006, p. 701. Italics in the text).

Not all the Illuminists shared d'Alembert's enthusiasm for mathematics: Diderot and Buffon are paradigmatic examples; but, certainly, mathematicians such as Monge, Carnot, Lacroix and Hachette, who had important roles during the Revolution and Napoleonic period, did acknowledge a crucial place to mathematics. Lagrange himself, who was less involved than Monge and Carnot in political tasks during that period, shared the idea of a humanistic, historical and

[5]There were other schools which trained engineers as the *École des Mines* founded in 1783 (mining engineers) and the *École de la Marine* founded in 1741, which took the name of *École des Ingénieurs-Constructeurs de Vaisseaux Royaux* since 1765 (naval engineers). However, for my aim the brief picture here outlined is sufficient.

educative value of mathematics. His engagement as a teacher at the *École Normal* and at the *École Polytechnique* is a clear evidence of his ideas. The historian of mathematics Montucla (1725–1799) also shared this view.

2) *The institutional centralization and rationalization*: these were two important aspects of revolutionary ideals. In pre-revolutionary France power was fragmented. Many local institutions existed which were dominated by the First and Second Order. Thus, the idea of centralizing and rationalizing power was consistent with a democratization of political and social life: since the central institutions were elected democratically, they were not subjected to the privileges of the clergy and nobility. It is not a coincidence that on 25 July 1793 the *Bureau de Longitude* was founded. Its aims were to solve the astronomical problems connected to the longitude's determination on board of travelling ships, to supply ephemerides of celestial bodies, to organize scientific expeditions useful for geodesy and astronomy, to offer scientific opinions on various problems. At the beginning, ten scientists composed the Bureau: Joseph-Louis Lagrange, Pierre Simon Laplace, Pierre Méchain, Joseph-Jérôme Lefrançais de Lalande, Giovanni Cassini, Jean-Baptiste Delambre, Jean-Charles de Borda, Louis Antoine de Bougainville, Jean-Nicolas Buache and Noël Simon Caroché. On 8 August 1793, the Academy of Science was suppressed and on 25 October 1793 the *Institut de France* was founded. Its scientific level had to be as high as that of the suppressed Academy, but no privilege connected to the nobility of birth was accepted. Only meritocratic criteria had to be adopted. On 9 September 1793 the military schools were closed.

3) *State organization and economy.* There were both ideal and practical reasons which induced the revolutionary government to found a national engineering school as the *École Polytechnique*. From an ideal standpoint, centralization, rationalization, democratization and meritocracy were important motivations. However, there were also urgent reasons: running the system of communication routes of a modern country such as France was a difficult task and, hence, well prepared engineers were needed to organize it. The preparation offered by the *École des Ponts et Chaussées* was adequate from a practical standpoint, but a modern engineer also needed a theoretical background which was not offered by the Schools of *Ancient Régime*. The revolutionary governments understood that a new global education was necessary for an engineer. This is especially true if we consider economy besides state organization. The French Revolution did not touch the private properties of the industries and industry in France was becoming important, at least in some regions and at least in some areas of productions. Thus, economy, which remained essentially in the hands of the private enterprise, needed well prepared engineers too.[6]

[6]I have remarked the differences between the ideals and the organization of the *Ancient Régime* and the Revolution because these are important aspects for my work. As a matter of fact, the situation is much more nuanced since strong elements of continuity also existed. Specifically with regard to the

In this context, Monge developed the idea to create a great School of engineering which could provide students with a profound mathematical and scientifical preparation.[7] Such School was not intended to replace the traditional *Écoles d'Applications*, such as the *Écoles du Génie* or the *École des Ponts et Chaussées*. Instead, it had to be propaedeutic for all the schools of applications. Carnot and Prieur de la Côte d'Or, members of the Committee of Public Safety and former students of Monge at Mézières, approved Monge's project. On 11 March 1794 the *Commission des Travaux publics* was founded. Among its tasks there was also the creation of the *École centrale des Travaux publics*. The idea of providing France with a centralized school of engineering useful for the entire nation was welcomed by the Committee of Public Safety. The logistic work was huge: a cabinet of physique was created at the Hôtel d'Aiguillon in Paris as well as a laboratory of chemistry and a library; a rich collection of imitation designs was produced. Since descriptive geometry needed a large amount of tables, the *Commission des Travaux publics* was allowed to employ 25 designers. This number could be increased, if necessary.

Antoine François Fourcroy (1755–1809) was chosen by the Committee of Public Safety to present the project of the School to the Convention. He was clear: the new School's purpose was not only to prepare engineers, but also to offer an advanced mathematical and scientific education to the most talented young people in France.

The *École centrale des Travaux publics* was opened on 25 September 1794 (7 vendémiaire of the third year). The following year the name was changed to *Ecole Polytechnique*.

The admission to the School was based on an examination that should testify the intellectual skills of the candidates rather than their preparation. Nonetheless, knowledge of elementary algebra and geometry was required. The candidates had to be between 16 and 20 years old (Fourcy, 1828, pp. 19–29). Initially the course of study lasted 3 years and was attended by 400 students per year, an enormous number.

With regard to the subjects taught, no discipline was neglected, but descriptive geometry, divided into stereotomy, architecture and fortification, was the leading one, because half the hours were devoted to this subject. In the following table I offer a synthetic view of the courses offered (Fourcy, 1828, pp. 51–52).

First year	Second year	Third year
Analysis' general principles and their application to solid geometry	Application of analysis to the mechanics of solids and fluids	Application of analysis to the mechanics of machines
Stereotomy	Architecture	Fortifications
General physics	General physics	General physics

(continued)

relation between the elements that the *École Polytechnique* inherited from the *Ancient Régime* Schools and the new ones, see, e.g., Belhoste (1989) and Langnis (1990).

[7]With regard to the discovery of the documents which testify the existence of a project developed by Monge to open a polytechnical school, see Belhoste (1994).

First year	Second year	Third year
Particular physics, namely chemistry. Saline substances.	Organic chemistry. Vegetables and animals.	Chemistry. Minerals.
Design (*étude du dessin*)	Design (*étude du dessin*)	Design (*étude du dessin*)

Many of the most important mathematicians and scientists were professors (*instituteurs*) at the *École*: Lagrange and Prony were two of the four professors of analysis; Monge, Lamblardie (who was also Director of the School) and Hachette were three among the eight professors of descriptive geometry; Neveu taught design (*dessin*); Hassenfratz and Barruel taught general physics; Fourcroy, Guiton and Berthollet were professors of chemistry (Belhoste, 2003, p. 79).

In 1795, the *Journal Polytechnique* was founded. The first article of the first issue is Monge's report of the course of stereotomy.

On the first September 1795, the Convention approved a law which established the new name of the School: *École Polytechnique*. It inherits most of the features connoting the *École centrale des Travaux publics*. The students were reduced to 360. On the 26th of May 1796 it was established that all the young people who aspired to enter the Public Service must have frequented the *École Polytechnique*.

From 1795 to 1800, the life of the *École* was not easy: the most illustrious French scientists were professors at the *École* (e.g. besides those I have already mentioned, Fourier became professor of analysis). The structure and the scientific level of the School were unequalled in Europe. Notwithstanding this, there were several criticisms: the *Écoles d'Applications* were not favourable to the fact that their students had to have attended the *École Polytechnique* and the teaching was considered too theoretical for future engineers. In 1798 the course of lectures was reduced to 2 years and the number of students to 260. However, this did not change the principal characteristics of the *École*: to use the appropriate expression of (Belhoste, 2003, Chap. 3), it was *une école de savants*. This is true under many points of view:

A) *The professors were illustrious scientists*. It was a novelty that advanced research and teaching were strictly tied. Between the end of the eighteenth century and the beginning of the nineteenth century, Gay Lussac, Ampère, Fourier, Poisson, Poinsot, Arago, J. Binet, P. Binet taught at the *École*. Some of them had been *élèves* of the *École* itself. Laplace strongly contributed to the organization of the *École*: in the late Napoleonic era and in the first years of the Restoration, his influence was fundamental. Progressively his influence surpassed that of Monge. A remarkable feature of the *École Polytechnique* was that the *savants* maintained the control of the School, at least in the period 1795–1816. This means that in the end, perhaps after long discussions and controversies, the management of the school reflected the will of the teaching staff.

B) *The School was not conceived only to train engineers, but also to create mathematicians and scientists*. This is confirmed by the fact that the particularly brilliant and motivated young people were allowed to frequent the *École* for 4 years if they wanted to improve their scientific knowledge. In 1804 Hachette founded the *Correspondance de l'École Polytechnique*, a journal in which the students of the School could publish their works. On this journal, between 1813 and 1816, Chasles published the initial results of his geometrical research.

C) *A cultural milieu around the École Polytechnique was created.* The *élèves* progressively began to recognize themselves as belonging to a cultural and scientific élite. This form of self-recognition was important in generating that *esprit de corps* of which Belhoste (2003) speaks: it was based on the recognition of mathematics as the leading subject for education and as a fundamental support for the state organization and administration. With the *École Polytechnique* modern technocracy was born. It was characterized by common cultural bases grounded on mathematics and science and by the elitist self-recognition. This milieu was also connoted by other aspects: a series of high-level handbooks were published for the students of the *École*. Many of these handbooks were also texts of advanced research. Probably the two most paradigmatic examples are Lagrange's *Théorie des fonctions analytiques* and Cauchy's *Cours d'analyse*. Other texts, such as the numerous handbooks published by Lacroix, though not devoted to research, were crucial for the teaching of mathematics and for the clarification of several concepts. The market of the handbooks for the *École Polytechnique* became an important economical enterprise. Finally, the *École* was recognized as a leading scientific and educative institution by many foreign scientists: in 1801, after the treatise of Lunéville, Volta, Brugnatelli and Rumford visited and appreciated the *École*. Alexander von Humboldt participated actively in the life of the School.

To conclude this brief excursus on the *École Polytechnique*, it is appropriate to point out that, starting from 1798, the hours dedicated to mathematical analysis and to mechanics were incremented while those devoted to descriptive geometry were reduced. This process, favoured by Laplace's ideas, continued in the nineteenth century. However, until Monge was active, namely until 1815, descriptive geometry remained a fundamental subject. It is not a coincidence that Belhoste names "École de Monge" the period which goes from the foundation of the School to 1816 and, from 1816, he speaks of "École de Laplace" (Belhoste, 2003, pp. 200–221).

Thus, Chasles was educated in the last years of the *École de Monge*, where the two dominant subjects were descriptive geometry and analysis (also applied to mechanics), but the basic "imprinting" was still geometrical.

1.4 A Hint to the Geometrical Context of the Period 1800–1825

Chasles wrote his first great memoirs starting from the second half of the 1820s. They belong to the field of pure or synthetic geometry. Therefore, it is useful to offer a brief picture of the development of geometry between the beginning of the nineteenth century and half the 1820s, because Chasles established and developed new bases for research in that period. There is a great deal of specialized literature on this subject. Thus, the picture I will provide should only be interpreted as a summary of already known and well-studied topics. However, it can be useful for the reader to have a general outlook of the mathematical environment in which Chasles began his research.

The mathematician who inspired the setting of geometry in the first 20 years of the nineteenth century was Gaspard Monge. Lagrange argued that Monge had made himself immortal with his geometry (Darboux, 1905, p. 518). As Darboux claims:

> And indeed descriptive geometry has not only enabled us to coordinate and perfect the process of every art 'where excellence and success in work and product are conditioned by precision of form', but also it has proved to be the graphic representation of a general and purely rational geometry whose fertility has been demonstrated by numerous and important investigations (ibid., p. 518).

Monge's method of projection considers the orthogonal projections of a three-dimensional object onto two mutually orthogonal planes. The projective radiuses are, thence, two bundles of parallel straight lines. After that a plane is rebated until it coincides with the other. The first plane is horizontal, the second is vertical. The rebatement or abatement takes place along the intersection straight line of the two planes. Monge's method offers a more complete view of an object with respect to a single orthogonal projection and can be developed through easier constructions than a perspective from an ordinary point of space. Through Monge's procedure, it is possible to infer spatial properties of a figure from its plane projection. For an exact description of the plane figure obtained after the abatement allows to reconstruct the form and position of the projected body (Monge, 1798, in Lorenat, 2015a, p. 285). At the same time, it is possible to discover the properties of plane figures which are difficult to obtain by the techniques of the ordinary synthetic and analytic geometry.[8] Therefore, Monge correlated spatial and plane geometry in a mutual relationship of dependence which represented an important novelty to reach a more general picture of the whole geometrical space. Chasles stressed this aspect of Monge's geometry (see, e.g., Borgato, 2006, pp. 125–128; Chemla, 2016, p. 64; Belhoste, 1998, p. 4). The systematic use of the method of projections posed the bases not only for the study of the relationships between space and plane developed by projective geometry, but also for an important didactic movement which was spread in the second part of the nineteenth century: the fusionism. I agree with Borgato (2006), which traces back to Monge's thought the remote origins of fusionism.

Jointly with the method of projection, Gino Fano identified the principles of contingent relations as the other pillar of Monge's geometry. This method was fruitful in the hand of Monge and his disciples:

> It is based on the idea to consider as fortuitous ("contingent") the presence or absence of certain conditions. Consequently, a proposition which is proved when these conditions take

[8] The deduction of plane theorems from spatial theorems in Monge's School is stressed, e.g., by Arana and Mancosu (2012, p. 300).

> place (for example, supposing a quadric to be cut by a straight line) is regarded as proved in the general case, that is also in the case that the straight line does not cut the quadric. [9]

This principle was also used by Carnot and was generalized and clearly stated by Poncelet in the celebrated principle of continuity, which was one of the bases of Poncelet's geometry. As we will see, Chasles reflected profoundly on this principle and the denomination of *principe de relations contingent* is due to him. Poncelet himself ascribed the origin of the principle of continuity to Monge (Nabonnand, 2015, p. 6). Fano explicitly claims that Monge's research was at the basis of the renovation of synthetic geometry: Monge's *Leçons de géométrie descriptive* is seen by Fano as the first text of the new synthetic geometry (Fano, 1907–1910, p. 230).

The principle of contingent relations is connected to the particular way in which Monge considered geometrical figures, that is in a genetic and dynamical way. This issue is crucial because Monge's view was shared by all the mathematicians of his School and, specifically, by authors such as Poncelet and Chasles who were among the founders of projective geometry. The dynamic conception of geometry is not an absolute novelty; one might think, for instance, of the generation of conics proposed by Kepler in his *Paralipomena ad Vitellionem* or of Newton's organic generation of conics in the *Principia*. However, in Monge the dynamic view concerned all the geometrical figures, it was not restricted to some curves or surfaces, but was extended to the entire geometrical space. Richards clearly explains this, also referring to (Dupin, 1819):

> Monge took a fluid view of geometry in which the boundary that separated the subject from dynamic processes was essentially blurred; he presented figures as generated from one another by continuous processes. Lines formed families as they rotated around fixed points. Ellipses became circles as their foci moved together; ellipses became parabolas as one focus moved away to infinity. As Charles Dupin (1784–1873), one of Monge's students, put it: "[In the study of descriptive geometry] the mind [*esprit*] learns to see internally and with perfect clarity, the individual lines and surfaces, [and the] families of lines and surfaces; it acquires a sense of the character of these families and individuals, ... it compares them, combines them and predicts the results of their interactions and their more or less intimate contacts (Richards, 2003, p. 455).

The "moving geometrical spectacle", to use an expression coined by Monge himself (ibid., p. 455), connotes the nature of the relations among figures. Glas points out that Monge conceived the idea that every surface is generated by the movement of a line which, during its motion, can change its shape and the generation can be produced in more than one way (Glas, 1986, p. 256). The surfaces, hence, have to be classified according to their generation. The two aspects of Monge's descriptive geometry, that is the dynamic generation of surfaces and the method of projections, are strictly connected because "[...] the method of projection applied to

[9]Fano (1907–1910, p. 230): " [...] nach welcher man das Auftreten oder das Nichtauftreten gewisser Umstände als zufällig („contingent") betrachtet, und folglich einen bei ihrem Auftreten bewiesenen Satz (z. B. in der Voraussetzung, daß eine Fläche 2. Grades von einer gewissen geraden Linie getroffen wird) stillschweigend als allgemein bewiesen ansieht und ausspricht (d. h. auch für den Fall, daß obige Linie die Fläche 2. Grades nicht trifft)".

surfaces needs a specific notion of surface, which is that every curved surface has to be considered as generated by the movement of a curved line" (Barbin, 2019, p. 9). Through the method of projections and his genetic-dynamical ideas concerning the generation of surfaces Monge gave fundamental contributions to the theory of developable surfaces and, in this context, he introduced the fundamental notion of edge of regression.[10]

Roughly speaking, the fathers of projective geometry can be divided among those who pursued a synthetic approach, such as Carnot, Brianchon, Poncelet, Chasles and Steiner, and those who also resorted to analytical methods, such as Gergonne, Möbius and Plücker. As is well known, Monge used extensively both synthetical and analytical procedures. He was not certainly a purist. Monge appreciated the generality offered to geometry by mathematical analysis. It is not a coincidence that, as Carrus states in the French translation of Fano (1907–1910): "In its general features, *G. Monge*'s work has shown us the happy alliance of geometry and analysis".[11] Lorenat (2015a, pp. 12–13) points out that "looming by early nineteenth century geometries, Gaspard Monge was invoked as a teacher among both synthetic and analytic practitioners". Granted the presence and importance of both methods in Monge's work, it seems to me that Glas' opinion may be shared: for Monge analysis is not a self-contained language, but the script of the geometrical spectacle, which is mobile (Glas, 1986, p. 257). This means that Monge believed in the existence of this mobile spectacle independently of its transcription into analytical terms. Analysis can, thus, be a part of the syntaxis of the geometrical language, which is synthetic in both its essence and semantics.

The further aspect of Monge's conception which influenced profoundly the ideas of the fathers of projective geometry is generalization. This is strictly connected to his dynamical view: the core of Monge's geometry are the figures. To use an appropriate expression used by Lorenat (2015c, p. 156) to connote Poncelet's geometry, but which can be extended to Monge: "the figure was never lost from view". However, the visualization of Monge's standpoint is that of a movie rather than that of a single figure. The movie starts with a photogram (a figure) and proceeds according to well-established geometrical-dynamical criteria. This conception intrinsically implies generality because a set of single figures of classical geometry are transformed into a single entity (the movie). Where, in classical synthetic geometry, each figure was studied separately, general methods to study the whole movie were devised by Monge. However, these methods were, in general, analytical. One of the greatest efforts developed by Poncelet and Chasles (and also

[10] On the contribution given by Monge to the theory of developable surfaces, a clear and informative paper is Lawrence (2011). See also Grattan-Guinness (1990, p. 262). The most complete work on Monge is the classical Taton (1951a).

[11] I will indicate the French translation of Fano (1907–1910) as Fano (1915). It presents some modifications with respect to the original German text due to Carrus, who was the translator. The passage mentioned in the running text does not exist in the German edition: "Comme caractère de l'ensemble de l'oeuvre de *G. Monge* nous avouns relevé l'alliance heureuse de la géométrie et de l'analyse". (Fano 1915 p. 193. Italics in the text).

Steiner) was to find purely geometrical methods which reached the same generality as the analytical ones. Thus, also with regard to the idea of generality, the fathers of projective geometry are indebted to Monge.

Nagel thinks, instead, that Monge was the founder of a synthetic geometry which goes beyond the figures: Monge—and before him, Desargues—introduced in geometry operations that are not restricted to intuitable figures. This led Monge to conceive three ideas which, appropriately developed by Poncelet, were at the base of the entire pure geometry: 1) the point at infinity; 2) the principle of continuity (contingent relations); 3) the imaginary in geometry. Nagel writes:

> Like Desargues he [Monge] therefore regarded parallel lines as having a common point "at infinity". Furthermore, whether two curves intersect in "real" or only in "imaginary" points was also of no consequence, because the occurrence of "real" or "imaginary" intersections depends entirely on the accidental positions of the bodies to be projected with respect to one another and to the plane upon which they are projected. (Nagel, 1939, p. 150).

Nagel sees in Monge the first step of a process in which the duality law plays a fundamental role: the process that privileges the structures instead of the objects. This is the base of the hypothetico-deductive approach to geometry typical of the end of the nineteenth century (Pasch, Hilbert and the Italian School with mathematicians such as Peano and Pieri). In this respect, Monge was fundamental for the creation of the mathematical order of ideas that influenced the founders of projective geometry. Nagel's exegesis is modernizing, but offers an interesting and stimulating perspective from which the history of the nineteenth century pure geometry can be seen. In his interpretation Monge is regarded as a precursor of abstract geometry.

The other mathematician whose influence was fundamental in shaping the "polytechnical" milieu in which Chasles studied was Lazare Carnot. In our context, Carnot is important for his mathematical ideas and for his impact on the teaching at the *École Polytechnique*. From 1790 onwards he developed a real foundational programme based on the following ideas and aims:

1) The true ontology of mathematics is given by geometry. The algebraic symbols do not distinguish between symbols for real objects and for fictitious ones. However, through some transformative techniques, algebra eventually allows to eliminate the symbols for fictitious entities. According to Carnot, algebraic and analytical methods are not appropriate to get the essence of geometry. Only synthetical methods are suited for this aim. Synthesis is not only a method which is preferable to analysis, it is something more: it is the correct way of thinking, not only in mathematics, but in the whole domain of human thought. As Schubring points out, the revaluation of synthesis at the expense of analysis became a "programmatic creed" (Schubring, 2005, p. 324) for Carnot. Schubring (2005, pp. 309–365) offers a valuable picture of the development of Carnot's ideas. The author also considers the possible influence exerted by the philosophical milieu contemporary to Carnot. He focuses on Condillac, Maine de Biran and Destutt de Tracy and, more generally, on the current of *idéologues*. During Carnot's scientific career, his opinions on the analytical objects and methods changed. I have referred only to his final position because it is that which

influenced the polytechnical milieu. Schubring offers the diachronic development of Carnot's ideas on the relations between analytical and synthetical methods. He argues that Carnot's reflection was prompted by the development of his ideas on negative quantities and numbers (ibid., p. 321). Therefore, Carnot was fundamental for the reappraisal of synthetic methods in France.

2) The advantages of the analytical methods in comparison with the synthetical ones typical of ancient geometry and of Newton's infinitesimal geometry too are their uniformity and generality. Therefore, the attempt to offer new geometrical methods which are as general and uniform as those analytical was one of Carnot's major tenets.[12] A principle became crucial in Carnot's geometry: the *principle of correlative figures*. In his *Géométrie de position* (Carnot, 1803), a text which was preceded by *De la Corrélation des figures de géométrie* (Carnot, 1801), where many of his ideas had already been expounded, Carnot explained: be given a figure *A* whose properties we search. Consider another figure *B* (the correlative one) of which the properties are known. If it is possible to determine a general correlation among the diverse parts of the two figures, then the properties of *A* can be deduced from those of *B*. As Nabonnand points out:

> Thus, the entire knowledge of a primitive system is implicitly transmitted to the correlative system. All that remains is to implement the procedures which allow to make this implicit knowledge explicit. Carnot justifies the use of correlative figures as an application in geometry of a general method which consists in reducing the analysis of what is unknown to known situations.[13]

Carnot's problem was, then, how to understand to which types of relationships the correspondence principle applies. The latter is not generally applicable to metric relations, but it is applicable to those positional. As Nabonnand clarifies:

> The starting point of Carnot's argument is the observation that the positional relations, though typical of a figure, can be applied—by changing the signs—to other figures which are called correlative, and which are obtained by modifying the relative positions of the initial figure's elements. The idea is that by expressing the properties of the figures through formulae and by specifying how the position changes affect these formulae, the positional relations can have a general treatment.[14]

---

[12] Without any claim to be exhaustive I mention the authors who address the problem of generality in Carnot: Chemla (2016, p. 53), Coolidge (1940, pp. 91–94), Lorenat (2015a, p. 168), Michel (2020a, pp. 59–61), Nabonnand (2011a, 2011b), Schubring (2005, pp. 318–365).

[13] Nabonnand (2011b, p. 3): "Ainsi, toute la connaissance que l'on a du système primitif se transmet sous forme implicite aux systèmes corrélatifs. Il ne reste plus qu'à mettre en place les procédures qui permettent de rendre explicite cette connaissance implicite. Carnot justifie l'utilisation des figures corrélatives comme l'application en géométrie d'une méthode générale qui consiste à ramener l'analyse de l'inconnu à des situations connues".

[14] Nabonnand (2011a, p. 4): "Le point de départ de l'argumentation de Carnot est le constat que les rapports de position bien qu'attachés à une figure peuvent être, moyennant des changements de signes, appliqués à d'autres figures dites *corrélatives* et obtenues en changeant les positions relatives des éléments de la figure initiale. L'idée est qu'en exprimant les propriétés des figures sous forme de formules et en précisant comment les changements de positions affectent ces

A figure and its correlative are seen by Carnot as a system. He had the idea that a system is given by a continuous series of figures. Each of them can be regarded as the correlative of the others. The concept of system was possibly inspired by Carnot's studies on the machines and on the passages from one state to another of the same machine. From a geometrical standpoint, each state could be regarded as a figure correlative to all the figures representing the other states.[15] Thus, Carnot's principle is connected with two fundamental characteristics of projective geometry: 1) the principle of continuity. For the correlation of figures can be seen as a particular formulation of the principle of continuity which, as is well known, became the pillar of Poncelet's projective geometry; 2) The non-metric properties. Carnot guessed that his principle holds for the positional properties of the figures, not necessarily for those metrical. This is a fundamental distinction for the birth of projective geometry, although Carnot did not explicitly distinguish between metric and graphical properties, as Poncelet did. However, he applied his principle correctly, so that we can certainly argue that he had more than one intuition of the two kinds of properties, although perhaps he did not have a complete view. It is not a coincidence that Grattan-Guinness claims:

> Carnot was concerned with some non-metrical properties in geometry, especially those which were preserved while a geometric configuration was continually changed (Grattan-Guinness, 1990, p. 254). [...] Like Carnot, Poncelet held that non-metrical properties of a geometrical figure could be preserved as the figure was gradually-that is, continuously-altered. (Grattan-Guinness, 1990, p. 709).

Thus, besides some important theories, as that of the transversals, Carnot reflected on and proposed a view which was important for the birth of projective geometry. In his contribution on this topic for the *Enzyklopädie der mathematischen Wissenschaften*, Schönflies pointed out the relevance of Carnot's theory of transversals and of his principle of correspondent figures in a continuous geometrical movement for the development of projective geometry (Schönflies, 1907–1910, pp. 393–395). While all scholars agree in recognizing Monge's great merits, it is worth pointing out that not all of them appreciate Carnot's approach and ideas. Coolidge explicit writes:

> He [Carnot] was possessed of the idea of overcoming the apparent increase in generality given by the algebraic methods of Descartes. How, for example, shall the geometer handle the question of positive and negative quantities? The idea developed in Carnot seems to be this. We establish certain geometrical relations in the simplest case where all quantities involved are positive but no further restriction is imposed. It is assumed as axiomatic that these relations are unaltered when the figure is replaced by what is called a 'correlative', which is essentially a figure obtained from the first by continuous deformation, such that one of the quantities involved has become negative. For instance, we might prove the law of cosines in the acute angle case where the projection involved has become negative. The

---

formules, les rapports de position peuvent être susceptibles d'un traitement général". (Italics in the text).

[15] On this aspect of Carnot's though see Gillispie (1971), Gillispie and Pisano (2014), Nabonnand (2011a, pp. 6–7), Schubring (2005, pp. 330–334).

whole idea seems to me vague and not worth of the esteem it was held by contemporaries (Coolidge, 1940, pp. 91–92).

Regardless of one's opinion on Carnot's mathematical work, undoubtedly, as Coolidge himself acknowledges, it was appreciated and influential between the end of the eighteenth century and the first 15 years of the following century. This is true both for Carnot's view of geometry and for his conceptions of the infinitely small quantities in analysis. For in the context of French mathematics and, specifically of the *École Polytechnique*, there was a discussion as to whether the method of limits or that of infinitesimals should be used in the teaching of mathematical analysis. At the time when Chasles was a student, analysis was taught by the method of infinitesimals, following a decision taken in 1811. Carnot's and Lacroix's ideas were fundamental as regard to this choice. It is significant that Lacroix, probably influenced by Carnot himself, changed his mind: if until the first 5 years of the nineteenth century he was convinced that the best way of teaching analysis was through the method of the limits, he claimed afterwards the necessity to adopt the method of infinitesimals (see, e.g., Belhoste, 2003, pp. 249–253). Schubring argues that the reappraisal of synthetic geometry is connected with the reintroduction of infinitesimals in the teaching of analysis.[16]

Chasles' opinion as to the role played by Carnot in the birth of projective geometry changed in the course of the years, as Michel appropriately points out: until the *Aperçu Historique* (1837), Chasles credited Monge for having introduced the main concepts from which this discipline arises. However, starting at least from his opening lecture in 1846 for the Chair of Higher Geometry at the Sorbonne, Chasles re-evaluated Carnot's work and fully recognized his scientific merits and profound influence on early nineteenth-century geometry (Michel, 2020a, pp. 54–62).

Daston stresses a further aspect that Poncelet and Chasles inherited from Monge and Carnot: geometry is strictly connected to mechanics. Carnot shared Locke's conviction that knowledge stems from experience and that the most abstract doctrines rely on an empirical base. The sharing of this empirical approach is the reason why Poncelet and Chasles considered the synthetical methods the most suitable for geometry: these methods enabled them to offer a palpable and almost-sensible interpretation of the ideal and imaginary entities. In Carnot and Monge the link between mechanics and geometry is guaranteed by the emphasis they placed on the concept of movement. Carnot introduced the geometrical movement. It should not be confused with the physical movement because the concept of velocity is not implied. However, the idea to apply extensively the notion of movement to geometry stems from the strong connection Carnot saw between the latter and physics. As we have seen, movement was a fundamental tenet of Monge's descriptive geometry too. Daston therefore traces Monge's and Carnot's profound influence on Poncelet and Chasles to the fact that the last two shared the ideas of the first two as to the relations

[16] Schubring (2005, pp. 371–410) offers a complete discussion on the way in which the teaching of analysis changed at the *École Polytechnique* from its foundation to the 1810s.

between geometry and mechanics (Daston, 1986). I am not as sure as Daston that Chasles can be considered an empiricist, but there is no doubt that the link with mechanics exerted a strong influence on his geometry.

The impact of Monge's and Carnot's on the conceptualization of pure modern geometry was remarkable. They were indeed two great mathematicians. It is, however, necessary to mention also a mathematician who, though less important than Monge and Carnot for the history of mathematics, played a fundamental role in spreading Monge's descriptive geometry at the *École Polytechnique*: Jean Nicolas Pierre Hachette (1769–1834). He gave several contributions to descriptive geometry (see, e.g., Hachette, 1817, 1822), to the theory of machines (Hachette, 1809) and to electricity, magnetism and optics. Moreover, his studies on the ruled quadrics are important for descriptive geometry. From a methodological point of view, Hachette was not a purist. Faithful to Monge, he used both analytical and synthetical methods. In 1788, only 1 year after his degree at the university of Rheims, Hachette was employed at the *École Royale du Génie* of Mézières as a draftsman and technician. Here he entered in contact with Monge and (to use a modern term) became his assistant. The spread and teaching of descriptive geometry became one of Hachette's main tasks. He was a teacher at the *École Polytechnique* from its foundation, and trained generations of young mathematicians until 1816, when he was forced out for his loyalty to Napoleon. Dupin and Brianchon were taught directly by Monge, but Poncelet and Chasles had Hachette as professor of descriptive geometry. He was also an organizer of cultural life. He was the editor of the *Journal de l'École Polytechnique* and in 1804 created the *Correspondance sur l'École Polytechnique*. This journal was particularly significant for the students of the *École Polytechnique* since many of them published their first scientific works on Hachette's journal. Chasles did the same thing, as we will see.

To summarize: when Chasles frequented the *École Polytechnique*, descriptive geometry was one of the leading subjects. It was taught both by synthetical and analytical methods, but the teaching approach was based on the implied idea that the true nature of descriptive geometry could be understood only through the extension of synthetic methods. Carnot was explicit in this sense. Monge and Hachette were less explicit. The teaching of mathematical analysis was based on the method of infinitesimals.

I offer here a picture of the staff working at the *École* when Chasles was a student there. The so-called *Directeur des Études* was Malus between 1811 and 1812, followed by Durivaux in the years 1812–1816. With regard to Analysis and Mechanics: Prony and Poisson (from 1815) worked as professors; Ampère, Reynaud, Lefebure de Fourcy, Paul Binet, Poinsot until 1815 (afterwards, in the same year, he became professor), Stainville and Cauchy from 1815 (in 1816 Cauchy became professor) were *répétiteurs* (lecturers); Lacroix (until 1815), Legendre and Poisson (from 1815) were examinators. As to descriptive geometry: Hachette (until 1815) and Arago were professors; Jacques Binet (until 1814), Duhays, Lefebure de Fourcy (from 1815) were *répétiteurs*. Hassenfratz was professor of physics until 1814 and Petit from 1815. The examinators of descriptive geometry, who acted possibly also as examinators of physics sometimes, were Ferry, until 1814, and Jacques Binet in 1815.[17]

The milieu around the *École Polytechnique* was the most influential in Chasles' mathematical education. However, it is necessary to mention the work and activity of a mathematician who was not directly connected to the *École*: Joseph Diaz Gergonne (1771–1859). He was deeply involved in the foundation of projective geometry. As we will see, he first used the term "polar" in 1813, while the term "pole" had been employed 2 years before by Servois. In 1813, Gergonne won the prize awarded by the Bordeaux Academy with a paper on analysis and synthesis. This prize had been offered "to characterize synthesis and analysis and to determine the influence these two methods had on the rigour, the progress and the teaching of exact sciences". The text with which Gergonne won the prize was never published and only a shorter version appeared (Gergonne, 1817).[18] Gergonne was the inventor of the double-column writing for the dual theorems and had a long and well-known priority dispute with Poncelet on the duality principle.[19] Though particularly skilled also in synthetic geometry, Gergonne claimed the superiority of the analytical methods. His influence on the French and, more generally, European mathematical environment was enormous. Basically, it was exerted through the journal he founded in 1810: the celebrated *Annales de mathématiques pures et appliquées*, also known as *Annales de Gergonne*. This journal published the articles of the most important French geometers: Servois, Brianchon, Poncelet, Bobillier, Chasles and Gergonne himself, only to mention some of the most celebrated ones. Starting from the second half of the 1820s, the *Annales* also published (in French) papers written by German-speaking geometers, such as Plücker and Steiner. Gergonne's journal was therefore an important means for the ideas and methods of such geometers to be known and discussed in France.

Among the mathematicians of the same generation as Chasles, Poncelet—who was 5 years older—was by far the most important and influential. His work was the reference point for Chasles and for all the mathematicians who cultivated pure geometry. In a very simplified—but basically truthful picture of the history of mathematics—Poncelet can be considered as the creator of modern projective geometry. In this book, I will develop a comparison between Poncelet's and Chasles' view on many aspects of geometry, especially on the principles of duality and of continuity. Therefore, now I will only offer a brief picture of Poncelet's conceptions which influenced Chasles.

Poncelet's programme is clearly explained in the Introduction of Poncelet (1822): Monge's great merit was to create descriptive geometry. However, in the works of

---

[17] These indications are drawn from Grattan-Guinness (2005, p. 235).

[18] With regard to Gergonne's original memoire, see Dahan-Dalmedico (1986), where the topic of the price is also referred to: "Caractériser la synthèse et l'analyse mathématique et déterminer l'influence qu'ont eue ces deux méthodes sur la rigueur, les progrès et l'enseignement des sciences exactes". (Dahan-Dalmedico, 1986, p. 97).

[19] I will deal in detail with the question of duality. Therefore, in this Introduction a hint to such issue is sufficient.

Monge himself and of his School (Hachette, for example) the necessary generality was obtained by using algebra and analysis. Poncelet, instead, *aims at creating a pure geometry as general and methodical as analytic geometry, but independent of algebraic analysis* (Poncelet, 1822, pp. xvii–xx). This is the main tenet of Poncelet's programme. Algebra is general because of the abstract symbols it uses. The ancients' analytical method had some similarities with algebra since the unknown data were treaded formally as the known ones and this allowed a generality that synthetic methods lack (ibid. pp. xxi-xxii). This statement is followed by two very dense pages in which Poncelet succinctly sets out the cornerstones of his method: he distinguishes explicitly between *graphical* and *metric* properties of the figures. Afterwards he introduces the principle of continuity: consider a figure and suppose it moves continuously ("par degrés insensibles", ibid., p. xxii). It is evident that many properties will be conserved during the movement, whereas some metric properties (such as the length or the sign of the segments) will change. The problem is, hence, how to determine which properties are conserved. The answer offered by Poncelet is that all the graphical properties are conserved as well as some of those metric, which will be called metric-graphical. Poncelet claimed that the principle of continuity is a general case of Carnot's principle of the figures' correlation and that it is used in the infinitesimal calculus, theory of limits and general theory of equations (ibid., p. xxiii). Through the principle of continuity, it is possible to introduce the ideal and infinite elements in geometry in a natural and general way (ibid., pp. xxiii-xxiv). The main difficulty in the concrete application of the principle of continuity consists in determining the general properties of a system which remains unmodified during the continuous movement and the particular properties, which, instead, change (ibid., p. xxv). It seems to me rather obvious that Poncelet identified the general properties with the graphical ones. To summarize: Poncelet's goal is to find a new synthetic-pure geometry. Since such a geometry will be based on a precise characterization of continuous movement, it is necessary to determine a transformation or a set of transformations which concretely transcribe the movement into geometrical terms. Poncelet's idea is to use the central projection ("projection relief", ibid., p. xxxi). The projections were the most general known transformations. Specifically, the central projection offers several advantages in comparison with Monge's method of projections and with the theory of coordinates. For in these two methods, in one case the lines and the surfaces are reduced to figures having only a dimension, and in the other case the surfaces are represented as flat areas. This is the best choice for Monge's aims, but for a general foundation of the pure methods, central projection is more suitable since it does not suffer the mentioned limitations (ibid., p. xxxi). Poncelet mentions Dupin and Chasles among the mathematicians who used the method of projections (ibid., p. xxxii). All properties that are preserved through a central projection, whether graphical or metric, are called "projective properties" (*propriétés projectives*) by Poncelet (ibid., p. xxxiii).

In his *Traité* Poncelet would develop the bases of pure projective geometry. It should finally be added that his concept of duality was reached specifically in relation to the theory of reciprocal polars, a topic he developed admirably and that he posed as a cornerstone of projective geometry. As we will see in detail, Chasles' celebrated memoir on duality stems from the whole of Poncelet's work and seeks to overcome it by showing that the duality relation has a more general value than the reciprocal polars theory.

The importance of Poncelet's work has been, of course, acknowledged: Fano claims that the passage from ancient to modern synthetic geometry is due to Poncelet's *Traitè*. Poncelet understood that all the positional relations are preserved in a projection, but that only some metric properties are conserved. Though Gergonne had a wider view than Poncelet as to duality, the latter was the first to apply concretely the duality law in order to study in depth a fundamental transformation: polarity. With regard to Poncelet's specific discoveries, Fano points out that the French mathematician first understood that all the circles of plane intersect in two imaginary points at infinity (Fano, 1907–1910, pp. 231–234). Schönflies stresses Poncelet's discovery that a projection conserves not only positional properties, but also some particular metric properties. He highlights that the main elements of homology theory are due to Poncelet, who also conceived the idea of the imaginary circle at infinity. Finally, the theory of reciprocal polars is also due to him (Schönflies, 1907–1910, pp. 397–401). Enriques underlines a particularly remarkable aspect of Poncelet's work: he realized that the metric properties can be considered as particular projective properties when some elements are specified and when the cyclic points are added to the ordinary ones. Particularly, in plane, the metric properties can be expressed as descriptive properties in relation to the cyclic points and, in space, in relation to the improper circle at infinity. Chasles was the most important geometer who prosecuted this line of thought (Enriques, 1907–1910, pp. 82–83). We will see that Chasles went far beyond Poncelet and added new transformations (the parabolic ones) which Poncelet had not regarded as particularly relevant for the reduction of metric to graphical properties. Kötter (1901, p. 93) and Coolidge (1940, p. 93) point out that Poncelet clearly distinguished the metric properties connected with the measures of segments and angles from the graphical ones, which concern the figures' position and are conserved in a projection. Both authors highlight the role played by the principle of continuity in Poncelet's geometry and its derivation from Carnot's principle of correlation. Among Poncelet's many merits, Darboux (1905, pp. 398–399) notes that he was the first to study in any detail a transformation in which a point does not correspond to a point: such a transformation is polarity. Poncelet was not the first to consider a polar transformation, but was the first to analyse its properties in every detail and with new projective instruments.

Nagel points out that, according to Poncelet, geometry does not have the same generality as algebra because of its dependence on the figure. Therefore, through his principle of continuity, Poncelet interpreted the geometrical figure as an abstract symbol subject to a series of operations as in algebra. Though being true that Poncelet refers his results to figures, they are no longer the sensible figures, but varying diagrams which can become ideal figures according to the specific conditions of the problem one is studying. Poncelet was so setting a new geometrical calculus, although he was not fully aware of all the potentialities of his method. The principle of continuity is a heuristic means to construct a formal system of signs. It can be interpreted as the genesis of an axiomatic system that defines implicitly the improper and imaginary elements and their properties. With Poncelet, geometry loses, at least potentially, its spatial bases because in our spatial world nothing as a point at infinity or an imaginary element exists. Thus, Nagel sees in Poncelet's work a fundamental step towards the modern axiomatization of geometry, though he highlights that, probably, Poncelet would have not accepted a formal interpretation of his work (Nagel, 1939, pp. 152–164). These observations are fundamental to placing Chasles in his cultural context because, as we shall see in the chapter on duality, Nagel interprets Chasles' work as the following step towards the abstract, hypothetico-deductive view of geometry.

Gray (2008, p. 41) remarks that Poncelet discovered a geometry concerned with the properties of the figures that do not regard the notion of distance and he called it non-metrical geometry. Gray defines Chasles as Poncelet's "successor as the leading projective geometer in France". Grattan-Guinness (1990, p. 262) also highlights that Carnot and Poncelet studied the conservation of non-metric properties in a continuous movement. Nabonnand (2015, pp. 18–22) underlines an important aspect of Poncelet's thought: he extends the property of a figure to a class of figures. As a matter of fact, Poncelet identified the peculiarity of algebra not in the use of its symbolism, but, rather, in its capability to reason on indeterminate variable quantities. The principle of continuity, if appropriately used, can allow pure geometry to reach the same generality as analytic geometry with the further important advantage that, by means of continuity, it is possible to identify exactly when a solution becomes imaginary or goes to infinity. The variable unknowns are, in this case, the geometrical figures assumed by the initial configuration in its continuous movement. Lorenat (2015c, p. 176) explains a similar concept while claiming that "the intermediate figure was Poncelet's purely geometrical equivalent to indeterminate coefficients in a coordinate equation". An important concept, expressed directly by Poncelet, as Lorenat emphasizes, is that the figure is never lost of view (ibid., p. 156ff.). Poncelet analyses mobile configurations rather than a single figure. Therefore, it is not possible to draw on a sheet of paper the entire situation described by Poncelet. However, the visual reference to the geometric movement is always present in Poncelet, so that one can speak of virtual figures. Furthermore, when some elements become imaginary the figure in the ordinary meaning does not exist and when they become infinite, it is irrepresentable. Nonetheless, the reference to visual intuition is the basis of Poncelet's dynamical geometry. Poncelet's connection to the figures is also emphasized by Nabonnand. For although Poncelet looked for generality in geometry, which he attempted to obtain through the principle of continuity, and despite his fundamental results on polarity, Poncelet's is still a geometry of figures, not yet a geometry of forms (as Steiner's) and of transformations (Nabonnand, 2011a, 2011b). It seems to me that, in this case, Nabonnand tends to underestimate the novelty of Poncelet's approach, especially with regard to transformations. I would like to conclude these considerations on Poncelet with some remarkable observations of Belhoste (1998): Poncelet was certainly a genius, but he was not an isolated genius; he worked within a research programme that traces back to the *École Polytechnique*. More particularly, it seems that Poncelet reached the principle of continuity while studying some issues discussed at the *École*. In this sense, a line of continuity Monge-Carnot-Poncelet can be traced. Chasles was the fourth "giant" within this line and he developed the original conceptions I am going to explain in this book.

In these pages I have tried to briefly describe the milieu in which Chasles was educated and worked as a student of the *École Polytechnique* and I have attempted to retrace the path that, starting from the late 1820s, made him one of the most eminent geometers in Europe. Obviously, this picture is lacking in many respects: for example, I have considered only geometry and, among the geometers, only the most illustrious ones; I have referred only to the French environment, while, starting from the 1820s, the German geometers also began to give important contributions to projective geometry. However, taking into account that my aim has been to offer only the general framework in which Chasles began his scientific career and that, as I will explain, he was scarcely influenced by the German geometers until the second half of the 1830s, I believe this brief overview to be sufficient for the scope of my work. It is perhaps appropriate to conclude with the observation that the Belgian mathematician Lambert-Adolphe-Jacques Quetelet (1796–1874) was, as will be considered, an important stimulus for Chasles. Several of Chasles' initial papers were actually published on the journal directed by Quetelet: *Correspondance mathématique et physique, publiée par A. Quetelet*. Furthermore, Quetelet invited Chasles to develop his studies in projective geometry so as to also include the properties of curves and surfaces with a degree bigger than two. Finally, the two crowned memories by Chasles on duality and homography were written to participate to the Price of the Royal Academy of Brussels in 1829. Thus, Belgium too had a role in the initial phases of Chasles' scientific career.

1.5 A Hint to Chasles' Biography and Works

Michel Chasles was born on November 15, 1793, in Épernon, chief town of the homonymous canton of the current department of Eure-et-Loir.[20] He belonged to an upper middle-class family: his grandfather traded in construction wood. His son, Chasles's father, continued this business and became president of the chamber of commerce in Chartres. The "boulevard Chasles" in Chartres is entitled to Michel, to his brother Adelphe, who was for a long-time mayor of the city, to his father, and to his uncle, who was curate of the cathedral. Chasles studied at the *collège* in Chartres and afterwards at the Imperial lyceum in Paris. According to Bertrand, Chasles was already very interested in geometry when he was a student in the lyceum. Since his passion for mathematics was shared by other classmates, a small academy was organized in which the best students exchanged ideas and exercises with the

[20]This general information on Chasles' life is taken from the following texts: Bertrand (1892), Koppelman (1971), Michel (2020a), Riccardi (1881).

solutions each had found (Bertrand, 1892, pp. XL–XLI). A very gifted classmate of Chasles's was Gaetano Giorgini. Upon the completion of his studies at the lyceum, Giorgini obtained the *prix d'honneur* and achieved the first place in the entrance examination at the *École Polytechnique*. Chasles gained the second position and entered the *École* in 1812. Based on the interests he had already been cultivating for some years, stimulated by the environment he found at the *École* and, specifically, by Hachette, Chasles wrote three notes when he was still a student (Chasles, 1813, 1814a, 1814b). They concerned the one-sheeted hyperboloid. In 1814, he was mobilized and took part in the defence of Paris. After the fall of Paris, the *École* remained closed for several weeks. During this period Chasles came back to his paternal house in Chartres and invited many of his companions to spend that difficult period there. The friendship with Giorgini became particularly close at that time. Upon resumption of the *École*'s activity Chasles graduated in 1815 and was selected among the ten students chosen by the *École du Génie*. However, Chasles did not accept and left his position to a companion, Coignet. For some years, Chasles chose not to work and to focus on geometry. Besides studying modern geometry, he developed a profound interest in Greek geometry. He worked on Apollonius and Archimedes and, particularly, he tried to interpret some difficult fragments by Pappus. For several years, Chasles' father accepted that his son dedicated most of his time to the study of geometry. However, when Michel was about 28, that is around 1820–1821, his father induced him to find a job as a stockbroker in Paris. However, a disastrous liquidation made Chasles insolvent. His father came to Michel's aid, and the latter thus returned to geometry. Until the first half of the 1820s, Chasles did not publish many works. But from 1827 onwards the situation changed completely: Michel began to publish an impressive number of long and significant papers on the most important mathematical journals, especially on the *Annales de mathématiques pures et appliquées* and the *Correspondance mathématique et physique, publiée par A. Quetelet*. The subjects dealt with by Chasles included all the most important topics of modern geometry and his ideas were very original: he gave significant contributions to the study of conics' systems; to the research on the second-degree cones and on the quadrics; he faced the parabolic transformations, which, as we will see in detail, are useful for the relations between descriptive and metric properties, and studied the spherical conics, a new and complex field of geometry. He applied geometry to the study of the movement of a rigid body proving what today is called Mozzi-Chasles theorem and offered a new general picture of the treatment of the system of forces and of couples of forces. Starting from these early works, where difficult geometrical figures and transformations are treated with extraordinary ability and with an astonishing capability to see spatial configurations, Chasles displayed the propensity for research that characterized his entire production and that is well summarized by Koppelman:

> Chasles's work was marked by its unity of purpose and method. The purpose was to show not only that geometry, by which he meant synthetic geometry, had methods as powerful and fertile for the discovery and demonstration of mathematical truths as those of algebraic analysis, but that these methods had an important advantage, in that they showed more clearly the origin and connections of these truths (Koppelman, 1971).

In 1837 Chasles published the *Aperçu Historique Sur l'Origine Et Le Développement Des Méthodes En Géométrie: Particulièrement de Celles Qui Se Rapporte à la Géométrie Moderne, suivi d'un Mémoire de Géométrie sur Deux Principes Généraux de la Science, la Dualité et l'Homographie* (about 800 pages). This seminal work appeared in Bruxelles for the types of the Belgian Royal Academy of Sciences. It is composed of an improved and enlarged version of the two memoirs with which Chasles won the prize of the Belgian Academy of Sciences in 1829. The two memoirs are preceded by a historical introduction on the development of geometry and by a long series of detailed notes where Chasles offers profound historical insights on the subjects he deals with in the book as well as in-depth analyses on the use of the anharmonic ratio (our cross ratio), probably his most important discovery in geometry. The study of transformations became one of the main tenets of Chasles' geometry.

From 1837 to 1845, Chasles set another milestone in his foundational research, aiming to build much of geometry and science on the bases offered by projective geometry. For in a series of long memoirs he dealt with the complex problem of the ellipsoid attraction, offering a completely original and synthetic solution.

In 1839, Chasles was elected a corresponding member of the French Academy of Sciences and became a full member in 1851. In 1841, he obtained the chair of Geodesy and Mechanics at the *École Polytechnique*, which he held until 1851.

From 1843, Chasles began to publish a series of works on the infinitesimal and finite geometrical movements with a style which would also characterize his research on enumerative geometry:[21] he poses few principles and some historical insights on the topics and afterwards a long series of theorems without demonstration or with only a sketch of demonstration. As Koppelmann recalls, in 1846 the chair of Higher Geometry was created for him at the Sorbonne, and he remained there until his death. Poinsot also played an important role in the creation of a chair of Higher Geometry conceived specifically for Chasles.

In 1852, the *Traité de géométrie supérieure* was published. It was an influential book, in which Chasles used systematically the cross ratio to offer a foundation to the whole of projective geometry. This work developed a series of results which were potentially included in the previous research by Möbius, Steiner and Chasles himself, but the numerous new acquisitions, the conciseness and precision of the style, as well as the breadth of the treatment make the *Traité* an important contribution to the inquiry on the bases of projective geometry. Another important text was the *Traité des sections coniques* (1865), where Chasles presented the complete projective theory of conics, which he and Steiner had founded some 30 years before.

In 1864 Chasles founded a new branch of mathematics: enumerative geometry (Chasles, 1864a, 1864b), a topic to which he dedicated a large number of contributions from 1864 to 1867. Some enumerative problems had been already faced by mathematicians—it is sufficient to think of Steiner's incorrect claim that the number of conics touching five given conics is 7776, which was a source of inspiration for

[21] On Chasles' enumerative geometry the fundamental text is Michel (2020a).

Chasles—but Chasles was the first to treat enumerative geometry in a systematic way grounded on few principles, basically the method of characteristics and the principle of correspondence. The discussion deriving from his work was broad and involved many great mathematicians such as Jonquières, Cremona, Halphen, Study, Schubert and Zeuthen. It also addressed crucial questions on mathematical methods, generality and existence, as Michel's complete account shows (Michel, 2020a).

In the course of his research, Chasles never neglected contributions to the history of mathematics: besides other minor contributions, in 1843 he wrote "Histoire d'arithmétique. Développements et détails historiques sur divers points du système de l'Abacus", a 28-page memoir which appeared in the *Comptes rendus de séances de l'Académie des Sciences*. He claimed the thesis that the decimal system had a Pythagorean rather than a Hindu origin. His most important historical contribution, after the publication of the *Aperçu* and before the *Le rapport sur le progrès de la géométrie* (Chasles, 1870), was *Les Trois Livres de Porismes d'Euclide* (Chasles, 1860a). Here, he offered a modernizing picture claiming that many results of the porisms had been obtained by the concept of cross ratio. This interpretation is not accepted nowadays (Koppelmann, 1971). This notwithstanding, "the historiographical tenets and practices involved in Chasles's restoration of the porisms as well as the philosophical and mathematical claims tentatively buttressed therewith" (Michel & Smadja, 2021, p. 1) are interesting, as Michel-Smadja's work points out.

Chasles was member of several important scientific institutions. As Koppelmann recalls:

> Chasles was elected a corresponding member of the Academy of Sciences in 1839 and a full member in 1851. His international reputation is attested to by the following partial list of his affiliations: member of the Royal Society of London; honorary member of the Royal Academy of Ireland; foreign associate of the royal academies of Brussels, Copenhagen, Naples, and Stockholm; correspondent of the Imperial Academy of Sciences at St. Petersburg; and foreign associate of the National Academy of the United States. In 1865 Chasles was awarded the Copley Medal by the Royal Society of London for his original researches in pure geometry. (Koppelmann, 1971).

In 1867, Chasles was also elected as the first foreign member of the London Mathematical Society, an institution founded in 1865.

Between 1861 and 1869, he was the victim of a fraud as famous as unbelievable: the forger Denis Vrain-Lucas sold Chasles numerous letters he passed off as a real correspondence between Pascal, Boyle and Newton. From these letters it resulted that Pascal had anticipated Newton in the discovery of the law of universal gravitation. It should be pointed out that one letter was dated 1654, when Newton was 12! The whole affair is truly grotesque since among the letters Vrain-Lucas presented to Chasles were not only those between Pascal, Boyle and Newton, but even a letter from Caesar to Vercingetorix written in mediaeval French! From 1867 to 1869, Chasles defended the authenticity of the letters in front of the Academy. When

Vrain-Lucas was condemned in 1869, Chasles was still not completely convinced that he had been a victim of a fraud.[22]

In 1867, the Ministry of Education asked Chasles to draw a historical and conceptual picture of geometry from the end of the eighteenth century to the 1860s. The result was *Le rapport sur le progrès de la géométrie* (Chasles, 1870). The author focused mainly on the achievements of French mathematicians—which was of primary interest to the Ministry—but also offered a good overview of the research developed abroad. In fact, Chasles did much more: he analysed the development of all branches of science that were connected, even indirectly, to geometry. The text is therefore also an excellent source for the study of the history of various branches of physics.

In 1872, Chasles became the first president of the *Société mathématique de France*. He was active in research and teaching until the last period of his life. He is one of the 72 scientists whose name is written in raised gold letters 60 cm high on the first floor of the Eiffel Tower. He is the 11th, on the face facing North.

His contributions to mathematics are highly original and creative. As we will see in this book, he did not consider geometry and physics as two separate doctrines. There was a discipline which was their conceptual basis and this was projective geometry. He therefore developed a research programme aimed at proving such a thesis. The exposition of such a programme is the purpose of my book. I would like to conclude this Introduction with Riccardi's assessment—with which I agree—of Chasles' main contribution to mathematics:

> Although he [Chasles] was versed in every branch of these sciences [mathematics], nevertheless the synthesis of the greatest number and most interesting of his scientific works was to raise the study of geometry to the point where it joined the highest theories of analysis; thus attempting to reveal that unity that exists between these two branches of mathematical disciplines in the science of infinity.[23]

[22]On this affair a well-informed and funny book is Vecchioni (2012).

[23]Riccardi (1881, p. 38): "Imperocchè sebbene versato in ogni ramo di queste scienze, tuttavia la sintesi del maggior numero e dei suoi più interessanti lavori scientifici fu il rialzare lo studio della geometria fino al punto nel quale essa congiungesi alle più elevate teorie dell'analisi; tentando così di rivelare quella unità che nella scienza dell'infinito pur deve esistere tra questi due rami delle matematiche discipline".

Chapter 2
Chasles' Foundational Programme for Geometry

Abstract After an introductory section where I summarize the most important contributions given by Chasles to projective geometry in the period 1827–1850, which support my interpretation concerning the existence of his foundational programme, this chapter presents four sections, divided into several subsections. One of my main theses is that Chasles tried to explain the metric properties within a projective context. He performed this task through a particular polarity he called "parabolic transformation". Therefore, in the first section I expand and comment on the contributions given by Chasles to the theory of polarity and his new reading of this transformation. The second section is dedicated to the development of Chasles' projective geometry and, specifically, to the notion of anharmonic ratio, today better known as cross ratio. The explanation of the way in which this concept enriched and specified his foundational programme is offered. The Swiss mathematician Jakob Steiner developed a programme that, to some extent, was similar to Chasles'. On the other hand, there are also remarkable differences in the approach of these two mathematicians to projective geometry and to the relation between projective and metric geometry. Therefore, in the third section, a close comparison between the two authors' conceptions is proposed. In the Conclusive considerations I develop a comparison between my interpretation of Chasles' works and those proposed by other scholars.

In 1837, Michel Chasles published his monumental *Aperçu historique sur l'origine et le développement des méthodes en géométrie*. In 1829, the Royal Academy of Sciences in Brussels had proposed a question concerning "a philosophical examination of the different methods in modern geometry, in particular the method of reciprocal polars". Chasles submitted two long memoirs on the subject, which won the Academy's prize. The Academy proposed Chasles to publish his two contributions. He accepted their offer but decided to enrich his work with an in-depth historical introduction and a huge series of annotations. The result of this effort was the *Aperçu historique*. The roughly 800 pages that compose the book are almost equally divided into three sections: the first part is devoted to the history of geometry, the second one contains the notes to the first section, notes which concern

P. Bussotti, *Chasles and the Projective Geometry*,
https://doi.org/10.1007/978-3-031-54266-4_2

several historical and conceptual questions, and the third and final part is made up of the two memoirs sent by Chasles to the Academy in 1829.

In 1837, Chasles was a mathematician who had already published an impressive series of contributions.[1] The context in which he was working is well defined. It is the context of what was then called modern or pure geometry. The principal reference point was Poncelet's masterpiece *Traité des propriétés projectives des figures: ouvrage utile à ceux qui s'occupent des applications de la géométrie descriptive et d'opérations géométriques sur le terrain* (Poncelet, 1822), where the author, by making clear distinctions between graphical and metric properties, defined a modern and broad field of research for pure geometry, which one might call projective geometry developed by synthetic methods. Nabonnand points out that, in fact, Poncelet had already clearly distinguished between graphic and metric properties in 1818. Poncelet referred to the metric properties as those which concern "les relations existantes entre les grandeurs mesurées des parties des figures" and to the graphic properties as those which are independent from such measures and which "n'ont trait qu'aux affections relatives à leur configuration, à leur manière d'être réciproque".[2] Furthermore, in this context—as pointed out in the introduction—Poncelet developed a series of brilliant, although not always precise, ideas: the principle of continuity, ideal secants and chords, the theory of poles and of reciprocal polars for conic sections, the theory of centres of similitude for the conics. Basically, Poncelet tried to reduce several properties of conic sections to those of circles by projections. In this setting, he introduced ideal elements and elements at infinity, the principle of continuity and the theory of polar reciprocity.

He was not the first to conceive projective concepts in the period under consideration. After Monge's descriptive geometry and before Poncelet's book, other mathematicians such as Carnot, Servois, Brianchon, Hachette and Gergonne (to mention only some of the most important ones) had grasped and proved several theorems of projective geometry. Servois (1810–1811) first used the word "pole" and Gergonne (1812–1813) the term "polar", though these concepts had been in use for a long time. Servois defined as follows the pole of a straight line with respect to a conic: "Given a straight line and a second degree line, I call *pole* of the straight line the point of the plane of this straight line and of the curve, around which all the chords of the contact points of the pairs of tangents drawn from different points of the straight line to the curve rotate". With regard to the definition of polar, Gergonne proved the following theorem and, immediately afterwards, he offered the definition of polar. He wrote: "If, through any point [P] belonging to the plane of a second degree line, a series of secants to this curve are drawn and if, from the two

[1] I mention the main contributions given by Chasles until 1837. Many of them will be analysed during this research. Chasles (1813; 1814a, 1814b; 1816a, 1816b; 1827–1828a, 1827–1828b, 1827–1828c; 1828a, 1828b; 1828–1829a, 1828–1829b, 1828–1829c; 1829a, 1829b, 1829c, 1829d, 1829e, 1829f, 1829g, 1829h; 1829i; 1830a, 1830b, 1830c, 1830d, 1830e, 1830f, 1830g, 1830h; 1831; 1832a, 1832b, 1832c; 1835a, 1835b; 1836a, 1836b; 1837a, 1837b, 1837c, 1837d).

[2] Poncelet (1818, p. 299). See also Nabonnand (2011a, p. 18). Reference to the free online version. The page 18 corresponds to the page 18 of the file.

intersection points of each of them with the curve, you draw two tangents to this curve—terminated to their intersection points—, the tangents of the same couple will form a series of circumscribed angles whose vertices will be on the same straight line [*Q*]. . . [Definition] Because of the relation between the point (*P*) and the straight line (*Q*), this point has been called the *pole* of this straight line; vice versa, it is possible to call the straight line (*Q*) the *polar* of the point (*P*)".[3]

However, Poncelet's *Traité* can be considered as the reference work for a great flourishing of studies concerning pure geometry that involved—between the 20s and the 30s of the nineteenth century—such important mathematicians as Gergonne, Hachette, Bobillier, Dupin, Dandelin, Quetelet, Chasles, Poncelet himself, Steiner, Plücker, Möbius and others.

Chasles was one of the protagonists of this many-faceted and multiform movement, and in this period, he had a plurality of ideas.[4] Many of these ideas represent a truly *foundational programme* concerning geometry and the relations between geometry and other branches of mathematics and science. His *foundational programme for geometry* entails the following issues:

A. *The general foundation of plane projective geometry insofar as graphic properties are concerned.*
B. *The general foundation of spatial projective geometry insofar as graphic properties are concerned.*
C. *The dependence of plane projective geometry on spatial projective geometry.*
D. *The reduction of the metric properties within a projective context.*

[3]Servois (1810–1811, 337): "Une droite et une ligne du second ordre étant assignées, j'appelle *pôle* de la droite, le point du plan de cette droite et de la courbe autour duquel tournent toutes les cordes des points de contact des paires de tangentes à la courbe issues de différens points de la droite". Gergonne (1812–1813, 297): "Si, par un point [*P*] pris arbitrairement sur le plan d'une ligne du second ordre, on mène à cette courbe une suite de sécantes; et que, par les deux points d'intersection de chacune d'elles avec la courbe, on mène à cette même courbe deux tangentes, terminées à leur points de concours, les tangentes de mêmes couples formeront une suite d'angles circonscrits dont les sommets seront tous sur une même ligne droite [*Q*]. . . [Definition] A cause de la relation qui existe entre le point (*P*) et le droite (*Q*), ce point a été appelé le *Pôle* de cette droite; et on peut, à l'inverse, appeler la droite (*Q*) la *polaire* du point (*P*)". Italics in the text. With regard to the history of the concepts of pole and polar, see also Chasles (1837a, 370–371) and Haubrichs dos Santos (2015, 124–126).

[4]The literature on projective geometry in the first 40 years of the nineteenth century is vast. I mention here some important works without any claim to be exhaustive. Furthermore, the texts referred to in this note concern general questions and are not specifically dedicated to the contributions of a single mathematician: Biosemat-Martagon (2010), Coolidge (1934, 1940, in particular chapters 1, V, VI; 2, III, VII); Delcourt (2011), Gerini (2000), Gray (2010), Haubrich dos Santos (2015), Kötter (1901). The text by Kötter, though dating at the beginning of the twentieth century, is fundamental. It offers a broad and clear picture of the synthetic geometry in the nineteenth century. It is a mine of information for everyone who is interested in the history of this discipline; Lennes (1930), Lorenat (2015a), Loria (1896, chapter 1), Nabonnand (2006, 2011a, 2011b), Obenrauch (1897), Pedoe (1975), Simon (1906), Voelke (2008, first section; 2010).

E. *The derivation of the projective properties from the conservation under projective transformations of a fundamental invariant: the anharmonic ratio (today called cross ratio*[5]*).*

To be more precise:

1) *Attempt to reduce the metric properties to the graphical ones.* This purpose can be realized by a thorough study of the *concept of transformation.* Chasles developed and realized these ideas in an impressive series of papers published between 1827 and 1830. In particular, he studied the polar transformations between conics and quadrics in an attempt to identify clearly those metric properties which are conserved under a particular polar transformation, which he called *parabolic*,[6] and which, based on few principles, allows one to obtain many metric-projective properties concerning conics and quadrics. At the same time, he recognized that polarity is a particular transformation of a more general class, the reciprocities, which, jointly with homographies, determine the projective transformations. The two memoirs from which the *Aperçu* arises are, in fact, dedicated to the two principles of duality and homography.[7] Chasles' basic idea was, therefore, to demonstrate that the metric properties arise from the graphical ones when particular specifications are considered concerning the entity with respect to which the projectivity is developed.
2) In the same period as the German mathematician August Ferdinad Möbius (1827) and the Swiss mathematician Jakob Steiner (1832),[8] Chasles was the first to use extensively the concept of cross ratio as a projective invariant. Möbius used the term *Doppelschnittverhältnis*, Steiner used *Doppelverhältnis* and Chasles called the cross ratio *rapport anharmonique*. The concept of cross ratio was already known; it is present in some theorems due to Pappus and was utilized in geometry before Chasles (e.g. by Poncelet). However, Möbius, Chasles and Steiner were the first to define the cross ratio and to consider it as one of the main objects of projective geometry. They also showed that the cross ratio can be the basis on which to develop research on the metric-projective properties. The history of the

[5]From a linguistic standpoint, the expression cross ratio was introduced by Clifford in 1878 in the first volume of his *Elements of dynamics* (Clifford, 1878, p. 42). See also Milne (1911) p. 2, note.

[6]A parabolic transformation is a polarity with respect to a parabola, if one works in the plane, and with respect to a paraboloid, if one works in the three-dimensional space.

[7]With regard to these memoirs, I will refer to the *Aperçu* (Chasles, 1837a, 573–848).

[8]The history of cross-ratio is long and dates back, at least, to Pappus, but my aim is not to trace such a history, rather to analyse Chasles' use of cross-ratio. Therefore, in this note, I mention some significant works on Steiner since he was the mathematician who shared with Chasles the full comprehension of how to use this concept. Moreover, his view was closer to Chasles' than to Möbius', as I will clarify in Sect. 2.3. Literature on Steiner, with no claim to be exhaustive: Baldus (1923), Blåsjö (2009), Graf (1897), Hesse (1863), Lange (1899), Lampe (1900), Rowe (1997). When addressing the way in which Chasles introduced the anharmonic ratio, I will draw a comparison between his conception and Steiner's one, because, under several viewpoints, Chasles' and Steiner's ideas present significant similarities.

cross ratio begins with these three mathematicians. Before them, one can only speak of a protohistory.

With regard to the possibility that projective geometry is the basis of several branches of other disciplines, this is the situation: many parts of optics (especially with regard to the aplanatic lines and to the caustics of refraction[9]) and mechanics (systems of forces, the theorem on the instantaneous centre of rotation of a rigid body, the attraction of an ellipsoid) can be obtained by means of considerations inherent to projective geometry.[10]

This programme implies, therefore, that projective geometry constitutes one of the most general[11] *and fundamental branches of mathematics and science, to which others can be reduced. This is a foundational programme.*

The bases of *Chasles' foundational programme for geometry* were constructed from the end of the 1820s until the end of the 1830s. Therefore, I will essentially focus on Chasles' contributions published in this period. Since, however, Chasles (1852) can be interpreted as a summary, development, specification and clarification of his thought insofar as projective geometry and its bases are concerned, I will also address some aspects of this work, which was well known at that time and also influenced the ideas of several mathematicians and scientists, as we will see.

Three considerations are necessary:

1) The mathematical context and milieu in which Chasles worked until the publication of the *Aperçu* is quite complex and interesting. The basic theorems of projective geometry were proved in this period and projective geometry as a fundamental branch of mathematics was created. The subjects under discussion were numerous: there was the concept of duality and the subsequent polemic between Gergonne and Poncelet, which was not only a problem of priority, but concerned the whole conception of duality; Plücker's profound studies on the abbreviated notation and on the so-called paradox of duality; Steiner's deep conception of projectivity; the problem of polarity for curves and surfaces of higher degrees to which Bobillier made important contributions; the general

[9]An optical system is defined "aplanatic" if it is free of spherical aberration and coma. Ernst Abbe (1840–1905) added a further condition, but this happened in the second part of the nineteenth century, not at the end of the 1820s. See Southall (1922, p. 610). A caustic is the envelope-surface of the rays reflected (caustic of reflexion) of refracted (caustic of refraction) which a not-stigmatic optical system makes to correspond to a homocentric beam of incidence rays. The reflected or refracted wave is not anymore spherical. There is spherical aberration.

[10]Chasles' idea that projective geometry might be the basis of several branches of science, and in particular, of physics, will be the subject of the following chapters.

[11]With regard to the problem of generality in Chasles, a valuable paper is Chemla (2016). The author assumes her conception of generality in Chasles as a guide to explain the relations seen by Chasles between analytical and pure methods in geometry. She also addresses the way in which Chasles introduced and interpreted the principle of "contingent relations", which is a generalization of Poncelet's principle of continuity. Finally, Chemla compares an axiomatic approach to geometry with Chasles' one. Chemla's analysis is based upon Chasles' *Aperçu historique*.

theory of projective transformations; and the construction of coordinate systems suitable for study projective properties (Plücker, Möbius). Many other subjects could be added.

2) Chasles' production after the publication of the *Aperçu* was very broad. To give only an incomplete idea:

   a) He continued to study the projective geometry of second-degree surfaces (Loria, 1896, 94, 97).
   b) He contributed to the development of algebraic geometry (ibid., pp. 59, 71, 89, 106, 133, 137, 247), also making important contributions to the correspondence principle (ibid., p. 65), the study of the curves of genus 0 (ibid., 76) and the developable surfaces (ibid., pp. 105, 120).
   c) Enumerative geometry (ibid., pp. 259–280). Chasles was one of the founders of this discipline and his contributions were seminal.
   d) He also wrote two huge treatises, with a partially didactical scope, but which can also be considered as a *summa* of his thought on projective geometry. One of these treatises concerns the conic sections (Chasles, 1865); the other is devoted to higher geometry and deals with the foundations of projective geometry (Chasles, 1852, second edition 1880).
   e) Chasles was also an important historian of mathematics. Among his historical works after the *Aperçu*, let us refer to his reconstruction of the three books of Euclid's porisms (Chasles, 1860a).
   f) In 1870, at the request of the Educational Minister, he wrote an interesting essay on the development of science in France from the beginning of the nineteenth century (Chasles, 1870).

3) The third question concerns the debate on the use of pure or analytical methods in geometry. The whole discussion is wide, and before describing Chasles' position, it is appropriate to give at least the basic coordinates of such discussion. In most cases, the term whose meaning is closer to our expression "synthetic geometry" was "pure geometry" ("géométrie pure") or "modern geometry" ("géométrie moderne), but there is not a perfect identification. Lorenat (2015a) argues that the application of the categories "analytic geometry" and "synthetic geometry" does not grasp the essence of the way in which projective geometry was conceived and developed between the 1810s and the 1830s. She writes:

> Similarly, 'analytic and synthetic' provide a means of association as well as dissociation to existent results and techniques. Rather than precise adjectives, the terms analysis and synthesis in geometry often acted as substitutes among historians for a perceived methodological bifurcation that might emerge under different names when applied by earlier geometers. Our thesis thus does not simply argue against a clear division between analytic and synthetic geometry in early nineteenth century geometry, since this could be easily shown by pointing the absence of 'synthetic geometry' as a common category among the authors themselves. Rather we will show the absence of a uniform dichotomy under any pair of contrasting adjectives (Lorenat, 2015a, p. 7).

The author also adds a rich bibliography. Lorenat's work is valuable and focuses on many useful nuances of mathematical terminology of that period. However, as I will try to show in the course of this book, my view is more traditional: I think that, beyond terminology, basically the geometers of that period clearly perceived the distinction between what nowadays we call analytic and synthetic geometry and that the two categories of "analytic" and "synthetic" make sense if applied to that epoch. With regard to the way in which Chasles used the term "synthèse", "synthétique" Lorenat writes:

> Chasles discussed 'synthesis' by name only rarely and in historical contexts, not to describe the works of his contemporaries [. . .]. Both of these references seem to imply a synonym between synthetic methods and so-called ancient geometry. (Ibid., p. 25).

This is true while referring to the *Aperçu* and to the memoirs analysed by Lorenat because, in that context, Chasles wanted to show how much "modern geometry" was richer than ancient geometry. Thus, possibly to avoid any ambiguity, he referred only to ancient geometry as "synthetic". However, in his mind ancient and modern geometry were both synthetic and conceptually well distinguished from analytic geometry. There is no doubt on this. Chasles' seminal memoir on the attraction of the ellipsoid (Chasles, 1837, 1846) [1837: presented at the Académie; 1846: published] is entitled "Mémoire sur l'attraction des ellipsoïdes. Solution synthétique pour le cas général d'un ellipsoïde hétérogène et d'un point extérieur". The term "synthétique" is here opposed to "analityque" in our meaning. In Chasles (1840, p. 466), the author speaks of Ivory's solution as "synthetic". In a note, he expresses a different opinion from that of Poisson as to the value of the analytic and synthetical methods, from which it follows that both of them opposed the analytic and the synthetical methods as we do and with the same terminology. Furthermore, in his *Systematische Entwicklung* Steiner explicitly used the expression "syntetische Methode" in contraposition to "analytische Methode" (see, e.g. Steiner, 1832, p. 5). Lorenat dedicates a remarkable and conceptually rich paper to Steiner's and Plücker's solutions of the Apollonius' and Lamé's problems (Lorenat, 2016). Her argumentative line is similar to that of her previous work. Her aim is to show that at the end of the 20s–beginning of the 30s of the nineteenth century, the distinction between what we call synthetic and analytical methods was not clear. She claims that Steiner and Plücker developed new and personal methods that can hardly be classified as analytical or synthetic. For example, though Steiner called his own geometry *synthetisch*,

> Steiner likewise distinguished his form of synthetic geometry from that of the ancients, as more general and complete, but still employing a "rigorously genetic path" [streng genetischen Gang], developing increasingly complex concepts from common constructions. In these qualities of generality and completeness, Steiner observed the connection and possible contributions between his method and analytic geometry" (Lorenat, 2016, p. 430).

There is no doubt that Steiner's and Plücker's methods were personal (which is also true for Gergonne's, Poncelet's or Chasles'), but there is also no doubt that Steiner's geometry is poor of calculations, that he rarely used systems of coordinates and that he imagined the situations from a visual point of view. Actually, Plücker's geometry is often based on equations written in abridged notation, which allowed him to simplify or, even, to avoid the calculations, but it is very significant that he started from equations. On the other hand, it is not appropriate to identify "calculation" with "typical of what can be called analytical" and "lacking of calculations" with "typical of what can be called synthetic". If we think of the extremely long series of proportions which connote the synthetic solutions of many difficult problems of Euclidean geometry or of the indefatigable Apollonius' work on proportions developed in his *Conics*, it will be patent that synthetic geometry is not necessarily associated with a lack of calculations. This is often the case, but not necessarily. The real difference is that in analytical geometry one starts from equations with some unknown variable quantities, which is not the case in synthetical methods. Therefore, though recognizing, as Lorenat points out (ibid., p. 418), a certain malleability of the methods as well as the fact that a mathematician can, obviously, use both methods and mix them also within the same problem, I think it makes perfectly sense to distinguish between a synthetic and an analytical approach in a relatively traditional meaning in reference to the kind of geometry developed in the period under examination.

As for Chasles' position, the situation is rather complex: he was surely sensitive to the problem of the purity of methods within geometry. As we shall see, several of his works belong to the field of pure geometry. Part of his research is dedicated to proving theorems or solving problems by means of "pure geometry". Therefore, the purity of methods was a fundamental issue for Chasles. However, in his view, there was an even more important priority, namely finding a set of concepts that might represent the true foundations of geometry. This means that it is necessary to separate the methodological aspect from the ontological one: first of all, it is necessary to determine the set of concepts which are fundamental within projective geometry (this concerns its ontology); secondly, it is appropriate to use synthetical methods to treat these concepts. However, at least in an initial phase, such concepts can also be introduced through analytical means. The study of the graphical properties of figures has to be reduced to these concepts. The metric properties have to be expressed in terms of graphic properties when the elements of a figure assume particular positions, which can be formulated in function of specific values assumed by the cross ratio. Thus, a limited set of notions should be the common basis of the entire geometry. The most important of these concepts is that of anharmonic ratio. Chasles developed this programme between 1827 and 1837, but the years 1827–1830 were particularly intense. The memoirs on duality and homography that appeared in the *Aperçu* in 1837—although originally written 8 years before—must be considered Chasles' milestones on these problems.

The situation with regard to Chasles' foundational programme for other than geometry is slightly different: in this case, as we will see in the next chapters, the methodological aspect is preponderant since Chasles tried to found such disciplines—and particularly several branches of physics—on concepts deriving from projective geometry and using synthetic methods in the broadest possible manner. Therefore, Chasles' foundational programme might be summarized in this way: there is a set of few concepts drawn from projective geometry which are more primitive than any other concept usable in geometry and in a great part of mechanics and optics. The nature of these concepts relies on the properties of particular geometrical configurations whose nature is independent of the equations through which it is possible to treat the mutual relations among such concepts. In other terms, their nature is synthetic, not analytical. However, if in an initial phase, the introduction of these concepts within geometry is easier by resorting to analytical methods, this is acceptable, provided that, afterwards, one is also able to reach a synthetic proof of the properties firstly demonstrated analytically.

With regard to the application of the synthetic method to physics, the situation is more complicated because in this case, not only is analytical geometry used, but infinitesimal calculus as well, as one might expect. This is particularly conspicuous when Chasles faced the problem of the ellipsoid attraction. This is how he solved the problem. First step: he identified a series of projective properties proved synthetically which represent basic concepts for the physical problems he was dealing with. Second step: insofar as this is possible, the reasoning was developed synthetically; when necessary, the data were transcribed into equations. Third and fundamental step: the infinitesimal elements necessary in a certain reasoning were introduced as infinitesimal geometrical elements, which are passible of an analytical transcription, but which, from an ontological point of view, have their foundation in geometry. One might claim that Chasles tried to extend as much as possible Newton's infinitesimal geometry on the basis of projective geometry.

These are the complex relations between methodology and ontology of mathematics and science in Chasles.

With particular regard to projective geometry, which I am discussing in this chapter, the *Aperçu* can be considered the text in which all the results of the first epoch, one might say "the heroic age", of projective geometry are summarized and generalized. It is an identity card of this epoch. Although there is not a complete caesura between ~~a~~ geometry before the *Aperçu* and ~~a~~ geometry after this publication, such book marks a strong periodization in the history of projective geometry. Since most of the foundational concepts created by Chasles were developed before and in the *Aperçu*, I will basically concentrate on Chasles' researches before his monumental work and in it. My aim is to clarify in what sense projective geometry is foundational for the whole geometry: for, in Chasles' opinion, there is the possibility to reconduct the metrical properties within a projective context.

2.1 Chasles and His Use of Polarity

The theory of the reciprocal polars plays a decisive role within Chasles' foundational programme because a particular polarity, the parabolic transformation, allowed him to establish the metric-graphical properties on the basis of those that are purely graphical.

Therefore, after an analysis of the first contributions by Chasles to geometry, to which I devote Sect. 2.1.1, four themes have to be addressed:

a) Chasles' foundational programme for graphical properties in the plane
b) The dependence of the graphical properties in the plane on the graphical properties in space
c) Metric-graphical properties and parabolic transformations
d) The extension of the foundational programme to algebraic curves and surfaces of any degree

I will concentrate on these subjects in Sects. 2.1.2–2.1.5.

2.1.1 Antecedents

Although it is always dangerous to claim that a specific work is the first one published by an author, given that some preceding contributions can perhaps be found, Chasles' first publication is likely a very brief note belonging to the theory of transversals and concerning a property of a skew quadrilateral in connection with the one-sheeted hyperboloid. This work appeared in the second volume of the *Correspondance sur l'École Royale Polytechnique* in 1813 (Chasles, 1813). In the following issue of the same journal, Chasles published four contributions (Chasles, 1814a, 1814b, 1816a, 1816b). The first and the fourth ones belong respectively to the theory of transversals and to some theorems on second-degree surfaces enunciated by Monge. The most interesting (and the longest) of these contributions is the third one, which also has significant connections with those two theories. Here, with a classical analytical treatment (it is not by chance that his work appeared under the rubric *Géométrie Analytique*), Chasles discovered some properties of the diameters of an ellipsoid. The most remarkable section of this paper is an appendix entitled "Application à l'ellipsoïde de différentes propriétés de la sphere" (Chasles, 1816a, 326–328).

Given a three-dimensional orthogonal coordinate system (x,y,z), Chasles developed a transformation (he spoke of *deformation*, ibid., 326) of this kind:

$$X = \frac{ax}{\theta}; Y = \frac{by}{\theta}; Z = \frac{cz}{\theta},$$

so to have

$$x = \frac{\theta X}{a}; y = \frac{\theta Y}{b}; z = \frac{\theta Z}{c}.$$

Chasles formulated one of his leading ideas: the idea to use the transformations to transfer some properties of a figure to another figure.

As he claimed: "This transformation permits one to apply to a surface some theorems characterising a different surface of the same degree" (ibid., p. 326). The surface in the coordinates (X, Y, Z) is called *derived* (*dérivée*).

Therefore, given the sphere $x^2 + y^2 + z^2 = \theta^2$, its derived surface is the ellipsoid

$$\frac{X^2}{a^2} + \frac{Y^2}{b^2} + \frac{Z^2}{c^2} = 1.$$

At this point, Chasles identified a series of properties concerning the contacts between figures (i.e. graphical properties) which can be transferred from a sphere to the derived ellipsoid. In particular:

a) The derived curve of a circle on the sphere is a plane curve.
b) The derived surface of a cone tangent at the sphere is a cone tangent at the ellipsoid. This means that the contact curve between a cone and an ellipsoid is a plane curve.
c) Given two circles in the sphere, two cones pass through them. Therefore, given two plane curves on an ellipsoid, two cones pass through them.
d) The derived curve of a great circle of a sphere belongs to a diametral plane of the ellipsoid.

Then Chasles argued that several properties concerning the theory of transversals, which hold for the sphere, are applicable to the ellipsoid (ibid., 326). In particular, he proved that all the spheres have as derived surfaces similar and similarly positioned (in what follows s.s.p.) surfaces. Thence, he continued: "From this, it follows that the theorems relative to the contacts of spheres are also valid for s.s.p. ellipsoids" (ibid., p. 327).

This memoir was located by Chasles within Carnot's theory of transversals and within Monge's descriptive geometry.

In this chapter, there is no reference to the theory of pole and polars. However, the idea to study the properties of the s.s.p. quadrics and conics considering them in the light of a transformation, which transforms a sphere in an ellipsoid, exists. We will see that the idea of transformation, already crucial in this early memoir, pervades many of the results—and many of the foundational results—obtained by Chasles in the years 1827–1830. This is the reason why this early memoir is significant in the context analysed here.

2.1.2 *The Great Memoirs of the Period 1827–1829: Pure Geometry and the Theory of Reciprocal Polars*

For more than 10 years Chasles did not write any other essays.[12] In contrast to this, from 1827 onwards, he began to write a series of profound memoirs that show how thoroughly he was following the progress of modern geometry. The panorama was different from what it had been in the period 1815–1817: projective geometry was no longer a discipline in an almost gestational phase; it had become the leading branch

[12] As far as I know, the only publication of Chasles in the period between 1816 and 1826 is an appendix to Hachette (1817); see Chasles (1817).

of modern geometry. If Poncelet can be considered as one of the main inspirational sources of the new movement, the train of thought developed mostly by French and German mathematicians, some of whom were mentioned in the Introduction, was so vast that a summary of the results obtained in these years is impossible here. The theory of polar reciprocity was one of the leading subjects. Chasles enriched it with new ideas which concern several topics: contributions to the theory of polar reciprocity; an extension of the theory of transformations beyond polar reciprocity; and the foundation of metric properties on graphical properties.

Chasles' return to mathematical research is attested by three papers published in the 18th issue of the *Annales de Mathématiques pures et appliquées* (Chasles, 1827–1828a, 1827–1828b, 1827–1828c). The last two memoirs are more interesting and hence I will focus on them.

The former (1827–1828b) is fundamental because it marks a series of research lines that influenced a great part of Chasles' subsequent production. At the beginning of his essay, Chasles reminds the reader of his former contribution—analysed above—on s.s.p. quadrics and conics. This subject remained one of his principal fields of research. Its insertion within the theory of reciprocal polars was aimed at shedding new light on the subject and on the potential of the theory. Chasles pointed out that the directions of this research have been marked by Poncelet's works (Chasles, 1827–1828b, 277–78). Chasles' idea is to construct a precise frame within which it is clear which properties of the circles can be extended to the conic sections. It is well known that Poncelet developed this field of research, but it is also a well-established fact that his brilliant and new results are not always completely precise, although almost always correct.

In his essay *Mémoires sur les propriétes des systemes de sections coniques situées dans un même plan* (Chasles, 1827–1828b) purely graphical properties of systems of conics are analysed. There is no reference to metric-graphical properties. This contribution is significant within Chasles' foundational programme because it represents a generalization and a systematization of the results obtained by other mathematicians. The entire situation can be interpreted this way: before dealing with the reduction of metric properties to graphical ones, Chasles tried to offer a general and satisfying treatment of purely graphical projective geometry. He aimed at supplying a standard form to this discipline so as to create a mathematical field with clear, well-defined and general properties, within which it would have been possible to develop his foundational programme. It is therefore necessary to explain the main concepts, methods and results introduced by Chasles in this memoir.

He explained that his work belongs to the context of Poncelet's theory of reciprocal polars[13] (ibid., 277–278).

The basic idea is to extend some properties of the systems of circles to the other conics.

[13]The theory of reciprocal polars plays a fundamental role in Poncelet's *Traité* (Poncelet, 1822). Poncelet (1827, 1828b, 1829a) offers a further refinement of this subject.

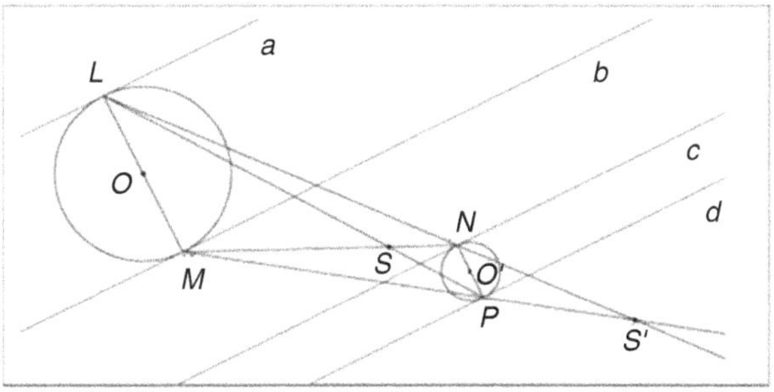

Fig. 2.1 Straight lines *a, b, c, d* are the four parallel tangents. The contact points *L, M, N, P* form the quadrilateral whose opposite sides are two parallel diameters of the circles. The points *S* and *S′* are the centres of similitude

Within this perspective, relying on a series of results obtained by Gaultier and other mathematicians (see Gaultier, 1813; Anonymous, 1822–1823; Steiner & Gergonne, 1826–1827), he referred to the most important features of a system of circles. The fundamental concepts on which Chasles' treatment is based are four:

1) The centre of similitude of two circles
2) The radical axis of two circles
3) Similitude polars of two circles[14]
4) The radical centre of three circles.

The concepts 1), 2) and 4) are well known, but it is interesting to examine the way in which they were introduced at that time because it is a mark of the particular climate of the then-pure geometry, a great part of which was dominated by the theory of polars. Chasles considered two circles in any position and any four parallel tangents (see Fig. 2.1).

The contact points form a quadrilateral of which two opposite sides are the diameters of the two circles. It is possible to prove that the contact points of the diagonals and of the two other opposite sides are constant for two given circles. These two points are the *centres of similitude* of the two circles. When the two circles are mutually external, the two centres of similitude can also be found as two intersection points of the four tangents common to the two circles (see Fig. 2.2).

However, the definition used by Chasles is more general. If, from one of the centres, a mobile secant is drawn to the two circles, the four tangents drawn from the four points where the secant cuts the two circles form a parallelogram. The opposite vertices *A* and *C* of the parallelogram are, of course, the poles of the secant relative to the two circles. Therefore, because of the reciprocity pole–polar, the points *A* and *C* belong to the same straight line as the centre of similitude *S* (see Fig. 2.3).

Furthermore, the locus described by the point *A*, while moving the secant, is the polar of *S* relative to the circle *c*. Analogously, the locus of *C* is the polar of *S* relative to *c*. Finally, the two other vertices of the parallelogram will describe a fixed line, independent of the similitude centre. This line is called the *radical axis* of the two circles. Chasles pointed out that, when the two circles intersect, the radical axis is

[14] Given two circles, their similitude polars are the four polars of the two centres of similitude with respect to the two circles.

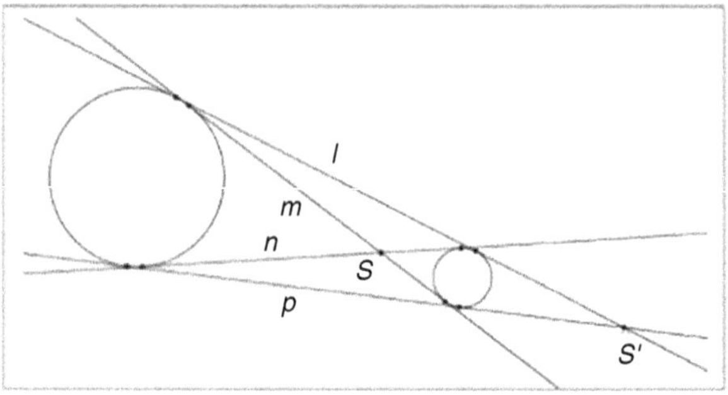

Fig. 2.2 Straight lines *l, m, n, p* are the four common tangents to the two circles, whose contact points with the circles are indicated by a dot. The points S and S' are the centres of similitude

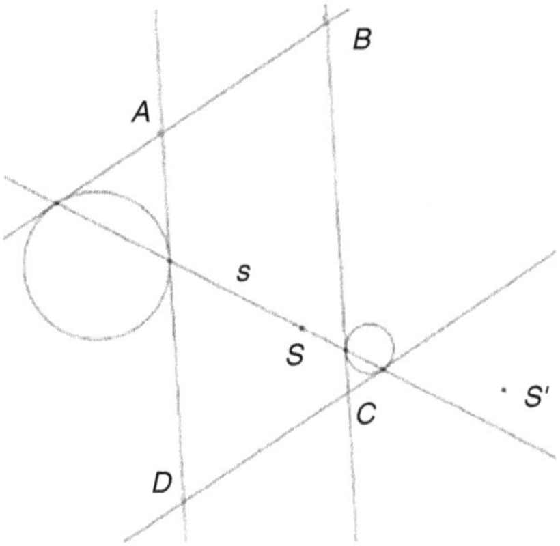

Fig. 2.3 Point S is the centre of similitude from which the (variable) secant s is drawn. The points A, B, C, D are the vertices of the parallelogram constructed by Chasles. The point S' is the other centre of similitude

their common chord, but the definition holds for any position of the circles (ibid., p. 279).[15]

The other fundamental concept is that of *similitude polars*. As explained, they are the four polars of the two centres of similitude with respect to the two circles (see Fig. 2.4).

These four straight lines are mutually parallel and are symmetric with respect to the radical axis. The two other introductory concepts concern a system of three circles for which the *radical centre* and the *similitude axes* are defined.

After these considerations, Chasles clarified the aim of his work: *to prove that the graphical properties of the systems of two or three circles can be extended to systems of two or three s.s.p. conics*. He developed his reasoning by means of a *purely geometrical method*, resorting to the stereographic projection (ibid., 280). To point out the *foundational character* of his work, Chasles coined a new word to indicate a system of similar and similarly positioned figures; this word is *homothetic*. Thus, two homothetic conics have two centres of similitude, which are the intersection points of two opposite sides and of the diagonals of a quadrilateral whose vertices are

[15]In the explanation presented in the running text, I have directly referred to Chasles' argumentation, but I have added the figures and the connected letters for the mathematical objects.

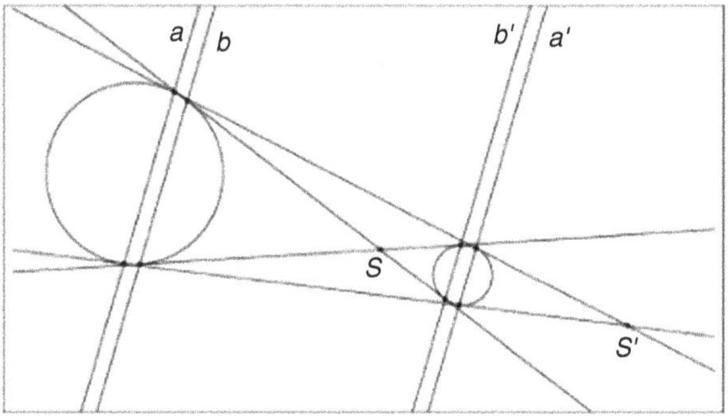

Fig. 2.4 Similitude polars: they are the four parallel lines a, b, b', c'. They are the polars of the two centres of similitude S and S' with respect to the two circles. In this diagram, the case in which the two circles are mutually external is drawn

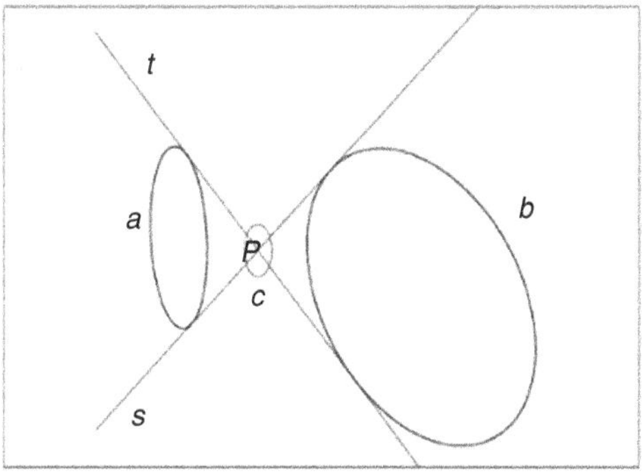

Fig. 2.5 The situation described by Chasles is here represented: the curves a and b are the two coplanar conics; the straight lines s and t are their common tangents intersecting in the point P; the conic c with centre P is that with respect to which Chasles will construct the polarity

the intersection points with the two conics of any four parallel tangents. The centres of similitude are also the intersection points of the four common tangents to two conics. These points are always real, although some of the common tangents might be imaginary. Analogously, the radical axis, which is parallel to the similitude polars, is always real, though the intersections between the conics can be imaginary. Chasles stressed the fundamental property that, although two coplanar conics have four real or imaginary intersection points, if they are homothetic, two of these points are at infinity (ibid., 281). In the language of projective transformations, which was developed a few years later and to which Chasles himself contributed, this means that a homothety is a homology whose axis is the line at infinity of the plane and whose centre is a proper point.

The concepts just expounded represent the first element in Chasles' method, the second one being the *theory of reciprocal polars*. If one considers (see Fig. 2.5):

a) Two coplanar conics a and b.
b) Two common tangents to the two conics, t and s, and their intersection point P.
c) A centred coplanar conic c, whose centre is P.

Then it is possible to construct the polar figure to that given with respect to the conic c. Such a figure is made up of two conics whose points are the poles of the tangents at a and b with respect to c. The contact points of a and b will have as polars the tangents at different points of the second system of conics.[16] Any planar polarity is an involutory correspondence, i.e. a bijective correspondence of points with straight lines. Therefore, the intersection points of *a* and *b* have as their polars the common tangents to the conics in which *a* and *b* are transformed by the polarity. Vice versa, the intersection points of these tangents are the poles of the common tangents of *a* and *b*. Since, by hypothesis, two common tangents pass through the centre of the directive conic[17] *c*, their poles with respect to *c* are at infinity. This means that the conics polar to *a* and *b* have two of their four intersection points at infinity. From which this theorem, which Chasles calls *fundamental*, follows:

> Any two coplanar conics placed in any position and considered with respect to a directive conic whose centre is at the intersection point of two common tangents to the two conics have two homothetic conics as reciprocal polar figures.[18]

To introduce the further essential concept Chasles created in this memoir, let us refer to a theorem he demonstrated, whose proof is quite easy. Chasles reported this theorem in two columns, according to the custom required by Gergonne for his *Annales* to indicate propositions to which duality is applicable. I will refer only to one of the two dual theorems because this is relevant to my aim (ibid., p. 284, see Fig. 2.6):

Rotate a mobile straight line in the plane of any two conics around the intersection point of two common tangents [*s* and *t*] of the two curves, so as to cut them in four points [*A, B, C, D*]. Then, the tangents [*a, b, c, d*] to the conics at these points are such that the four intersection points of the two tangents to a curve with the two tangents to the other curve *describe a system of two fixed straight lines*. These lines are the common chords of the two curves if they mutually intersect.

Chasles highlighted that when the two conics do not intersect the two fixed lines of the theorem are those called by Poncelet "common ideal chords" (ibid., p. 285). On the other hand, we have seen that the definition of radical axis is independent from the fact that two circles mutually intersect. The reader can easily check that the conditions of the theorem are exactly those which define the radical axis for two circles or for two homothetic conics. Given this analogy, Chasles coined a new locution for these two fixed straight lines and called them *axes de symptose*.

For the intersection points of the common tangents of two conics chosen on the basis of the stated criteria, Chasles continued to use Poncelet's (as well as the current) locution: *centres of homology*. Referring to the polarity described to introduce the fundamental theorem, Chasles called *primitive system* the system of the two arbitrarily placed conics, and he called *derived system* its polar system made up of the two homothetic conics. Therefore, the centre of homology of the primitive

[16] In this case, too, I have introduced letters to denote mathematical objects, which are not present in Chasles, but which are helpful for the reader.

[17] The directive conic is the conic with respect to which the polarity is constructed.

[18] Ibid., p. 282: "Deux coniques quelconques, situées d'une manière quelconque dans un même plan, et rapportées à une conique directrice ayant son centre au point de concours de deux tangentes communes aux deux courbes, ont pour polaires réciproques deux coniques homothétiques".

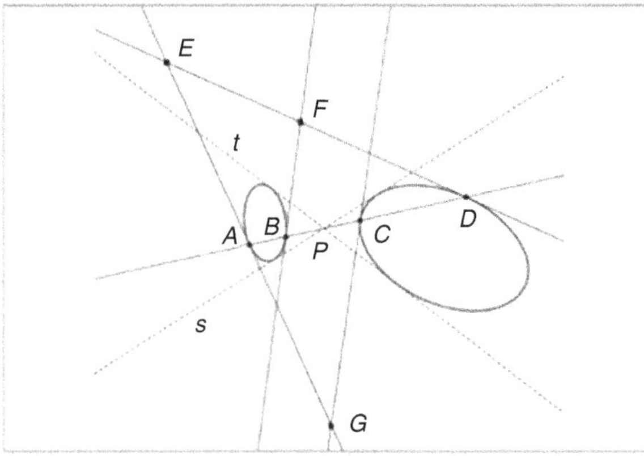

Fig. 2.6 The straight lines s and t are two common tangents at the two conics, being P their intersection point. The straight line through A, B, C, D is the (variable) secant to the two conics. The tangents passing through the points A, B, C, D are, respectively, a, b, c, d. (These letters are missing in the figure to avoid confusion.) The points E and G represent the intersection points of the tangent a with the tangents c and d. Therefore, the line EG is one of the fixed straight lines of which Chasles spoke. The point F is the intersection between the tangents b and d. As often in projective geometry, the intersection point of b and c is out of the diagram. (It might be at infinity.) I have represented a case in which the two conics do not mutually intersect

system is the pole of the radical axis of the transformed system and the *axes de symptose* of the primitive system are the polars of the two centres of similitude of the second system.

At this point, the quadrilateral theorem for two conics assumes the following form:

> Any quadrilateral, whose sides touch two conics in the four points where they are cut by a straight line passing through their centres of homology, has its four vertices on the two *axes de symptose* of these curves, provided that no vertex coincides with the intersection point of two tangents to the same curve.[19]

It is easy to prove that if two conics intersect in four real points, they have six centres of homology and six *axes de symptose*. The centres of homology are the six intersections (considered two by two) of the four common tangents, while the *axes de symptose* are the six lines which join two by two the four intersection points of the two conics (see Figs. 2.7 and 2.8).

Conjugate centres of homology are those which do not belong to the same tangent and *conjugate axes de symptose* are those which do not concur in a point common to the two curves. Therefore, there are three systems, each one made up of two conjugate centres of homology, and two conjugate *axes de symptose*.

[19]Ibid., pp. 285–286: "Tous les quadrilatères dont les côtés touchent deux coniques aux quatre points où elle sont coupées par un droite menée arbitrairement par leur centre d'homologie ont leurs quatre sommets sur les deux axes de symptose de ces deux courbes, pourvu qu'on ne prenne, pour aucun sommet, le point de concours de deux tangentes à la même courbe".

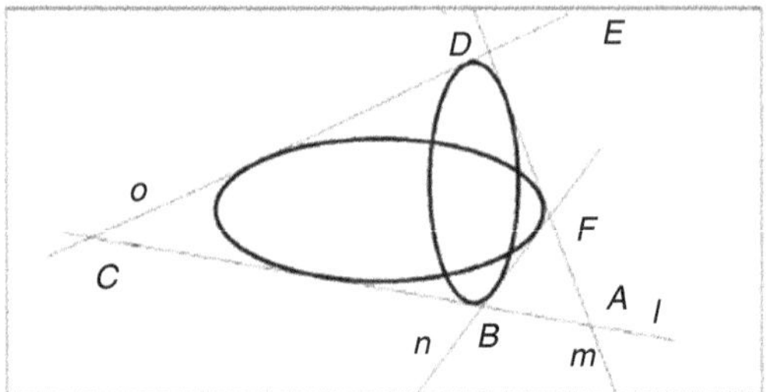

Fig. 2.7 Centres of homology of two conics intersecting in four real points. The straight lines *l, m, n, o* are the common tangents at the two conics. The points *B* (intersection of *n* and *l*) and *D* (intersection of *m* and *o*) are two conjugate centres of homology, so as *C* (intersection of *l* and *o*) and *F* (intersection of *m* and *n*). The other two conjugate centres of homology are *A* (intersection of *l* and *m*) and the intersection of *n* and *o*, which does not appear in the diagram

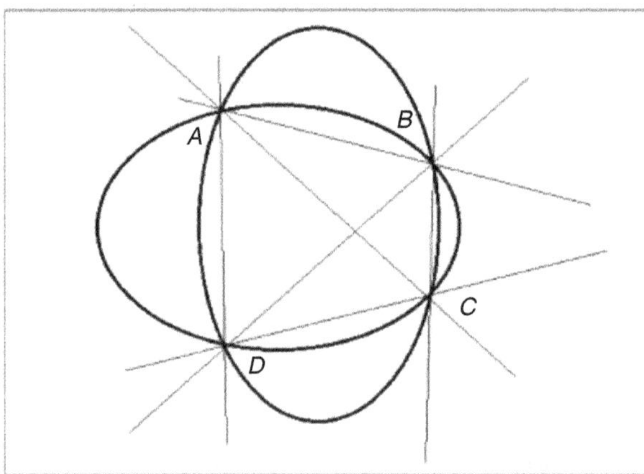

Fig. 2.8 *Axes de symptose* of two conics intersecting in four real points. The three couples of conjugate *axes de symptose* are (1) lines *AB* and *CD*; (2) lines *AC* and *BD*; and (3) lines *AD* and *BC*. For they do not mutually intersect in a point belonging to any of the two conics

Chasles showed that in any position of the two conics, at least two centres of homology and two *axes de symptose* are real. He also illustrated very briefly the analytical reasons why this is the case (ibid., p. 286).

Chasles clarified why the *axes de symptose* are so important. His explanation shows his very general view of geometry: the *axes de symptose* allow one to know the intersection points of the two curves. Chasles proved that these points permit one to solve all the determined problems having three or four solutions. Therefore, if it were possible to construct the *axes de symptose* by means of straight edge and compass, this would mark a considerable progress in geometry.

It is appropriate to highlight that one of the results obtained by Chasles through the expounded apparatus extends an important property proved by Steiner (Steiner & Gergonne, 1826–1827, 309). He had demonstrated that the straight lines joining the radical centres of three circles to the poles of one of the similitude axes relative to these circles cut the three circles at their contact points with a fourth circle, which touches all of the three. Chasles claimed that this property holds if one replaces the locution "three circles" with "three homothetic conics". Thus, it is possible to solve the following problem (Chasles, 1827–1828b, p. 294), which I refer to in the dual notation used by Chasles:

Given three conics having a common centre of homology, describe a fourth conic touching all of the three and having with them the same homology centre.	Given three conics having a common *axe de symptose*, describe a fourth conic touching all of the three and having the same *axe de symptose*.

In this contribution, Chasles placed the bases of his foundational research programme in pure projective geometry. As a matter of fact, he began to develop a line of thought based on projective transformations, which would comprise a great part of his subsequent productions. Furthermore, he was able to reduce most of the graphical properties concerning the conic sections to a few concepts, basically those of the centre of homology and *axe de symptose*. He would later extend this kind of research to spatial geometry, to graphical-metric properties and, in some cases, to curves and surfaces of degree greater than 2. Certainly, he was not an isolated thinker, but entered into the great field of research that was projective geometry.

2.1.3 Dependence of the Graphic Properties in the Plane from the Graphic Properties in Space

Chasles' foundational programme continues by showing the dependence of graphical properties in the plane on spatial properties and constructions. The context is once again that of pure geometry, for the memoir I will analyse appears under the rubric "Géométrie pure" in volume 18 of Gergonne's *Annales* (Chasles, 1827–1828c). In his previous contribution, Chasles had shown that important graphical properties of a system of two or three conics can be reduced, by polar reciprocity, to those of a system of two or three homothetic conics and these, in turn, to those of a system of two or three circles. But this reduction is not the fundamental one, for this depends on the fact that both circles and conics can be obtained as a section of a cone. Of course, this property was well known from Greek mathematics onwards, but Chasles illustrated all the necessary steps that connect the section of a cone to the specific graphical properties concerning the homothetic conics.

He observed that an important contribution to clarifying the situation was due to Dandelin (Dandelin in Quetelet, 1825; Dandelin & Gergonne, 1825–1826; Dandelin, 1827), but that the latter had examined only the stereographic projection of circles belonging to spheres, whereas a complete analysis should start with a generic surface of the second degree (Chasles, 1827–1828c, p. 305). Chasles considered three properties: let us suppose that a second-order surface E (e.g. an ellipsoid; see Fig. 2.9) be given.

Then:

a) If we cut E by two planes α and β, two conics C and C' will be obtained. Two conic surfaces S and S', which are the projections of the two conics from two

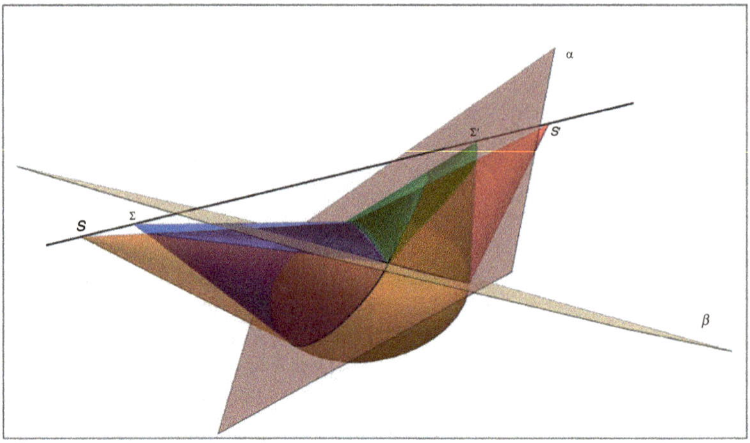

Fig. 2.9 Planes α and β cut the ellipsoid in two conics C and C'. (A part of their perimeter is indicated in bold.) The cones whose vertices are Σ and Σ' are the cones touching the ellipsoid along C and C'. The cones, whose vertices S and S' are collinear with Σ and Σ', cut the ellipsoid along C and C'. The line $S\ \Sigma\ \Sigma'\ S'$ is the polar with respect to the ellipsoid of the intersection line of α and β

points of space can be constructed. On the other hand, two conic surfaces Σ and Σ' exist that are tangent to E along the conics C and C'. If the vertices of the so-constructed cones belong to the same straight line, this line is the polar in relation to E of the line in which the planes containing C and C' mutually cut.

b) If the plane γ containing the vertex of one of the two cones, e.g. S, and the intersection line of the planes containing C and C' is drawn, then the polar of this plane with respect to S is exactly the straight line of the four vertices. Furthermore, it is possible to prove that this implies that all the sections made in S parallel to γ have their centres on the line joining the four vertices.

c) It is evident that all the sections made in E and in the cones S or S' parallel to the planes of the conics C and C' will produce similar and similarly placed conics as C and C' (ibid., 306).

These were known theorems.

Chasles' reasoning continued like this: suppose that one of the two curves C' and C, e.g. C', becomes a point P. Its plane will become tangent to E in P. Therefore, only the cone S will remain. The vertex of this cone will coincide with P.[20]

My explanation: Chasles imagined that the plane containing C' moves until it becomes tangent to E. During this movement, the vertices of the cones S and S'—which have to be imagined to be reciprocally constrained—converge to P. At the end, the cone S' has disappeared and become the tangent plane, while the cone S has remained a cone with its vertex at P. Therefore, in this new position all the planes

[20] In this theorem, the letters are used by Chasles. They are not an addition of mine.

parallel to the plane of the curve C' (the tangent plane in P) will continue to cut the cone S in homothetic conics.

Chasles now reconsidered (see previous item b) the plane passing through the vertex of S' and the intersection line of the planes containing C and C'. It has become the tangent plane and its polar with respect to S, which contains the four vertices, is the line joining P to the vertex of Σ. This line contains all the centres of the sections made in S parallel to the tangent plane in P.

Therefore, a theorem follows, which Chasles expressed in a language typical of descriptive geometry and which he calls "[...] a theorem that one can regard as fundamental":

> Let the eye be placed at any point of a second-degree surface. Let the plane of the table be parallel to the tangent plane of this surface at such a point:
>
> 1. All the plane curves drawn on the second-degree surface are projected on the table in mutually similar and similarly placed curves. They are also s.s.p. in relation to the intersection of the surface with the plane of the table.
> 2. The projections of these different curves on the table have as their centres of projection on the table itself the vertices of the cones circumscribing the second-order surface according to these same curves.[21]

Then, Chasles proved the reciprocal theorem:

> Reciprocally, given in a plane any number of homothetic conics, they can be considered as the stereographic projection of the same number of plane curves drawn on a second-degree surface. Their centres will be the projections of the vertices of the cones circumscribing the surface along these curves.[22]

These two propositions are fundamental for showing that *the graphical properties of the systems of conics are a direct consequence of general graphical properties of the surfaces of second degree, their mutual intersections, projections from points, intersections with planes and so on.*

These two theorems, and especially the former, were considered seminal by Chasles. In an important memoir that appeared in tome 19 of Gergonne's *Annales*

[21] Ibid., p. 307: "L'oeil étant placé en un quelconque des points d'une surface du second ordre, et le plan du tableau étant parallèle au plan tangent à cette surface en ce point; 1. Toutes les courbes planes, tracées sur la surface du second ordre dont il s'agit, se projeteront sur la tableau suivant des courbes semblables et semblablement situées, tant entre elles que par rapport à l'intersection de la surface du second ordre avec le plan du tableau; 2. Les projections de ces diverses courbes, sur le plan du tableau, auront respectivement pour centres les projections, sur ce tableau, des sommets des cônes circonscrits à la surface du second ordre suivant ces mêmes courbes". Chasles reminded the reader that he had proved the first part of this theorem in Chasles (1814b) and that he had proved analytically the second part (Chasles, 1817, in Hachette, 1817).

[22] Ibid., pp. 308–309: "Réciproquement, des coniques homothétiques étant tracées dans un même plan, en tel nombre qu'on voudra, on pourra toujours les considérer comme les projections stéréographiques d'autant de courbes planes tracées sur une même surface du second ordre; et leurs centres seront alors les projections des sommets des cônes circonscrits à cette surface suivant ces même courbes".

(Chasles, 1828–1829b), he gave a slightly more general formulation of the former theorem, stated like this:

> Let a series of second degree surfaces be given. Let these surfaces be inscribed in the same second degree surface and the eye placed at any point [P] of the circumscribing surface. Let the plane of the plane table be parallel to its tangent plane in [P], then.
>
> 1. The projections of apparent contours of the inscribed surfaces onto the table are homothetic conics.
> 2. The centres of such conics are the projections of the poles of the planes[23] of the contact lines of the inscribed surfaces with the circumscribing one, when the polarity is considered with respect to this surface or with respect to any other inscribed surface. [Chasles proved that such poles are the same in the two polarities.][24]

The proof, developed by purely synthetical means, is a remarkable example of the projective methods applied to this kind of propositions (Fig. 2.10).

For Chasles observed: be the second-degree surfaces s, s', s'', ... inscribed in the second-degree surface S. Thence, both the cone C which determines the apparent contour of s and the surface S are circumscribed to s. Therefore, they mutually cut along two plane curves, whose planes pass through the intersection straight line of the plane of the curves along which C and S touch s.[25] According to the hypothesis of the theorem, the vertex of C, which is the eye, is in a point of S. Ergo, one of the intersections between C and S is a point and the plane of this intersection is nothing but the tangent plane to S at the vertex of C. The cone C will cut S according to a further plane curve whose plane, as well as the tangent plane, will pass through the straight line along which the planes of the two contact lines of s with S and C pass. The cone C cuts the surface S along a plane curve, whose section will be a conic homothetic to the section of the surface S with this plane.[26] But this section is the perspective of the apparent contour of the surface s. Consequently, the perspective of all the apparent contours of the surfaces s, s', s'', ... are homothetic with the section of the surface S with the plane of the table. Thence, they are mutually homothetic. In this manner, the first part of the theorem is proved.

With regard to the second part of the proof Chasles argued: as seen in the first part, the plane α of intersection between the cone C with the surface S, the plane β of the contact line of the two surfaces S and s and the plane γ tangent to S at the eye

[23] It is worth recalling here the concept of pole of a plane with respect to a quadric: two points A and B are called conjugate or reciprocal with respect to a quadric when they divide harmonically the two intersections of the straight line AB with the quadric. Given a quadric, the locus of the conjugate points of a given fixed point P with respect to the quadric is a plane which is called *polar plane* of the point P (which is its *pole*) with respect to the quadric.

[24] Ibid., p. 158: "Plusieurs surfaces du second ordre étant inscrites à une même surface de cet ordre, l'oeil étant placé en un quelconque des points de cette dernière, et le plan du tableau étant parallèle à son plan tangent en ce point; 1. Tous les contours apparens des surfaces inscrites seront, en perspective, des coniques homothétiques; 2. Les centres de ces coniques seront les projections des pôles des plans des lignes de contact de ces surfaces avec celle à laquelle elles sont inscrites, pris par rapport à cette surface, ou respectivement par rapport à chacune des autres".

[25] This property had been proved in Chasles (1816b, p. 339).

[26] This property had been proved in Chasles (1827–1828c, p. 307).

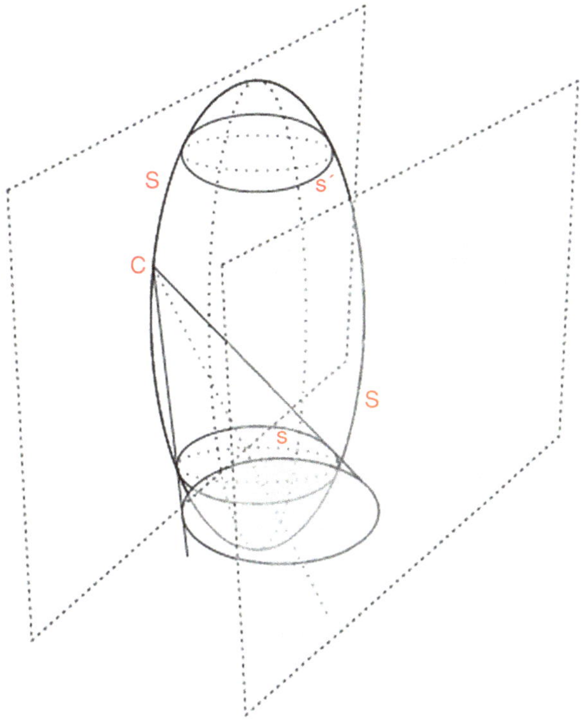

Fig. 2.10 Reconstruction of the figure described by Chasles. To make the construction sufficiently clear, two of the second-degree surfaces s, s', s''... inscribed in the second-degree surface S are drawn. They are s and s'. The cone with vertex C is drawn only for the surface s

(which is the vertex of C) are collinear. Therefore, because of the reciprocal character of the polar transformation, the poles of the first two planes α and β with respect to the surface S are on a straight line passing through the eye. The centre of the section of the cone C with the plane of the table is on the straight line joining the eye[27] with the pole of α. Therefore, such a centre is also on the straight line joining the eye with the pole of β, i.e. the pole of the plane of the contact line of the surfaces S and s. Evidently this pole is the same that is considered with respect to the surface S or to the surface s. Consequently, the second part of the theorem is also proved.

The proof of this theorem is evidence of the edifice Chasles was founding, an edifice concerning the fundamental notions and constructions of projective geometry, based on the results of other mathematicians as well as on the refinement and extension of his own previous results.

With regard to Chasles' foundational programme, there is another important consideration in this contribution: he began to develop a train of thought that will converge in his memoirs on duality and homography published in the *Aperçu*. This research would lead him to consider projective transformations from a more general perspective in which the polarity itself is included. Therefore, the idea of reducing any projective consideration to the concept of polar reciprocity is not necessary. This was one of the most significant results obtained by Chasles in his memoirs on the

[27] For the proof of this property, see Chasles (1827–1828c, p. 308).

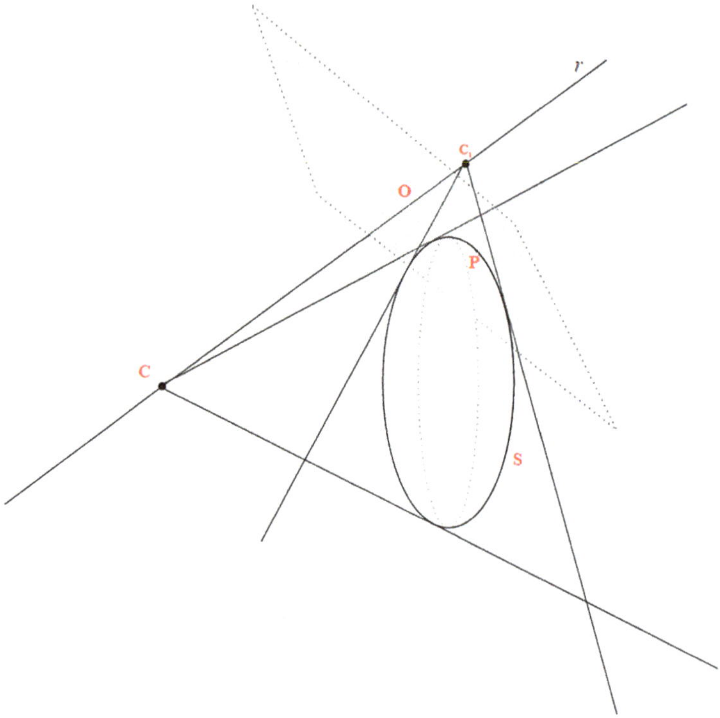

Fig. 2.11 The points C and C_1 are the vertices of the two cones circumscribing the second-degree surface S. The plane α is tangent to S at P. The intersection point O between the straight line CC_1 and α is the second centre of homology considered by Chasles

polarity and homography, granted that the theory of reciprocal polars was considered as the cornerstone of projective geometry. At this stage, Chasles proved a theorem by means of which the reduction of a homological transformation to the method of reciprocal polars is not necessary anymore. The theorem, defined as "remarkable" by Chasles, states that:

> Let several cones, whose vertices are collinear, be circumscribed to a second-order surface. All the planes tangent to this surface cut these cones in conics having two common centres of homology. Such conics, hence, will fulfil all the properties characterizing the conics inscribed in a quadrilateral.[28]

The proof runs like this: (Fig. 2.11) given two cones circumscribed to a second-degree surface, the plane tangent to this surface at any point P cuts the two cones

[28]Ibid., p. 165: "Si l'on circonscrit à une même surface du second ordre plusieurs cônes dont les sommets soient situés sur une même droite quelconque, tout plan tangent à cette surface coupera ces cônes suivant des coniques qui auront deux centres d'homologie communs, et qui jouiront conséquemment de toutes les propriétés d'une série de conique inscrites à un même quadrilatère".

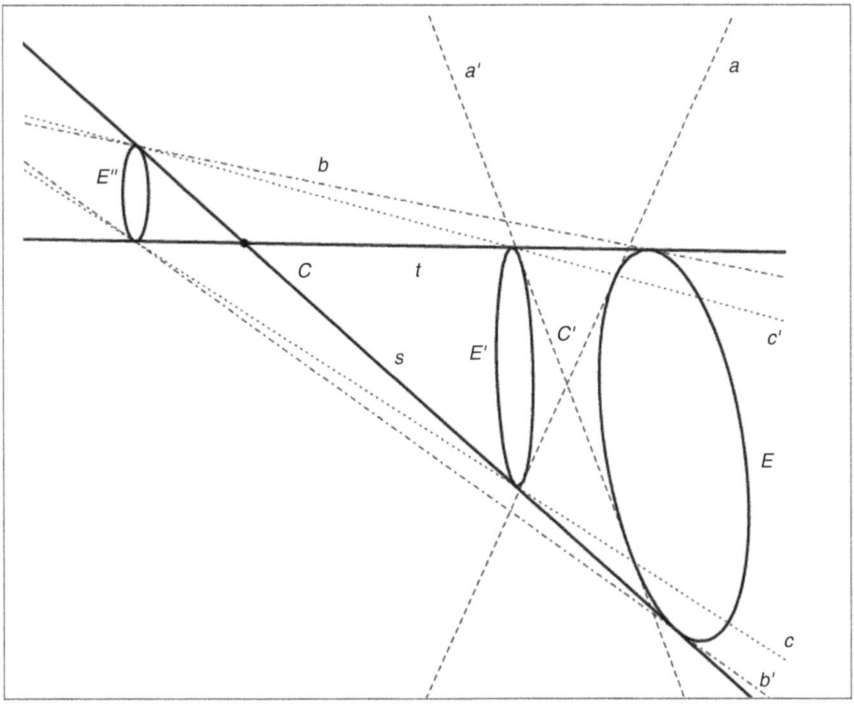

Fig. 2.12 The curves E, E', E'' are the three conics to which Chasles referred. The straight lines s and t are the tangents common to the three conics. Their intersection C is the centre of homology common to the three conics. The point C', which is the intersection of the other two tangents to the conics E and E', is one of the three centres of homology conjugate to C. The other two are obtained as the intersection of the straight lines b and b', which are the other two tangents common to the conics E and E'' (centre of homology C'') and as the intersection of the straight lines c and c', which are the other two tangents common to the conics E' and E'' (centre of homology C'''). The four centres of homology are collinear

along two conic sections which have P as their centre of homology.[29] The second centre of homology of the two conics is the intersection between the tangent plane and the straight line joining the vertices of the two cones, because, through this straight line, it is possible to draw two planes tangent to both cones. Ergo, from the point where such a straight line cuts the plane of the two conics, it will be possible to draw the common tangent to the conics.

Chasles claimed that this theorem offers a new method for deriving the general properties of any two conics and those of three conics with the same centre of homology. He offered an example of this: let there be given three conics having the same centre of homology C. Then, consider the conics two by two; they will have a second centre of homology conjugate to C (Fig. 2.12). The system of these four

[29] The proved theorem is indicated as Proposition 13 in Chasles (1827–1828c). The property to which this note is referred is indicated by Chasles as Proposition 12 (ibid., p. 165).

centres of homology is collinear. For it is enough to consider these conics as the sections made in three cones circumscribing a second-order surface by a plane tangent to the surface. Their common centre of homology C is the contact point of the tangent plane, and the centres of homology conjugate with C are the points where this plane is cut by the straight lines joining two by two the vertices of the three cones. These lines belong to the plane determined by the vertices of the three cones. Therefore, the three centres belong to the intersection of this plane with the tangent plane; i.e. they are collinear.

This method, whose elements are:

(a) A second-degree surface;
(b) A plane tangent in a give*n* point of this surface;
(c) A series of cones intersecting the surface;
(d) Appropriate plane sections of these cones,

works independently of the theory of polar reciprocity, as Chasles' example makes clear. It can be the basis for the theory of homology. When using this method, the duality typical of the polar reciprocity cannot be applied and each theorem has to be proved independently (ibid., p. 166). In his next contributions, Chasles found a form of duality applicable to this method that enabled him to offer a more general and comprehensive formulation of some results obtained by Poncelet, Quetelet and Dandelin (ibid., pp. 167 and 168).[30]

2.1.4 Towards the Metric-Graphical Properties: The Theory of Reciprocal Polars and Parabolic Transformations

Chasles continued his foundational programme by including the metric-graphical properties within his plan for projective geometry. This project was realized in two fundamental memoirs on the transformations of metric relations of figures (Chasles, 1829g, 1830a). In the brief and clear exposition which opens the first memoir, Chasles specifies that the graphical properties of the figures are deducible from the general rules of the theory of reciprocal polars. In his *Traité* (Poncelet, 1822) and in the celebrated memoir (Poncelet, 1829a) on the reciprocal polars, Poncelet had studied the metric properties of figures concerning angles and length of segments insofar as the properties included within the theory of transversals are considered. Chasles' intention was to extend the sphere of the metric properties treatable by means of specific graphical instruments. He reminds the reader that this programme is realizable using the theory of polar transformations, but also by means of a different theory (Chasles, 1829g, p. 282). Chasles refers specifically to the theory

[30]Poncelet (1822, p. 261), Dandelin and Gergonne (1825–1826). With regard to Quetelet, Chasles did not clarify the work to which he was referring. It is likely that he was referring to Quetelet (1826).

of the involution of six points. He had already outlined this theory in a previous memoir (Chasles, 1827–1828a), but in that work, too, after all, he had resorted to the theory of the reciprocal polars. That brief memoir did not have a general and foundational character, whereas the two contributions I am analysing are seminal works, because in them numerous properties are derived from a very limited number of principles, in fact, only from one principle.

In the first part of the first memoir, Chasles analyses the parabolic transformations of the plane figures. The necessary presuppositions are three well-known theorems of the theory of polars:

(1) The polar with respect to a parabola of a point at infinity is parallel to the parabola's axis.
(2) The poles of mutually parallel lines belong to a diameter of the parabola, i.e. to a line parallel to the axis.
(3) Any line at infinity has as its pole the point at infinity at the extremity of the axis of the parabola (Chasles, 1829g, 283).[31]

A remark on the condition 3) is necessary here: Chasles spoke of "Toute droite située à l'infini", but, obviously, any plane has its line at infinity. Therefore, until only plane geometry is analysed, it is sufficient to say "the line at infinity", not "any line at infinity".[32] The line at infinity is tangent to the parabola in the point at infinity of its axis, and, as for any tangent, its pole is the point where the tangent touches the curve. The fundamental principle of transformation is the following one:

> The polars referred to a parabola of any two points cut out on the parabola's axis a segment that is as long as the orthogonal projection on the axis of the segment straight line joining the two points.[33]

The proof runs like this (see Fig. 2.13):be given two points A and B. Draw through them the two perpendiculars c and d to the axis. The polars a and b of these two points will pass through the poles C and D of the two perpendiculars. The poles C and D are as distant from the vertex of the parabola as the lines c and d.[34] The distance between these two poles is, then, equal to the distance between the two lines. The distance of the two lines is exactly the orthogonal projection of the line

[31] From an intuitive point of view, the situation can be represented in this way: let us consider the vertex and the axis of a parabola. On the axis, take into account the points situated in the zone of the plane where there is no parabola. The polar of any of these points tends to the line at infinity when the point tends to the point at infinity of the axis.

[32] In this subsection, I will clarify Chasles' conception of the elements at infinity.

[33] Ibid., p. 283: "Les polaires de deux points quelconques, prises par rapport à une parabole, interceptent sur l'axe de cette courbe un segment qui est égal an longueur à la projection orthogonale sur cet axe de la droite qui joint les deux points".

[34] My explanation: this depends on the property that, given a point P belonging to the parabola's axis, if the two tangents to the parabola from P are drawn, the line connecting the two contact points, which is the polar of P, saws the axis in a point Q, which is as far from the vertex as P. This polar is perpendicular to the axis.

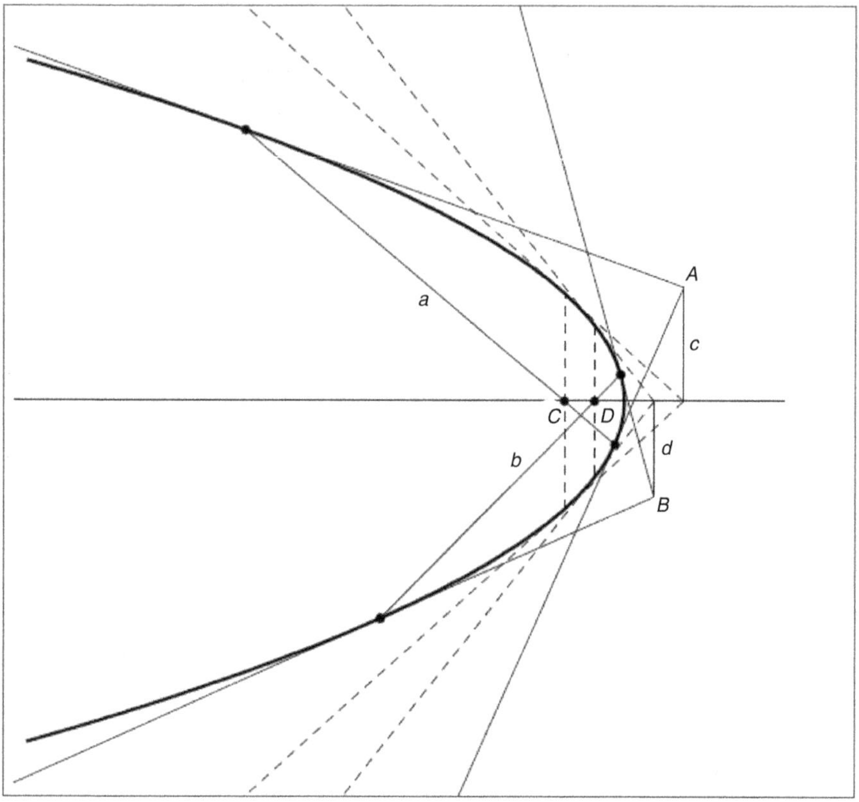

Fig. 2.13 Segments c and d, traced from the points A and B, are perpendicular to the axis. The straight line a is the polar of A (constructed by tracing the two tangents from A to the parabola). The straight line b is the polar of B. The pole C of the perpendicular c is determined as the intersection point of the polar a of A, point belonging to c, and the polar of another point belonging to c. I have chosen the point belonging to the axis. The same construction is applied to determine the point D, pole of d

joining A and B. Their polars pass through C and D, which proves the theorem (ibid., 283).

This simple principle is the key point of Chasles' reasoning, which is based on the following argument: be given a series of points A, B, C, D, E,..., belonging to a plane figure. Let $F(AB,CD, EF,\ldots) = 0$ be a relation between the distances of these points. (AB indicates the distance between A and B.) Let us make the polarity of this figure with respect to a parabola ("a parabolic transformation" in Chasles' language). Let $a,b,c,d,e,\ldots$, be the polars of A, B, C, D,.... Let α, β, γ, δ, ε, ..., be the points where the polars meet the parabola's axis X. Then, the following relations hold:

$$\alpha\beta = AB\cos(AB, X)$$
$$\gamma\delta = CD\cos(CD, X)$$
$$\dots$$

The relation $F(AB, CD, EF, \dots) = 0$ is transformed into the following one:

$$F\left(\frac{\alpha\beta}{\cos(AB, X)}, \frac{\gamma\delta}{\cos(CD, X)}, \frac{\varepsilon\varphi}{\cos(EF, X)}, \dots\right) = 0 \tag{2.1}$$

If in this equation, according to the problem one is dealing with, the cosines disappear, a relation expressing a general metric property of the polar figure will be obtained (ibid., 284). As Chasles claimed, this type of transformation is applicable to a vast class of problems.

Chasles provided several remarkable theorems proved with this uniform method. I will mention one of them with Chasles' proof, so that the reader is offered a clear picture of the way in which Chasles applied his method.

This theorem concerns the parabolic transformation of a theorem Chasles had proved in a previous memoir (Chasles, 1828b; see Fig. 2.14).

He had demonstrated that if a conic is inscribed in a quadrilateral *ABCD* and a tangent *t* is drawn cutting the opposite sides *AB* and *CD* in two points *M* and *N*, then, for any given tangent the relation

$$\frac{AM}{BM} = K\frac{DN}{CN} \tag{2.2}$$

holds, where *K* is a constant. By means of a parabolic transformation, the points *A,B, C,D* will be transformed into four tangents *a,b,c,d*. To the tangent *t* of the first conic, a point *T* corresponds in the transformed figure, and to the points *M* and *N*, the straight lines *m, n* drawn from *T* to the intersection points of *a* and *b*, *c* and *d* correspond. Let μ and ν be the points where the parabola's axis cuts *m* and *n*.

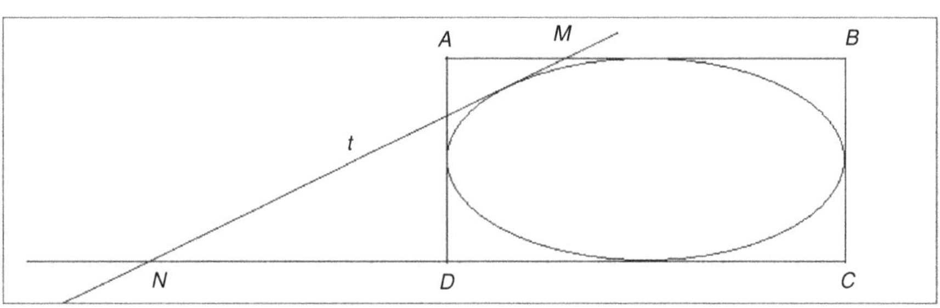

Fig. 2.14 Straight line *t* is the tangent to the conic inscribed in the quadrilateral *ABCD*. Such a tangent cuts the opposite sides *AB* and *CD* of the quadrilateral in the points *M* and *N*, respectively

Let $\alpha,\beta,\gamma,\delta$ be the points where the axis cuts the sides a,b,c,d. According to the reasoning just given, it is the case that

$$\begin{cases} \alpha\mu = AM\cos(AM, X) \\ \beta\mu = BM\cos(AM, X) \\ \gamma\nu = CN\cos(CN, X) \\ \delta\nu = DN\cos(CN, X) \end{cases}$$

Therefore, from Eq. (2.1), it follows that[35]

$$\frac{\alpha\mu}{\beta\mu} = k\frac{\delta\nu}{\gamma\nu}$$

This relation expresses the following theorem:

> Consider a conic circumscribing a quadrilateral. Let a fixed transversal be drawn. From any point of the curve let two rays be drawn to two opposite vertices of the quadrilateral. The ratio of the segments lying on the transversal and included between the first ray and the two sides of the angle from which the ray is drawn and the ratio of the segments lying on the transversal and included between the second ray and the two other sides of the quadrilateral are in a constant ratio, whatever is the point of the curve from which the two rays are drawn.[36]

Chasles claimed that this theorem is susceptible to many applications and developments as one can verify by a direct reading of his memoir.

The proof of this proposition provides an example of what Chasles means when asserting that the cosines disappear from the equation $F(AB, CD, EF, \ldots) = 0$. In particular, when, according to the problem analysed, the points α and δ coincide, the points β and γ also coincide. Chasles examined several cases (see, e.g., ibid., pp. 293–294).

The general and foundational character of the parabolic transformations is further demonstrated by Chasles through the extension of this method to spatial constructions and problems. In this case, the figure, with respect to which the polarity has to be developed, is a paraboloid and the principles are the following ones:

[35] Because of a typo, the letter δ is written in the denominator of the expression on the left, but the correct letter is β.

[36] Chasles (1829g, p. 293): "Quand une conique est circonscrite à un quadrilatère, si l'on tire arbitrairement une transversale fixe, puis, que d'un point quelconque de la courbe, on mène deux rayons aboutissans à deux sommes opposés du quadrilatère, le rapport des segments compris sur la transversale entre le premier rayon et les deux côtés de l'angle au sommet duquel est mené ce rayon, et le rapport des segments compris sur la transversale entre le second rayon et les deux autres côtés du quadrilatère, seront entre eux dans une raison constante, quel que soit le point de la courbe d'où l'on a mené des deux rayons".

1) The polar plane relative to a paraboloid of a point at infinity is parallel to the paraboloid's axis.
2) The polars of mutually parallel straight lines are coplanar. Their plane is parallel to the axis of the paraboloid.
3) The poles of mutually parallel planes belong to a straight line parallel to the paraboloid's axis.
4) The polar of any line at infinity is parallel to the paraboloid's axis.
5) The pole of each plane at infinity is the point placed at infinity on the axis of the paraboloid.

A remark concerning item 5) is necessary. This observation is analogous to that concerning item 3) in reference to plane geometry. In the final version of classical projective geometry, only a plane at infinity exists in space. But Chasles had a different view, a different kind of intuition. It is as if he thought that space was limited at infinity by a series of planes. This is the reason why he spoke of "[…] tout plan situé à l'infini". Notwithstanding this different kind of intuition, of visual perspective, the final principles and theorems are correct. Thus, the fundamental proposition for the parabolic transformations in space reads as follows:

> The polar planes of two points with respect to a paraboloid cut out on the paraboloid's axis a segment equal to the orthogonal projection on the axis of the straight line joining the two points.[37]

The proof is analogous to that expounded for the plane case.

By means of these principles and theorems, Chasles proved several propositions concerning the theory of transversals and skew quadrilaterals inscribed in or circumscribed to quadrics. However, the most interesting theorems concern propositions that go beyond the theory of transversals and that regard three subjects:

a) Metric-graphical properties of systems of solid angles or of points having a particular disposition in space.
b) A theorem on the centre of gravity of a system of planes analogous to that concerning the system of lines expounded in the plane case (this is only natural if one reflects on the fact that the duality in the plane is point-straight line, whereas in space it is point-plane).
c) Metric-graphical properties of second-degree surfaces.

I shall give only one significant example of the application of the parabolic transformation to second-degree surfaces. Chasles considered a hyperboloid (see Fig. 2.15).

If a secant cuts the hyperboloid and its asymptotic cone in four points *A, B* (belonging to the hyperboloid) and *C, D* (belonging to the cone), then $AC=BD$ (ibid., pp. 321–322). A parabolic transformation transforms the hyperboloid into

[37] Ibid., p. 306: "Les plans polaires de deux points, pris par rapport à un paraboloïde, interceptant sur l'axe de paraboloïde un segment égal en longueur à la projection orthogonale sur cet axe de la droite qui joint les deux points".

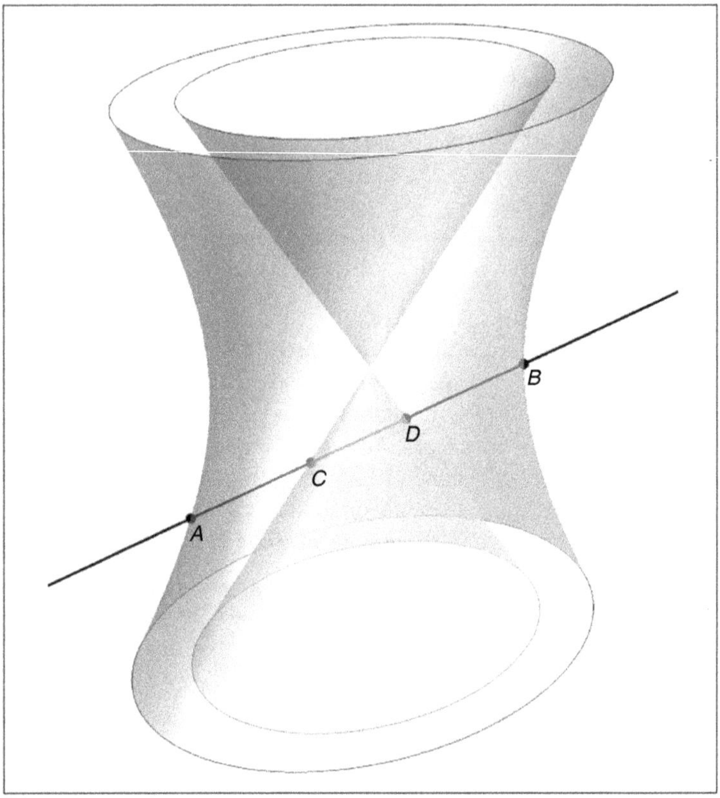

Fig. 2.15 Hyperboloid and its asymptotic cone are represented as well as the secant *ABCD*, cutting the hyperboloid in *A* and *B* and the cone in *C* and *D*. It is $AC = BD$

another second-degree surface, and the asymptotic cone corresponds to a plane section of such surface.[38] This plane curve is the contact curve of the transformed surface with the circumscribing cylinder having its edges parallel to the paraboloid's axis.[39] To the secant and to the points *A*, *B* and *C*, *D*, a line and four planes *a*, *d* and *c*, *d* drawn through this line tangentially to the transformed surface and to its plane section correspond. If α, β, γ, δ are the points where these planes cut the paraboloid's axis, then

[38] My explanation: this depends on the fact that the lines generating the cone converge in a point; thence, by duality, the lines corresponding to the generating lines of the cone are coplanar.

[39] My explanation: this depends on the fact that the plane at infinity is secant to the hyperboloid; hence in the polar parabolic transformation, according to the principle 4), the lines at infinity cutting the hyperboloid are transformed into lines parallel to the paraboloid's axis. They form the cylinder, to which Chasles refers.

$$\alpha\gamma = AC\cos(AC, X)$$

$$\beta\delta = BD\cos(AC, X)$$

Therefore, it is $\alpha\gamma = \beta\delta$. Hence:

> Consider a second-degree surface. Be given its plane section made with a diametral plane. If, through any line, the planes tangent to the surface and to the plane section are drawn, and if, furthermore, a transversal parallel to the diameter conjugate to the diametral plane is drawn, the segment cut out on this transversal between a plane tangent to the surface and a plane tangent to the curve will be equal to the segment intercepted between the two other tangent planes.[40]

These represent only a part of the remarkable metric-graphical properties discovered by Chasles. His programme has a further extension because he wrote a second memoir on the metric-graphical properties that are not limited to second-degree curves and surfaces but also concern some features of other algebraic curves.

2.1.5 Beyond the Problems of Second Degree: Parabolic Transformations and Algebraic Curves and Surfaces

In 1829, several extracts from the letters sent by Chasles to Quetelet were published in the *Correspondance Mathématique et Physique* (Chasles, 1829c, 1829f, 1829i). In the first two letters, while dealing with problems concerning aplanatic lines, Chasles illustrated some properties of fourth-degree curves as the projection on a plane of the intersection of two second-degree surfaces. In Chasles (1829i) additional properties of the curves of third and fourth degree are obtained from those of the second-degree curves and surfaces. In his answer to the second letter written by Chasles (1829f), Quetelet developed some considerations concerning the general condition of geometry (Quetelet, 1829, 191): knowledge of the properties of the second-degree curves and surfaces was quickly improving, whereas research on algebraic curves, considering only those of the third and fourth degree, was proceeding slowly and without the support of general methods. Quetelet invited Chasles to devote a part of his research to enriching the theory of the third- and fourth-degree curves. Quetelet's idea was that such a theory should be presented as a consequence of the properties of second-degree curves and surfaces (ibid., p. 191). Chasles shared this view and developed it far beyond Quetelet's *desiderata*. Importantly, he discovered several general metric properties of algebraic curves and surfaces relying upon those of the second-degree entities. The paper where he presented his results is

[40] Ibid., p. 322: "Si l'on a une surface du second degré et une section plane faite par un plan diamétral, et que par une droite quelconque on mène des plans tangents à la surface et à sa section plane; puis, que l'on tire une transversale parallèle au diamètre conjugué au plan diamétral, le segment intercepté sur cette transversale entre un plan tangent à la surface et un plan tangent à la courbe, sera égal au segment intercepté entre les deux autre plans tangens".

the second memoir on parabolic transformations (Chasles, 1830a). The basic idea is that all the characteristics of the algebraic curves and surfaces could be obtained as a consequence of the properties of second-degree curves and surfaces.

This perspective belongs to an initial phase of algebraic geometry, and, in a few years, such a conception turned out to be too simplistic. However, it was important in this initial phase since it gave rise to a series of investigations. Chasles' researches were among the most significant studies in this subject, and they led to remarkable and correct results. Those obtained in Chasles (1830a) are part of Chasles' foundational programme because the idea of reducing a wide number of properties belonging to different mathematical objects (such as the algebraic curves and surfaces) to a few features of more elementary objects (such as the second-degree curves and surfaces) is typical of a foundational approach.

In his contribution, after explaining a series of introductory concepts, Chasles analyses algebraic plane curves, surfaces, skew curves, developable surfaces and systems of algebraic entities having common points of tangency.

In an attempt to give a precise idea of some of his methods, I will focus on three subjects:

a) General concepts.
b) Algebraic plane curves.
c) Algebraic plane surfaces.

It is first of all necessary to introduce the concept of the centre of *the mean distances* (*centre des moyennes distances*). It is defined like this: let different points $a,b,c,\ldots$ be given belonging to a straight line. A point g on this line exists such that, being m any other point of the line and n the number of the points $a,b,c,\ldots$, the following equation holds:

$$ma + mb + mc + \ldots = n \cdot mg$$

where mi indicates the distance between the points m and i. In practice, the centre of the mean distances g is the barycentre of the given system of points if one supposes that all these points have the same mass.

Chasles' inquiry starts from a theorem proved by Newton in his *Enumeratio linearum tertii ordinis*, which can be expressed as follows: let an algebraic curve of order n be given in the plane, as well as a series of lines parallel to a fixed axis. These lines will cut the curve in n real or imaginary points. On each line consider the centre of the mean distances of such intersection points. All these points belong to a straight line. This line is called a diameter of the curve. Therefore, every curve has an infinity of diameters, one for each direction. Such a diameter is said to be *conjugate to the original direction*. The proof is very easy and relies on the algebraic equation of the curve (Chasles, 1830a, p. 2). In the demonstration, it is also proved that the centre of the mean distances is always a real point, although all or some of the intersection points might be imaginary.

A completely analogous theorem can be formulated for algebraic surfaces: if a series of parallels to a fixed axis is drawn to an algebraic surface and on each of them the centres of the mean distances of the intersection points are considered, all these centres belong to a plane. This plane is said to be the conjugate-diametral plane to the axis or to the direction. A similar concept is expressed for the skew curves (ibid., p. 3).

A very important concept introduced by Chasles is the following one: a set of straight lines in a plane can be considered as an algebraic curve. Just like "normal" algebraic curves, these lines will have an infinity of diameters, each of which is *conjugate* to a given direction. For it is enough to determine the points where a line parallel to a fixed axis cuts the given lines and to find the centre of the mean distances of these points. An analogous concept is defined for algebraic surfaces and skew lines.

Let us consider the fundamental lemma, which allowed Chasles to extend the parabolic transformation to algebraic curves and surfaces. First, I refer to the lemma for plane curves, whose proof, which is similar to the lemmas on parabolic transformations explained in Chasles (1829g), I omit:

> Let there be given a system of points belonging to a straight line as well as the centre of their mean distances. If a parabolic transformation is performed, one will obtain a system of lines intersecting in a point [dual of the original points belonging to a line] and the diameter of these lines conjugate to the direction of the parabola's axis.[41]

Chasles applied this series of concepts to the plane curves. If the parabolic transformation is applied to the figure of Newton's theorem, a second algebraic curve will correspond to the given one. Given a system of mutually parallel transversals, to any transversal p, a point P on a fixed line parallel to the parabola's axis will correspond. This depends on the fact that in a plane parabolic transformation, the polar with respect to a parabola of the point at infinity is parallel to the parabola's axis (see the first of the three presuppositions on the parabolic polarities explained at the beginning of the Sect. 2.1.4). To the intersection points between p and the curve, the tangents to the second curve drawn from P will correspond. Finally, because of the lemma, to the centres of the mean distances of the points where the parallels cut the first curve, the diameters to these tangents will correspond, which are conjugate to the direction of the parabola's axis. These diameters will pass through a fixed point that corresponds to the diameter containing the centres of the mean distances of the first figure. Thus, the following theorem holds true:

> Given a plane algebraic curve and a fixed coplanar line l, if, from each point of l, a pencil of tangents to the curve is drawn as well as the diameter of the pencil, conjugate to l, all these diameters will pass through a point L.[42]

---

[41] Chasles (1830a, p. 6): "Quand on a un système de points en ligne droite, et leur centre des moyennes distances, si l'on fait la transformation parabolique, on aura un système de droites concourant en un même point, et le diamètre de ces droites, conjugué à la direction de l'axe de la parabole auxiliaire".

[42] Ibid., p. 7: "Si l'on a une courbe plane géométrique et une droite fixe tracée dans son plan, et que par chaque point de cette droite, on mène un faisceau de tangentes à la courbe, et le *diamètre* de ce faisceau, *conjugué* à la droite fixe, tous ces diamètres passeront par un même point". Italics in the text.

Chasles added two definitions, which make it clear that he was trying to develop a theory of algebraic curves relying upon the concepts related to second-degree curves. He claimed that if the curve is a conic, then the point through which the diameters of the pencils of tangents (drawn from the different points of the fixed straight line) pass is the pole of a fixed line with respect to the conic. Therefore, extending the concept of pole and polar, the point L will be called the *pole* of l with respect to the given algebraic curve. This granted, the diameters are *conjugate* to l. If the straight line l is the line at infinity, the preceding theorem still holds true because of its projective character. But, if l is at infinity, the tangents of each pencil will be mutually parallel. Hence:

> If all the tangents parallel to a fixed line are drawn to an algebraic curve, the diameters of these tangents will pass through a fixed point, for any direction of the fixed line.[43]

This fixed point is called by Chasles the *centre of the curve*. Given this definition, every algebraic curve has a centre. Relying upon these concepts, Chasles discovered a series of metrical-projective properties for algebraic curves (ibid., 8–12).

By means of a lemma analogous to that for plane curves, Chasles addressed the problem of the algebraic surfaces. The lemma is:

> Let there be given a system of collinear points and the centre of their mean distances. If the transformation with respect to a paraboloid is developed, a system of planes will be obtained, which pass through the same straight line and their diametral plane will also be obtained, which is conjugate to the direction of the paraboloid's axis.[44]

After this lemma, analogous concepts and results to those obtained for algebraic plane curves are achieved.

Commentaries: the theory of polars for algebraic curves is not an original achievement of Chasles. Between 1827 and 1829, Bobillier, for example, had already developed several concepts and theorems on the successive polars to an algebraic curve.[45] In his *Recueil*, Chasles actually mentions Bobillier for his researches on the successive polars to the algebraic curves and surfaces (Chasles, 1870, 65–68) and claims that such researches have been a rich source of inspiration for the other geometers (ibid., 67). Interestingly enough, Chasles includes himself among these geometers. However, it seems to me that his aims were, at least partly, different from Bobillier's.

Chasles was intending to found the basis of a projective doctrine, which could include the metric-graphical properties, relying on a few concepts reducible to the following:

---

[43] Ibid., p. 8: "Si l'on mène à une courbe géométrique toutes ses tangentes parallèles à un même droite, le diamètre de ces tangentes passera par un point fixe, quelle que soit la direction de cette droit". The term "geometrical curve" was then used to denote what today we call an algebraic curve. In the translation proposed in the running text, I have used the modern locution.

[44] Ibid., p. 11: "Quand on a un système de points en ligne droite et leur centre des moyennes distances, si l'on fait la transformation par rapport à un paraboloïde, on aura un système de plans passant par une même droite et leur plan-diamètre, conjugué à la direction de l'axe du paraboloïde".

[45] See Haubrichs dos Santos (2015, 186–197).

1) The intersections of first-degree entities (straight lines and planes), considering imaginary elements and elements at infinity
2) Polarities with respect to conics and quadrics

He was interested in studying the widest possible extension of these very basic concepts. It is not by chance that, to use Bobillier's language, Chasles considered the $(m-1)$th order polars of an algebraic curve of order m, i.e. straight lines. Furthermore, he investigated the possibility of extending the synthetic method, which was certainly not Bobillier's aim.

As explained above, Chasles was interested in establishing the power of the pure methods, although this did not prevent him either from using analytical methods in some papers and circumstances or from considering the foundational problems strictly connected to these methods.

In the papers analysed here, formulas are almost completely absent. In particular, the equations of the curves are almost all missing, which is a clear mark of Chasles' intention to use synthetical methods as much as possible. It also emerges that he was trying to separate clearly the metric from the graphical properties. Chasles' researches in geometry after the publication of the *Aperçu* did not go either in the direction of studying the properties of the higher polars for an algebraic curve or in that of developing an analysis of the relations between graphical and metric properties *à la* Von Staudt. I mean that, after having introduced the anharmonic ratio, he did not deem problematic the fact that the anharmonic ratio—though being a projective invariant—was defined in terms of metric notions. As a matter of fact, Chasles was neither involved in an examination of the concept of cross ratio insofar as it is based on metric elements nor in an attempt to define a projective metric *à la* Cayley–Klein. However, his work certainly supplied the initial bases for these researches.[46]

A particular commentary is necessary with regard to the transformation used by Chasles to obtain metrical properties by means of a projective structure. He used a parabolic transformation where the polarity is developed with respect to a parabola in the plane or with respect to a paraboloid in space. The problem is that, from a purely projective point of view, no parabola and no paraboloid exist. For it is today known that the projective classification of the conics admits only five objects whose equations, if an appropriate frame of homogeneous coordinates is used, assume the following forms in the projective real plane:

1) $X_0^2+X_1^2+X_2^2=0$, which is a not degenerate conic having no real point.
2) $X_0^2+X_1^2-X_2^2=0$, which is a not degenerate real conic.
3) $X_0^2+X_1^2=0$, which is a degenerate conic given by a real point.

[46]Three Chasles' important and long memoirs are Chasles (1829h, 1830h, 1831). He applied his foundational ideas to several geometrical questions. Though these memoirs are very significant from a mathematical point of view, they do not add new elements insofar as Chasles' foundational programme is concerned. Thence, I will not analyse their content.

4) $X_0^2 - X_1^2 = 0$, which is a real degenerate conic given by two straight lines.
5) $X_0^2 = 0$, which is a degenerate conic given by two coincident straight lines.

With regard to the quadrics and considering only those that are not degenerate, there are three kinds of projective quadrics:

1) $X_0^2 + X_1^2 + X_2^2 + X_3^2 = 0$, which is a quadric having no real point.
2) $X_0^2 + X_1^2 + X_2^2 - X_3^2 = 0$, which is a quadric composed of elliptical points.
3) $X_0^2 + X_1^2 - X_2^2 - X_3^2 = 0$, which is a quadric composed of hyperbolic points.

Therefore, from a purely projective standpoint, one cannot speak either of parabola or paraboloid. For the concepts of parabola and paraboloid are affine, not projective concepts, which are produced if one considers the behaviour of a conic with respect to the line at infinity. Therefore, if the line at infinity is regarded as a particular entity different from the ordinary straight lines—which implies a departure from a purely projective approach—the equations of the not degenerate conics assume the following form, in an affine coordinate system (x,y):

1) $y^2 = x$, which is a parabola.
2) $x^2 + y^2 + 1 = 0$, which is an ellipse having no real point.
3) $x^2 + y^2 - 1 = 0$, which is a real ellipse.
4) $x^2 - y^2 - 1 = 0$, which is a hyperbola.

Other concepts, such as those of equilateral hyperbola or circumference, pertain to the metric classifications of the conics, which, obviously, is even more specific than the affine one.

Therefore, what Chasles did from a mathematical standpoint was to prove that several metrical properties are reducible to projective ones provided that an affine object as a parabola for the plane or a paraboloid for space is introduced. Obviously, he was aware of the fact that "parabola" and "paraboloid" are not projective concepts. Nonetheless, his results are extremely significant because they show that also without the development of a projective metric, it is possible to reduce many metric properties to projective ones, if you are allowed to introduce an affine object.

Although Chasles' main interest was to develop his foundational programme, his researches on the parabolic transformations, if analysed from another perspective, concern the results that can be obtained in a certain field of mathematics if you introduce one and only one object which is extraneous to that field. As to Chasles, his intention was to obtain some metrical properties within a projective context, and he introduced an object, a parabola/paraboloid, which is not a projective figure. There are several cases of procedures which formally resemble this approach: for example, consider the famous Poncelet–Steiner theorem[47] whose idea is to develop

[47] The basic ideas of the Poncelet–Steiner theorem exist in Poncelet's *Traité* (Poncelet, 1822). However, the rigorous demonstration that all the constructions by ruler and compass can be developed using only a ruler and a single circle of given centre is offered in Steiner (1833). Among the texts dedicated to the Poncelet–Steiner theorem, see Martin (1998), p. 97–105.

all the constructions by straight line and compass using only a straight line. This is possible, provided that a single circle and its centre are given. The circle in this case plays a role analogous to that of the parabola/paraboloid by Chasles: as the parabola/paraboloid are not projective objects, so the circle is non-constructible through a straight line. A similar though not identical example is provided by the constructions that Mascheroni developed only by the compass (Mascheroni, 1797). Mascheroni actually proved that all the constructions which can be obtained by straight line and compass can be obtained only by compass, provided that you admit that a straight line is constructed when two of its points are given. In this case what is added to the constructions allowed by the sole compass is not a further object which cannot be constructed with the compass, but a further hypothesis: a straight line is geometrically constructed through two of its points (as a matter of fact, Mascheroni used the compass also to transport distances, whereas in Euclid—see the third postulate—it is admitted only to draw a circle). The examples might be multiplied, but what expounded so far is enough to clarify the situation. From a conceptual point of view, this also has a connection with abstract algebra, because when, given a field of rationality, a single new entity is added to the field, one is studying the set of operations which are not allowed by the initial field, but which become possible with the addition of that single new element. Therefore, Chasles' use of parabolic transformation is an interesting source for several branches of the historical-mathematical research.

Another noteworthy observation concerns the use of the concept of cosine. In itself, the concept of cosine is a metrical concept since it is tied to the measure of an angle. However, as we have seen, Chasles used the concept of cosine to define a certain relation (see Eq. 2.1), which he employed within projective geometry showing that, for his specific applications, the cosines disappear, so that this problem does not subsist. However, it is clear that the use of trigonometric functions implies a departure from a purely projective context. This problem persisted also when the concept of anharmonic ratio was developed and was overcome only thanks to Von Staudt's work: in the latter's study the construction of a harmonic group of four points was obtained without resorting to the concept of cross ratio. In contrast to this, the very concept of cross ratio was founded on purely projective constructions.

A remarkable historical observation concerns the way in which Poncelet reacted to Chasles' publication of the two memoirs on parabolic transformations. This testifies to the two scholars' different approaches to the concept of transformation: in 1832, Poncelet published the important memoir *Note sur quelques principe généraux de transformation des relations métriques des figures* [...] (Poncelet, 1832, in Poncelet, 1866, pp. 332–345). In this work he claims that Chasles' parabolic transformations are not general, but, rather, they represent a particular case of the transformations which Poncelet himself called "projective". The programmatic aim of this memoir was, in fact, to show that Chasles' parabolic transformations are a subset of those transformations Poncelet called "relations projective orthogonalement" (Poncelet, 1832, 1866, p. 333). In substance, Poncelet projected orthogonally a segment or an area. He wrote:

> Suppose to have replaced, for any distance and flat area entering the proposed relation [a relation in which cosines appear], its value in function of the distance or the area which is the orthogonal projection [...].[48]

After that Poncelet developed a series of reasonings and showed that Chasles' parabolic transformations belong to the *relations projective orthogonalement*, which are, in turn, specific projections whose general theory he had offered in the *Traité* of 1822 and in the memoirs on the centres of the harmonic means. What to say with regard to Poncelet's reaction? Kötter's interpretation is the following one: Poncelet was more profound than Chasles. He was the inventor of the theory of projective transformations. Chasles was the greatest mathematician to develop Poncelet's results, but, essentially, Kötter agrees with Poncelet's stance: Chasles' theory of parabolic transformations is an important application of Poncelet's results, but it does not imply any substantial novelties from a conceptual point of view, though being a significant piece of projective geometry. As Kötter writes:

> It is perhaps possible to claim that Poncelet understood the essence of the transformation more deeply than Chasles [...]. Poncelet's personality is far more stately. His force consists in the broad viewpoints, from which it is possible to guess the relations among things. It seems to me impossible to deny that Chasles did not reach the same profoundness in the conception of the theory as a whole—at least until that moment [...].[49]

Therefore, according to Kötter, Poncelet's geometry is the essential step from a geometry in which the basic elements are the figures towards a geometry where the study of the transformations becomes the theory's focus. Chasles' theory of transformations would be derived from Poncelet's and is included within it.

Nabonnand expresses a completely different opinion, according to which Poncelet's is still, basically, a geometry of figures. Nabonnand comments thus:

> Poncelet's theory remains a geometry of figures, of problems, and of theorems. Certainly, the nature of the problems and of the theorems changes because they concern the projective properties (invariant by a central projection) of the figures [...]. The particular figure remains eventually the proof's centre, but insofar as it is considered in the net of the projective figures or in the set of figures whose elements glide continuously the one in relation to the other, the properties relative to a particular figure acquire a general character and are extended to other figures.[50]

---

[48] Poncelet (1832, 1866), p. 333: "Supposant donc qu'ayant substitué, pour chacune des distances et des aires planes qui entrent dans la relation proposée, sa valeur ci-dessus en fonction de la distance ou de l'aire qui en est la projection orthogonale [...]".

[49] Kötter (1901), pp. 182–183: "Man wird vielleicht zugeben, dass Poncelet das Wesen des Transformation tiefer erfasst hat, als Chasles. [...] Die weitaus wuchtigere Persönlichkeit ist Poncelet. Seine Stärke liegt in der Ausmittelung des grossen Gesichtspunkte, von denen aus man den Zusammenhang der Dinge zu überblicken vermag. Es kann, scheint mir, nicht geleugnet werden, dass Chasles in der Auffassung der grossen Ganze – wenigstens zunächst – nicht dieselbe Tiefe erreicht [...]".

[50] Nabonnand (2011b, p. 9): "La théorie de Poncelet reste une géométrie de figures, de problèmes et théorèmes. Certes la nature des problèmes et des théorèmes changent de nature puisqu'ils concernent les propriétés projectives (invariantes par projection centrale) des figures [...]. La figure particulière reste en dernière instance le lieu de la démonstration mais comme celle-ci est prise dans

Thus, according to Nabonnand, Poncelet's is, so to say, a geometry of "general figures", something between the geometry of specific metrical figures—as Euclid's—and a geometry in which the central point is represented by the transformations which conserve some invariant properties.

The judgement on how much innovative Chasles' conception was depends, in great part, on the consideration of Poncelet's geometry. As outlined in the Introduction, I am inclined to think that Poncelet's was not only a geometry of figures. It seems to me that works as Poncelet (1832) clearly show that he was also thinking in terms of general projective transformations. It is true that Poncelet also thought of transformations, as the polarities, always in relation to a figure; on the contrary, Chasles and Steiner, as we will see, inserted the theory of reciprocal polars within a broader theory of transformations. However, this does not mean that Poncelet did not think in terms of transformations, but that his view was still anchored to a concrete view of the transformations in which the transformation is not separated from the figures to which it is applied, whereas Chasles and Steiner represented a further step towards a general theory of transformations, but it is true that the seed of the transformation theory is to be found in Poncelet. On the other hand, Poncelet's idea—which seems to be also shared by Kötter—that Chasles' results on the parabolic polarities are a part of Poncelet's theory of projections is true from a mathematical point of view, because they can be deduced within this theory. Chasles certainly could not deny this truth. However, their intentions were profoundly different. Poncelet first distinguished clearly between metric and graphic properties and, throughout his scientific career, he deduced many metric-projective theorems. This notwithstanding, the idea to study systematically the metric properties of segments and angles and of ratio of segments and angles in function of a particular projective transformation—the parabolic polarity—is typical of Chasles. I mean that the reductionist approach characterizes Chasles' point of view rather than Poncelet's, who, among his theorems, proved many graphic metric properties, but whose main scope was not to reduce the metric properties to those graphic; rather Poncelet's aim was to offer a general basis for projective geometry made up of new principles and conceptions. Therefore, I conclude that Chasles' theory of parabolic polarities is original and that it is a crucial step in his foundational programme as well as an essential step in the development of his conception, according to which the specific study of the transformations is a basic element of the new geometry.

Let us now analyse a further step in Chasles' foundational programme: the overcoming, in projective geometry, of the phase in which the theory of polar reciprocity was the leading theory. This phase can be associated with the period between the beginning of the nineteenth century and Chasles' *Aperçu*. After that, the theory of polarity came to be included within a more general theory of transformations.

des réseaux de figures projectives ou dans des déploiements de figures dont les éléments glissent continument les uns par rapport aux autres, les propriétés relatives à une figure particulière acquièrent un caractère de généralité en s'étendant à d'autres figures".

2.2 Beyond the Theory of Polar Reciprocity: Chasles' Memoir on Duality and Homography

To conclude the explanation of Chasles' ideas on the foundation of geometry up to the *Aperçu*, some foundational aspects contained in the memoir on duality and homography inserted in his masterpiece will be analysed.

This memoir is a work containing about 300 pages in which several ideas are introduced and a plurality of theorems is proved. Only the main concerns can be described here. The foundational ideas on which the memoir is based are:

a) The concept of cross ratio as the cornerstone of projective geometry. It is not only a useful element with which to solve specific problems; it is the foundational concept. Chasles had already called the cross ratio the "anharmonic ratio" and the cross ratio whose value is -1 "harmonic ratio", but he had not yet placed this concept at the basis of pure geometry. He therefore shares with Steiner and Möbius the merit of having fully grasped the generality of this concept. Reciprocal influences can be excluded.
b) Cross ratio represents the basis for the theory of transformation in pure geometry. This theory includes that of reciprocal polars, a theory which thus becomes an application of a more general concept. Chasles fully understood that finding the logical basis of a transformation means determining something invariant with respect to this transformation. The cross ratio is the invariant with respect to projective transformations, those analysed by Chasles.
c) The two fundamental principles of transformation are *duality* and *homography*. By means of them the purely graphical and the metric-graphical properties of the figures can be established.
d) The two principles can be proved in an analytical and in a purely geometrical form. With regard to the method, a remark is necessary: as already noted, Chasles had pointed out the importance of purely geometrical methods, but when he had to found geometry, he was more interested in finding the foundational concepts rather than concentrating on methods. In the memoir under examination, the analytical method (to be more explicit: the resort to a coordinate system) is also used. Chasles stressed the genetic importance of geometrical methods, but he also used analytical ones insofar as they are more comfortable, although less simple (Chasles, 1837a, pp. 575–577).

2.2.1 *The Concept of* Rapport Anharmonique

As soon as Chasles introduced the concept of *rapport anharmonique* and his applications in the *Aperçu*, he highlighted its revolutionary novelty and its generality.[51] The context is that of the analysis concerning the doctrines of the deformation and transformation of the figures, namely the doctrine of duality and homography. In the sixth chapter of the *Aperçu*, which should be interpreted as an introduction to Chasles' two memoirs on duality and homography (though written afterwards), he claimed explicitly:

> We will prove these two principles [deformation and transformation] in a direct way, which will show a series of absolute and abstract truths completely independent of particular methods and which will be suitable to justify and facilitate the applications in some particular cases.
>
> Thence, we will present them, as we have already remarked, in a more general form than [that allowed by] any other method. Our extension will find its main utility in a principle inherent to the relations of magnitude, which is very simple and which will make them applicable to numerous new questions.
>
> This principle relies upon a lone relation, to which it will be always sufficient to refer all the others. This relation is the one we have named *rapport anharmonique* of four points or of a pencil of four straight lines. This is the sole type of all the relations *transformable* through the two principles which we will prove. The law of correspondence between a figure and its transformed one consists in the identity of the corresponding anharmonic ratios.
>
> The simplicity of this law of the anharmonic ratio makes this form of relation particularly suitable to play an important role in the science of extension.[52]

---

[51] In the *Aperçu* Chasles mentioned the expression "*rapport* or fonction *anharmonique*" (italics in the text) on p. 35 in reference to a theorem by Pappus, where Chasles claimed that such a function might produce a remarkable simplification in several geometrical theorems. The value of the anharmonic ratio had been introduced on p. 34. He then mentioned the anharmonic ratio in reference to Pascal's hexagram (ibid., p. 72). With regard to Desargues' theory of involution, Chasles spoke of the case in which the anharmonic ratio becomes a harmonic ratio (ibid., p. 77), claiming that Desargues introduced the notion of anharmonic ratio (ibid., p. 80). De la Hire is mentioned as the mathematician who also used the notion of harmonic ratio (ibid., p. 120). Before the quotation I mention in the running text, the anharmonic ratio (also considering the harmonic one) is named some other times by Chasles (e.g. ibid., p. 159, pp. 211, 218n), but the quotation referred to is the first one in which he made it completely explicit the value that, according to his opinion, the concept of anharmonic ratio holds in projective geometry. Several considerations on the concept of anharmonic ratio in Chasles are present in Chemla (2016, pp. 73–85). Milne (1911) also presents many remarks on the important contributions given by Chasles to the cross ratio theory. I am facing the different aspects of the way in which Chasles introduced, used and interpreted such notion in this section and in Sect. 2.3 where a comparison with Steiner's notion of *doppelverhältnis* is developed.

[52] Ibid., pp. 254–255: "Nous démontrerons ces deux PRINCIPES d'une manière directe, qui en fera vérités absolues et abstraites, dégagées et indépendantes de toutes méthodes particulières, propres à les justifier on à en faciliter les applications dans quelques cas particuliers.

Nous les présenterons, ainsi que nous l'avons déjà dit, dans une plus grande généralité qu'aucune de ces méthodes. L'extension que nous leur donnerons trouvera sa principale utilité dans un principe de relations de grandeur, extrêmement simple, qui les rendra applicables à de nombreuses questions nouvelles.

Ce principe repose sur une relation unique, à laquelle il suffira toujours de ramener toutes les autres. Cette relation est quelle que nous avons appelée *rapport anharmonique* de quatre points ou d'une faisceau de quatre droites. C'est là le type unique de toutes les relations *transformables* par les deux principes que nous démontrons. Et la loi de correspondance entre une figure et sa transformée, consiste dans l'égalité des rapports anharmoniques correspondants.

While reading this passage, a question arises: we have seen that Chasles was careful to highlight the distinction between metric and graphic properties from the beginning of his scientific career and that he had tried to reduce the former to the latter. This was one of the bases of his foundational programme. This notwithstanding, the concept that, starting from the beginning of the 30s, he considered as the paramount notion to develop a unitary picture of projective geometry, namely the concept of anharmonic ratio, was based on metric notion as those of sinus of an angle and length of a segment. Chasles clearly recognized this feature of the cross ratio because, in the just mentioned passage, he spoke of "un principe de relations de grandeur". Then: why did he not problematize this aspect of the cross ratio's nature? The answer is not easy and cannot be but conjectural. Chasles (jointly with Möbius and Steiner) discovered the invariance of the cross ratio by projections and sections; it is likely that he was impressed by the simplicity of this relation, by its extensive use to solve several problems connected to the theory of projections and by the general panorama it offered. This possibly induced him to accept the notion of anharmonic ratio as a primitive projective concept. It should be highlighted that he did not accept the notion of length of a segment or of sinus of an angle and either the ratio of two segments or of two sinuses of two angles as primitive projective concepts. On the contrary, he considered the cross ratio as a unitary entity which, from a projective standpoint, is more "primitive", more fundamental than the ratios and the single elements of the ratios which compose it. This statement does not solve the problem from a mathematical and logical point of view, but it can explain Chasles' order of ideas, which was not unique to him, but which was also shared by Steiner (and independently from Chasles, as we will see) and which Chasles himself maintained also after Von Staudt published his *Geometrie der Lage* (1847). In fact, I doubt that Chasles knew Von Staudt's work since the latter is mentioned neither in Chasles (1852) nor in Chasles (1870). It therefore seems to me that one can exclude the idea that Chasles knew von Staudt's results but decided not to mention them, because Von Staudt, at the beginning of his Preface to *Geometrie der Lage*, wrote explicitly:

> In these last times, the geometry of position has been distinguished from the metric geometry. This separation is correct. Nonetheless, propositions which are not involved with any concept of magnitude are, in general, proved through the consideration of ratios. In this work, I tried to develop the geometry of position as an independent science, where the concept of measure is not necessary.[53]

La simplicité de cette loi et celle du rapport anharmonique rendent cette forme de relations éminemment propre à jouer un rôle si important dans le science de l'étendue". Capitals and italics in the text.

[53] Von Staudt (1847), p. III: "Man hat in den neueren Zeiten wohl mit Recht die Geometrie der Lage von der Geometrie des Masses unterschieden, indessen gleichwohl Sätze, in welchen von keiner Grösse die Rede ist, gewöhnlich durch Betrachtung von Verhältnissen bewiesen. Ich habe in dieser Schrift versucht, die Geometrie der Lage zu einer selbstsändigen Wissenschaft zu machen, welche des Messens nicht bedarf".

The idea expressed by von Staudt is so clearly connected to the concept of cross ratio that Chasles could not have ignored von Staudt's work if he had known it. The only logical—though conjectural—explanation is that he was not aware of his work, or at least that he did not know it in depth, also considering that it was written in German.[54] Von Staudt can be regarded as the geometer who completed and overcame Chasles' programme insofar as geometry is concerned.

Chasles discussed in depth the mathematical properties of the concept of *rapport anharmonique* in the long Note IX of the *Aperçu* (ibid., pp. 302–308), which is entitled *Sur la fonction anharmonique de quatre points, ou de quatre droits* claiming that, given four points a, b, c, d, the function

$$\frac{ac}{ad} : \frac{bc}{bd}$$

is named *rapport anharmonique* of a, b, c, d.

This note is important because in it Chasles explained the most significant features of the cross ratio. He referred to Proposition 129 of the seventh book of Pappus' *Collections* translating its enunciation in modern terms: given four straight lines passing through a point and a series of transversals cutting these straight lines, the cross ratio of the four contact points has the same value for any transversal (ibid., p. 302) (Fig. 2.16).[55] He comments like this on the previous property:

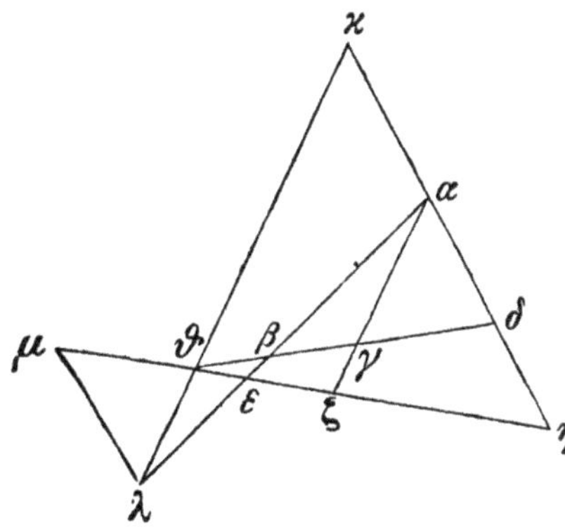

Fig. 2.16 The figure used by Pappus in the statement and demonstration of Proposition 127, seventh book of his *Collections*

[54] In a note of the *Aperçu*, while referring to the works by Steiner, Plücker and Möbius, Chasles claimed, with regard to the fact they were written in German: "I regret much the fact I cannot quote their works, that I do not know because of my ignorance of the language in which they are written" (Chasles, 1837a, p. 215, n.1). In successive works, Chasles mentioned the results of these three mathematicians, but it is absolutely plausible that he always had difficulties in reading German.

[55] Proposition 129 of Pappus' *Collections*'s seventh book claims that (see Fig. 2.16), given three straight lines $\alpha\beta$, $\gamma\alpha$, $\delta\alpha$, and two other straight lines $\theta\varepsilon$, $\theta\delta$, the following relation holds: $\theta\epsilon \cdot \eta\zeta : \theta\eta \cdot \zeta\varepsilon = \theta\beta \cdot \delta\gamma : \theta\delta \cdot \beta\gamma$, where η is the intersection between $\delta\alpha$ and $\theta\varepsilon$ and ζ is the intersection between $\theta\varepsilon$ and $\gamma\alpha$ (Pappus, 1877, II, pp. 871–873). Since Pappus' relation can be written as $\frac{\theta\epsilon}{\theta\eta} : \frac{\epsilon\zeta}{\eta\zeta} = \frac{\theta\beta}{\theta\delta} : \frac{\beta\gamma}{\delta\gamma}$, this establishes the invariance of the cross ratio by projection. In this case, the centre of projection is α, the two projective lines are $\theta\varepsilon$ and $\theta\beta$ and the points corresponding in

This property of the *anharmonic function* of four points distinguishes it from any other function which is possible to obtain with the segments" that these four points form.[56]

Chasles continued with what might be called "the fundamental theorem of the cross ratio". It reduces the cross ratio of four collinear points to the cross ratio of the sinuses of the angles formed by four straight lines drawn from a point external to the straight line on which the four points lie. Chasles stressed the projective character of this theorem. His words are very clear, for he wrote:

> But the anharmonic function holds an even more important property from which the former derives, namely:
>
> *If, from an arbitrary point, four lines are drawn to four collinear points, the anharmonic function of these four points will hold precisely the same value of this function when, to the four segments determined by the four points, lie the sinuses of the angles among the straight lines including the four segments are replaced.*
>
> This function among the sinuses of the angles of four straight lines drawn from the same point will be called *anharmonic function* of the four straight lines.
>
> This theorem proves that the anharmonic function of four points is of *projective* nature, namely that this function maintains the same value when you make the projection or the perspective of the four points to which it is referred.
>
> This theorem can be generalized to space, drawing any four planes through the four points, provided that they are collinear: *the anharmonic function of the four points will maintain the same value, if the segments are replaced by the sinuses of the dihedral angles formed by the planes which include these segments.*[57]

In this way, Chasles established the main feature of the cross ratio in the forms we call of first species, namely the dotted straight line, the pencil of straight lines and the sheaf of planes. He then clarified that the cross ratio of four points a,b,c,d has three different values:[58]

---

the projection from α are (θ,θ); (ε,β); (ζ,γ); (η,δ). I do not enter the difficult question how modern Pappus' conception is. It seems that Chasles gave a modernizing interpretation of Pappus' work.

[56] Ibid., p. 302 : "Cette propriété de la *fonction anharmonique* de quatre points la distingue de toute autre fonction qu'on pourrait former avec les segments que ces quatre points font entre eux". Italics in the text.

[57] Ibid., p. 302: "Mais la fonction anharmonique jouit d'une autre propriété encore plus capital, et dont cette première dérive, c'est que: *Si, d'un point pris arbitrairement, on mène des droites aboutissant à quatre points situés en ligne droite, la fonction anharmonique de ces quatre points aura précisément pour valeur ce que deviendra cette fonction quand on y substituera, aux quatre segments qui y entrent, les sinus des angles que feront entre elles les droites qui comprendront ces segments.* Cette fonction entre les sinus des angles de quatre droites issues d'un même point, sera dite *fonction anharmonique* des quatre droites. Ce théorème prouve que la fonction anharmonique de quatre points est de nature *projective*, c'est-à-dire que cette fonction conserve la même valeur quand on fait la projection ou la perspective des quatre point auxquels elle se rapporte. On peut généraliser ce théorème, en menant par les quatre points, quatre plans quelconques, pourvu qu'ils se coupent suivant une même droite, prise arbitrairement dans l'espace: *la fonction anharmonique des quatre points conservera la même valeur, si l'on y substitue, à la place des segments, les sinus des angles diédres que les plans qui comprennent ces segments font entre eux*". Italics in the text.

[58] We know that, given four points, there are six different values of the cross ratio and not only three. I will face this problem in Sect. 2.3.

$$a)\ \frac{ac}{ad}:\frac{bc}{bd};\quad b)\ \frac{ac}{ab}:\frac{dc}{db};\quad c)\ \frac{ab}{ad}:\frac{cb}{cd}$$

He pointed out that, if the point d is at infinity, the three cross ratios assume respectively the forms:

$$a')\ \frac{ac}{cb};\quad b')\ \frac{ca}{ab};\quad c')\ \frac{ba}{bc}$$

He then proved that, given two groups of four point $a, b, c, d;\ a', b', c', d'$ such that one of the three equations of the following system holds:

$$\begin{cases} \dfrac{ac}{ad}:\dfrac{bc}{bd}=\dfrac{a'c'}{a'd'}:\dfrac{b'c'}{b'd'} \\ \dfrac{ac}{ab}:\dfrac{dc}{db}=\dfrac{a'c'}{a'b'}:\dfrac{d'c'}{d'b'} \\ \dfrac{ab}{ad}:\dfrac{cb}{cd}=\dfrac{a'b'}{a'd'}:\dfrac{c'b'}{c'd'} \end{cases} \tag{I}$$

then the other two also hold.

After having proved some elementary features of the cross ratio for plane configurations, Chasles claimed that the anharmonic ratio can be very useful in spatial geometry too (ibid., p. 305). He offered an interesting example concerning the projective generation of the one-sheeted hyperboloid. As he writes:

> [1] The surface generated by a mobile straight line A, which lays on three fixed straight lines can also be generated in a second manner by a mobile straight line which lies on three positions of A. [2] This surface has the property that any plane cuts it along a conic.[59]

As a matter of fact, this proposition consists of two theorems whose proof offers the projective generation of the one-sheeted hyperboloid. The proposition [1] is, in its turn, divided into two reciprocal statements. The proofs of both [1] and [2] are a clear example of the manner in which Chasles used the cross ratio to reach the projective generation of the ruled quadrics. The first part of [1] states that:

> When each of four straight lines lays on three fixed straight lines, which are posed in any position in space, the anharmonic ratio of the segments they form on one of these three straight lines is equal to the anharmonic ratio they form on each of the other two.[60]

---

[59]Ibid., pp. 305–306: "La surface engendrée par une droite mobile, qui s'appuie sur trois droites fixes, peut être engendrée, d'une seconde manière, par une droite mobile qui s'appuie sur trois positions de la première droite génératrice; et cette surface jouit de la propriété, que tous plan la coupe suivant une conique".

[60]Ibid., p. 306: "Quand quatre droites s'appuient chacune sur trois droites fixes, situées d'une manière quelconque dans l'espace, le rapport anharmonique des segments qu'elles forment sur l'une de ces trois droites, est égal au rapport anharmonique des segments qu'elles forment sur l'une quelconque des deux autres".

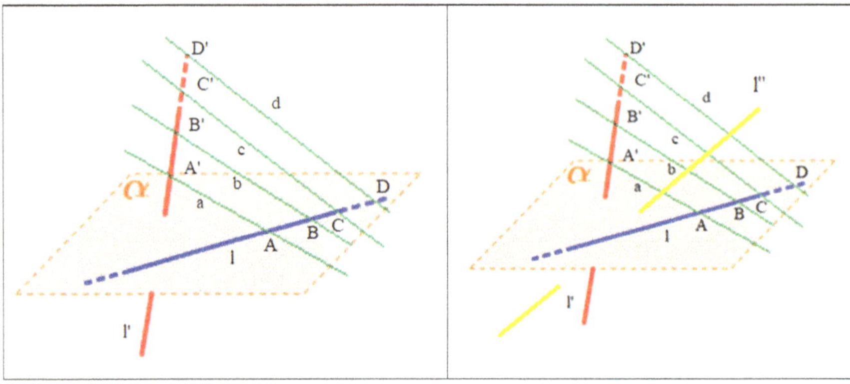

Fig. 2.17 A reconstruction of the situation described by Chasles in his theorem. On the left: the straight lines D and D' lie on the four straight lines a,b,c,d. On the right: the straight line l'' which lies on three of the straight lines a,b,c,d also lies on the fourth one

The proof runs like this: Be L, L', L'' the three fixed lines in space and a, b, c, d the four points where the straight lines A, B, C, D which lie on L, L', L'' cut L and a', b', c', d'; a'', b'', c'', d'' the points where they cut respectively L' and L''. The anharmonic ratio of the four points a, b, c, d is equal to that of the four points a', b', c', d' because each of them is equal to that of the four planes passing through L'' and respectively through A, B, C, D. These two cross ratios are, thence, equal. Obviously a completely analogous reasoning can be repeated for the group of points a, b, c, d and a'', b'', c'', d''.

The reciprocal proposition claims that:

> If four straight lines lie on two fixed straight lines in space, so that the anharmonic ratio of the segments they form on one of these straight lines is equal to the anharmonic ratio of the segments they form on the other, every straight line which lies on three of these four straight lines will lie on the fourth, as well.[61]

The proof runs like this (see Fig. 2.17)

The hypothesis of the theorem states that the following relation holds:

$$\frac{AC}{BC}:\frac{AD}{BD}=\frac{A'C'}{B'C'}:\frac{A'D'}{B'D'}$$

It is to prove that any straight line l'' lying on a,b,c, also cuts d.

Chasles argues as follows: through the point D of l, draw the straight line d' which lies on l' and l''. Be E' and E'' the points where d' cuts l' and l''. The four straight lines a,b,c,d' lie on l,l',l'', thus determining the following anharmonic ratios:

[61] Ibid., p. 306: "Si quatre droites s'appuient sur deux droites fixes dans l'espace, de manière que le rapport anharmonique des segments qu'elles font sur l'une de ces droites soit égal au rapport anharmonique des segments qu'elles font sur l'autre, toute droite qui s'appuiera sur trois de ces quatre droites s'appuiera nécessairement sur la quatrième".

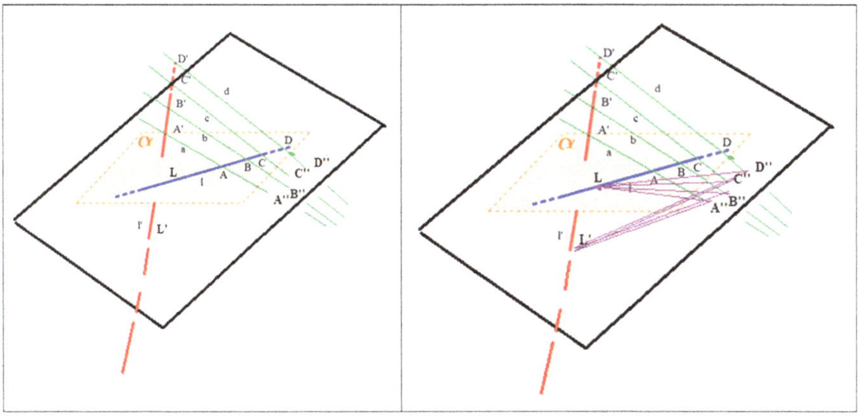

Fig. 2.18 Reconstruction of the situation described by Chasles in his theorem

$$\frac{AC}{BC}:\frac{AD}{BD}=\frac{A'C'}{B'C'}:\frac{A'E'}{B'E'}$$

This equation, compared with the previous one, proves that D' and E' coincide and, hence, also d and d' do. Thus, the straight line l'', which lies on the three straight lines a,b,c, also lies on the fourth d. This proves the theorem.

An easy consequence of this proposition is that, if l,l',l'' are three skew straight lines and $a,b,c,d,\ldots$ is seen as the position of a mobile straight line which lies on l,l',l $''$, each straight line which lies on three of the four lines a,b,c,d also lies on the fourth one.

The proof of the second part of this theorem, namely that any planar section of the surface generated by such a mobile straight line is a conic, is further evidence that shows the profound interconnections of the edifice Chasles was constructing, an edifice whose milestones are composed of the anharmonic ratios among its elements (see Fig. 2.18).

Be L and L' the points where a plane cuts l and l', while A'', B'', C'', D'' are the points where the same plane cuts a,b,c,d. It is required to prove that the points L,L', A",B",C",D" lie on a conic section. Because of the projective generation of conics (proved by Chasles in Note XV of his *Aperçu*, as we will see in the next pages) it is enough to prove that the four straight lines LA'', LB'', LC'', LD'' of the pencil L and L' A'', $L'B''$, $L'C''$, $L'D''$ of the pencil L' have the same anharmonic ratio. The anharmonic ratio of the first quatern of straight lines is the same as that of the four planes through L and through each of such four lines, which, in turn, is the same as that of the four points A'',B'',C'',D", but this is exactly the same anharmonic ratio of the second quatern of straight lines. Thus, the two quaterns of straight lines have the same anharmonic ratio, and, hence, they generate a conic to which the six points L,L', A'',B'',C'',D'' belong.

These theorems state that the projective generation of a one-sheeted hyperboloid as this quadric is given by the array of all the straight lines which lie on three skew straight lines that form an array reciprocal to the former. Chasles' theorem proves that it is enough that the lines of an array lie on two lines of the other array and the enunciated relation among the anharmonic ratio holds. Furthermore, the projective generation of the conics allowed him to prove that each section of the hyperboloid is a conic, so that, independently of any other knowledge, he was also able to prove that the hyperboloid is a ruled second-degree surface. He stressed the importance of this projective generation of the hyperboloid when writing:

> Thus, the theorem of the double generation [that is with two arrays of skew straight lines] of the one-sheeted hyperboloid by means of a straight line is completely proved and this has happened through absolutely elementary geometrical considerations.[62]

Chasles continued stating that, as a matter of fact, the projective generation of the hyperboloid also allows to reach a general theorem from which the projective generation of the asymptotic cone of the hyperboloid also arises. As he wrote:

> It is proved in Analysis that the straight lines parallel to the generatrices of a hyperboloid drawn from a point of space form a cone of second degree. The theory of the anharmonic ratios offers once again an extremely easy proof of this proposition. It is enough to apply to the section of a cone through a plane the reasoning we have shown for a planar section of the hyperboloid. You see that such a section is, once again, a conic.[63]

If the point from which to draw the lines parallel to the generatrices of the hyperboloid is the centre of the hyperboloid, you get the asymptotic cone.

Therefore, the projective generation of the hyperboloid allowed immediately Chasles to also determine some of the basic features of the hyperboloid such as the existence of its asymptotic cone.

Among the first results which can be obtained through the concept of anharmonic ratio Chasles dealt with the theory of the involution of six points (ibid., Note X, pp. 308–327)[64] and the projective properties of the conic sections (ibid., Notes XV and XVI pp. 334–341 and 341–344, respectively).

Note XV is important because Chasles, thanks to the concept of anharmonic ratio, reached the projective generation of conics. The theorem can be enunciated like this: the intersection point of two plane projective but not perspectives pencils of radiuses is a conic. Basing on the concept of anharmonic ratio and on its conservation by projections and sections the proof is extremely easy, *d'un simplicité extreme* (ibid., p. 335). Chasles argued that: consider two points A and B on a circumference; from A and B draw respectively four straight lines L, M, N, O and L', M', N', O' cutting the circumference in four points l,m,n,o. The angles formed by any pair of lines of the

[62]Ibid., p. 307: "Ainsi, le théorème de la double génération de l'hyperboloïde à une nappe, par une droite, est démontré complétement, et par des considérations géométriques tout à fait élémentaires".

[63]Ibid., p. 307: "On démontre, en Analyse, que les droites menées par un point de l'espace, parallèlement aux génératrices de l'hyperboloïde, forment un cône du second degré. La théorie du rapport anharmonique donne encore une démonstration extrêmement facile de cette proposition. Il suffit d'appliquer, à la section du cône par un plan, le raisonnement que nous venons de faire pour une section plane de l'hyperboloïde; on voit que cette section est encore une conique".

[64]See also Chemla (2016, pp. 75–77).

first pencil are equal to those of the corresponding lines of the second pencil because these angles subtend the same arcs. Thus, the anharmonic ratio of the sinuses of the first angles is equal to the anharmonic ratio of the sinuses of the second angles. Since each conic can be seen as the projection of a circumference, so that the cross ratio is conserved, the theorem is proved.

I contend that in this demonstration, besides the cross ratio, another metric element is used, namely the circumference. The extension of graphic properties from the circumference to the other conics was a procedure adopted by Poncelet in his *Traité* and, after him, it came into use.

From a historical point of view, it should be recalled that Poncelet was not the first to deduce systematically properties of the conics considering them as a projection of a circumference. Maclaurin also did in his *A treatise of fluxions* (Maclaurin, 1742). As a matter of fact, at the beginning of the 14th chapter of his work Maclaurin claimed and demonstrated that several properties proved for the circle by Euclid, Pappus and other mathematicians also hold for the ellipse, regarding this conic as a section of a circular right cylinder. For Maclaurin considered the ellipse as the projection of a circle by straight lines parallel to the generatrices of the cylinder on a plane oblique with respect to the plane of the circle. The projection of the circle's centre determines the centre of the ellipse. Maclaurin was able to determine several properties of the circle which, through the described parallel projection, can be extended to the ellipse. Among these properties, the most significant ones are metric-projective. Poncelet and Chasles knew very well Maclaurin's work. It is not to exclude that, in the same way, they were influenced by him, though the geometrical context in which Poncelet and Chasles worked was incomparably broader than Maclaurin's. It should be added that the metric-projective properties proved by Maclaurin are fundamental in the treatment of the problem concerning the ellipsoid's attraction he developed at the end of the *Treatise*. We will see that, as to this issue, Maclaurin—jointly with Newton—was Chasles' essential reference source.

From a conceptual standpoint, an interesting question arises: is the use of the circumference, which is a metric object, correct in this projective context? The answer is not obvious: from a certain standpoint it is correct, because it is common in mathematics to prove a theorem which holds for a particular object having specific properties and to show that those properties are not significant for the extension of the theorem one was thinking of. On the other hand, Chasles exploited metric properties of the circumference, so that from an absolutely purist point of view, Chasles' proof is debatable. I am inclined to think that his demonstration is not subject to the criticism of having used a circumference. In this case, it seems to me that the metric properties of the circumferences are an instrument, which, after the use, disappears: something similar to the use of the imaginary numbers to solve the third-degree equations with three real roots. However, it should be highlighted that a strong purist might have some doubts as to the use of the circumference's metric properties in a projective context.

After the proof, Chasles also offered a beautiful dynamical image of the projective generation of the conics. For he considered three straight lines of a pencil and

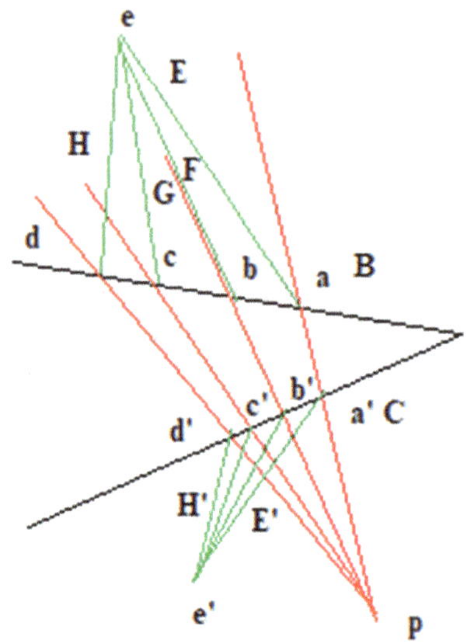

Fig. 2.19 The situation described by Chasles. The pencils generating the conic are those whose centres are e and e'. To avoid confusion in the picture I do not indicate the letters denoting the straight lines F', G'

their correspondent straight lines belonging to another pencil as fixed elements, while a fourth straight line of the first pencil and their correspondent in the second pencil were assumed to rotate in all the possible positions around the centres of their respective pencils. The intersection points of these variable fourth straight lines generate a conic determined by the two centres of the pencils and the three intersections of the fixed straight lines. Chasles proved several classical theorems as applications of the projective generation of conics based on the cross ratio. Two of the most remarkable ones concern Pascal's mystic hexagram theorem and Newton's organic generation of conics. For example, with regard to Pascal theorem (see Fig. 2.19),

Chasles argued that, given a fixed angle, a pole p and a transversal A rotating around the pole, such transversal in each of its positions will cut the two sides of the angle B,C in two points. Thus, if four points on one side of the angle a, b, c, d are determined jointly with their correspondent points a', b', c', d' on the other side, the two anharmonic ratios will be equal. Thence, if from a fixed point e four straight lines E, F, G, H are drawn to the points signed on a side of an angle and from a second fixed point e' four straight lines E', F', G', H' are drawn to the correspondent points on the other side of the angle, the intersections of the lines passing through the corresponding points generate a conic also passing through e and e'.

Therefore, Chasles concludes that:

> When the three sides of a triangle of variable shape rotate around three fixed points [in our case p, e and e'] and two vertices of the triangle run across two fixed straight lines [in our

case *B* and *C*], the third vertex generates a conic which passes through the two points around which the two sides adjacent to such a vertex rotate.[65]

Chasles claimed that this construction is nothing but Pascal's theorem expressed in another manner. As a matter of fact, this configuration was first introduced by William Braikenridge (1700–1762) around 1726, though his work was published only in 1733 and Maclaurin proved that from this configuration Pascal theorem is easily deducible (ibid., p. 336).

Relying on the concept of cross ratio, in Note XVI Chasles proposed the projective generation of the conics as envelope of tangents. In this case, the proposition which states such a generation reads as follows:

> When each of two coplanar straight lines is divided into four segments and the points of division of one line are made to correspond one to one to the points of division of the other line, if the anharmonic ratio of the first four points is equal to the anharmonic ratio of the other four points, the four lines joining the correspondent points and the two given straight lines will be six tangents to a conic.[66]

Thus, the cross ratio allows us to prove theorems for configurations of points with theorems for configurations of straight lines and vice versa. For example, to Pascal theorem, Brianchon theorem corresponds. To Desargues theorem stating that, given a quadrangle inscribed in a conic and a straight line *L* passing through two points of the conics, these two points are in involution with the three couples of points where the diagonals and the four sides of the quadrangle cut *L*, Sturm's theorem, according to which, given a quadrilateral circumscribed to a conic, the straight lines drawn from any point to its four vertices and the two tangents drawn from *l* to the conic form a pencil in involution corresponds, and so on (ibid., p. 341–342). It should be pointed out that, according to Chasles, the properties of the cross ratio which hold for quatern of points as well as for quaterns of straight lines belonging to a pencil are the basis of duality in the plane. We will see how he justifies duality in space.

Therefore, in these notes Chasles offered the basic properties of the cross ratio and some significant applications. The notes do not have a systematic character, whereas the two memoirs on duality and homography have. In these two memoirs, the anharmonic ratio plays a fundamental role.

A brief commentary is appropriate: in the works of several mathematicians before Chasles' *Aperçu*, it is possible to find segmentary relations which are equivalent to the cross ratio. Chasles himself, as we have seen, mentioned Pappus and De La Hire. More recently, many relations of Carnot's theory of transversals are based, from a

---

[65]Ibid., p. 336: "Quand les trois côtès d'un triangle, de forme variable, tournent autour de trois points fixes, et que deux des sommets du triangle parcourent deux droites fixes, le troisième sommet engendre une conique qui passe par les deux points autour desquels tournent les deux côtès adjacents à ce sommet".

[66]Ibid., p. 342: "Quand deux droites, situées dans un même plan, sont divisées chacune en quatre segments, et que les points de division de la première droite correspondent, un à un, à ceux de la seconde; si le rapport anharmonique des quatre premiers points est égal au rapport anharmonique des quatre autres, les quatre droites qui joindront un à un les points correspondants, et les deux droites données, seront six tangentes d'une même conique".

conceptual standpoint, on the concept of anharmonic ratio, as well as on diverse identities proved by Poncelet in his *Traité*. However, Chasles (and at the same time Möbius and Steiner) was the first to realize that the anharmonic ratio is a universal basis which permits to develop a unified view on projective geometry, being a projective invariant. It is the foundation of the theory of transformations, on which I will comment after having presented Chasles' treatment of the two most important projective transformations: duality and homography.

2.2.2 Duality

The first part of Chasles' memoir concerns duality. His intention was to prove the validity of the duality law.

Such proof is based upon four theorems from which the dual character of the analysed properties is conspicuous. The first theorem claims that if, in three-dimensional space, a plane with coordinates (x,y,z) is given such that the parameters of its equation contain to the first degree the coordinates of a point, called the director [*directeur*] by Chasles, then:

1) When the point moves on a plane, the mobile plane rotates around a fixed point.
2) When the point moves on a straight line, the mobile plane rotates around a straight line.
3) When the point moves on a curved surface, the mobile plane rotates on another curved surface. If the first surface is of degree 2, then the second surface also is. If the first surface is algebraic of degree m, then the second is algebraic, too, and given any line, it is possible to draw m tangent planes to it (Chasles, 1837a, p. 577).

Chasles pointed out that the fixed point in 1) is the pole of the plane traversed by the director point.

The language of movement used by Chasles is perfectly equivalent to the projective language: issue 1) claims that, in space, the correlative dual figure of the pointed plane is the bundle of planes. Issue 2) claims that the correlative dual figure of a straight line is a sheaf of planes. Issue 3) deals with the distinction between order and class of a curve.

I will refer to the proof of property 1) so that the reader can achieve a precise idea of Chasles' way of reasoning. Chasles named x', y', z' the coordinates of the director point; the equation of the mobile plane takes then the form

$$Xx' + Yy' + Zz' = U \tag{2.3}$$

where X, Y, Z, U are polynomials of the form $ax + by + cz - d$.

Be

$$Lx + My + Nz = 1 \tag{2.4}$$

the equation of the plane on which the director point moves (L, M, N are constant). If, in Eq. (2.4) you write the coordinate of the mobile point, you get the equation

$$Lx' + My' + Nz' = 1 \tag{2.5}$$

Comparing Eqs. (2.3) and (2.5), the following relations hold:

$$X = LU; \quad Y = MU; \quad Z = NU \tag{2.6}$$

Equations (2.6) have two features: a) they determine a point because L, M, N, U are constant values and b) this point belongs to the mobile plane whatever the value of the triple (x', y', z') on the plane of Eq. (2.5) be. Thence, Chasles can conclude that the mobile plane, in all its positions, passes through a fixed point whose coordinates are determined by Equations (2.6). Thus, the first part of the theorem is proved.

The second theorem proves that if the director point moves on a curved surface A, the surface enveloped by the mobile plane is the geometrical locus of all the planes tangent to A. The point where the mobile plane—in one of its positions—touches such an enveloped surface is the pole of the plane tangent to the surface A, drawn through the director point, to which this position of the mobile plane corresponds (ibid., pp. 578–579).

The third theorem claims that if the point moves at infinity, the mobile plane rotates around a fixed point, as if the director point traversed a plane (ibid., pp. 580–581).

The enunciation of this theorem is extremely significant: it makes clear that Chasles' conception of the elements at infinity is correct, but it is not yet included within a formalized geometrical structure because, for later projective geometry, the object at infinity of space is the plane at infinity, which has exactly the same projective properties as any other plane. Hence, there is no need of such a theorem as Chasles' third one.

The fourth theorem concerns the cross ratio: if the director point assumes four collinear positions a,d,c,d, the mobile plane assumes four corresponding positions A, B,C,D. The four planes belong to a sheaf and their anharmonic ratio (cross ratio) is equal to that of the four points (ibid., pp. 582–584). Namely:

$$\frac{\sin(C,A)}{\sin(C,B)} : \frac{\sin(D,A)}{\sin(D,B)} = \frac{ca}{cb} : \frac{da}{db} \tag{2.7}$$

This theorem expresses that a correlative projective transformation preserves the cross ratio.

Given the importance of this theorem, it is appropriate to refer to its proof (ibid., pp. 582–583) also considering that it offers some interesting food for thought.

The demonstration runs like this: since the points *a,b,c,d* are collinear (be *s* their straight line), according to Theorem 1), item 2) the planes *A,B,C,D* also are. Therefore, drawn any transversal cutting the four planes respectively in the points α, β, γ, δ, the following equation holds:

$$\frac{\gamma\alpha}{\gamma\beta} : \frac{\delta\alpha}{\delta\beta} = \frac{\sin(C,A)}{\sin(C,B)} : \frac{\sin(D,A)}{\sin(D,B)} \tag{2.8}$$

It is, thence, enough to prove that the first member of (2.8) is equal to the second member of (2.7).

If ξ, ν, ζ are the coordinates of a fixed point of *s* and r', r'', r''', r^{IV} the respective *distances* of the points *a,d,c,d* from *K*, their coordinates can be respectively written as

$$\left.\begin{aligned} x' &= \xi + lr' \\ y' &= \nu + mr' \\ z' &= \zeta + nr' \end{aligned}\right\}, \quad \left.\begin{aligned} x'' &= \xi + lr'' \\ y'' &= \nu + mr'' \\ z'' &= \zeta + nr'' \end{aligned}\right\}, \quad \left.\begin{aligned} x''' &= \xi + lr''' \\ y''' &= \nu + mr''' \\ z''' &= \zeta + nr''' \end{aligned}\right\}, \quad \left.\begin{aligned} x^{IV} &= \xi + lr^{IV} \\ y^{IV} &= \nu + mr^{IV} \\ z^{IV} &= \zeta + nr^{IV} \end{aligned}\right\} \tag{2.9}$$

where *l,m,n* are the director cosines of *s*.

Taking into account Eq. (2.3), the equation of the mobile plane *A* corresponding to the director point *a* will be

$$X\xi + Y\nu + Z\zeta - U + (lX + mY + nZ)r' = 0 \tag{2.10}$$

If ϱ' indicates the *distance* between *K* and the point where the plane of Eq. (2.10) cuts *s*, the coordinates of such a point will be

$$\left.\begin{aligned} x &= \xi + l\varrho' \\ y &= \nu + m\varrho' \\ z &= \zeta + n\varrho' \end{aligned}\right\} \tag{2.11}$$

This point belongs to *A*. Thence, obtaining the values of ξ, ν, ζ from (2.11) and replacing them in (2.10), the equation of such plane will be of the form

$$P + Q\varrho' + (R + S\varrho')r' = 0 \tag{2.12}$$

The expressions, *P, Q, R, S* are functions of ξ, ν, ζ, l, m, n.

Analogously, indicating by ϱ'', ϱ''', ϱ^{IV} the respective distances of *K* from the points where *B,C,D* cut *s*, the planes *B,C,D* will assume equations of the respective forms:

$$P + Q\varrho'' + (R + S\varrho'')r'' = 0$$

$$P + Q\varrho''' + (R + S\varrho''')r''' = 0$$

$$P + Q\varrho^{IV} + \left(R + S\varrho^{IV}\right)r^{IV} = 0$$

From these equations, the following identities are easily deducible:

$$\frac{\rho''' - \rho'}{\varrho''' - \varrho''} = \frac{r''' - r'}{r''' - r''} = \frac{Q + Sr''}{Q + Sr'}$$

$$\frac{\rho^{IV} - \rho'}{\varrho^{IV} - \varrho''} = \frac{r^{IV} - r'}{r^{IV} - r''} = \frac{Q + Sr''}{Q + Sr'}$$

And finally

$$\frac{\rho''' - \rho'}{\varrho''' - \varrho''} : \frac{\rho^{IV} - \rho'}{\varrho^{IV} - \varrho''} = \frac{r''' - r'}{r''' - r''} : \frac{r^{IV} - r'}{r^{IV} - r''} \tag{2.13}$$

But the first member of (2.13) is equivalent to the first member of (2.8) and the second member of (2.13) is equivalent to the second member of (2.7), which proves the theorem.

This demonstration is a prototype of several kinds of proofs in which the cross ratio is used. I point out that the concept of distance is employed and the term distance is explicitly mentioned by Chasles (I have indicated in italics when Chasles used the terms "distance", "distances"). This is the well-known problem with the cross ratio: it is a projective invariant, but it is constructed on a metrical basis. In my opinion, at least from a conceptual point of view, if not by means of a direct influence, Chasles foundational programme represents the first and crucial step towards the reduction of metric properties to graphic properties, whose following and conclusive steps were Von Staudt's theorem on the complete quadrangle and the Cayley–Klein metric.

The demonstration of the duality law for space is divided into two parts. In the first part, Chasles gave the definition of two correlative figures and established the duality points-planes-straight lines/planes-points-straight lines. What is very interesting is the second part of the theorem:

> In two correlative figures, to four collinear points of the former, four planes belonging to a sheaf correspond in the latter. Their anharmonic ratio is the same as that of the four points.
>
> To four planes belonging to a sheaf of the former, four collinear points correspond in the latter. The anharmonic ratio of the four points is the same as that of the four planes.[67]

[67] Ibid., p. 588: "Dans deux figures corrélatives, à quatre points la première, situés en ligne droite, correspondent, dans la seconde, quatre plans passant par un même droite, et dont le rapport anharmonique est égal au rapport anharmonique des quatre points; Et, à quatre plans de la première

This principle is an immediate application of this theorem.

Chasles clarified how the cross ratio can be used to deal with metric-graphical properties, which is one of the main purposes of his foundational programme: let $a,b,$ c,d be four points of one of the two correlative figures and A,B,C,D the correlative planes in the other figure. If a transversal is drawn cutting the four planes A,B,C,D respectively in α, β, γ, δ, the relation

$$\frac{\gamma\alpha}{\gamma\beta}:\frac{\delta\alpha}{\delta\beta}=\frac{ca}{cb}:\frac{da}{db} \tag{2.14}$$

holds. If d, for example, is at infinity, the relation is simplified to

$$\frac{\gamma\alpha}{\gamma\beta}:\frac{\delta\alpha}{\delta\beta}=\frac{ca}{cb}. \tag{2.15}$$

If the two points at infinity mutually correspond, the relation is further simplified to

$$\frac{\gamma\alpha}{\gamma\beta}=\frac{ca}{cb}. \tag{2.16}$$

The application to metric-graphical properties runs as follows: suppose there is a relation between several points of a figure. Form the correlative figure. An equation similar to (2.7) will be achieved. One will try to eliminate the distances between the points of the first figure from Eq. (2.7) (something similar to the elimination of the cosines in the parabolic transformation). A relation between sinuses will remain, or between the segments such as those indicated by Greek letters in Eqs. (2.14)–(2.16). If it is possible to introduce relations concerning elements at infinity, the whole proposed problem will be simplified. Chasles claimed that the metric applications of the projective theory are the most important part of the duality principle because in most of the researches concerning extension, metric properties are involved (ibid., pp. 595–596).

As always in mathematics, an example is useful for clarifying the application of a general concept or procedure. Chasles offered a large number of applications of his method to several properties of algebraic curves and surfaces and obtained more specific results for the second-degree curves and surfaces. I will present one example concerning the already analysed concept of "centre of the mean distances".

Let us reconsider the centre of the mean distances g of a series of collinear points $a,b,c,\ldots,g,m$ and the definitional identity (ibid., pp. 616–618)

figure, passant par un même droite, correspondent, dans la seconde figure, quatre points situés en ligne droite, dont le rapport anharmonique est égal au rapport anharmonique des quatre plans".

$$ma + mb + mc + \ldots = n \cdot mg$$

which can also be written

$$\frac{ma}{mg} + \frac{mb}{mg} + \frac{mc}{mg} + \ldots = n.$$

Given the system of collinear points *a,b,c,* ..., the correlative figure will be given by a sheaf of planes *A,B,C,...,G,M*. Be *I* the plane corresponding to the point at infinity i_∞ of the line *ab*. The general principle of correlation, expressed by Eq. (2.7), then becomes

$$\frac{\sin(M,A)}{\sin(M,G)} : \frac{\sin(I,A)}{\sin(I,G)} + \frac{\sin(M,B)}{\sin(M,G)} : \frac{\sin(I,B)}{\sin(I,G)} + \ldots = \frac{ma}{mg} : \frac{i_\infty a}{i_\infty g} + \frac{mb}{mg} : \frac{i_\infty b}{i_\infty g} + \ldots = \frac{ma}{mg} + \frac{mb}{mg} + \ldots = n.$$

This equation can be written in the form

$$\frac{\sin(M,A)}{\sin(I,A)} + \frac{\sin(M,B)}{\sin(I,B)} + \ldots = n \frac{\sin(M,G)}{\sin(I,G)}.$$

Now draw a transversal cutting the planes *A,B,C,... G,M* respectively in the points α, β, γ, ..., θ, μ, *i*. Given the properties of the cross ratio, it is the case that

$$\frac{\mu\alpha}{\mu\theta} : \frac{i\alpha}{i\theta} = \frac{\sin(M,A)}{\sin(M,G)} : \frac{\sin(I,A)}{\sin(I,G)}.$$

Therefrom, it follows that:

$$\frac{\mu\alpha}{i\alpha} + \frac{\mu\beta}{i\beta} + \frac{\mu\gamma}{i\gamma} + \ldots = n \frac{\mu\theta}{i\theta}. \tag{2.17}$$

The point *m* is arbitrary. Therefore, the plane *M*, which in the initial reciprocity, corresponds to *m*, and the point μ, which, in the reciprocity between the planes *A,B, C,* ... and the points α, β, γ, ..., corresponds to the plane *M*, are arbitrary as well. This means that, in order to grasp some metrical properties, it is legitimate to consider μ as the point at infinity on the straight line $\alpha\beta$. Given this condition, Eq. (2.17) takes the form

$$\frac{1}{i\alpha} + \frac{1}{i\beta} + \frac{1}{i\gamma} + \ldots = n \frac{1}{i\theta}.$$

This equation expresses the property of a concept that had been studied deeply at that time. For the distance $i\theta$ is the harmonic mean of the distances $i\alpha$, $i\beta$, ... (the locution "harmonic mean" associated with segments and not only with numbers had been used by McLaurin) and Poncelet had called the point θ "the centre of the harmonic means among the points α, β, γ, ... " and had dedicated important investigations into the use of this concept in a geometrical context.[68]

These considerations have a geometrical interpretation in the following theorem:

> Be given the planes *A*,*B*,*C*,... passing through a straight line and be *I* the last of them. A plane *G* that passes through this line exists such that, any transversal being drawn, the centre of the harmonic means of the points where the transversal cuts the given planes in relation to the point where it cuts the plane *I* belongs to *G*.[69]

The reasoning used by Chasles includes:

1) The use of the projective transformations to obtain general graphical properties.
2) The particularization of some of the elements of a transformation (in this case by introducing elements at infinity) to get metric-graphical properties from the cross ratio.
3) The intrinsic compositionality of the projective transformations. Namely the product of two projectivities is a projectivity.

Issue 3) is particularly significant. It is not made explicit by Chasles. Nonetheless, such a concept of compositionality is clear and, in the perspective of the history of mathematics, it shows that the acquisition of the concept of group of transformations passes through a series of steps, of which these researches developed by Chasles can be considered as an initial phase.

What is particularly interesting is to fully understand the reason why Chasles considered that the foundation of the projective doctrine on the concept of anharmonic ratio overcame and generalized the theory of polar reciprocity. As a matter of fact, he pointed out the well-known truth that in two polar correlative figures, to a point of space considered as belonging to the first figure and to the second figure, the same plane corresponds. This stresses the involutory character of the polarities, which is not true anymore for two general correlative figures which are not polar reciprocals (ibid., 659), as he had proved analytically in the XVIIIth section of this memoir (ibid., 639–643). Therefore, the profound invariant property of correlations is given not by the polar reciprocity but by the conservation of the anharmonic ratio, which holds for any reciprocal projective transformation and, in general, for any projective transformation. It follows that the theory of reciprocal polars is a part of the theory of anharmonic ratios. Chasles wrote explicitly that no

[68] See, e.g., Poncelet (1828a).

[69] Ibid., p. 618: "Étant donnés plusieurs plans *A*, *B*, *C*,... et un dernier plan *I*, passant tous par une même droite; il existera toujours un certain plan *G*, passant aussi par un cette droite, et jouissant de cette propriété, que, si l'on mène une transversale quelconque, le centre des moyennes harmoniques des points où elle percera le plans *A*,*B*,*C*,... par rapport au point où elle percera le plan *I*, sera toujours dans ce plan *G*".

one before him had thought of placing such relation at the foundation of the whole doctrine of projectivity. We read:

> [...] although the anharmonic relation is projective, no one has thought of assuming it as the sole prototype of the projective relations or of the relations transformable through the principle of duality. This means that the anharmonic ratio is the unique form to which all the other relations have to be compared and reduced. This gives a precision and a generality to the transformation methods, which, beforehand, were limited and had something vague and uncertain in their applications.[70]

These words express Chasles' thought very clearly and can be regarded as a sort of synthesis of his foundational programme. That said, they explain the most significant part of his convictions, but not all of them. He actually expounded another conviction which shows that his mentality was still tied to the idea that internal properties of the figures existed which can be proved by duality, but which hold for more intrinsic reasons than duality. In practice, the duality law—based on the conservation of the cross ratio—explains the ὅτι, but not the διότι, to use the Aristotelian distinction; i.e. the duality law explains "that" something happens, but not "why" it happens. This idea is so clearly expressed by the following quotation that no commentary is necessary:

> We consider the transformation methods a precious means for the discovery of new theorems and the proof of some partial truths. But, insofar as truths belonging to an already formed theory are concerned, the proofs produced by these artificial methods do not seem completely satisfactory. This theory has to find in itself the resources necessary for the direct demonstration of its truths, without relying upon the corresponding truths in the correlative theory. So, for example, if we seek to enter the treatment of the second-degree surfaces using the new properties found in the previous paragraphs, it will be necessary that we prove these properties directly, without exploiting the principle of duality. These direct proofs will bring a notable improvement in the geometrical theories.[71]

---

[70]Ibid., p. 661: "[...] bien que la relation anharmonique soit projective, on n'a pas songé à la prendre pour le type unique des relations projectives, ni des relations transformables par le principe de dualité, c'est-à-dire pour la forme unique à laquelle devaient être comparées et ramenées toutes les autres relations; ce qui donne un caractère de généralité et de précision aux méthodes de transformation, qui auparavant étaient restreintes, et avaient quelque chose de vague et d'incertain dans leurs applications". Actually, though several geometers before Chasles had considered expressions mathematically equivalent to what he called *rapport anharmonique*, no one before him thought of defining such expressions as "single objects". In this case, Chasles seems to concede to his predecessors a—though partial—consciousness of the cross ratio's importance they did not have.

[71]Ibid., p. 615: "Car nous considérons les méthodes de transformation comme des moyens précieux pour la découverte de théorèmes nouveaux, et la démonstration de quelques vérités partielles; mais, quand il s'agit de vérités appartenant à une théorie déjà formée, les démonstrations que procurent ces méthodes artificielles ne nous paraissent pas complétement satisfaisant: cette théorie doit trouver en elle-même les ressources nécessaires pour la démonstration directe des vérités qui lui appartiennent, sans qu'on soit obligé de s'appuyer sur les vérités correspondantes, dans la théorie corrélative. Ainsi, par exemple, si nous faisons entrer, dans un Traité des surfaces du second degré, les propriétés nouvelles que nous avons trouvées dans les paragraphes précédents, telle que celle des *axes conjugués relatifs à un point*, ce serait directement que nous démontrerions ces propriétés, et

The theory of homography should be the final step towards an intrinsic and convincing foundation of projective geometry and towards the reduction of metric properties to graphical ones.

2.2.3 Homography

With regard to the way in which Chasles introduced and used the principle of homography, I will follow the same line of explanation adopted for duality. Therefore, the genesis of this principle and an example of its utilization will be proposed.

From a conceptual point of view, Chasles considered the homographies as deriving from the product of two reciprocities (ibid., 695). Consider three figures F,F',F'', being F' the reciprocal of F and F'' the reciprocal of F', then, F and F'' are homographic. The following relations between F and F'' hold:

A. To each plane, line, point of F, a plane, a line, a point of F'' correspond. Therefore, in particular, to the points at infinity of F, coplanar points in F'' correspond.
B. The anharmonic ratios of four collinear points are conserved in the transformation.
C. The anharmonic ratios of four planes belonging to a sheaf are conserved in the transformation.

By using the concept of anharmonic ratio, Chasles proved three important metric properties of two homographic figures. Let us see the proof of the first property: let a, b,c,d be four collinear points of F and a',b',c',d' their homologous points in F''. The identity of anharmonic ratios

$$\frac{ca}{cb}:\frac{da}{db}=\frac{c'a'}{c'b'}:\frac{d'a'}{d'b'}$$

holds. This equation can be written in the form

$$\frac{ca}{cb}:\frac{c'a'}{c'b'}=\frac{da}{db}:\frac{d'a'}{d'b'}$$

The second member does not depend on the position of c in the straight line $abcd$. This means that the ratio

non par le principe de dualité. Ce sont ces démonstrations directes qui, nécessairement, amèneront un perfectionnement notable dans les théories géométriques".

$$\frac{ca}{cb}:\frac{c'a'}{c'b'}$$

is constant for any position of c. Draw a plane M through c and consider it as belonging to the first figure. Let M' and c' be homologous to M and c. Be p and q the distances of the plane M from the points a and b respectively, then:[72]

$$\frac{p}{q}=\frac{ca}{cb}$$

Analogously, the following identity holds (the meaning of the symbols is evident):

$$\frac{p'}{q'}=\frac{c'a'}{c'b'}.$$

Therefore

$$\frac{p}{q}:\frac{p'}{q'}=\text{const.}$$

This means that:

> In two homographic figures, the ratio of the distances of any plane belonging to the first figure, to two fixed points of such figure, is to the ratio of the distances of the homologous plane in the second figure to the two fixed points that correspond to those given in the first figure, in a constant ratio.[73]

The second fundamental proposition is the following one, which is the dual of that just proved:

> In two homographic figures, the ratio of the distances of any point belonging to the first figure, to two fixed planes belonging to this first figure, as well, is to the ratio of the distances of the homologous point in the second figure to the two fixed planes that correspond to those given in the first figure, in a constant ratio.[74]

The third fundamental property needs the distinction between the finite elements and the elements at infinity. Its enunciation is the following:

---

[72] My explanation: consider the orthogonal projection a_1 of a on M and the orthogonal projection b_1 of b. The triangles aa_1c and bb_1c are similar and $aa_1=p$, $bb_1=q$.

[73] Ibid., p. 698: "Dans deux figures homographiques, le rapport des distances d'un plan quelconque de la première, à deux points fixes appartenant à cette figure, est au rapport des distances du plane homologue, dans la seconde figure, aux deux points fixes qui correspondent à ceux de la première figure, dans une raison constante".

[74] Ibid., p. 699: "Dans deux figures homographiques, le rapport des distances d'un point quelconque de la première, à deux plans fixes appartenant à cette première figure, est au rapport des distances du point homologue dans la seconde figure, aux deux plans fixes qui correspondent aux deux premiers, ceux de la premiers, dans une raison constante".

> In two homographic figures, the distance of any point belonging to the first figure, to a fixed plane belonging to this first figure, as well, is to the ratio of the distances of the homologous point in the second figure to the two planes that correspond, in this figure, the former to the fixed plane and the second to the plane at infinity of the first figure, in a constant ratio.[75]

At this point, Chasles concluded the introductory section concerning homography with a series of considerations which make his foundational vision clear.

He claimed that the principle of homography or of descriptions of figures ("description de figures") includes two parts, the former concerning the descriptive relations and the latter the metric relations. The descriptive relations consist in the fact that to any point and plane of a figure, a point and a plane of the other figure correspond. The metric relations correspond to the conservation of the anharmonic ratios. Chasles concluded by expressing the clear idea that the metric relations can be reduced to the graphical relations, something proved by Cayley in his celebrated *A Sixth Memoir upon Quantics* (Cayley, 1859), even though Chasles expressed this concept several years before Cayley. So, Chasles paved the way for the following researches concerning the reduction of metric properties to graphical ones. He wrote:

> From here, it follows that, in the correlative figures, the metric relations are a consequence of the descriptive relations. But, if we present directly—that is, without resorting to the principle of duality—the theory of the homographic figures, we will refer to the definition,[76] which depends only on graphical properties, and from this definition itself, we will derive the metric relations of the figures and all their properties.[77]

Chasles added that the entire theories of plane perspective and of homology enters the theory of homography as particular cases.

These ideas pervade Chasles' thought. For they are reproposed in a substantially analogous manner in Chasles (1852), where an interesting chapter dedicated to the metric properties of the homographic figures is inserted (Chasles, 1852, pp. 342–347) and where the properties of the homological figures are also treated after having explained the spatial genesis of the plane homology (ibid., pp. 352–356).

Commentary: in Chasles' mind *almost* all the most important features of the relations between metric and graphic properties were clear. He had understood that

---

[75] Ibid., p. 700: "Dans deux figures homographiques, le distance d'un point quelconque de la première, à un plan fixe de cette première figure, est au rapport des distances du point homologue, dans la seconde figure, aux deux plans qui correspondent, dans cette figure, l'un au plan fixe e l'autre à l'infini de la première, dans aux deux premiers, ceux de la premiers, dans une raison constante".

[76] The definition is simply that if two figures satisfy the condition that to any point and any plane of the former, a point and a plane of the latter correspond, then the two figures are *homographic* (ibid., 701).

[77] Ibid., p. 701: "Cela résulte de ce que, dans les figures *corrélatives*, les relations métriques sont aussi une conséquence des relations descriptives. Mais quand nous présenterons directement, et sans le secours du principe de dualité, la théorie des figures homographiques, nous nous renfermerons dans la définition que nous venons de faire reposer sur leurs relations descriptives seules, et nous conclurons, de cette définition même, les relations métriques des figures et toutes leurs propriétés".

the cross ratio is an invariant quantity per projections and sections and that it is usable to reduce the metric properties to those graphic. What was still missing is the idea that the cross ratio could be replaced by a completely graphic construction—namely the one concerning the harmonic group of four points and four planes—and that all the projective properties could be obtained studying the conservation of the harmonic groups in the graphic transformations of two forms of first species. This proved, the cross ratio can be associated with the projective transformations when metrical properties have to be obtained, but in this case all the graphic properties are obtained without using the cross ratio, which, therefore, becomes a useful application when particular metric properties have to be obtained, and not the concept through which the graphic properties themselves are achieved. This important step was done by Von Staudt, but Chasles' train of thought is not only the initial but a fundamental passage towards the reduction of the metric properties to those graphic.

The applications of the theory of homography are numerous and Chasles laid the bases for a series of new fundamental concepts: in the 13th chapter (ibid., 739–745), he introduced a new system of coordinates, which he defined as a generalization of the usual systems of coordinates. They are the projective coordinates for spatial geometry. Chasles referred to the coordinates of a point to a fundamental tetrahedron. But this is not all: in the seminal 15th chapter (ibid., 754–764) he proved what is known as the "fundamental theorem of projectivity for projective space", i.e. given five independent elements in a homography, any sixth element is determined automatically. In modern terms, in the form of third species,[78] five elements determine univocally a homography. The proof is developed by relying on the properties of the anharmonic ratio. It is well known how difficult it is to reach a rigorous proof of this theorem. All the subtleties—in great part connected to the concept of continuity—involved in the demonstration were extraneous to Chasles' basic reasoning.[79] Nonetheless, his ideas were clear and he deserves a place in the history of the fundamental theorem.

Let us see Chasles' reasoning to prove this fundamental theorem. He argued like this (Fig. 2.20):be given a figure in space and five points a, b, c, d, e belonging to it. Let us suppose that these five points correspond to other five points, respectively, a', b', c', d', e', in a homography. Then the homography is determined, i.e. the figure homographic to the given one is determined, so that it is possible to find the homographic transformed m' of any point m and M' of any plane M. For given the two tetrahedra $abcd$, $a'b'c'd'$, draw the planes ebc, mbc of the first figure which cut the edge ad respectively in ε and α and in the second figure the planes $e'b'c'$, $m'b'c'$

[78]The projective forms of first species are 1) dotted straight line; 2) pencil of lines; 3) sheaf of planes. In the language of movement, they are generated by a *simple movement* of their generating elements (points, lines and planes, respectively). The projective forms of second species are 1) dotted plane; 2) ruled plane; 3) bundle of straight lines; 4) bundle of planes. These forms are generated by a *double movement* of their generating element. The projective forms of third species are 1) space of points; 2) space of planes. These forms are generated by a *triple movement* of their generating element.

[79]See Voelke (2008).

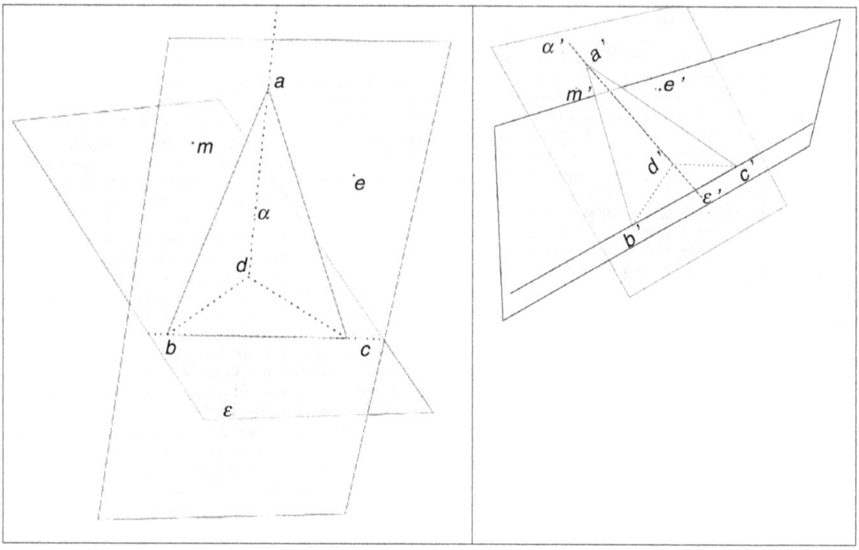

Fig. 2.20 Figure which reconstructs the situation described by Chasles

(the position of the point m' is, for the moment, unknown), which cut the edge $a'd'$ respectively in ε', α'. The four points a', d', ε', α' are, in the second figure, homologous to a, d, ε, α in the first figure. Therefore, the following relation between the cross ratio holds:

$$\frac{a\alpha}{d\alpha} : \frac{a\varepsilon}{d\varepsilon} = \frac{a'\alpha'}{d'\alpha'} : \frac{a'\varepsilon'}{d'\varepsilon'}$$

The three points a', d', ε' are given. This equation will, therefore allow us to determine the point α'. Thence, the plane passing through the edge $b'c'$ of the tetrahedron $a'b'c'd'$ and through the required point m' will be determined. By means of two similar equations, the planes passing through the edges $a'c'$; $a'b'$ and through the required point m' will be determined, so that the point m' is obtained as the intersection of three planes.

The proof concerning the determination of the plane M' homologous of M could have been obtained by Chasles through duality. Instead, he presents a direct demonstration. In the next commentaries, I will explain why.

An interesting metrical application of the train of thought just expounded concerns the relations between segments, areas and volumes of particular homographic figures. As a matter of fact, Chasles proved that:

> Given any two tetrahedra $abcd$, $a'b'c'd'$, for any point of a given figure draw three planes passing through the three edges bc, ca, ab of the first tetrahedron. These planes cut the opposite edges of the tetrahedron respectively in the points α, β, γ. On the three edges $a'd'$, $b'd'$, $c'd'$ of the second tetrahedron choose three points α', β', γ' such that

$$\frac{\alpha a}{\alpha d}=\lambda\frac{\alpha' a'}{\alpha' d'};\ \frac{\beta b}{\beta d}=\mu\frac{\beta' b'}{\beta' d'};\ \frac{\gamma c}{\gamma d}=\nu\frac{\gamma' c'}{\gamma' d'} \tag{2.18}$$

> where λ, μ, ν are any three constants. Under this condition, the intersection point of the three planes $\alpha' b' c'$; $\beta' c' a'$; $\gamma' a' b'$ belongs to a second figure which is homographic to the first one.[80]

The relations (2.18) referring to homographic figures do not appear particularly significant from a metric point of view. They become significant if some elements are specified like Chasles did: suppose (using a modern language) that we have a homography in which the plane at infinity is fixed. Chasles hypothesized, hence, that to the plane at infinity *abc* of the first figure corresponds to the plane at infinity $a'b'c'$ of the second figure and vice versa.[81] Then, the following relations hold (ibid., pp. 761 and 811):

$$\alpha d=\lambda\cdot\alpha' d';\beta d=\mu\cdot\beta' d';\gamma d=\nu\cdot\gamma' d' \tag{2.19}$$

Vice versa, if these relations hold, the two figures are homographic and the planes at infinity are reciprocally homologous. Therefore, *from an affine point of view*, to two parallel straight lines or planes of the first figure, two parallel straight lines or planes of the second figure correspond. From *a metric standpoint*, two homologous lines are divided into proportional parts by homologous points (ibid., p. 812), i.e. if the points *a,b,c* belong to a straight line and a',b',c' to its homologous, then it holds that

$$\frac{ba}{bc}=\frac{b'a'}{b'c'} \tag{2.20}$$

In conclusion, in this kind of homographies the following theorems are valid:

1. *The ratio of two segments considered on two parallel lines in the first figure is equal to the ratio of the corresponding segments in the second figure.*
2. *The ratio of the areas of two plane polygons placed between parallel planes and belonging to the first figure is equal to the ratio of the areas of the two corresponding polygons in the second figure.*

---

[80] Ibid., p. 757: "Étant donnés deux tétraèdres quelconques *abcd*, $a'b'c'd'$; si, par chaque point d'une figure donnée, ou mène trois plans, passant par les trois arêtes *bc, ca, ab* du premier tétraèdre, et rencontrant respectivement les arêtes opposées en α, β, γ et que, sur les trois arêtes $a'd'$, $b'd'$, $c'd'$ du second tétraèdre, on prenne trois points α', β', γ', déterminés par les trois équations $\frac{\alpha a}{\alpha d}=\lambda\frac{\alpha' a'}{\alpha' d'};\ \frac{\beta b}{\beta d}=\mu\frac{\beta' b'}{\beta' d'};\ \frac{\gamma c}{\gamma d}=\nu\frac{\gamma' c'}{\gamma' d'}$, λ, μ, ν, étant trois constants, prises arbitrairement: Le point d'intersection des trois plans $\alpha' b' c'$; $\beta' c' a'$; $\gamma' a' b'$, appartiendra à une seconde figure qui sera HOMOGRAPHIQUE à la première". Capitals in the text.

[81] Note that Chasles continued to think of different planes at infinity for any figure. But this does not compromise the correctness of his reasoning.

3. *The volumes of two corresponding parts of the two figures are in a constant ratio.*[82]

This is interesting because these features denote the properties of the transformation which we call an affinity or affine transformation.

Commentary: the way in which Chasles interpreted the duality law is rather subtle: he wanted to establish homography as a principle which holds independently of duality, a very foundational principle. The question is rather complex, since, as we have seen at the beginning of this subsection on homography, Chasles introduced this concept basing on two reciprocities and the notion of reciprocity is tied to that of duality. Then, his idea was to develop the theory of the anharmonic ratio independently of duality and to construct reciprocities and homographies—thence duality itself—on the theory of cross ratio which will be the foundational cornerstone of his programme. This is noteworthy because, as we will see in the fourth section, Chasles developed an authentic philosophy of duality. However, the basic duality, that born in the context of projective geometry, should be founded on an even more primitive concept: that of anharmonic ratio. Chasles' train of thought is clearly expressed at the end of chapter XV (ibid., pp. 754–764), the same in which the previous theorems are proved, where he demonstrated the two following statements, which are a re-proposition of the theorem in which Eq. (2.18) appears (ibid., pp. 762–763):

1) Be given a tetrahedron *SABC* and any plane π (Fig. 2.21). For any point Q of π three planes passing through the three edges of the basis *ABC* of *SABC* be drawn. Cut these planes the edges opposite to the one belonging to them respectively in α,β,γ. Consider the ratios

$$\frac{\alpha S}{\alpha A}, \frac{\beta S}{\beta B}, \frac{\gamma S}{\gamma C}.$$

Consider, then, a second tetrahedron $S'A'B'C'$ and assume three points α',β',γ' on its edges so that the three ratios

$$\frac{\alpha' S'}{\alpha' A'}, \frac{\beta' S'}{\beta' B'}, \frac{\gamma' S'}{\gamma' C'}$$

are to the previous ratios in given and constant ratios, then the planes drawn respectively through the point α' and the edge $B'C'$; through the point β' and

[82]Ibid., p. 813: "1. Le rapport de deux segments, pris sur deux droites parallèles quelconques, dans le première figure, est égal au rapport des deux segments correspondants, dans la seconde figure. 2. Le rapport des aires de deux polygones plans quelconques, situés dans deux plans parallèles, appartenant à la première figure, est égal au rapport des aires des deux polygones correspondants, dans la seconde figure. 3. Les volumes de deus parties correspondantes des deux figures sont dans un rapport constant".

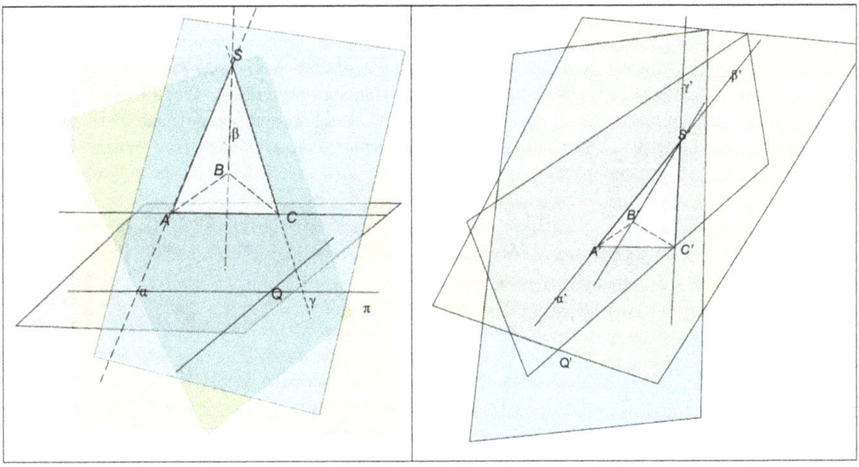

Fig. 2.21 Diagram representing the situation described by Chasles

the edge $A'C'$; and through the point γ' and the edge $A'B'$ mutually cut in a point whose locus is a plane.

2) Be given a tetrahedron *SABC* and any point *P*. Rotate a plane π around *P*. For a given position, cut π the edges *SA*, *SB*, *SC* of the tetrahedron respectively in three points α,β,γ. Consider the following ratios of the segments made by the plane with the edges

$$\frac{\alpha S}{\alpha A}, \frac{\beta S}{\beta B}, \frac{\gamma S}{\gamma C}.$$

Consider, then, a second tetrahedron $S'A'B'C'$ and a transversal plane π' cutting its edges $S'A'$, $S'B'$, $S'C'$ respectively in three points α',β',γ' such that the three ratios

$$\frac{\alpha' S'}{\alpha' A'}, \frac{\beta' S'}{\beta' B'}, \frac{\gamma' S'}{\gamma' C'}$$

are to the previous ratio in constant ratios, the plane π' will pass, in all its positions, through a point.

Chasles' commentary is the most evident litmus paper of his ideas. As he wrote:

> This theorem and the previous one, which we have deduced from the theory of homographic figures, contain the whole doctrine. If one of these two theorems were proved *a priori* [namely without any reference to duality], we would deduce from it our principle of homographic deformation including the descriptive and metrical relations of the figures.

> One or the other of these two theorems is that of which we have spoken in our *Aperçu historique sur les méthodes en géométrie* (Fifth Epoch, § 28) claiming that the whole doctrine of *the transformation of the figures in other figures of the same kind* is based on a lone and unique theorem of geometry. We will give, in another writing, which will deal with the *anharmonic ratio* and its numerous applications, the direct and geometrical proof of this theorem, so that the principle of *homographic deformation* is proved directly and independently of the duality principle.[83]

Chasles wrote the text of which he spoke in this quotation, though several years after 1837. It is the *Traité de Géométrie supérieure* (Chasles, 1852) in which the whole theory of the cross ratio is completely expounded.

Thence, Chasles foundational programme is absolutely clear: the basic concept is that of anharmonic ratio.

Therefore, in the *Aperçu*, Chasles came up with the idea of constructing the foundations of geometry by relying on the concepts of transformation and anharmonic ratio. We have seen that this conclusion represents the highest point in a coherent itinerary that he had been thinking of, starting as early as 1827. One of Chasles' main concerns in this itinerary is the attempt to consider the metric properties as specific cases of graphical configurations and related transformations. In the first phase, the 4 years 1827–1830, the specified element is the figure with respect to which a polarity is considered. This figure is a parabola or a paraboloid. In the second phase, the specified element is the value of the anharmonic ratio of projective transformations which are required to yield the figures' metric properties. Thus, it is clear in what sense the metric properties can be interpreted as particular graphic properties. What is still missing is the projective definition of the size of an angle and of the distance between two points.

The whole picture just described shows that Chasles had a truly foundational programme in his mind and realized it over the course of several years. Certainly, he was not an isolated thinker. Other mathematicians had similar ideas, but no one developed a complete and original foundational programme.

[83] Ibid., pp. 763–764: "Ce théorème et le précédent, que nous venons de déduire de la théorie des figures homographiques, en renferment, l'un et l'autre, toute la doctrine. Si l'un ou l'autre de ces deux théorèmes était dèmontré *a priori* et directement, nous en conclurions notre principe de déformation homographique, comprenant les relations de description et les relations de grandeur des figures.

C'est l'un ou l'autre de ces deux théorèmes dont nous avons voulu parler dans notre *Aperçu historique sur les méthodes en géométrie* (cinquième Époque, §28) en disant que toute la doctrine de *transformation des figures en d'autres du même genre*, reposait sur un seul et unique théorème de Géométrie. Nous donnerons dans un autre écrit, qui traitera du *rapport anharmonique* et de ses nombreuses applications, la démonstration directe et géométrique de ce théorème. De sorte que le principe de *déformation homographique* se trouvera démontré directement, et indépendamment du principe de dualité". Italics in the text.

2.3 A Comparison with Steiner's Conceptions

The situation concerning the studies on projective geometry between the 20s and the 30s of the nineteenth century is extremely multifaceted and one might think of different foundational approaches which converge in a sole monumental foundational programme. To cite just a few mathematicians, Poncelet (1788–1867) in his *Traité* (1822) had the enormous merit to found projective geometry (which he called modern geometry) on a set of new synthetic concepts and principles which—though not always completely rigorous—gave a uniform vision of the results obtained before the publication of the *Traité* and allowed him to achieve a strong development of the researches concerning the projective properties of the figures; Poncelet and Gergonne (1771–1859) developed the conception of duality which is one of the milestones of the edifice of projective geometry. Poncelet was only 5 years older than Chasles, while Gergonne 22 years older. They were active, with important publications, until the second half of the nineteenth century, so that their work intersected that of Chasles during many years. However, their publications preceding the 1830 belong, so to say, to the first phase of the birth of modern geometry, a very productive period in which the ground concepts were created, sometimes at the expense of an absolute perspicuity. This first phase also has a protohistory represented by Monge's descriptive geometry and by Carnot's theory of transversals. In the following phase of projective geometry (which corresponds to the end of the 20s and the 30s), other protagonists came on the scene. The most important are Chasles, Steiner (1796–1863), Plücker (1801–1868) and Möbius (1790–1868).

Plücker, who was 8 years younger than Chasles, gave some fundamental contributions, the most important of which were a) Plücker (1828–1831), where he introduced the abridged notation and explored the possibility to consider the straight lines rather than the points as the basic elements of geometry. This allowed him to develop an elegant formulation of the duality principle; b) Plücker (1835), where he introduced the system of coordinates which are nowadays known as Plücker coordinates; and c) Plücker (1839), where the celebrated formula which connects the number of singularities of an algebraic curve to that of its dual curve is proposed and proved.

Möbius, 3 years older than Chasles, in his *Der barycentrische Calcul* (1827) gave profound contributions to the foundation of projective geometry: a) he invented a new method to determine the barycentre of any system of points on the basis of the position of three points he called *fundamental points* of the barycentric coordinates; b) he taught how to pass from barycentric to Cartesian coordinates and vice versa; c) he fully understood the importance of the affine transformations (*Verwandtschaf*) within the theory of projections; and d) in the second part of his work, he introduced the concept of *Doppelschnittsverhaeltniss* (cross ratio) and applied it to the geometrical nets, from which the definition of complete quadrangle follows. The aim of Möbius was to connect, through the barycentric calculus, the evidence of the synthetic method with the generality of that analytic (Möbius, 1827, p. 12). All these authors contributed greatly to the foundation of projective geometry.

It should be noted that Chasles' foundational programme differs from the ideas of Plücker and Möbius for the following reasons: 1) Chasles developed the problem of the relations between metric and graphic properties in a more profound manner than the two German mathematicians;[84] 2) he reached the conclusion that the cross ratio is the fundamental projective invariant which can also explain the duality law;[85] 3) Chasles' aim was to offer a new generality to the synthetic methods. This aspect represents a development of Poncelet's train of thought; and 4) his foundational programme was not restricted to geometry, but, as we will see, his goal was to prove that the basic concepts of projective geometry also represent the foundation of several sections of physics. These characteristics granted, the author whose ideas are more consonant with Chasles' is the Swiss mathematician Jakob Steiner, 3 years younger than Chasles, who shared items 2) and 3) of the latter's foundational programme. Since Steiner wrote most of his works in German but also published several works in French, Chasles, who could not read German, was nevertheless familiar with some of Steiner's ideas thanks to the latter's publications in French. Furthermore, in the course of the years Chasles possibly became a good connoisseur of the *Systematische Entwicklung* (Steiner, 1832), the text in which Steiner presented his foundation of projective geometry basing on the concept of cross ratio. However, it is noteworthy that in the *Aperçu* only the works Steiner wrote in French are mentioned and Chasles wrote explicitly:

> Several German geometers: Steiner, Plücker, Möbius, etc., who are respectable collaborators of the famous analysts Gauss, Crelle, Jacobi, Lejeune-Dirichlet, etc. write, in this last journal,[86] on the new doctrines of rational Geometry. We regret not to be able to mention their works, which we do not know, for our ignorance of the language in which they are written.[87]

This means that when he wrote the *Aperçu*, he knew only the general conceptions proposed by Steiner in the *Systematische Entwicklung*, but, for sure, not the details. Thence, the consonance of their ideas is even more significant as reciprocal influences can be excluded. It is quite likely that Chasles improved his knowledge of Steiners' *Systematische Entwicklung* along his scientific career because in the preface of his *Traité de Géométrie supérieure* (Chasles, 1852) he mentioned Möbius and Steiner as the mathematicians who made vast use of the concept of cross ratio to found projective geometry. Möbius' *Der barycentrische Calcul* and Steiner's

[84] I am not claiming that Chasles' approach is, generally speaking, more profound than Plücker's and Möbius', but that it is focused on slightly different features of projective geometry.

[85] Möbius too was completely aware of the importance of the cross ratio, but he did not consider it as the fundamental element of his construction. It was just one important element among others.

[86] Chasles is referring to the *Journal für die reine und angewandte Mathematik* (also known as *Crelles Journal*).

[87] Chasles (1837a, p. 215): "Plusieurs géomètres allemands: MM. Steiner, Plücker, Möbius, etc., dignes collaborateurs des célèbres analystes Gauss, Crelle, Jacobi, Lejeune-Dirichlet, etc., écrivent dans ce dernier recueil, sur les nouvelles doctrines de la Géométrie rationnelle. Nous éprouvons un vif regret de ne pouvoir citer leurs ouvrages, qui nous sont inconnus, par suite de notre ignorance de la langue dans laquelle ils sont écrits".

Systematische Entwicklung are mentioned in Chasles (1852, pp. XXII–XXIII). It is, however, curious that in his treatise on the conic sections (Chasles, 1865), where he presented the projective generation of the conics as the fundamental approach to face the properties of these curves, Chasles did not mention Steiner, who more than 30 years before, had shared with him the merit for this discovery, while he mentioned Möbius more than once and Plücker once. In contrast to this, in the *Rapport* (Chasles, 1870), while commenting a property proved by Abel Transon, Chasles pointed out that it had already been proved by Steiner in his *Systematische Entwicklung* (Chasles, 1870, pp. 166–167) and he expressed his high esteem for Steiner:

> Steiner, who consecrated his tenacious and powerful mind only to pure geometry, had published numerous memoirs in the *Journal de Mathématiques* de Crelle and two volumes entitled *Développement systématique de la dépendence reciproque des figures géometriqués, avec citation des travaux de géomètres anciens et modernes sur les Porismes, les methods de projection, la Géométrie de situation, les transversales, la dualité, et la reciprocité,* etc., Berlin 1832—*Constructions géométriques par la ligne et le cercle.* Berlin 1833.
>
> These works and the lectures of this celebrated professor of the Berlin University contributed to spread the taste and the study of the methods of pure geometry.[88]

Thus, it can be argued that Chasles fully recognized Steiner's merits.

A comparison between Chasles' foundational programme and Steiner's ideas insofar as they are expressed in the *Systematische Entwicklung* is useful to guess the consonances and the differences between the conceptions of these two great geometers, granted that Steiner did not develop such a vast foundational programme, extended to various branches of science and philosophy, as Chasles did.

In this section, Steiner's ideas will be analysed. The commentaries represent an occasion to compare Steiner's and Chasles' line of thought and, in particular, to compare Steiner's conceptions with those developed by Chasles in his *Aperçu* as well as in his later *Traité de Géométrie Supérieure*, which illustrates the line of continuity of Chasles' thought in the course of the decades.

[88]Chasles (1870, pp. 217–218): "Steiner, qui consacrait exclusivement à la Géométrie pure ses facultés tenaces et puissantes, avait publié de nombreux Mémoires dans le *Journal de Mathématiques* de Crelle, et deux volumes intitulés: *Développement systématique de la dépendence reciproque des figures géométriques, avec citation des travaux de géomètres anciens et modernes sur les Porismes, les méthodes de projection, la Géométrie de situation, les transversales, la dualité, et la reciprocité,* etc., Berlin 1832—*Constructions géométriques par la ligne et le cercle*. Berlin 1833.

Ces ouvrages et ces leçons du célèbre professeur de l'université de Berlin contribuèrent a répandre le goût et l'étude des méthodes de pure Géométrie ". Italics in the text. In a note Chasles also referred to the original German titles of the two mentioned Steiner's works.

2.3.1 *Steiner's* Systematische Entwicklung

As the title clearly states, Steiner's work is, jointly with Chasles' Notes to the *Aperçu* and the two memoirs on duality and homography, the first systematic work on the use of cross ratio in projective geometry. The way in which the cross ratio is presented by the two authors is different: we have seen that Chasles introduced the main properties of the anharmonic ratio and immediately applied them to the proof of significant theorems and to duality and homography. On the contrary, Steiner, before applying the cross ratio to classical or new theorems, developed a systematic treatment of this concept. One might say that Steiner's *Systematische Entwicklung*, which was then a book of advanced research, appears in retrospect, as a handbook where the reader can follow all the steps through which his theory was constructed, whereas Chasles' Notes of the *Aperçu* and two memoirs do not give the impression to be a handbook.

2.3.1.1 The Basic Elements of Steiner's Theory

Steiner's Preface clearly illustrates his programme. For he began his work with the following words:

> This work contains the final results of a research I am developing from many years on the fundamental spatial properties which contain the seed of all propositions, porisms and problems of geometry, which the ancient and the new epoch donate us so generously. For the set of these mutually separated properties, it was necessary to find a guiding thread and a common root, from which it was possible to obtain a global and clear panorama of the propositions and to glance more freely the particularity of any proposition and its position in the rest of the edifice.[89]

Steiner continued claiming that it is necessary to put order in the foundation of geometry taking into account that neither the synthetical (*synthetische*) nor the analytical (*analytische*) method, namely the classical methods used in geometry, reaches the core of the problem; rather, it is necessary to develop a new synthetic conception of the mutual dependence of figures starting from some initial and simple configurations, from some initial ground elements (*Grundelemente*, ibid., p. VI). His new foundation will allow the mathematicians to obtain a new view on two very debated questions, which were then the basis of modern geometry: a) Gergonne's and Poncelet's different conceptions of duality as well as their priority dispute and b) the profound nature of the theory of reciprocal polars (ibid., pp. VII–VIII). Steiner

[89] Steiner (1832, p. V): "Das vorliegende Werk enthält die Endresultate mehrjähriger Forschungen nach solchen räumlichen Fundamentaleigenschaften, die den Keim aller Sätze, Porismen und Aufgaben der Geometrie, womit uns die ältere und neuere Zeit so freigebig beschenkt hat, in sich enthalten. Für dieses Heer von auseinander gerissenen Eigenthümlichkeiten musste sich ein leitender Faden und eine gemeinsame Wurzel auffinden lassen, von wo aus eine umfassende und klare Uebersicht der Sätze gewonnen, ein freier Blick in das Besondere eines jeden und seiner Stellung zu den übrigen geworfen werden kann."

claimed that duality follows immediately from his basic principles, whereas the theory of reciprocal polars will be derived through a series of deductions. He recognized the clever (*scharfsinnige*) Möbius as the mathematician who, in his *Barycentrische Calcül*, first guessed the correct way to found modern geometry (ibid., p. VIII). After that Steiner presented the steps of his programme: 1) projective line, pencil of lines, sheaf of planes; 2) projective planes and star of straight lines in space; 3) projective space; 4) systems of correlations and nets: theory of involution; and 5) projective construction and properties of the second-degree curves and surfaces (ibid., p. VIII).

After the Preface, in the section "Introductory Concepts" (*Einleitende Begriffe*, ibid., pp. XIII–XVI) Steiner referred to the following fundamental forms:[90]

a) The straight line (*Gerade*), which is a set (*Menge*) of points whose quantity is innumerable (*unzählige*) and which follows immediately (*unmitteblar*) one to the other.
b) The plane pencil of radiuses (*ebene Strahlbüschel*), which is given by the innumerable straight lines through a point which is their centre.
c) The sheaf of planes (*Ebenenbüschel*), which is composed of the infinite (*unendlich*) planes passing through a straight line, their axis.
d) The plane, which is composed of the innumerable (*zahllose*) straight lines and points (or pencils of straight lines) it contains.
e) The star of straight lines, or to be more adherent to Steiner's language, the pencil of radiuses in space (*Strahlbüschel in Raume*). It is given by all the straight lines passing through a point in space, which is its centre. Steiner pointed out that such a configuration contains not only infinite (*unendlich*) straight lines but also innumerable (*zahllose*) plane pencils of radiuses. The star of planes also exists, given by all the planes passing through a point.

After that, Steiner analysed the basic relations between such forms: a plane pencil of radiuses and a straight line are correlated because to each point of the straight line a radius corresponds, e.g. the radius passing through that point, if the straight line

[90]Steiner was the first mathematician to have a clear conception of the objects of projective geometry regarded as forms of three different species. A precious source to valuate the meaning of the concept of Steiner's forms within geometry is Obenrauch (1897). The author of this remarkable work defines Steiner as the greatest and most clever geometer of Germany born in the town of Utzsensdorf in Berna Canton (ibid., p. 246) and dedicates to Steiner the pages 246–287 of his book, where you can find not only important observations on Steiner's research in geometry but also on his didactical activity as well as his relations with great men of that period such as Alexander von Humboldt and Carl Gustav Jacob Jacobi. As to the concept of geometrical form in Steiner, Obenrauch highlights its importance on pp. 256–257.

does not belong to the pencil. Analogously a straight line and a sheaf of planes are correlated considering, for example, the points where the straight line cuts the planes of the sheaf. Thence, also a pencil of radiuses and a sheaf of planes are correlated. Planes and stars are correlated because to each point of a plane a straight line of the star radiuses corresponds and to each straight line of a plane the plane of the star, for example that passing through such a straight line, corresponds. What Steiner claimed with regard to space is a clear litmus paper of the cleverness and clarity of his views. As a matter of fact, Steiner wrote that space corresponds to itself, so that from a geometrical point of view, space is twofold: there are two spaces. Hence, to each element of a space an element of the other space corresponds.

Duality derives from the nature itself of these fundamental forms. As Steiner wrote:

> The essence of such duality of the properties and propositions derives necessarily from the fundamental forms themselves, namely from the complete representation of the space-elements.[91]

Commentary:

1) Steiner's distinction in different forms is very advanced. He perfectly identified the forms we call of first species, namely dotted straight lines, pencils of straight lines and pencil of planes, and stated that a correspondence exists among them. It is clear that he is referring to a projective correspondence, a projective transformation between forms. Analogously he identified the forms of second species, namely the dotted and lined plane, the star of straight lines and the star of planes as correspondent projective forms. It is evident that he also considered space as dotted space and planned space. This is the meaning of his assertion according to which space is twofold. Therefore, Steiner introduced his forms and spoke of the transformations among them. With regard to the relations between forms and transformations, Nabonnand's opinion seems convincing to me: according to the latter's view, it is appropriate to stress the importance of the forms. However, rather than in themselves, the forms are crucial as they are the elements which allowed Steiner to construct a whole theory of the transformations, i.e. of the dependence among figures. This was his more profound aim (Nabonnand, 2006, p. 80). The classical figures of the Euclidean synthetic geometry did not permit the construction of a general theory of the functional dependence. Thus, Steiner introduced the forms.

   With regard to the distinction in different projective forms, we have seen that in his Note IX of the *Aperçu* Chasles proved some basic projective features and properties of the first species forms basing on the concept of cross ratio. It is, thence, clear that both mathematicians fully understood the difference in three kinds of projective forms. Steiner's treatment is more systematic, but conceptually it is close to Chasles'.

[91] Ibid., p. XVI: "Das Wesen dieser Dualität von Eigenschaften und Sätzen ist also durch die Grundbilde selbst, d.h. durch die umfassende Vorstellung der Raumelemente, notwendig bedingt".

2) As to duality there is a difference: Steiner thought that duality was intrinsic in its definitions of forms and in the correspondences between forms. Thence, he did not feel the need to prove the duality law relying upon something more basic. Under this respect, Chasles' conception is partially different because—as we have seen—he felt the need to prove the duality law. His demonstration is grounded on the identity of the cross ratio between four points of a dotted line and four radiuses belonging to a pencil of straight lines, considering the correspondence point-straight line of the pencil passing through such a point (and analogously for the duality dotted line-sheaf of planes and pencil of straight lines-sheaf of planes). Such an identity—as we will explain—is the conceptual basis of Steiner's idea of projective forms. However, it is also true that Chasles introduced three theorems on the director point of a plane. Thence, according to Chasles, duality is a derivable truth, whereas for Steiner, it is intrinsic in the definition of the correspondence between geometrical forms. Although this is not a huge difference, it is nonetheless a nuance that should be highlighted. According to a more modern approach it is to consider duality for space as given axiomatically, whereas duality for the forms of second species (and afterwards, and consequently, for those of first species) is deduced as a theorem. Furthermore, the resort to the cross ratio is avoided because of the presence of metric elements in the definition of this concept.[92] If one adopts an even more modern approach based on abstract algebra—for example resorting to the notion of duality in a lattice—it is possible to prove the duality law both in space and in the plane.
3) With regard to the methods, there is also a consonance between Chasles and Steiner. In this aspect, both of them shared Poncelet's opinions and thought that the synthetic methods are the intrinsic basis of geometry. However, in some circumstances the help of analytical methods can be useful. It is appropriate to point out that Steiner's approach in the *Systematische Entwicklung* is more faithful to the synthetic methods than Chasles' in the two memoirs on duality and homography because, as we have clarified, Chasles also resorted to the equations of the figures he introduced (though he claimed that the same properties proved analytically might be demonstrated also synthetically) whereas Steiner used no equation in the work I am analysing and no Cartesian system of coordinates. The only element which might be defined analytical is the cross ratio. However, one might wonder whether it is correct to define such a concept as "analytical". The answer is that it is not completely correct. Certainly, it indicates a number, but a number which represents an invariable intrinsic projective property. If one considers the cross ratio in itself as an analytical element, then

[92]The *Lezioni di geometria proiettiva* (first edition 1898 second improved edition 1904) by Enriques, though being a handbook, also represents a summary of the projective geometry developed through synthetic methods in the nineteenth century. It is not a coincidence that Corrade Segre defined Enriques' work as a book of research rather than a mere handbook. The conception of duality adopted at the end of the nineteenth century is explained in the second chapter entitled "Duality law – preliminary theorems" (*Legge di dualità – teoremi preliminari*, Enriques, 1898, second edition 1904, pp. 31–56).

one is obliged to consider the simple ratio used by Euclid in the theory of similarity as an analytical element, too. Therefore, the methods developed in the *Systematische Entwicklung* have to be regarded as purely synthetic, though not purely graphic. This is an aspect on which a reflection is necessary: if you aim at founding a purely graphic doctrine within which the metric properties have to be considered as particular cases of the general objects and relations you are dealing with, then the cross ratio can represent, so to say, an intermediate phase in your foundational programme, because the cross ratio is involved with metric objects. However, if your aim is to found the old synthetic geometry on new unitary bases, but you are not interested in founding the metric properties on a graphic ground, then the cross ratio presents no problem. This is exactly what Blåsjö claims with regard to Steiner: unlike the majority of interpreters, he argues that Steiner's aim is to found the classical metric geometry on more solid and general bases, not to determine a purely graphic theory. Thence, the use of the cross ratio is, in the case of Steiner, not susceptible to any criticism (Blåsjö, 2009, pp. 21–22). Thus, one might argue that Steiner's aim is more restricted than Chasles' because the reconduction of the metric properties to those graphic is not part of his foundational programme. As a matter of fact, if the previous interpreters were too prone to the idea that Steiner intended to found a purely graphic doctrine, Blåsjö, as we will see, is too inclined to deny this idea. Steiner was less interested than Chasles in founding such a doctrine and in reducing the metric properties within a graphic domain, but "less interested" does not mean "not interested, at all" as Blåsjö argues.

2.3.1.2 The Concept of *Doppeltverhältnis*

The first section is entitled "Consideration of the straight lines, the plane pencil of radiuses and the sheaf of planes in reference to their mutual projective relations" (*Betrachtung der Geraden, der ebenen Strahlbüschel und der Ebenenbüschel in Hinsicht ihrer projektivischen Beziehungen unter einander*). In the first chapter, the basic projective elements are introduced. Steiner considered (Fig. 2.22) a pencil of radiuses whose centre is $\mathfrak{B}$ and a straight line A not passing through $\mathfrak{B}$.

Given a radius a belonging to $\mathfrak{B}$, it is possible to establish a correspondence between $\mathfrak{A}$ and the point $\mathfrak{A}$ where a cuts A. Suppose now that a rotates—for example—counterclockwise. In its various positions b, c, ... it will cut A in different points $\mathfrak{b}$, $\mathfrak{c}$, ... until reaching the position q in which it is parallel to A. The straight line q cuts the point at infinity (*unendlich entfernt Punkt*, ibid., p. 2) $\mathfrak{q}$ of A. Steiner felt the need to clarify that there is a sole point at infinity because there is a sole parallel and two straight lines cut mutually in a sole point.

With regard to the mutual correspondence between A and $\mathfrak{B}$, when it takes place as in Fig. 2.22, the forms A and $\mathfrak{B}$ are called *perspective*. It is also possible to think of different projective correspondences between A and $\mathfrak{B}$, so that the straight line a does not correspond anymore to the point $\mathfrak{a}$ where it cuts A. For such a correspondence to be possible it is sufficient that it is ordered, so that, if a rotates, e.g.,

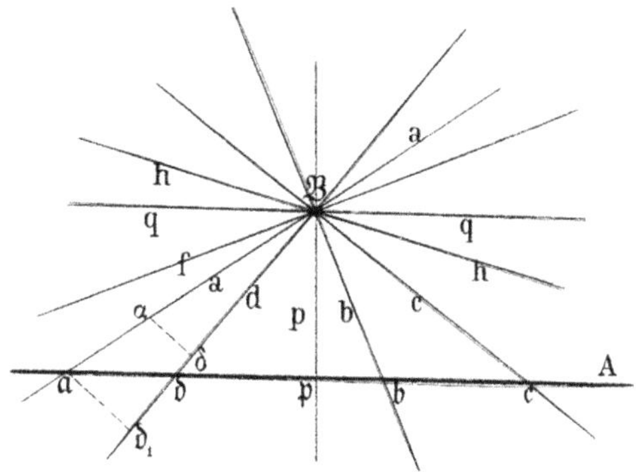

Fig. 2.22 Steiner's figure to explain what is described in the running text

clockwise, its correspondent point $\mathfrak{a}$ rotates always either clockwise or counterclockwise (see Fig. 2.23). Steiner called the projective position of two figures which are not perspective "oblique position" (*schiefe Lage*, ibid., p. 3).

After giving these basic elements of the projective positions of a straight line and of a pencil of radiuses, Steiner began the process that led him to define the concept of cross ratio. He considered the perspective position of Fig. 2.22: to each segment of A an angle of $\mathfrak{B}$ corresponds. From any point α of a draw the perpendicular $\alpha\delta$ to d and from $\mathfrak{a}$ the perpendicular $\mathfrak{a}\mathfrak{d}_1$ to d. Furthermore, be p the perpendicular from $\mathfrak{B}$ to A. The two right triangles $\mathfrak{B}\mathfrak{p}\mathfrak{d}$ and $\mathfrak{a}\mathfrak{d}_1\mathfrak{d}$ as well as $\mathfrak{B}\alpha\delta$ and $\mathfrak{B}\mathfrak{a}\mathfrak{d}_1$ are similar, so that the two proportions

$$\frac{\mathfrak{B}\mathfrak{p}}{\mathfrak{B}\mathfrak{d}} = \frac{\mathfrak{a}\mathfrak{d}_1}{\mathfrak{a}\mathfrak{b}} \text{ and } \frac{\mathfrak{a}\mathfrak{d}_1}{\mathfrak{B}\mathfrak{a}} = \frac{\alpha\delta}{\mathfrak{B}\alpha}$$

hold, from which it follows:

$$\mathfrak{B}\mathfrak{p}\cdot\mathfrak{a}\mathfrak{d} = \mathfrak{B}\mathfrak{a}\cdot\mathfrak{B}\mathfrak{b}\cdot\frac{\alpha\delta}{\mathfrak{B}\alpha} \tag{2.21}$$

The ratio $\frac{\alpha\delta}{\mathfrak{B}\alpha}$ is the sinus of the angle (a,d), so that relation (2.21) can be written as

$$\frac{\mathfrak{a}\mathfrak{d}}{\sin\,(ad)} = \frac{\mathfrak{B}\mathfrak{a}\cdot\mathfrak{B}\mathfrak{b}}{\mathfrak{B}\mathfrak{p}} \tag{2.22}$$

This expression, Steiner claimed, represents the relation between the angle (ad) of the pencil $\mathfrak{B}$ and the correspondent segment of A (ibid., p. 6). If four straight lines of $\mathfrak{B}$ such as a,b,c,d and the correspondent points on A, namely, $\mathfrak{a},\mathfrak{b},\mathfrak{c},\mathfrak{d}$ are considered, you will obtain five other expressions analogous to (2.22). From the combination of such six expressions, it follows easily the relation

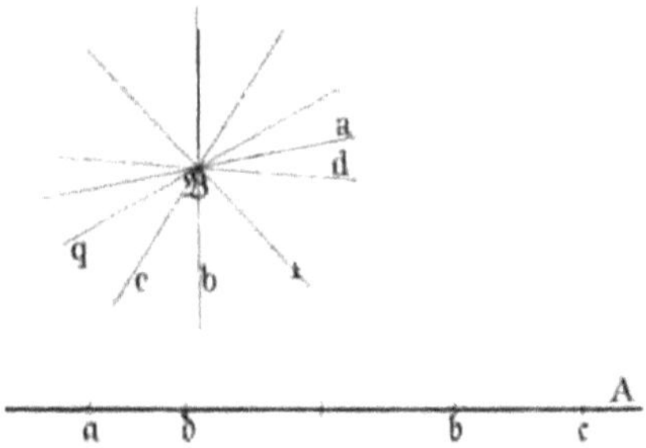

Fig. 2.23 Steiner's figure—with a slight modification—used to expound an oblique projective position of *A* and *B*

$$\frac{\mathfrak{ad}}{\mathfrak{bd}}:\frac{\mathfrak{ac}}{\mathfrak{bc}}=\frac{\sin(ad)}{\sin(bd)}:\frac{\sin(ac)}{\sin(bc)} \tag{2.23}$$

Equation (2.23) is obtained relying upon the perspective position of *A* and $\mathfrak{B}$, but since no reference to the perspective centre $\mathfrak{B}$ is present in such an expression, it holds for any position of the two projective forms, not only for that perspective. Steiner named the left side of (2.23) the "double ratio" (*Doppeltverhältnis*) of four points and the right side the double ratio of four straight lines. Thus, he stated the fundamental relations of the cross ratio:

> Given two groups of four corresponding elements *a,b,c,d* and $\mathfrak{a},\mathfrak{b},\mathfrak{c},\mathfrak{d}$, the double ratio $[(\mathfrak{ad}:\mathfrak{bd}):(\mathfrak{ac}:\mathfrak{bc})]$ of four segments lying in the straight line *A* is equal to the double ratio [(sin(*ad*) : sin (*bd*) : (sin*ac*) : sin *bc*)] which is formed, in a correspondent manner by the sinuses of those angles of the radiuses' pencil $\mathfrak{B}$ which correspond to such segments.[93]

Steiner (Fig. 2.22) continued claiming that given a group of four elements *a,b,c,d*, it is possible to obtain three different double ratios (ibid., p. 8), namely (*abcd*), (*acbd*), (*adbc*), where (*abcd*) means $\frac{ad}{bd}:\frac{ac}{bc}$, and the other expressions consequently.

The next step is another fundamental acquisition because Steiner distinguished the case in which two couples of elements mutually separate or do not separate. In a disposition as (*abcd*) they separate, while in dispositions as (*acbd*), (*adbc*) they do not separate.

It seems that, according to Steiner, what follows is a full justification of the duality law for the points of a dotted line and the radiuses of a pencil: if you consider the four intersection points $\mathfrak{a},\mathfrak{b},\mathfrak{c},\mathfrak{d}$ of *A* with the corresponding straight lines of $\mathfrak{B}$ and you move the point $\mathfrak{B}$, the angles between the lines *a,b,c,d* vary, but the double ratio of their sinuses remains unchanged. Analogously, if $\mathfrak{B}$, and hence *a,b,c,d* remain unchanged, while *A* varies, the distances between $\mathfrak{a},\mathfrak{b},\mathfrak{c},\mathfrak{d}$ will vary, but their cross ratio will not vary. Thence, the following propositions, which Steiner

[93] Steiner 1832, p. 7: "Dass bei irgend vier entsprechenden Elementarpaare *a,b,c,d* und $\mathfrak{a},\mathfrak{b},\mathfrak{c},\mathfrak{d}$, ein gewisses Doppelverhältniss $[(\mathfrak{ad}:\mathfrak{bd}):(\mathfrak{ac}:\mathfrak{bc})]$ gebildet aus vier Abschnitten der Geraden *A*, gleich ist dem Doppelverhältniss [(sin(*ad*) : sin (*bd*) : (sin(*ac*) : sin *bc*)], welches auf entsprechende Weise aus den Sinussen derjenigen Winkel des Strahlbüschels $\mathfrak{B}$, die jenen Abschnitten entsprechen, gebildet ist".

wrote in the typical double column notation invented by Gergonne to indicate two dual statements, hold:[94]

In any pencil of radiuses, four of which lines pass through four given points $(\mathfrak{a}, \mathfrak{b}, \mathfrak{c}, \mathfrak{d})$ of a straight line A, the three double ratios which are obtained by the sinuses of the angles of such four lines have well-determined values. Precisely, these values are always equal to the values of the three double ratios which are obtained from the distances of the four fixed points.	In any straight line, which cuts four given radiuses (a,b,c,d) of a radiuses' pencil $\mathfrak{B}$, the three double ratios which are obtained by the distances of the four intersection points $(\mathfrak{a}, \mathfrak{b}, \mathfrak{c}, \mathfrak{d})$ have well-determined values. Precisely, these values are always equal to the values of the three double ratios which are obtained from the sinuses of the angles of the four fixed radiuses.

Through a series of not difficult considerations based on a purely synthetic reasoning, Steiner proved the fundamental theorem of projectivity, namely that given three elements of a projection in the form of first species, the fourth element is automatically determined. As he wrote:

> In a system of corresponding couples of elements of two projective forms A and B, as soon as three couples are given, that is, if any three couples of elements are given, it is possible to find through them the fourth element of a form which corresponds to any given fourth element of the other form and, if the configuration is oblique, it is possible to pose it in its original or perspective position.[95]

Along his argumentation, Steiner added, in a note, an interesting observation: in a projective correspondence the reciprocal position of any element should be deduced only by the sign of the cross ratio, without resorting to the use of a figure, which,

[94] Ibid., p. 10:

Bei allem Strahlbüschel, von welchen vier Strahlen durch die nämlich vier bestimmten Punkte $(\mathfrak{a}, \mathfrak{b}, \mathfrak{c}, \mathfrak{d})$ einer Geraden A gehen, haben die drei Doppeltverhältnisse, die sich aus den Sinussen der von den jedesmaligen vier Strahlen eingeschlossenen Winkel zusammensetzen lassen einerlei Werthe; nämlich diese Werthe sind jedesmal den Werthen der drei Doppeltverhältnisse gleich, welche aus den Abständen der vier festen Punkte von einander zusammengesetzt sind.	Bei allem Geraden, welche die nämlich vier bestimmten Strahlen (a,b,c,d) eines Strahlbüschel $\mathfrak{B}$ schneiden, haben die drei Doppeltverhältnisse, die sich aus den Abständen der jedesmaligen vier Durchschnittspunkte $(\mathfrak{a}, \mathfrak{b}, \mathfrak{c}, \mathfrak{d})$ von einander zusammensetzen lassen einerlei Werthe; nämlich diese Werthe sind jedesmal den Werthen der drei Doppeltverhältnisse gleich, welche aus den Sinussen der von den vier festen Strahlen eingeschlossenen Winkel zusammengesetzt sind.

[95] Ibid., pp. 10–11: "Dass das ganze System der einander entsprechenden Elementenpaar zweier projectivischen Gebilde A, B bestimmt sei, sobald irgend drei Paare gegeben sind, d.h. wenn irgend drei Elementpaare gegeben sind, so kann mittelst derselben zu jedem gegebenen vierten Element des einem Gebildes das entsprechende Element des anderen Gebildes gefunden werden, und die Gebilde lassen sich dadurch, wenn sie sich in schiefer Lage befinden, in die ursprüngliche oder perspectivische Lage zurückbringen."

hence, becomes an explicative support, but the foundation of the projective relations might be obtained through the pure analysis of the cross ratio (ibid., p. 12).

Steiner analysed then some specific cases of the cross ratio: in particular a) when in the straight line *A* one of the four considered points lies in the middle between two of the other three or b) when one of the points is at infinity and a′) when a straight line of $\mathfrak{B}$ bisects two of the other three lines or when b′) two of the four straight lines are perpendicular (ibid., p. 15). Propositions a) and a′) are dual. Probably Steiner also considered b) and b′) to be dual.

In case a), if $\mathfrak{d}$ (Fig. 2.22) lies in the middle between $\mathfrak{a}$ and $\mathfrak{b}$, so that $\mathfrak{ad} : \mathfrak{bd} = 1$, then the double ratio on the left side of Eq. (2.23) is a simple ratio so that such an equation is transformed into

$$\mathfrak{ac} : \mathfrak{bc} = \frac{\sin{(ac)}}{\sin{(bc)}} : \frac{\sin{(ad)}}{\sin{(bd)}}$$

On the other hand, if $\mathfrak{q}$ is the point at infinity of *A*, the three segments $\mathfrak{aq}, \mathfrak{bq}, \mathfrak{cq}$ will be equal, so that their ratio is equal to 1, and hence, the cross ratio $(\mathfrak{abcd})$ is transformed into

$$\mathfrak{ac} : \mathfrak{bc} = \frac{\sin{(ac)}}{\sin{(bc)}} : \frac{\sin{(aq)}}{\sin{(bq)}} \tag{2.24}$$

where q is the parallel to *A* from $\mathfrak{B}$. Analogous is the situation for the other two cross ratios obtained by the group of four points $\mathfrak{a}, \mathfrak{b}, \mathfrak{c}, \mathfrak{d}$.

The reason why Steiner seemed to consider 2′) dual to 2) is noteworthy. For when two straight lines of a group of four radiuses *a,b,c,d* belonging to $\mathfrak{B}$, for example *a* and *b*, are perpendicular, then sin(*bc*) = cos (*ac*) and sin(*bd*) = cos (*ad*), or sin (*ac*) = cos (*bc*) and sin(*ad*) = cos (*bd*), and it is

$$\left\{ \begin{array}{l} \dfrac{\mathfrak{ac}}{\mathfrak{bc}} : \dfrac{\mathfrak{ad}}{\mathfrak{bd}} = \tan{(ac)} : \tan{(ad)} \\ \dfrac{\mathfrak{ac}}{\mathfrak{bc}} : \dfrac{\mathfrak{ad}}{\mathfrak{bd}} = \tan{(bd)} : \tan{(bc)} \end{array} \right. \qquad \text{or}$$

whose form is, *mutatis mutandis*, the same as (2.24). It seems that Steiner considered two perpendicular straight lines as functionally analogous to the point at infinity of a straight line because in both cases it is possible to reduce one of the two double ratios of the equation expressing a projectivity to a simple ratio, so simplifying the treatment. This is the reason why Steiner put 2′) in dual position with respect to 2).

Commentary: Steiner's *Systematische Entwicklung* is framed as a foundational and systematic work on the cross ratio from the very beginning. There are several consonances with Chasles' way to introduce the anharmonic ratio: both authors proved the invariant projective character of the cross ratio; both individuated three different values for the cross ratio of a group of four elements; both thought that duality might be deduced from the application of the concept of cross ratio. According to Steiner, this deduction was an immediate consequence; according to Chasles, conversely, an argument was necessary. However, both Chasles and Steiner thought that duality—which, as we will see, Chasles posed at the basis of a philosophical way of thinking, as well—relies upon the projective nature of the double ratio. Both authors analysed the peculiar positions of a group of four elements, particularly when one of the four elements, specifically a group of four points in a dotted straight line, is at infinity. This is useful because, in this case, the double ratio becomes a simple ratio. There is a difference in the presentation of the subject: if Steiner expounded the properties of the cross ratio in a systematic manner, Chasles mentioned and proved the most important of such properties and applied them immediately to several questions of projective geometry, whereas Steiner offered a more complete treatment of such properties. One might say that, from a structural point of view, Chasles' text which corresponds to Steiner's *Systematische Entwicklung* is the *Traité de Géométrie Superieure* (Chasles, 1852) some aspects of which will be analysed in the next sections. It is, however, clear that in 1852 the concept of cross ratio was well known and accepted by the mathematical community and that Von Staudt had already proved that projective geometry can be founded on a pure graphic basis without resorting to the cross ratio.

In conclusion, in the period 1832–1837 there were two mathematicians, one in France and the other in Swiss-Germany, who, independently of one another, developed the theory of cross ratio in the belief it could offer a basis for projective geometry. As we have seen, the literature is divided whether Steiner intended to reduce the metric properties to the projective ones, whereas this was the core of Chasles' programme. Nonetheless, it is clear that Steiner, too, thought of projective properties as the most profound of geometry. A particular observation regards the fundamental theorem of projectivity: we have seen that Chasles proved it for the forms of third species (and implicitly also for those of first and second species). Steiner demonstrated this theorem separately for the form of first, second and third species. This is a further consonance. Obviously, what has been said for Chasles is also valid for Steiner: he resorted to the concept of the double ratio and he did not develop any considerations on the problems of continuity. A proof of the theorem independent of the cross ratio is due to Von Staudt (1847) and the considerations on continuity date from the end of the '70 to beginning of the '80 and rely upon Dedekind's work.

2.3.1.3 The Concept of Harmonic Ratio

While a theory of the double ratio did not exist before Steiner and Chasles, the concept of harmonic division of a segment or a pencil of straight lines already existed. In a note Steiner recalled that this notion was used by De La Hire and Brianchon (ibid., p. 19). Steiner defined four elements to be harmonic if their cross ratio is equal to 1. In this case, given four harmonic points (Fig. 2.22) $\mathfrak{a},\mathfrak{d},\mathfrak{b},\mathfrak{c}$ and four corresponding harmonic straight lines *a,b,c,d* the two dual identities hold:

$$\frac{\mathfrak{ac}}{\mathfrak{bc}} = \frac{\mathfrak{ad}}{\mathfrak{bd}}; \qquad \frac{\sin(ac)}{\sin(bc)} = \frac{\sin(ad)}{\sin(bd)}$$

That is, the two points $\mathfrak{a}$ and $\mathfrak{b}$ divide the segment $\mathfrak{dc}$ in proportional parts and vice versa (analogously for the sinuses of the corresponding straight lines).

Some propositions that after Steiner became classical follow, the most important of which are:

a) if in two projective forms *A*,*B* four elements of a form are harmonic, the correspondent elements of the other form are also harmonic (ibid., pp. 19–20). Within the commentaries of this proposition Steiner also stated that, given the 24 permutations of four elements, in the case of the harmonic ratio the equivalence classes are composed of eight equal double ratios; i.e. there are only three classes of equivalence (whereas normally the classes are six for the other cross ratios).
b) Steiner clarified that the couple of points $(\mathfrak{a},\mathfrak{b})$ and $(\mathfrak{d},\mathfrak{c})$ mutually separate.
c) It is $\mathfrak{ac} : \mathfrak{bc} = (\mathfrak{ab}+\mathfrak{bc}) : \mathfrak{bc} = 1 + \mathfrak{ab}/\mathfrak{bc}$. Let us suppose $\mathfrak{c}$ to be the point at infinity of the straight line $\mathfrak{ab}$, while $\mathfrak{a}$ and $\mathfrak{b}$ are two given ordinary points. In this case $\mathfrak{ac} : \mathfrak{bc} = 1$. Since the harmonic ratio $(\mathfrak{a},\mathfrak{b},\mathfrak{c},\mathfrak{d}) = 1$, it follows that $\mathfrak{ad}/\mathfrak{bd} = 1$, so that the point $\mathfrak{d}$ is the middle point of the segment $\mathfrak{ab}$. This is the proof of the theorem that, in a harmonic ratio, the middle point of a couple corresponds to the point at infinity of the considered dotted line and vice versa (ibid., pp. 22–23). An interesting consideration concerns the dual property: consider two couples of straight lines (*a*,*b*) and (*d*,*c*) belonging to a pencil and composing a harmonical group. Furthermore, suppose that *d* bisects the angle between *a* and *b*. Because of this last hypothesis, it is $\sin(ad) : \sin(bd) = 1$, and, since the group is harmonic, it is also $\sin(ac) : \sin(bc) = 1$, so that also the angles (*ac*) and (*bc*) are equal. Therefore, *d* and *c* are perpendicular. Thus, given a pencil through $\mathfrak{B}$ the harmonic conjugate of a straight line *l* which bisects the angle between two other given straight lines is the perpendicular to *l* through $\mathfrak{B}$ and vice versa (ibid., pp. 22–23). Thence, two perpendicular straight lines behave, in this respect, as the middle point of a segment and the point at infinity of the straight line to which the segment belongs. Through this property Steiner proved a beautiful theorem according to which, given four harmonic points $\mathfrak{a},\mathfrak{d},\mathfrak{b},\mathfrak{c}$ and, through them, four straight lines belonging to a pencil whose central point $\mathfrak{B}$ moves in the plane, if one couple of those lines always bisects the angle formed by the other couple, then $\mathfrak{B}$ moves on the circumferences of two circles, one of which has $\mathfrak{ab}$ as a diameter and the other one $\mathfrak{cd}$. Dually, if four given harmonic straight lines *a*,*d*,*b*,*c* cut a variable straight line *A* in four points $\mathfrak{a},\mathfrak{d},\mathfrak{b},\mathfrak{c}$ such that $\mathfrak{d}$ bisects $\mathfrak{ab}$, so that $\mathfrak{c}$ is the point at infinity of *A*, then *A* is parallel to one of the fixed straight lines and vice versa (ibid., pp. 24–25).

These are the most important of the initial properties connoting the harmonic ratio introduced by Steiner.

Commentary: in his systematic construction of projective geometry, Steiner expounded the basic properties of the harmonic ratio. Chasles dedicated some specific and interesting considerations on the harmonic ratio in Chasles (1852): first of all, he clarified the origin of the name *rapport harmonique*. This relies on the fact that, considered a point of a harmonic group, the three segments having as extremes the given point and each of the other three points are in harmonic proportion (Chasles, 1852, p. 37). This proportion is characterized by the fact that the ratio between the first and the third segment is as the ratio between the first minus the second and the second minus the third. Since in a harmonic proportion of four points a,e, a', f, it is $\frac{af}{ae} = \frac{a'e}{a'f}$, if one takes into account the three segments af, aa', ae, the previous equation becomes

$$\frac{af}{ae} = \frac{af - aa'}{aa' - ae}$$

which is a harmonic proportion.

It is apropos of the equation $\frac{ae}{af} = \frac{a'e}{a'f}$ that Chasles made an important observation: for he pointed out that this expression is correct only if the absolute value of the ratio is considered, but to give a general meaning to the cross ratio, it is necessary to take into account the sign too. Since, in the harmonic ratio, the couples mutually separate, this means its sign is minus. Thence, the correct manner to express the harmonic ratio is

$$\frac{ae}{af} : \frac{a'e}{a'f} = -1$$

rather than $\frac{ae}{af} : \frac{a'e}{a'f} = 1$ (Chasles, 1852, pp. 38–39). This last expression presents no ambiguity since the anharmonic ratio of four points can be +1 only if two points coincide. Therefore, from a rigorous point of view the value of the harmonic ratio is −1, not +1.

As Steiner had highlighted, in a harmonic group of four points, three different values of their cross ratio exist[96] and Chasles proved that these other two values are ½ and 2. Specifically if $\frac{ae}{af} : \frac{a'e}{a'f} = -1$, then (ibid., p. 40)

$$\frac{af}{aa'} : \frac{ef}{ea'} = \frac{1}{2}; \qquad \frac{aa'}{ae} : \frac{fa'}{fe} = 2.$$

Afterwards the properties of the harmonic ratio, most of which already existing in Steiner, are presented. Important applications of the harmonic ratio were developed

[96] We know that, given four forms of first species, there are, in general, six classes of equivalence of their cross ratios and that, as Chasles proved, if the ratio is harmonic, the classes are three. In 1852, this was recognized, but at the beginning of the 30s, this question was not so plain, as we will see in Sect. 2.3.1.4.

by Chasles as to the imaginary elements in geometry (ibid., Chapter V, pp. 54–63). This chapter is particularly interesting because Chasles introduced the imaginary elements in a purely geometrical manner and presented the properties of the harmonic ratio when some elements are imaginary. Thus, it can be regarded as a step of his foundational programme. It is necessary to remark that, as Chasles himself pointed out, his theory concerns only couples of conjugate imaginary points and not a complete projective doctrine including all the imaginary quantities. Therefore, the couples of imaginary points became objects of Chasles' projective theory, but not a single imaginary element.[97]

In particular: given a straight line l, it is possible to determine two of its points a and a' given their middle point α and the product of their distances from a point M of l. If such product $Ma \cdot Ma'$ is indicated by ν, it is easy to prove that the distances Ma and Ma' are a function of the expression $\sqrt{\overline{M\alpha}^2 - \nu}$. If ν is positive and, in absolute value, bigger than $\overline{M\alpha}^2$, the two points a and a' are imaginary; otherwise they are real (ibid., pp. 56–57). Chasles had proved (ibid., p. 43) that, given the middle point α of two points a and a' as well as two points e,f with their middle point O, the equation which expresses the fact that $(a, a', e, f) = -1$ has the form

$$ma \cdot ma' + me \cdot mf = 2m\alpha \cdot mO$$

where m is the origin of the reference system of coordinates. Since he had demonstrated that the product of the distances between a real point and two imaginary points is a real number (ibid., p. 56), then it is possible the harmonic conjugates of two real points e and f to be two imaginary points a and a'. But there is something more: Chasles had showed that, given two imaginary points a, a' as well as the middle point O of two other points e,f, these last two points can be constructed, which means that they are real. Hence, when two couples of points a,a' and e,f are in harmonic ratio, only one of the two couples can be composed of imaginary points. Given two imaginary points a,a', their mean point α and a real point e, the fourth harmonic f after a,a',e can be found by the equation

$$ea \cdot ea' = e\alpha \cdot ef$$

After showing that the concept of harmonic ratio can be extended to the case in which two of the four elements are imaginary, Chasles cared about the interpretative aspect: he claimed that, in these conditions, the classical equations expressing the harmonic ratio of four points such as

[97] Chasles (1852, pp. 55–56). See also, e.g., Nabonnand (2011a). The problem of the imaginary in projective geometry is difficult and a decisive step was performed by the profound work (Von Staudt, 1856–1860). This text developed the connections between an elliptic involution and a single imaginary element. A fundamental contribution which completed Von Staud's work is Lüroth (1875, 1877). The theory was completed by Kötter (1887). A good paper on the imaginary in geometry is Coolidge (1900).

$$\frac{ae}{af}=-\frac{a'e}{a'f} \quad \text{and} \quad \frac{2}{ef}=\frac{1}{ea}+\frac{1}{ea'}$$

cannot have the same geometrical interpretation as in the case in which all the points are real (ibid., p. 61). As far as this is concerned, Chasles developed an interesting reasoning: if you recognize that the same algebraic operations of sum, division, etc. which are valid for the real equations are also valid for the equations in which imaginary elements occur, then, for, example, the equation $\frac{2}{ef}=\frac{1}{ea}+\frac{1}{ea'}$ will be transformed into

$$\frac{2}{ef}=\frac{ea+ea'}{ea\cdot ea'}=\frac{2e\alpha}{ea\cdot ea'}\rightarrow ef\cdot e\alpha=ea\cdot ea'$$

In this equation all the elements are real because e,f,α are real points and the product $ea\cdot ea'$ is real, though being a and a' imaginary.

Chasles interprets the imaginary segments as elements of the symbolism, as mere symbols which allow us to reach some properties of the real figures which might be hidden if working only with real elements. So to say: something similar to what happens when in an equation of third degree with three real solutions, an imaginary quantity appears, which is necessary exactly to obtain such solutions. He was clear. As we read:

> Then the segments in these relations such as $\frac{ae}{af}=-\frac{a'e}{a'f}$ have to be considered as *symbols* through which you allude to the case where the points are real and which, reciprocally combined, as in this special case, guide to relations where only the *elements* of the two points enter. Thus, the primitive symbolic relation is, after all, nothing but an expression of this relation among *elements*, which are always real.[98]

Thence, Chasles gave personal contributions to the theory of the harmonic ratio and these contributions are tied to his previous work as well as to Steiner's studies and they were particularly important to specify the nature of the harmonic ratio and the extension of projective geometry to imaginary elements, which belong to Chasles' foundational programme. It should be observed that Chasles' novelty is not the idea of constructing an imaginary geometry. As a matter of fact, he extended the theory of the cross ratio to imaginary elements but considered them as instruments to study the properties of the real figures. Therefore, once again, his main point is to show the power of the basic concepts of projective geometry, as that of cross ratio, through which it is possible to unify the treatment of the real and

---

[98] Ibid., p. 65: "Mais alors les segments qui entrent dans ces relations, comme dans $\frac{ae}{af}=-\frac{a'e}{a'f}$ doivent être considérés comme des *symboles*, au moyen desquels on fait allusion au cas où les points seraient réels, et qui, combinés entre eux, comme dans ce cas spécial, conduisent à des relations où n'entrent que les *éléments* des deux points; de sorte que la relation symbolique primitive n'est, au fond, qu'une expression de cette relation entre des *éléments* toujours réels". Italics in the text. On Chasles' introduction of the imaginary elements in his *Traité*, see also Nabonnand (2006, pp. 166–169).

imaginary elements, though, after all, only the real ones are susceptible of an authentic geometrical interpretation.

2.3.1.4 Projective Forms of the Same Name

Until this moment Steiner considered the projective correlation between the points of a dotted straight line and the straight lines of a pencil. In the following section, entitled "Two and more straight lines; two and more plane pencils of straight lines" ("Zwei und mehrere Gerade, und zwei una mehrere ebene Strahlbüschel"), he took into account the projectivity between forms of the same name. In particular two dotted straight lines A and A_1 which are projective to the same pencil $\mathfrak{B}$ are mutually projective, so that among their points a correspondence exists. Two dotted straight lines are perspective if and only if they are perspective to the same pencil (Fig. 2.24); otherwise they are oblique (Steiner, 1832, pp. 29–30).

To the point $\mathfrak{r}$ where the line r of $\mathfrak{B}$ parallel to A_1 cuts A, the point at infinity of A_1 corresponds, and to the point $\mathfrak{q}_1$ where the line q parallel to A cuts A_1, the point at infinity of A corresponds. Nowadays, we call these two points "limit points", while Steiner called them "intersection of the parallel lines" (*Durchschnitte der Parallelstrahlen*, ibid., p. 30). Steiner highlighted that, when two straight lines are perspective, their common point is a double point. Dual considerations hold for two projective pencils of rays (ibid., pp. 30–32).

Now there is a section which is important from a historical point of view because you can deduce the reason why only three and not six classes of equivalence of cross ratios, given four forms of first species, were then taken into account: Steiner remarked that in two projective straight lines or pencils of straight lines, four corresponding elements have the same double ratio. Restricting to two groups of four corresponding points $\mathfrak{a},\mathfrak{b},\mathfrak{c},\mathfrak{d}$ and $\mathfrak{a}_1,\mathfrak{b}_1,\mathfrak{c}_1,\mathfrak{d}_1$ (but a completely analogous relation holds among the sinuses of the angles of two groups of corresponding straight lines), the three relations hold:

$$\begin{cases} \frac{\mathfrak{ac}}{\mathfrak{bc}} : \frac{\mathfrak{ad}}{\mathfrak{bd}} = \frac{\mathfrak{a}_1\mathfrak{c}_1}{\mathfrak{b}_1\mathfrak{c}_1} : \frac{\mathfrak{a}_1\mathfrak{d}_1}{\mathfrak{b}_1\mathfrak{d}_1} \\ \frac{\mathfrak{ab}}{\mathfrak{cb}} : \frac{\mathfrak{ad}}{\mathfrak{cd}} = \frac{\mathfrak{a}_1\mathfrak{b}}{\mathfrak{c}_1\mathfrak{b}_1} : \frac{\mathfrak{a}_1\mathfrak{d}_1}{\mathfrak{c}_1\mathfrak{d}_1} \\ \frac{\mathfrak{ab}}{\mathfrak{bd}} : \frac{\mathfrak{ac}}{\mathfrak{bc}} = \frac{\mathfrak{a}_1\mathfrak{b}}{\mathfrak{b}_1\mathfrak{d}_1} : \frac{\mathfrak{a}_1\mathfrak{c}_1}{\mathfrak{b}_1\mathfrak{c}_1} \end{cases} \tag{I}$$

Steiner claimed that it is essential to remark (*wesentlich zu bemerken*, ibid., p. 34) that the projective correspondences are ordered. This granted, he claimed:

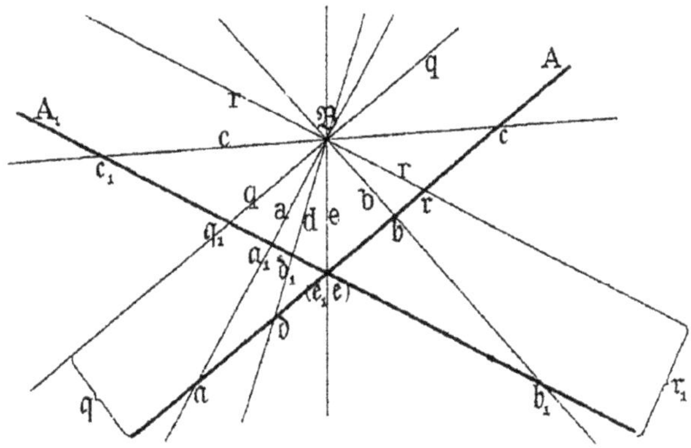

Fig. 2.24 The figure used by Steiner to illustrate the projection (in this case a perspectivity) between two dotted straight lines

> Because of this concordance [the order] and because of expression I) it follows that if, of the eight elements to which expression I) is referred to, seven are given, the eight element is completely determined.[99]

So that:

> α) In two projective forms [...] the double ratios determined by four elements of one form are equal to the double ratios determined by the four elements of the other form; furthermore, the reciprocal position of four elements of the first form agrees with that of the four elements belonging to the other form.
>
> β) Therefore, the entire system of corresponding pairs of elements is given when three pairs are given.
>
> γ) And certainly such three pairs can be given arbitrarily and vice versa.[100]

Commentary: To summarize, in these two last quotations, Steiner argued that the fundamental theorem of projectivity holds for three pairs of elements of two projective forms of first species having the same name. Besides this, there is a consideration which allows us to understand why Steiner, and also Chasles in his *Aperçu*, considered only three cross ratios and not six, given a group of four projective elements. What Steiner claimed is that the cross ratio in itself is not yet sufficient to individuate a projectivity in a complete way, since it is also necessary to take into account the fact that the projectivities are ordered correspondences. He clearly

[99]Ibid., p. 35: "Vermöge dieser Uebereinstimmung und vermöge der obigen Ausdrücke (I) selbst folgt also, dass wenn von den 8 Elementen, auf die sich einer dieser Ausdrücke bezieht, irgend 7 gegeben sind, dann das achte Element dadurch ganz unzweideutig bestimmt sei".

[100]Ibid., p. 37: "α) Bei zwei projektivische Gebilden [...] sind die Doppeltverhältnisse, welche durch irgend vier Elemente des einen Gebildes bestimmt werden, gleich den Doppeltverhältnisse, welche durch die vier entsprechenden Elemente des anderen Gebildes bestimmt werden; ferner ist die gegenseitige Lage der vier Elemente des einen Gebildes übereinstimmend mit der der vier Elemente der anderen Gebildes.

β) Daher ist das ganze System des entsprechenden Elementpaare zweier projektivische Gebilde bestimmt, wenn irgend drei Paare gegeben sind.

γ) Und zwar können solche drei Paare ganz nach Willkür angenommen werden. Und umgekehrt."

claimed: "Because of this concordance [the order] and because of the expression I)". For let us consider a group of four projective elements A,B,C,D. In our perspective the 24 permutations of these elements are divided into 6 classes of equivalence, precisely, told $(A,B,C,D)=k$, the six classes are:

1) $(A,B,C,D)=(B,A,D,C)=(C,D,A,B)=(D,C,B,A)=k$
2) $(A,B,D,C)=(C,D,B,A)=(B,A,C,D)=(D,C,A,B)=1/k$
3) $(A,C,B,D)=(D,B,C,A)=(C,A,D,B)=(B,D,A,C)=1-k$
4) $(A,D,B,C)=(C,B,D,A)=(D,A,C,B)=(B,C,A,D)=(k-1)/k$
5) $(A,D,C,B)=(B,C,D,A)=(D,A,B,C)=(C,B,A,D)=k/(k-1)$
6) $(A,C,D,B)=(B,D,C,A)=(C,A,B,D)=(D,B,A,C)=1/(1-k)$

If we look at Steiner's Eq. (I) and at Chasles' system (I) in Sect. 2.2.1, we see that only three cross ratios are represented, namely: Steiner referred to the classes I have indicated by 1), 3), 4) and Chasles to 1), 3), 5). This clearly means that both of them unified, from the point of view of the double ratio, a certain class C with the class C', whose elements have the cross ratio which is the reciprocal of those belonging to C. Obviously, this does not mean that they identified these classes from a projective standpoint because they are different insofar as the order of the elements is concerned. Therefore, the cross ratio plus the order determines a projective class, according to the conception expressed by Steiner and Chasles. As it is easy to see, in the case of the harmonic groups, also adopting Steiner's and Chasles' view, the only cross ratio is sufficient to identify a projectivity.

In the rest of the section I am analysing, Steiner showed the transitive character of the projectivities and proved the projective character of the harmonic groups. Interesting considerations concern the concept of similarity: though Steiner did not use this expression, his treatment of the similarities can be interpreted as a metric application of the projective theory. For he claimed that in a perspectivity, it can happen that a) the projection point $\mathfrak{B}$ is at infinity, so that the projection-lines $a,b,c,d,\ldots$ are parallel, or b) the two projective lines A and A_1 are parallel and in this case the centre of projection can lie between the two lines or be external to their plane-strip (see Fig. 2.25a–c).

In these cases, Steiner claimed (ibid., pp. 42–43), the following identities between the simple ratios hold:

$$\frac{\mathfrak{ab}}{\mathfrak{a}_1\mathfrak{b}_1}=\frac{\mathfrak{ac}}{\mathfrak{a}_1\mathfrak{c}_1}=\frac{\mathfrak{ad}}{\mathfrak{a}_1\mathfrak{d}_1}=\frac{\mathfrak{bc}}{\mathfrak{b}_1\mathfrak{c}_1}=\cdots$$

The two straight lines A and A_1 are, then, called "similar" (*ähnlich*, ibid., p. 43). These properties of the cross ratios hold independently of the perspective position of A and A_1. Steiner identified the main properties of two similar projective straight lines, namely that the similarity is given as soon as two corresponding couples are given. From a metric-projective standpoint Steiner identified the following fundamental property of two projective straight lines: in two projective straight lines the points at infinity mutually correspond; conversely, if in two projective lines the points at infinity mutually correspond, then the two lines are similar (ibid., p. 44).

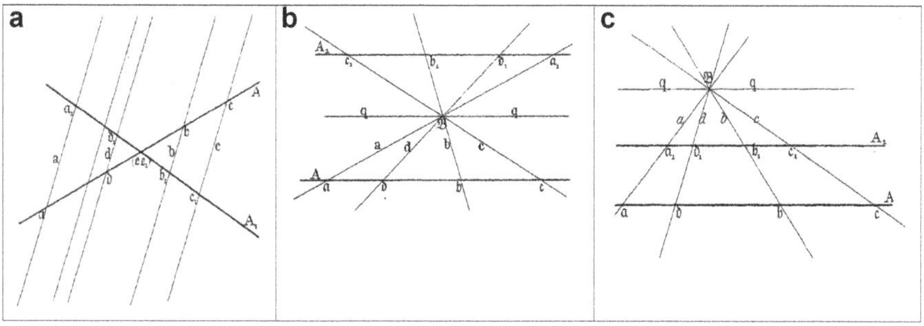

Fig. 2.25 (**a**–**c**) The particular projectivities (similarities and isometries) analysed by Steiner

Steiner eventually defined the concept of two equal or congruent straight lines: this is the case if, in the disposition of Fig. 2.25a, the two angles $\mathfrak{b}\mathfrak{a}\mathfrak{a}_1$ and $\mathfrak{b}_1\mathfrak{a}_1\mathfrak{a}$ are equal, or if in the disposition of Fig. 2.25b, the locus of the point $\mathfrak{B}$ lies in the line parallel to A and A_1 which is in the middle of the plane-strip AA_1 or if the straight lines A and A_1 are parallel as well as the projection-straight lines $a,b,c,d,\ldots$ (ibid., pp. 44–45). In all these cases, it is $\mathfrak{a}\mathfrak{b}=\mathfrak{a}_1\mathfrak{b}_1;\mathfrak{a}\mathfrak{c}=\mathfrak{a}_1\mathfrak{c}_1;\mathfrak{b}\mathfrak{c}=\mathfrak{b}_1\mathfrak{c}_1$ and the straight lines are equal or congruent. When two straight lines are congruent, a pair of elements is sufficient to construct the projectivity. Steiner also distinguished between direct and indirect congruence (ibid., p. 46–47). The concept of identity can also be extended to the pencils of straight lines.

Commentary. The part just analysed is important in a comparison with Chasles because both mathematicians interpreted the similarities and the isometries as projective transformations which satisfy specific metric properties. Therefore, Blåsjö's idea that Steiner was not trying to found a projective theory within which the metric-graphic properties are seen as particular cases of the graphic ones is not correct, even if it is true that the reduction of the metric properties to those graphic was not Steiner's main aim. Thence, the idea that projective geometry was the basis of the whole of geometry was shared by both authors. As already claimed, Chasles analysed the reduction of the metric properties to those graphic in a deeper way and Steiner examined the concept of cross ratio in all its nuances and properties. Chasles, at least until his *Traité*, offered the basic notions concerning the concept of the cross ratio and, after that, showed a remarkable amount of applications. With regard to the reduction of the similarities and isometries within a projective context, in the commentaries to Sect. 3.2. I will argue that Chasles, though not resorting to the concept of anharmonic ratio, considered the similarities and the isometries as particular homologies. This idea is confirmed by the material referred to in Sect. 3.3. This means that, even before the composition of Steiner's *Systematische Entwicklung* Chasles thought of the reduction of metric transformations within a projective context. This train of thought is developed coherently by Chasles until his *Traité*, where he clearly identified an affine transformation between two figures lying in two planes (also considering two superimposed planes) as that homography in which the lines at infinity of the two planes are homologous, so that parallel straight lines of a figure are transformed into parallel straight lines of the homographic figure and the areas of the parallelograms lying in a plane have a constant ratio with the areas of the homologous parallelograms (Chasles, 1852, pp. 359–364). The theory of affinity is original by Möbius and not by Chasles, but it has so to say "its natural place" within Chasles' foundational programme, in which affinities, similarities and isometries are seen as projective transformations when particular affine or metrical elements are specified. The similarities can be seen as particular cases of affinities.

With regard to the use of the anharmonic ratio in the theory of two projective figures we have seen that the properties identified by Chasles in his theory of homographic figures (Sect. 2.2.3) are exactly the same as those identified by Steiner. The former analysed immediately the transformations in the form of second species, whereas the latter started from the forms of first species, but this is due to the fact that the two memoirs on polarity and homography were not conceived to develop a systematic approach *à la* Steiner, but were intended as a way to apply the concept of cross ratio to geometry, granted some basic properties of the double ratio. This is only a difference in the way to present the subject. As a matter of fact, in those years (it should be considered that Chasles' memoirs were written in 1829 even though they were published in 1837) there was a perfect consonance between Chasles and Steiner on the fact that projective geometry had to be found on the anharmonic ratio.

2.3.1.5 The Complete Quadrilateral

After the section I have analysed, in the *Systematische Entwicklung* there is another long section which is dedicated to projective dotted lines and pencils of radiuses set in particular positions: perspective, overlapping, oblique. It is entitled "Respective position of the forms: propositions and problems connected to it" (*Von der gegenseitigen Lage der Gebilde und den durch sie bedingten Sätzen und Aufgaben*, Steiner, 1832, pp. 47–71). This section contains much interesting material, although it is not so important for the aim of my study, unlike the following section entitled "Propositions and porisms which arise from the disposition of the forms" (*Sätzen und Porismen, die aus Zusammenstellung der Gebilde entspringen*, ibid., pp. 71–97) which is instead extremely important because the theorem of the complete quadrangle together with other significant incidence theorems is dealt with.

Steiner began his analysis by pointing out that his foundation of projective geometry will allow him to treat in an easy manner a class of problems whose solution is, otherwise, rather complex. In particular, Carnot often used the figure he called "complete quadrilateral" (*vollständiges Vierecke*, *quadrilatère complet*, in French, ibid., p. 72). Steiner clarified that, under this denomination, two dual figures are included: 1) the complete quadrilateral (*vollständiges Vierseit*) and the complete quadrangle (*vollständiges Vierecke*). The former (Fig. 2.26a) is given by the four straight lines *A,B,C,D* and by the six points $\mathfrak{a},\mathfrak{b},\mathfrak{c},\mathfrak{d},\mathfrak{e},\mathfrak{f}$. This figure has three diagonals $\mathfrak{af},\mathfrak{bc},\mathfrak{cd}$, joining the opposite vertices. It entails three simple quadrilaterals $\mathfrak{abfea},\mathfrak{acfda},\mathfrak{bcedb}$. Dually, the latter (Fig. 2.26b) is composed of the four points $\mathfrak{a}$, $\mathfrak{b},\mathfrak{c},\mathfrak{d}$ and of the six straight lines joining them. The three points $\mathfrak{e},\mathfrak{f},\mathfrak{g}$ joining the

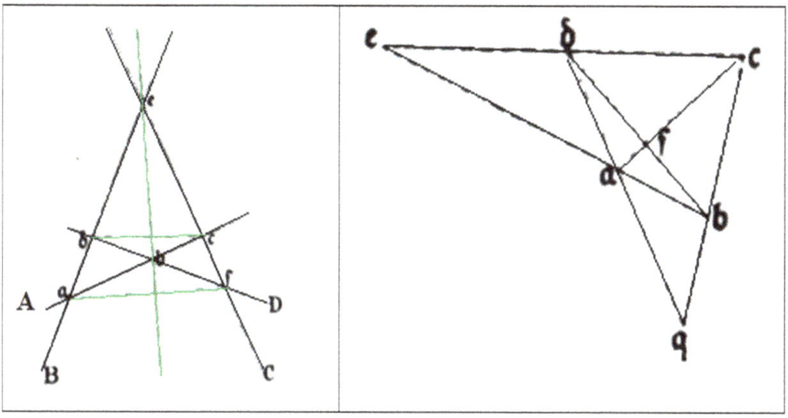

Fig. 2.26 (**a**) The figure used by Steiner to describe the complete quadrilateral. I have used the green colour to make the figure clearer. (**b**) The figure used by Steiner to describe the complete quadrangle

opposite sides are the diagonal points. Three simple quadrangles $\mathfrak{abdca}, \mathfrak{acdba}, \mathfrak{acbda}$ are included in one complete (ibid., p. 72).

Steiner claimed that complete n-laterals and n-angles with any number $n>3$ of sides and vertices can be constructed.

The fundamental proposition proved by Steiner is the theorem of the complete quadrangle-quadrilateral. He proved that in a complete quadrangle the straight lines connecting the intersecting points of two pairs of opposite sides are harmonic to the last pair. A dual proposition holds for the quadrilateral (ibid., p. 75). Steiner argued that (see Fig. 2.27)

Be a, b, a_1, b_1 a complete quadrangle. Cut the three diagonals $\mathfrak{AB}, \mathfrak{ab}_1, \mathfrak{a}_1\mathfrak{b}$ in $\mathfrak{C}, \mathfrak{B}, \mathfrak{E}$. Be d the fourth harmonic straight line after a,b,c and d_1 the fourth harmonic after a_1, b_1, c_1, then because of the conservation of the harmonic groups in a projection, the straight lines d and d_1 cut the straight line $\mathfrak{ab}_1\mathfrak{C}$ in $\mathfrak{D}$, which is the fourth harmonic point after $\mathfrak{a}, \mathfrak{b}_1, \mathfrak{C}$ and which is the conjugate of $\mathfrak{C}$. Analogously, d and d_1 pass through the point of $\mathfrak{a}_1\mathfrak{b}\mathfrak{E}$ which is the fourth harmonic after $\mathfrak{a}_1, \mathfrak{b}, \mathfrak{E}$. Thence, this point coincides with $\mathfrak{D}$. Thus, the two groups of four points $\mathfrak{a}, \mathfrak{b}_1, \mathfrak{C}, \mathfrak{D}$ and $\mathfrak{a}_1, \mathfrak{b}, \mathfrak{E}, \mathfrak{D}$ are harmonic. From these results, it is easy to prove that $\mathfrak{C}, \mathfrak{E}, \mathfrak{A}, \mathfrak{B}$ is a harmonic group of four points, too. Since $\mathfrak{aba}_1\mathfrak{b}_1$ is any quadrangle, the explained reasoning shows that the straight lines d,c are the harmonic conjugate of a,b; the straight lines c_1, d_1 are the harmonic conjugate of a_1, b_1, and d, d_1 are the harmonic conjugate of $\mathfrak{ab}_1$ and $\mathfrak{a}_1\mathfrak{b}$. This proves the theorem.

Steiner's demonstration is based on the conservations of the harmonic groups in a projectivity, which, in turn, relies upon his use of the cross ratio.

Steiner continued his analysis claiming that his projective approach allows scholars to see in a new unitary perspective a "set of consequences" (*eine Menge Folgerungen*, ibid., p. 77) tied to the theory of triangles, transversals and, in general,

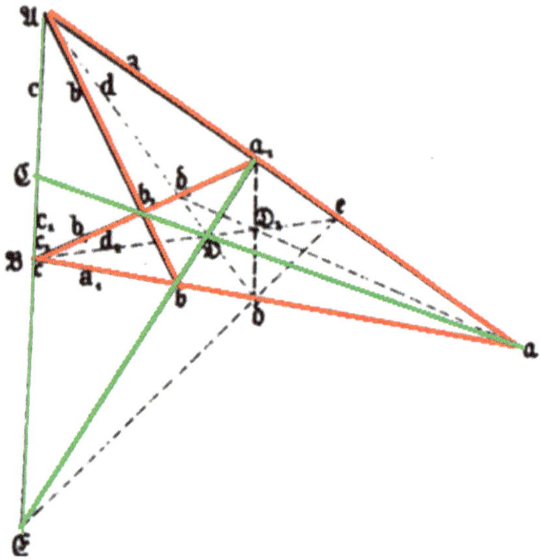

Fig. 2.27 The figure used by Steiner to prove the quadrangle theorem. I have used the green and red colours to make the figure clearer

problems of incidence. Among the propositions he proved, the most important is the following one:[101]

Given three mutually projective straight lines A, A_1, A_2, which concur in a point where three corresponding points converge, so that two any of such lines are perspective, then the three projection-points B, B_1, B_2 are collinear.	If three pencils of rays B, B_1, B_2 are mutually projective and lie so that three corresponding straight lines are so set that their centres of projection are collinear, so that each two pencils are perspective, then the three collinear straight lines A, A_1, A_2 cut in a point.

As Steiner remarked, an immediate consequence of these propositions is the fundamental Desargues theorem of the homological triangles and its dual proposition (ibid., pp. 78–79).

[101] Ibid., p. 78:

Sind die drei Gerade A, A_1, A_2 unter einander projectivisch und liegen sie so, dass sie sich in einem Punkte schneide, un dass in demselben drei entsprechende Punkte vereinigt, und mithin je zwei Gerade perspectivisch sind, so liegen die drei Projectionspunkte B, B_1, B_2 in eiener Geraden.	Sind drei Strahlbüschel B, B_1, B_2 unter einander projectivisch und liegen sie so, dass drei entsprechende Strahlen aufeinander fallen, also ihre Mittelpunkte in einer Geraden liegen, und mithin je zwei Strahlbüschel perspectivisch sind, so treffen sich die drei perspectivischen Durchschinitte A, A_1, A_2 in einem Punkt.

Steiner proved a plurality of other propositions relative to the n-laterals and the n-angles, but what expounded is enough for the reader to get a precise idea of his foundational programme.[102]

Commentary. 1) With regard to the properties of the complete quadrangle, Steiner pointed out that they were also known by Carnot and by Pappus (*Collections* seventh book). With regard to Carnot, the complete quadrangle is one of the most important configurations in his theory of transversals. In Carnot (1801), the IV problem (Carnot, 1801, pp. 117–138) consists in finding the ratios among the sides of any quadrilateral, its diagonals, the segments and the angles which result. In this context, Carnot introduced the name of *quadrilatère complet* (ibid., p. 122). The methods he used rely on classical geometry (e.g. the concept of similarity is used), on trigonometrical properties and on segmentary relations, to which he arrived through his theory of transversals. The projective methods are implicit in Carnot's work, but they are not explicitly used. His results concerning the quadrangles are obtained through a long series of algebraic manipulations in turn obtained through an appropriate use of classical synthetic geometry and of the aforementioned methods. Many of the segmentary and angular identities are expressible through a cross ratio or product of cross ratios, but Carnot did not use such a concept directly. Therefore, though several projective results and the concept of cross ratio are implicit in his work, the process towards the explication of these notions and their use is a long and difficult one. Steiner, clearly relying upon the work of other mathematicians such as Poncelet, carried out this step and his foundational programme sheds a new light on the whole of geometry.

Apropos of the problem of determining the fourth harmonic point after three given points in a straight line and the dual problem of determining the fourth harmonic straight line after three given in a pencil (problems which can be very easily solved through his complete quadrangle), Steiner claimed that the former was solved by De La Hire in his *Sectiones Conicae* (De La Hire, 1685). As a matter of fact, the first book of De La Hire's work begins with the definition of harmonic division of a straight line.[103] De La Hire defined a line AD

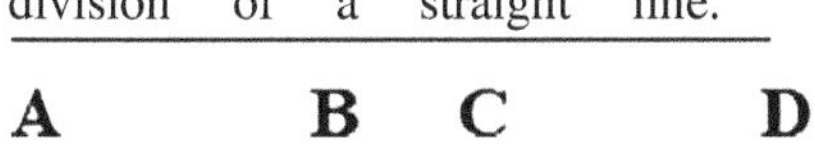

to be divided harmonically if $AD : AB = CD : BC$. In Proposition I, he illustrated how to divide harmonically a straight line. In the proof there is no use of the complete quadrangle. Whereas in Proposition XX where he required to solve the same problem with a different method (*methodo diversa*, De La Hire, 1685, p. 9), the complete quadrangle is used in the construction. However, it should be remarked that only a mathematician who, in retrospect, had guessed the importance of the complete quadrangle could interpret De La Hire's construction as one in which the quadrangle is used, because De La

[102]See, e.g., Nabonnand (2006, pp. 93–99).

[103]We have seen that Chasles in his *Traité* remembered the origin of the locution "division harmonique".

Hire did not pose any particular emphasis on this figure. It appears as a means to solve an important but specific problem.

2) As to the way in which Chasles treated the properties of the complete quadrangle—excluding, for the moment, their connections with the theory of conic sections—in the already mentioned Note X of the *Aperçu*, concerning the theory of involutions,[104] he claimed that the theorem of the involution of six points, stating that the three opposite sides of a complete quadrangle cut a straight line, which does not belong to any vertex of the quadrangle, in six points in involution, was known by Pappus (Proposition 130 of the seventh book of the *Collections*). He also claimed that, if the projective character of the involution is known, the proof of the theorem can be simplified considering a parallelogram rather than a generic quadrangle, as Brianchon did. The merit for the full understanding of the equations concerning the involution of six points was ascribed to Desargues (Figs. 2.28 and 2.29).[105] In what follows in Note X, Chasles used the anharmonic ratio to find new properties of the involution. In particular, he proved that when six points are in involution, the ratio of four any of them is the same as that of the corresponding points (Chasles, 1837a, p. 320). Among the new properties proved by Chasles with regard to the involution there is a remarkable one, from which he deduced several consequences and which is considered the main feature of the hyperbolic involutions: be AA', BB', CC' three pairs of correspondent points in an involution on a dotted straight line and be α, β, γ the central points of the three respective segments, it is rather easy to prove that the following equation holds:

[104]On the theory of involution in Chasles' *Aperçu*, see, i.e., Chemla (2016, pp. 73–77).

[105]The history behind Pappus' Proposition 130 is very interesting: in the second volume of Hultsch's edition (Pappus, 1877, pp. 873–875) the proposition sounds like this: given the figure $\alpha\beta\gamma\delta\varepsilon\zeta\eta\vartheta\lambda$, if it holds

(1) $\alpha\zeta \cdot \beta\gamma : \alpha\gamma \cdot \beta\zeta = \alpha\zeta \cdot \delta\varepsilon : \alpha\delta \cdot \varepsilon\zeta$, the points ϑ, η, ζ belong to a straight line (see Fig. 2.28).

However, Eq. (1) does not represent an involution of six points. In his book on Euclid's porisms (Chasles, 1860a, p. 102) Chasles changed the orders of the letters in $\alpha\varepsilon\beta\zeta\gamma\delta$. With this disposition, Eq. (1) is that of an involution. Hultsch (Pappus, 1877, II, p. 873) in a note refers to Chasles' disposition and to other four different ones existing in the Commandinus' edition. In Chasles (1860a, p. 102), Chasles used a figure with Latin letters, apart from ρ (see Fig. 2.29). The transcription between the order of the letters on the line *AB* and that referred to the Greek letters is obvious. Ver Eecke also refers to the disposition used by Hultsch (Pappus, 1933, pp. 675–677). For a more modern edition of Pappus' seventh book, see that edited by Alexander Jones (Pappus of Alexandria, 1986). Proposition 130 is on p. 264. With regard to Brianchon, Chasles referred to Brianchon (1817). As to Desargues, it is well known that his seminal work (Desargues, 1639) was appreciated by few mathematicians until the nineteenth century. One of these was Fermat. Great part of the merit for the rediscovery of Desargues as the mathematician who had already guessed many of the concepts which are the basis of projective geometry, is due to Poncelet and to Chasles himself. On Desargues, see, without any claim to be exhaustive, Andersen (1991), Anglade and Brien (2017, 2019) (see also my review of this very good paper, Bussotti, 2020), Catastini (2004), Catastini and Ghione (2005), Field (1987), Field and Gray (1987), Hogengijk (1991), Ivins (1943, 1947), Le Goff (2005), Schneider (1983), Taton (1951b), Viola (1946), Vita (1974), Zacharias (1941).

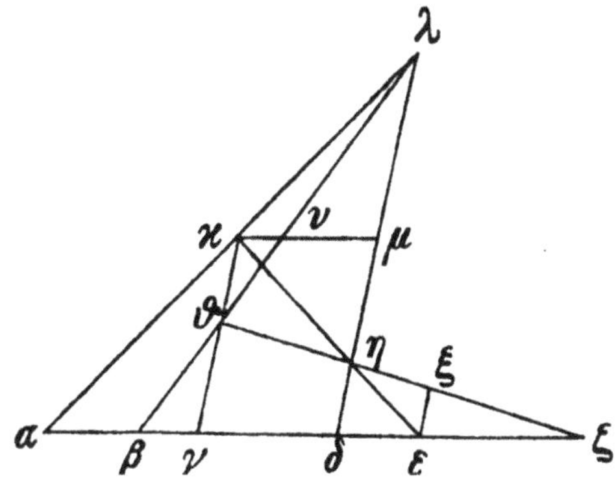

Fig. 2.28 The figure referred to by Hultsch in reference to Pappus' Proposition 130, seventh book of *Collections*

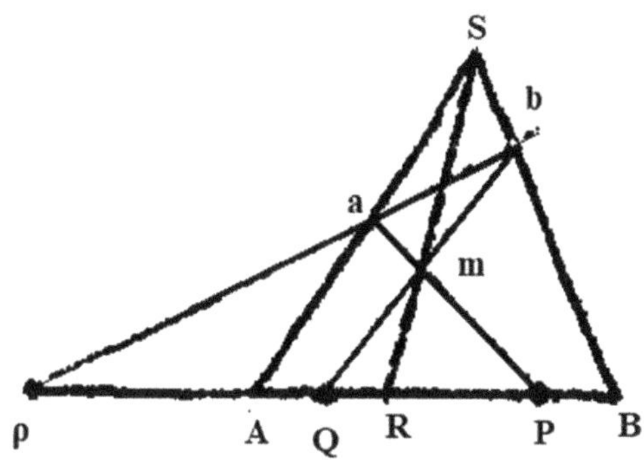

Fig. 2.29 The figure used by Chasles in reference to Pappus theorem

$$\overline{\alpha A}^2 \cdot \beta\gamma - \overline{\beta B}^2 \cdot \alpha\gamma + \overline{\gamma C}^2 \cdot \alpha\beta = \alpha\beta \cdot \beta\gamma \cdot \gamma\alpha$$

If the points C, C' converge in E, and if B, B' converge in F, namely if E and F are two double points of the involution, the equation is transformed into

$$\alpha A^2 = \alpha E \cdot \alpha F.$$

But this formula expresses the fact that the points A and A' are harmonic conjugate of E and F (ibid., pp. 321–322).[106] In this way, to use a modern language, applicable to Chasles' theory after 1852, Chasles proved that the absolute invariant of a hyperbolic involution is -1.

In Chasles (1852, Chapter IX, pp. 119–161), a complete theory of involution is expounded. The whole theory is based on the properties of the anharmonic ratio. Among the numerous properties proved by Chasles, the most important in our context are the following: i) the rigorous proof that any involution has two double

[106] A complete proof of the known fact that this equation represents the equation of four harmonic points can be found in Chasles (1852, pp. 41–45, particularly pp. 44–45).

points, either imaginary—we call elliptic such an involution—or real—we call it hyperbolic (ibid., §II, pp. 127–129); ii) as already proved in the *Aperçu*, the double points divide harmonically each other pair of corresponding points (ibid., p. 128); iii) the two double points are imaginary when two pairs of corresponding points mutually separate; they are real when they do not separate (ibid., § III, pp. 134–136). In Chapter XVIII (ibid., pp. 231–241), Chasles connected the theory of the involution and the properties of the complete quadrangle. This chapter is opened by the theorem of the six points (already analysed in the *Aperçu*, as we have seen) of which Chasles offered a proof completely based on the concept of cross ratio. Let us see his demonstration (Fig. 2.30).

The theorem states that each transversal belonging to the plane of a complete quadrilateral cuts its four sides and the two diagonals in six points in involution.

Proof: be *ABCD* the quadrilateral and *L* the transversal cutting the two opposite sides *AB*, *CD* in a,a' and *AD* and *BC* in b,b' and the two diagonals in c,c'. The four straight lines *AB*, *AC*, *AD*, Ac' from *A* cut the diagonal *BD* in the same points as the four straight lines *CB*, *CA*, *CD*, Cc' from *C*. Thence, the anharmonic ratio of the first four straight lines is equal to that of the second four. Ergo, the anharmonic ratio of the points a, c, b, c' in which the straight lines from *A* cut *L* is the same as that of the points b', c, a', c' where the straight lines from *C* cut *L*. The order of the last group of four points can be changed in a', c', b', c without changing their anharmonic ratio. Consequently, the two groups a, c, b, c' and a', c', b', c give rise to the three systems of corresponding points a, a'; b, b'; c, c', but the cross ratio of the points a, b, c, c' is the same as that of the corresponding points a', b', c', c; thence, the three pairs of points are in involution (ibid., p. 231).[107]

The proof is completely based on the property of the anharmonic ratio, namely on its conservation in sections and projections.

From this theorem Chasles deduced the harmonic theorem of the quadrilateral, namely that in any quadrilateral the two diagonals divide harmonically the straight line *D* joining the intersection points d,d' of the opposite sides. He argued that this property derives directly from the theory of the involution because, in this case, the transversal *L* is the straight line *D*. The points d,d' are, thus, the double points of an involution, so that, according to what already proved in the *Aperçu*, they divide harmonically the segment included between the two diagonals (ibid., p. 232).

In the rest of the chapter, Chasles proved several theorems concerning segmentary and angular properties of the quadrilateral/quadrangle through the features of the involution and through the basic theorems I have explained. In the next chapter, dedicated to the triangles (ibid., pp. 242–264), he proved numerous segmentary and angular properties of the triangles, some of which known and others new. For example, the theorem according to which in a triangle *ABC* whose sides are cut by a transversal in the points a,b,c (a lies on *BC*, b on *AC* and c on *AB*) the relation $\frac{aB}{aC} \cdot \frac{bC}{bA} \cdot \frac{cA}{cB} = +1$ holds (ibid., p. 242); or the proposition which states that the

[107] The identity of the cross ratios (a, b, c, c') and (a', b', c', c) is one fundamental condition for three couples of points to be in involution (ibid., p. 119).

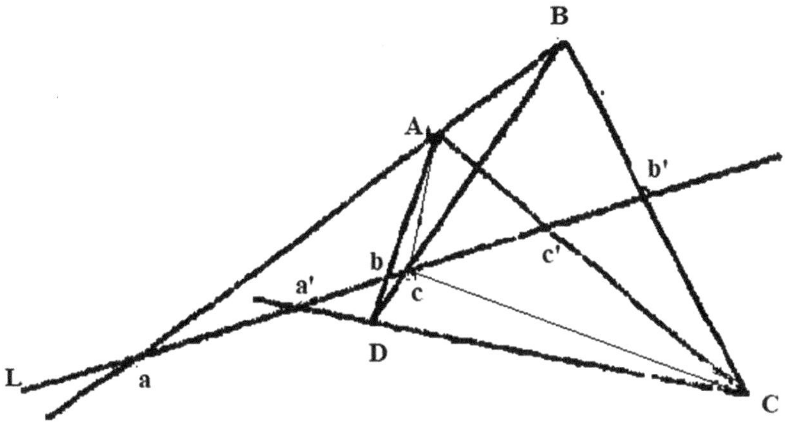

Fig. 2.30 The figure used by Chasles to prove the theorem concerning the involution of six points

straight lines drawn from a point to any vertex of a triangle cut the straight lines of the opposite sides in three points a,b,c, such that $\frac{aB}{aC} \cdot \frac{bC}{bA} \cdot \frac{cA}{cB} = -1$ (ibid., p. 246); or the one which establishes that if, through the vertices of a triangle ABC three lines Aa, Bb, Cc are drawn which form a triangle abc circumscribed to ABC, the relation $\frac{Aa}{Ac} \cdot \frac{bB}{Ba} \cdot \frac{Cc}{Cb} = \frac{\sin \widehat{aAC}}{\sin \widehat{aAB}} \cdot \frac{\sin \widehat{bBA}}{\sin \widehat{bBC}} \cdot \frac{\sin \widehat{cCB}}{\sin \widehat{cCA}}$ holds, and so for other theorems. What should be observed is that these propositions were proved by Chasles through a uniform method, based on the properties of the involution and of the quadrangle, which are, in turn, founded on the conservation of the anharmonic ratio by sections and projections. Chasles always cared about his foundational programme and his methods. As a matter of fact, in the chapter on the triangles, subsection entitled "Reflexions on the character of the proofs based on the theories explained in this work and other demonstrations of the previous theorems" (*Réflexions sur le caractère des démonstrations fondées sur les théories exposées dans cet Ouvrage – Autres démonstrations du théorème précèdent*, ibid., pp. 250–252), he claimed explicitly that the concept of anharmonic ratio seems destined to become the cornerstone of modern geometry and seems suitable to offer geometry a generality comparable with that of the analytical methods. As he wrote:

> It is the notion of *anharmonic ratio* and the theories naturally deduced from it which seem destined to supply these appropriate demonstrations. Under this standpoint, these theories hold, in modern geometry, a character which distinguishes them essentially and makes them suitable to give the geometrical conceptions all the generality that connotes the results of analysis.[108]

[108] Ibid., p. 250: "C'est la notion du rapport anharmonique et les théories qui s'en sont déduites naturellement, qui paraissent destinées à procurer ces démonstrations adéquates. Sous ce point de vue, ces théories ont dans la Géométrie moderne un caractère qui les distingue essentiellement, en les rendant propres à donner aux conceptions géométriques toute la généralité dont sont empreints les résultats de l'analyse".

Therefore, there is a line of continuity in Chasles' conceptions. The *Traité* was, in a sense, inscribed in Chasles' researches of the period 1828–1837, whose culminating point is represented by the use of the cross ratio in the duality and homography theory.

In a comparison with Steiner, it is only natural to highlight that he, too, had a foundational programme for geometry based on the concept of cross ratio. His programme was less wide than Chasles' because it was restricted to geometry and because Steiner was not as interested as Chasles in showing that the metric properties can be reduced to those graphic. This notwithstanding, the idea of founding the projective transformations on the concept of cross ratio was an important point shared by the two authors. It should be recalled that Chasles proved the harmonic theorem of the quadrangle including it in the general theory of involution, whereas Steiner gave a proof independent of the theory of involution. In 1852, Chasles saw such a theorem as an application of a broader theory.

3) It is to remember that in the modern and purely graphic approach to projective geometry, the one developed by Von Staudt in 1847, a group of four harmonic points is not defined by the fact that their cross ratio is -1, but by their construction through the complete quadrangle. Von Staudt in § 8 of his *Geometrie der Lage* wrote:

> If, given in a straight line three points $A,B,C,$ a quadrangle is constructed so that a diagonal passes through the second of the given points, and, in each of the two remaining points, two opposite sides of the quadrangle mutually cut, then the other diagonal will cut that straight line in a fourth point D, which is determined by the three points and which is called their fourth harmonic.[109]

Von Staudt proved that any other quadrangle such that a diagonal passes through B, a couple of opposite sides through A and the other couple through C has the feature that its other diagonal will cut ABC in D. Through a brilliant reasoning based on spatial considerations he also proved that the points A and C are separated by B and D. The whole argumentation is independent of the concept of cross ratio because his aim was exactly to offer a purely graphic foundation to projective geometry. The concept of harmonic ratio can then be seen as an application of Von Staudt's construction, and since the whole theory of the cross ratio can be found on the harmonic ratio, the whole theory of cross ratio can be interpreted as an application of Von Staudt's theory.

[109] Von Staudt (1847, p. 43): "Wenn in einer Geraden drei Punkte A, B, C gegeben sind, und alsdann ein Viereck so construirt wird, dass eine Diagonale durch den zweiten der gegebenen Punkte geht, in jedem der beiden übrigen aber zwei einander gegenüberliegende Seiten sich schneiden, so schneidet die andere Diagonale des Vierecks jene Gerade in einem vierten Punkte D, welcher durch die drei gegebenen Punkte bestimmt ist und zu denselben der vierte harmonische Punkt heist".

2.3.1.6 Sheaf of Planes

The second chapter of the *Systematsche Entwicklung* is entitled "Projective straight lines, plane pencils of radiuses and sheaves of planes in space" (*Von projectivischen Geraden, ebenen Strahlbüscheln und Ebenenbüscheln in Raume*, Steiner, 1832, pp. 97–127). In this chapter, Steiner analysed the projectivities between the named forms. From a conceptual point of view, this is an important section because Steiner showed that the traditional distinction between geometry in the plane and geometry in space is not suitable to grasp the essential features of projectivities since there is a spatial configuration, the sheaf of planes, which behave, from a projective standpoint, as the plane configurations dotted straight line and pencil of planes. The fundamental distinction depends on the movement of the element which produces the form as the three mentioned forms are produced by a simple movement of the generating elements, respectively, the point, the straight line and the plane. One might also say that, if the difference between the geometrical dimension of support and of the generating element is 1 then the form is of first species. For, in the dotted line the support is the straight line (dimension 1) and the element is the point (dimension 0); in the pencil of radiuses the support is the plane (dimension 2) and the element is the straight line; in the sheaf of planes, the support is space (dimension 3) and the element is the plane.

Steiner began the chapter by claiming that, since the new configuration is spatial, the drawing of figures is complicated. On the other hand, the reader who intends to become an expert in the new geometry has to imagine the construction of the figure through his force of representation (*Vortsellungskraft*, ibid., p. 98), without resorting to the help of the senses (*Versinnlichungsmittel*, ibid., p. 98). He seems to claim that logic and representative faculty are sufficient for geometry; the resort to the visual support is only a means to facilitate the comprehension, but, in itself, it is not necessary.

Steiner analysed then the reciprocal positions and incidence relations between a straight line A and a sheaf of planes $\mathfrak{A}$. The straight line will cut the planes $\alpha, \beta, \gamma, \delta, \ldots$ of the sheaf in a series of points $\mathfrak{a}, \mathfrak{b}, \mathfrak{c}, \mathfrak{d}, \ldots$ The plane parallel to the straight line or which cuts it in its point at infinity was called by Steiner "parallel plane" (*Parallelebene*, ibid., p. 99). Today we say that the point at infinity of the straight line belongs to the line at infinity of the plane. Steiner also analysed the case in which A cuts the axis of $\mathfrak{A}$ and its particular subcase in which A is parallel to the axis. With regard to the intersection of a plane B with the sheaf, it is, in general, given by a pencil of radiuses whose centre is in the intersection between B and the sheaf's axis. In particular if B is parallel to the axis, the intersection is the point at infinity of the axis so as to obtain a pencil of parallel lines. If B is perpendicular to the axis, then the angle between the planes $\alpha, \beta, \gamma, \delta, \ldots$ is the same as the angle between the straight lines $a, b, c, d, \ldots$ of the pencil lying respectively on those planes (ibid., pp. 100–101). Steiner also considered the sheaves of parallel planes.

It is possible to refer projectively the points $\mathfrak{a}, \mathfrak{b}, \mathfrak{c}, \mathfrak{d}, \ldots$ of the straight line A as well as the straight lines $a, b, c, d, \ldots$ of the pencil to the planes $\alpha, \beta, \gamma, \delta, \ldots$ of the sheaf. In particular, if they are referred so that each plane is related to the elements belonging to it, then the straight line and the sheaf as well as the pencil and the sheaf are correlated in a perspectivity (ibid., p. 101).

Steiner developed this noteworthy projective observation (Fig. 2.31): consider two pencils of radiuses belonging to B and B_1. They generate two planes which, without any ambiguity, can also be named B and B_1. Be $A = B \cap B_1$. Consider now the sheaf of planes whose axis is the straight line BB_1. Indicate this sheaf by $\mathfrak{A}$. The plane α belonging to $\mathfrak{A}$ cuts B and B_1 in the straight lines a,a_1 respectively. Be $\mathfrak{a}$ $=a \cap a_1$. This point belongs to A. The plane β belonging to $\mathfrak{A}$ cuts B and B_1 in the straight lines b,b_1 respectively. Be $\mathfrak{b}$ $=b \cap b_1$. This point belongs to A. Analogously for the plane γ (indicated in Fig. 2.31) and so on. It is possible to correlate the pencil B and the straight line A, so that the straight line a corresponds to $\mathfrak{a}$, the straight line b corresponds to $\mathfrak{b}$, the straight line c corresponds to $\mathfrak{c}$ and so on. Furthermore, it is possible to correlate the pencil B_1 and the straight line A so that so that the straight line a_1 corresponds to $\mathfrak{a}$, the straight line b_1 corresponds to $\mathfrak{b}$, the straight line c_1 corresponds to $\mathfrak{c}$ and so on. Consequently, the two pencils B and B' are projectively correlated so that a corresponds to a_1, b to b_1, c to c_1. Moreover, the intersection points of two homologous straight lines lie on the straight line A. Thence, even though the two pencils are not collinear they are in perspective position. Steiner claimed that this result is obtained basing on "intuition" (*Anschauung*, ibid., p. 102). Thus, he concluded that two pencils of radiuses B and B' which lie (namely which are perspective to) in a sheaf $\mathfrak{A}$ are perspective and the incidence straight line of the planes in which the two pencils lie is the perspectivity axis. Vice versa, if two projective pencils of radiuses B and B' have a perspectivity axis A, then they lie in a sheaf of planes $\mathfrak{A}$ (ibid., pp. 104–105).

Completely analogous theorems hold for the dotted straight lines, so that if two straight lines A and A' referred to the same sheaf are projective and if they mutually cut, then they are perspective and their perspectivity centre lies on the axis of the sheaf (ibid., p. 105). It is the point where the plane AA' cuts the axis.

Steiner also considered particular metric applications: all the straight lines which are perspective to a sheaf $\mathfrak{A}$ and which are parallel to the same plane are projectively similar. Vice versa all the straight lines perspective to a sheaf $\mathfrak{A}$ and similar are parallel to a plane. In particular if the sheaf is composed of parallel planes, the two dotted straight lines are equal. Similar theorems hold for the pencils of straight lines (ibid., pp. 105–106).

Given the three projective forms $\mathfrak{A}$=sheaf of planes with elements the planes $\alpha, \beta, \gamma, \delta, \ldots$; B=pencil of straight lines, with elements the straight lines $a,b,c,d,\ldots$; A=pointed straight line, with elements the points $\mathfrak{a},\mathfrak{b},\mathfrak{c},\mathfrak{d}, \ldots$ the fundamental theorem on the identity of the double ratios of four corresponding element holds:

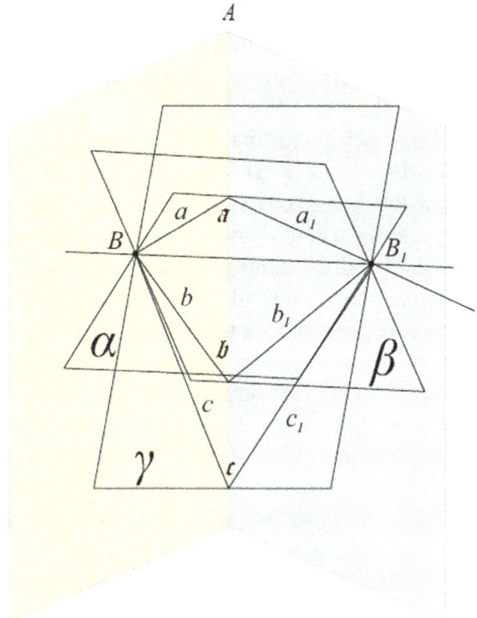

Fig. 2.31 Reconstruction of the figure describing Steiner's reasoning. The plane generated by the pencil B is the pink one and that generated by the pencil B_1 is the blue one

$$\begin{cases} \dfrac{\sin(\alpha\gamma)}{\sin(\beta\gamma)} : \dfrac{\sin(\alpha\delta)}{\sin(\beta\delta)} = \dfrac{\sin(ac)}{\sin(bc)} : \dfrac{\sin(ad)}{\sin(bd)} = \dfrac{\mathfrak{ac}}{\mathfrak{bc}} : \dfrac{\mathfrak{ad}}{\mathfrak{bd}} \\ \dfrac{\sin(\alpha\beta)}{\sin(\gamma\beta)} : \dfrac{\sin(\alpha\delta)}{\sin(\gamma\delta)} = \dfrac{\sin(ab)}{\sin(cb)} : \dfrac{\sin(ad)}{\sin(cd)} = \dfrac{\mathfrak{ab}}{\mathfrak{cb}} : \dfrac{\mathfrak{ad}}{\mathfrak{cd}} \\ \dfrac{\sin(\alpha\beta)}{\sin(\delta\beta)} : \dfrac{\sin(\alpha\gamma)}{\sin(\delta\gamma)} = \dfrac{\sin(ab)}{\sin(db)} : \dfrac{\sin(ac)}{\sin(dc)} = \dfrac{\mathfrak{ab}}{\mathfrak{db}} : \dfrac{\mathfrak{ac}}{\mathfrak{dc}} \end{cases}$$

Conversely: if the elements of two forms $\mathfrak{A}$ and B or $\mathfrak{A}$ and A are referred so that among each four pairs of elements the identity of the double ratios holds, the forms are projective (ibid., pp. 106–107). Consequently, according to what proved in the first chapter, Steiner concluded that the whole system of the corresponding pairs of elements of two projective forms $\mathfrak{A}$ and B or $\mathfrak{A}$ and A is given as soon as three pairs of corresponding elements are given.

Steiner analysed then the specific case of the harmonic ratio. In particular, from what proved on the conservation of the harmonic ratio in the previous sections, it follows that if four planes α, β, γ, δ of a sheaf $\mathfrak{A}$ are cut by a straight line A in four harmonic points $\mathfrak{a}, \mathfrak{b}, \mathfrak{c}, \mathfrak{d}$ or by a plane B in four harmonic straight lines a,b,c,d they will be cut by any other straight line and by any other plane in four harmonic points/ straight lines. These planes will, thence, be called "harmonic". For them the same relations holding for harmonic points and harmonic straight lines are valid. From the perfect analogy between the properties of the cross ratio in a straight line, in pencil of radiuses and in a sheaf of planes, several general theorems are deducible. The first one (Theorem I) claims that if three planes α, β, γ of a sheaf $\mathfrak{A}$ are cut by any straight line A in three points $\mathfrak{a}, \mathfrak{b}, \mathfrak{c}$ or by a plane B in three straight lines a,b,c, the locus of the points $\mathfrak{d}$ or of the straight lines d which are the fourth harmonic after $\mathfrak{a}, \mathfrak{b}, \mathfrak{c}$ and a, b,c is a plane δ of $\mathfrak{A}$.

The most interesting of the theorems proposed by Steiner in this context is the following: from the property of the complete quadrangle, it is easy to prove that, given two perspective straight lines a and a', their collineation axis is the fourth harmonic of a and a' with respect to the straight line joining the intersection point of a and a' and the centre of perspective. A dual theorem holds for two perspective pencils of radiuses (ibid., p. 76). Combining Theorem I and this last proposition, the following property can be inferred: given three planes α, β, γ of a sheaf $\mathfrak{A}$ and a point $\mathfrak{b}$ (belonging—e.g.—to β) from which any two straight lines A, A' are drawn cutting α, γ in $\mathfrak{a}, \mathfrak{c}$ and $\mathfrak{a}', \mathfrak{c}'$ respectively, the locus of the intersection points of the straight lines $(\mathfrak{a}\mathfrak{c}')$, $(\mathfrak{a}'\mathfrak{c})$ is the plane δ which is the fourth harmonic of β with respect to α and γ. Namely, if $\mathfrak{b}$ moves on β, the intersection points of $(\mathfrak{a}\mathfrak{c}')$, $(\mathfrak{a}'\mathfrak{c})$ move on δ. The dual proposition for three points belonging to a straight line holds (ibid., p. 110).

These propositions are general and include several theorems whose proof was beforehand not based on such general principles of projective theory. Steiner mentions a theorem by Carnot which states that: be given two tetrahedra $\mathfrak{b}\mathfrak{a}\mathfrak{a}_1\mathfrak{a}_2$ and $\mathfrak{b}\mathfrak{c}\mathfrak{c}_1\mathfrak{c}_2$, named α the plane $\mathfrak{a}\mathfrak{a}_1\mathfrak{a}_2$, γ the plane $\mathfrak{c}\mathfrak{c}_1\mathfrak{c}_2$ and $\mathfrak{A}$ their sheaf. Be β the plane of $\mathfrak{A}$ passing through $\mathfrak{b}$, then the intersection points of the diagonals of the three quadrangles $\mathfrak{a}\mathfrak{a}_1\mathfrak{c}\mathfrak{c}_1, \mathfrak{a}\mathfrak{a}_2\mathfrak{c}\mathfrak{c}_2, \mathfrak{a}_1\mathfrak{a}_2\mathfrak{c}_1\mathfrak{c}_2$ lie in a further plane δ of $\mathfrak{A}$ which is the harmonic correlate of β in the group of four planes α, β, γ, δ (Fig. 2.32).

The dual proposition also holds (ibid., pp. 111–112). This theorem can be proved as an application of the previous propositions.

Steiner continued his inquiry analysing the projectvities between two sheaves of planes. They are projective if they are projective to the same straight line A or pencils of straight lines $\mathfrak{B}$. The identity of the double ratios between the sinuses of the corresponding groups of four planes is shown as well as the inverse theorem (ibid., p. 112). All the projective theorems valid for the other forms of first species are extensible to the projective sheaves of planes. With regard to the perspectivities, two sheaves of planes are perspective if the straight lines along which the correspondent planes mutually cut are coplanar. Given this property, two sheaves of planes are perspective if two corresponding planes coincide or if the three intersections straight lines of three corresponding pairs of planes are coplanar.

As a conclusion of the second chapter Steiner added a section of problems and porisms, as well as two annotations. For my aim, the first annotation is interesting (ibid., pp. 121–124). It concerns the projective forms which lie in a spatial pencil of straight lines. The initial paragraph of this annotation clarifies what Steiner meant: he is going to analyse some properties of the set of all the straight lines passing through a point D, namely, in our language, the properties of the stars of straight lines[110] and

[110] In what follows, to simplify the language, I will use this expression rather than "spatial pencil of straight lines". Thence, the expression "pencil of straight lines" will be referred to the planar configuration.

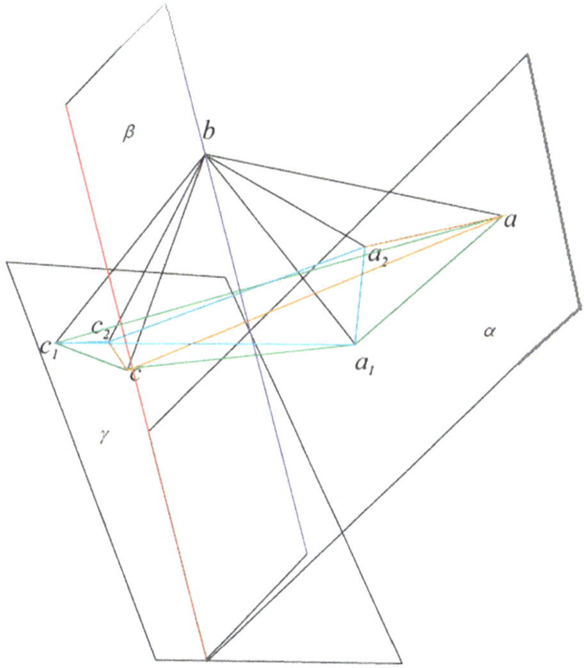

Fig. 2.32 Partial reconstruction of Carnot's theorem mentioned by Steiner. I have used Latin letters rather than Gothic letters, but in this context there is no ambiguity. I did not draw the diagonals of the three quadrangles because, otherwise, the figure would become too complicated and hardly legible

of the forms belonging to them. It is clear that any pencil of straight lines will belong to a star and a sheaf of planes will also, if its axis passes through the centre of the star. Interestingly enough, Steiner claimed that, this granted, the projective relations between a pencil of radiuses and a sheaf of planes belonging to *D* are the same as those between a dotted straight line and a pencil lying in the same plane. For if *D* is cut by a plane *E*: 1) each sheaf of planes belonging to *D* is cut in a pencil of radiuses; 2) each pencil of radiuses is in a dotted straight line and 3) each dotted straight line is cut in a point. Thence, if two forms in *E* are projective; their corresponding forms in *D* also are and vice versa. What Steiner is saying is interesting: it is clear, first of all, that while referring to the star of straight lines, he is also referring to the stars of planes. It is not a coincidence that he claimed that the sheaves of planes belong to a star. Therefore, Steiner is explaining a duality in which, in a star, to each plane a straight line corresponds. This derives from the previous item 1). In the plane to each straight line a point corresponds. This derives from the previous items 2) and 3). This is exactly the duality law for the forms of second species star of planes-star of straight lines and lined plane-dotted plane. The previous items 1), 2), 3) also allowed Steiner to correlate the star and the plane. For example, given a trihedron in the star, a triangle in the plane corresponds where the three planes of the trihedron correspond to the three sides of the triangle and the wedges of the trihedron to the vertices of the triangle. In other words, the star of planes is correlated to the lined plane through the duality plane-straight line and the star of straight lines is correlated to the dotted plane through the duality straight line-point. Thence, though Steiner had not yet completely developed the whole language and concepts of the projective forms, the fundamental notions were absolutely clear to him.

Commentary.

1) A particular remark has to be dedicated to the role of the figures in the then projective geometry. As we have referred to, Steiner explicitly claimed that the reader has to imagine the spatial configurations, so that, also considering the difficulty to draw spatial diagrams, he will avoid to draw them. Nabonnand clearly explains the reason why Steiner's can be considered as a "géométrie sans figures", or, at least, where the figures are not necessary elements. As he writes:

> With Steiner the figure is not anymore at the centre of the cares of a geometer. It is replaced by the projective relations among forms. The research of fundamental propositions on which geometry will be based to solve its problems and to develop its theories implies a subordination of the figure. The means available to the geometer do not concern anymore the figures directly, but the fundamental forms which become the basic-objects. On the other hand, the notion of projective relation relies on that of cross ratio, which is metric. Thus, it depends on the notions of angle and of length of a segment, used by ancient geometry".[111]

Certainly, Steiner is not an exception. Chasles, in his memoirs and in the *Aperçu*, drew almost no figure. In the *Traité* (Chasles, 1852) there are figures, but without any doubt Chasles shared Steiner's conviction that the reader had to construct the figures he was describing. With some different nuances, it is possible to claim that projective geometry in the first 40–50 years of the nineteenth century was a geometry, not without figures, but a geometry with an extremely moderate use of figures. Not only does this represent a clear contraposition to Greek geometry, but also to the treatises of the seventeenth and eighteenth centuries where the resort to figures is extensive. It is enough to think of Newton's *Principia* or of De La Hire's *Conics*. This might appear paradoxical if we think that projective geometry represents the theory on which, retrospectively, the methods of perspective used in painting are to be founded as well as particular techniques such as the cut of the stones and specific disciplines as the graphic statics. Thus, one might expect that the founders of projective geometry ought also offer a visual support to their theories. However, besides the reason of which Steiner spoke explicitly, there is a more profound one, which can be explained through an anecdote referred to Von Staudt, but which also holds for the other fathers of projective geometry. This anecdote is narrated by Corrado Segre in his introductory study to the Italian translation of the *Geometrie der Lage* (Segre in Staudt, 1889, pp. XIX–XX): Culmann applied Von Staudt's theory in his courses of graphical static at the Zürich Polytechnique starting from 1860, and in 1866, he published the important text *Die graphische Statik* (Culmann,

[111] Nabonnand (2011b, p. 14): "Avec Steiner, la figure n'est plus au centre des préoccupations du géomètre: elle est remplacée par les relations projectives entre formes. La recherche de propositions fondamentales sur lesquelles le géomètre s'appuiera pour résoudre ses problèmmes et développer ses théories induit une subordination de la figure. Les outils dont dispose le géomètre ne concerne plus directement les figures mais les formes fondamentales qui deviennent les objets de base. Par contre, la notion de relation projective repose sur celle, métrique, de birapport, et donc dépend des notions d'angle et de longueur utilisées dans l'ancienne géométrie".

1866). In the Easter of 1865 Culmann visited Von Staudt in Erlangen. During a conversation he expressed to Von Staudt the desire that his masterpiece *Geometrie der Lage* could be reprinted enriched with the diagrams. Von Staudt refused this proposal as, he claimed, a figure shows only a particular case, whereas he should have imagined simultaneously a series of figures, all relative to the same proposition. Therefore, there are two intrinsic reasons to explain why, in most cases, the figures are missing or almost missing: the first one concerns the generality that the founders of projective geometry ascribed to their works. Independently of what was the real meaning of the diagrams in Greek geometry,[112] there is no doubt that authors such as Poncelet, Chasles, Steiner and Von Staudt considered the theorems of Euclidean geometry as being confined to specific situations, while their aim was to offer a new general approach based on more general transformations than those purely metric (this is valid for Steiner, as well). Therefore, even though the theorems of classical geometry refer to an infinity of situations (e.g. the theorem of Pythagoras holds for any right triangle, not only for that specific Euclid drew in his proof), these situations can be represented in paradigmatic figures far more easily than in projective geometry, whose generality makes this operation more difficult. As it is often the case, Chasles is the most explicit in this respect because in his *Aperçu* he claimed more than once that classical geometry is confined to the analysis of specific configurations, situations and transformations, whereas his purpose was to offer a synthetic theory where any theorem includes in itself a plurality of situations which are separated in a metric approach.[113] In his Introduction to Poncelet (1822) Poncelet also underlined the new character of modern geometry, but Chasles addressed this subject in a more extensive way.

The second aspect which makes it more difficult to draw figures in projective geometry is the dynamical character of the theorems connoting this discipline. This is absolutely clear in Chasles. We have seen that in his memoir on duality the first and fundamental theorem is expressed in terms of a mobile plane and of a mobile point. This dynamical view on the nature of projective geometry is typical of all the fathers of such discipline. When the theory of transformations was completed and organized in an abstract algebraic doctrine, the dynamical view lost part of its importance, but during the 30s and 40s of the nineteenth century, a phase in which one sees a transition from a geometry of figures to a geometry of transformations, the dynamical approach was a fundamental aspect for the geometers to visualize with their mind the continuous action of a certain transformation, duality, homography,

[112]The problem of how the Greeks considered the diagrams within their geometry is rather complex. Since, in itself, it is not strictly connected to the subjects I am dealing will, I will restrict to mention some important works, without any claim to be exhaustive: Acerbi (2011), Catton and Montelle (2012), Dimmel and Herbst (2015); Ferreirós (2016), specifically chapter 5 entitled "Ancient Greek Mathematics: A Role for Diagrams"; Hanna and Sidoli (2006a, b), Macbeth (2010), Manders (2008), Miller (2007), Mumma (2008, 2012), Netz (1998, 1999), Panza (2012), Saito (2012), Saito and Sidoli (2012).

[113]Chemla (2016, section 2.2.1, pp. 50–56) offers a satisfying view of Chasles' opinion on the way of proceeding of classical geometry and on its limits of generality.

etc. I am not claiming that a dynamical approach to geometry was an absolute novelty; for it is enough to think of the way in which Kepler saw the conics in his *Paralipomena ad Vitellionem* or of Newton's organic generation of conics,[114] and the examples might be multiplied. However, the projections are naturally associated with the change of the viewpoint, continuous transformations (it is not a coincidence that the principle of continuity was introduced by Poncelet in projective geometry) and the passage of elements from "normal" positions to "particular" positions as it is the case for the elements at infinity and for those imaginary. Thus, dynamicity is consubstantial to projective geometry. It is more difficult to represent dynamical situations than statical ones. When these situations are spatial, the enterprise becomes very arduous. If to all these aspects—which far from being secondary reveal the profound nature of the researches the projective geometers were developing in the first 40 years of the nineteenth century—we add that the construction of the figures is seen as a good "exercise" for the reader, you can guess why projective geometry was then a "geometry without figures".

Of course, it is not impossible to draw spatial figures which represent projective theorems. The handbooks of projective geometry published in the second half of the nineteenth century often do. The reader can check how much, sometimes, these figures are intricate, though very useful to get an intuitive view of the theorem's meaning.

Finally: before the period we are analysing, the almost complete lack of figures was a prerogative of the analytic approach to geometry and science. What Lagrange wrote at the end of the "Avertissement" of his *Méchanique Analitique*'s first edition is paradigmatic. For we read:

> No figure will be found in this work. The methods I present require neither constructions nor geometrical or mechanical arguments, but solely algebraic operations subject to a regular and uniform procedure. Those who appreciate mathematical analysis will see with pleasure mechanics becoming a new branch of it and, hence, will recognize that I have enlarged its domain.[115]

The reasons why the supporters of the analytic approach refuse the use of figures and the reasons why there is a moderate use of diagrams in the founders of modern geometry have, in part, analogous grounds, but the final goals are opposite: the analysts think that their methods, which transform a geometrical or a mechanical

---

[114]On Kepler's conception of conics in the *Paralipomena* (Kepler, 1604), a good paper is Del Centina (2016a). The well-known organic generation of conics was offered by Newton in *Enumeratio Linearum Tertii Ordinis*. A dynamical generation of conics is also supplied by Newton in the fourth and fifth sections of the *Principia*.

[115]Translation retrieved from Lagrange (1997, p. 7). The translation concerns the 1811 edition of Lagrange's masterpiece, but the "Avertissement" is referred to in this edition exactly with the same words as in the first one: "On ne trouvera point de Figures dans cet Ouvrage. Les méthodes qui j'y expose ne demandent ni constructions, ni raisonnemens géométriques ou méchaniques, mais seulement des opérations algébriques, assujetties à une marche régulière & uniforme. Ceux qui aiment l'Analyse, verront avec plaisir la Méchanique en devenir une nouvelle branche, & me sauront gré d'en avoir étendu ainsi le domaine" (Lagrange, 1788, p. VI).

problem into an algebraic-analytical one, are far preferable because they are general and uniform. The fathers of projective geometry had the same purpose: to make their methods uniform and general. But their aim was to compete with the purely analytical procedures. This aspect stresses a further important nuance of modern geometry which is also connected to the problem of generality: the creation of uniform methods to prove the entire set of theorems so as to make synthetic geometry as systematic as analytic geometry was one of the goals of the projective geometry's founders. Therefore, the representation of their results through diagrams was not interesting for them because a figure photographs only a single situation. Nonetheless, their aspiration was to find general laws for the "world of geometrical figures": according to them, the figures were particularly important and belonged to such an abstract and general world—be this world a platonic one, or created by our mind; this makes no difference in this context—that a drawn diagram was, in general, not able to reproduce the richness of the properties relative to the world of the figures. In contrast to this, the analysts considered that figures had to be eliminated because the most profound as well as more malleable, general and easily treatable structures of mathematics and science are algebraic-analytical.

In the transition between classical geometry and projective geometry, an important author is Lazare Carnot because he considered that the real objects were represented by the figures and that the algebraic-analytical language expressed mere symbols which did not correspond to a reality in themselves. However, he felt the need both to generalize the approach and results of classical geometry and to introduce the algebraic-analytical symbolism within geometry in an ontologically satisfying manner. Therefore, though not arriving at conceiving a projective doctrine of geometry, in the works written between 1800 and 1806 (Carnot, 1801, 1803, 1806) he introduced the concept of configuration, which is not correlated with projection, but which is more general than the concept of figure. To give an example: Menelaus theorem represents a configuration rather than a figure, because it is applicable to flat triangles, to other flat figures belonging to the same plane, to spherical triangles and to other figures composed of arcs of great circles belonging to a sphere. Geometry has to study configurations rather than figures. The theorems of the figures have to be deduced as particular cases of configurations. Within this context a geometrical meaning to the algebraic quantities can be assigned. Though Carnot's approach is not projective, he introduced several ideas which were improved, generalized and made more perspicuous by the fathers of projective geometry. For example, in Carnot you can find a sort of continuity principle—the principle of correlation among figures—which was then specified by Poncelet who also criticized some aspects of Carnot's principle.[116] In this context, it is worth recalling the valuable second chapter of Lorenat (2015a), entitled "The role of the figure: a case study in methodological differences". She refers to a locution used by Tournès (2012, p. 272), that of "virtual figures". They are figures which are not

[116]A good paper from which I have drawn several of the ideas expressed on Carnot is Chemla (1998).

drawn in the text, but whose construction is described. Lorenat's perspicuous analyses show how in Poncelet (1817) and Plücker (1826)[117] there is a recurrent use of virtual figures. The author examines a single problem consisting in the construction of "a second-order curve sharing a third order contact with a given planar curve, whose multiple solutions included conic sections, tangents, secants and points of intersection constructed with the ruler alone" (Lorenat, 2015a, pp. 62–63). In Poncelet (1817), this problem is solved at pp. 153–154. However, as Lorenat herself points out Poncelet, Plücker and Gergonne (we might add almost all the geometers of that period) resorted to virtual figures on many occasions. I think the use of virtual figures to be strictly connected with the fact that the then geometry was, in any case, a visual geometry, namely a geometry in which the visual intuition of the geometers played a fundamental role in the heuristics of their results (Lorenat points out this aspect) and to be also connected with the fact that they imagined dynamical configurations rather than static figures. Thence, they provided the reader with all the indications through which to construct a figure, whose elements can be changed according to their described movement, which is typical of projective geometry. Therefore, they drew virtual configurations corresponding to several actual figures (had they drawn). Of course, as already pointed out, this does not mean that they did not draw actual figures, but certainly the virtual-dynamical configurations were more suitable for their approach. Finally, Lorenat also stresses a practical-economical aspect: to draw numerous figures was expensive (ibid., p. 76).

2) In the *Aperçu* Chasles introduced the concept of anharmonic ratio among four planes in an *Addition* (Chasles, 1837a, pp. 550–552) while dealing with a problem posed by De La Hire in his *Gnomonique* (De La Hire, 1698) (Fig. 2.33). In this context, he claimed that the anharmonic ratio of four straight lines is equal to that of four planes of which such straight lines are the traces on a certain plane, so that, if A, B, C, D are the planes and a, b, c, d their traces, it holds

$$\frac{\sin(ca)}{\sin(cb)} : \frac{\sin(da)}{\sin(db)} = \frac{\sin(CA)}{\sin(CB)} : \frac{\sin(DA)}{\sin(DB)}$$

The example examined as well as Chasles' commentaries are interesting: De La Hire in the *Gnomonique* posed the problem of drawing all the hour lines of a sundial, given some conditions. The most significant case addressed by De La Hire concerns that in which seven consecutive hour lines are given. De La Hire solved the problem like this: be given the lines X,XI, XII, I, II, III, IV. From a point of the line IV, draw a transversal l parallel to the line X. It will cut the lines III, II, I, XII, XI in the points a, b, c, d, e. Draw the segments oa', ob', oc', od', oe' equal respectively to the segments oa, ob, oc, od, oe on l from the opposite part of o. The points a', b', c', d', e' belong to

[117] The original version of this paper was published in Plücker and Schoenflies (1904), whereas a modified version was published—in two papers—by Gergonne in his Annales with the name Pluker (1826). Following Lorenat's way of writing I will indicate the 1826 versions as Plücker (1826a, 1826b) and the original one as Plücker and Schoenflies (1904).

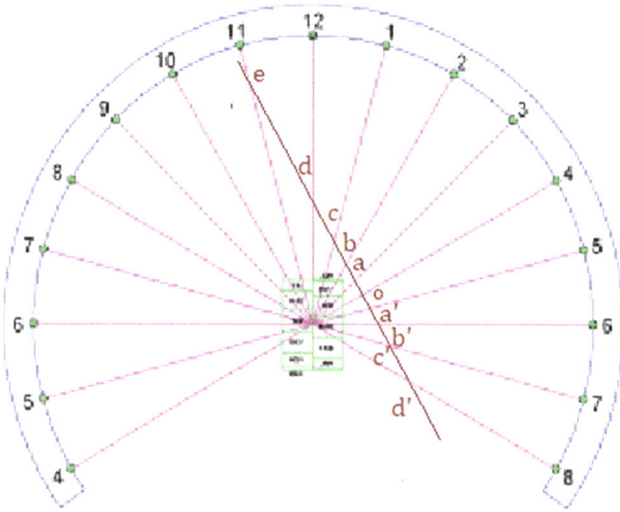

Fig. 2.33 My representation of De La Hire's diagram described by Chasles to construct the hour lines, given seven of them

the searched hour lines. Thence, the problem is solved. For, the two hour planes[118] X and IV are perpendicular and the planes III and V are equally inclined on the plane IV. Therefore, Chasles explains, these two planes are harmonic conjugate of the two planes X and IV. Ergo, the hour lines III and V are harmonic conjugate of X and IV. Thus, each transversals cut these four straight lines in four harmonic points. Consequently, if this transversal is parallel to the straight line X, the two points where it cuts the straight lines III and V are equidistant from that where it cuts the straight line IV, which proves the theorem (Chasles, 1837a, p. 550). Now, Chasles posed a typical foundational question: are the conditions of De La Hire minimal or is it possible to obtain the same results with a smaller number of conditions? Generally, what is the least number of conditions for the problem to be solved? Thanks to the concept of anharmonic ratio among four planes and four straight lines this problem can be solved when only three hour lines are given. For be a,b,c the given hour lines. Suppose that the line d is the one to construct. The anharmonic ratio of these four straight lines will be the same as that of the four hour planes of which they are the traces on the quadrant. If A,B,C,D are these four planes it holds

$$\frac{\sin(ca)}{\sin(cb)} : \frac{\sin(da)}{\sin(db)} = \frac{\sin(CA)}{\sin(CB)} : \frac{\sin(DA)}{\sin(DB)}$$

[118] An hour plane is the plane which, at any instant, contains the sun, the observer and the north celestial pole. The gnomon and the appropriate hour line lie in the hour plane.

Since the angles among the planes A,B,C,D are given because D is determinable through observation, the second member is a given quantity and, hence, the position of d can be found.

Therefore, the concept of anharmonic ratio allowed Chasles to see in a vast and general perspective a series of problems which, beforehand, were isolated problems that could be solved with different methods, or, at least, through methods that, unlike the cross ratio, were not absolutely uniform. As we have seen in Sect. 2.2.2 on duality, the cross ratio of four planes in comparison to that of four points and straight lines became a fundamental element, or better, probably the fundamental element, in Chasles' programme from the beginning of his two crowned memoirs. In the *Aperçu*, he did not offer a systematic treatment because this work was not conceived as a research-handbook like Steiner's *Systematische Entwicklung*, but certainly Chasles had arrived at the same results as Steiner's. The way in which he used the cross ratio in his two memoirs leaves no doubt. In the second chapter of Chasles (1852), which, instead, has the structure of a systematic work on the recent projective geometry, the author, after having defined the anharmonic ratio of four points and straight lines, defined that of four planes (Chasles, 1852, pp. 7–8). He proved that if four planes belong to a sheaf, a plane not belonging to the sheaf cuts them in four straight lines whose anharmonic ratio is the same as that of the four planes and that if they are cut by a transversal they are cut in four points whose anharmonic ratio is the same as that of the planes (ibid., pp. 13–14). This shows the consonance between Steiner's and Chasles' thought. With regard to the stars of planes and straight lines, we will see in the next chapter that they will play an important role in Chasles' studies on the movements of a rigid body.

2.3.1.7 The Projective Generation of Conics

The third chapter of Steiner's masterpiece is entitled "Erzeugung der Linien und der geradlinigen Flächen zweiter Ordnung durch projectivische Gebilde" ("Generation of the lines and ruled surfaces of the second order through projective forms", Steiner, 1832, pp. 127–295). This is the longest and more profound section of his work. The style of this section is, at least in part, different from that of the previous ones. For the concept of cross ratio was used to found the basic properties of the projective forms—therefore, it is the basis also of the chapter I am analysing—but the considerations and the proofs expounded here by Steiner are purely synthetical. The cross ratio enters directly in a few circumstances, though, as it has been observed, it is the basis of the previous theory of projective forms used here. Steiner clearly pointed out the foundational character of his work while referring to the "necessary generation of the conics from the geometrical basic forms" ("nothwendige Entstehung der Kegelschnitte aus dem geometrischen Grundgebilden", ibid., p. 128). More than this: he was going to show a remarkable ("merkwürdige", ibid., p. 128) double projective generation both of conics and ruled surfaces so that the properties of these figures appear under a new, unified light.

First Steiner considered a double oblique circular cone and the star of planes passing through its vertex. Some planes will not contain any generatrix of the cone (planes of kind A), some will contain one (kind B) and some two (kind C). All the planes of space parallel to planes of kind A cut only one sheet of the cone in ordinary points. The set of these section points is called ellipse (ibid., pp. 131–132). Each plane α parallel to those of kind B cuts only one sheet of the cone and cuts all the generatrices in ordinary points apart from that generatrix belonging to the plane of kind B to which α is parallel. This is touched by α in its point at infinity. The set of these section points is a parabola. Each plane β parallel to the planes of kind C cuts the two sheets of the cone and it cuts each generatrix in ordinary points apart from the two generatrices cut by the plane through the vertex parallel to β. These are cut in their respective points at infinity. This section is a hyperbola (ibid., p. 133). Thus, the ellipse has no point at infinity, the parabola one and the hyperbola two. The parabola has a tangent at infinity (ibid., p. 134).

Steiner then claimed that, since the conic sections are generated by section of a cone with a plane, they share their own projective properties. Thence, for the study of such properties, it is appropriate to analyse the projective generation of the circumference, the easiest and well-known conic section.

Steiner (see Fig. 2.34) observed that: be $\mathfrak{B}\mathfrak{B}_1$ a secant of the circle C. From any point $\mathfrak{a}, \mathfrak{b}, \mathfrak{c}, \ldots$ of the circumference draw the straight lines $a,b,c,\ldots a_1,b_1,c_1,\ldots$ respectively to the points $\mathfrak{B}$ and $\mathfrak{B}_1$ where the secant cuts the circumference. Therefore, it is $\left(\widehat{ab}\right) = \left(\widehat{a_1b_1}\right); (\widehat{ac}) = (\widehat{a_1c_1}); \left(\widehat{bc}\right) = \left(\widehat{b_1c_1}\right), \ldots$. This means that the projective pencils of radiuses $\mathfrak{B}$ and $\mathfrak{B}_1$ are equal. Furthermore, to the line $\mathfrak{B}\mathfrak{B}_1$ considered as belonging to the pencil $\mathfrak{B}$, the tangent to the circumference at $\mathfrak{B}_1$ corresponds and, considered as belonging to the pencil $\mathfrak{B}_1$, the tangent at $\mathfrak{B}$ corresponds. Ergo, since the straight line $\mathfrak{B}\mathfrak{B}_1$ is not self-corresponding, the two projective pencils $\mathfrak{B}$ and $\mathfrak{B}_1$ are not perspective, so that a circumference is generated by two projective not perspective equal pencils of radiuses.

Steiner then proved (see Fig. 2.35) that, given any two tangents A, A_1 to a circumference, the set of all the other tangents produces a projectivity between the points of A, A_1 such that to each point in which a tangent a cuts A, the point in which it cuts A_1 corresponds. In the figure, the points $\mathfrak{q}_1, \mathfrak{r}$ represent respectively the limit points of A_1 and A, being q parallel to A_1 and r to a. It is important to remark that Steiner's proof relies on properties of elementary geometry and on a property deriving from the invariance of the cross ratio in a projectivity, namely that if, in two projective straight lines a and b, you consider two corresponding points A and B and the limit points I of a and J of b, it is $AI \cdot BJ = \text{const}$ (ibid., § 12.1, pp. 38–40). In the case under examination, Steiner proved that $\mathfrak{a}\mathfrak{r}.\mathfrak{a}_1\mathfrak{q}_1$ is constant (ibid., p. 136). After a brief further reasoning he was, thence, able to state the two dual following fundamental propositions:[119]

[119] Ibid., p. 137:

Irgend zwei Tangenten (A,A_1) eines Kreises sind in Ansehung der entsprechenden Punktpaare, in whelchen sie von den übrigen Tangenten geschnitten werden, projectivisch, und zwar entsprechen dem in ihrem Durchschnitte vereinigten Punkten $\mathfrak{d}, \mathfrak{e}_1$, ihre wechselseitigen Berührungspunkte $\mathfrak{d}_1, \mathfrak{e}$.	Irgend zwei Punkte $(\mathfrak{B},\mathfrak{B}_1)$ eines Kreises sind die Mittelpunkte zweier projectivischen Strahlbüschel, deren entsprechende Strahlen sich in den übrigen Punkten der Kreislinie schneiden, und zwar entsprechen den vereinigten Strahlen d, e, die wechselseitigen Tangenten d_1, e in jenen Punkten $(\mathfrak{B},\mathfrak{B}_1)$.

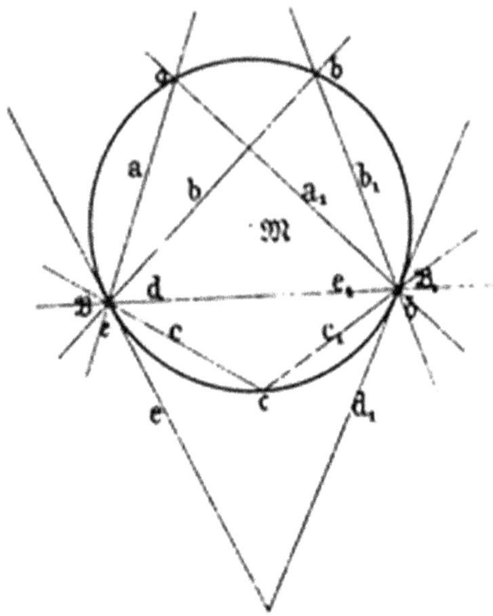

Fig. 2.34 The figure used by Steiner to explain the projective generation of a circumference

Proposition *P*. Any two tangents A, A_1 to a circle are projective if you consider as corresponding the points where they are cut by the other tangents. To their intersection point $\mathfrak{d}, \mathfrak{e}_1$ their reciprocal tangent points $\mathfrak{d}_1, \mathfrak{e}$ correspond.	**Proposition *P′*.** Any two points $\mathfrak{B}$, $\mathfrak{B}_1$ of a circle are the centres of two projective pencils of straight lines. The corresponding radiuses mutually cut in the other points of the circumference. To the common straight line d, e_1 the reciprocal tangents d_1,e at those points $\mathfrak{B}$, $\mathfrak{B}_1$ correspond.

Since these propositions P and P' concern only incidence and position properties, their formal structure continues to be valid if to the configurations they express, others which are their projections or sections are replaced. In particular: a circumference is the section of a second-degree cone. Therefore, if we project the configuration of Proposition P from a point not belonging to the plane of the circle, we obtain that: "In any two planes α, β tangent to a second-degree cone, there are two projective pencils of radiuses, whose corresponding couples of straight lines lie in the other tangent plane. In particular, to the common radius (d, e_1) of the two planes, the two straight lines (d_1,e) in which β and α touch the plane correspond. To Proposition P' dual of P, the following one corresponds:

> Each two straight lines of a second degree conic surface are the axes of two projective sheaves of planes, whose corresponding couples of planes cut in the other straight lines. In particular, to the common plane (δ, ε_1) of those two straight lines, the planes (δ_1, ε) which touch the cone along the two given straight lines correspond" (ibid., p. 138).

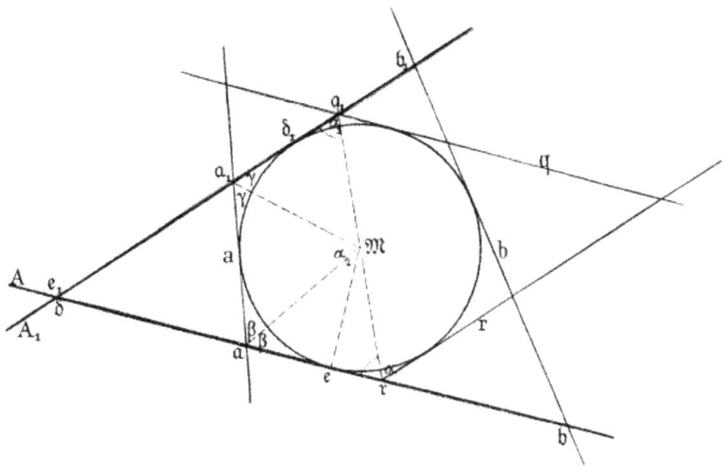

Fig. 2.35 The figure used by Steiner to explain the projective properties of the tangents to a circumference

Two propositions which express the converse property also hold.

It is remarkable the fact that Propositions P and P' show a duality in the plane between points and straight lines, whereas the propositions derived from them show a duality in the star between planes and straight lines.

Now the chain of deductions can be continued and Steiner inferred a third group of propositions. Since the conics are obtained as a section of a second-degree conic surface, the four previous propositions imply this other quatern of propositions, which express Steiner's famous projective generation of the conic sections:

Prop. A. Each two tangents A, A_1 of a conic section are projective considering the points in which they are cut by the other tangents. In particular to their common point $\mathfrak{d}, \mathfrak{e}_1$, their reciprocal tangent points $\mathfrak{d}_1, \mathfrak{e}$ correspond.	**Prop. A′.** Each two points $\mathfrak{B}, \mathfrak{B}_1$ of a conic are the centres of two projective sheaves of radiuses whose corresponding straight lines mutually cut in the other points of the conic. In particular, to the common straight line (d, e_1) the tangents (d_1, e) in the reciprocal points $(\mathfrak{B}, \mathfrak{B}_1)$ correspond.

Vice versa	
Prop. B. Any two coplanar projective, but not perspective, straight lines A,A_1 generate a conic section, which they touch. Namely, they and all their projection-straight lines are all the tangents of a determined conic. In particular, this conic section touches A,A_1 in those points $(\mathfrak{e}, \mathfrak{d}_1)$, whose correspondent points $(\mathfrak{e}_1, \mathfrak{d})$ are united.	**Prop. B′.** Any two projective, but not perspective, coplanar pencils of radiuses $\mathfrak{B}, \mathfrak{B}_1$ generate a conic section passing through their centres. Namely these and the intersection points of the corresponding couples of radiuses are all the points of a determined conic. In particular, this conic is touched in those

(continued)

Vice versa	
	projective centres by those lines (e, d_1) whose correspondent straight lines (e_1, d) are united. (Ibid., p. 139).

Steiner stated that such propositions are those fundamental ("die eigentlichgen wahren Fundamentalsätze") for the research of the conics' properties, because almost all of them can be deduced from these propositions (ibid., p. 140).

After having explained the projective generation of conics Steiner analysed some significant particular cases (ibid., pp. 140–147), namely he analysed what nowadays we call the affine and metric properties of the conics. First of all, he examined the projective generation of the parabola: since two dotted similar straight lines have a projection-straight line at infinity and vice versa two dotted lines are similar if their points at infinity mutually correspond and, since, as Steiner had just proved, among the conics, only the parabola has a tangent at infinity, then it follows that two projective but not perspective similar pointed straight lines generate a parabola. Vice versa, each two tangents at a parabola are cut by all other tangents in two projective similar straight lines. This projective generation explains why the parabola does not have parallel tangents: this depends on the fact that, given two similar projective dotted straight lines, no parallel projective radiuses exist (ibid., p. 141).

Steiner offered other interesting properties concerning the generation of particular conics. For example (Fig. 2.36), if, given two intersecting straight lines A, A_1, their common point is the limit point both for A and A_1, namely the point at infinity of A corresponds to $A \cap A_1$ regarded as belonging to A_1 and the point at infinity of A_1 corresponds to $A \cap A_1$ regarded as belonging to A, then A and A_1 are the asymptotes of a hyperbola.

Therefrom, since, given two corresponding points A and B on two projective straight lines a and b and the limit points I of a and J of b, it is $AI \cdot BJ = \text{const}$, then the well-known feature of the hyperbola follows: "the rectangles $\mathfrak{r}\mathfrak{a}, \mathfrak{q}_1\mathfrak{a}_1, \mathfrak{r}\mathfrak{b}, \mathfrak{q}_1\mathfrak{b}_1$ that any tangent *a,b,...* cut from the asymptotes A, A_1 have a constant magnitude" (ibid., p. 142). This means that such metric property of the hyperbola also has a projective basis.

Steiner developed a remarkable observation concerning the projective movement of the two straight lines A and A_1 (Fig. 2.37).

He observed that, when their position becomes perspective, the conic enveloped by their projective radiuses degenerates into the straight line ee_1 connecting the projection point B with the intersection point $\mathfrak{e}\mathfrak{e}_1$ of A and A_1. In this case, the ellipse of which A and A_1 are the tangents degenerates into the segment B-$\mathfrak{e}\mathfrak{e}_1$ and the hyperbola into the two other infinite parts of the straight line B-$\mathfrak{e}\mathfrak{e}_1$ and the parabola, obtained when the projection point B is at infinity, into the half-line B-$\mathfrak{e}\mathfrak{e}_1$ posed in the part of plane containing the parabola. Today we see this question in a slightly different manner. For we say that two perspective dotted straight lines generate a couple of points given by the perspectivity centre and by the intersection of the two straight lines (degenerate envelop conic), whereas the intersection locus of two perspective pencils of straight lines is a couple of straight lines given by the

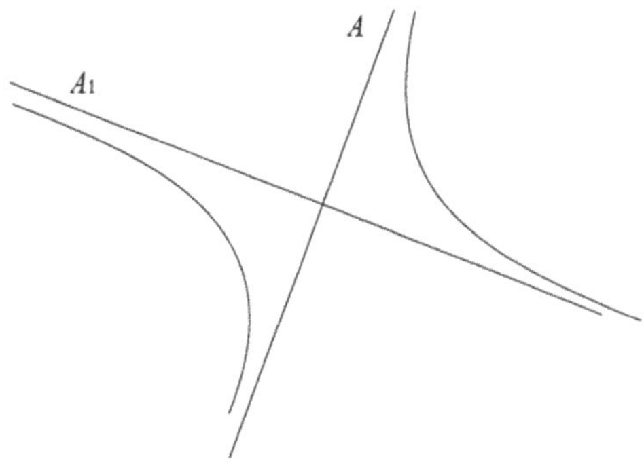

Fig. 2.36 A sketch of the figure described by Steiner

perspectivity axis and by the united common radius of the two pencils (degenerate conic locus).[120]

After the projectivity between two dotted straight lines, Steiner considered that between two coplanar pencils of straight lines. He developed this brilliant reasoning: as in two projective and concentric pencils of radiuses $\mathfrak{B}$ and $\mathfrak{B}'$ either two or one or no double straight line can exist, so in two pencils of projective coplanar radiuses, posed in any position of their plane, either two or one or no parallel corresponding straight lines can exist. This means that two pencils of projective straight lines can generate any of the three kinds of conics.

If the projectivity between the two pencils $\mathfrak{B}, \mathfrak{B}'$ is direct, all the three kinds of conics can be generated. When the intersection points of the corresponding radiuses are included in a finite space, an ellipse is generated; if, as Steiner claimed, at the limit of this space, a couple of corresponding straight lines are parallel, then a parabola is generated. This is a visual and intuitive image of how a parabola can be seen as the separation line between the series of the ellipses generated by the two pencils $\mathfrak{B}, \mathfrak{B}'$ and the series of hyperbolas. For Steiner stated, hyperbolas are generated if the intersection lines of two corresponding straight lines intersect beyond such limit.

If the projectivity is inverse, then only hyperbolas can be generated because, in this case, since the inverse projections of two concentric pencils have two double straight lines, when the pencils are not concentric, they have two couples of parallel straight lines.

Steiner showed then that two directly equal pencils of radiuses generate a circle, whereas two inversely equal pencils generate an equilateral hyperbola (ibid., pp. 143–145).

Finally, he analysed the various kinds of cones generated by sheaves of planes, so also introducing the projective generation of the elliptic, parabolic and hyperbolic cylinders (ibid., pp. 145–147).

[120] See, e.g., Enriques (1904, pp. 223–224).

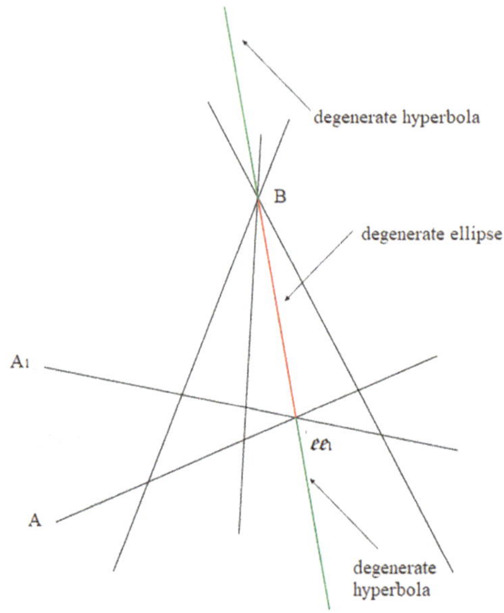

Fig. 2.37 Reconstruction of the figure described by Steiner where the degenerate hyperbola and the degenerate ellipse are drawn

The next step connoting Steiner's foundational programme consists in showing that the conics' properties can be explained within a new unitary picture if their projective generation is taken into account.

One of the most classical theorems on the conics claims that a conic is univocally determined by five tangents or five points. This theorem enters easily the structure constructed by Steiner. For given two projective dotted straight lines A and A', the projectivity is univocally determined if three couples of corresponding points are given, or, which is the same, three projective radiuses. Therefore, granted Proposition B concerning the projective generation of the conic sections, assuming the straight lines A and A' as the tangents to a conic section, it will be enough that three of their projective radiuses are given to determine univocally the projectivity and, hence, the conic section determined by it (ibid., p. 148). Through an easy dual consideration, it is possible to prove that five points determine univocally a conic. Other not difficult arguments allow us to prove projectively any of the five conditions determining a conic.

Two other beautiful considerations allowed Steiner to prove that the conics are curves of the second class and the second order. For since, according to a fundamental theorem proved in § 18 of the *Systematische Entwicklung* (ibid., p. 70), given in a plane a projectivity between two dotted straight lines, which is not a perspectivity, through any point of the plane at most two projective radiuses pass. Then in the plane of a conic, at most two tangents to the conic can pass. Thus, the conics are curves of second class; the dual reasoning allowed Steiner to conclude that they are of second order (ibid., p. 149).

The classical theorems of Pascal and Brianchon (ibid., p. 150), as well as their reciprocal propositions, are also easily deducible from the projective generation of the conics and from the properties of the hexagons generated by two projective, but not prospective, dotted straight lines and by four of their projective radiuses or by two points assumed as centres of projective, but not perspective, pencils of radiuses and by the intersection points of four corresponding radiuses (ibid., § 24, p. 90). Steiner claimed that the French mathematicians Gergonne, Poncelet, Chales (!!!), Sturm, Bobillier, the Belgian Dandelin and the German Moebius and Plücker proved these important theorems with various methods. However, he underlined the novelty of his own foundational approach, in which such theorems are not the basis of the theory; rather they rely upon a more fundamental ground, given by the projective theory he constructed. Steiner highlighted explicitly:

> This deduction of such propositions clarifies their nature from a new standpoint. It shows that they are not the real foundation for the researches on the conics. Rather, they together with many other properties, derive in an easy and clear way from a wide and comprehensive source, namely from the relation of the projective forms.[121]

From the properties concerning the complete n-angles and n-sides (ibid., p. 74–82) and from the projective generation of the conics, Steiner was able to prove several theorems and solve several problems connected to the tangents to and to the points of a conic. To give some significant examples: "If five tangents to a conic are given, it is required to find the contact points using only the ruler"; "If five points of a conic are given, to find the tangents through them using only the ruler" (ibid., p. 152); "In each triangle circumscribed to a conic, the three straight lines joining a vertex with the point in which the opposite side touches the conic concur in a point"; dually: "In each triangle inscribed in a conic, the three points where the sides are cut by the tangents at the opposite vertex are collinear" and vice versa (ibid., p. 156).

Afterwards, Steiner introduced the important concepts of "four harmonic tangents to a conic" and "four harmonic points of a conic". Four tangents to a conic which cut another tangent of a conic in four harmonic points cut every other tangent in four harmonic points (as it is very easy to prove). They are named "harmonic tangents" and the conic inscribed in their quadrilateral is called "the inscribed harmonic conic". The four contact points of the four harmonic tangents are the "four harmonic points". The straight lines joining them form a quadrangle called by Steiner "the inscribed harmonic quadrangle" (ibid., pp. 157–158). These harmonic figures fulfil several properties (see Fig. 2.38):

For, be A, A_1, A_2, A_3 four harmonic tangents to a conic and A_4 be a fifth tangent. Thence, the four points $\mathfrak{h}, \mathfrak{J}, \mathfrak{f}, \mathfrak{l}$ are harmonic. Consider the points in which, e.g., A_1 and A_4 are cut by the other tangents. Then, to the common point $\mathfrak{h}$ regarded as belonging to A_4, the point $\mathfrak{a}$ in which A touches the conic corresponds. Therefore, to the points $\mathfrak{h}, \mathfrak{J}, \mathfrak{f}, \mathfrak{l}$, the points $\mathfrak{a}, \mathfrak{b}, \mathfrak{c}, \mathfrak{d}$ correspond in A. Thus, this quatern of points is

[121] Ibid., p. 151: "Die gegenwärtige Ableitung der Sätze beleuchtet sie von einer neuen Seite, sie zeigt, dass dieselben nicht die eigentliche Grundlage für die Untersuchung der Kegelschnitte sind, sondern dass sie vielmehr, mit vielen andern Eigenschaften zugleich, aus einer umfassenderen Quelle, nämlich aus der Beziehung projectivischer Gebilde, sehr leicht und klar hervorgehen."

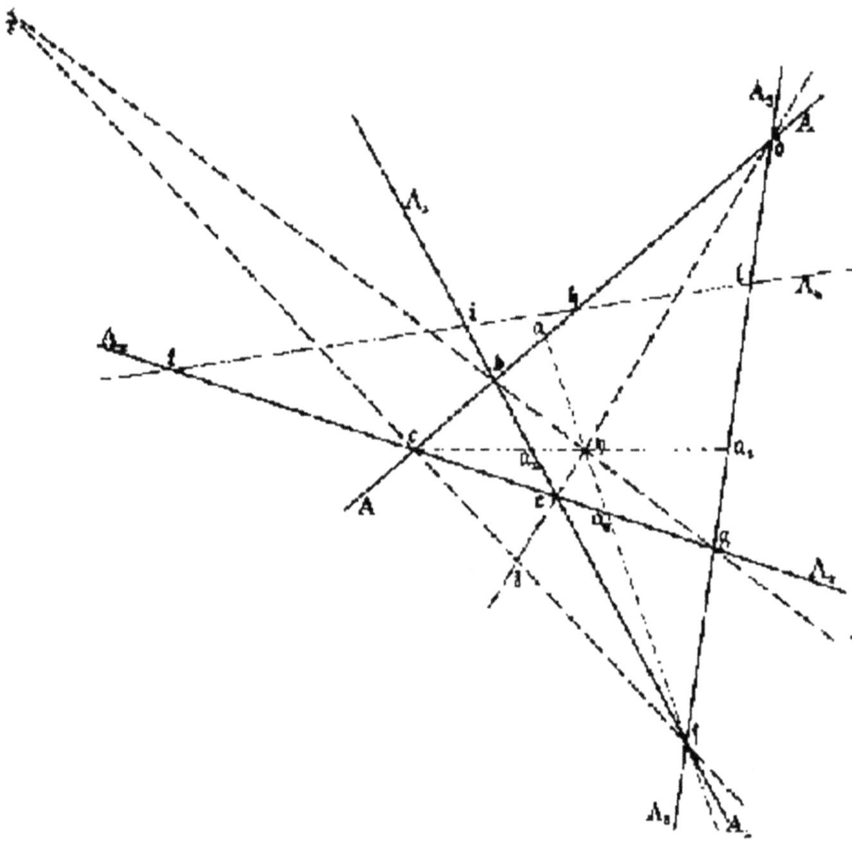

Fig. 2.38 The figure used by Steiner to prove the theorems on the harmonic tangents

harmonic. The same reasoning can be applied to the other three tangents. If the points $\mathfrak{c},\mathfrak{b},\mathfrak{a},\mathfrak{d}$ as well as $\mathfrak{c},\mathfrak{e},\mathfrak{a}_2,\mathfrak{g}$ are harmonic, then the three straight lines $\mathfrak{be},\mathfrak{aa}_2,\mathfrak{bg}$ mutually cut in a point $\mathfrak{f}$ (this is a well-known property of the harmonic groups proved by Steiner in §§ 12 and 14 of the *Systematische Entwicklung*). Analogously the points $\mathfrak{a},\mathfrak{a}_1$ where the harmonic tangents A_1, A_3 touch the conic are collinear with the intersection point $\mathfrak{c}$ of the two tangents A, A_2. From these considerations, Steiner concluded that, given four harmonic tangents to a conic: a) the contact point of each tangent is the fourth harmonic of the triplet of points in which such tangent is cut by the three other tangents and b) the contact point of two harmonic tangents are collinear with the intersection point of the other two. Vice versa: a′) if four tangents of a conic fulfil one of the two previous conditions they are harmonic and b′) four harmonic tangents touch a conic in four harmonic points (ibid., pp. 158–159).

Through this theorem Steiner solved some other problems connected to conics (ibid., pp. 160–161).

Commentary

All the affine and the metric properties of the conic sections are seen by Steiner as particular cases of the projective generation of conics. These cases, as we have seen, are obtained when the elements which remain unspecified while dealing with the general properties of the projectivities are, instead, specified. In this way affine objects as the parabola, the hyperbola and the ellipse and metric objects as the equilateral hyperbola and the circle are obtained. Thus, the metric properties are framed as particular cases of projective properties. Ergo, Steiner's conception has several convergence points with Chasles'. Steiner's line of thought can be summarized like this: a) the concept of cross ratio is defined; b) all the main features of the projectivities are reducible to the relations among cross ratios; c) in particular, when some elements of the cross ratios are specified, this concept allows us to also face, through a projective treatment, several cases of affine and metric properties; and d) the conic sections are generated projectively; thence their affine and metric properties are subject to the same features as those connoting the cross ratios when some of their elements are particularized.

Chasles had two approaches: the former developed in those works I have called "the great memoirs of the period 1827–1829". Here, from a mathematical point of view, he showed that assuming an object as a parabola or a paraboloid, the most important properties of the metric transformations can be obtained through a polarity with respect to such two figures. There is no reference to the cross ratio. The second approach, developed in the same years in which the *Systematische Entwicklung* was conceived and published (1830–1832), is more similar to Steiner's insofar as it is based on the concept of cross ratio. There are some differences in the approaches of the two mathematicians: Chasles had a specific interest in showing the reduction of the metric properties to the graphic ones in both phases of his thought (1827–1829; 1830–1837). We have seen that he dedicated specifically some memoirs to this problem, and in his works devoted specific sections to the metric properties seen as particular cases of projectivities. Steiner's approach is slightly different: he had the intention to found projective geometry on the concept of cross ratio and to determine a projective generation of conic sections. Within this picture, he also pointed out some metric properties connected to projectivities, which showed the generality of his theoretical construction. But he did not develop the dependence of the metric properties from the graphic ones in such a wide manner as Chasles did. Whereas he developed in a systematic way the foundation of the concept of cross ratio and the foundation of projective geometry on the cross ratio, which Chasles did, with a slightly different approach, in his two crowned memoirs on duality and homography and in his *Traité de géométrie supérieure*. This text is Chasles' closest work, in style as well, to Steiner's *Systematische Entwicklung*. As already pointed out, it should be considered that it was published 20 years after Steiner's work, which is a "geologic era", given the celerity with which projective geometry was developed in those 20 years.

Granted these significant differences, the similarities are more conspicuous and significant: in Sect. 2.2.1 I have shown how Chasles reached the projective generation of the conics and how, through this generation, the classical theorems on the conic sections, as Pascal's theorem, are deducible. In this case, the difference is that Chasles used directly the concept of cross ratio, whereas Steiner preferred to introduce the conics starting from the circle and the circular oblique cone of second order, as we have seen, and to develop his reasoning without using extensively, in this case, the concept of cross ratio, but geometrical constructions. It is worth pointing out that Chasles, relying on the use of cross ratio, developed systematically all the consequences of the projective generations of the conics in his treatise on conic sections (Chasles, 1865). Here, at the very beginning, Chasles wrote:

> A sole property will serve as the basis for the whole theory of such curves, that is: THEOREM. "If, through four points of a conic, the tangents and four other straight lines towards any other fifth point of the curve are drawn, the anharmonic ratio of these four straight lines will be equal to that of the four points in which any other tangent cuts the given ones.[122]

However, since the projective properties of Steiner's generation of the conics rely on the cross ratio, the approaches of the two mathematicians are rather similar. Both of them proved that most of the known theorems on the conic sections depend on their projective generation. We have seen this in the case of Pascal and Brianchon theorems, but several other propositions might be added. Therefore, this is a further confirmation of the foundational character that both Chasles and Steiner ascribed to their works.

Another fundamental point of convergence is the conception that Steiner and Chasles had of duality: both of them thought that, in spite of the fact that duality is the most important and specific property of the projective transformations, it is a derived property. We have seen that Chasles, in his memoir on duality, proved that the duality law in space is deducible from three theorems and a fourth theorem, which states the relations between the director point and the mobile plane used by Chasles. It is evident that Steiner, as well, conceived duality as a derived property. In the specific but significant case of the conic sections (but in most of other cases), he used the dual notation due to Gergonne in which a theorem and its dual are written in two columns, but he never wrote something as "since I have proved the theorem for a configuration, then it also holds for the dual configuration". He always offered a new and complete demonstration, after which he wrote the theorems according to the dual notation. Therefore, neither Chasles nor Steiner accepted the duality law as a postulate but thought that it derived from the properties of the cross ratio.

A further contact point is represented by the idea to use the projective properties of the easiest conic section, namely the circle, and to deduce from them the projective properties of other conics: we have seen that Steiner used the projective generation of the circumference to deduce the projective properties of all the conic sections. Chasles did not use this approach to reach the projective generation, but, as

[122] Chasles (1865, pp. 1–2): "Une seule propriété servira de base à toute la théorie de ces courbes. La voici: THÉORÈME.- Si par quatre points d'une conique, on mène les tangentes et quatre autre droites aboutissant à un cinquième point quelconque de la courbe: le rapport anharmonique de ces quatre droites sera égal à celui des quatre points de rencontre des quatre tangentes et d'une cinquième tangente quelconque".

already highlighted, he used such an approach to prove that the graphical properties of the systems of two or three circles can be extended to systems of two or three s.s.p. conics. The idea to use some properties of the circles which can be transferred to the conics in order to study such sections was expressed by Poncelet who wrote:

> [...] all the reasonings we have used to establish the different properties of the coplanar circles can be directly applied, with some restrictions, to the general case in which these circles are replaced by any s.s.p. coplanar conic sections [...]. It is evident that the restrictions are referred to what concerns, explicitly or implicitly, absolute and determined magnitudes, namely to those properties reducible only to what has been said as to the relations of equality of the rectangles corresponding to the secants of a circle and to the orthogonality of two circles having reciprocally as radiuses the equal tangents drawn from their respective centres.[123]

In this quotation, it is evident that Poncelet is writing that only those properties which are exclusively graphic can be extended from the circles to the other conics, which was what Chasles and Steiner did.

2.3.1.8 Polarity

The section on what Steiner named "Harmonische Pole und Gerade" with respect to a conic (ibid., pp. 162–170) is important because the author proved that all the results obtained in the theory of reciprocal polars have their foundation in his own theory. He recognized that the initial and decisive impulse to this part of geometry was given, in modern times, by Monge.

It is worth following the whole reasoning proposed by Steiner because it is emblematic of a purely projective approach where the concept of harmonic ratio plays the decisive role.

He considered four tangents A, A_1, A_2, A_3 to a conic and their contact points $\mathfrak{a},\mathfrak{a}_1,\mathfrak{a}_2,\mathfrak{a}_3$ (see Fig. 2.39).

The four tangents form a complete quadrilateral circumscribing a conic and the four contact points a complete quadrangle inscribed. Consider the quadrangle $\mathfrak{a}\mathfrak{a}_1\mathfrak{a}_2\mathfrak{a}_3$. The point $\mathfrak{f}$ is the intersection of the two opposite sides $\mathfrak{a}\mathfrak{a}_1,\mathfrak{a}_2\mathfrak{a}_3$. Therefore, the straight lines $\mathfrak{f}\mathfrak{a},\mathfrak{f}\mathfrak{p},\mathfrak{f}\mathfrak{a}_3,\mathfrak{f}\mathfrak{x}$ are harmonic, being $\mathfrak{p}$ the intersection point of the diagonals connecting the two pairs of opposite vertices and $\mathfrak{x}=\mathfrak{a}_1\mathfrak{a}_2\cap\mathfrak{a}\mathfrak{a}_3$. This is a property of any complete quadrilateral proved by Steiner, as we have seen in Sect. 2.3.1.5. Therefore, every straight line intersected by these four radiuses is cut in four

[123]Poncelet (1822, 1865, art. 288, pp. 147–148): "[...] tous les raisonnemens dont nous nous sommes servi pour établir les diverses propriétés des cercles situés sur un même plan s'appliquent directement, à quelques restrictions près, au cas général où l'on remplace ces cercles par des sections coniques quelconques, s. et s.p. sur un plan. [...] Les restrictions ne portent évidemment que sur ce qui concerne explicitement ou implicitement des grandeurs absolues et déterminées, c'est-à-dire que cela se réduit uniquement à ce qui a été dit sur la relation d'égalité des rectangles correspondans aux sécantes du cercles et sur l'orthogonalité de deux cercles qui ont réciproquement pour rayons les tangentes égales issues de leurs centres respectifs".

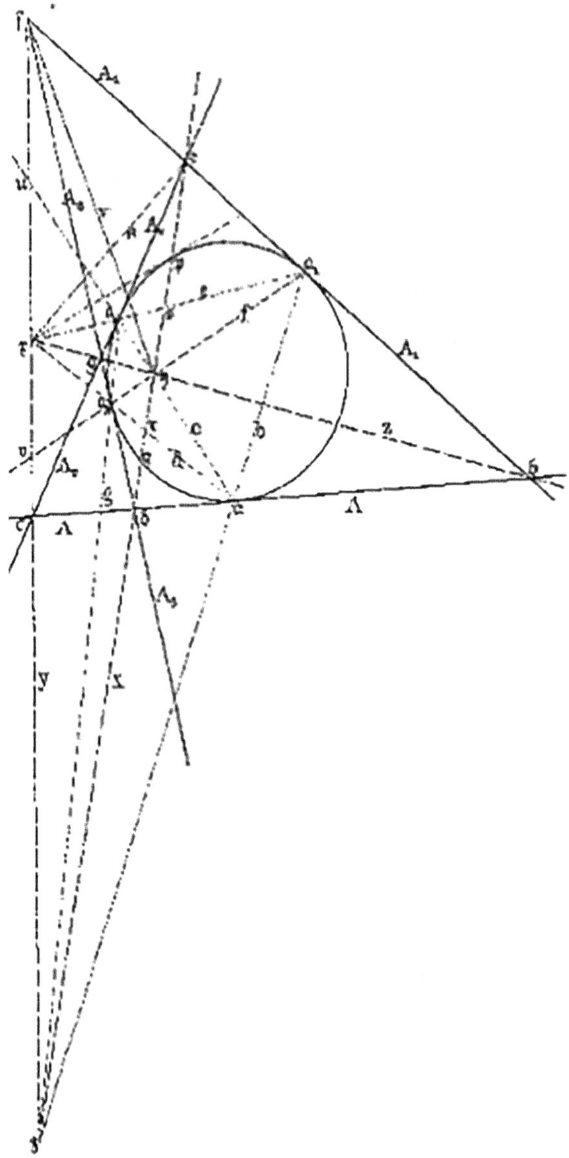

Fig. 2.39 The figure used by Steiner to develop the theory of polarity with respect to a conic

harmonic points. Thus, the four quaterns of points $\mathfrak{a}, \mathfrak{r}, \mathfrak{a}_3, x; \mathfrak{a}, \mathfrak{p}, \mathfrak{a}_2, \mathfrak{u}; \mathfrak{a}_1, \mathfrak{p}, \mathfrak{a}_3, \mathfrak{u}; \mathfrak{a}_1, \mathfrak{s}, \mathfrak{a}_2, \mathfrak{r}$ are harmonic, so that the two quaterns of straight lines $\mathfrak{f}\mathfrak{a}_1, \mathfrak{f}\mathfrak{h}, \mathfrak{f}\mathfrak{a}_3, \mathfrak{f}\mathfrak{u}$ and $\mathfrak{e}\mathfrak{a}_1, \mathfrak{e}\mathfrak{s}, \mathfrak{e}\mathfrak{a}_2, \mathfrak{e}\mathfrak{r}$ also are.

This granted, Steiner developed the following reasoning: since, given three points as $\mathfrak{a}, \mathfrak{a}_2, \mathfrak{u}$, the fourth harmonic (homologous of $\mathfrak{u}$) is a unique point $\mathfrak{p}$, if in the conic, the three straight lines A, A_2 and $\mathfrak{c}\mathfrak{u}$ are assumed as fixed, the chord $\mathfrak{a}_1\mathfrak{a}_3$ connecting the two contact points of the tangents A_1, A_3 passes always through the point $\mathfrak{p}$ which is the fourth harmonic with respect to $\mathfrak{a}, \mathfrak{a}_2, \mathfrak{u}$. It may also be assumed that the point $\mathfrak{f}$ where the variable tangents mutually cut moves on the straight line y. An analogous reasoning holds if the two straight lines A, A_3 and the straight line $\mathfrak{dr}$ are assumed as fixed. In this case the fixed point functionally equivalent to $\mathfrak{p}$ is $\mathfrak{x}$ and the straight line functionally equivalent to y is x. Furthermore, since the contact points of two harmonic tangents are collinear with the intersection point of the other two, the straight line $\mathfrak{de}$ passes through the intersection points $\mathfrak{p}, \mathfrak{q}$ of the two tangents $\mathfrak{xp}, \mathfrak{xq}$ mutually cutting in $\mathfrak{x}$.

Ergo, Steiner could conclude:[124]

If a straight line ($\mathfrak{a}_1\mathfrak{a}_3$ or $\mathfrak{a}_1\mathfrak{a}_2$) cutting a conic section rotates around a fixed point ($\mathfrak{p}$ or $\mathfrak{x}$) belonging to it, then: α) the locus of the point ($\mathfrak{u}$ or $\mathfrak{s}$) which is the fourth harmonic with respect to the two contact points ($\mathfrak{a}_1, \mathfrak{a}_3$ or $\mathfrak{a}_1, \mathfrak{a}_2$) and the fixed point and which is homologous of the fixed point, belongs to a determined straight line (y or x); β) in this straight line also moves the intersection point ($\mathfrak{f}$ or $\mathfrak{e}$) of the two tangents (A_1, A_3 or A_1, A_2) through whose contact points that movable, intersecting straight line passes.	If a point ($\mathfrak{f}$ or $\mathfrak{e}$) moves in a fixed straight line (y or x) in the plane of a conic: α) that straight line (v or s), which is the fourth harmonic of the tangents (A_1, A_3 or A_1, A_2) passing through that point and of the fixed straight line, passes through a determined point ($\mathfrak{p}$ or $\mathfrak{x}$); β) around this point that straight line ($\mathfrak{a}_1\mathfrak{a}_3$ or $\mathfrak{a}_1\mathfrak{a}_2$) passing through the contact points ($\mathfrak{a}_1, \mathfrak{a}_3$ or $\mathfrak{a}_1, \mathfrak{a}_2$) of those two respective tangents passes.

Steiner called the straight line y "die Harmonische" of the point $\mathfrak{p}$ and named the point $\mathfrak{p}$ "der harmonische Pol der Geraden" y with respect to the given conic. He pointed out that the French mathematicians use the names of "Polaire" and "Pol" (ibid., p. 164). Steiner was able to show easily the reciprocal character of the pole–polar relation, namely that if a straight line p passes through a point P, the polar q of P passes through the pole Q of p, and if a point P lies on a straight line p, the pole Q of p lies on the polar q of P (ibid., p. 165). The two straight lines p and q are the reciprocal polars and P and Q the reciprocal poles.

[124]Ibid., p. 163:

Dreht sich eine Gerade ($\mathfrak{a}_1\mathfrak{a}_3$ oder $\mathfrak{a}_1\mathfrak{a}_2$), die einen Kegelschnitt schneidet, um irgend einen (in ihr liegenden) festen Punkt ($\mathfrak{p}$ or $\mathfrak{x}$): α) so ist der Ort desjenigen Punkts ($\mathfrak{v}$ or $\mathfrak{s}$) welcher zu den zwei Durchschnittspunkten ($\mathfrak{a}_1, \mathfrak{a}_3$ or $\mathfrak{a}_1, \mathfrak{a}_2$) und dem festen Punkte der vierte, und zwar dem letzteren zugeordnete, harmonische Punkt ist, eine bestimmte Gerade (y oder x); und β) in dieser Geraden bewegt sich zugleich der Durchschnitt ($\mathfrak{f}$ oder $\mathfrak{e}$) derjenigen zwei Tangenten (A_1, A_3 oder A_1, A_2) durch deren Berührungspunkte jene bewegliche schneidende Gerade geht.	Bewegt sich ein Punkt ($\mathfrak{f}$ oder $\mathfrak{e}$) in einer festen Geraden (y oder x) in der Ebene eines Kegelschnitts: α) so geht diejenige Gerade (v oder s), welche zu den zwei durch den Punkt gehenden Tangenten (A_1, A_3 oder A_1, A_2) und der festen Geraden die vierte der letzteren zugeordnete, harmonisch Gerade (Strahl) ist, durch einen bestimmten Punkt ($\mathfrak{p}$ oder $\mathfrak{x}$); und β) um diesen Punkt dreht sich zugleich diejenige Gerade ($\mathfrak{a}_1\mathfrak{a}_3$ oder $\mathfrak{a}_1\mathfrak{a}_2$), welche durch die Berührungspunkte ($\mathfrak{a}_1, \mathfrak{a}_3$ oder $\mathfrak{a}_1, \mathfrak{a}_2$) der jedesmaligen zwei Tangenten geht.

This clarified, Steiner taught how to draw the polar of a given point with respect to a given conic and reciprocally, and how, through this rule, to draw the tangents through a given external point (ibid., p. 166). He claimed that the propositions he had proved by his construction explain the fundamental properties of the theory of reciprocal polars, which the French mathematicians were successfully developing (ibid., p. 167).

Now, Steiner claimed that the theory he had expounded here, though more general than the one developed until that moment, was not yet the most fundamental one: rather a common original source ("gemeineschaftliche Urquelle", ibid., p. 167) to the entire projective geometry exists, which will show the internal nature of this discipline and its unitary character. He affirmed this "Urquelle" would have been explained in the fourth chapter of his work, but this chapter was never written.

To prove the richness of his theory, he showed how easily important propositions of the theory of reciprocal polars can be proved. He offered a significant example: suppose two conics K, K' to be given in a plane. Consider the tangents of K'. What is the locus of poles of such tangents with respect to K? Steiner argued that: be *a,b,c,d, e,f* six tangents of K' and be $\mathfrak{a}, \mathfrak{b}, \mathfrak{c}, \mathfrak{d}, \mathfrak{e}, \mathfrak{f}$ their poles with respect to K. The diagonals of the hexagon (*Sechsseit*) *abcdef* circumscribed to the conic cut reciprocally in a point. Therefore, since the intersection points of the three opposite sides of the hexagon (*Sechseck*) $\mathfrak{abcdef}$ are the poles of the diagonals of the hexagon *abcdef*, they are collinear.[125] Hence, the hexagon $\mathfrak{abcdef}$ is inscribed in a conic K''. Since five points are sufficient to determine a conic, it is possible to conclude that the required locus is a third conic K'' (ibid., p. 169). This reasoning solves, thus, the following problem: given two conics in a plane, to determine the law fulfilled by the poles of the tangents of one conic with respect to the other one. Steiner pointed out that, in general, the following question is raised: if a set of straight lines in the plane fulfilling a certain law is given, how is it possible to determine the law satisfied by the set of the poles of such straight lines with respect to a conic lying in the same plane as the straight lines? The dual problem can also be posed. Steiner claimed that this question—of which the just analysed one is a specific case—is the basis of the "Théorie des polaires réciproques" of the French mathematicians. The general law behind the theory of reciprocal polars has been given by Möbius (1827, § 287).

Through some further not particularly difficult passages Steiner proved a famous proposition by Brianchon: "If two variable tangents to a conic K move so that the straight line passing through their contact points always touches a second conic K', their intersection point passes through a third conic K'' (ibid., p. 170). The dual theorem also holds.

[125] In § 44 II (p. 164) Steiner had proved that the poles with respect to a conic of all the straight lines passing through a point lie on a determined straight line, namely in the polars of this point, and dually the polar with respect to a conic of all the points lying in a straight line passes through a determined point, namely through the pole of this straight line.

Commentary.

1. *Comparison with the theory of reciprocal polars before Steiner's and Chasles' work.* We have seen how Steiner proved that if the point $\mathfrak{f}$ moves on the straight line $\mathfrak{cu} \equiv y$, then the fourth harmonic of the variable point $\mathfrak{f}$ with respect to two points of a conic A,B aligned with $\mathfrak{f}$ is a fixed point Y, namely if the chord of a conic rotates around Y the fourth harmonic of the two extremes A and B of the chord is a straight line. In this reciprocal involutory relation, Y is called by Steiner the harmonic pole of y and y the harmonic polar of Y. The whole proof of the fact that the two loci are respectively a point and a straight line as well as of the involutory character of the relation relies on the harmonic properties of the quadrangle inscribed in and circumscribed around a conic.

This definition of pole and polar is something new in projective geometry. It is very important because it shows that the properties of the poles and polars with respect to a conic derive directly from a sole common root, namely from the concept of cross ratio and, in this case, from the specific properties of the harmonic ratio connected to the hexagons and quadrangles inscribed in and circumscribed around a conic.

The original definition of pole given by Servois is different because Servois defined the concept of pole of a straight line with respect to a conic independently of any reference to the harmonic relation and afterwards he proved that some harmonic properties of points and straight lines are connected to the notion of pole and polar. For, given a straight line and a coplanar conic, he named pole of the straight line the point, belonging to the plane of the line and the conic, around which all the chords connecting the contact points of pairs of tangents to the conic drawn from any point of the straight line rotate (Servois, 1810–1811, p. 337). Obviously, the straight line is the polar of the point, though, as already highlighted, this name was given by Gergonne. Servois connected several properties of the poles and polars to those of the conics, but the idea of posing the harmonic ratio as the basis of the theory of polarity is missing. Rather the harmonic properties of the poles and polars are seen as an application of Servois' (and Gergonne's) definition and of a series of techniques which were enormously developed in the first 30 years of the nineteenth century. The theory of reciprocal polars reached the status of the fundamental doctrine within projective geometry, as Steiner himself pointed out. However, the set of techniques composing the theory of reciprocal polars had not been reduced to a single concept as Steiner's double ratio and, in particular, in this case, to that of harmonic ratio.

To clarify the whole picture, it is useful to summarize the steps through which Steiner arrived at the theory of reciprocal polars: 1) definition of cross ratio and analysis of its properties; 2) the harmonic ratio is simply a particular value of the cross ratio; 3) all the classical theorems on the harmonic ratio (as the quadrangle theorem) are deducible from the consideration that the harmonic ratio is a particular case of cross ratio; 4) distinction between n-laterals and n-angles; 5) projective definition of the conics and its consequences; 6) definition of pole and polar with respect to a conic through the harmonic ratio; and 7) deduction of the properties of the poles and polars, already proved by the French mathematicians with other methods, from the theoretical structure developed in the items 1)–6) and demonstrations of new properties of the poles and polars.

The logic behind the theory of reciprocal polars before Steiner's and Chasles' work—namely before the introduction of the concept of cross ratio and of the projective generation of conics—can probably summarized in the best way following the steps through which Poncelet in his *Traité* (Poncelet, 1822) arrived at this theory.[126] This work can actually be seen as the one which offers the most complete foundation of projective geometry before the Chasles–Möbius–Steiner era, with the introduction of the principle of continuity, of the ideal elements, of the projective properties of conics deduced from those of the circle. These are the basic elements through which Poncelet gave a unitary view on projective geometry, and by means of which he summarized the results of his predecessors. To fully appreciate the novelty of Steiner's and Chasles' approach, it is appropriate to offer some details of Poncelet's view on the theory of reciprocal polars. I will explain such a view through 16 steps:[127]

---

[126]It is not exactly true that, before Steiner's *Systematische Entwicklung* and Chasles' *Aperçu*, a sole approach to the concept of pole and polar with respect to a conic existed, because such an approach is also connected to the way in which duality is interpreted and we know that the discussion Poncelet–Gergonne on duality was not only a priority dispute; it also concerned a different conception of duality. Under this respect, Kötter's position is very interesting because, apropos of duality, he summarized like this Gergonne's idea of duality: 1) apart from the metric properties, in any truth concerning the plane geometry, the term "point" can be replaced by the term "straight line" *salva veritate* (Kötter, 1901, p. 162); 2) though Gergonne highlighted the properties of duality through those of polarity with respect to a conic, he considered duality a general law of nature which goes beyond polarity (ibid., p. 164); 3) Poncelet had the priority on duality and he was outstanding in solving problems and demonstrating theorems, but he saw duality always in connection with polarity (Poncelet also considered metric-graphical properties which are transmissible through duality); thus, Gergonne's conception on duality is more profound than Poncelet's (ibid., pp. 164–167). Coolidge is inclined to share Kötter's opinion (Coolidge, 1940, p. 95). With regard to further bibliographical indications on Poncelet's idea of duality and polarity, see next note). I will address Gergonne's conception of duality in more detail in Chap. 6. However, no one before Steiner and Chasles saw the polarity as a specific application of the properties of a general concept as that of cross ratio. I am referring to the way in which Poncelet interpreted the polarity with respect to a conic in his *Traité*, given the enormous importance and influence of this text.

[127]I will offer the elements of Poncelet's theory of reciprocal polars only insofar as they are necessary to develop a comparison with Chasles and Steiner. For more specific properties of such theory, see: Gray (2010, chapter 5, pp. 53–61), where Poncelet's conception of polarity is connected to that of duality and a reference to the polemic on duality with Gergonne also exists; Gray also develops interesting considerations on the relation between Poncelet's polarity theory and duality in the *Traité* in Gray (2005); Haubrichs dos Santos (2015, Section 3.1, pp. 114–137) is a good text on Poncelet's theory of reciprocal polars; Kötter (1901, pp. 129–135), where the theory of polar reciprocity of Poncelet (1822) is referred to. The whole first section (pp. 121–160) of the second chapter is dedicated to Poncelet's *Traité*; Kötter (1901, pp. 169–172) regards Poncelet's general theory of reciprocal polars (Poncelet, 1829a); you can find interesting considerations on poles, polars and polarity in Ostermann and Wanner (2012). This book presents geometry referring to the single subjects (Euclid, Euclidean geometry, trigonometry, Descartes' geometry and so on) according to a historical approach. In chapter 7 "Cartesian Coordinates", pp. 185–240, and in chapter 11 "Projective Geometry", pp. 319–344, there are numerous theorems on polarity, among

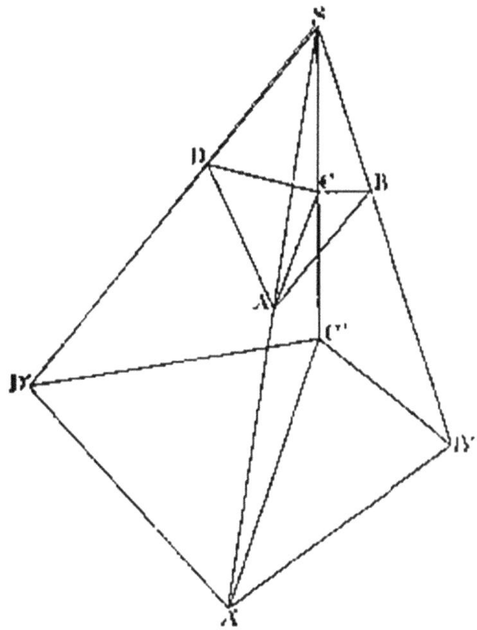

Fig. 2.40 The first figure used by Poncelet in his *Traité* to distinguish the features of the graphic properties from those metric

1) Identification of the projective relations as those properties which subsist independently of the length of the projective radiuses. For example in Fig. 2.40

the two projective figures $ABCD$ and $A'B'C'D'$ share their graphic properties independently of the length of the projective radiuses SA, SB, SC, SD; SA', SB', SC', SD' (Poncelet, 1822, art. 8–10, pp. 6–8).

2) The harmonic ratio, defined *à la* De La Hire, is a projective invariant (ibid., arts. 21–22, pp. 12–13).
3) Properties of the harmonic groups of four elements (ibid., arts. 34–39, pp. 18–20).
4) Extension of the concept of harmonic ratio in relation to a conic (Fig. 2.41).

Consider the centre O of the chord MN. Be AB the diameter conjugate to the direction of MN. The point O' to which the tangents drawn at the extremities of MN concur is the fourth harmonic with respect to A,B,O. It is $\frac{O'A}{O'B} = \frac{OA}{OB}$. Among the straight lines m,m' and the points O,O', a remarkable relation, hence, subsists such that, when one element of the quatern is given, the other three also are. Because of this relation, Poncelet named the two points and the two straight lines "harmonic conjugate" with respect to the direction of the diameter AB. The point O' is the pole of the straight line mn and the point O is the pole of $m'n'$ with

which many are due to Poncelet, so that the reader can get an idea—though obtained through a series of single theorems and not by a complete historical endeavour—of Poncelet's results on polarity.

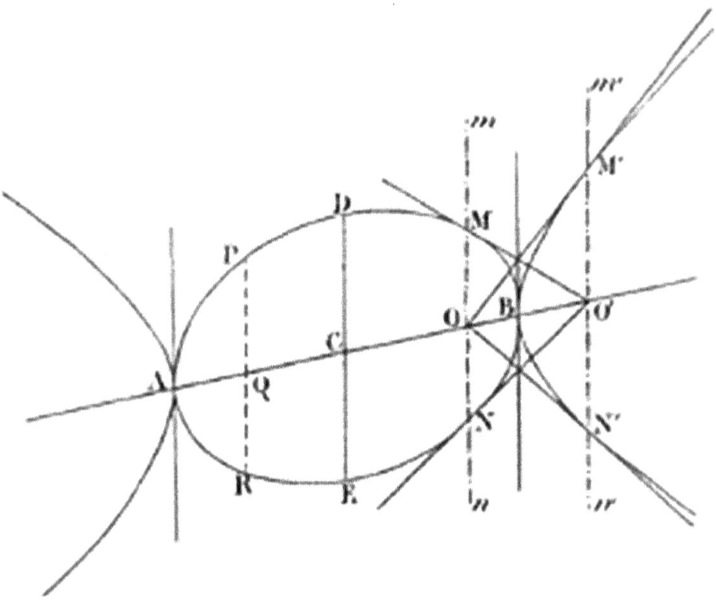

Fig. 2.41 The figure used by Poncelet to prove the properties explained in item 4)

respect to the conic. These straight lines are reciprocally the polars of such points (ibid., arts. 48,49, pp. 26–27).

5) Introduction of the concept of ideal chord of a conic in connection with the already analysed pole–polar relation (ibid., art. 50, p. 27).[128]

Consider a real conic C and a real straight line cutting C in P_1 and P_2. The segment P_1P_2 belonging to the straight line l (Fig. 2.42a) is a real chord of C. Be d the diameter conjugate to l, which cuts C in the points $Q1$, $Q2$. Name Q the intersection between l and d. Then, for a basic property of the conics, it is

$$(QP_1)^2 = k(Q_1Q)(Q_2Q)$$

being k a real constant, which does not vary for any line parallel to l intersecting C. If l' is parallel to l, but does not intersect C, you can, anyway, consider the intersection point M between l' and d, as well as two points R_1, R_2 of l' such that

[128] A clear and synthetic explanation of Poncelet's concept of ideal chord is offered by Del Centina (2016b, pp. 22–23).

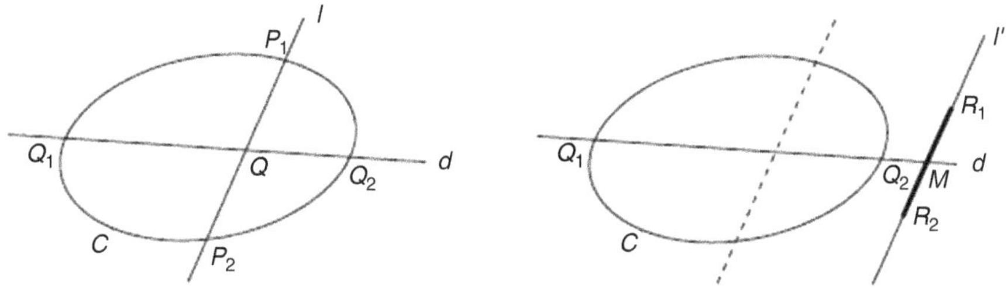

Fig. 2.42 (**a**) (on the left) and (**b**) (on the right). The diagrams used by Del Centina to explain Poncelet's concept of ideal chord of a conic. Retrieved from Del Centina (2016b, p. 23)

$$(MR_1)^2 = k(Q_1M)(Q_2M)$$

with the further condition $MR_1 = MR_2$ (Fig. 2.42b). The segment R_1R_2 is called by Poncelet an "ideal chord" of *C*. Poncelet developed in the *Traité* all his theory of the ideal chords without the algebra of complex numbers, but his considerations can be transcribed in an algebraic form. For example, R_1, R_2 are the conjugate complex points where the straight line l' cuts the conic *C*. But Poncelet did not use such a language.

6) Properties of the simple and complete quadrilaterals. In particular: a) in any complete quadrilateral each diagonal is divided harmonically by the other two (ibid., art.155, pp. 79–80), in any complete quadrilateral the middle points of the diagonals are aligned (ibid., art. 164, pp. 83–84).
7) Quadrilaterals inscribed in conic sections (ibid., arts. 170–184, pp. 87–95).
8) Fundamental properties of the quadrilateral inscribed in a conic and circumscribing a conic proved transforming the conic section in a circle, the inscribed quadrilateral in a rectangle and the circumscribed quadrilateral in a parallelogram. The projective properties can be extended from this particular configuration to any conic (ibid., art. 185, pp. 95–96). Poncelet proved the following fundamental properties:

a) Given a conic, the four diagonals of the inscribed quadrilateral *ABCD* and of the circumscribed one *abcd*, posed as in the figure (Fig. 2.43) mutually cut in a point *P*; b) the four intersection points *L, M* and *l, m* of the opposite sides of the inscribed and circumscribed quadrilaterals belong to the polar of *P*; c) the diagonal *bd* of the circumscribed quadrilateral and the opposite sides *AB, CD* of the inscribed one belong to a pencil whose centre is *M*. Analogously the diagonal *ac* and the sides *AD* and *BC* belong to a pencil whose centre is *L*; d) each of the two points *M* and *L* is the pole of the diagonal passing through the other point and through *P*; e) each segment of any straight line through *P*, whose extremes belong to the conic or to the opposite sides of one of the two quadrilaterals, is divided harmonically in *P* and in the point where the straight line cuts the polar *LM* of *P* (ibid., art. 186, p. 96).

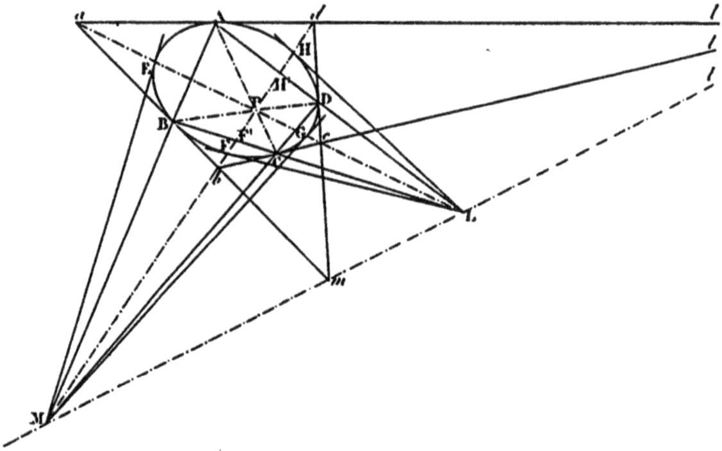

Fig. 2.43 The figure used by Poncelet to prove item 8)

9) In any quadrilateral inscribed in a conic, the intersection points of the diagonals and of the two couples of opposite sides are three points such that each of them is the pole of the straight lines joining the two other points; in any complete quadrilateral circumscribing a conic, each of the three diagonals is the polar of the intersection point of the other two (ibid., art. 192, p. 99).
10) From the properties proved in art. 186, the following fundamental ones derive:

Be $AB, A^{'}B^{'}, A^{''}B^{''}, \ldots$ a series of chords of a conic mutually cutting in P (Fig. 2.44): a) the points $C, C^{'}, C^{''}, \ldots$ which are the fourth harmonics of the extremes of each chord and of P belong to the polar of P; b) the points $L, M; L^{'}, M^{'}, \ldots$ where the chords A^iB^j, A^jB^i and A^iA^j, B^iB^j mutually cut belong to the polar of P; c) all the intersection points $T, T^{'}, \ldots$ of the couples of tangents drawn at the extremities of the chords $AB, A^{'}B^{'}, \ldots$ belong to the polar of P; d) reciprocally: if from different points $T, T^{'}, \ldots$ of a straight line p belonging to the plane of a conic, couples of tangents are drawn to the curve, the contact chords $AB, A^{'}B^{'}, \ldots$ of these tangents mutually cut in a point P which is the pole of p (ibid., art. 194, pp. 100–101). All these properties establish the reciprocal and involutory character of the relation pole–polar, until the property that if a point belongs to a straight line in the plane of a conic, its polar passes through the pole of such a straight line and vice versa (ibid., art. 196, pp. 101–102).

11) Fundamental properties of the inscribed and circumscribed hexagons with respect to a conic. In particular (see Fig. 2.45):

If *ABCDEF* is a hexagon inscribed in a conic and *abcdef* is circumscribed so that the contact points of the tangents which are the sides of *abcdef* are the vertices of *ABCDEF*, the opposite vertices *a* and *d* of the circumscribed hexagon have as their polars the two opposite sides *AF* and *CD* of the inscribed hexagon. The

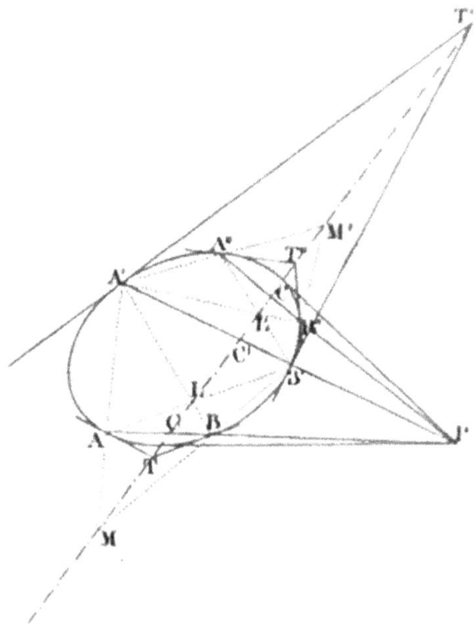

Fig. 2.44 The figure used by Poncelet to prove item 10)

diagonal *ad* is the polar of the intersection point *I* of *AF* and *CD* and contains the pole *P* of *IKL* which contains the intersection point of the three pairs of opposite sides of *ABCDEF*. Therefore, *P* is also the intersection point of the three diagonals of the circumscribed hexagon. From these considerations on the poles and polars, the proof of Brianchon theorem follows easily (ibid., art. 208, pp. 107–108).

12) Through the principle of continuity, the properties of the hexagons inscribed in and circumscribing a conic are extended to pentagons and quadrilaterals (ibid., arts. 214–221, pp. 109–113).
13) Particular metric cases. Concept of indicatrix ("indicatrice", ibid., arts. 222–226, pp. 113–116). This concept is very interesting. For consider a conic, a pole *P* and its polar *p* with respect to the conic. Consider, then, the pencil of straight lines through *P*. They will intersect the polar in points Q_i, thus producing an infinity of segments PQ_i. The mean points of such segments are on a straight line *i*. Poncelet proved *i* to be a tangent if the conic is a parabola. If the conic is a hyperbola, then *i* cuts the curve in two points. If you join both points with the pole, you obtain two new straight lines, which are parallel to the hyperbola's asymptotes. If the conic is an ellipse, the straight line is external. Poncelet called *indicatrice* the straight line *i* because many properties of the conic are deducible from it. Let us consider a straight line *PQ* of the pencil *P* and suppose that it intersects the conic in the points *A* and *B*, then the segment *PQ* is the harmonic means between *PA* and *PB*. This property connotes the relation pole–polar. Whereas if *R* is the point where *PQ* cuts the indicatrix, then *PR* is the

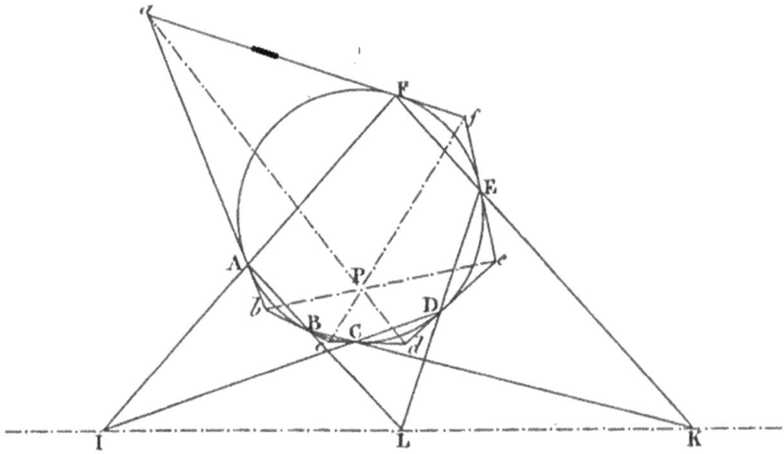

Fig. 2.45 The figure used by Poncelet to prove item 11)

geometrical means between *PA* and *PB*. This connotes the relation pole-indicatrix (ibid., art. 226, p. 116).

14) Reciprocal character of the relation pole–polar. Fundamental theorem: if two hexagons *H* and *H′* lying in the plane of a conic section are such that the vertices of *H* are the poles of the sides of *H′*, reciprocally the vertices of *H′* are the poles of the sides of *H* (ibid., art. 227, p. 116). If one of the two polygons is inscribable in a conic section, the other one is circumscribable. These theorems are applicable to polygons of any numbers of sides, which will be called reciprocal polar with respect to the conic named directrix (ibid., arts. 228–229, pp. 116–117).
15) Since each conic can be determined by five of its elements (points or tangents), it is possible to deduce that, given two conic sections in a plane and five points or tangents in one of them, their polars or poles with respect to the other lie on a third conic section to which also the polar or the pole of a sixth point or tangent assumed in the first conic belongs. Thence, if a polygon in a plane of a conic is inscribed in another conic, its reciprocal polar polygon is circumscribed at such curve. Ergo, if a point or pole in the plane of a conic moves on another conic, its polar envelops a third conic and reciprocally (ibid., art. 230–231, pp. 117–118).
16) Finally, Poncelet shows that many theorems on the reciprocal polars with respect to a conic also hold for two reciprocal curves of any degree. In particular, if two any curves in the plane of a conic are such that the points of one of them are the poles of the tangents to the other, reciprocally the points of the second one are the poles of the tangents to the first curve (ibid., art. 232, p. 118).

Therefore, before Steiner and Chasles' work the concept of harmonic ratio played already a fundamental role in projective geometry, but it was not seen as a particular case of a more general projective invariant, the anharmonic or cross ratio, which is conserved in the projective transformations. Rather, it was still seen as the specific kind of proportionality between the four points of a segment defined by De La Hire.

Furthermore, the projective generation of the conics relies on the invariance of the cross ratio in projective transformations. Thus, before the identification of the double ratio relation as a projective invariant, a projective generation of the conics would have been inconceivable. In the development of the theory of polar reciprocity, as we have shown expounding the steps followed by Poncelet, the polygons inscribed in and circumscribing a conic play a significant role. Steiner fully clarified this role proving that the properties of the complete quadrangle rely on the invariance of the harmonic ratio by projection, a property which derives, once again, from the invariance of the double ratio. Moreover, the dual distinction between n-ogons and n-laterals was not fully grasped in its great importance by the mathematicians preceding Steiner. This distinction spreads a new and unitary light on the relation between poles and polars with respect to a conic and, in general, to the relation of duality which was the cornerstone of projective geometry. For, e.g., in the Pascal theorem you will speak of hexagon, in the dual Brianchon theorem of hexalateral. This means that the elements: discovery of the invariant character of the cross ratio by sections and projections—distinction between n-ogons and n-laterals—projective generations of conics offer a new unitary picture to the theory of reciprocal polars which had been invented by the French mathematicians and to which Poncelet had given the first unitary view just synthetically expounded. Steiner and Chasles showed that the duality law and the theory of reciprocal polars, the two milestone of projective geometry, are, in fact, founded on a more basic concept, that of cross ratio, enriched with the projective generation of the conics.

This second phase in the foundation of projective geometry will be furtherly overcome by Von Staudt, through his proof that all the projective properties can be obtained without any reference to the concept of cross ratio, relying on the definition of the harmonic groups of four points, two of which are those where the two diagonals of a complete quadrangle cut the straight line joining the intersection points of the opposite sides and the other two points are such intersections points. An analogous theorem holds for a group of four planes and a complete tetrahedron.

2. *Comparison with Chasles*. As shown in Sect. 2.1.2 Chasles gave fundamental contributions to the theory of reciprocal polars before his *Aperçu*, namely before the introduction of the concept of cross ratio and of the projective generation of the conics. At the beginning of his memoir on duality inserted in the *Aperçu* he analysed the properties of duality in space. In his demonstration the use of the anharmonic ratio plays a crucial role. Therefore, he considered the projective properties of the plane geometry as a particularization of the projective properties of the spatial geometry. He fully understood that the forms of first species, namely dotted straight lines, pencils of radiuses and sheaves of planes are analogous from a projective standpoint. However, his mathematical references are not the forms of the three species—whose character he fully recognized—but the plane and space without further specifications. In a sense, in the *Aperçu* Chasles imposed the whole new structure he was constructing—based on the concept of anharmonic ratio—on space, deduced a series of general properties and saw the properties of the plane projective

geometry as a particularization of such spatial projective properties. Of course, he knew perfectly and—as we have seen—also developed the projective properties of the plane without resorting to that of space, but his main concern is such a reduction. Chasles was perfectly aware that the duality he based on the anharmonic ratio which he invented allows us to pass from plane to spatial configurations and vice versa and that, hence, the distinction between straight line, plane and space in projective geometry is not the one which captures the essence of such discipline. This notwithstanding, I think that the basis of his results and way of proceeding the idea that space is the "natural environments" of projective geometry is present and that, hence, the planar properties derive from the spatial ones. On the contrary, Steiner worked directly with the projective forms. He was less interested than Chasles in showing the reduction of the planar graphic properties to those spatial and was more interested than Chasles in highlighting the new character of his forms. In other terms, though both authors developed similar methods and approaches to projective geometry, my impression is that the basic mentality of Chasles might be expressed by the tendency to think in terms of plane and space, while Steiner thought in terms of forms. As a matter of fact, the latter's analysis of polarity in the *Systematische Entwicklung* starts from planar figures as the conics, while the first theorem on the poles and polars proved by Chasles in the *Aperçu* is a theorem of spatial geometry presented in an interesting paragraph (the sixth one) entitled *Sur les poles et les plans polaire des surfaces du second degree* (Chasles, 1837a, pp. 596–597). Here Chasles considered a centred surface of second degree C, two tangent parallel planes α and β. The chord joining the contact points a and b passes through the centre o of the surface, so that $oa=ob$. In the correlative figure, according to the duality law, just proved by Chasles, to the surface of second degree C another second-degree surface C' corresponds; to the plane at infinity of the first surface, the point i corresponds; a straight line l through i corresponds to the intersection line at infinity of α and β. If a' and b' are the points where l cuts C', and A and B the planes tangent to C' in a' and b $'$, these planes correspond respectively to the points a and b. Thence, their intersection straight line m belongs to a fixed plane O corresponding to the centre o of C. Consider now the plane I through m and i. This point corresponds to the point at infinity of ab. Thence, because of the conservation of the anharmonic ratio in the projective transformations, the following identity holds:

$$\frac{\sin\left(\widehat{OA}\right)}{\sin\left(\widehat{OB}\right)}:\frac{\sin\left(\widehat{IA}\right)}{\sin\left(\widehat{IB}\right)}=\frac{oa}{ob}=1$$

If γ, δ, ω are the points where a transversal through i cuts the planes A,B,O, the identity

$$\frac{i\gamma}{i\delta}:\frac{\omega\gamma}{\omega\delta}=1$$

is satisfied. From this, it follows that if a transversal t cutting a second-degree surface rotates around a fixed point, the pairs of planes tangent to the surface at the pairs of intersection points between each of the straight lines generated by the rotation of t and surface mutually cut on straight lines belonging to a fixed plane. The transversal cuts this plane in a point which is the harmonic conjugate of the fixed point in relation to the two points of the surface. Chasles stated that this is the fundamental property of the poles and the polar planes for the second-degree surfaces (Chasles, 1837a, p. 597). The preceding theorem can also be expressed claiming that the correlative figure of a second-degree surface C is a second-degree surface C' in which the pole, considered with respect to C', of the plane corresponding to the centre of C is the point corresponding to the plane at infinity of C. Here Chasles introduced the concept of polar of a straight line in space with respect to a second-degree surface. For the intersection line of two tangent planes to a second-degree surface is called the *polar* of the chord joining the two contact points, so that the just proved theorem states that all the straight lines through a point have their polars on a plane (ibid., p. 597).

Chasles observed then that his new approach to duality can be of help to solve many well-known problems of the theory of poles and polars. Therefore, there is a strong similarity between Steiner's and Chasles' approach to polarity because both of them resorted to the use of the anharmonic ratio and, through it, managed to show that some objects in two correlative figures fulfil a harmonic relation. These objects are called pole and polars with respect to a conic. In the old approach to the theory of polarity, instead, the relations pole–polar were defined independently of the reference to the harmonic ratio. It was then proved that such objects satisfy the harmonic ratio, which was not seen as a particular case of a more general relation as in Steiner from the *Systematische Entwicklung* and in Chasles from the *Aperçu* onwards. The difference is that Chasles began speaking directly of a spatial situation, while Steiner constructed his theory from the planar polarity.

A possible and reasonable objection in my view, according to which, besides many important similarities between Chasles' and Steiner's approach, there is the slight difference I have pointed out, is that the purposes of the *Systematische Entwicklung* and of the two crowned memoirs which constitute the second part of the *Aperçu* were not exactly the same. For, Chasles had to present a research on the two principles of duality and homography, whereas Steiner presented a research on the basic concept which found such a principle. This objection can be substantiated by the fact that in the notes of the *Aperçu*, written after the two crowned memoirs, Chasles founded the concept of anharmonic ratio and, as we have seen, applied it to the conics without any reference to spatial geometry. Furthermore, in the *Traité de géométrie supérieure*, he introduced the anharmonic ratio and its properties for quaterns of points, straight lines belonging to a pencil and planes belonging to a sheaf (Chasles, 1852, Chapter II, pp. 7–27), but in the following chapters he developed the theory basically for planar configurations. However, this does not contradict my idea because the *Traité* is a late work in comparison with those of the period 1830–1837 and, therefore, an approach based on the idea of expressing the basic properties of the cross ratio and of showing their applications from elementary cases to more complex ones was catching on, and the most elementary cases are, in general, planar ones. Hence, in my opinion, after the introduction of the notion of cross ratio, Steiner immediately began to think in terms of forms of different species, whereas Chasles, in this faithful to his foundational programme, was still inclined to think in terms of the distinction plane-space. Steiner's awareness of the importance of projective forms is confirmed by an "Anmerkung" where he explicitly posed the duality between the dotted-ruled plane and the star of planes-radiuses. For he wrote that the following dualities hold (Steiner, 1832, p. 182):

Star	Plane
Straight line	Point
Pencil of straight lines	Straight line
Sheaf of planes	Pencil of straight lines
Solid n-ogons figure	n-ogons
Solid n-edges figure	n-lateral
Cone	Conic section

Chasles was perfectly aware of such dualities and, as we will see in the next chapters, he developed more complex forms of duality, but no such specific references to the projective forms exist in Chasles: rather he applied the properties of the projective forms directly to spatial and planar geometry, which was done by Steiner as well, but after having developed the specific properties of the projective forms in themselves. We will consider now exactly Steiner's application of his theory to the projective generation of the quadrics.

2.3.1.9 The Generation of Projective Forms in Space

Steiner faced systematically the forms' projective generation in space. Namely: given two projective forms of first species, what configuration is generated by their corresponding elements? Steiner analysed the six possible cases:

1) A straight line A and a sheaf of planes $\mathfrak{A}$;
2) Two pencils of radiuses B,B';
3) A sheaf of planes $\mathfrak{A}$ and a pencil of radiuses B;
4) A pencil of radiuses B and a straight line A;
5) Two straight lines A, A';
6) Two sheaves of planes $\mathfrak{A}$, $\mathfrak{A}'$.

Steiner claimed that only in the last two cases new and interesting figures are generated. He treated quickly the first four cases. Cases 1) and 2) produce immediately no remarkable figure. As to the former, Steiner clarified that this specific construction can be developed to obtain not an immediate generation but a mediate generation "mittelbares Erzeugnis" (Steiner, 1832, p. 183) of the hyperbolic paraboloid: be A skew with respect to sheaf's axis. From any point a of A draw the perpendicular to the plane α corresponding to *a*. All these straight lines will generate a hyperbolic paraboloid.[129] In the third case, a conic section is generated because the plane of the pencil *B* cuts the sheaf in a pencil *B*′ which is projective to *B*. As to the fourth case, each pair of corresponding elements generates a plane, namely the plane passing through a point of the straight line and through its corresponding straight line of the pencil. Steiner also identified the particular positions of the so generated planes. For he imagined a further pencil *B*′, concentric (and, obviously, not coplanar with *B*, so that *B* and *B*′ belong to a star) with *B* and *B*′ perspective to *A*. Therefore, *B* ′ is projective to *B* and the planes generated by two corresponding straight lines of *B* and *B*′ are the same as those generated by the points of *A* and the corresponding straight lines of *B*. Since in § 38, II (ibid., p. 138) Steiner had proved that two projective, but not perspective, sheaves of planes belonging to a star envelop a second-degree cone, then the planes generated by the pairs of corresponding elements in the projection between a straight line *A* and a pencil of radiuses *B* envelop a second-degree cone whose vertex is the centre of *B* (ibid., pp. 184–185).

This clarified, Steiner faced the two most difficult and interesting cases: those of two projective straight lines *A* and *A*′ and of two projective sheaves of planes $\mathfrak{A}, \mathfrak{A}'$. He considered all the possible subcases, the easiest of which concerns the figure generated by any point *D* of space and by the projective radiuses of the two dotted straight lines $\mathfrak{A}$ and $\mathfrak{A}'$, and dually by any plane *E* and by the intersection straight lines of the homologous planes in $\mathfrak{A}$, $\mathfrak{A}'$. Through a reasoning similar to that analysed in the previous paragraph, he was able to demonstrate that the planes generated by *B* and by the projective radiuses connecting the corresponding points of *A* and *A*′ are the tangent planes to a second-degree cone whose vertex is *D*. Dually the figure generated by a plane *E* and by the intersection points of the homologous planes in $\mathfrak{A}, \mathfrak{A}'$ is a conic section (ibid., pp. 186–187).

Steiner introduced the next subcase—whose steps I am summarizing—which is the most interesting one, through the following reasoning: consider two skew straight lines *A* and *A*′ perspective to a sheaf of planes whose axis is *A*″. Therefore, *A* and *A*′ are projective. The pairs of corresponding points $\mathfrak{a}, \mathfrak{a}'; \mathfrak{b}, \mathfrak{b}'; \mathfrak{c}, \mathfrak{c}'; \ldots$of *A* and *A*′ belong to the planes α, β, γ of *A*″. Therefore, their projective radiuses $\mathfrak{a}\mathfrak{a}' \equiv a$; $\mathfrak{b}\mathfrak{b}'$ $\equiv b; \ldots$ also belong to these planes. Hence, they cut *A*″, so that they are a system of radiuses cutting all of the three straight lines *A*,*A*′,*A*″ (see Fig. 2.46).

Dually, given two sheaves of planes $\mathfrak{A}, \mathfrak{A}'$—whose axes *A* and *A*′ are skew—perspective with a straight line *A*″, they are projective and the intersection straight line of two homologous planes cuts each of the straight lines *A*,*A*′,*A*″.

Vice versa, if in space three straight lines *A*,*A*′,*A*″, which are skew two by two, are given, an array composed of an infinite (*unzhälig*) number of straight lines *a*,*b*,*c*,

[129] Steiner did not present this proposition in the analysis of the six cases proposed at the beginning of § 49 I am examining entitled "Projective generation of the forms in space", rather as a specific application in the "Appendix", §60, 26 (ibid., pp. 302–303), where theorems proposed without demonstration to the reader exist. An indirect construction can also be obtained for case 2), but it is not interesting for my aim, which is to present Steiner's general way of thinking and his most relevant results comparing them with Chasles'.

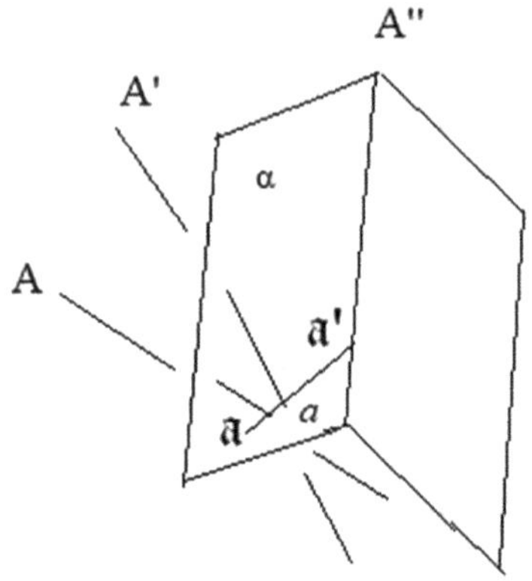

Fig. 2.46 My reconstruction of the figure described by Steiner in reference to the sole straight line ***aa'***$\equiv a$

d,... exists such that each of them cuts *A*,*A'*,*A*" (ibid., pp. 187–188, 190). Furthermore, each two of the straight lines *A*,*A'*,*A*" are cut projectively by the straight lines of the array and are the axes of two projective sheaves of planes whose correspondent planes mutually intersect in straight lines of the array. Steiner proved that, reciprocally, the projective radiuses *a*,*b*,*c*,*d*,... of any two skew projective straight lines *A*,*A'* or the intersection straight lines *a*,*b*,*c*,*d*,... of the couples of correspondent planes belonging to two projective sheaves whose axes *A*,*A'* are skew are cut by straight lines with the characteristics of *A''*. To prove such a proposition, he applied the fundamental theorem of projectivity, according to which in the projective forms of first species a projectivity is determined when three pairs of corresponding elements are given (ibid., pp. 191–192). Hence, Steiner summarized his results as follows:

1) The projective radiuses *a*,*b*,*c*,*d*,... of any two skew projective straight lines *A*,*A'* are cut by an infinite number (*unzählig*) of straight lines *A''*, *A'''*,*A''''*,... such that each of these straight lines cuts all the projective radiuses *a*,*b*,*c*,*d*,... if it cuts three of them.	1') The intersection straight lines *a*,*b*,*c*,*d*,... of the corresponding planes, which belong to two projective sheaves whose axes *A*,*A'* are skew, are cut by an infinite number of straight lines *A* ",*A'''*,*A''''*,... such that each of these straight lines cuts all the lines *a*,*b*,*c*,*d*,... if it cuts three of them.
Therefore there are two arrays of straight lines *a*,*b*,*c*,*d*,... and *A*,*A'*, *A''*,*A'''*,... which satisfy perfectly reciprocal relations. Specifically: 2) Each two straight lines belonging to each of the two arrays are projective and the straight lines of the other array are their projective radiuses.	2') Each two straight lines belonging to each of the two arrays are the axes of projective sheaves of planes, whose corresponding planes cut mutually along straight lines of the other array (ibid., pp. 193–194).

It was well known that two arrays of skew straight lines mutually cutting generate a one-sheeted hyperboloid. Therefore, this set of reasonings show the *projective generation* of this second-order surface, so that Steiner could conclude this part of his argumentation claiming that

Each two skew projective but not perspective straight lines *A*,*A*′ generate a one-sheeted hyperboloid; namely they and their projective radiuses, jointly with the array of straight lines cut by these radiuses, lie in a one-sheeted hyperboloid.	Each two projective but not perspective sheaves of planes *A*,*A*′ generate a one-sheeted hyperboloid; namely the intersection straight lines of two homologous planes, jointly with the array of straight lines which cut them, lie in a one-sheeted hyperboloid (ibid., p. 194).

The projective generation of the hyperboloid implies numerous properties explained by Steiner. I will focus on those most remarkable. The first property shows the projective generation of the hyperboloid's asymptotic cone. Steiner observed that: each plane cutting the hyperboloid in a straight line *l* of an array also cuts it in a straight line *L* of the other array. The point where *l* and *L* mutually cut was called by Steiner "contact point" (*Berührungspunkt*) and the plane *lL* "contact plane" (*Berührungsebene*), so that all the contact planes of a hyperboloid passing through the same point envelop a second-degree cone and dually the intersection points of a hyperboloid and of any plane *E* generate a conic section (Fig. 2.47).

Since: i) each two projective straight lines of an array have the straight lines of the other array as projective radiuses, and ii) given a projectivity which is not a similarity, a projective radius exists parallel to one of the projective straight lines (ibid., § 9, pp. 32–33), it follows that for any straight line of an array generating the hyperboloid, a straight line parallel to it in the other array exists. Therefore, according to what we have just seen, all the planes passing through the parallel straight lines of the two arrays generate a second-degree conic surface passing through the centre of the hyperboloid. Such a cone is the *asymptotic cone* of the hyperboloid of which Steiner has so shown the projective generation (ibid., pp. 195–196).

Steiner highlighted several properties of the one-sheeted hyperboloid. I will focus on the following one because it allows us to comprehend perfectly the concept of projective generation of the ruled quadrics and shows that, from a projective point of view, each ruled quadric is equivalent since the hyperbolic paraboloid is a limit case of the one-sheeted hyperboloid when the projectivity between the two arrays of straight lines (I) *A*, *A*′, *A*″, *A*‴, ... and (II) *a*,*b*,*c*,*d*,... assumes a particular form. I will refer below to the steps of Steiner's reasoning.

1) Suppose that the projectivity between *A* and *A*′ to be a similarity. In this case, the two points at infinity of *A* and *A*′ are homologous. Thence, one of the projective radiuses, namely one of the straight lines of (II), is at infinity. Be it indicated by *e*.
2) Since *e* is a projective radius for all the straight lines of (I), this means that all such straight lines are similar.
3) Consider a sheaf of planes having a straight line of (I)—for example *A*″—as axis. Be α'', β'', γ'', ... its planes, which pass through the straight lines *a*,*b*,*c*,... of (II) respectively. The plane *E* which passes through *e* will be parallel to all of the lines belonging to (I) because it passes through their points at infinity.
4) This means that a plane parallel to *E* passes through each straight line of (I). Thence, the sheaf of parallel planes *E*, *E*′, *E*″,... cuts the projective straight lines of (II). Therefore, the projectivity between such lines is also a similarity because

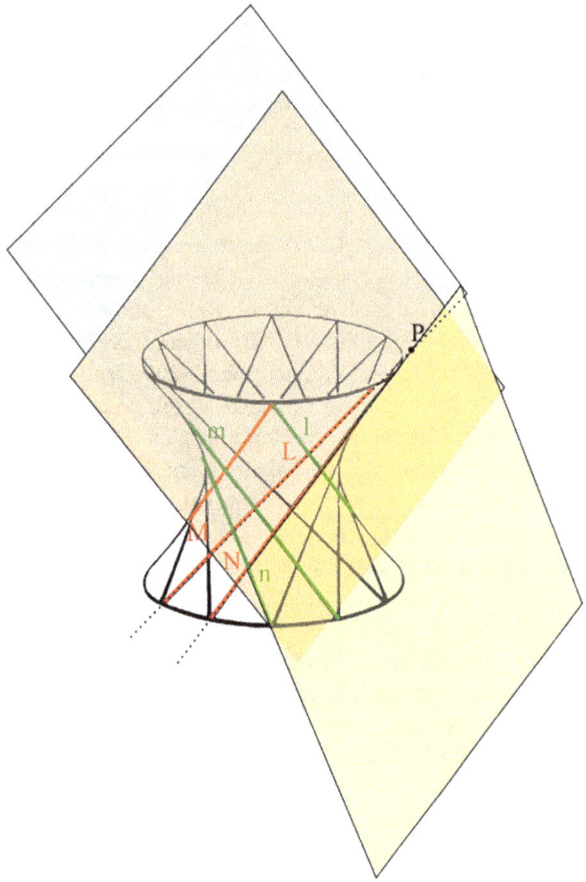

Fig. 2.47 In this diagram I explain the situation described by Steiner. The planes *Ll*, *Mm*, *Nn* are contact planes. They meet in the point *P*, which is the vertex of the second-degree cone enveloped by the three planes

the parallel planes cut these lines in segments having the same ratios. Therefore, a straight line $A^{(n)}$ of (I) will be at infinity; all the straight lines of (II) will be parallel to a plane and a sheaf of parallel planes $\alpha^{(n)}, \beta^{(n)}, \gamma^{(n)}, \ldots$ exists through which the straight lines of (II) respectively pass.

5) Consider once again the sheaf of axis A'', whose planes $\alpha'', \beta'', \gamma'', \ldots$ pass through the straight lines *a*,*b*,*c*,... of (II) respectively (item 3). This sheaf will be cut by each or the parallel planes $\alpha^{(n)}, \beta^{(n)}, \gamma^{(n)}, \ldots$, e.g. $\alpha^{(n)}$, in a pencil $\alpha_1^{(n)}$,of straight lines $a^{(n)}, b^{(n)}, c^{(n)}, \ldots$. Each one of such straight lines will be parallel to one of the array (II), as can easily be proved. Therefore, this array contains all the directions of the straight lines belonging to a pencil or to a plane. An analogous reasoning holds for the straight lines of the array (I).
6) It is possible to prove that the straight lines of $\alpha_1^{(n)}$ are projective with those of (I).

Analogously, the straight lines of (I) are parallel to a pencil projective to the straight lines of (II).

7) Since (I) and (II) are parallel to the planes having the position of E and $\alpha^{(n)}$ and since they are parallel to the straight lines of the asymptotic cone (ibid., § 51, IV, 8, pp. 197–198), such a cone degenerates, in this case, in two sheaves of parallel planes. Their line at infinity touches the figure. Thence, they are called by Steiner *asymptotic planes* (ibid., p. 203).[130]
8) With regard to the form of the surface generated by two arrays, it should be highlighted that, considered the intersection straight line x of a couple of two asymptotic planes, in each of the arrays (I) and (II), there is a sole straight line perpendicular to x. The contact point X of these two straight lines is the vertex of the figure. The intersection line of the two asymptotic planes passing through X is the axis of the figure (ibid., p. 204). The general intersection between a plane π and the surface is a hyperbola, because π cuts the two asymptotic planes passing through the vertex in two lines which cut the surface at infinity and which are the asymptotes of the hyperbola. If the plane π is parallel to any asymptotic plane, the hyperbola degenerates in a parabola (ibid., pp. 205–206).[131]

Therefore, given these features, the figure produced by the degeneration of a one-sheeted hyperboloid is a *hyperbolic paraboloid*, of which Steiner has shown this projective generation. This means that, from a projective standpoint, all the ruled second-degree surfaces belong to a sole class and the hyperbolic paraboloid is a one-sheeted hyperboloid tangent at the plane at infinity. Steiner added a series of specific features of the ruled quadrics: for example, he explained how, through the specifications of some projective elements, it is possible to obtain a metric figure as the equilateral hyperbolic paraboloid. This is the case when the two asymptotic planes are perpendicular (ibid., pp. 210–211). Steiner also showed that the generation of a ruled quadric through the movement of some elements is reducible to its projective generation (ibid., pp. 222–223). For my aims, this presentation of Steiner's spatial theory of projectivity, insofar as the quadrics are concerned, is sufficient.

In the next section entitled "Zusammengesetztere Sätze und Aufgaben" ("Collected propositions and problems", ibid., pp. 234–250) Steiner proved several interesting propositions on the n-ogons and n-wedges, but I will not consider this subject which can be seen as an extension to the spatial case of the properties connoting the flat n-ogons and n-laterals.

The final section of Steiner's masterpiece entitled "Allgemeine Anmerkung: Ueber Abhängigkeit einiger Systeme verschiedenartiger Figure von einander" ("General annotation: on the mutual dependence of some system of various figures", ibid., pp. 251–295) is important. Here he analysed several transformations of two configurations and in each of them individuated a dual correspondence between two figures of each configuration. This section is a litmus paper of the broadness of his view on geometry and of his profound analysis of the geometrical transformations.

[130] The planes are usually called "director planes".

[131] Steiner did not mention the cases in which the intersection of the figure with a plane are a couple of straight lines or a straight line, but they are easily deducible as limits cases of the previous ones.

Only to mention some of Steiner's acquisitions referred to in this section, the following should be recalled:

a) He explicitly distinguished the double projective nature of the straight line in space: as set of its points or as axis of a sheaf of planes. For he wrote:[132]

> According to the previous researches a straight line $\mathfrak{A}$ in space can be considered in a double manner, namely, as a real straight line $\mathfrak{A}$ (that is containing an infinite set of points) or as a sheaf of planes $\mathfrak{A}$ (that is as axis of the sheaf). Under this double standpoint, a straight line satisfies the following relation with all the points and all the planes in space:[...]

It (the axis $\mathfrak{A}$), with every point (not lying in it), determines a plane.	It, with every plane (in which it does not lie), determines a point.

These few lines entail several interesting statements. First of all, the double nature of the straight line in space as well as the dual character of the points and the planes is formulated in a manner which was the one used in the handbooks of projective geometry until around the 40'–50' of the twentieth century, namely until the period in which this discipline was still taught in the universities and some kind of high school following the classical approach. Furthermore, it is also appropriate to focus on the expression used by Steiner to denote the points of a straight line. He spoke of *unendliche Menge Punkte*. The expression, but also the concept behind it, has a Cantorian flavour. For "unendliche Menge" is exactly the expression Cantor used to indicate the infinite sets. It seems to me also clear that no ambiguity can exist: Steiner is thinking of the points composing a straight line as an actually infinite quantity. Finally, it is worth pointing out that Steiner, while referring to the dotted straight line, spoke of "straight line in a proper sense" ("eigentliche Gerade"), which is something like survival in his new conception of the classic way of conceiving a straight line.

b) Steiner presented a beautiful transformation whose basic elements are two planes E and E' belonging to a sheaf and two skew straight lines A and A' which cut the two planes in four points. If from each point a of E the straight line which cuts A and A' is drawn, a point a' on the plane E' corresponds to a. It has remarkable properties: in general, the homologous of a point in E is a point in E', but there are three points (whose identification is not difficult), he called "fundamental", in

[132] Ibid., p. 252: "Bei den vorhergehenden Untersuchungen wurde eine Gerade $\mathfrak{A}$ im Raume auf doppelte Weise, d.h., in Hinsicht zweier Gebilde, betrachtet, nämlich entweder als eigentliche Gerade $\mathfrak{A}$ (d.i. als seine unendliche Menge Punkte enthaltend), oder als Ebenenbüschel $\mathfrak{A}$ (d.i. als Axe des Ebenenbüschels). In dieser doppelten Hinsicht steht daher eine Gerade $\mathfrak{A}$ mit allen Punkten und allen Ebenen in Raume in folgender Beziehung[...]

Sie (die Axe $\mathfrak{A}$) bestimmt mit jedem Punkt (der nicht in ihr liegt), eine Ebene.	Sie bestimmt mit jeder Ebene (in der sie nicht liegt), einen Punkt.

E and three in E' to which a whole straight line corresponds. They are the vertices of the "fundamental triangle" (*Haupdreiecke*, ibid., p. 255). It is possible to prove that, in general, in this transformation a straight line in one of the two planes corresponds to a conic section in the other one and vice versa (ibid., p. 256), but if the straight line passes through one of the fundamental points, then its correspondent is a straight line and not a conic (ibid., p. 257). Steiner developed all the properties of this correspondence (ibid., pp. 253–270).

c) This issue in connection with the previous one is noteworthy because through them Steiner showed that his theory of transformations is so wide as to include the theory of reciprocal polars. Now he considered the system of all straight lines passing through the two skew straight lines A,A' and indicated it by $[A,A']$. He constructed the transformation between two stars D and D' like this:[133] given the plane $\alpha \in D$ and passing through $a \in [A, A']$ its corresponding plane $\alpha' \in D'$ is that passing through a (Fig. 2.48).

Be the planes r, $s \in D$ those which cut the two axes A,A' and be $x = r \cap s$. Analogously be z', y' the planes of D' passing through A,A' respectively and $t' = z' \cap y'$. Therefore, since x,t' cut A,A', they belong to the system $[A, A']$. Be t the plane of D passing through t' and $t \cap r \equiv y$; $t \cap s \equiv z$. Be x' the plane of D' passing through x and $x' \cap z' = s'$; $x' \cap y' = r'$. Steiner named the planes $r, s, t; z', y', x'$ "principal" or "fundamental planes" ("Hauptebenen", ibid., p. 272) and $z, y, x; r', s', t'$ "Hauptstrahlen"; the three-surfaces rst; $z'y'x'$ form the "Hauptdreiflache" ("principal three-surface") of the star of planes D,D'. Finally, be $ee' \equiv D \cap D'$.

This stated, it is not difficult to prove that to all the planes of the sheaves z, y, z the principal planes $z',y'x'$ correspond respectively, and to all the planes of the sheaves r $',s't'$, the principal planes r,s,t correspond. Each plane of the sheaf ee' corresponds to itself.

Steiner demonstrated (ibid., pp. 273–274) that, in general, to all the planes of the two stars D,D' which belong to a sheaf correspond, in the other star, the planes which are tangent to a second-degree conic surface. It is also possible to say that to a straight line in D,D' a conic surface in the other star corresponds. This surface is inscribed within the principal three-surface. There are exceptions to this general correspondence. For if the axis of one of the considered sheaves of planes in D belongs to one of the principal planes r,s,t, its correspondent in D' is another sheaf of planes whose axis belongs to the corresponding principal plane.

Steiner also proved that, in general, to each system of planes belonging to D which envelop a n-th-degree conic surface, a conic surface of class $2n(n - 1)$ in D' corresponds. However, since through the principal straight lines in D, $n(n - 1)$ planes pass enveloping the conic surface, the corresponding conic surface in D' will touch the principal planes z',y',x', $n(n - 1)$ times, so that if the conic surface in D is of second degree and touches any two of the three principal planes, its correspondent

[133] Steiner used the expression "Strahlbüschel D,D'", but it is obvious that he is considering the system of all the planes and straight lines passing through a point, namely what we call a star.

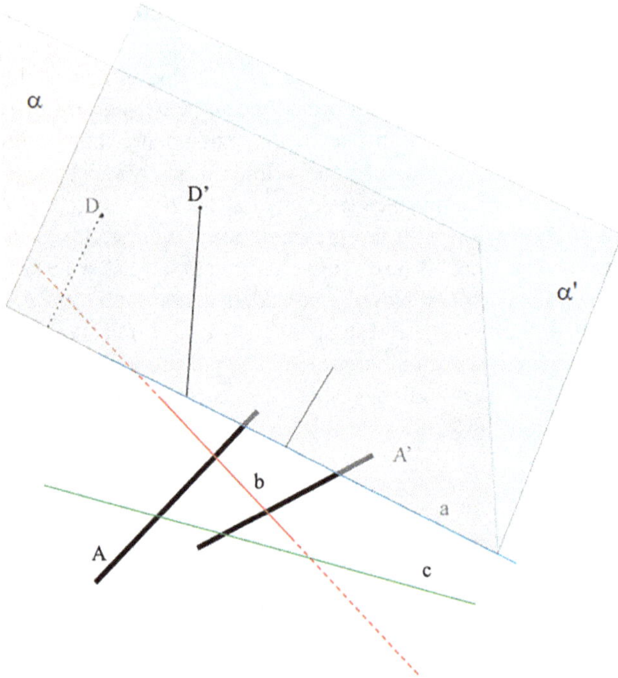

Fig. 2.48 Diagram representing the situation described by Steiner

in D' is of second degree and touches two of the three principal planes and vice versa (ibid., pp. 275–276).

Steiner added a further element to the transformation just described: he considered any two planes C and C' cutting the previous configuration [I indicate it by (i), while I indicate the new configuration by (ii)], namely the star of planes D and D' correlated through the system $[A,A']$. In this section (ii) the principal three-surfaces rst, $z'y'x'$ of (i) produce three principal trilaterals $\mathfrak{rst}, \mathfrak{z}'\mathfrak{y}'\mathfrak{x}'$, the principal planes r,s,t; z',y',x' produce the sides $\mathfrak{r},\mathfrak{s},\mathfrak{t},\mathfrak{z}',\mathfrak{y}',\mathfrak{x}'$ of the trilaterals and the principal straight lines x,y,z; r',s',t' produce the three principal points $\mathfrak{x},\mathfrak{y},\mathfrak{z};\mathfrak{r}',\mathfrak{s}',\mathfrak{t}'$ (ibid., pp. 276–277). Steiner identified 22 reciprocal (dual) correspondences (and other might be), divided into four groups, among the elements of C and C' which, obviously, are deducible from the properties of (i) from which they derive (ibid., pp. 277–280). I restrict to refer to the most important ones:

Plane C	Plane C'
Straight line a	Straight line a'
A straight line R	The straight line at infinity R'
Single sides $\mathfrak{r},\mathfrak{s},\mathfrak{t}$ of the principal threeside	All the straight lines of the stars $\mathfrak{r}',\mathfrak{s}',\mathfrak{t}'$
Point B not lying in one of the three principal straight lines	Conic section $[B]$ inscribed in the principal threeside[a]

(continued)

Plane C	Plane C'
Points belonging to the three principal straight lines	Points belonging to the three principal straight lines
Array of the points of R	Array of all the parabolas inscribed in the principal threeside[b]
Curve of degree n. Since $n(n-1)$ tangents to C pass through each of the principal points	Curve of class $2n(n-1)$. Each of the principal straight lines will touch $n(n-1)$ times the curve

[a]My explanation: This depends on the fact that, in (i) to each straight line a conic second-degree surface corresponds.
[b]My explanation: This depends on two facts: 1) to a point a conic corresponds; 2) to R the line at infinity corresponds. Therefore, the conic corresponding to each point of R is a parabola because it is tangent to the line at infinity.

Now Steiner summarized this complex situation. There are three configurations he analysed: a) the planes *E,E'*; b) the stars D, D'; c) the planes C, C'. Within each configuration a dual and reciprocal law allows us to connect two by two the elements, as we have seen. However, since in all the three cases the relations are obtained through the use of the system [*A,A'*], a duality also holds between the elements of a)-b); a)-c); b)-c) (ibid., pp. 280–282).

Steiner expounded the dual relations between a) and b), whose most important are (ibid., pp. 282–283):

Plane E	Star D'
Point	Plane
Straight lines A—considered as the totality of its points—not passing through the three principal points $\mathfrak{r},\mathfrak{s},\mathfrak{t}$	Conic surface [A'] considered as the totality of its tangent planes—inscribed in the principal three-surface z',y',x'
Pencil of planes whose centre is not one of the three fundamental points	Array of second-degree conic surfaces touching a plane which is not one of the three principal planes
Conic circumscribed to the principal triangle	Straight line (or sheaf of planes) not lying in any of the three principal planes
Conic passing through two principal points (e.g. r,s)	Second-degree conic surface touching the two correspondent principal planes (z',y')

Later on, it should be considered the dual relation between a) and c) whose main correspondences are (ibid., pp. 283–285):

Plane E	Plane C'
Point	Straight line
All the points in one of the three principal straight lines	One of the three principal straight lines
Principal points r,s,t	All the straight lines of the three principal pencils $\mathfrak{r}',\mathfrak{s}',\mathfrak{t}'$
a) Straight line passing through none of the three principal points; b) set of its points	a) Conic inscribed in the principal triangle; b) array of its tangents
Line at infinity	A certain conic

(continued)

Plane E	Plane C'
a) Straight line through one of the principal points; b) set of its points	a) Point in one of the three principal straight lines; b) pencil of straight lines through it
Pencil of straight lines whose centre is not one of the fundamental points	Array of conic sections touching the straight line corresponding to the point (such a line is not one of the principal ones)
Limit straight line	Array of parabolas inscribed in the principal triangle
a) Conic circumscribed to the principal triangle; b) sets of points belonging to it; c) sets of straight lines touching it	a) Point; b) pencil whose such a point is the centre; c) array of conics through the point and inscribed in the principal triangle

Now there is the culminating point of this complex construction: as already pointed out the two straight lines—as every straight line—A,A' which are the bases of the just analysed systems of relations can be considered either as a set of their points or as axes of two planes' sheaves. Suppose that, instead of being skew, such straight lines are coplanar. Then, if regarded as dotted straight lines they generate a plane E_2', and as axes of two sheaves of planes, they generate the star of planes passing through the intersection point of the planar projection of A and A'. Suppose now that the two straight lines A,A' be, once again, skew and that their system be indicated by R. We will have three systems of straight lines: R, E_2, D_2 (ibid., p. 287). These three systems of straight lines are correlated by specific relations in which to a straight line in one of them a straight line in the other one corresponds. The fundamental principle of such relations, on which the others are based, is the following one: given a straight line a in R, which passes through the points $\mathfrak{a} \in A$ and $\mathfrak{a}' \in A'$ and in which the two planes $\alpha \in A$, $\alpha' \in A'$ mutually cut, then, to a in E_2 a straight line $\mathfrak{a}\mathfrak{a}'$ corresponds which is the projection of a on E_2 and to a in D_2 a straight line $\alpha\alpha'$ corresponds which is the intersection straight line of the two planes α and α', projection of a on D_2 (ibid., pp. 287–288). For example, given a system of straight lines in R which generates an one-sheeted hyperboloid, its corresponding figure in E_2 will be a conic section enveloped by the projections on E_2 of the straight lines generating the hyperboloid and which touches the straight lines A,A', while in D_2 the corresponding figure is a second-degree conic surface passing through the two axes A,A'.

An explanation is here appropriate. Consider, as Steiner did, a straight line which lies on the two skew straight lines A and A' cutting them in $\mathfrak{a}$ and $\mathfrak{a}'$. Be a the straight line connecting $\mathfrak{a}$ and $\mathfrak{a}'$. Each couple of straight lines A and a as well as A' and a generate a plane. The two planes mutually cut along the straight line a. Now, move the whole configuration so that A and A' become coplanar. The new position of the straight lines A and A' determines, hence, the straight line which corresponds to a. Steiner named $\mathfrak{a}\mathfrak{a}'$ such a straight line, which might create an ambiguity, but, in fact, this new straight line is that transformed of a after the movement of A, A'. Such a straight line, obviously, belongs to the plane E_2 produced by the straight lines which are those transformed of A and A'. *Mutatis mutandis*, the same reasoning applies to the relations between R and D_2.

It is possible to prove that the following relations hold between the system E_2 and the other two systems, being *e* the straight line that passes through the two points where *A* and *A′* cut E_2 (ibid., pp. 290–291). Let us name Π′ the first system of relations and Π″ the second one:

Plane E_2	System *R*
Straight line	Straight line
Point *B* considered as central point of a pencil	One-sheeted hyperboloid passing through *A,A'*
Straight line considered as set of its points	An array of a one-sheeted hyperboloids passing through the three straight lines *A,A′,e* and a fourth straight line
Straight line at infinity	All the hyperbolic paraboloids passing through the straight lines *A,A′,e.*
...	...

Plane E_2	Star D_2
Straight line	Straight line
Point	Plane
Straight line considered as the set of its point	A sheaf of planes
Conic touching the principal straight lines *A,A'*	Second-degree surface passing through the principal straight lines *A,A'*
...	...

Finally, Steiner sectioned the star of planes E_2 with a plane C_2, thus obtaining the following relations (I name Π such a system) (ibid., p. 291):

Plane E_2	Plane C_2
Point	Straight line
Straight line	Point
Point and straight line through it	Straight line and point lying on it
Curve; point of the curve; tangent to the curve	Curve; tangent to the curve; contact point curve-tangent
Curve of degree *n*	Curve of degree $n(n-1)$ or of class *n*
...	...

If the two planes E_2 and C_2 coincide, the system of relations Π establishes a correlativity among points and straight lines of a plane and if such a transformation is involutory it is a *polarity*. Therefore, the whole theory of polarity is included as a specific case in the far more general theory presented by Steiner, who explicitly highlighted this feature of his works. For we read:

> The preceding system of relations (Π) contains the fundamental propositions on which the so called "*Théorie des polaires reciproques*" is founded, theory that is usually presented through an auxiliary conic section, where necessarily the two systems of figures lay in a plane (namely the planes E_2 and C_2 are superimposed). Here these properties are presented in

a general manner, that is independently of a conic section and they are a special case of the previous system of relations Π''.[134]

Therefore, the theory of projective transformations developed by Steiner allowed him to treat as a specific case the theory of polar reciprocity, namely the basic theory on which, in French, projective geometry was developing.

Commentary: The fundamental section of the *Systematische Entwicklung* analysed above highlights several features of Steiner's work and is useful in a comparison with Chasles because it allows us to clearly detect the common ideas, discoveries and techniques of the two mathematicians as well as their differences:

1) Steiner perfectly understood that the basic concept of geometry is that of *transformation*. For his work can be interpreted as that in which all the characteristics of the projective transformations are examined. The projective generation of the conics and of the ruled quadrics is a clear indication that he was trying to identify all those properties of the figures whose common root relies on their projective nature and which remain unmodified through a projective transformation. It is clearly identified as a transformation in which the cross ratios are conserved. He had a clear idea of the distinction between a) *projective transformations*, where, in the case of the ruled quadrics, you can speak only of the one-sheeted hyperboloid; b) *affine transformations*, which, in the case of the quadrics, allow you to consider the hyperbolic paraboloid as a limit case of the one-sheeted hyperboloid. You reach such transformations specifying the nature of some specific objects, typically the elements at infinity; c) *metric transformations*, which are even more specific than those affine and, for example specifying some conditions of perpendicularity, allow you to obtain figures as the equilateral one-sheeted hyperboloid. This research on the transformations is particularly important insofar as space geometry is concerned. Steiner inaugurated, hence, a new era in projective geometry, that in which the projective transformations are identified as the most general ones, whereas the others are obtained by specifying the nature of some elements. He made this in a systematic manner. Therefore, although he was less interested than Chasles in basing the metric properties on those graphic, nonetheless, at the end of his work Steiner was driven, almost naturally, to identify the projective transformations as those most fundamental and general within geometry.

   If one compares Chasles' and Steiner's conceptions, it should be noted that Chasles' too considered the concept of transformation as that fundamental in

[134] Steiner named β and γ the systems of relations I have preferred to indicate by Π'' and Π. Ibid., p. 292: "Das vorstehende Beziehungssystem (Π) enthält übrigens die Fundamentalsätze auf denen die sogennante "*Théorie des polaires reciproques*" beruht, welche Theorie gewöhnlich mittelst eines Hülfkegelschnitts dargestellt wird (§ 44), wobei nothwendigerweise beide Systeme von Figuren in einer und derselben Ebene liegen (d.h. die Ebenen E_2 and C_2 liegen aufeinander). Hier stellen sich diese Eigenschaften auf allgemeinere Weise, unabhängig vom Kegelschnitt, dar, und zwar, wie schon bemerkt worden, nur als besonderer Fall des obigen Beziehungssystems (IV, 2, Π")". Italics in the text. Steiner also pointed out that Möbius first in his *Barycentrische Cälcul* had recognized the possibility to found the theory of polarity independently of the reference to a conic.

geometry. We have already seen that, as Steiner did, he identified the projective transformations as those in which the cross ratio is conserved. Specifically with regard to his generation of the one-sheeted hyperboloid, the genesis he proposed is, in substance, analogous to Steiner's. For, as we have seen, the latter claimed that "the projective radiuses $a,b,c,d,\ldots$ of any two skew projective straight lines A,A' are cut by an infinite number (*unzählig*) of straight lines A'',A''',A'''',... such that each of these straight lines cut all the projective radiuses $a,b,c,d,\ldots$ if it cuts three of them". Chasles wrote: "If four straight lines lay on two fixed straight lines in space, so that the anharmonic ratio of the segments they form on one of these straight lines is equal to the anharmonic ratio of the segments they form on the other, every straight line which lay on three of these four straight lines will lay on the fourth, as well" (Chasles, 1837a, p. 306). They also proved the reciprocal proposition because Steiner demonstrated that "Each two straight lines belonging to each of the two arrays are projective and the straight lines of the other array are their projective radiuses". Chasles expressed exactly the same concept when he stated and proved that: "When each of four straight lines lays on three fixed straight lines, which are posed in any position in space, the anharmonic ratio of the segments they form on one of these three straight lines is equal to the anharmonic ratio they form on each of the other two" (ibid.).

Therefore, in both cases, the projective transformations are assumed as the basic elements to generate the figures. The two authors analysed completely the projective nature of the second-degree curves and surfaces, but they also spread the seeds to deal with the projective properties of algebraic curves and surfaces of higher degree. As already pointed out, Steiner worked on the forms of the three species, whereas Chasles was a little bit less sensible to develop a complete classification of such forms as Steiner did, but the approach of these two great mathematicians was, in substance, analogous and quite innovative.

A further fundamental convergence point of Chasles' and Steiner's conception regards their ideas on the theory of reciprocal polars. As it has been recalled, this theory was the milestone of projective geometry until the beginning of the 30s and, due to its remarkable applications, it remained important also afterwards. However, from a foundational standpoint, Chasles and Steiner proved that the theory of reciprocal polars enters as a specific case within the theory of projective reciprocal transformations or reciprocities. Chasles was extremely direct and clear in showing that reciprocities can be reduced to the study of the anharmonic ratios of the reciprocal figures and that polarities are involutory reciprocities. He expounded the situation in his general and correct terms. I have explained it in Sect. 2.2 and will come back on the subject in Sect. 3.3 *Explanations and commentaries* 2. In this respect, Steiner was less clear. For he developed the transformation I have referred to immediately before these commentaries and proved that the theory of reciprocal polars is a particular case of such a transformation, within which the concept of fundamental triangle of a polarity is also included. This transformation was developed by Steiner to show the power of his projective method, but, at first reading, it might appear as an example among many others—and indeed it is. It is only by reading carefully such an example that one sees one of its more significant aims: to show that the theory of reciprocal polars can be included within a particular composition of projective transformations, as those showed by Steiner. In this case, Chasles was far clearer and Steiner probably more profound, but the aim of the two authors was analogous.

2) A particular remark concerns the concept of *foundation*. In this context, it is necessary to distinguish two meanings of this term: a) Steiner developed a foundational programme for projective geometry because he created the concept of cross ratio; he proved that in the projective transformations it is conserved and developed his considerations on this basis, with all the enclosed characteristics: projective forms of the three species, theory of conics, theory of quadrics, general theory of projective transformations and so on. Within this picture, the idea of considering the affine and metric properties as a specification of those projective also enters; b) Chasles, in the same period as Steiner, also developed a similar programme, as we have seen. However, his programme was far broader than Steiner's and far more radical: for Chasles tried to reduce the metric properties to those projective—both in reference to metric transformations and to the fundamental metric concept of measure of a segment—before inventing the concept of anharmonic ratio, as seen with his theory of parabolic transformation. One of the main points of his programme was exactly to find a projective characterization of the metric notion of distance, which does not exist in Steiner or in any other mathematician until Cayley and Klein. Furthermore, as we will see, he extended his foundational ideas on projective geometry until considering several parts of physics as reducible, ultimately, to projective considerations. He also developed a philosophy of duality which relies on a projective ground and which should be the basis of a new natural philosophy neither monistic, nor pluralistic, but dualistic. In this sense, Chasles' programme was radical, unique and different from that of any other mathematician and, thence, worth being known in detail.

2.4 Conclusive Considerations

In these conclusive considerations, I would like to discuss three significant aspects related to the foundations of a mathematical discipline; the former concerns Chasles' idea of the demonstrative bases of geometry. The second one regards the notion of mathematics' foundation itself. The third one deals with Chasles' general view of geometry. Finally, I will develop a comparison between my view of Chasles' foundational programme and the view expressed by Michel in his remarkable work (Michel, 2020a).

Chasles and his idea of demonstration. To face Chasles' concept of demonstration, a good guide is Chemla (2016), where the author highlights the value of generality in Chasles in an effective manner. In particular, it is true that one of Chasles' main concerns was to show that different propositions and parts of a theory should derive from a single or from few sources (Chemla, 2016, p. 51). This implies that, given some fundamental theorems, many other truths are deducible as their corollaries (ibid., p. 61). We have seen that the cross ratio was one of such main sources. As Chemla herself points out, the notion of transformation, which is not completely separated from that of cross ratio, is a further source from which Chasles derived many of his results and which, more generally, structures the whole of his view of geometry. Granted this general correct view, the author states that:

> For Chasles, the roots of mathematical knowledge are provided by "general propositions" from which the others are derived as mere "transformations" or "easy consequences." In contrast, an axiomatic-deductive organization places emphasis on another kind of starting point and stresses the rigor of the derivations rather than their generality (ibid., p. 55, n. 15).

In my opinion, it is true that "For Chasles, the roots of mathematical knowledge are provided by 'general propositions' from which the others are derived as mere 'transformations' or 'easy consequences'", but it is false to argue that this represents a contrast with an axiomatic-deductive approach.

In the Conclusion of her paper, Chemla speaks of the difference between Chasles' standpoint and a deductive-axiomatic one. Therefore, a substantial difference would exist between an axiomatic approach to geometry and an approach based on the transformations of few basic propositions through given criteria. Actually, it seems to me that things are more complicated than Chemla argues: it is true that, from a geometrical point of view, the idea of working on the transformations is new and far more general than an approach *à la* Euclid. But this concerns the fact that projective geometry is more general than metric geometry and, hence, it allows more easily than metric geometry to focus on the importance of the transformations. However, there is no contradiction and no opposition between the two approaches. Chasles belongs to the first phase of the geometry of transformations. It is, hence, understandable that he did not develop a precise axiomatic theory of transformations. As the future theory of the transformation groups developed by Lie, Klein and Poincaré shows, the axiomatic approach is perfectly consistent with the idea that the transformations are the bases of geometry. Chasles did not develop an axiomatic approach because he was in an initial phase of such a new theory and no theory (at least until the twentieth century) is born on an axiomatic basis; the axioms are introduced only after the main truths of the theory have been discovered and one aims at offering a precise deductive structure to the theory itself. It is only natural that in an initial phase intuition and general views guide the development of the new theory. Afterwards, the axiomatization takes place. For, e.g., Euclid's *Elements* with their axiomatic structure were written after that Greek geometry spent a long phase in which the theorems were likely proved without explicitly resorting to an axiomatic system. On the other hand, the fragmentation of the truth in an axiomatic system has the aim to make the deductions possible, but the guiding truths of a theory are independent of axioms, which concern their formal structure. For example, Enriques claims that, among the two guiding truths of the first book of the *Elements*, there are the property of sum of the angles of a triangle and the Pythagoras theorem, whereas other propositions are the connection-elements of the chain with leads to the guiding truths (Enriques, 1922, reprint 2003, p. 6). Furthermore, in the period in which Chasles wrote, the axiomatic method was not of general—or at least of exclusive—

use in mathematics. If you look at the production of that time, not all of the mathematicians formulated explicitly or implicitly a clear system of axioms on which to rely. For example, if you consider mathematical analysis, the 20s of the nineteenth century were the period in which a more rigorous approach to its foundations began, but certainly one cannot claim that the basic axioms of analysis were clear, well-stated and shared by the entire community of mathematicians. On the other hand, if one also considers Hilbert's abstract axiomatic approach to geometry, the deductive aspect is only one of those which are behind such an approach. The other significant aspect is that the idea of mathematical structure is privileged with respect to that of mathematical object and, for sure, the idea of founding a discipline on the transformations and the concept itself of duality is absolutely coherent with that of privileging the structures rather than the objects, a position which Chemla herself seems to share. Hilbert's analyses of the role of Desargues and Pascal theorems within geometry[135] are clear evidence of the fact that he did not deny the existence of truths which are more fundamental than others to connote the structure of a certain geometrical universe. The abstract axioms become the advanced means through which the role of such truths can permeate theory. It is a normal process of mathematics' progress that a set of propositions which appeared separated and needing of diverse and sometimes complex demonstrations are unified in a more comprehensive view so that many of such propositions appear as simple corollaries of new, more comprehensive ideas and theories. There are several examples: a remarkable one is given by the tensor algebra and the exterior algebra which allow us to obtain a unitary vision of the whole electromagnetism in the picture of Minkowski spacetime so as to obtain a covariant form of the Maxwell equations as well as the variational formulation of electrodynamics.[136]

In conclusion, I do not see a contrast between an approach to geometry based on axioms and one based on transformations. Certainly, Chasles tried to reduce several truths to more basic and less numerous ones (this is the basis of the foundational programme I am expounding) from which the proofs of the other propositions should follow almost as corollaries. It is also true that Chasles had a new idea of generality. However, this is not in conflict with an axiomatic approach, because in the axiomatic geometry there are also some propositions, principles and transformations which are recurrent in the proofs of the others. They represent, so to say, the conceptual skeleton of the theory. In the plane Euclidean geometry, for example such a skeleton might be represented by the properties of the parallel lines, the Pythagoras theorem, the theorem of the secant and the tangent, the properties of the angles at the centre and at the circumference and the theory of similarity (not by

[135] The fifth and the sixth chapters of Hilbert (1899) are dedicated respectively to Desargues' and Pascal's theorems and to their mutual interrelations as well as to their connections with plane and spatial axioms of projective geometry. Hilbert added many interesting considerations on the value and interpretations of these two theorems which he regarded as seminal truths for projective geometry.

[136] On these subjects, see, e.g., Sonego (2020, chapters 1 and 3). The author also offers an extensive list of references.

chance a transformation). In projective geometry, you can consider the cross ratio, its properties and its specific cases as the conceptual apparatus from which you can deduce many of the basic truths, if you are not interested in a purely graphic approach. I add that the two memoirs on duality and homography as well as Chasles' works on the ellipsoid attractions—only to mention some examples of his works—are full of demonstrations. This means that Chasles did not underestimate at all the role of the deductive process. It is true that he was convinced it might be remarkably simplified in the light of his general notions. It is also true that in the theory of the rigid body's movement and in enumerative geometry Chasles wrote several papers in which the basic principles are stated and lists of unproved theorems follow. However, this does not mean that he underestimated the role of demonstrations, but that he considered—I do not enter the question whether he was right—such propositions almost an immediate consequences of the principles and, thus, not needing a proof.

Foundations of mathematics. The debate on the foundations of mathematics does not characterize only the last 30 years of the nineteenth century and the twentieth century.[137] When a branch of mathematics reaches a broad series of results, it is only natural that the mathematicians wonder what the relations among the concepts they are introducing and the theorems they are demonstrating are. Foundational problems arise always after a discipline has passed through an inventive and innovative phase. They represent the reflexion of the mathematicians and, in a broader sense, of anyone interested in philosophy of mathematics and in its ontology, on mathematical activity. Probably the difference between the foundational debate at the end of the nineteenth century and the other ones is that in the former an absolute foundation of the entire mathematics was sought for; mathematics was considered as a whole, whereas the other debates concern some sections of mathematics. It is enough to think of the numerous discussions associated with the tumultuous development of the infinitesimal calculus in the seventeenth and eighteenth centuries. Part of the activity of mathematicians such as Bolzano, Cauchy and Dirichlet (just to mention some of the most important) in the nineteenth century was a foundational activity. Projective geometry between the end of the eighteenth century and the 1830s had a febrile development. The genesis of many projective concepts dates back to the Greeks (especially Apollonius and Pappus), and, at the beginning of modern mathematics, Desargues introduced many of the most important concepts of projective geometry and proved several of the main theorems. De la Hire was also important. However, projective geometry as a phenomenon which involved many of the great mathematicians dates from the beginning of the nineteenth century. This discipline had a remarkable philosophical or, at least, methodological character from the beginning. It was important not only to prove some theorems but also to consider

[137] Grattan Guinness, for example, points out that the mathematicians of the late eighteenth century were interested in foundational problems. The focus of their interest was to understand whether the approach based on infinitesimals or that based on limits was preferable for infinitesimal calculus. They did not restrict to this topic, but also reflected on the algebraic and geometrical aspects of calculus (Grattan-Guinness 1990, pp. 137–145).

the way in which they were proved. The dissatisfaction of Poncelet with the analytical methods, as well as his attempt to create a pure geometry as powerful as or even more powerful than analytical geometry, is well known. But the dichotomy pure/analytical methods is certainly not the only methodological aspect connected to projective geometry. The interpretation of duality is another one, which is strictly tied with the role of transformations in geometry. The attempt to construct systems of coordinates that expressed graphical properties better than ordinary Cartesian coordinates is also a liminal subject between mathematics and methodology of mathematics. Not to speak of the interpretations of the point at infinity and imaginary elements. Last but not least, the relations between graphical and metric properties are an important aspect, too.

Given this picture, I have argued that Chasles was one of the first mathematicians who tried to offer a foundational programme for geometry, where projective geometry was the basis for the rest of this discipline. He was deeply interested in grasping the fundamental concepts. It is not by chance that in the *Aperçu* he resorted to analytical methods, though he himself considered them less suitable for geometry than pure methods, but first of all, it was important to establish a precise foundation. Methodological purity is desirable, and, as in part we have seen and as we will observe in next chapters, Chasles sought for it in many parts of his production such as the ellipsoid's attraction. However, establishing the basic concepts of a mathematical discipline is even more important. It seems that Chasles believed in the existence of a geometrical universe founded on some basic concepts (one might speak of "conceptual atoms"), which have to be discovered. This universe is written in the language of the pure geometry, but, if analytical geometry is a useful means for grasping the nature of such a universe, in a first foundational phase it is a precious instrument to utilize. Chasles was not the only one who had foundational intentions in the period analysed in this book. Surely, at least Poncelet, Steiner, Möbius and Plücker had profound foundational interests. Steiner was the mathematician whose way of thinking was closer to Chasles'. Furthermore, the way in which Chasles and Steiner faced a foundational programme is different from Poncelet's. Though generalizations are always dangerous and criticizable at a deeper analysis, it is perhaps legitimate to claim that, starting from the end of the eighteenth century and during the entire nineteenth century, the development of projective geometry was characterized by different phases: in the first one, it is possible to insert Monge's descriptive geometry and Carnot's theory of transversals. Carnot's results have often a projective character, but they are obtained through the use of classical geometry, algebra and trigonometry; to the second phase belongs the set of researches which found their coagulation point in Poncelet's *Traité*. This phase is connoted by the introduction of purely projective methods, by the discussion concerning the use of analytical or synthetical methods within projective geometry and by an enormous set of results. The works by Chasles, Möbius, Steiner and Plücker belong to the third phase of projective geometry characterized by the beginning of a deeper analysis of the projective concepts with respect to the metric ones, by the predominant use of the cross ratio and by the introduction of the projective systems of coordinates. All these aspects can be considered as foundational and Chasles, as we have seen, gave essential contributions to all of them, making them to converge into an authentic foundational programme. Of course, a huge amount of new results was obtained in this phase as well, but a rational systematization of old results and of the classical concepts within projective geometry represents an aspect of this phase which is as important as the achievement of new results. The fourth phase is represented by the great work of Von Staudt, in which the projective concepts were made completely independent of any metric references, which still existed in the notion of cross ratio. The fifth phase is that represented by the work of Cayley and Klein with regard to the projective metrics, also including the projective metric of not Euclidean geometries. These mathematicians brought to a conclusion an order of idea along which Chasles had taken several fundamental steps, though not fully recognized by Cayley and Klein. Finally, a sixth phase can be identified: that of the explicit axiomatization of projective geometry. It seems to me that the two fundamental works of this last phase of the productive period of projective geometry are the *Lezioni di geometria proiettiva* by Enriques (Enriques, 1898, second edition 1904) and "A Set of Assumptions for Projective Geometry" by Veblen and Young (1908). Here the authors, following, formalizing and generalizing an order of ideas which dates back to Von Staudt, Dedekind, Enriques, Vailati, Fano and Huntington, arrive at the formulation of a set of axioms and assumptions for projective geometry functionally analogous to those developed by Hilbert for metric geometry. Specifically, the authors determine a system of numbers constructed on the basis of projective concepts which is isomorphic with the system of real numbers (ibid., pp. 368–369), so that a system of projective coordinates expressed by real numbers is fully justified. They also face the problem of the projectivities in the complex space and that concerning the mutual independence of their assumptions.

Chasles' view of geometry. What was Chasles' view of geometry? Was it the science of extension or a more abstract discipline? The answer to this question is connected with his foundational conception but is not identifiable with it. My idea is that, according to Chasles, geometry is the science of an extension E which is broader than the visual extension E_v and which includes objects such as imaginary elements and elements at infinity. The visual space E_v is a subset of E and the ordinary metric space E_m is a subset of E_v when the nature of some elements is specified. As Nagel writes:

> He [Chasles] distinguished between two types of geometrical configurations subject to general conditions: in the first, certain parts and positions of the figure, upon which the construction of the figure does not however depend, are "real" and "palpable"; in the second, these parts no longer occur, so that with respect to the first case they have become imaginary, even though the general conditions for constructing the figure have remained the same (Nagel, 1939, p. 164, Note 35).

As to visual space, it has a double nature: it can be seen either as a space of points, which is the traditional view, or as a space of planes, which is the new view opened by Chasles. With regard to the plane, it can be seen either as dotted plane or as lined plane (Chasles, 1837a, pp. 408–409; Nagel, 1939, p. 187). What allows to pass from the dotted space to the planned space is duality, which, however, has a broader range of applicability than that to extension. As Nagel recalls:

> The most general transformations in the sense that CHASLES thought of them are the "linear" or projective ones; and he pointed out that these transformations entail dual theorems not only in geometry but in pure algebra as well. The obvious consequences of this analysis are that GERGONNE's belief in the principle of duality as based upon the alleged nature of extension is certainly mistaken, since the principle could be shown to follow from considerations which did not apply exclusively to space. (Nagel, 1939, p. 186).

We shall see this to be fully coherent with Chasles' idea that duality is a universal law which goes beyond projective geometry. In particular, with regard to geometry, duality is applicable also in cases in which some elements become imaginary and, thence, it is not restricted to E_v. It is extended to the whole E. This is clear in the analysis made by Chasles of the imaginary in geometry (Chasles, 1852). Therefore, duality represents a meta-level with respect to the distinction between real and imaginary elements because it transcends such a distinction.

Duality is the property of a transformation. What is its role? It allows us to connect propositions of different spaces, e.g. E and E_v, or, within the latter, dotted space and planned space, provided that these propositions have the same syntaxis. Therefore, from this perspective Chasles' geometry is not only a geometry of figures, but it is a more abstract discipline. It is a geometry of transformations. They permit to interconnect the figures and they need a theory separated and more abstract than the usual theory concerning the figures they interconnect. The new theoretical approach is based on the concept of cross ratio which can give geometry the same abstractness and generality as algebra.

With regard to the imaginary objects, it seems that they have a double role in Chasles' view: when one works in E_v, they can be employed as symbols which are used during the demonstrative process, which are contingent elements and disappear in the final result. On the other hand, at least in Chasles (1852) they live in the extended spatial environment E. In this case, the imaginary elements do not belong exclusively to the contingent parts of a figures, but, at least with regard to two conjugate elements, they are a constitutive part of the geometrical world E. Perhaps one might hazard the guess that, while single imaginary elements are only object belonging to the symbolism of geometry without a real ontological value, two mutually conjugate imaginary elements belong to E. A pair of conjugate imaginary elements is a single object, a single molecule in E, composed of two atoms.

These observations confirm that, basically thanks to the notion of cross ratio, Chasles progressed in the theory of transformations compared to his predecessors, notably Poncelet. His geometry was more abstract and more structures—and transformations—centred than that of Poncelet. However, the structures and the transformations were still a means to develop the mathematical research on a world of objects which exists independently of them. After having analysed Chasles' conception of geometry and of its foundation in this chapter, one can claim that Nagel's continuist vision, according to which the abstract view of geometry as a hypothetico-deductive system was reached at the end of the nineteenth century through a process of which Chasles is an important step can be confirmed. Nagel tends to overemphasize the road of geometry towards the modern approach and sometimes tends to interpret the acquisitions of the early nineteenth century in too modern terms. For this reason, I have pointed out that the geometrical object plays still a fundamental role in Chasles, though such an object is not anymore that of classical geometry. This notwithstanding, Nagel's conception captures significant elements of the historical evolution of geometry, which is an important aspect because, otherwise, no synthesis work is possible, and the historical reconstruction is too compartmentalized. It is only natural that in general syntheses such as that of Nagel, not all the tiles of the mosaic are in order. Nonetheless, the general view can be interesting, stimulating and basically correct even if some of the single steps of the narration are debatable.

I would like to add a brief comparison between Chasles and Plücker's conception of geometry. It is well known that Plücker developed his ideas on duality and his system of coordinates in the same years as Chasles. He conceived the line as a locus of points and the point as a locus of lines. From his view on the relations between systems of coordinates, immediately an advanced conception of spaces follows: the dimensionality of a space depends on the constitutive element one considers. For example, the plane is two-dimensional if the elements are points or lines, but it is five-dimensional with respect to conics because five independent coordinates are necessary to specify a conic (Nagel, 1939, pp. 189–191). Therefore, Plücker's view is more abstract and, in our perspective, more advanced than that of his contemporaries. However, a consideration is necessary: Plücker achieved his results on duality and, more in general, on projective geometry, without the constraint of being tied to methods that are as synthetic as possible. He had an analytical-algebraic approach. In contrast to this, Chasles had a double purpose, not a single one: 1) like Plücker, he aimed at generalizing and specifying (it is not a contradiction) the results of the previous geometers. In particular, both Chasles and Plücker had the idea that the element of space is not necessarily the point and that duality has to be framed in a more general theory of which that of reciprocal polars is a particularization; 2) however, the former also had the goal of producing a synthetic theory that met the mentioned requirements. This is a difficult challenge and possibly it is one of the reasons why the level of abstractness and precision achieved by Plücker seem superior. He had less constraints.

Finally, with regard to Steiner, I have already pointed out that his view of geometry is less tied than Chasles' to the concepts of space and extension and is connected to the more abstract notion of "forms of a certain species", which opens a new perspective in geometry. The transformations act between forms and they represent the basis of his new geometry. The forms as objects exist independently of the transformations. Thus, like Chasles: 1) the theory of transformation is the basis of geometry; 2) however, the objects (spaces for Chasles, forms for Steiner) exist independently of transformations. The forms of Steiner are more abstract and surely more suitable for projective geometry than Chasles' spaces. On the other hand, Chasles offered a systematic treatment of the imaginary objects, especially of the conjugate elements, whereas this aspect is not present in Steiner's *Systematische Entwicklung*. Traditionally this has been interpreted as a limit of Steiner's approach. Blåsjö claims, instead, that this is due only to the fact that, among Steiner's purposes, the treatment of the imaginary elements was not included (Blåsjö, 2009, pp. 21–22). I do not deal with this question because it is not crucial for my research.

Comparison with Michel's Of Words and Numbers. In his work (Michel, 2020a) the author speaks of a programme developed by Chasles. On some important issues, the opinion of Michel does not collide with the one I expressed in Bussotti (2019) and which I am generalizing and deepening in this book. However, in other equally important respects, my interpretation of the development of projective geometry in the nineteenth century does not coincide with his. This difference is partially due to the fact that my approach is that typical of a historian of mathematics, while I would define Michel's work as belonging to historical epistemology. Furthermore, the focus of my research covers, for the most part, a period prior to the foundation of the enumerative geometry by Chasles in 1864–1867. Instead, Michel focuses exactly on Chasles' enumerative geometry, though he dedicates a long series of profound considerations to Chasles' work before this period, especially concerning the *Aperçu* and the *Traité de Géométrie supérieure*, and he concludes his work describing a wide picture of enumerative geometry until the end of the nineteenth–beginning of the twentieth century. However, even taking into account the differences that arise from our two approaches and the different periods of Chasles' activity that Michel and I have considered, as well as the diversity of topics addressed, there are some important issues on which Michel and I disagree. Thus, I would like now to offer a picture where initially I discuss Michel's arguments on which I agree and afterwards those on which I do not agree.

Before embarking on this discussion, I would like to recall that one of the most important theses of Michel's text—argued right from the abstract—is that one of the main purposes of Chasles' foundational programme is to offer pure geometry the same generality as algebraic language. Michel traces this research programme to Chasles' lectures of Higher Geometry at the Sorbonne in 1846, although he also finds many aspects of it in earlier works and, especially, in the *Aperçu.*

Michel claims that one of Chasles' main tenets—as is transpires from the *Aperçu*—is to offer pure methods so general that each one can apply them to prove new theorems without an excessive effort, namely without any necessity to be a genius. These methods have the advantages, with respect to the analytical ones, to be intrinsic (namely independent of system of coordinates) and more easily applicable. For the analytical methods are not always direct and the transcriptions of geometrical objects into equations might be not as automatic as one could think. Chasles' focus on generality also derives from his education at the *École Polytechnique,* where the value of generality was inculcated in students (Michel, 2020a, pp. 20–21).

Michel focuses on the idea that generality was an epistemic value in the polytechnical education also while referring to the role of descriptive geometry and mathematical analysis as taught at the *École* (ibid., pp. 27–29). The author refines furtherly the concept of generality typical of the polytechnical milieu and highlights that Lagrange's, Chasles' and Comte's historical epistemologies express three different conceptions of generality:

> For Lagrange, this generality was attained through the unification of historical results within a single formula and principle [the principle of virtual velocities]; for Comte, history served the identification of the general and fundamental principles of a given science. Similarly the *Aperçu historique* opens with what Karin Chemla called a 'diagnosis about the limits of ancient geometry'; and these limits consist mainly in the lack of generality that can be found in both the mathematical statements and the geometrical methods used by ancient Greek geometers" (ibid., p. 32).[138]

What Michel claims is absolutely true. It seems to me appropriate to add only an observation: in the *Aperçu*, while referring to geometrical methods which can be as general as those analytical, Chasles is thinking of the procedures based on the anharmonic ratio. As we have seen, they have considerable generalizing and unifying power, but when applied to concrete mathematical problems they also require a certain amount of discernment and acuity. That is, the concept of having such powerful methods that can solve difficult problems through relatively easy and mechanical steps is an ideal which rarely takes place in the practice of mathematics, although, obviously, some methods permit to solve given classes of problems more easily than other methods.

Michel stresses that Chasles criticized Greek geometry because it lacked of generality under two respects: 1) Greek geometry was too figure-bound. Although the Greek geometers saw their figures as diagrams applicable to all the configurations a certain theorem deals with and not only to the physically drawn one, the need of a visual support was still a limitation; 2) the methods of the ancients (as the exhaustion), though being based on a directive idea, lacked systematicity because any application needed a particular and specific intuition, not included a priori in the method (ibid., pp. 35–37). This is a correct exposition of Chasles' ideas on Greek geometry.

A further important aspect of Chasles' epistemology in the *Aperçu* is the concept of naturality: geometry has methods which are natural and which allow simple and general proofs. In the case of projective geometry such methods are those of duality and homography, in turn based on the concept of anharmonic ratio. Michel writes:

> [...] nature provides the germs for simple methods, which therefore must be general. Conversely, by following natural methods, one will walk upon simple paths, thereby reaching generality (ibid., p. 39).

Therefore, generality and naturality are the two main tenets of Chasles' epistemology. This also explains why he re-demonstrated several theorems: they were not located in their natural place, which Chasles' proofs, instead, did (ibid., pp. 43–44).

What claimed by Michel is true. Only an observation has to be added: of course that of "nature" and "natural" is not a precise concept. It seems to me that it can be divided into two sub-concepts: the first one is that of intrinsicity. It is perhaps here appropriate to add a historical observation: the problem of the mathematical methods' intrinsicity did not originate in geometry, but in arithmetic. In a letter sent in February 1657 to the English mathematicians as a challenge to solve the

---

[138]The reference to Chemla is Chemla (2016, p. 50).

so-called Pell equation (in fact, Fermat equation), and significantly entitled *Second défi de Fermat aux mathématiciens*, Fermat wrote:

> Hardly anyone proposes purely arithmetical questions, hardly anyone understands them. Doesn't this perhaps depend on the fact that arithmetic has so far been treated geometrically rather than arithmetically? In fact, most of the volumes of the ancients and moderns show this. Diophantus himself shows this. He departed from geometry slightly more than the others insofar as he restricted analysis to rational numbers only; Viète's *Zetetica* are a sufficient proof that this part of arithmetic is not free from geometry. For, in such work Viète extends Diophantus' methods to continuous quantity, namely to geometry".[139]

Fermat is lamenting the fact that number theory is still treated with geometric methods and not, instead, with methods that are its own (see Bussotti, 2006a, p. 25, pp. 177–180, pp. 187–213). In a sense, Chasles develops the same observation, but in an almost opposite direction: in his epoch geometry was not treated with geometrical method because of the analytical approach. Fermat vindicated a heritage of arithmetical methods for number theory. These methods had to be neither geometrical nor analytical, at least in modern terms, namely tied to the continuum. It can be added that Archimedes' double demonstrations, by exhaustion and by the mechanical method, are connected to the most appropriate way to offer a mathematical demonstration. However, the reasons why Archimedes did not consider geometrical the proofs given by the mechanical method are still under discussion, while Fermat is crystal clear in his methodological assertion.

Coming back to our main subject, a geometrical method is intrinsic if the proofs based on it must not resort to system of coordinates. The notion of "being intrinsic" is, hence, one component of Chasles' concept of naturality. But it seems to me that there is something else which is connected to the metaphysics, to the ontology of geometry rather than to its epistemology: Chasles imagined the existence of a geometrical world with its objects (the figures), its relations and its transformations. A method and a set of symbols (including figures) are the more natural the more the symbols mirror the objects of the geometric world and the more the method induces on the symbols a language whose syntaxis reflects the relations and transformations of the geometric world. According to Chasles, the pure methods are more natural than the analytical ones in this sense too. Therefore, "naturality" is both an epistemic and a metaphysical concept in Chasles.

According to Michel the programme of generalization of geometrical methods by Chasles continues in the *Traité de géométrie supérieure* where he treats anharmonic ratio, homographic division and involution with a more perspicuous language. An important principle is that of signs, which allows Chasles to introduce the couples of

---

[139] Fermat (1891–1922, 2, pp. 334–335). "Questiones pure arithmeticas vix est qui proponat, vix qui intelligat. Annon quia Arithmetica fuit hactenus tractata geometrice potius quam arithmetice? Id sane innuunt pleraque et Veterum et Recentiorum volumina; innuit et ipse Diophantus. Qui licet a Geometria paulo magis quam caeteri discesserit, dum Analyticen numeris tantum rationalibus adstringit, eam tamen partem Geometria non omnino vacare probant satis superque *Zetetica* Vietaea, in quibus Diophanti methodus ad quantitatem continuam, ideoque ad Geometriam porrigitur".

conjugate imaginary elements (ibid., p. 61–69). After further analyses of the methods used by Chasles in his *Traité*, Michel concludes that Chasles' higher geometry is a language which reaches the same generality as algebra without losing intrinsicity (ibid., pp. 81–82). These opinions can be shared.

Michel points out that the idea of generality pervades Chasles' production from his first works until his contributions to enumerative geometry. Two principles were fundamental in order to offer general methods of construction and demonstration. Chasles named both of them "principle of correspondence", but, in fact, they are two different assumptions: the former concerns projective geometry and the latter enumerative geometry. The principle of anharmonic correspondence for projective geometry is expressed by two propositions: 1) be given a problem or a theorem where no transcendental function is involved. Consider two series of points on two straight lines or on a single straight line. If to a point of the first series a single point of the second series corresponds and, conversely, one is allowed to conclude that the two series are homographic. Consequently, the anharmonic ratio of four points of the first series is the same as that of the corresponding points of the second series; 2) be given a problem or a theorem where no transcendental function is involved. Consider two series of points on two straight lines or on a single straight line. If it is proved that to a point of the first series only one point of the second series corresponds, but to a point of the second series two points of the first series correspond simultaneously and indistinctly, then all these couples of points are in involution and correspond anharmonically to the single points belonging to the second series (Michel, 2020a, pp. 103–104; Chasles, 1855, pp. 1098–1100). As for the correspondence principle devised by Chasles to solve questions of enumerative geometry, it is based on two dual lemmas. I refer to the first which reads as follows "Let x and u be two series of points, on a straight line L, between which there exists a correspondence which to a point x match α points u, and to a point u match β points x, then the number of the correspondence's fixed points is (α+β)" (Michel, 2020a, p. 152).[140] The other lemma can be obtained by mutually replacing the terms "straight line" and "point". Michel stresses that, although the two principles are different, "this method of proof [the principle of correspondence applied to enumerative geometry] is a generalized version of the principle of anharmonic correspondence" (ibid., p. 152). In the course of his work, Michel shows how this principle was applied in enumerative geometry and that it is the basis of Chasles' research programme based on the epistemological tenet of generality. These descriptions of Chasles' conceptions and methods that Michel offers can be shared.

Now I will consider the aspects of Michel's work on which I do not agree. These aspects are important because the value and the nature of Chasles' foundational programme depend on their interpretation. Generally speaking it is seems to me that

[140] Chasles (1864b, p. 1175): "Lorsqu'on a sur une droite L deux séries de points x et u, tels, qu'à un point x correspondent α points u, et à un point u, β points x: le nombre des poinys x qui coincident avec des points correspondants u, est (α+β)".

Michel underestimates the importance of projective geometry in the nineteenth century as well as the continuity of its history in this period. Michel claims:

> Much has been written on the renewal of pure (or synthetic) Geometry in early 19th century France. This historical episode is one which took place during the two or three first decades of the nineteenth-century, and it saw an effort from various geometers to broaden the scope of geometrical methods by introducing concepts such as transformations and projections, often in order to enable geometry to compete with analysis. The main actors of this historical narrative are typically considered to be Gaspard Monge and Lazare Carnot, but also Charles Dupin (1784-1873), Jean-Victor Poncelet (1788-1867), Charles Brianchon (1783-1864), and of course Chasles. In first approximation, we shall refer to this collective as the "French tradition of pure geometry", a label whose relevance we shall discuss shortly. Already at the end of the nineteenth century, mythical retellings would present this historical episode as the birth of a new branch of mathematics, namely projective geometry. A key text responsible for this historiographical narrative is Klein's famous Erlanger Programm. This French tradition, Klein suggested, "provided a sound foundation for that distinction between properties of position and metrical properties". By subordinating the study of metrical properties (such as lengths and areas of figures) to the study of properties which are not altered by projection (such as the fact that two figures intersect, or are in certain specific configurations), Poncelet, Chasles, and their colleagues allegedly discovered a new geometry, which would later be made entirely free from metrical concepts (such as lengths or angles) by German geometers such as Von Staudt (Michel, 2020a, pp. 23–24).

In order to justify the strong thesis that Klein's historical reconstruction—which I consider correct—is a myth, Michel adds:

> However, recent scholarship has definitively put the lie to this reading, and revealed it as decidedly anachronistic. None of the authors of this French tradition suggested an exclusion of metric properties from geometry (ibid., p. 24).

It is certainly true that the metric properties were not excluded. However, this observation—if interpreted correctly—is favourable to Klein's thesis that the French mathematicians founded projective geometry between the end of the eighteenth century and the 1830s. First of all, there is no doubt that Poncelet, as we have seen, explicitly distinguished between graphic and metric properties and that he was interested in those properties which are conserved by sections and projections. With regard to Chasles, his study of the parabolic transformations is an emblematic example of precisely what Michel thinks did not happen: the subordination of the study of metrical properties to projective properties when the elements of a transformation are specified (e.g. when the ordinary elements are distinguished from the elements at infinity, or when one specifies the type of conic to which a polarity is referred). I have analysed all the details of this idea. Another example drawn from Chasles' work is his treatment of the problems connected to the centre of the mean distances. We have seen that some metric properties can be obtained when the point μ goes to infinity. Thus: 1) the distinction between metric and graphic properties was well established in Poncelet's and Chasles' works; 2) given the set of metric properties, it was known which of them are conserved by sections and projections; 3) the attempt to reduce the metric properties to particular graphic ones dates at least to Chasles' memoirs on parabolic transformations. This seems to me sufficient to claim that Michel's statement against Klein's interpretation is not acceptable.

Furthermore, while claiming that "However, recent scholarship has definitively put the lie to this reading, and revealed it as decidedly anachronistic. None of the authors of this French tradition suggested an exclusion of metric properties from geometry", Michel refers to Chemla (2016, p. 66). But, as far as I read, Chemla on p. 66 of the mentioned work does not deal with the problem of metrical properties and, more generally, in her paper the terms "metric" and "metrical" never appear. Thus, this reference seems inappropriate to me.

My observations on Michel's work, when I agree with him as well as when I do not, have the aim to highlight our different approaches to Chasles' foundational programme. Michel points out the values of generality and simplicity subtended to such a programme and does so with refinement, presenting the reader with a broad and profound view of the picture of Chasles' enumerative geometry. On the other hand, the attempt to reach generality and simplicity connotes every foundational programme. Certainly, it is necessary to specify the meaning of these two polysemous terms in any single context, which Michel makes very clearly as to Chasles. However, I claim that Chasles also had a more specific purpose than just to obtain maximum generality and simplicity from its methods: he aimed at posing projective geometry as the conceptual centre of the whole geometry and of large part of physics. Obviously, Michel cannot reach my conclusion because it is largely based on the acceptance of Klein's view which Michel considers mythical and false. It is perhaps possible to add that, in fact, the "imprinting" deriving from projective geometry also pervades Chasles' enumerative geometry. The two fundamental lemmas which give rise to the principle of correspondence in enumerative geometry are, as we have seen, two dual propositions in the projective sense of this term: they are obtained the one from the other by replacing the term "point/s" with the term "line/s". Furthermore, as Michel himself recognizes, the principle of correspondence of enumerative geometry has connections to the principle of homographic correspondence. Thus, the basic principles of Chasles' enumerative geometry have a projective syntaxis, though the meaning is not projective. Hence, in a methodological sense, projective geometry is important for enumerative geometry, too.

This general picture can be useful to give an idea of the broadness of Chasles' foundational programme with respect to the conceptions of the great mathematicians who were the protagonists of the development of projective geometry.

However, his programme was far broader because it included the idea that several fundamental parts of physics as well as a whole philosophical conception of the universe and, consequently, of the way in which we know the universe were dependent on projective geometry. In what follows, I will analyse the main features of this programme.

Chapter 3
Displacement of a Rigid Body

Abstract This chapter is devolved to the analysis of the way in which Chasles' foundational programme was extended beyond geometry. The problem of the movement of a rigid body is addressed. An initial brief historical introduction is followed by five sections, which explore the work of Chasles by dividing it into three main periods. Section 3.1 is dedicated to the first one, which corresponds to the year 1830. Section 3.2 regards, instead, the second period, which dates from 1831 to the publication of the *Aperçu* in 1837. The third section concerns a comparison between Chasles' conceptions and those developed by other mathematicians on this same subject as well as a contribution he gave in 1843. The influence exerted by his ideas and methods is also addressed in this section, which is useful to understand the complex relations between synthetic and analytical methods in geometry at that time. Chasles offered a complete and advanced treatment of the movement of a rigid body in a series of papers published between 1860 and 1861. This connotes the third phase of his thought on such a topic, to which the fourth section is devoted. Chasles offered almost no demonstrations of the propositions he conceived in this third phase, but other mathematicians embarked in such an enterprise. In this section we also provide an overview of their work. The Conclusive considerations follow as fifth section.

With regard to the movement of a rigid body, it is necessary to distinguish five aspects:

1) Distinction between infinitesimal and finite displacements. The latter problem is more difficult than the former.
2) The existence of formulas which, given the position of a rigid body, allow us to obtain the position after an infinitesimal displacement.
3) The same problem for finite displacements.
4) Beyond the formulas: to find general principles which permit to understand the displacement of a rigid body as a unitary phenomenon. In this case, within a foundational order of ideas, it is appropriate to distinguish three questions: a) the only geometrical properties of the displacements, namely the geometrical transformations characterizing the movement; b) the kinematical properties which regard the motion's velocity independently of its causes; c) the dynamical

P. Bussotti, *Chasles and the Projective Geometry*,
https://doi.org/10.1007/978-3-031-54266-4_3

properties connected to the forces which cause the motion. Though acceleration, in itself, is a kinematic quantity, given its connection with the forces, it has been treated, as a matter of fact, as a dynamical concept.
5) The method (analytical or synthetic) by which the results are obtained.

Chasles gave crucial contributions to all these aspects relying upon synthetical methods. His first contribution on this subject is Chasles (1830i). This was fundamental and thank to such work nowadays the main theorem on the infinitesimal displacements of a rigid body is called "Chasles theorem". However, before—and in one case contemporarily (Giorgini)—to Chasles (1830i), other scholars dealt with this question. The most important are four: Euler, Mozzi, Lagrange, Carnot, Giorgini.

In the course of his long and intense scientific career, Euler gave significant contributions to the representation and the displacement of a rigid body in space. Although in one circumstance, Euler claimed the utility of separating the geometrical elements of the movement of a rigid body from those in which dynamical quantities are involved, as a matter of fact, in all his studies, he did not treat the movements independently of their causes. In other terms, his contributions concern only indirectly that branch of physics which in 1834 Ampère named "cinématique".[1] In our prospective the most remarkable results obtained by Euler concern: 1) the invention of the so-called Euler angles which allow us to describe the movement of rotation of a rigid body; 2) the discovery that if a body rotates around a fixed point, it also rotates around an instantaneous axis of rotation; 3) for every rotation, the existence

[1] Ampère introduced the term "cinématique" in the *Preface* of Ampère (1834, p. xlii). In this work the section 3, §1, the first part is entitled *Cinématique*, pp. 50–53. Here Ampère clarified that, through this term, he was referring to the study of the movement independently of its causes. As several physicists, mathematicians and engineers of his period, Ampère was interested in the movements of the different parts of a machine. In this context, he defined *cinématique* as follows: "A treatise where you consider all the movements independently of the forces which produce them will be very useful in education because it presents the difficulties inherent in the interplay of diverse machines. In this way the spirit of the student will not be obliged to deal with the difficulties arising from considerations concerning the forces' equilibrium. I call *kinematics*, from κίνημα, *movement*, this science which considers in themselves the movements of the bodies of our environment and especially of the machines' machineries. After these general considerations on what *movement and velocity* are, kinematics has to address the relations among the velocities of different points of a machine". Ampère (1834, p. 52): "Un traité où l'on considérerait ainsi tous les mouvemens indépendamment des forces qui peuvent les produire, serait d'une extrême utilité dans l'instruction, en présentant les difficultés que peut offrir le jeu des diverses machines, sans que l'esprit de l'élève eût à vaincre en même temps celles qui peuvent résulter des considérations relatives à l'équilibre des forces. C'est à cette science où les mouvemens sont considérés en eux-mêmes tels que nous les observons dans les corps qui nous environnent, et spécialement dans les appareils appelés machines, que j'ai donné le nom de *cinématique*, de κίνημα, *mouvement*. Après ces considérations générales sur ce que c'est que *mouvement et vitesse*, la cinématique doit surtout s'occuper des rapports qui existent entre les vitesses des différens points d'une machine [...]". Italics in the text. Ampère clarified that statics studies forces independently of movements. Dynamics studies forces as causes of movements, so deducing the general laws of movement (ibid., pp. 53–54).

of an instantaneous axis of rotation, 4) the expressions to determine the momenta of the forces acting on the body with respect to the three principal inertia axes which he used to determine his three equations of motion which establish "[. . .] the changes in the position of the axis of rotation through the center of gravity and the changes in the angular velocity about this axis".[2] As Koetsier points out, the geometrical proof concerning the existence of an instantaneous axis of rotation given by Euler might also have been applied to two different positions of a finite displacement. However, Euler did not develop this consideration. The concepts and the methods he introduced in his studies on the movement of a rigid body are crucial for dynamics: it is enough to think of the concept of axis of inertia, of his idea to consider a fixed orthogonal reference frame in the absolute space and an orthogonal reference frame comoving with the body, of his genial method to project the instantaneous movement of a rigid body on the surface of a sphere so to reduce the problem of the movement of a rigid body to that on a surface of a sphere. Nonetheless, though several of the results obtained by Euler can be exploited in a kinematical context, they concern dynamics. He did not develop an independent kinematic and did not reach the result that any infinitesimal motion and any finite displacement of a rigid body can be described by the movement of a screw in its nut. Euler's treatment is almost entirely analytical; complicated formulas are often involved, which Euler in the course of the years tried to improve and simplify.

Giulio Mozzi in 1763 proved that any infinitesimal displacement can be considered as the movement of a screw in its nut. This proposition is one of the first proved by Mozzi in his treatise. For Mozzi in the fourth Corollary of the first Lemma at page 5 of his work clearly stated that any infinitesimal movement of any particle of a rigid body can be decomposed into two movements: 1) a rectilinear movement common to all the parts of the body and parallel to the axis of rotation passing through the gravity centre; 2) the rotation around an axis parallel to the mentioned one. Mozzi named this axis "Asse spontaneo di rotazione" ("Spontaneous axis of rotation"). He proved as first Lemma a proposition already known to Euler:

[2]Koetsier (2007, p. 185). The Euler's angles were introduced by Euler in the Appendix to the second volume of his *Introductio in analysin infinitorum*, entitled *Appendix de Superficiebus* (Euler, 1748, vol. 2, pp. 321–398), chapter IV, entitled *De immutatione coordinatarum* (ibid., pp. 365–373). With regard to property 2), see Euler (1752). A very good explanation of the main results obtained by Euler in this work is offered by Koetsier (2007, pp. 182–184). The result 3) was also reached in Euler (1752). For the property 4) see Euler (1752, 1765a, 1765b, 1767). A remarkable work on the movement of a rigid body in Euler is Langton (2007). Euler did not prove that any displacement can be reduced to a rotation around an axis to which the translation is parallel, that is he did not reach Chasles theorem. It is to remember that D'Alembert also had an important role in the study of the rigid body's movement. In D'Alembert (1749) the author writes the second order differential equations relative to the angles which describe the movement of the Earth's axis in reference to the Ecliptic. See also Magri (2013).

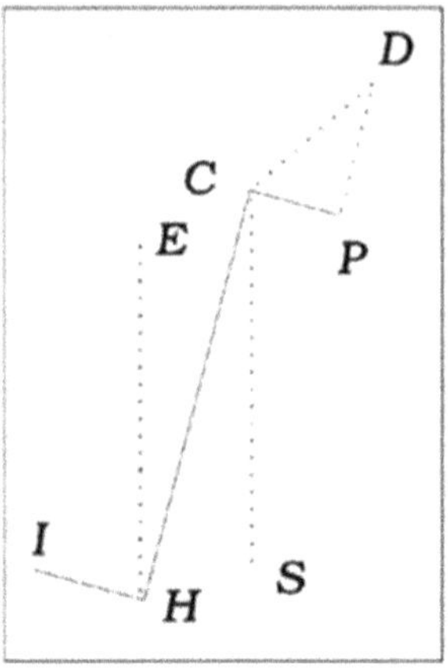

Fig. 3.1 The diagram used by Mozzi to prove Corollary III. 6 (Chasles-Mozzi theorem). Legenda: *C*=gravity centre of a rigid body; *CS*=line of rotation through *C*; *CD*=displacement of *C*; *CD* is decomposed in a component *PD* parallel to *CS* and a component *CP* in a plane π perpendicular to *CS*. *CH* is orthogonal to *CP*. The displacement *HI* is equal and opposite to *CP*. With these elements, it is possible to prove Chasles-Mozzi theorem (Ceccarelli, 2007, pp. 282–283)

> If a sphere is in movement, but its centre is at rest, in any instant of its motion the sphere will rotate around an immobile axis, which is a diameter of the sphere.[3]

From this lemma Mozzi deduced two corollaries which state the theorem of the instantaneous movement of a rigid body, so that the name Mozzi is sometimes associated with that of Chasles to indicate this fundamental theorem. We read:

> Corollary III. 5. Therefore, every moving body in each instant of its motion will be animated only by two movements: a rotation around its gravity centre; a progressive motion, which is common to all of its parts, along a straight line. 6. Thus, it will be possible to deduce that these two movements are reducible to two others: a rectilinear movement parallel to the axis of rotation passing through the gravity centre, which is shared by all the bodies' parts, and a rotation movement whose axis is parallel to the mentioned one.[4] (Figs. 3.1 and 3.2).

Mozzi offered several interesting applications of his result. Many of them concern the action of instantaneous forces on the modification of the state of motion of a rigid body. All the applications are based on analytical reasoning, but the demonstration of the fundamental theorem is developed (if one excludes the concept of centre of mass) by means of a descriptive geometrical reasoning, as it is evident by Mozzi's

---

[3] Mozzi (1763, p. 1): "Se una sfera si muove, mentre il di lei centro rimane in quiete, dico che in ciasched'un istante del moto essa dovrà rivolgersi intorno ad un asse immobile: che sarà uno de' suoi diametri".

[4] Ibid., pp. 4–5: "Corollario III. 5. Adunque qualsivoglia corpo, che si muova, in ciascun istante del moto due soli movimenti potrà avere, uno di rotazione intorno al centro di gravità, e l'altro progressivo in linea retta comune a tutte le parti sue. 6. Quindi ancora si potrà dedurre, che i suddetti due movimenti si riducono a due altri, uno de' quali sarà rettilineo e comune a tutte le parti del corpo, e parallelo all'asse di rotazione, che passa per il centro di gravità, e l'altro pure di rotamento, che avrà un asse di rotazione parallelo all'asse mentovato".

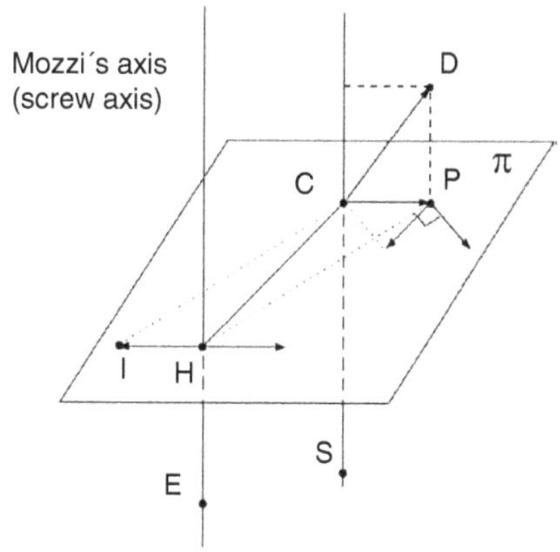

Fig. 3.2 Modern image to describe Chasles-Mozzi theorem (Ceccarelli, 2007, p. 283)

argumentation and as Ceccarelli has pointed out (Ceccarelli, 2000, 2007). Mozzi did not deal with finite displacements. His works were still unknown to Chasles and Giorgini when the latter reached their results. They became acquainted with his treatise only afterwards.

In his *Mécanique Analytique*, Lagrange offered the equations for the movement of a rigid body by means of the principle of virtual velocities. While dealing with the instantaneous movement of a rigid body, Lagrange considered the infinitesimal time as a constant and derived the infinitesimal changes of the coordinates under the condition that the distances between two given points of the body must be the same before and after the movement (condition of rigidity). Lagrange in § 3 of *Mécanique Analytique*'s third section offers the formulas for the infinitesimal motion of translation of a rigid body and in §§ 4–7 those for the rotational infinitesimal motion. He exploited many of Euler's results and gave them a general form thanks to the principle of virtual velocities. The method used by Lagrange is purely analytical. His results do not belong to the geometrical study of the movement because the concepts of force and momenta are used.

Lazare Carnot was an important reference point for Chasles because, in his *Géométrie de position* he clearly expressed the concept of a geometrical movement in which only the positions of the bodies are considered independently of movement's cause. As Carnot wrote:[5] "Geometry could consider the movements which do not result from the mutual action and reaction of the bodies". He also identified the reversibility of any geometrical movement, which is not always the case with a physical movement. We read:

[5]Carnot (1803, pp. 336): "La géométrie [. . .] pourroit embrasser les mouvemens, qui ne resulte pas de l'action et de la reaction des corps les uns sur les autres". See also Atzeni (2014–2015, p. 49).

> [. . .] given a movement in a system of bodies, the opposite movement is possible at any moment, which is not the case when the movement is not geometrical.[6]

Carnot had, hence, the clear idea of a geometrical movement. The opinion expressed by Charles Gillispie and Adolf Youschkevitch can be shared. Gillispie claims that there are elements in Carnot which will also be present in the vector analysis and that in his work the distinction between reversible and irreversible movements is clear. We read:

> There are considerations which allow us to consider the concept of geometrical movement as the precursor if not as the common ancestor of the reversible processes and of the vector analysis. In effect, reversibility is the criterion which permits to recognize the independence of such movements in respect to the rules of dynamics. Furthermore, these movements are determined only by the system's geometry.[7]

In the first 30 years of the nineteenth century, Lazare Carnot was an important source of inspiration for the development of projective geometry and for the study of the geometrical movement.

One of the scholars who also drew inspiration from Carnot's work was Gaetano Giorgini. He was an important author in our context because in his memoir entitled "Intorno alle proprietà geometriche dei movimenti di un sistema di punti di forma invariabile", published in 1836, but presented in 1830 (see Giorgini, 1836) he proved the main properties of the geometrical movement and in the Appendix to this work he had the aim to develop Carnot's ideas. Giorgini stressed the importance of the theory of projections for statics as well as the necessity—pointed out by Carnot—to separate the geometrical and the dynamical elements of any rigid movement. He also made an interesting remark: in order to have a completely satisfactory geometry of all the movements, geometry should be developed beyond the old and known limits. He did not have such an ambitious aim. Rather he concentrated on "[. . .] the examination of the movement of a body or of a system of points which always maintain the same reciprocal distances",[8] which might be the first step of the new geometrical science of the movements. Giorgini's order of ideas is close to Chasles': he dealt with the geometry of the infinitesimal movement, reaching the same results as Chasles and providing explicitly a series of formulas which are also deducible from Chasles' work, but which the latter did not write explicitly. Nonetheless, Giorgini reached his results by means of mathematical analysis applied to analytical geometry, not by synthetic methods. He proved the

[6]Ibid., pp. 336–7: "[. . .] lorsqu'un mouvement existe dans un système du corps, le movement contraire à chaque instant est toujours possible; ce qui n'a pas lieu lorsque le mouvement n'est pas géométrique". See also Atzeni (2014–2015, p. 49).

[7]Gillispie and Youschkevitch (1979, p. 103): "Ce sont ces considérations qui nous permettent de considérer la concept de mouvement géométrique comme le précurseur, sinon l'ancêtre commun, et des processus réversibles et de l'analyse vectorielle. D'une parte, en effet, la réversibilité est le critère permettant de reconnaître l'indépendance de tels mouvements par rapport aux règles de la dynamique; et, d'autre part, ces mouvements sont déterminés par la seule géométrie du système".

[8]Giorgini (1836, p. 2): "[. . .] nell'esame del movimento di un corpo, o di un sistema di punti che tutti si muovono conservando tra loro le medesime distanze".

"Mozzi-Chasles theorem". In the Appendix of Giorgini (1836), he explicitly referred to Mozzi (1763) claiming that when he developed his order of ideas he did not know Mozzi's work. Giorgini also remarked a slight mistake in Mozzi's proof (Giorgini, 1836, pp. 47–48). He continued mentioning Chasles (1830i), in which the same results as Giorgini's are expounded by synthetic methods and claimed—which is true—that he and Chasles reached these results independently one from the other and independently of Mozzi's work. Giorgini stated in these terms the fundamental theorem:

> A rigid body or a rigid system of points can be driven from a position to another by means of a continuous movement analogous to that of a screw. This means by a translation in the direction of a determined straight line and of a simultaneous rotation around the same line.[9]

This was the state of the art when Chasles entered the scene.

The works specifically dedicated by Chasles to the movement of a rigid body are Chasles (1830i, 1843, 1860–1861) (this is a long paper published in five parts). Fundamental observations are to be found in the *Aperçu* (Chasles, 1837a); some useful considerations are also present in the *Rapport* (Chasles, 1870) and in the treatise on higher geometry (Chasles, 1852).

My thesis is the following one:

> *The foundational programme consisting in the reduction of the most basic sections of mechanics (statics, kinematics and part of dynamics) to geometry and, hence, to projective geometry was already present in the paper written in 1830. This means that the idea to found the whole of geometry and fundamental parts of physics on projective geometry was already present in an early stage of Chasles' scientific career. With regard to the specific problem of the relations geometry-mechanics, the following papers were specifications and additions to concepts, which, in nuce, were already present in Chasles (1830i).*

3.1 The Results Obtained in 1830

The succession in which the theorems are expounded in this early work leaves no doubt that Chasles interpreted the movement of a rigid body as a geometrical transformation and, specifically, as a particular kind of isometry. For the first proposition stated by Chasles is that, given in a plane two congruent polygons, a point of the plane exists which is equidistant from two homologous vertices of the polygons. The second theorem reads as follows: given in a plane two similar polygons, posed in any position, a point P of the plane exists such that its distances from two homologous vertices of the two polygons are in a constant ratio, which is the same as the ratio of the polygons' homologous sides. Chasles stressed that it is enough to rotate one of the two polygons around P for the sides of the polygons to

[9]Ibid., p. 48: "Un corpo solido, o un sistema rigido di punti, può sempre esser condotto da una posizione in un'altra qualsivoglia diversa, mediante un movimento continuo analogo a quello della vite; cioè a dire di traslazione nella direzione di una determinate retta, e di rotazione simultanea attorno alla retta medesima".

become parallel—that is for the polygons to be similar and similarly posed. The point P is, hence, the centre of similitude and, if from P the perpendiculars on the lines joining any vertex of the first polygon to its correspondent are traced, the perpendiculars' feet are the vertices of a polygon similar to the given two.

After stating these two theorems connoting the movement in a plane, Chasles enunciated the *fundamental theorem* for the movements in space:

> Given in space two similar bodies located in any positions: 1) a point O of space exists whose distances from the homologous vertices of the two bodies are in a constant ratio. This point is unique and it is similarly posed in respect to the two bodies. This means that if one considers this point as belonging to one of the two bodies, it is self-homologous in the other body; 2) a straight line D exists whose distances from two any homologous points of the two bodies are in a given constant ratio. This straight line in unique and is similarly posed in respect to the two bodies, that is, as considered belonging to the first body, it is self-homologous in the second body; 3) a plane P exists such that the distances of two homologous points from this plane are in a constant ratio. This plane is unique and is similarly posed in respect to the two bodies, that is, if one considers it as belonging to the first body, it will be self-homologous in the second body; 4) finally such a plane and the straight line D are perpendicular and pass through O.[10]

Chasles added that, if through D two planes are drawn, which pass, respectively, through two homologous points of the two bodies, and these planes are rotated around D until they overlap, the two bodies will become similarly posed. From this it follows that, given in space two similar bodies, a straight line always exists, rotating around which the two bodies become similarly posed.

Now from the case of two similar and similarly posed bodies, Chasles passes to the specific one in which the bodies are also equal, *the point* O *and the plane* P *go to infinity*. The theorem thus reads as follows:

> Given in space two perfectly equal bodies, placed in any manner, a straight line exists, which considered as belonging to the first body, is self-homologous in the second body [thence. . .]: given in space a free solid body, if it is subject to any finite displacement, a straight line will always exist in this body, which, after the displacement, will be in the same place as beforehand.[11]

[10] Chasles (1830i, pp. 322–323): "Quand on a dans l'espace deux corps semblables entr'eux, et situés d'un manière quelconque, l'un par rapport à l'autre: 1° il existe toujours dans l'espace un certain point O, dont les distances à 2 sommets homologues quelconques des 2 corps, sont entr'elles dans un rapport constant; ce point, qui est unique, est semblablement placé par rapport aux a 2 corps; c'est-à-dire que si on le considère comme appartenant à un des 2 corps, il est lui-même son homologue dans le second; 2° il existe toujours un certaine droite D, dont les distances à 2 points homologues quelconques des 2 corps, sont entr'elles dans un rapport constant; cette droite, unique, est semblablement placée par rapport aux 2 corps; c'est-à-dire que, considérée comme appartenant au Ier, elle est elle-même son homologue dans le seconde; 3° il existe toujours un certain plan P, tel que les distances de 2 points homologues quelconques des 2 corps à ce plan, sont entr'elles dans un rapport constant; ce plan, qui est unique, est semblablement placé par rapport aux 2 corps; c'est-à-dire que si on le considère comme appartenant à l'un d'eux, il sera lui-même son homologue dans le second ; 4° enfin, ce plan et la droite D sont à angle droit, et passent par le point O".

[11] Chasles (1830i, pp. 323–324): "Quand on a dans l'espace 2 corps parfaitement égaux, et placés d'une manière quelconque, l'un par rapport à l'autre, il existe toujours dans l'espace une certaine droite indéfinie, qui, considérée comme appartenant au Ier corps, est elle-même son homologue dans

And now the fundamental theorem on the displacement of a rigid body is stated:

> If the second body (namely the body in its second position) is made to rotate around such a straight line, it will become similarly posed as the first one. Furthermore, if a translational movement is given to the body in the direction of the straight line, it will overlap the first one. Hence: *it is always possible to move a free solid body from a position to any other determined position through the movement of a screw to which the body will be invariably fixed.*[12]

If the displacement is infinitesimal, in the enunciation of the theorem, it is possible to replace the term "displacement" with "continuous movement", so that Chasles concluded that, *when a solid body is subject to an infinitesimal movement, a straight line exists in this body which translates on itself while the body is rotating around such line, so that the movement of the body is nothing but that of a screw in its nut.*

Chasles added two further conclusions: 1) the principle of virtual velocity can be deduced from these theorems because each virtual movement is nothing but the movement of a screw in its nut. Thus, it is enough to prove this principle for the movement of a screw; 2) if a body is constrained to a fixed point, its movement is reducible to a rotation around an axis passing through such point.

In a final and important theorem, Chasles stated that given two equal bodies in any position, if the homologous points of the two bodies are joined by straight lines, the mean points of the joining segments form a further solid body. It is possible to give this solid body an infinitesimal movement such that all its points move in the direction of a given straight line.

Commentaries:

1) The fundamental theorem is related to the properties of any similarity in three-dimensional space. This means that Chasles considered the similarity as the object of his research. He was beginning to study the properties of the geometrical transformations. The finite displacement of a rigid body is based on the idea of considering the positions of two congruent rigid bodies, which, of course, can be considered as a sole rigid body, initially in a given position and later in its position after the displacement. The geometrical transformation in this case is an isometry composed of two isometries, but the characteristics of this isometry are grasped by Chasles interpreting it as a particular case of a spatial similarity. The first isometry is a rotation around the straight line D. By means of this transformation, the two congruent bodies become similarly posed. The second isometry

le second [...] quand on a dans l'espace un corps solide libre, si on lui fait éprouver un déplacement fini quelconque, il existera toujours dans ce corps une certaine droite indéfinie, qui, après le déplacement, se retrouvera au même lieu qu'auparavant". Italics in the text.

[12]Ibid., p. 324: "Si on fait tourner le second corps (c'est-à-dire le corps pris dans sa seconde position) autour de cette droite, il deviendra semblablement placé au Ier ; et si, ensuite, on lui donne un mouvement de translation dans le sens de cette droite, il viendra se superposer sur le Ier corps; ce qui prouve que: *l'on peut toujours transporter un corps solide libre d'une position dans un autre position quelconque déterminée, par le mouvement continu d'une vis à laquelle ce corps serait fixé invariablement* ". Italics in the text.

is a translation, which is produced by the similarity when the point O and the plane P go to infinity. The elements at infinity are the feature of projective geometry, or better, they have to be specified and separated from the elements at finite when projective geometry is applied to affine or metric contexts. Therefore, it is not an over-reading to claim that Chasles clearly interpreted the translation as a homology whose centre and plane of homology are both improper. Furthermore, the spatial rotation can be interpreted as a particular homology in the star of planes and the spatial similarity also has an interpretation in terms of homology.[13] Therefore, *Chasles considered the transformations necessary to explain the geometrical movement of a rigid body as an application of projective geometry to a metric and mechanical context.* Obviously, his ideas on the geometrical transformations in 1830 were not yet completely clear, but the train of his thoughts was.

2) Chasles claimed that the straight line D might be considered as belonging to one or to the other of two congruent bodies. This means that he was not considering the body as an object in itself independent of the space in which it is inserted. He was beginning to consider space itself jointly with the objects "immersed" in it as a subject of study. The line D is in space and, hence, if this implies no mistake, we are free to consider it as belonging to one or to the other body. Chasles was beginning to consider the movements of space on itself, in these movements two congruent objects can overlap and this geometrical feature is useful to face the motion of a rigid body. This is also typical of projective geometry where one can study the correspondences between two overlapped forms of first, second or third species—to use more modern terms. In this case, Chasles studied two spaces as two overlapped forms of third species. Though this might appear a modernization of his thoughts, I think that it grasps his real way of thinking.
3) A question related to the following works of Chasles on the movement of a rigid body concerns the demonstrations. Chasles gave almost no explicit proof of his assertions. He only added some words of clarifications between a theorem and the following one which are not at all real demonstrations, but only sentences which explain the logic of the successive steps of Chasles' work. His propositions are all true and other mathematicians, as we will see, gave explicit synthetic demonstrations of his theorems. Chasles (1830i) is not an exception. There is no explicit proof. This work is followed by a brief appendix by Hachette where the existence of an equidistance point between two congruent polygons is proved. Through a similar method the existence of the centre of similitude might be proved and the case of two congruent polygons might be deduced as a particular case.[14]
4) As a final geometrical consideration, I point out that Chasles spoke, in the last-mentioned theorem, of the existence of a third body obtained joining the middle points of the segment connecting two homologous points. As we will see,

[13] In the next section, dedicated to the *Aperçu*, we will see Chasles' ideas in regard to these two transformations.

[14] See Hachette in Chasles (1830i, p. 326).

30 years later he named this segment "chord" and this body "mean body" and analysed all their properties of which he made no mention in Chasles (1830i). This is a clear litmus paper that in the course of the years he developed an order of ideas and a foundational programme whose bases were already solid enough in 1830.

5) Finally: today the name of Chasles is associated with the "Chasles-Mozzi theorem" on the infinitesimal movements. This fundamental theorem is deduced very easily by Chasles as a sort of corollary of all the expounded geometrical considerations. Therefore, from his point of view, this result was an application of projective geometry. With regard to the principle of virtual velocities, I do not address here all the interesting discussions concerning the doubts it created, I just observe that Chasles reduced the problem of a general demonstration of this principle to its proof for the case of infinitesimal movements of a screw. In his following works he will be clearer on this point. Thence, it will be addressed while dealing with such works.

3.2 Transformations and Displacement of a Rigid Body in the *Aperçu*

My thesis is that, in the course of the years, Chasles developed the foundation of mechanics based on projective geometry through a clarification and, at the same time, a generalization of ideas he had already expounded in Chasles (1830i). Since similarity plays a fundamental role in the context of that paper, it is first of all necessary to clarify how Chasles interpreted similarity in the panorama of projective geometry. Later on, we will consider his new results regarding the infinitesimal movements of the rigid body in the light of his progresses in the theory of transformations.

Chasles' results 1
In Sect. 3.1, we have seen how Chasles considered a translation in terms of projective transformations, but the fundamental transformation for the movement of a rigid body is the similarity because Chasles considered the isometries as a particular case of similarities. In the Additions to the historical section of his *Aperçu*, he clarifies the way in which similarity can be expressed by means of projective concepts. He wrote:

> *If in a plane, we consider two figures, one of which is the perspective of the other, and if they are placed mutually in any position: 1) each point of one of the two figures will have its homologous in the other one; 2) three points exist in one of the two figures, which will be overlapped their homologous in the second figure. One of these points is real, the others can be imaginary.* It follows that there are also three straight lines which are overlapped their homologous in the second figure. These are the lines joining the three points two by two. One of these lines is always real, the other two can be imaginary. When the figures are similar, which is a particular case of perspective, two of the three points and two of the three lines are always imaginary. The third point is real. The third straight line is also real, but it is at infinity. This is also the case when two figures are equal. These properties of the flat

figures have their analogous in the three dimensional figures for which I have already enunciated some theorems referred to this theory (See *Bulletin universel des Sciences*, t. XIV, p. 321, year 1830).[15]

Several explanations are necessary: a) by the term "perspective", Chasles intended two homographic figures, not necessarily perspective; b) item 1), the biunivocity of homography, is not a problematic item;[16] c) if we exclude the presence of imaginary elements, which is a fundamental item for projective geometry in the whole of the nineteenth century, but which is not so important with regard to the movement of a rigid body, similarity is characterized by Chasles as a homography between two overlapped planes in which the line at infinity is self-corresponding. It should be pointed out that this feature is a necessary, but not a sufficient condition for a transformation to be a plane similarity, because what Chasles claimed characterizes all the affinities, not only the similarities. It should also be added that the given homography transforms the absolute involution of one plane in the absolute involution of the overlapped plane. In this case, the two lines at infinity, beyond being self-corresponding, are also congruent and, hence, to any plane angle, a congruent plane angle corresponds. This order of ideas was developed thanks to the contribution given by Von Staudt to the doctrine of the involutions and thanks to the projective definition of the measure of an angle provided by Laguerre. Cayley gave its final form to this doctrine with regard to Euclidean geometry. However, Chasles' quotation clearly shows that he was trying to express the metric concepts and transformations by projective means. The homotheties and the isometries enter easily this picture; d) With regard to the extension to space of the notion of similarity, analogous considerations have to be added; e) it is remarkable that Chasles connected his considerations on similarity to the paper of 1830 previously analysed. It is, hence, clear that he considered the movement of a rigid body as a section of projective geometry. Therefore, this is a confirmation that the ultimate

[15]Chasles (1837a, p. 549): "*Si l'on conçoit dans un plan deux figures qui ont été primitivement la perspective l'une de l'autre, et qui se trouvent actuellement placées d'une manière quelconque l'une par rapport à l'autre;*

Chaque point de l'une des figures aura son homologue dans l'autre figure;

Il existera généralement trois points dans l'une des figures, qui se trouveront superposés respectivement sur leurs homologues dans la seconde figure;

L'un de ces trois points sera toujours réel; les deux autres pourront être imaginaires.

Il résulte de là qu'il y a aussi trois droites, dans l'une des figures, qui se trouvent superposées sur leurs homologues dans la seconde figure; ce sont les droites qui joignent deux à deux les trois points.

L'une de ces droites est toujours réelle, et les deux autres pourront être imaginaires.

Quand les deux figures sont semblables, ce qui est un cas particulier de la perspective, deux des trois points et deux des trois droites sont toujours imaginaires; la troisième point est réel; la troisième droite est aussi réelle; mais elle se trouve située à l'infini.

Cela a lieu pareillement quand les deux figures son égales entre elles.

Ces propriétés des figures planes ont leurs analogues dans les figures à trois dimensions, pour lesquelles j'ai déjà énoncé quelques théorèmes qui se rapportent à cette théorie (Voir *Bulletin Universel des Sciences*, t. XIV, p. 321, année 1830)". Italics in the text.

[16]I postpone slightly the commentary on the united points and straight lines of a homography.

foundation of mechanics relies on projective geometry and, particularly, on the doctrine of projective transformations.

Finally, with regard to the proof that, given a homography (in fact a non-homological homography), three points exist in one of the two figures, which will be overlapped their homologous in the second figure, I report the beautiful and easy proof by Chasles (1837a, pp. 834–835). This is useful to give an idea of his way of reasoning. It is based only on the theorem (proved by Steiner and by Chasles himself) that, given in a plane two projective pencils of straight lines, the intersection point of two homologous lines belongs to a conic section. Chasles reasoned like this: in the (overlapped) planes of the two homographic figures, let us consider two pencils of straight lines whose centres are O and O'. The intersection point of two homologous lines belongs to a conic. Let us consider two other pencils with centres Ω and Ω'. In this case too, two homologous straight lines will mutually cut in a point belonging to a conic. The two conics will pass through the intersection point of the two lines $O\Omega$ and $O'\Omega'$ because these rays reciprocally correspond. Be A,B,C the three other intersection points of the two conics. Each of these points considered as belonging to the first figure is self-homologous in the other figure. Indeed, the straight lines $O'A$ and $\Omega'A$ of the second figure correspond, respectively, to OA and ΩA in the first figure. Therefore, the intersection point of $O'A$ and $\Omega'A$ corresponds to the intersection point of OA and ΩA. This means that A considered as belonging to one of the two figures is self-homologous in the other one. Since the intersection point of $O\Omega$ and $O'\Omega'$ is real, it is (and was) well known from the theory of conics that the two conics have another real intersection point. Hence, one among A,B,C is a real point, the other two can be imaginary.

In the *Aperçu* Chasles offered an analysis of the infinitesimal motion more precise than in Chasles (1830i). Such analysis is strictly connected to the projective transformations, in particular to the reciprocities. He wrote:

> The properties of a system of two equal bodies and, analogously, of two similar bodies, posed in any position in space are, hence, a consequence of this same theory [the theory of reciprocity and homography]. These properties, on which no research has yet been developed, are numerous and lead to several curious theorems on the infinitely small movements and on the finite displacement of a solid body, as well.[17]

The foundation of the theory of the rigid body's infinitesimal movements is developed in the memoir on duality. It relies on the first fundamental theorem proved by Chasles. The method is that analytical, but Chasles claimed that such proposition is provable in a synthetic way. He used the analytical method for sake of brevity.

Such theorem (Theorem 1) claims that if, in three-dimensional space, a plane with coordinates (x,y,z) is given such that the parameters of its equation contain the

[17] Chasles (1837a, p. 266): "Les propriétés que présente le système de deux corps parfaitement égaux, et même de deux corps semblables situés d'une manière quelconque dans l'espace, sont aussi des conséquences de cette même théorie. Et ces propriétés, qu'on n'a point encore cherchées, sont nombreuse et conduisent à divers théorèmes curieux sur le mouvement infiniment petit, et même sur le déplacement fini quelconque d'un corps solide".

coordinates of a point called the director [*directeur*] point by Chasles at the first degree, then:

1) When the point moves on a plane, the mobile plane rotates around a fixed point.
2) When the point moves on a straight line, the mobile plane rotates around a straight line.
3) When the point moves on a curved surface, the mobile plane rotates on another curved surface. If the first surface is of degree 2, then the second surface also is. If the first surface is algebraic of degree *m*, then the second is algebraic too, and given any line, it is possible to draw *m* tangent planes to it.

Chasles pointed out that the fixed point in 1) is the *pole* of the plane traversed by the *director point* (Chasles, 1837a, p. 577).

Let us now see the elements of the theory of the rigid body's infinitesimal movement developed by Chasles in the *Aperçu*. The § XXIV of the memoir on duality is dedicated to this subject. The first theorem proved by Chasles (Theorem 2) in this context is based on the following reasoning: he considered a body in an orthogonal system of coordinates. According to formulas due to Euler and Lagrange, the infinitesimal variations of the coordinates of any point are given, by

$$\begin{aligned} \delta x' &= \delta l - y'\delta N + z'\delta M \\ \delta y' &= \delta m - z'\delta L + x'\delta N \\ \delta z' &= \delta n - x'\delta M + y'\delta L \end{aligned} \tag{3.1}$$

The six coefficients δl, δm, δn, δL, δM, δN are constant for all the body's points. Let us now consider the equation of the plane passing through (x', y', z') perpendicularly to the rectilinear element described by this point during the infinitesimal movement of the body. The equation of this plane is:

$$(x - x')\delta x' + (y - y')\delta y' + (z - z')\delta z' = 0$$

i.e.,

$$x\delta x' + y\delta y' + z\delta z' = x'\delta x' + y'\delta y' + z'\delta z'$$

And, after a very easy calculation, one has

$$x\delta x' + y\delta y' + z\delta z' = x'\delta l + y'\delta m + z'\delta n$$

The functions $\delta x'$, $\delta y'$, $\delta z'$ are linear functions of x', y', z' so that the equation of the plane contains these coordinates only at the first degree. Thence, Chasles concluded, Theorem 1 can be applied so that:

> When a solid body is subject to an infinitesimal movement, the planes normal to the trajectories of the body's points belonging to a plane pass through a point.

The planes normal to the trajectories of the points posed on a straight line pass through a straight line.

The planes normal to the trajectories of the points posed on a second-degree surface are tangent to another second degree surface.

In general, the planes normal to the trajectories of the points of a surface of degree m, will envelop a second geometrical [today we say algebraic] surface, to which from a given line it is possible to draw m tangent planes.[18]

More generally (that is independently of being the surface algebraic) when any figure is subject to an infinitesimal displacement, the planes normal to the trajectories of its points will envelop a figure which is correlative to the given one.

Explanations and Commentaries 1

a) A hint to the genesis of the system of equations (3.1) is useful to grasp Chasles' reasoning. Lagrange's explanations in the second edition of his *Méchanique Analytique*, first tome, chapter IV dedicated to the equilibrium of a solid body are particularly clear. Lagrange had proved in the previous chapter of his work that the rigidity condition implies the validity of the following system of equations which connect the variations with the differences:

$$dxd\delta x + dyd\delta y + dzd\delta z = 0$$

$$d^2xd^2\delta x + d^2yd^2\delta y + d^2zd^2\delta z = 0$$

$$d^3xd^3\delta x + d^3yd^3\delta y + d^3zd^3\delta z = 0$$

$$\cdots\cdots\cdots\cdots\cdots\cdots$$

In the specific conditions of these equations, it is possible to assume a first difference, for example dx, to be constant so that the differences of higher order are equal to 0. Lagrange replaced the values $d^2x = 0$; $d^3x = 0$ in the second and third equations of the previous system. From the second equation, he obtained a second order differential equation, which he differentiated so that a third order differential equation is obtained. He integrated this equation reducing it to a new second order equation and indicated by δL the constant of integration. He integrated the so obtained second order differential equation indicating by $-\delta$

[18] Chasles (1837a, pp. 675–676): "Quand un corps solide éprouve un mouvement infiniment petit, les plans normaux aux trajectoires des points du corpssitués dans un même plan, passent tous par un même point;

Les planes normaux aux trajectoires des points situés sur un même droite, passent tous par un même droite, passent tous par une même droite;

Les planes normaux aux trajectoires des points situés sur une surface du second degré, sont tous tangens à une autre surface du second degré;

Et, en général, les plans normaux aux trajectoires des points d'une surface du degré m, enveloppent une seconde surface géométrique, à laquelle on peut mener m plans tangents par une même droite".

Mdx and by δNdx the two constants of integration which appear in the first order equation relative to the differences of the variations δy, δz he had obtained. By means of a third integration and indicating through δl, δm, δn the new constants of integration, Lagrange achieved the system referred to by Chasles in which the variations are free from any differentiation.

b) *The idea to reduce the movement of a rigid body within a projective context is here clearly realized*: for the mobile plane of Theorem 1 is, in the specific situation concerning the infinitesimal movement of a rigid body, the plane normal to the instantaneous trajectory of a point moving on a plane (which is not, obviously, the mobile plane itself). The director point is the moving point and the pole of the plane is a point to which Chasles gave, for the moment, no name, but that, starting from his memoir of 1843, he named the *foyer* (*focus*) of the plane. This means that the focus is the pole of the mobile plane in the particular situation in which the plane is perpendicular to the trajectory of the point.

In Theorem 1 there is no specification with regard to the positions of the mobile plane and the trajectory of the director point. Here the interesting relation for the instantaneous movement of a rigid body is specified: it is a relation of perpendicularity.

The second proposition of the theorem, according to which the planes normal to the trajectories of the points posed on a straight line *a* pass through a straight line *b* is also important because it can be exploited to detect the role of the two lines within an infinitesimal movement. There is thus a perfect analogy with the proposition concerning the relations between the planar trajectory of a point and the focus of the plane which can be used to prove that any infinitesimal movement can be reduced to the movement of rotation around a fixed line and a translation parallel to such a line, that is to a screw movement. Chasles enunciated the following theorem, which he called *metrical*:

> In a body, an axis [*X*] exists which moves only along its own direction.
>
> The planes normal to the trajectories of any two points *a*, *b* of the body will meet such axis in two points α, β which are the feet of the perpendiculars drawn on the axis from *a* and *b*, so that it always holds $\alpha\beta = ab \cos(ab.\, X)$.[19]

Chasles claimed to have only enunciated this theorem whose proof he was going to postpone to a work specifically dedicated to the movement of a rigid body. The first proposition of this theorem is exactly the conclusion of Chasles (1830i); the announced work will be Chasles (1843). It is, hence, clear that he was continuing to develop a foundational programme, whose bases he had given in Chasles (1830i).

[19]The symbols $\alpha\beta$ and *ab* indicate, respectively, the distances between the two points α-β, and *a*-*b*. Chasles (1837a, p. 676): "Il existe dans le corps un certain axe qui n'a de mouvement que dans sa propre direction;

Les plans normaux aux trajectoires de deux points quelconque *a*,*b* du corps rencontrent cet axe en deux points α, β qui sont les pieds des perpendiculaires abaissées sur lui, des points *a*, *b*;

De sorte que l'on toujours $\alpha\beta = ab\cos(ab.\, X)$".

Chasles' Results 2

The results explained by Chasles with regard to the finite displacement of a rigid body are a confirmation of what I claimed in the last sentence. As it was the case for the infinitesimal movement, he inserted such results within the theory of reciprocal figures and used them, in turn, to construct reciprocal figures. On this subject, the main theorem concerning the construction of correlative figures by the finite displacement of a rigid body expounded in the *Aperçu* is the following one of which Chasles claimed to be provable by simple geometrical considerations or by analysis. He also stated that such proofs will be provided by him in a later contribution. The theorem is the following:

> Given in space two equal figures in any reciprocal position, if the points of one figure are joined to the corresponding in the other figure [by a segment] and if, from the middle points of such segment the normal planes to the segment itself are drawn, all these planes will envelope a figure, which will be correlative to each of those given and will also be correlative to the third figure made up by the middle points of the segments joining the homologous points of the first two figures.[20]

In a note (Chasles, 1837a, pp. 677–678) Chasles also offered the formulas by which, given two equal bodies in two different positions in an orthogonal system of coordinates (*système de coordonnées rectangulaires*), it is possible to express the coordinates of a point in the second position as a function of those in the old position (he claimed: after and before the displacement). His formulas needed six coefficients, but the value under a square root is one single value, whereas the transformation formulas given by Monge needed six different radical expressions.[21] Chasles stressed that his formulas could also be used for the transformation of a system of orthogonal coordinates into another system.

He concluded his series of theorems on the movement/displacement of a rigid body analysing a property of the screw which derives from the fact that the infinitesimal movement of any rigid body can be considered like that of a screw in its nut. This property derives directly from Theorem 2 and it is notable that, though its proof relies on the infinitesimal movement, in its enunciation there is no reference to the movement. For the theorem establishes a property of the screw as an object. It reads as follows:

> Be given a screw posed in any position in space. Suppose that from any point of a given figure, some helices of the screw pass:
>
> 1. The planes normal to the helices, drawn through the figure's points, will envelope a second figure which will be correlative to the first one.

[20]Ibid., p. 677: "Si l'on a dans l'espace deux figures parfaitement égales, et placées d'une manière quelconque l'une par rapport à l'autre;

Que l'on joigne par des droites les points de la première figure aux points correspondans de la seconde; et que par le milieu de chacune de ces droites on lui mène un plan normal;

Tous ces plans envelopperont une figure qui sera corrélative de chacune des deux figures proposées, et corrélative, aussi, d'une troisième figure formée par les points milieux des droites qui joignent les points homologues dans les deux premières".

[21]Chasles was referring to Monge (1784–1785).

2. The segments on the screw's axis included between two normal planes will be equal to the orthogonal projection of the segment straight line joining the two corresponding points.[22]

Chasles continued by claiming that if the surface of a screw is cut by a plane and if, through the different points of the intersection curve, the planes normal to the helices passing through these points are drawn, all these planes will mutually intersect in a point belonging to the cutting plane. This will offer new and interesting properties useful in the methods of descriptive geometry and in the arts.

Explanations and Commentaries 2
Some brief explanations are useful with respect to this last theorem: item 1., according to which the planes normal to the helices, traced through the figure's points, will envelop a second figure which will be correlative to the first one, derives directly from the consideration that when any figure is subject to an infinitesimal displacement, the planes normal to the trajectories of its points will envelop a figure which is correlative to the given one. Item 2. was also proposed by Chasles in Chasles (1843). He never gave an explicit demonstration—probably deeming it to be trivial—but we will see that a group of scholars offered explicit and synthetic demonstrations of Chasles' statements. Chasles' final observation on the fact that the planes normal to the helices mutually intersect in a precise point depends on the fact that the helix can be considered as the trajectory of the infinitesimal movement of a point belonging to the cutting plane, so that Theorem 2 can be applied.

As a general comment: the *Aperçu* is a coherent development of the train of thought Chasles expressed in 1830: the infinitesimal movements and the finite displacements are reduced to particular projective transformations. Specifically, after having clarified the entire theory of the correlative transformations in his memoir on duality, he included his results within such a projective theory, which, hence, becomes the real foundation for the study of the rigid body's movement. Furthermore, there is another aspect which should be highlighted: Chasles saw this part of mechanics as a section of projective geometry; at the same time, the results obtained in such a section of mechanics—precisely because they rely upon projective geometry—can be legitimately used to prove projective and projective-metric properties of the figures. They also become a means to discover and to prove new geometrical properties. This is already conspicuous in some theorems of the *Aperçu*, but it will become one of the preponderant aspects of Chasles' work starting from Chasles (1843). With regard to the application of the rigid body's infinitesimal movement to geometrical problems, Chasles' way to draw the tangents to some curves is particularly noteworthy: he (Chasles, 1837a, p. 548) referred to a method

[22]Chasles (1837a, p. 678): "Soit une vis, placée d'une manière quelconque dans l'espace; concevons que par tous les points d'une figure proposée passent des hélices de la vis;

1° Les plans normaux à ces hélices, menés par les points de la figure, envelopperont une seconde figure qui sera corrélative de la première;

2° Le segment compris sur l'axe de la vis, entre deux plans normaux, sera égal à la projection orthogonale de la droite qui joint les deux points correspondans".

invented by Descartes useful to draw the normal—and hence the tangent—to the ordinary cycloid from one of its points. Such method, as Chasles wrote, is different from those used by Fermat, Roberval and Pascal: Descartes considered both the straight line and the circle generating the cycloid as polygons having an infinite number of sides, which, during the infinitesimal movement[23] of the circle on the straight line, share an infinitesimal side and, hence, two vertices. During such movement, the first polygon rotates around one of the two contact points. This means that the point generating the cycloid produces an infinitesimal arch of circle whose centre is one of the two vertices. Thence, the normal to this infinitesimal arch is the normal to the cycloid. Chasles highlighted that Descartes' method is extendible to all the cycloids and to many other curves. He gave the example of Nicomedes' conchoid (ibid., p. 549). Chasles showed the dependence of Descartes' method from the following general theorem:

> When a plane figure is subject to an infinitesimal movement remaining in its plane, a fixed point exists during this movement. The straight lines drawn from the diverse points of the figure and perpendicular to the trajectories described by the points during the infinitesimal movement pass through this point.[24]

This theorem is nothing but a specification of the first theorem in Chasles (1830i), according to which, as we have seen, given two equal polygons in a plane, a point—that is similarly posed with respect to the two polygons—exists which is equidistant from two homologous vertices. On that occasion Chasles also claimed that this theorem is the basis of a method to trace the normal to many curves (Chasles, 1830i, pp. 321–322). What stated in Chasles (1837a) is, hence, a clarification and specification of an order of ideas already existing in Chasles (1830i).

Therefore the panorama devised by Chasles is the following: projective geometry found mechanics, mechanics in turn—insofar it has a geometrical foundation—can be used to derive further geometrical truths. Chasles was absolutely explicit with regard to the role of geometry. For we read:

> Geometry has so been able to regain easily its generality and intuitive evidence on mechanics and on the physical-mathematical sciences. [...] Descriptive geometry has fortified and developed the power of our conception. It has given more cleanness and certainty to our judgement as well as precision and clearness to our language. Under this respect it has been infinitely useful to mathematical sciences in general.[25]

[23]It is appropriate to recall that Roberval's method is also based on the generation of the cycloid by motion. But it is not connected to the consideration of the infinitesimal motion and with the idea that a curve is a polygon made up of infinitesimal sides (Chasles, 1837a, p. 549).

[24]Chasles (1837a, p. 548): "Quand une figure plane éprouve un mouvement infiniment petit dans son plan, il existe toujours un point qui, pendant ce mouvement, reste fixe;

Les droites menées par les différens points de la figure, perpendiculairement aux trajectoires qu'ils décrivent pendant le mouvement infiniment petit, passent toutes par ce point fixe".

[25]Chasles (1837a, p. 190): "La Géométrie devint aussi en état de répandre plus aisément sa généralité et son évidence intuitive sur la mécanique et sur les sciences phisico-mathématiques. [...] La Géométrie descriptive, en un mot, fut propre à fortifier et à développer notre puissance de conception ; à donner plus de netteté et de sûreté à notre jugement; de précision et de clarté à notre

The doctrine by which geometry can play a foundational role is the theory of projective transformations. Chasles dedicated a series of interesting considerations (Chasles, 1837a, pp. 190–252) to such a doctrine which he founded on the concept of anharmonic ratio in the two memoirs on duality and homography. Since this work is dedicated to the foundation of mechanics on projective geometry, I will deal with Chasles' theory of geometrical transformations only insofar as it is connected to the present aim, but it is at least necessary to point out that he developed the theory of reciprocity beyond the theory of the reciprocal polars, which was one of the cornerstones of the projective geometry of that time and framed this last theory into a more general picture. He developed the theory of homography beyond that of homological figures and interpreted homologies as particular homographies. He offered a profound interpretation of the so debatable Poncelet's principle of continuity in terms of principle of contingent relations and showed the heuristic as well as the demonstrative power of synthetic methods in projective geometry. A particularly remarkable section of Chasles' work is the study of the metric-graphical properties. Therefore, Chasles' foundations of mechanics belong to a general foundational picture based on projective transformations which he was developing.

3.3 The Period 1840–1859: Specification and Spread of Chasles' Ideas

With regard to the movement of a rigid body the 20 years 1840–1859 are characterized by the publication of Chasles (1843). This is a paper in which the author proposed a long series of concepts and theorems concerning the infinitesimal movement of a rigid body. These propositions were, in part, already present in Chasles (1830i, 1837a), but in Chasles (1843) they were framed in a new context, from which it was clear that Chasles' ideas could open a new way to see mechanics as a branch of projective geometry and as a discipline, in turn, useful to geometry. This was already the case in Chasles (1830i, 1837a), but the quantity and the profoundness of the theorems enunciated in Chasles (1843) are far superior than those connoting the two previous works. Chasles gave no explicit proof of his assertions. However, some scholars, basically in France, but also in Italy and in Germany, adhered to Chasles' ideas and, for a relatively brief period, founded a "School" of, so to say, "projective mechanics" based on Chasles' works. The publications of these authors are, in great part, dedicated to clarify Chasles' theory and to provide the readers with the synthetic proofs of his theorems. However, before Chasles (1843), Olinde Rodrigues published a paper on the displacement of the rigid body (Rodrigues, 1840). This paper is extremely advanced because Rodrigues, who mentioned the previous works by Chasles, offered a clear picture of the basic

langage; et, sous ce premier rapport, elle fut infiniment utile aux sciences mathématiques en général".

geometrical properties of the rigid body's movement as well as a complete series of formulas in which the previous results by Euler, Lagrange and Monge were included. The concepts introduced by Chasles starting from his paper in 1843 are tied to his own previous contributions and seem scarcely influenced by Rodrigues' essay, though Chasles knew for sure this work,[26] which, probably, induced him to offer a more extended picture of his ideas. Given such panorama, this section will be divided into three subsections: in the first, I will present a brief hint of Rodrigues (1840)'s results; in the second one, I will discuss Chasles (1843) and in the third one, I will consider the spread of Chasles' ideas and the synthetic demonstrations which other authors gave of his theorems.

3.3.1 A Hint to Rodrigues' Results

In the period 1840–1880, Rodrigues (1840) was a relatively well-known paper within the milieu of the scholars dealing with the theory of the rigid body's movement. However, Rodrigues' ideas and contributions were progressively forgotten or neglected. It is a great merit of Jeremy Gray to have rediscovered the importance and originality of Rodrigues (1840). After the pioneering (Gray, 1980), a series of publications on Rodrigues appeared.[27] Gray (1980) points out how advanced Rodrigues' conception of geometrical transformations was, until including most of the features which connote the concept of transformation group. For the context I am dealing with, it is important to refer to Rodrigues' results which are directly connected to Chasles' work. Rodrigues (1840) is divided into a synthetic (pp. 380–397) and an analytical part (pp. 398–440), in which several kinds of transformation formulas for the movements of a rigid body are given. I will focus on the synthetic part because it is more directly connected to Chasles' results.

In the introductory section of his paper, Rodrigues claimed that in an infinitesimal rotation, the arc described by any point of a rigid body can be identified as the chord connecting the two positions of the point, before and after the movement. On the other hand, any (finite or infinitesimal) translation can be considered as a rotation of infinitesimal amplitude around an infinitely far axis which is perpendicular to the translation's direction (Rodrigues, 1840, p. 381). After having explained the way in which different translations have to be composed, a purely geometrical demonstration follows concerning the property according to which if a rigid body rotates around a point it also rotates around a straight line. As we have seen, this feature was already proved by Euler, but Rodrigues' demonstration is far more instructive in

[26] In the fourth paper of Chasles (1860–1861), p. 194, n. (1) the author mentions Rodrigues (1840). In particular, he refers to a theorem proved by Rodrigues, that he demonstrated with a different method and remembered that Rodrigues analysed the composition of two finite rotations around two incident axes.

[27] Besides Gray (1980), see Altmann (1989, 2007); Altmann and Ortiz (2005), Garza and Quintanilla (2011), Williams and Fyfe (2010).

order to guess the geometrical properties of a rigid body's movement (ibid., pp. 383–384). After these introductory theorems, the fundamental part of Rodrigues' works follows. Rodrigues considered a rigid body in two different positions. Through a point P in the initial position, he imagined to draw a series of straight lines equal and parallel to those which, in the final position, join the point P' corresponding to P to all the points of the body in its final position. In this manner, a body C will be constructed which is identical to the original one, but whose position is derived from the original position of the body rotated around a fixed axis. If C is now opportunely translated in the direction which joins P to its correspondent P', then C will overlap with the final position of the displacement (ibid., p. 384). Though conceived in a slightly different manner, this reasoning is the same used at the beginning of Chasles (1830i)—see Sect. 3.1—to show that any two positions of a rigid body can mutually overlap by a rotation around a fixed axis and a translation. After an additional, brief and easy reasoning, Rodrigues came to state what he called "Théorème fundamental" which he expressed in these terms:

> Be a solid body transported from a place to another place, it is always possible to consider such displacement as resulting from two consecutive displacements, one of which is a rotation and one is a translation. The rotation takes place around a fixed axis through any point of the solid in the first position. Such axis is drawn parallel to a certain direction, which is determined by the two considered positions of the solid body. The two positions also determine the sense and the amplitude of the rotation. The translation takes place parallel to the straight line joining a point of the axis to its correspondent in the second position of the system. The length of this line measures such translation.[28]

In general, for any given axis of rotation, the axis of translation is not parallel to that of rotation. Rodrigues added that, in this situation, the rotation and the translation commute. After that, he proved a series of results, many of which had been enunciated by Chasles without demonstration: in particular, given a displacement of a rigid body, the straight line segment joining any point to its correspondent after the displacement can be interpreted as the third side c of a triangle of which the first side a varies according to the chosen point of the body and is perpendicular to the rotation axis. The second side b is constant and measures the translation of the system relative to the origin of the displacement (Fig. 3.3).

To be more precise: the projection of c along the rotational axis is constant for any chosen axis, but it varies from axis to axis because such a projection is the sum of two projections. One is that of the chord of the rotation's arc, and it is variable according to the chosen axis of rotation; the other one is due to a pure translational

[28] Rodrigues (1840, p. 385): "De quelque manière qu'un solide ait été transporté d'un lieu dans un autre, on peut toujours considérer ce déplacement comme résultant de deux déplacements consécutif en rotation et en translation, la rotation s'effectuant autour d'un axe fixe mené par un point quelconque du solide dans le première situation, parallèlement à une certaine direction, invariablement déterminée par les deux situations considérées du solide, aussi bien que le sens et l'amplitude de la rotation; la translation s'opérant ensuite parallèlement à la droite qui joint un point de cet axe à son correspondant dans le seconde situation du système, la grandeur de cette droite mesurant celle de la translation".

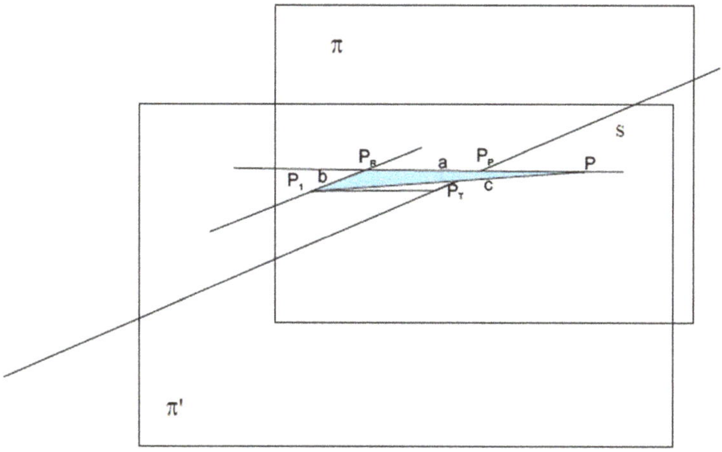

Fig. 3.3 Reconstruction of the figure described in the running text with all the steps: s is the axis of rotation. The considered point is P. Its projection on the axis is P_p, which lies in the plane π orthogonal to the axis. In π the point P_r symmetric of P_p with respect to P is determined. The rotation around s moves P in P_r. The chosen translation b (case in which the translation is parallel to the axis of rotation, but the general case is easily constructable on the basis of this specific one) moves P_r in P_1, which is the point corresponding to P after the whole movement

movement, and it is constant for any axis of rotation. A direction exists which makes the first component of the projection null. This direction is that parallel to the axis of rotation. Therefore, if one chooses such an axis, the translation will be parallel to it and the projection on it will be a minimum. Rodrigues called this projection "the absolute translation of the system" (ibid., p. 386). The so determined axis is the *axe central du déplacement* (central axis of displacement, ibid., p. 386). Rodrigues also taught how to construct the central axis (ibid., pp. 386–387). In a note he claimed that this theorem was due first—probably—to Chasles (1830i). After that, until the section entitled *Composition des rotations infiniment petit* ("Composition of infinitesimal small rotations", ibid., p. 391), Rodrigues specified a series of results, many of which had already been obtained by Chasles. For example, while dealing with the displacement of a flat figure in its plane, he "rediscovered" the mean plane, of which Chasles had already spoken, as we have seen, and that he explicitly named "mean plane" in 1860. Rodrigues discovered that any displacement of a rigid body can also be obtained by the rotation around two axes, one of which is arbitrary and the other is determinable. These two axes are nothing but the straight lines discovered by Chasles in what I have indicated as Theorem 1) item 2 (Sect. 3.1 *Chasles' results 1*). The nature of such axes, specifically in reference to the movement of a rigid body, was specified by Chasles in Theorem 2), item 2 (Sect. 3.1 *Chasles' results 1*). Chasles would furtherly study the features of these two straight lines. Therefore, until the named section *Composition des rotations infiniment petit* Rodrigues might have been seen, in fact, as an epigone of Chasles who specified his ideas on the foundation of mechanics on projective geometry. However, in the second longer and more important part of his paper, Rodrigues developed a personal theory, that of the composition of movements. In this theory, as Gray claims, there are many elements of group theory also concerning the distinction between commutative and not commutative movements with the fundamental theorem that the infinitesimal rotations commute, whereas the finite ones do not. Chasles developed his theory in a different direction. Though in his following papers on the movement of a rigid body there are some theorems concerning the composition of rotations, this was certainly not the focus of his research. Rather, he tried to further specify his basic concepts and to connect them to some properties of the algebraic surfaces, basically, but not exclusively, to the quadrics. We will now follow such developments.

3.3.2 Chasles' Theory in 1843

1843 is an important year for Chasles' foundational programme because he published the 15-long pages paper "Propriétés géométriques relatives au movement infiniment petit d'un corpe solide libre dans l'espace" (Chasles, 1843). On this occasion he:

a) Named many objects and specified many concepts which were already present in Chasles (1830i) and, above all, in Chasles (1837a). Naming the object and the concepts is a meaningful operation because it indicates that in the mind of the author such objects and concepts have reached a definiteness which they did not have before.
b) Offered an enormous series of theorems concerning the movement of a rigid body, the most important of these theorems have a dual character.
c) Showed that many geometrical properties are deducible from the objects and concepts he had introduced.
d) Gave some trigonometric formulas concerning the infinitesimal movement of a rigid body.

The demonstrations are missing.

The three basic concepts from which the whole of Chasles' treatment arises are those of "foyer d'un plane" ("focus of a plane"), "caractéristique d'un plan" ("characteristic of a plane") and "droits réciproques" ("reciprocal straight lines"). Considering an infinitesimal movement, Chasles wrote:

> Given a plane regarded as a part of a body, the planes normal to the trajectories of its points will pass for a point of this plane itself. I will call this point *focus* of the plane.[29]

The focus of a plane is exactly the object he had introduced in Theorem 2) (see Chasles, 1837a, pp. 675–678), item 2., of Chasles (1837a) without naming it. After

[29] Chasles (1843, p. 1420): "Un plan étant considéré comme faisant partie du corps, les plans normaux aux trajectoires de ses points passeront tous par un même point de ce plan. J'appellerai ce point *foyer* du plan". Italics in the text.

that he defined the concept of straight line *characteristic* of a plane in the following manner:

> In a plane an infinity of points whose trajectories belong to the plane itself exists. All of them belong to a straight line. I call this line the *characteristic* of the plane. I will explain why I am using this denomination.[30]

Finally:

> When several planes pass through a straight line D, their *foci* are on a second straight line Δ; reciprocally, if the planes pass through this straight line Δ, their foci will lay on the first straight line D, so that these straight lines have reciprocal properties.[31]

Chasles specified that, considered a straight line D as belonging to the body, the planes normal to the trajectories of its points will pass through the straight line Δ. Reciprocally, the planes normal to the trajectories of the points of Δ, considered as belonging to the body, will pass through the straight line D. Chasles named D and Δ "droites conjugées" ("conjugated straight lines", ibid., p. 1421) and claimed that they allow us to discovery many properties of the infinitesimal movements.

Three commentaries are necessary: a) the straight lines D and Δ had been already introduced by Chasles in the *Aperçu* (item 2 of the proposition I have called Theorem 2, Chasles, 1837a, p. 675) as a particularization of Theorem's 1 item 2 (Chasles, 1837a, p. 577). This marks a noticeable continuity in Chasles' way of thinking and shows that he, in the course of his scientific career, often reworked and specified conceptions already present in the previous stages of his thought; b) the theory of projective transformations is the basis for the entire branch of physics concerning the movement of a rigid body. When metrical properties like the perpendicularity are necessary, this does not constitute a problem for Chasles since he also had the idea to express the transformations connoting the metrical properties as particular reciprocities or homographies. This means that Chasles' foundational programme is internally interconnected. It is a real programme, not a series of isolated ideas; c) the assertion that a generic straight line can be considered as belonging to the body means that Chasles was not referring to the body as a single object but to two different positions of space considered as overlapped to itself. That is: he was considering the entire space and its projective transformations as the object of his research. The body is seen as a subset of space.

The next propositions by Chasles stated that, given some planes passing through a point P, their foci belong to the same plane π having its focus in such a point (Chasles, 1843, p. 1421).

---

[30] Ibid., p. 1420: "Dans le plan, il existe une infinité de points dont les trajectoires seront comprises dans le plan même; tous ces points sont situés en ligne droite. J'appellerai cette droite *caractéristique* du plan; je dirai plus loin la raison de cette dénomination". Italics in the text.

[31] Chasles (1843, p. 1420): "Quand plusieurs plans passent par une même droite D, leurs *foyers* sont sur une deuxième droite Δ; réciproquement, si plusieurs plans passent par cette droite Δ, leurs *foyers* seront sur la première droite D. De sorte que ces deux droites jouissent de propriétés réciproques". Italics in the text.

The dual character of these propositions is conspicuous: in space the two conjugated straight lines as well as the point P and the plane π are dual entities of that particular reciprocity which is the infinitesimal movement of a rigid body.

With the clarification of the concepts already explained, Chasles was able to insert the screw-theorem within a complete theory. These are the steps:

First of all, he claimed that when some planes are parallel (i.e., they meet on the line at infinity), their foci are on a straight line which is always parallel to an axis. This axis has a particular property which is a direct consequence of the way in which it has been determined: for since the points of the axis are the foci of the given parallel planes, then the instantaneous trajectories of all these points are mutually parallel because they are normal to the planes of the given bundle. This means that, during the body' displacement, the axis only translates on itself without any rotational motion. The reduction of the infinitesimal movement of a rigid body to that of a screw in its nut is immediate (ibid., p. 1421). Chasles claimed that the theory of the infinitesimal movements is a part of the theory of the finite displacements, which he intended to complete in a following contribution. This contribution is Chasles (1860–1861).

Next, he claimed that the infinitesimal movement of a rigid body can also be determined by means of two rotations around two conjugated straight lines D and Δ (ibid., p. 1421). Though the proofs of several of these theorems are not completely trivial, Chasles did not give them. We will see that his epigones would provide these proofs. A commentary seems appropriate: Chasles was framing all his concepts within a general theory. This is an absolutely original theory. An indirect proof of this fact is that some basic demonstrations—as that of the screw-theorem—are founded in Chasles on concepts which are slightly different from those used by Rodrigues, though both proofs are synthetic.

At this point, Chasles established the main properties of the *characteristic* straight line. He claimed that the tangent to the trajectory of a point is the characteristic of a plane and, vice versa, the characteristic of a plane is tangent to the trajectory of one of its points. Furthermore, the tangent to the trajectory of a point P has as conjugate the characteristic of the plane whose focus is P. This means that, if, through the focus of a plane, the normal to the plane is drawn, then the conjugated straight line to such a normal is the characteristic of the plane. Therefore, the movement of a plane is reducible to a rotation around the characteristic, while the characteristic rotates around the plane's focus. Thence, the infinitesimal movement of a flat figure in space is reduced to the rotation of the figure's plane around a straight line belonging to the plane, while this straight line rotates around the focus of the plane, without exiting the plane. The characteristic is, therefore, the intersection of two infinitely close positions of the plane (ibid., pp. 1421–1422).

This completes the treatment of both spatial and plane infinitesimal displacements. However, Chasles' research is far from being concluded: he proceeded by showing the profound interconnections between the concepts already introduced. As a matter of fact, he explained some features of the conjugated straight lines, which implies unsuspected geometrical properties. The most interesting of such properties are the following ones: a) two couples of conjugated straight lines D, Δ and D', Δ'

are the generators—belonging to the same family—of a one-sheeted hyperboloid, that is, if a straight line meets three of them, it also meets the fourth one; b) the segment measuring the smallest distance of two conjugated straight lines cut the axis of rotation and is perpendicular to the axis; c) all the planes perpendicular to the axis X cut D, Δ and X in three collinear points. That is, through two conjugated straight lines D and Δ and through X, it is possible to construct a hyperbolic paraboloid whose generatrices are perpendicular to X; d) when some lines $D, D', \ldots$ are incident in a point P, their conjugates $\Delta, \Delta', \ldots$ belong to a plane which is normal to the trajectory of P. Reciprocally, when some lines belong to a plane π, their conjugated lines pass through the focus of π.

Commentaries

i) These theorems show the inner interconnections of Chasles' theory, which goes far beyond the "Mozzi-Chasles theorem". This proposition is only the peak of an iceberg, where the most interesting propositions show the dependence of mechanics from projective geometry and, at the same time, the possibility to deduce geometrical theorems from propositions concerning the movement, but only when the doctrine of the movement has been geometrically founded. If one analyses Chasles' train of thought in depth, one will realize that his fascinating foundational programme shows all its power and inner coherence. The basic concepts of projective geometry were the continuous reference point for Chasles. For example, the theorems concerning the focus of a plane show a beautiful example of duality straight line-straight line and point-plane (plane-point) obtained by the concepts Chasles introduced to address the motion of a rigid body.
ii) The concepts he created in this theory were new at that time and today they are not used anymore, or they are only marginally used. It is true that, following the scarce indications left by Chasles, it is possible to achieve the proofs of his theorems. Nonetheless, the proofs given by some mathematicians of Chasles' assertions are very useful to grasp the demonstrative tissue connoting this theory. For this reason, I will dedicate the next section to this subject. Before dealing with the works of Chasles' epigones, it is, however, useful to refer briefly to another series of Chasles' results which connect the theory of the rigid body's movement and geometry.

One of these results also has a terminological value. For Chasles claimed that if a straight line is tangent to the trajectory of one of its points, the tangents to the trajectories of its other points are coplanar and envelop a parabola whose focus is the focus of the plane (ibid., p. 1424). Hence, the name of focus of a plane derives from that of parabola's focus.

The theory of the movements is also connected with some elements of algebraic geometry. Chasles actually claimed that if a curved surface makes an infinitely small movement, the planes normal to the trajectories of its points envelop a second surface such that the planes normal to its trajectories will envelop the first surface in turn. The two surfaces are reciprocal. If the first one is algebraic, the second is algebraic too, though, generally, the degrees of the two surfaces are different (ibid.,

p. 1425). These theorems are a specification of items 3) of Theorem 1) and of Theorem 2) (see Chasles, 1837a, p. 577 and p. 675, respectively).

Chasles also provided the reader with some metrical formulas concerning the infinitesimal movements (Chasles, 1843, pp. 1426–27): be, as before, D and Δ two conjugated straight lines; X the rotation axis; ν the rotation around X; e the translation of the axis along its own direction; r the smallest distance between D and X; ρ the smallest distance between Δ and X (we recall that r and ρ are segments of the same straight line perpendicular to the axis); (D,X) and (Δ,X) be the angles between the straight lines D and Δ and the axis X, respectively, then it holds

$$r \tan(\Delta, X) = \rho \tan(\Delta, X) = \frac{e}{\nu}$$

Let us now consider the infinitesimal movement as decomposed into the rotations around D and Δ, then, if D is directed along the trajectory of one of its points, Δ will be in the plane normal to such trajectory and it will be:

$$\tan D \cdot \tan \Delta = 1; \qquad r\rho = \frac{e^2}{\nu^2} = \text{constant},$$

If Ω and ω denote, respectively, the rotations around D and Δ, the following relations hold:

$$\Omega^2 = \frac{e^2\nu^2}{[r\nu \sin(D,X) + e\cos(\mathrm{D},\mathrm{X})]^2}; \qquad \omega^2 = \frac{\nu^2(e^2 + r^2\nu^2)\sin^2(D,X)}{[r\nu \sin(D,X) + e\cos(D,X)]^2}$$

from which, with relatively easy calculations, it follows:

$$\frac{\Omega}{\omega} = \frac{\sin(\Delta,X)}{\sin(D,X)}, \qquad \Omega^2 + \omega^2 + 2\Omega\omega\cos(D,\Delta) = \nu^2, \qquad \Omega\omega \sin(D,X) \times (\varrho + r) = e\nu$$

A further remarkable result, with which I close the exposition and commentary of Chasles (1843), is that, if one projects on a straight line D the trajectories of its points, all the projections are equal. Their length p is inversely as the body's rotation around D. Precisely:

$$\Omega p = \nu e.$$

As Chasles pointed out, the projection p expresses the quantity of which each point of the straight line D moves in the direction of this line itself. Thus, it indicates the movement of the straight line in its own direction. The previous equation shows that the rotation of a body around any straight line is inversely as the movement of this straight line considered in its own direction (ibid., p. 1427).

Chasles also obtained other metrical results, but these are enough for my aim.

3.3.3 The Spread of Chasles' Theory

From a historical standpoint, a remarkable question concerns the spread of Chasles' ideas on the treatment of the movement of a rigid body. As a matter of fact, the approach to kinematics adopted by most scholars was analytical, whereas in Chasles' works the formulas are scarce and almost all the reasonings are developed through geometrical methods. Therefore, one may wonder how much his results as well as his mathematical methods were spread and known. The answer is that a certain, though not a great, number of mathematicians and physicists followed Chasles' approach. In spite of the fact that the analytical methods were preponderant, then, he had some epigons.

The most reasonable manner to understand the situation is to refer to Chasles' works themselves because he was concerned with the spread of his ideas. Hence, the historical annotations in Chasles (1860–1861, fifth part, pp. 499–500, notes) are a good reference point to grasp how much his ideas were spread in the mathematical environment. Among the authors who were inspired by his work, Chasles mentioned Breton 1838 (solution of some geometrical problems basing on the displacement of a rigid body); Transon 1845 (a paper on the curvature radius of some classes of curves); Salmon 1852 (a very famous treatise on plane curves); Bresse 1853 (a paper on the application of the infinitesimal movement to the determination of the curvature radiuses); Gilbert 1857 (an important memoir on the geometry of the plane movements which won the price of the Real Belgian Academy and was also published as separated volume in 1861, see Gilbert 1861), Mannheim 1858 (paper on the use of the rigid movement to determine the curvature centres of a plane figure which rotates in its plane).

Let us see the most important of these works in order to understand Chasles' influence on these scholars.

Breton (1838) is an interesting paper. The author moves from the principle that every infinitesimal movement of a rigid body constrained at a fixed point is a rotation around an axis passing through that point. Immediately afterwards, Breton ascribed to Chasles the discovery of the instantaneous centre of rotations and claimed to have been inspired by Chasles' method to draw the tangents to a vast class of curves (Breton, 1838, pp. 488–489). This method is based on the instantaneous movement of a curve on another. Descartes had also given a single example of its utilization. Breton studied the instantaneous centre of rotation in connection with the theory of envelopes. He faced problems such as how to construct the envelope of the successive positions of a segment which glides on the sides of a right angle (ibid., p. 490). In the third section Breton used the properties of the instantaneous centre of rotation to solve some geometrical problems. Therefore, Breton's work is close to Chasles' way of thinking and is profoundly influenced by his ideas and results.

Salmon mentioned Chasles five times, but his reference to Chasles' works on the displacement of a rigid body is marginal within the structure of Salmon's treatise.

Bresse (1853) mentioned the results on the instantaneous centre of rotation claiming that they can be used to draw the tangents to several curves, but he did not mention Chasles and ascribed the merit to Euler. Chasles wrote a polemic note in Chasles (1860–1861) against Bresse on this issue, thus claiming credit for the discovery of the instantaneous centre of rotation and for the method of drawing the tangents basing on the infinitesimal movements (Chasles, 1860–1861, fifth part, p. 499, n. (1)). Bresse, in an interesting appendix to his work (Bresse, 1853, pp. 109–115), also quoted other authors as Arnoux, Duhamel and Rivals in connection with these ideas, but he did not mention Chasles.

This shows that the above-mentioned works used some of Chasles' results to develop a series of independent considerations and, which is still more important in my perspective, did not consider Chasles' results as an organic doctrine. A remarkable exception is Breton (1838).

However, there is a series of other works, which, *in toto* or partially, are dedicated to the spread of Chasles' doctrine. The authors of these treatises recognized Chasles as the founder of a School based on geometrical methods and on the conviction that projective geometry is the basis to deal with the fundamental sections of mechanics. Chasles (1860–1861, fifth part, p. 500, note (4)) mentions Steichen (1855), Jullien (1855), Jonquières (1856), Lamarle (1859) and Bellavitis (1860). After 1861, it is at least necessary to remember Chelini (1862b), Brisse (1870, 1874, 1875).

Steichen (1855) is a noteworthy introduction to all the geometrical properties necessary in the study of mechanics. The author recognized Chasles as a prominent scholar in this field. First of all, he named Chasles in the Introduction (ibid., p. 13), essentially because of the concept of anharmonic ratio. He referred to Chasles' *Traité de géométrie supérieure* as an important reference point. In the course of his treatise, Steichen dedicated the §§ 38–40 to the geometrical infinitesimal movement of a rigid body and to its finite displacement. He mentioned Euler and Lagrange (ibid., p. 151), but Chasles is by far the most mentioned author. The § 38 is entitled "Thèoréme de Chasles". Steichen quoted and proved the initial theorem of Chasles (1830i). He also mentioned Bobillier. He expounded the theorem in the following manner:

> When two equal figures, which are overlapping without an inversion are posed in any manner on a plane, a point equidistant from two any homologous points of the figure exists. This point is the intersection of the perpendiculars which join two any vertices of a figure to their homologous points of the other figure: this is the centre of equidistance, which is self-homologous.[32]

[32]Steichen (1855, p. 119): "Quand deux figures égales et superposables sans retournement sont placées d'une manière quelconque dans un plan, il existe toujours un point du plan également distant de deux points homologues quelconques des deux figures. Ce point est à l'intersection des perpendiculaires, aux milieux des droites qui unissent deux sommets quelconques de l'une aux deux points homologues de l'autre : c'est le centre d'équidistance, qui est son propre homologue".

Following the contributions given by Euler, Lagrange and Chasles himself, Steichen arrived at demonstrating the screw-theorem, he called Chasles' theorem ("découvert par M. Chasles", ibid., p. 131). In an interesting remark (ibid., p. 160–162), after having expounded the geometrical and the analytical features (Moebius, Poinsot, Stegmann and Rodrigues are mentioned) of the infinitesimal movements and of the finite displacements, Steichen underlined that both methods are useful and that one cannot be neglected in favour of the other. In particular, he pointed out that all the analytical formulas discovered until that moment were, anyway, referred, at least with regard to their origin, to geometrical considerations. They were not, so to say, "purely analytical". Furthermore, the principles behind Euler's and Lagrange's formula were not yet clear. This was not true anymore for the formulas discovered after Chasles (1830i). Thence, his work sheds a great light on the general understanding of the problem of rigid body's movement. Steichen wrote:

> One has to say that M. Chasles spread a new light on the advanced theory of the two geometers [Euler and Lagrange], which made such a theory complete and more explicit thanks to the enunciation of his two theorems [on the infinitesimal and finite displacements], given without a proof. These theorems make any secondary consideration on the rotation around parallel axes useless and can be applied to the virtual and finite displacements of the invariable bodies.[33]

Chasles is also named because of the consequences of his considerations on the displacement of a rigid body as to the problem of drawing the tangent to a curve c from a point of C (ibid., p. 164). The other mentions concern the concept of anharmonic ratio. It is worth highlighting that Steichen mentioned only Chasles (1830i, 1852).

Among the authors profoundly influenced by Chasles' results and methods one of the most important is Jullien. Jullien (1855) is a text basically conceived to solve the problems of rational mechanics proposed at the universities or at the polytechnic schools. It is an advanced handbook. Between the sections dedicated to statics and kinematics, Jullien inserted a section entitled "Étude géométrique du movement ou cinématique". It is divided into two chapters, the former is a theoretical one, the latter is applicative. The theoretical chapter (ibid., pp. 163–179) is entirely dedicated to Chasles' results and to the demonstration of the theorems he had only enunciated.[34] The name of Poinsot and his celebrated work *Théorie nouvelle de la rotation des corps* (Poinsot, 1851) are associated with Chasles, and the way in which Jullien referred to the propositions and his geometrical proofs are clearly in Chasles' style.

---

[33] Ibid., p. 161: "On peut dire que M. Chasles a jeté une nouvelle lumière sur la théorie savante de ces deux géomètres, et qu'il l'a rendue plus complète et plus explicite par l'énoncé de ses deux théorèmes, donnés sans démonstration, qui rendent même inutile toute conception secondaire de rotation autour d'axes parallèles, et qui se rapportent aux déplacements virtuels et finis à la fois des corps invariables".

[34] Jullien referred to two works by Chasles. The former is clearly Chasles (1830i). With regard to the latter, Jullien spoke of a paper published "[...] dans les *Comptes rendus de l'Acadèmie des Sciences* pour 1853" (ibid., p. 163). The date is clearly a typo. The reference and the content leave no doubt that the correct indication is Chasles (1843).

Jullien, rightly, considered Chasles and Poinsot as the representatives of a conception of science in which the geometrical aspects of mechanics are regarded as essential.

Lamarle's work on the instantaneous centres and axes of rotation of various rigid figures in plane and in space starts from presuppositions which are slightly different from Chasles' because Lamarle considered the field of velocities of an instantaneous motion. For example, his first theorem reads as follows:

> When the simultaneous velocities of different points of a straight line are transported in the same point, the locus of their extremities is a straight line normal to the first one.[35]

In contrast to this, Chasles did not introduce the notion of velocity in his contributions on the geometrical motions because, in his point of view, the geometry of the motion should provide the real geometrical bases of the movement. Velocity is a kinematical concept, but not a geometrical one. This notwithstanding, Chasles was a constant reference point for Lamarle. He referred to and proved Chasles' proposition according to which the minimal distance segment between two conjugated straight lines cuts in its central point the axis of rotation and is perpendicular to it (ibid., p. 58). He mentioned Chasles' results on the conjugated straight lines obtained by Lamarle with slightly different concepts and demonstrative methods (ibid., pp. 108–109). Lamarle also dealt with the straight line called by Chasles "characteristic"—as Lamarle himself reminded the reader—(ibid., p. 129). Lamarle recalled that Chasles (1843) was one of his sources of inspiration (ibid., pp. 130–131). He referred to several of Chasles' propositions as "[. . .] une suite de propositions curieuses" ("[. . .] a series of curious propositions", ibid., p. 131). This shows that the mathematical environment perceived the novelty of Chasles' approach and way of thinking. At the same time, the adjective "curieuses" seems to indicate some clever and interesting discoveries, but whose real importance within mathematics is debatable. Almost surely Lamarle was not referring to Chasles' propositions directly concerning the geometry of the rigid body's movement, but to that very long series of propositions inherent to the geometrical figures (I have offered the example of some of them regarding the one-sheeted hyperboloid and the paraboloid) in which Lamarle, probably, perceived an exoteric character. He also addressed the results obtained by Chasles on the focus of a plane as well as the metrical formulas expounded in Chasles (1843). All in all, this contribution was the fundamental reference point for Lamarle.

Bellavitis in his treatise published in 1860 shows to be well informed on Chasles' results because he mentioned Chasles (1830i, 1837a, 1843), but Chasles' influence on this author does not seem preponderant with respect to that of other mathematicians. It is enough to refer to the huge literature posed by Bellavitis at the end of every chapter of his work to understand that Chasles' influence, though significant, was not crucial. Bellavitis was, however, well aware of Chasles' results. He actually

---

[35] Lamarle (1859, p. 3): "Lorsque les vitesses simultanées des différents points d'une droite sont transportées , en un même point, le lieu de leurs extrémités est une droite normale à la première".

referred to the latter's theorem according to which an instantaneous movement can be reduced to a rotation around two axes and to the consequences drawn by Chasles (Bellavitis, 1860, p. 126). He also remembered that Chasles had introduced the notions of focus and characteristic of a moving plane (ibid., p. 127).

Chelini (1862b) is an important text in the contest I am analysing because this author, besides referring to all the mathematicians who dealt with the displacement of a rigid body and ascribing a particular merit to Euler, Moebius and Olinde Rodrigues for the analytical part of this topic, added that he himself had arrived at the initial theorems explained without the demonstration in (Chasles, 1860–1861) which are dedicated to the finite displacement of a rigid body. Chelini decided to publish the results of his research because they are a useful introduction—enriched with all the demonstrations—to Chasles (1860–1861), and, in fact, of Chasles (1843) too (Chelini, 1862b, p. 362). His memoir is divided into two parts, the first one is geometrical, the second analytical. The former—to which I will refer after having examined (Chasles, 1860–1861)—can be interpreted as the demonstration of the initial theorems given by Chasles (1860–1861). Therefore, Chelini clearly recognized the latter as the father of the new geometrical doctrine of movements.

Brisse (1870) and Brisse (1874) are two papers in which the author proved, respectively, the theorems enunciated in Chasles (1843) and Chasles (1860–1861). These works are useful, but, in my opinion, several proofs, though correct, are not given in the most clear and transparent possible manner. In this respect Chelini's work is better conceived, but since Brisse's concerns most of Chasles' theorems, it is more important than Chelini's from a historical standpoint, though it is inferior in a purely mathematical perspective.

I have postponed the analysis of Jonquières (1856) because this is the work in which the author showed to have profoundly understood and shared the reasons which induced Chasles to formulate his theory. All the works I have mentioned so far regard as fundamental Chasles' contributions to the study of the movement of a rigid body, but Jonquières went beyond these considerations. He actually recognized Chasles as the founder of a School in which a new way to conceive synthetic projective geometry and his relations to mechanics is proposed and developed. Furthermore, Jonquières' proofs of the theorems enunciated in Chasles (1843) are so clear and well argued that, probably, Chasles himself could not have done better! Jonquières also spoke of a philosophy devised by Chasles (Jonquières, 1856, pp. 44–48). I will refer to this part of his work in the chapter on philosophy of duality. The other reference point of Jonquières' treatise was Poinsot. Jonquières showed a precise knowledge of Chasles' works, in particular of Chasles (1830i, 1837a, 1843, 1852). In the next section, I will report and comment on some of the proofs given by Jonquières of the theorems enunciated in Chasles (1843). This is necessary to grasp exactly what these mathematicians meant by geometrical methods and why the reader cannot have a precise opinion of Chasles' way of thinking and work without the proofs of his theorems. Probably he was so used to face these questions daily that the proofs were evident to him, but, for sure, they were not for his contemporaries, who offered them explicitly and they are not evident for most of us.

All in all, it can be argued that in the period 1830–1870, Chasles' doctrine was certainly known, though only by a restricted group of mathematicians. Jonquières himself regretted the fact that the methods of pure geometry were neglected for a long time in France, even if France was the homeland of these methods. He claimed to be glad that thanks to the works of some illustrious scholars the situation was changing. Jonquières was afraid that in England, Belgium, Italy or Germany these methods would be cultivated more than in France (Jonquières, 1856, pp. v–vi). Chasles' results on the displacement of a rigid body and his foundational ideas were spread in a relatively small group of mathematicians because of the progressive replacement of the synthetic methods with those analytical. Furthermore, the geometrical concepts by Chasles were replaced by vector concepts, which have a geometrical origin, but whose applications do not need the whole set of notions invented by Chasles.

3.3.4 *Jonquières and His Demonstrations of the Theorems in Chasles (1843)*

Chasles used these terms to define the demonstrations given by Jonquières of his theorems:

> All these theorems have been proved with a rare simplicity and elegance through the geometrical method by M. Jonquières in his *Mélanges de géométrie pure* [. . .].[36]

Jonquières' proofs are referred to Chasles (1843), therefore, as the author himself pointed out, they concern the geometrical properties of a rigid body's infinitesimal movement. The first chapter (pp. 1–54) of Jonquières (1856) is dedicated to this question. I will focus on some demonstrations which are particularly significant to grasp the geometrical bases of the reasonings behind Chasles' doctrine.

Jonquières, following Chasles (1843)'s train of thought, analysed, first of all, the infinitesimal movement of a plane (see Fig. 3.4).

He considered a plane Q in a given initial position and its position R after an infinitesimal movement. The two planes will cut along a straight line $a'b'$. This is Chasles' *characteristic*.[37] Be ab the original position of $a'b'$ on Q. After the movement the point a has reached the position a' and b the position b'. Now Jonquières considered the mean point α of aa' and erected the perpendicular αO in the plane R. He carried out the same operation with the segment bb' which has two homologous points as extremes, so determining the position of the point O.

[36] Chasles (1860–1861, 4, p. 500): "Tous ces théorèmes ont été démontrés avec une facilité et une élégance de méthode géométrique rare, par M. Jonquières, dans ses *Mélanges de géométrie pure* [. . .]". Italics in the text.

[37] Though the properties of the "characteristic of a plane" were profoundly studied by Chasles with a synthetic approach, Jonquières (ibid., p. 2) pointed out that this concept was introduced by Monge (see, i.e., Monge, 1807).

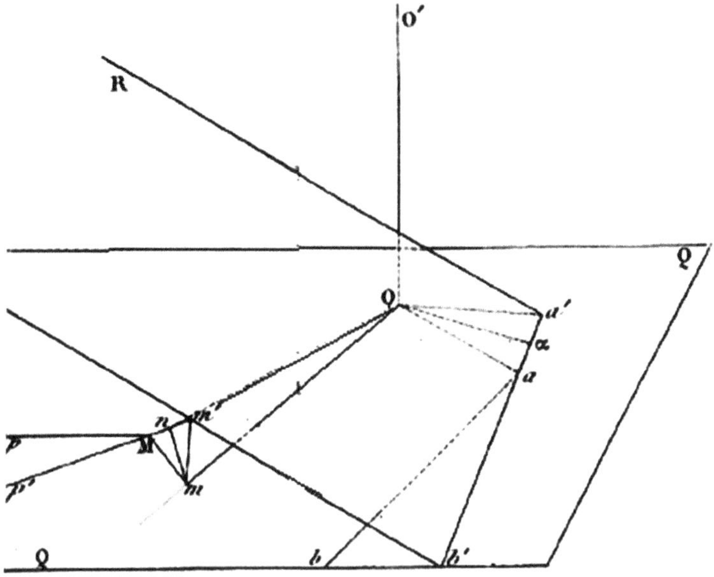

Fig. 3.4 The figure used by Jonquières to analyse the infinitesimal movement of a plane (from Jonquières, 1856, Planche I, Fig 1)

Furthermore, let us draw the perpendicular OO' to the plane Q. Given these constructions, a rotation of the plane Q around the axis OO' will drive the straight line ab on the characteristic $a'b'$. Finally, a rotation of the plane Q around $a'b'$ will drive the plane Q in the position R. The two rotations of Q around OO' and around $a'b'$ can be imagined successive or simultaneous. Therefore, Jonquières concluded:

> The movement of a plane is reduced to a rotation around the characteristic, while this straight line rotates, in the primitive position of the plane, around a point which can be considered a fixed point and which we call the focus of the plane.[38]

The point O is the focus of Q.

As Jonquières pointed out, any infinitesimal movement of a plane in space is reducible to a rotation of the plane around a straight line belonging to the plane itself, while this straight line rotates around a point remaining in the plane.

Since the motion is infinitesimal and since the focus of Q is on the centre of rotation of Q during its rotation around the characteristic, the movement of the focus will be an infinitesimal arc perpendicular to Q (ibid., p. 3).

[38]Jonquières (1856, p. 2): "Le mouvement du plan se réduit à une rotation autour de la caractéristique, pendant que cette droite tourne, dans la position primitive du plan, autour d'un point qu'on peut considérer comme un point fixe, et que nous nommerons le foyer du plan". Italics in the text.

Brief explanation: the focus does not move on Q, but only participates to the rotation of Q around the characteristic. Since the movement is infinitesimal, the instantaneous tangent to the trajectory of the focus is orthogonal to Q.

Each other point M belonging to Q has a double movement: it describes an arc Mm around the focus and an infinitesimal arc mm' around the characteristic; mm' is perpendicular to Q. Thence, M describes the side Mm' of the triangle Mmm'. Therefore, Om is perpendicular to the plane of the infinitesimal triangle Mmm'. Consequently, if from m, the perpendicular mn to the trajectory is drawn, the plane Omn is perpendicular to the trajectory.

In this way, Jonquières has proved in a purely geometrical manner the existence of the basic concepts created by Chasles while dealing with the movement of a plane considered as a part of a rigid body moving in space. Jonquières thus reached these initial results:

> Therefore, given a plane considered as a part of a solid body, the planes normal to the trajectories of its points pass through the same point of this plane, which is the point that we have called the focus of the plane.[39]

The other fundamental concept introduced by Chasles (1843) was that of "conjugated straight lines". To prove their existence Jonquières proceeded like this: he considered two planes π and Π cutting along a straight line D. The planes perpendicular to the trajectories of their points will pass, respectively, through the foci F and F' of π and Π. Therefore, all the planes normal to the trajectories of the points belonging to D will pass through another straight line Δ because D is the intersection of the two planes. That is: the locus of the foci of a straight line considered as the axis of a bundle of plane is another straight line. Jonquières also proved the validity of the reciprocal proposition: the locus of the foci of the points belonging to the straight line Δ is the straight line D (ibid., pp. 4–5). The two straight lines D and Δ are Chasles' conjugated straight lines.

Jonquières considered the two conjugated straight lines D and Δ as belonging to the moving body. Then, the planes normal to the trajectories of the points belonging to Δ pass through D. This implies that the movement of Δ is nothing but a rotation around D. Vice versa, the movement of D is a rotation around Δ. The consequence is that in order to transport a body in its position after the infinitesimal movement, it is enough to impart the body two successive or simultaneous rotations around two conjugated straight lines. The two conjugated straight lines can be interpreted as simultaneous axes of rotation (ibid., p. 5).

The case of the star of planes is interesting, too. For if M is the centre of the star, the plane normal to the trajectory of M passes through the foci of the planes belonging to the star because M belongs to each of these planes. Therefore, all the foci of the star belonging to M belong to the plane having M as focus. Thence, when

[39] Jonquières (1856, p. 3): "Donc un plan étant considéré comme faisant partie du corps, les plans normaux aux trajectoires de ses points passent tous par un même point de ce plan, qui est le point que nous venons d'appeler le foyer du plan". Italics in the text.

several planes pass through a point, their foci belong to a plane having its focus in such a point.

These theorems show the geometrical reason of the duality straight line-straight line in the case of the two conjugated straight lines and the duality point-plane in the case of the planes' star when you consider the isometry determining the movement of a plane regarded as a part of a rigid body.

Jonquières is, thus, proving step by step all the theorems enunciated in Chasles (1843) and he is making this with the geometrical method already devised by Chasles.

The "Chasles-Mozzi theorem" is proved by Jonquières through this clear geometrical reasoning: let us consider a plane π as belonging to the instantaneously moving body. Be D the line at infinity of π and let us consider D as the intersection line of parallel planes. If Δ is the conjugate of D, its movement is a rotation around D, but since D is at infinity this rotation is, in fact, a translation of Δ parallel to itself. The translation moves Δ to the position X. If a series of parallel planes, whose intersection is a line at infinity D' different from D, is considered, the conjugate Δ' of D' will be moved by the translation exactly to the position X because the body is rigid. Therefore all the straight lines Δ, Δ',. . .conjugated of the lines at infinity belonging to all the planes regarded as part of the rigid body are mutually parallel. The trajectories of all the points of Δ are mutually parallel because these points are the foci of mutually parallel planes. Let us now consider the trajectories of the points of Δ'. The trajectories' directions of the points of Δ and Δ' make an angle equal to that of the two series of planes to which they are referred (ibid., pp. 6–7). To explain even more clearly the situation, one might replace Jonquières' expression "to that of the two series of planes to which they are referred" with "to that of the lines at infinity D, D', . . . of which Δ, Δ', . . . are, respectively, the conjugates". Therefore, Jonquières concluded:

> When some planes are mutually parallel, their foci are on a straight line which is always parallel to an axis, independently of the planes' common direction. This straight line has the property that the all of its points are mutually parallel, so that in the body's displacement the straight line has only a movement of translation parallel to itself. [. . .] If all the planes are perpendicular to the direction of this straight line, their foci will lay on a certain straight line X, which is parallel to that straight line. The trajectories of all the points will be directed precisely along this line X; so that this line will translate on itself during the movement of the body. During this translation the body cannot do nothing but rotate around X.[40]

[40] Jonquières (1856, p. 7): "Quand plusieurs plans sont parallèles entre eux, leurs foyers sont sur une droite qui est toujours parallèle à un même axe, quelle que soit la direction commune des plans. Cette droite jouit de la propriété que les trajectoires de tous ses points sont parallèles entre elles, de sorte que, dans le déplacement du corps, la droite n'a qu'un mouvement de translation parallèlement à elle-même. [. . .] Si tous les plans sont perpendiculaires à la direction de cette droite, leurs foyers seront sur une certaine droite *X*, parallèle à celle-là, et dont les trajectoires de tous les points seront dirigées précisément suivant cette droite *X*; de sorte que cette droite glissera sur elle-même pendant le mouvement du corps. Pendant ce glissement, le corps ne pourra que tourner autour de *X*". Italics in the text. Jonquières adds that in Poinsot (1851) *X* is named «axe spontané glissant».

This means that the instantaneous movement of the rigid body is reduced to a rotation around X and to a translation parallel to X.

With regard to the property that the tangent to the trajectory of a point is the characteristic of a plane and reciprocally, the characteristic of a plane is tangent to the trajectories of its points, the proof is not difficult (Jonquières, 1856, pp. 8–9).

Rather, it is worth noting the proof of Chasles' theorem according to which *the conjugate of the tangent to the trajectory of a point is the characteristic of a plane of which this point is the focus* (ibid., p. 9). For let us draw the normal n to a plane π through its focus F. All the points of n only rotate around the characteristic c of π in a plane Π normal to c. I add that this is the case because the movement of the focus around the characteristic is circular. Hence, the straight line n drawn through the focus perpendicularly to π is *tangent to the trajectory of the focus* and such a line belongs to a plane Π normal to π and, hence, to its characteristic. This means that n is the characteristic of Π. The planes normal to the trajectories of its points will pass through c and reciprocally. Thence, the straight lines c and n are conjugated. Considering that n is tangent to the trajectory of F, it follows exactly that *the conjugate of the tangent to the trajectory of a point is the characteristic of a plane of which this point is the focus*.

All these basic properties of the infinitesimal movement of a rigid body were proved by Jonquières only referring to geometrical concepts—thence without considering kinematical notions as velocity or acceleration—and through a pure geometrical method. Chasles in his *Aperçu* had claimed that the most appropriate method to prove such theorems was the geometrical one, but that in the *Aperçu* he had also resorted to the analytical method for the sake of brevity. Thence, Jonquières' work is an important step towards the full clarification of Chasles' foundational programme since he actually provided the geometrical demonstrations, thus proving what Chasles had claimed: all these truths are susceptible of a geometrical proof. We have seen that the first Chasles' step was to find geometrically the doctrine of the movement of the rigid body; the second was to apply, in turn, such a doctrine to geometry. In Chasles (1837a, 1843) he had offered several propositions either without giving their proofs or giving an analytical one. Let us see the Jonquières' geometrical proofs of the most important of Chasles' propositions.

As we have seen, Chasles in the *Aperçu* (Chasles, 1837a, pp. 675–6) stated that the planes normal to the trajectories of the points of an algebraic surface of degree m envelop a second algebraic surface to which it is possible to draw m tangent planes from a given line. Let us analyse Jonquirès' geometrical proof:[41] he considered an algebraic surface (as it was usual, he spoke of geometrical surface) of degree m. A transversal straight line D cuts such a surface A in m points. To each of these points, a plane normal to its trajectory will correspond, which touches a further surface A'. Since these m tangent planes are normal to D, they will pass through the straight line Δ, which is the conjugate of D. This means exactly that A' is an algebraic surface having m tangent planes passing through the straight line Δ. Jonquières added that

[41] In what follows the points and the planes can be real or imaginary.

the two surfaces belong to that kind of figures called "figures correlatives" by Chasles (ibid., p. 10). However, in the case of the figures generated by the infinitesimal movements, a particular dependence exists, which does not take place in the general correlative transformations because the planes of the generated figures pass through the points of the generating figure, which is not generally true in the reciprocities. In this case there is a perfect reciprocity between the order of the generating surface and the class of the generated surface, even though, in general, the two surfaces are of different order. In the specific case in which the first surface is of second degree, it is easy to notice that also the second is, so that when a quadric makes an infinitesimal movement, the planes normal to the trajectories of its point envelop another quadric and vice versa (ibid., pp. 11–12).

After that, Jonquières analysed the properties of two conjugated straight lines. Let us see how he proved two of the theorems stated in Chasles (1843): i) two couples of conjugated straight lines D, Δ and D', Δ' are the generators—belonging to the same family—of a one-sheeted hyperboloid, that is, if a straight line meets three of them, it meets also the fourth one; ii) all the planes perpendicular to the axis X cut D, Δ and X in three collinear points. That is, through two conjugated straight lines D and Δ and through X, it is possible to construct a hyperbolic paraboloid whose generatrices are perpendicular to X.

The first theorem proved by Jonquières states that:

> If the straight line D is normal to the trajectory of one of its points, all the other points will have their trajectories normal to this line, so that D will be self-conjugated.[42]

Jonquières considered such a straight line as belonging to the plane passing through it and through the trajectory to which it is perpendicular. Obviously, this plane normal to the trajectory will cut D along D itself. This second plane will pass through the focus of the first one.[43] Therefore, this focus will belong to D. Let us now consider the normal plane $a'OO'$. Its focus is the point a' because this plane is normal to the trajectory of a'. Therefore, two different planes, namely Q and $a'OO'$ passing through the straight line D, have their foci on D. Thence, D is self-conjugated. Afterwards Jonquières proved directly that D is normal to the trajectory of each of its points (ibid., pp. 14–15).

The next step to prove Chasles' propositions is this brilliant reasoning: Jonquières supposed that the straight line L lies on two conjugated straight lines D and Δ in the points m and n. Since the movement of a rigid body can be considered as a rotation around D and Δ, the point m will have no rotation around Δ because it belongs to such axis. This means that, during the infinitesimal movement, m describes the element of a circumference that is the basis of the cone of which D is the axis and

[42] Jonquières (1856, p. 14): "Si la droite *D* est normale à la trajectoire d'un de ses points, tous ses autres points auront leurs trajectoires normales à cette droite. De sorte que la droite *D* sera elle-même sa conjuguée". Italics in the text.

[43] As Jonquirès pointed out, in the figure the line D is Oa', which is normal to the trajectory aa' of the point a. The first plane is Q and the second plane is $a'OO'$, which passes through the focus O of the first plane, through the normal OO' and through the straight line Oa' (ibid., p. 14).

n the vertex. The generatrix nm of this cone is, hence, normal to the element of the circle, that is to the trajectory of the point m. *Mutatis mutandis*, the reasoning is the same considering n. Therefore, because of the previous theorem, every straight line which lies on two conjugated straight lines is normal to the trajectory of its points (ibid., pp. 15–16).

Let us now consider two pairs D, Δ and D', Δ' of conjugated straight lines and a straight line L lying on the first three of them. Since L lies on D and Δ, it is normal to the trajectories of all its points. Thence, the plane normal to the trajectory of the point where L touches D' passes through D' itself, but this plane also passes through the conjugated Δ' of D'. Thus, L lies on the four straight lines D, Δ and D', Δ'. This is exactly the general condition for four lines to belong to the same family of generatrices of a one-sheeted hyperboloid (ibid., p. 16). This proves the theorem in Chasles (1843) I have indicated by a).

Now Jonquirès' reasoning continues like this: be given the axis of rotation X and two conjugated straight lines D and Δ. Through any point M of X draw a plane π perpendicular to X. Be m and n the points where D and Δ cut π. The line at infinity of π is conjugated to X. Let us draw the straight line through M and m. It lies on two conjugated straight lines (X and the line at infinity of π). Since it lies on D, it also lies on Δ, which is the conjugated of D. Therefore:

> Each plane perpendicular to the axis of rotation cuts the two straight lines D, Δ and the axis in three points which are collinear.[44]

While varying the point M, the straight line mn changes its position in space, but it remains parallel to a director plane normal to X; and since such a line continues to lie on D, Δ, it generates a hyperbolic paraboloid.

This proves the previous theorem by Chasles which I have indicated by b). These are the geometrical proofs of the theorems by Chasles that were defined as "curious" by Lamarle and by Chasles himself. Jonquières proved all the theorems enunciated in Chasles (1843), but what shown so far is sufficient to highlight how articulated and refined the geometrical proofs of Chasles' theorems are.

An interesting question, which is both terminological and conceptual, concerns the name of "focus of a plane". Chasles (1843), p. 1424 explains the origin of the name and Jonquières gave the proof of this assertion. As a matter of fact, when a straight line moves on a plane, it is tangent to the trajectory of one of its points. It is the characteristic of the plane. It rotates around the focus of the plane. Let us now consider two infinitely close positions L and L' of this line. Be a,b,c,d, four points on L and a',b',c',d' their homologous on L'. The transformation from L to L' is a homography, hence the straight lines aa', bb', cc', ... envelop a conic section.[45] To the point at infinity of L the point at infinity of L' corresponds. This means that one of

[44] Jonquières (1856, p. 16): "Ainsi tout plan perpendiculaire à l'axe de rotation rencontre les deux droites D, Δ et l'axe lui-même, en trois points qui sont en ligne droite". Italics in the text.

[45] It is to remark here a typo in Jonquières' text (ibid., pp. 28–29) because Jonquières refers to Chasles (1852), Article 548 as that in which Chasles introduced the projective generation of conics, whereas such an article is numbered 558.

the straight lines enveloping the conic is at infinity. Hence, such a conic is a parabola. By construction, the focus of the plane is such that the straight lines Fa', Fb', Fc', Fd',... are perpendicular to the trajectories aa', bb', cc', ..., that is they are perpendicular to the tangents to the parabola. Therefore, *the focus of the plane is the focus of the parabola* (ibid., pp. 28–29).

I will not deal with the proof of the metric relations enunciated in Chasles (1843) and demonstrated in Jonquières (1856, pp. 31–53) because, though very interesting, they do not add any significant element in order to grasp the main points of Chasles' foundational programme. Such points were already known in the period 1850–1860. They were shared by the mathematicians who recognized the foundational character of Chasles' work. As we have seen, the number of such mathematicians was not great, but neither small. Let us summarize the main items of Chasles' foundational programme analysed so far.

1) The metrical properties can be interpreted in the light of some particular projective concepts and constructions. This means that projective geometry is the foundation of the whole geometry. The basis of projective geometry is the theory of geometrical transformations which was developed in the French and German mathematical environment between the 20' and the 40' of the nineteenth century and to which Chasles also gave significant contributions.
2) The study of the motion of bodies includes three disciplines: a) the geometry of the movement, which is independent of any concept including time; b) kinematics, which excludes the causes of motion, but which takes into account velocity and acceleration. Items a) and b) jointly make up what Ampère called "cinématique", but Chasles—item a)—distinguished the pure geometry of motion; c) dynamics, which studies the causes of motion, namely the forces. Projective geometry is the foundation of a), which, in turn, is the foundation of b) and c). Hence, projective geometry is the foundation of the whole doctrine of motion.
3) The infinitesimal movement (we have seen) and the finite displacement (we will see in Sect. 3.4) of a rigid body allow us to deduce several "curious" geometrical properties.
4) Geometry also concerns the foundation of several problems and principles connected to the equilibrium of forces' systems (we will see in Chap. 4) and in some problems connected to gravity (we will see in Chap. 7).
5) All the theorems dealt with are provable by geometrical methods. For the sake of brevity Chasles (1837a) provided the demonstration with a mixture of analytical and geometrical proofs. However, there is no doubt that he also had the geometrical demonstrations of his theorems, later on offered by other mathematicians. We have seen Jonquières' work, which, with regard to Chasles (1843), is the most extended one.
6) Projective geometry induces a philosophical ontology: the ontology of duality as well as a new epistemology (we will see in Chap. 6).

As a final step of this chapter I will refer to the last work that Chasles dedicated entirely to the movement of the rigid body. In Chasles (1843) he faced the infinitesimal movement; in Chasles (1860–1861), he developed the theory of the finite displacements.

3.4 1860–1861: The Final Structure of Chasles' Theory and the Problem of Finite Displacements

Chasles (1860–1861) is the last paper entirely dedicated by Chasles to the displacement of a rigid body. It is a rich and complicated work where the author faced the problem of the finite displacements. As we have recalled, this work was published in five parts. The main content of this text is composed of six sections, respectively, dedicated to: i) the displacement of plane figure moving in their plane; ii) properties of symmetric figures belonging to a plane; iii) displacement of a straight line in space; iv) displacement of a flat figure in space; v) displacement of a spherical figure on a spherical surface and of a rigid body fixed at a point; vi) displacement of a rigid body in space.

Chasles enunciated 150 theorems, of which those numbered 63–150 concern the displacement of a rigid body in space. He offered no demonstration. Charles Brisse in 1874, 1875 proved all of them. Most of the propositions enunciated by Chasles are original. I mean that they were enunciated for the first time by him either in this paper or in his previous works. These propositions (of which I will analyse the most significant) concern a plurality of subjects, also including several properties of the algebraic curves and surfaces deduced by means of the finite displacements' geometrical features.

Not less interesting than the propositions are the "Introduction" and the "Notice historique sur la question du déplacement d'une figure de forme invariable", added by Chasles as a conclusion of his paper. The foundational character of his work, which is apparent in the theorems he enunciated is explained in words in the course of the "Introduction" and in the "Notice historique".

In the "Introduction" Chasles explained that the geometrical properties of the displacements have to be the basis for any further analysis of the movement. Carnot had clearly understood that, without a geometrical study of the displacement, the entire problem of the rigid body's finite displacements would be obscure. Thought not explicitly written, Chasles let the reader understand that he was the mathematician who realized a coherent programme, an important part of which was based on Carnot's pioneering ideas (Chasles, 1860–1861, I, p. 856). Ampère distinguished kinematics—as the discipline which studies the motions independently of their causes—and associated kinematics with static. However, Chasles did not agree. The reason why he did not share Ampère's approach sheds a further light on the foundational character of his work: kinematics—in Chasles' meaning—is a branch of science which is not separated from geometry. It belongs to geometry, whereas statics is already a part of rational mechanics, though separated from dynamics. Kinematics is the conceptual basis of rational mechanics, but it is also the foundation of mechanic engineering (Chasles spoke of "Méchanique pratique") since it is applicable to the communication and transformation of movement (ibid., p. 856). Because of this, the treatises of kinematics begin with a section dedicated to the results obtained by Chasles himself (ibid., p. 857).

In the conclusive "Notice historique" Chasles offered a brief but interesting historical fresco of the results obtained with regard to the infinitesimal movement and finite displacement of a rigid body until "Chasles-Mozzi" theorem. It is indicative that his treatment is actually divided into three parts: the first one concerns the proto-history of the problem. This is a geometrical proto-history in which Chasles included few results by Greek geometers, Descartes and Newton (Chasles, 1860–1861, V, pp. 489–491). The second part regards the results obtained between the end of the seventeenth century and the beginning of the nineteenth. Here, great physicists like Bernoulli, D'Alembert, Euler and Poinsot (in a later period) dominate the scene. However, these authors do not separate the geometrical properties of the movement from the others. The third part (ibid., pp. 497–501) concerns the geometrical study of the movement. It is also verbally separated from the second because we read: "I return to the question of the displacement of a figure considered from a geometrical point of view".[46] There is no doubt that Chasles considered himself to be the creator of this foundational part of, so to say, geometrical mechanics. He mentioned other authors who addressed this subject. For example, Giorgini, Rodrigues and Stegmann are mentioned in note 1 at page 500. Mozzi is also mentioned. Nonetheless, he basically referred to his Chasles (1830i, 1837a, 1843) as the fundamental and foundational works of kinematics. In two long notes (ibid., note 1 p. 499, note 4, p. 500), he referred to contributions which were directly inspired by his ideas.

Therefore, he was perfectly aware of the novelty of his foundational programme.

Let us now see some of the properties concerning the most difficult part of such a programme: the finite displacement of a rigid body.

3.4.1 *The Displacements of Two Coplanar Superimposable Figures*

This section includes 32 propositions and concerns all the positions of two congruent flat figures which can be superimposed through a motion taking place in the plane of the figure. Therefore, the movements of two symmetrical figures (as the right and the left hand) which can be overlapped only rotating around a straight line of the plane are excluded because the figure during the motion exits the plane, though being the initial and final position in the plane.

[46] Chasles (1860–1861, V, p. 497): "Je reviens à la question du déplacement d'une figure, considérée au point de vue géométrique".

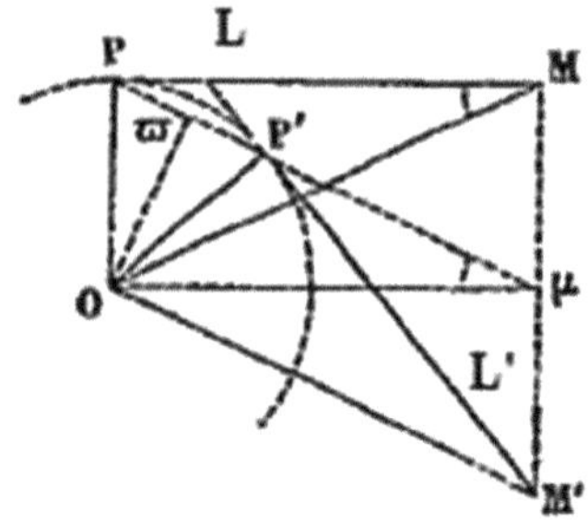

Fig. 3.5 Adaptation of the figure used by Brisse to prove Chasles' theorem 2 (Brisse, 1874, p. 222)

The first theorem is a generalization of the first proposition in Chasles (1830i). For he claimed that, given two congruent figures (in Chasles (1830i) he had referred only to polygons) in a plane, a point exists that is self-conjugated in the displacement which overlaps the two figures, so that it is enough to rotate the second figure around this point for the two figures to coincide. This means that, given two different positions of two superimposable figures, the movement which overlaps one figure on the other is a rotation around the self-conjugated point.

The second theorem is very interesting: in Chasles (1830i) (see my observation 4 in Sect. 3.1 *Commentaries*), he had considered two solid bodies in two different positions and had spoken of the third body obtained joining the middle points of the segment which connect two homologous points of the two given ones. Now Chasles offered the ultimate structure of his theory and also gave a systematic form to that observation starting from the movement of two superimposable plane figures. He stated:

> When in the two figures, two homologous straight lines L, L' and the straight lines (which we call *chords*) AA', BB', ... joining two by two the homologous points are considered, *the middle points of these chords belong to a straight line* Λ *which forms equal angles with the two straight lines* L, L'.[47]

Chasles named *chords* the segments joining two homologous points and, as already pointed out, naming a concept also means to fully recognize its importance and distinctness. Given the relevance of this proposition, let us see Brisse's proof (Brisse, 1874, pp. 222–223, see Fig. 3.5).

Be L and L' two homologous straight lines and P, P' the feet of the perpendiculars drawn from the centre of rotation O, respectively, on L and L'. Since OP is perpendicular to L and OP' to L', the two homologous straight lines are perpendicular to the circle of centre O and radius $OP=OP'$. Be M and M' the corresponding positions of any other two points. Let us draw the perpendiculars $O\varpi$ to PP' and $O\mu$ to MM'. These perpendiculars cut the segments PP' and MM' in their mean points.

---

[47] Chasles (1860–1861, I, p. 859): "Que l'on considère dans les deux figures deux droites homologues L, L' et les droites AA', BB', ... qui joignent deux à deux leurs points homologues, droites que nous appellerons des *cordes*:

Les milieux de ces cordes sont sur une droite Λ *qui fait des angles égaux avec les deux droites* L, L'". Italics in the text.

An explanation is here necessary to grasp Brisse's argument in any detail. His idea is to show that PP' is the mean straight line and that, hence, the points $PP'\mu$ are collinear. He will prove that this happens when the angle in ϖ is right. Since it is right, then those three points are collinear and, hence PP' is the mean straight line.

Brisse's reasoning continues like this: the two triangles POP' and MOM' are similar, so that $\frac{O\varpi}{OP} = \frac{O\mu}{OM}$. Consider that $O\varpi$ and $O\mu$ makes equal and equally directed angles with OP and OM, so that $\varpi\widehat{O}\mu = P\widehat{O}M$. This means that the triangles $\varpi O\mu$ and POM are similar. Therefore, when and only when $PP'\mu$ is a straight line, the angle in ϖ is right, but it is, so that PP' is the mean straight line. Furthermore, the angles $\mu P'M'$ and μPM are equal because their sides are, respectively, perpendicular to the sides of the two equal angles $PO\varpi$ and $P'O\varpi$, which proves Chasles' theorem in all its parts. Brisse's demonstration also teaches, hence, how to construct the mean straight line.

It should be observed that a trivial consequence of this proposition is Chasles' theorem 3, according to which the mean line relative to two homologous straight lines passes through the feet of the perpendiculars drawn from the centre to such lines.

A curious annotation is that, after some easily provable propositions concerning the chords, Chasles stated as theorem 6 a proposition which is false. This is the one of the very few cases with regard to the Chasles' woks I have seen.[48] For Chasles claimed that, given two homologous straight lines L and L', their chords envelop a parabola tangent to L and L', whose focus is the centre O of rotation and whose directrix is the mean line Λ of L and L' (Chasles, 1860–1861, I, p. 859).

As a matter of fact, Brisse proved that the theorem imposes that the term "directrix" is replaced by "tangent at the vertex" (Brisse, 1874, p. 224). For Fig. 3.5, $O\widehat{\mu}M'$ is a right angle of which one side passes through the fixed point O and whose vertex μ describes a fixed straight line (that is the mean line between L and L'). It is necessary to find the curve enveloped by the other side $\mu M'$ (namely MM'). Given a parabola whose focus is O and whose tangent at the vertex is Λ, it is well known that the straight line MM' is tangent to such a parabola. Therefore, it is the required envelope. But, according to Chasles' theorem 3, the mean line relative to two homologous straight lines passes through the feet of the perpendiculars drawn from the centre to such straight lines L and L' and since the locus of the perpendiculars' feet to the tangents of a parabola drawn from the focus is the tangent to the vertex and vice versa, the theorem follows.[49] I am inclined to think that Chasles' was only a *lapsus calami* and not a mathematical mistake.

---

[48] For a further case, see Brisse (1874, p. 242). Chasles wrote "rotation", whereas the right term is "half-rotation".

[49] Brisse did not explicitly mention this last property. It is a well-known metric-projective property of the parabola. For the reader interested in a beautiful and projective proof, see, e.g., Enriques (1898, pp. 310–311).

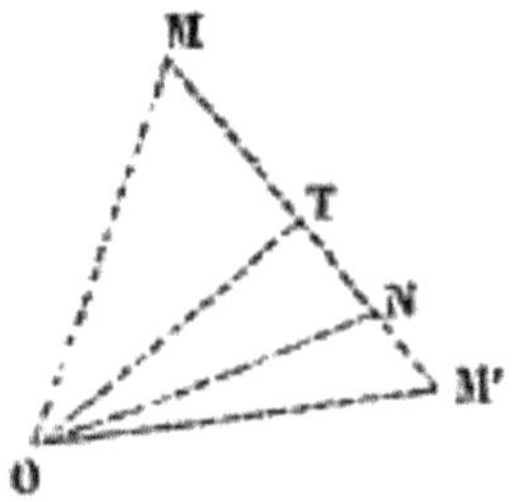

Fig. 3.6 The figure used by Brisse to prove Chasles' theorem 11, and also useful in the proof of theorem 19 (Brisse, 1874, p. 227)

Theorems 7–18 concern several properties of the conic sections provable by the means of the principles already expounded. I refer directly to Chasles' enunciations and Brisse's proofs.

It is also interesting to focus on some aspects of the subsection entitled by Chasles "Propriétés relatives à deux courbes géométriques égales". This is the case because such a section shows how vast Chasles' programme was. Here he highlighted some properties of the algebraic curves which are derivable from the displacement of two flat superimposable figures in their plane. This subsection includes the theorems 19–28. With regard to the proofs, I will deal only with that of theorem 19 and draw some considerations relative to the internal connections of Chasles' work.

The theorem sounds like this:

> When two equal curves of order m are posed in any manner in the same plane: the straight lines joining two by two the homologous points of these curves envelop a curve of class $2m$ and order $m(m+1)$. This curve has three multiple tangents of order $2m$, one of which is real at infinity, and the two others are imaginary. They are the asymptotes of the circle having its centre in the central point common to the two equal figures [the centre of rotation] to which the two curves of order m belong.[50]

It seems to me appropriate to point out that the geometrical proofs of Chasles' theorems concerning algebraic geometry and enunciated by him in this circumstance are anything but easy. Thence, Brisse's work is really precious. I will focus on the questions concerning the class of the enveloped curve and on the problems of the three multiple tangents of order $2m$.[51]

For the proof of this theorem, it is necessary to refer to Chasles' 11th proposition (See Fig. 3.6).[52] It claims that given two homologous (obviously equal in this context) figures, the chords joining the homologous points and incident in a point O considered as a part of the second figure, belong to circle's circumference which

[50]Chasles (1860–1861, I, pp. 860–861): "Quand deux courbes égales d'ordre m sont placées d'une manière quelconque dans un même plan: Les droites qui joignent deux à deux les points homologues de ces courbes enveloppent une courbe de la classe $2m$ et de l'ordre $m(m+1)$; Cette courbe a trois tangentes multiples de l'ordre $2m$, dont une, réelle, est à l'infini, et les deux autres, imaginaires, sont les asymptotes d'un cercle qui aurait son centre au point central commun aux deux figures égales auxquelles appartiennent les deux courbes d'ordre m". Italics in the text.

[51]For the order of the enveloped curve the reader can see directly Brisse's work. The whole proof of this theorem is in Brisse (1874, pp. 233–234).

[52]Brisse wrote Proposition 13, but, in fact, the correct number is 11.

passes through the centre of rotation C, the point O and the point O' that is the homologous of O in the first figure.

The proof runs as follows: to determine the class of the enveloped curve, Brisse searched how many chords pass through the point O. Since, because of theorem 11, the points of the first figure, which, if joined to their homologous in the second figure, pass through the point O, belong to the circle through C, O, O', the class of the enveloped curve is found taking into account the number of intersections between a circle and the curve of order m. Since these intersections are $2m$, this is the class of the enveloped curve.

Let us now consider the problem of the triple tangent: it is easy to prove that the circle COO' can be determined as the circle passing through O and C and whose tangent in C is a straight line making with CO—in the sense of the rotation—an angle equal to the complement of half the rotation. Through C, let us draw the straight line OC and the tangent to the circle at C, which will remain constant while O going to infinity. When O will be at infinity the circle is reduced to the line at infinity and to the tangent through C. The point O is on the line at infinity which, given that the curves are of order m, will cut one of them in m points. Joining O with these m points, we will obtain the tangents generating the enveloped curve.[53] Let us suppose now that O is one of the two circular points. In this case the straight line CO will be one of the asymptotes of the circle with centre C. Because of the properties of the isotropic straight lines, the straight line which makes with an isotropic straight line any finite real angle is such an isotropic straight line itself. The circle is so reduced to the line at infinity and to the imaginary straight line CO. The straight line at infinity will supply—through its intersections with the curve of order m—the already found tangent of order m. The imaginary straight line CO—which is tangent by construction—also touches the curve in m points. The same is true for the other circular point O'. This proves completely Chasles' theorem with regard to the class of the enveloped curves and with regard to the tangents of order m.

First of all, an observation with regard to the demonstrative method is necessary: I have referred to Brisse's proofs because they show that the whole of Chasles' construction has projective geometry as reference point. The concepts used by Brisse are typically projective notions: for to move a point to the line at infinity and to use the circular points are operations which find their historical and conceptual foundation in projective geometry. Obviously metric notions, as that of equiangularity, are used too, but all of Chasles' results and Brisse's proofs of the most difficult of Chasles' propositions have a clear reference to projective geometry. This confirms that projective geometry was seen by Chasles as the foundation of the doctrine of metrical-graphical properties and of that concerning the movement of a rigid body, which, in turn, are usable to find further geometrical properties. As to the method, it is remarkable that all these theorems are provable by the synthetic method, as

[53] Brisse added a further consideration on the simple and double tangents. Since it is not included in Chasles' theorem, I refer directly to Brisse (1874, end of p. 233).

Chasles had claimed. On the method of proof I will add a further annotation while dealing with the mean body.

The second observation concerns the inner connection of Chasles' theorems. He was creating an edifice in which he was trying to prove the properties of the movement, of the second-degree curves and surfaces as well as the properties of curves and surfaces of higher degree relying upon few projective principles. His theorems concerning the equal algebraic curves support my thesis. As always, a concrete example is useful to clarify the situation. In his Theorem 20, Chasles claimed (and Brisse wrote the proof) that, given two equal curves of class n, the intersection points of the homologous tangents belong to a curve of order $2n$ and class $n(n+1)$ (Chasles, 1860–1861, I, p. 861). This curve has three multiple points of order n of which one is real and two are imaginary. That real is the centre of rotation and the two others are at the infinity on a circle. This theorem can be seen as the dual of theorem 19, with the replacements explained in the following table:

Theorem 19	Theorem 20
Order m	Class n
Straight lines joining two by two the homologous points	Intersection points of two homologous tangents
Three multiple tangents of order $2m$	Three multiple points of order n
Line at infinity	Centre of rotation
Isotropic straight lines	Circular points

Given these two propositions, Chasles in his theorems 23 and 24 (ibid., pp. 861–862) analysed the situation when the movement is not finite, but infinitesimal, that is when the two equal curves are infinitely close. He deduced these two theorems: if the vertex of a right angle of which a side rotates around a fixed point moves on a curve of order m, the other side will envelop a curve of class $2m$ and order $m(m+1)$. This curve has three multiple points of order m. One is real and it is the centre of rotation of the first angle's side; the two others are imaginary at infinity on a circle (Theorem 23). Theorem 24 states that the perpendiculars' feet drawn from a fixed point on the tangents of a curve of class n lie on a curve of order $2n$ and class $n(n+1)$. This curve has three multiple tangents of order n. One is real and is at infinity; the other two are imaginary and are the asymptotes of a circle having its centre in the fixed point.

These propositions confirm that Chasles had in front of him a vast panorama: a) the theorems—or at least many of them—concerning the infinitesimal movements and the geometrical properties can be studied in themselves, but can also be seen, in a more mature perspective, as limit cases of theorems regarding the finite displacements; b) as theorems 19 and 20, those 23 and 24 can be interpreted as dual propositions; c) the first part of theorem 23 can be seen as a generalization of the already mentioned property according to which if the vertex of a right angle, whose one side rotates around a fixed point, moves on a straight line, the other side envelops a parabola.

Therefore: duality and the finite displacements are used as the basic instruments to connect several geometrical properties that generally pertain to the algebraic curves of which the conics are specific cases. These properties can be achieved dealing with infinitesimal displacements, but they are, in turn, limit cases of the finite displacements.

The final subsection concerns the composition of finite rotations and translations. In this regard a remark is paramount: the theorems enunciated by Chasles in this subsection were already present in Rodrigues (1840), or, at least, they were easily deducible from Rodrigues' work. However, the latter analysed the displacement of a plane figure in its plane relying upon the properties of the displacements in space, whereas in Chasles (1860–1861), the author—according to a foundational idea—had the intention to clearly separate what is deducible in the plane from what is deducible also taking into account the movement in space. It is not a coincidence that he separated the treatment of the flat superimposable figures from that of flat symmetrical figures. With regard to the composition of rotations and translations for superimposable figures, the fundamental theorem is that a rotation around a fixed point A can be replaced by an equal and equally directed rotation around any other point B and by a translation, which is equal to the double distance AB by the sinus of half the rotation. The opposite is also true (ibid., pp. 862–863).[54]

3.4.2 *The Displacement of Two Symmetric Coplanar Figures*

The section concerning the properties of two coplanar figures (Chasles, 1860–1861, II, pp. 905–909, theorems 33–45) which are symmetric and not superimposable is surely one of the most important to grasp Chasles' ideas, because: 1) he proved the existence of many properties which separate the displacement of two symmetric figures from that of two superimposable figures. He pointed out that, before his research, such properties were either unknown or not studied in a systematic manner. However, 2) he proved that both kinds of properties—those inherent to superimposable figures and those relative to symmetric figures—are included in the general doctrine of homography. Simply, they are different homographies. Therefore, this relatively brief section is a fundamental step for the explanations of the displacements and the geometrical properties deriving from them on the basis of projective transformations.

Chasles claimed, which is intuitive enough, that, given two symmetric figures in any position on a plane, they become superimposable through two rotations: one rotation of 180° around any straight line of the plane. This is equivalent to a rebatement of the plane on itself and makes the two figures superimposable. The other is an ordinary plane rotation around a point of the plane. Hence, with regard to

[54]The proof of this theorem is easy and does not offer new elements to develop my discussion, thence I refer to Brisse's demonstration (Brisse, 1874, pp. 240–241).

its genesis, the theory of symmetric figures is comprehended within spatial geometry, but since this case is so interesting, Chasles decided to consider its properties separately (Chasles, 1860–1861, II, p. 905).

If two figures have no common point, there is no point of the figures which is self-conjugated. This is the case of a general glide symmetry. If they share one point, they also share a straight line through this point, which is the symmetry line. In this case the transformation is simply a rotation around the symmetry line (ibid., p. 906). An important and intuitive property is that the two figures share a common line X such that it is enough to translate one of the two figures parallel to such a line for the two figures to become symmetrical with respect to X (ibid., p. 906).

As always, from a conceptual point of view, Chasles' treatment can be divided into two parts: 1) the geometrical properties which determine the symmetric displacement; 2) the geometrical properties which are determined by the symmetric displacement. This distinction is not made explicitly by Chasles, but it is evident that he thought of the ideal cycle: geometry explains displacements which, in turn, are usable to find further geometrical properties.

Among the properties of the first kind, the most remarkable is the following one whose projective character is clear:

> [Theorem 37] [1] Given a straight line L belonging to the first figure, a point O exists around which it is possible to rotate the second figure so as to posit it in a position such that the symmetry axis is L. In this case, the straight line L', homologous of L, will coincide with L. [2] Reciprocally, given any point O, two homologous straight lines L and L' exist in the two figures, such that through a rotation around O, the straight line L' coincides with L. Hence, the two figures are in a symmetric position with respect to such a line. [...].
>
> [Theorem 40] To any straight line L, a point O corresponds, and to any point O, a straight line corresponds (37). When the [considered] straight lines L are incident in a point, the points O belong to a straight line; reciprocally when the points O belong to a straight line, the straight lines L are incident. Furthermore, the anharmonic ratio of four [considered] points is equal to that of four [considered] straight lines.[55]

Brisse's proof of theorem [37, 2] is useful to grasp that of theorem 40. Let us refer to the following figure 3.7:

Be XX' the axis of the displacement. Let us draw the perpendicular OP from the point O to the axis. From P draw two segments $P\lambda$ and $P\lambda'$, respectively, equal to

[55]Chasles (1860–1861, II, pp. 906–907): "37. Une droite L étant prise arbitrairement dans le première figure, il existe toujours un point O autour duquel il suffit de faire tourner la seconde figure pour l'amener dans une position symétrique à la première, l'axe de symétrie étant la droite L, sur laquelle est venue se placer son homologue L' de la seconde figure.

Réciproquement: Un point O étant pris arbitrairement, il existe deux droites homologues L, L' dans les deux figures, telles, que, par une rotation de la seconde autour du point O, la droite L' vient se placer sur la droite L, et les deux figures se trouvent dans une position de symétrie par rapport à cette droite. [...].

40. À chaque droite L correspond un point O, et à chaque point O correspond une droite L (37): Quand des droites L passent par un même point, les points O sont sur une même droit; et réciproquement quand des points O sont en ligne droite, les droites L passent par un même point.

En outre, le rapport anharmonique de quatre points est égal à celui des quatre droites". Italics in the text.

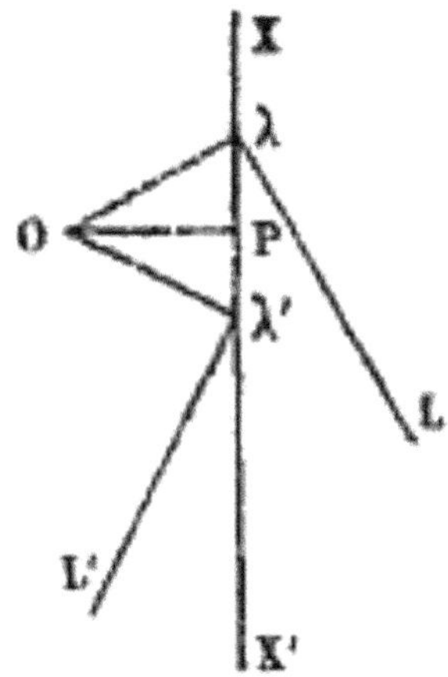

Fig. 3.7 Brisse's figure used to prove theorem [37, 2] (Brisse, 1874, p. 247)

half the translation (glissement) parallel to XX' necessary for the figure containing L to become symmetrically posed with respect to XX' to that containing L'. Draw $O\lambda$ and $O\lambda'$. Through these two points draw the perpendiculars L and L' to $O\lambda$ and $O\lambda'$, respectively. These are the required homologous lines of the displacement such that if L' rotates around O it overlaps the line L.

Brisse's proof of theorem 40 reads as follows: he considered a pencil of straight lines generally indicated by L passing through a point l and the straight lines L' passing through the point l', homologous of l. This means that, for any choice of L and its homologous L', the point O of theorem 37's first part belongs to the axis of the straight line ll'. Reciprocally, given a straight line and the point O belonging to it, it is possible to determine two pairs of homologous straight lines $L, L'; L_1, L_1'$, and, therefore, the two fixed points l, l'.

To prove that the anharmonic ratio of four collinear points O is equal to that of four lines belonging to the pencil l or l', let us refer to Fig. 3.7: if four points O are collinear, their anharmonic ratio is the same as their projections P on the displacement axis XX'. Since $P\lambda$ and $P\lambda'$ are equal and constant for any point P, the cross ratio of four points λ (and λ') is equal to that of four points P. But, since all the lines L pass through a point l (and the lines L' through l'), the cross ratio of four points λ (λ') is equal to that of four lines L (L').

On the basis of what Chasles enunciated, we can conclude that the glide symmetry is a projectivity which is the product of two plane homologies: a translation and an orthogonal symmetry with respect to an axis. The glide symmetry subordinates a reciprocity point-straight line which is shown by theorems 37 and 40. This is confirmed by theorem 41 where Chasles claimed that any two lines and their correspondent points form two correlative figures (Chasles, 1860–1861, II, p. 907). This is the projective structure of the glide symmetry.

With regard to the properties deducible from the nature of the glide symmetries, Chasles enunciated some theorems which resemble those stated for the infinitesimal displacements or for the finite displacement of two superimposable figures. For example, theorem 42 claims that given two equal symmetric figures in any position in a plane, if two homologous straight lines rotate around two homologous points, the intersection points of these lines describe an equilateral hyperbola. An asymptote of this hyperbola is the line common to the two figures. The chords joining two by two the homologous points of two homologous straight lines envelop a parabola tangent to the two straight lines and tangent in its vertex to the common straight line of the two figures (Chasles, 1860–1861, II, p. 907). Brisse offered the proofs of these theorems.

Commentary: Chasles thought of projective geometry as the foundation of the doctrine he was explaining, because, after having pointed out once again the different properties of the superimposable and the symmetric figures, he stressed that the transformations typical of the two cases have, in fact, a common profound conceptual basis. As he wrote:

> But the geometrical properties in these two systems [superimposable and symmetric figures] have a constant analogy. As a matter of fact, the two systems are nothing but particular cases of two homographic figures.[56]

Therefore, at a general and profound level, the displacement of two superimposable figures and that of two symmetric figures are both projectivities. When one descends from the projective level to that projective-metrical, the two isometries connoting the two displacements can be distinguished like this: since any transformation which overlaps two superimposable figures can be reduced to a rotation, this means that such a transformation is a homography in which of the three self-conjugated points, one is real (the centre of rotation, which is at infinity if the rotation is a translation) and the two others are imaginary (the circular points). Of the three self-conjugated straight lines, one is real and it is the line at infinity and the two others are imaginary (the isotropic straight lines). Whereas in the glide symmetry—Chasles claimed—two symmetric figures are two homographic figures having two common points and two lines. One of these points is at infinity and one of the two lines is at infinity.

I contend that the assertion according to which two symmetric figures have two common straight lines is perfectly clear. These two lines are the line at infinity and the displacement axis. Some brief explanations are necessary with regard to the two self-conjugated points: in the normal situation, that is, when the glide symmetry is not reducible to an orthogonal symmetry with respect to a line, both the two common points are at infinity, one of them corresponds to the line at infinity and the other to the displacement axis.[57] In the case of an orthogonal symmetry, the displacement is an affine homology in which the centre of homology is the self-corresponding point at infinity. It is posed on the perpendicular to the axis of homology.

[56] Chasles (1860–1861, II, p. 908): "Mais les propriétés géométriques, dans ces deux systèmes, ont une analogie constante. C'est qu'en effet les deux systèmes ne sont que des cas particulaires de deux figures homographiques quelconques".

[57] See, e.g., Enriques (1898, pp. 177–189), and, specifically, pp. 182–187.

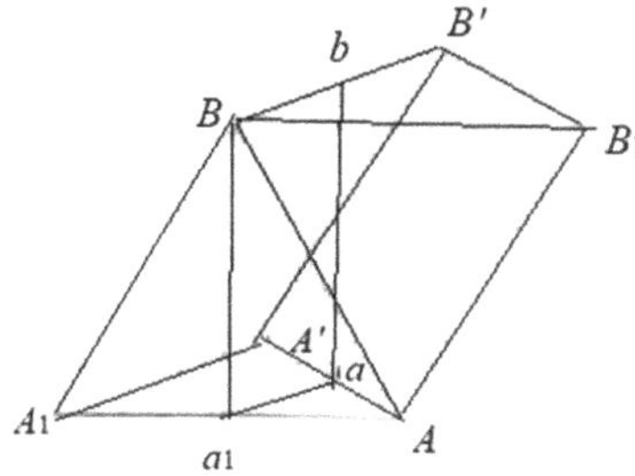

Fig. 3.8 The (adapted) figure used by Chelini to prove Chasles' theorem. Chelini (1862b), sheet with the figure, between p. 428 and p. 429

3.4.3 Displacement of a Straight Line and of a Plane Figure in Space

In his systematic construction of the geometrical doctrine of the rigid movement, Chasles, before dealing with the general displacement of a rigid body, faced the displacement of a straight line (Chasles, 1860–1861, II, p. 909) and of a flat figure (ibid., pp. 910–912) in space. With regard to the first problem, his final aim was to prove that any finite displacement of a straight line can be obtained through the rotation of the line around a fixed axis. Besides the concept of axis of rotation, the most interesting notion introduced by Chasles is that of mean straight line. We have already found this concept with regard to the planar movement of a figure. Chasles enunciated the following theorem: given two straight lines L and L' and the pairs of homologous points A, A'; B,B'; C,C',... (A,B,C,.. belong to L and A', B', C',... to L '), the chords AA', BB', CC', ... have their middle points a, b, c,... belonging to a straight line Λ—called the mean line. This line makes equal angles with the two straight lines and belong to a plane parallel to L and L' (ibid., p. 909). I will refer to the beautiful proof given in Chelini (1862b), pp. 377–378 (Fig. 3.8), which is clearer than Brisse's (1874), pp. 254–255.

Chelini reasoned like this: through A' draw the straight line segment $A'A_1$, which is equal and parallel to $B'B$. On the triangle $AA'A_1$ as a basis be constructed a prism with lateral edges AB_1 and A_1B parallel and equal to $A'B'$. Be the side AA_1 bisected in a_1. The segments aa_1, $A'A_1$, $B'B$, besides being parallel, have also the property

$$aa_1 = \frac{1}{2}\,A'A_1 = \frac{1}{2}B'B = bB$$

Therefore, the segments ab, a_1B are also parallel and equal and their direction bisects the angle ABA_1, which is equal to that of the straight lines. Let us now consider the two homologous points C and C' and the middle point c of the chord CC '. It is necessary that the two straight lines ab and ac bisect the angle of the directions AB, $A'B'$. Thence, they are segments of the same straight line, which is, in fact, the mean line.

After that, it is not difficult to prove that all the planes drawn through the points of the mean line perpendicularly to each chord AA', BB', CC', ... belong to the same straight line λ and that the two lines L and L' can be superimposed through a rotation around λ.

Therefore, in the displacement of two lines, the two main objects are the mean line and the axis of rotation.

The finite displacement of a flat figure in space has many interesting features which show connections both with the infinitesimal movement of a plane considered as belonging to a rigid body and with the finite displacement of a straight line in space. This is a further proof of the unity of Chasles' theory because he tried to determine the geometrical similarities of the different kinds of displacements. These similarities have the clear "flavour" of a duality because one passes from the features of one kind of displacement to those of another by replacing some words and maintaining unaltered the syntactical structure of the sentence (though these "dualities" are not always projective dualities), I propose the following table whose reading is evident.

Finite displacement of a plane figure in space	Finite displacement of a straight line in space	Infinitesimal movement of a plane considered belonging to a rigid body
The middle points $a,b,c,\ldots$ of the chords AA', BB', CC' belong to a plane Π (mean plane) which makes equal angles with the planes P, P' of the two figures.	The middle points $a,b,c,\ldots$ of the chords AA', BB', CC' belong to a straight line Λ which makes equal angles with the straight lines L, L' of which they are chords.	
The planes perpendicular to the chords drawn through their middle points pass through a point (focus) of Π.	The planes perpendicular to the chords drawn through their middle points pass through a straight line λ.	The planes perpendicular to the trajectories of any point pass through a point (focus) of the plane.
Only for the focus, it holds that the chord of which it is the middle point is perpendicular to Π.		The tangent to the trajectory of the focus is perpendicular to the moving plane.

In the case of the infinitesimal movement, it makes no sense to speak of mean plane because the positions of the moving plane are infinitely close and, hence, the mean plane tends to be the moving plane itself. For the finite movement of a straight line in space, the concept of mean plane and, hence, of its focus does not exist. This is the reason why I claimed that, in a sense, the movement of a plane figure in space sums up some properties of the two other movements.

The three properties on the table's first column are those fundamental with regard to the movement of a flat figure in space. From them, it is possible to deduce some beautiful and unexpected properties of this movement summarized by Chasles in his theorem 50 (ibid., p. 910):

1) the mean plane cuts the moving plane (one might speak of two planes) in its two positions along two straight lines L, L', which are homologous in the considered isometry;
2) the mean line of L and L' belongs to the mean plane Π;

3) the planes drawn through the middle points of the L and L''s chords perpendicularly to such chords pass through a straight line λ (this follows from what seen with regard to the displacement of two straight lines in space). I add this observation: the straight line λ is perpendicular to L and L' because L and L' belong to the mean plane and λ is orthogonal to such a plane;
4) This straight line has a segment which is a chord whose middle point coincides with the focus of Π (this follows from the property 3 expounded in the preceding table);
5) It is possible to overlap the straight lines L and L' through a rotation around λ. After that, through a rotation of the plane P around L, the two flat figures will be superimposed.

From this it follows that *any displacement of a flat figure in space can be carried out through two rotations around two perpendicular straight lines. The first of these two straight lines is perpendicular to the mean plane and passes through the focus of such plane. The other straight line is the trace on the mean plane of the figure's plane in its first position* (ibid., p. 910).

The finite displacement can also be produced in another manner: for given the two planes P and P', be D' their intersection straight line considered as belonging to the second figure. Be D its homologous in the first figure. Then, D belongs to P. Therefore, the plane P contains two homologous straight lines D and D'. They can be superimposed through a rotation around a fixed straight line perpendicular to P. After that, through a rotation around D, the two planes are made coincident. This means that every displacement of a plane figure in space can be obtained through two successive rotations. One around a straight line perpendicular to the plane of the figure; the other around a given straight line belonging to the plane itself.

These theorems determine the basic geometrical properties of a flat figure's finite displacement. Afterwards, Chasles enunciated his usual series of "curious" theorems which connect the geometrical properties of the displacement with those of several figures. We will see some of these theorems. But before dealing with them, it seems to me appropriate to present to the reader at least Brisse's proof of the fundamental theorem regarding the existence of the mean plane. This is important to make it clear that Chasles' work is unitary not only with regard to the results, but as for the methods of proof are concerned, as well.

Let us prove that the middle points $a,b,c,\ldots$ of the chords AA', BB', CC', ... belong to a plane Π, which makes equal angles with the plane P,P'.

I will focus on the existence of the mean plane. Brisse (1874), p. 256 proposes the following reasoning: three points A, B, C and their homologous A', B', C' are sufficient to determine the displacement. Obviously the middle points a,b,c of the three chords AA', BB', CC' determine a plane Π. Now Brisse considered the two straight lines BC and $B'C'$. The mean straight line L relative to them passes through the points b and c. Analogously, the mean straight line M of CA and $C'A'$ and the mean straight line N of AB and $A'B'$ pass, respectively, through a, c and a, b and belong to Π. Let us now consider any point D of P and its homologous D' in P'. Be d the middle point of the chord DD'. Through D let us draw a straight line cutting the triangle ABC. The point D', homologous of D, will cut $A'B'C'$. The mean line of these two straight lines passes through the middle points of the chords whose extremes are the points where D and D' cut, respectively, the triangles ABC and $A'B'C'$. Thence, this straight line belongs to Π. Consequently, the point d, which belongs to this line, also belongs to Π.

The "curious" theorems expounded by Chasles in this section are based on the following considerations: the straight line intersection of the two planes P and P' where the two equal figures lie is a chord. Thence, it contains two homologous points. Chasles indicated by D' this straight line if it is regarded as belonging to the first figure. In this case a straight line D'' in P' will correspond to D' and a point a'' lying on D'' will correspond to a point a' of D'. If D' is considered as belonging to the second figure, a point a belonging to the correspondent straight line D of D' in the first figure will correspond to a'. Hence, a straight line $a'a''$ of the second figure corresponds to the straight line aa' of the first figure. These two straight lines are, thus, homologous. According to an already explained theorem (Chasles, 1860–1861, I, p. 859, corrected and proved in Brisse, 1874, p. 224), they envelop two parabolas belonging to P and P'. The contact points of aa' and $a'a''$ with the parabolas are homologous and the two parabolas touch D' in the two homologous points lying on such a straight line.

All these properties are a direct consequence of the apparatus introduced by Chasles, but he went far beyond these considerations and stated some theorems of algebraic geometry connected to this context. The first theorem, which gives, so to say, the tone of Chasles' considerations is the following:

> The plane of the two straight lines aa' and $a'a''$, which are tangent to the two parabolas, envelops a developable surface. It is of fourth degree and its generatrix is the straight line joining the contact points of aa' and $a'a''$ with the two parabolas respectively.[58]

Brisse (ibid., pp. 261–261) reasoned like this: since the plane α of the straight lines aa', $a'a''$ passes from a position to another in a defined manner, it envelops a developable surface. Let us now take into account two infinitely close positions of this plane. Their intersection is a generatrix of the developable surface. To obtain the generatrix, it is enough to consider the traces of two positions of α on the planes P and P' of the two figures and to join the so obtained points. These intersections are the contact points of two infinitely close tangents to each of the two parabolas. Therefore, the generatrix is the straight line joining such a contact points.

Now, be L a straight line belonging to P; be L' its homologous in P'. They will meet the two parabolas in two points, respectively. Be Λ the mean line of L and L'. Thence, the mean line will meet the surface at least in two points. To prove that the surface is of fourth degree, it is now necessary to prove that such a line cuts the surface only in two other points. To develop this reasoning Brisse will prove that, in

[58] Chasles (1860–1861, p. 912): "Les plan des deux droites aa' and $a'a''$, tangentes à ces deux paraboles, enveloppe une surface développable du quatrième ordre dont la génératrice est la droite qui joint les points de contact des droites aa' et $a'a''$ avec les deux paraboles, respectivement". Italics in the text.

fact, each line L and L' can be associated with two points where Λ cut the developable surface.[59] Thence, the surface is of fourth degree.

Commentary: until this moment, all the theorems on the finite displacements proposed by Chasles were referred exclusively to the initial and final positions of two given flat figures in space. This theorem also has to take into account the whole movement between the initial and final positions because, according to different movements having the same initial and final positions, the plane of which Chasles speaks changes its position instantaneously. Therefore, it envelops different surfaces sharing the two tangent planes at the beginning and at the end of the movement. Thus, there is an infinity of surfaces enveloped by such planes having the same initial and final positions. It is necessary to add the condition that the trajectory is given, though it is not needed at all to specify its nature. In this case the enveloped surface is unique.

Chasles stated several other beautiful theorems of algebraic geometry, for example the edge of regression of the developable surface is a third degree curve with double curvature and, if the two movements concerns a curve of degree m, the chords generate a regular surface of degree $2m$ (Chasles, 1860–1861, II, p. 912).

This confirms the existence of a foundational programme which, starting from few elements of projective geometry, tends to include a prominent part of the basic elements of mechanics and, through them, of algebraic geometry. It is a scheme like this:

Basic elements of projective geometry→geometrical theory of the rigid movement→several properties of algebraic figures

I will now analyse the final step of this theory developed by Chasles: the geometrical properties of the movement of a rigid body as they are presented in Chasles (1860–1861, III, IV, V).

3.4.4 Displacement of a Rigid Body in Space

The section dedicated to the displacement of a rigid body is the longest of Chasles (1860–1861) because it includes the articles 63–150. Following—though not literally—the author's division into subsections, it is possible to enucleate the following conceptual elements: 1) introductory theorems to the subject; 2) properties of two homologous straight lines regarded as belonging to two homologous rigid bodies and of their chords; 3) properties of two homologous planes belonging to two rigid bodies; 4) concept of mean body and its properties; 5) features of a system of two conjugate rotations; 6) construction of the central axis of two equal bodies.

[59] I do not enter the details of this last part of Brisse's proof because they would add nothing important to the picture I am drawing.

With regard to item 1), Chasles referred to properties which in the 60s of the nineteenth century were well known. The most important one is the screw-theorem discovered by Chasles himself in 1830, according to which the displacement of a rigid body in space can be carried out by means of a rotation around a straight line (central axis of rotation) which translates on itself. That is: any displacement can be achieved by a screw at which the body is attached (Chasles, 1860–1861, III, p. 78). The other important theorem of this section states that each displacement of a solid body in space can be obtained through two rotations around two perpendicular straight lines, one of which can be freely chosen, while the position of the other is determined by such a choice (Chasles, 1860–1861, III, p. 79). This theorem resembles the way in which the displacement of a plane figure in space can be described (Chasles, 1860–1861, II, p. 910).

Items 2) and 4) are the most interesting because they present that series of "curious theorems" which connote the relations between the movement of a rigid body and several geometrical properties. These propositions are the very mark of distinction of Chasles' work in this field. Therefore, I will focus on the most important of the propositions presented in these items.

Item 5), as Chasles himself wrote,[60] refers to some results already obtained by Rodrigues 20 years before; therefore, it is not an original section and hence, I will not deal with it.

The most interesting properties of the homologous straight lines concern the geometrical figures they can generate. It is easy to prove that, given any point in space, two homologous straight lines pass through this point and that each of these lines is a chord (Chasles, 1860–1861, III, proposition 80, p. 80). Chasles had this idea: given two homologous bodies, he considered all the points belonging to a line as points through which two homologous straight lines pass and he reached the following beautiful result:

> If two homologous straight lines *D*,*D*' are drawn through any point of a straight line *L*, the straight lines *D* of the first body, will generate a hyperbolic paraboloid passing through *L* and through the line, which, in the first body, corresponds to *L* considered as belonging to the second body.[61]

In Brisse (1875), p. 81 the theorem is proved like this: be a' the point where a line D cuts L. If a' is considered as belonging to the second body, its homologous a in the first body will lie on the straight line D itself and on the straight line L_1, which is the homologous on L in the second body. Therefore, the segment aa' is a chord because the two points belong to the homologous lines L and L_1. If one considers another couple of homologous straight lines passing through another point of L, a different chord bb' connecting the points b and b' of L and L_1 is obtained. A previous theorem

[60] Chasles (1860–1861, IV, p. 194, footnote 1).

[61] Chasles (1860–1861, III, pp. 80–81): "Si par chaque point d'une droite *L* on mène les deux droites homologues *D*,*D*' qui se rencontrent en ce point, les droites *D* du premier corps forment un paraboloïde hyperbolique qui passe par la droite *L* et par la droite qui correspond dans le premier corps à cette droite *L* considérée comme appartenant au second corps".

by Chasles[62] stated that if a straight line L in which the points $a,b,c,\ldots$ are marked is moved in the position L' and $a',b',c'\ldots$ are the points corresponding to $a,b,c,\ldots$, the chords aa', bb', $cc',\ldots$ are parallel to a plane with respect to which the straight lines L and L' are equally inclined. But this is the feature of the lines generating a hyperbolic paraboloid. Hence, the theorem is proved.

The lines D' will form a second paraboloid whose features are completely analogous to those of the first paraboloid.

This is a typical metrical-projective theorem where one exploits the fact that the displacement of a rigid body is an isometry, hence a particular homography, so that the homologous of a straight line is a straight line.

The limit case of the proposition I have presented is a significant example which confirms the profound links connecting each theorem discovered by Chasles. For the limit case reads as follows: if the straight line L is the intersection of two homologous planes, the two homologous straight lines which one can draw from all the points of L belong to these two planes and the two paraboloids degenerate into two parabolas belonging to these two planes. From an intuitive and visual standpoint this theorem can be interpreted as the fact that the straight lines generating the two previous paraboloids become the straight lines enveloping Chasles' parabolas. From a rigorous point of view the proof of this proposition relies on Chasles' proposition 74 according to which if the chords connecting the homologous points of two bodies belong to a plane, they envelop a parabola (Chasles, 1860–1861, III, p. 79. Proof in Brisse, 1875, p. 146).

On the one-sheeted hyperboloid, there are two theorems. The former states that if around two homologous straight lines two homologous planes rotate, their intersection generates a hyperboloid (Chasles, 1860–1861, III, proposition 85, p. 81). The other proposition is the following theorem: given the planes passing through a straight line L, in each of them a system of two homologous straight lines D,D'—belonging, respectively, to the two bodies—exists, which holds the following three properties:

1) The straight lines D of the first body generate a hyperboloid and those of the second body a second hyperboloid;
2) Any pair of homologous straight lines D and D', each belonging to the same plane passing through L, are incident. The locus of their incidence-points is a skew curve of third degree;
3) For any straight line of space, it is possible to draw four planes tangent to the curve, so that the developable surface of which the curve is the edge of regression is of fourth degree (Chasles, 1860–1861, III, proposition 87, p. 82).

In his proofs of Chasles' theorems, Brisse (1875, p. 153) observes: consider a plane P through L, regarding L as belonging to the second position of the body. Be L_1 the homologous of L and P_1 the plane homologous of P in the first position of the body. The plane P_1 will pass through L_1. The intersection of P and P_1 is D. Therefore,

[62] Chasles (1860–1861, II, p. 909). For the proof, see Brisse (1874, pp. 253–254).

on the basis of the first theorem on the hyperboloid according to which if around two homologous straight lines two homologous planes rotate, their intersection generates a hyperboloid, Chasles' assertion 1) is proved.

Item 2) is proved by this interesting reasoning: Brisse considered now L as belonging to the first body and its correspondent straight line L_2 in the second body. This means that the hyperboloid generated by the straight lines like D passes through L and L_1, while that generated by the straight lines as D' (belonging to the second figure) passes though L and L_2. Hence, these two hyperboloids share the generatrix L. Ergo, as it is easy to prove, they mutually cut along a third degree curve, which is the locus of the incidence-points of the straight lines like D and D'.

Item 3) is proved considering that, given a point in space, only four planes tangent to the third order curve can be drawn. When a plane is tangent it contains one generatrix of the ruled surface. Therefore, there are only four generatrices which cut the straight line L. Thence, this straight line cuts the surface in four points, which proves the theorem in all its parts.

It is appropriate to point out that Brisse's demonstrations make it clear that each element of space (a point, a straight line and a plane) can be considered as belonging to the first or to the second solid body (or to the first and second position of the same solid body, as one prefers). Thence, what Chasles is considering is a rigid movement of the whole space in itself. Space overlaps to itself, but before and after the displacement, obviously, the elements of space, in general, do not mutually superimpose. Therefore, Chasles' theory is a general theory of the transformation of space in itself through isometries. If one considers solid bodies, that is limited regions of space, then one obtains the application to physics of Chasles' theory, which is, in fact, a general theory of transformations.

Chasles obtained several properties which he transcribed into a quantitative form by means of trigonometric relations (Chasles, 1860–1861, III, propositions 92–101, pp. 83–84). The first of these relations concerns the angle of two homologous straight lines. These results are important to develop the calculations and to solve problems with specific data. However, if one looks at the foundational and purely geometrical aspect of Chasles' programme, they can be neglected.[63] Therefore, I will now analyse the properties of the mean body.

The theory of the mean body is an important conclusion of Chasles' geometrical researches on the displacement of a rigid body. It includes theorems 104–118. After this section, the sections concerning the properties relative to the composition of two rotations and to some further properties of the chords follow. However, the ideas expressed in the mean body's theory represent the final step of the new series of concepts introduced by Chasles. The two other sections are, so to say, applications of such train of thought. The concept of mean body is so important because it allowed Chasles to connect the theory of the infinitesimal movements with that of the finite displacements and to solve some problems concerning the finite displacements

[63] The proofs of these properties can be found in Brisse (1875, pp. 155–160).

through the properties of the infinitesimal displacements which are easier to treat. As he wrote explicitly:

> The notion of *mean body* that can have an infinitesimal movement where the trajectories of the mean body's points are directed along the chords of which they are the middle points leads to numerous properties concerning the finite displacement of a body in space; these properties are deduced from those of the infinitesimal displacement, question which is easier to treat.[64]

Chasles began his treatment considering two equal bodies V and V' and the middle points $a, b, c, \ldots$ of their chords $AA', BB', CC', \ldots$. They form a new body which is homographic to V and V'. This is the *mean body* of V and V'.

The two initial and fundamental theorems of this doctrine imply the introduction of several concepts: let us suppose to give an infinitesimal movement to the mean body M so that its points $a, b, c, \ldots$ have their trajectories, respectively, directed along the chords $AA', BB', CC', \ldots$ of which $a, b, c, \ldots$ are the middle points. Be a', $b', c', \ldots$ the positions of $a, b, c, \ldots$, respectively, after the infinitesimal movement. Be M' the new position of M. Under these conditions, the central axis of the finite displacement of V and V' is the same as the axis of the infinitesimal movement by which M assumes the position M'. Therefore, there are two rotations and two translations: a finite and an infinitesimal rotation; a finite and an infinitesimal translation. Be U the finite rotation around the axis and E the finite translation parallel to the axis which leads V in V'. Be ω the infinitesimal rotation around the axis and h the infinitesimal translation which lead M in M'. Since all the chords have the same projections on the central axis of rotation (Chasles, 1860–1861, III, proposition 67, p. 78) and since—because the chords are tangent at the infinitesimal trajectories $aa', bb', cc', \ldots$ — aa' is collinear to AA'; bb' is collinear with $BB', \ldots$, the ratio between these chords and their orthogonal projection on the axis is the same. Therefore, if ε indicates any infinitesimal constant, the relation $\frac{aa'}{AA'} = \frac{bb'}{BB'} = \ldots = \frac{h}{E} = \frac{\varepsilon}{2}$ is satisfied.

This clarified, the following theorems hold:

1) the ratio between the infinitesimal rotation ω and translation h can be expressed like this:

$$\frac{\omega}{h} = \frac{tan\,\frac{1}{2}U}{\frac{1}{2}E},$$

so that

[64]Chasles (1860–1861, IV, pp. 190–191): "La notion de ce *corps-milieu* qui peut prendre un mouvement infiniment petit dans lequel ses trajectoires sont dirigées suivant les cordes dont les points de ce corps sont les milieux, conduit à de nombreuses propriétés relatives au déplacement fini d'un corps dans l'espace ; ces propriétés se concluent de celles du déplacement infiniment petit, question plus facile à traiter". Italics in the text.

$$\omega = \varepsilon \cdot tan \frac{1}{2}U; \;\; h = \varepsilon \cdot \frac{1}{2}E$$

2) The element aa' described in the infinitesimal movement by each point a of the middle body is the half-chord $aA\left(=\frac{AA'}{2}\right)$ multiplied by ε. It is $aa' = \varepsilon \cdot aA = \varepsilon \cdot \frac{AA'}{2}$.[65]

Chasles stated, as we have seen, that the planes drawn through the middle points of the chords belong to the mean straight line Λ and that the planes drawn through the middle points of the chords (which make up the mean line) perpendicularly to the chord themselves belong to a straight line λ (Chasles, 1860–1861, II, p. 909). Therefore, according to the concept of conjugated straight lines in an infinitesimal movement, the two lines Λ and λ are two conjugated axes of rotation for the infinitesimal movement of the mean body (Chasles, 1860–1861, IV, p. 191). Now, the reciprocal theorem also holds: if the straight line λ is considered as the mean line of two homologous straight lines l, l', the planes normal to the chords having their middle point on λ will pass through Λ.[66] This means that the straight lines Λ and λ considered in the finite displacement are conjugated and have all the properties characterizing the conjugated straight lines of the infinitesimal movements, already analysed in Chasles (1843) and on which Chasles came back in theorems 122–130 of Chasles (1860–1861, IV). The analogy between the geometrical features of the infinitesimal movement and those of the mean body (or plane or line) offers further properties connoting the finite displacements. As a matter of fact, we have seen that two homologous straight lines L,L' in a plane Π exist which can be superimposed by a rotation around a fixed point F which is the focus of the plane. Now, as Chasles claimed, if Π is considered as the mean plane relative to two homologous planes P,P', the point F will be the focus of this plane in the infinitesimal movement of the mean body.[67]

Another important proposition which allowed Chasles to transfer the properties connoting the conjugated straight lines, the focus and the characteristic in an infinitesimal movement to the finite displacements is the following one: we have seen the theorem according to which the homologous straight lines belonging to the mean plane Π are the intersections of this plane with the two homologous planes P,P'. From this, it follows that the characteristic of the plane Π in the infinitesimal movement is the mean line Λ of the straight lines L and L' because the chords corresponding to the points of these straight lines belong to Π, hence Λ belongs to Π (Chasles, 1860–1861, IV, p. 191).

[65] The two theorems I have mentioned are Chasles' propositions 105 and 106 (Chasles, 1860–1861, IV, p. 190). For their demonstrations, see Brisse (1875, pp. 162–164).

[66] For the proof of this theorems, see Brisse (1875, pp. 164–65).

[67] Chasles (1860–1861, IV, p. 191) referred for this property to Chasles (1843, p. 1420). We have seen in detail the features of the planes' focus in the infinitesimal movement in Sect. 3.4.2.

With these theorems the properties of the infinitesimal movements involving the described concepts can be extended to the finite displacement, for which we refer to Chasles (1860–1861, IV), pp. 191–192 and 195–197. I will not consider these properties because my aim is to offer the fundamental concepts and correlations of Chasles' work.

Commentary

A specific observation concerns the value indicated by Chasles through ε. He considered it an infinitesimal constant. This means an actually infinitesimal value. It is worth pointing out that it cannot be seen as an ordinary potential infinitesimal of mathematical analysis because it represents the ratio between an infinitesimal and a finite chord. Therefore, both the infinitesimal chord and the value ε are actual quantities. At the same time, Chasles clarified that the value of ε, and hence, of the infinitesimal chords as aa' is arbitrary. This might seem strange: how can a quantity be given and, at the same time, be arbitrary? It seems that Chasles meant something like this: the ratio between a finite and an actually infinitesimal quantity cannot be specified by something like an infinitesimal number. However, we are entitled to consider an actually infinitesimal chord. Its ratio with a finite actual chord will be a constant, but since this constant is not a part of the numerical *continuum* of the real numbers, we can consider it as an arbitrary value less than any given finite number. Therefore, Chasles is not thinking of something like a calculation with infinitesimal small quantities whose magnitude can be represented by infinitesimal numbers. Giuseppe Veronese thought of such a theory and developed a part of it and Robinson's not standard analysis relies on this kind of ideas.[68] Chasles did not arrive at thinking of an algebra of infinitesimal quantities, but undoubtedly he deemed legitimate to consider infinitesimal geometrical objects and infinitesimal constants. Their ratios with the finite quantities are constant, but these ratios are not differentiable one from the other. This means that, for example, the quantities $2\varepsilon, \varepsilon, \frac{\varepsilon}{2}, \frac{\varepsilon}{3}$ and so on are different from a signic standpoint, but are not from an actual point of view. They simply indicate actually infinitesimal quantities. Therefore, Chasles did not foresee an algebra of infinitesimals and did not take into account the problems deriving from the insertions of actually infinitesimal quantities in the structure of the real numbers' *continuum*, because, in fact, he did not operate a separation and a distinction between these quantities. However, he admitted the existence of infinitesimals. In a sense the situation is similar to that of the points, the lines and the plane at infinity: these entities are actual ideal objects in projective geometry and it makes perfectly sense to consider, e.g., the cross ratio of four points, some of which are at infinity. This notwithstanding, the distance between a point at infinity and a point at finite, though being a given segment, has not been represented by an infinite number conceived as an element of an algebraic system of infinite numbers. It is not by chance that Cantor, though referring to the point at infinity of

[68] Veronese (1891): *Introduzione. Principi fondamentali delle forme matematiche astratte*, pp. 1–205. See also Bussotti (1997) and Cantù (1999). The classical text on non-standard analysis is Robinson (1966).

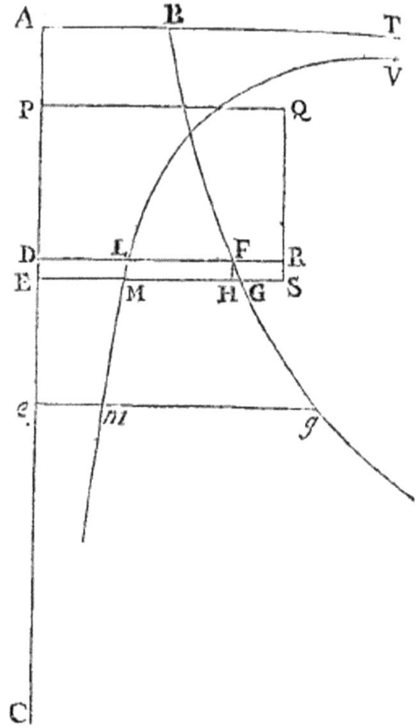

Fig. 3.9 The figure used by Newton to prove Proposition XXXIX, section VII, Book I of *Principia*

projective geometry and of the function theory as an actually infinite element, used a completely different road to construct his systems of transfinite numbers.[69] It is appropriate to recall that, in the course of the history of mathematics and science—beyond the well-known logical mistakes concerning the actual infinitesimals—some scientists considered—without making any error—infinitely small given magnitudes. This example drawn from Newton's *Principia* is paradigmatic: while dealing with the famous *Inverse problem of central forces*, solved in Proposition XLI of the *Principia*, section VIII, Book I, Newton, given a centripetal force and granted the quadrature of the curvilinear figures, searched the trajectory of the bodies and the times along those trajectories. A necessary presupposition to solve this problem is Proposition XXXIX, section VII, Book I of *Principia*, which reads as follows:

> Supposing a centripetal force of any kind, and granting the quadratures of the curvilinear figures; it is required to find the velocity of a body, ascending or descending in a right line, in the several places through which it passes; as also the time in which it will arrive at any place; And vice versa (Newton, 1729, p. 163).[70] (Fig. 3.9)

[69] See Cantor (1883, § 1).

[70] I will use Mottes' English translation of the third edition of Newton's *Principia* referring to it as Newton (1729).

During his reasoning, Newton referred to the segment *DE* as:[71]

> For in the right line *AE* let there be taken the very small line *DE* of a given length, and let *DLF* be the place of the line *EMG*, when the body was in *D* [...] (ibid., p. 164).

Therefore: what kind of entity is this *lineola* (*linea quam minima datae longitudinis*)? Is it an actual infinitesimal? Newton spoke of it as a given quantity smaller than any given real quantity. Thence, the *lineola* might be thought of as an actual infinitesimal. But, in fact this is not the case. Newton's *lineola* is a *physical fiction*: in a phase of his reasoning, it is convenient to consider an infinitesimal given straight line segment in order to develop the argumentation. This is an expedient, a device, which allowed Newton to develop the proof. This does not mean that Newton believed in the physical existence of an actually infinitesimal segment. Rather, it is an instrument to develop the reasoning. It is a useful mathematical and physical device to express the concept that the *lineola* is a given finite quantity, which is so small with respect to the other finite quantities involved in the reasoning that its length is negligible at all. In this sense, it is an infinitesimal. Therefore, so to say, the *lineola* is an infinitesimal in the context Newton is analysing. It is not a real actually infinitesimal quantity.

Newton, and Leibniz too, referred to given infinitesimal elements more than once. But, probably they are not interpretable as actually infinitesimal objects. Rather, their nature is analogous to that of the *lineola*.[72]

Can Chasles' infinitesimal chords and the infinitesimal constant value ε be interpreted as Newton's *lineola*? There are some analogies because an infinitesimal chord in an infinitesimal displacement is, in fact, a *lineola*. However, I tend to exclude this interpretation exactly because Chasles spoke of the infinitesimal quantity ε as the ratio between an infinitesimal and a finite chord, whereas, though Newton developed a series of operations with its *lineola*, a specific element as ε does not exist. This means that the chord is an actual infinitesimal, not a finite quantity which, in a given context, can be considered infinitely small. Nonetheless, it is not easy to grasp exactly what Chasles meant. I interpret his entities as actually infinitesimal ideal objects, which, nonetheless, do not produce a system of infinitesimal numbers.

---

[71] Newton (1726, p. 120): "Etenim in recta *AE* capiatur linea quam minima *DE* datae longitudinis, sitque *DLF* locus lineae *EMG*, ubi corpus versabatur in *D* [...]".

[72] See Bussotti (2015, pp. 50–53) and Pisano and Bussotti (2017). With regard to Leibniz, there is a certain amount of literature on his infinite and infinitesimal concepts. I mention Esquisabel and Raffo Quintana (2021), where the reader can also find abundant references to the literature.

3.5 Conclusive Considerations

The theory of the mean body represents, from a conceptual standpoint, the final step of Chasles' theory of the displacements because it connects the results obtained for the infinitesimal displacements with those connoting the finite displacements, explicitly showing the unity of Chasles' construction. The metric-graphical properties characterizing the isometries which generate the displacements are the fundamental and foundational elements of Chasles' theoretical developments. The concept of focus of a plane as a point whose trajectories are perpendicular to the plane itself, the notion of conjugated straight lines, the analysis of the conjugated straight lines in the specific case in which they are orthogonal, the study of the orthogonal projections on the axis of rotation, the homography between the mean body and the two positions of the moving body as well as all the results obtained for the algebraic figures, whose bases are those achieved for the hyperbolic paraboloid and for the one-sheeted hyperboloid, belong to the context of the metric-graphical geometry. Since the basic step of Chasles' foundational programme was to explain the metric-graphical properties within a projective context, the conceptual itinerary I propose shows that he considered, in the whole of his scientific career, projective geometry as the foundational discipline for the entire geometry and mechanics. This will become even more apparent in the following sections of this book, but the situation is already clear.

The problem of method deserves a particular consideration. On several occasions Chasles claimed that the synthetic methods are those intrinsic to geometry and that in his *Aperçu* he had also resorted to analytical means because, in establishing the fundamental concepts, they are quicker. There is no doubt that he considered the concept of cross ratio as belonging to the geometrical entities and, actually, as the most mature and rich geometrical notion. For he thought of this concept as the unifying basis of projective geometry, also including the duality law. One passes to analytical methods only if the figures are posed within a system of coordinates, but insofar as ratios or cross ratios are considered the context is synthetic. Therefore, the methods are important for Chasles and he took into account this problem on several occasions; but he maintained separated the problem of methods—which is an epistemological problem—from that of the foundation of geometry and mechanics, which is an ontological problem, at least in Chasles' perspective. I mean: projective transformations are the foundation of all geometry and of the doctrine of movements. This is the primary and ontological question. The epistemological question is that such transformations can be studied by intrinsic synthetic methods (which is preferable) or by analytical methods, that is by equations expressed in a system of coordinates. These equations are the models of the geometrical figures and transformations, they are not the figures and transformations themselves. However, given the isomorphism between geometrical world and model, the analytical methods are absolutely legitimate, though the synthetic ones are consubstantial with the authentic nature of the studied problems. Therefore, Chasles was a purist, but not a fanatic purist. For given the whole of his activity as a geometer, he was perfectly aware of the power of the analytical methods.

Chasles' methodological refinement is highlighted by what he wrote almost at the beginning of his treatment of the displacement of a rigid body in space. We read:

> The previous theorem [screw theorem] is one of those which we proved several years ago in the *Bulletin des Sciences mathématiques du baron de Férussac* (t. XIV, p. 324, year 1830) as deriving from the more general consideration of two *similar* bodies posed in any manner in space.[73]

What Chasles meant is clear: each method opens new perspectives in the research. The general method used in Chasles (1830i) is perfect from a foundational point of view because it shows that the whole doctrine of the displacement of a rigid body, as based on an isometry, is included in the more general doctrine of the similar bodies. Nonetheless, if one researches specific properties of such particular isometry, it is preferable to consider the isometry in itself, not its genesis as a particularization of similarities.

Finally, it seems to me appropriate to add a further consideration: Chasles' doctrine concerning the movement of a rigid body must be considered in its globality: the concepts introduced and their direct application to kinematics are interesting, but the geometrical consequences drawn by Chasles are even more significant. The "curious" theorems concerning the algebraic curves and surfaces show the profoundness and the broadness of his conception. They are his mark of originality. He had the merit to have discovered these theorems connecting them to the concepts he introduced in kinematics. These theorems are the litmus paper of his foundational programme and of his way of conceiving the relations between geometry and the rest of mathematics and science. As we have told, these ideas by Chasles had a certain circulation in the mathematical environment, but they never became "popular", probably because algebraic geometry was taking a general road, to which Chasles himself gave important contributions, that was completely different from that connoting his works on the displacement of the rigid body. For a historical work dedicated to this great mathematician, the "curious" theorems are unavoidable reference points because they represent the philosophical cipher of his way of conceiving foundations of mathematics.

[73]Chasles (1860–1861, III, p. 78): "Le théorème précédent est un de ceux que nous avons fait connaître anciennement dans le *Bulletin des Sciences mathématiques du baron de Férussac* (t. XIV, p. 324, année 1830) comme dérivant de la considération plus générale de deux corps *semblables* placés d'un manière quelconque dans l'espace". Italics in the text. Obviously, Chasles was referring to Chasles (1830i).

Chapter 4
Chasles and the Systems of Forces

Abstract The way in which a system of forces modifies the kinetic state of a physical system was a serious issue in the first 30 years of the nineteenth century. Mathematicians and physicists knew how to solve single problems, but the foundations of the theory were not solid. In 1830, Chasles offered a remarkable contribution to this question. His results were not new from a physical standpoint, but they are expressed in a way which can supply a precise geometrical foundation to the study of the system of forces. Therefore, in the first of the two sections in which this chapter is divided, I provide the reader with the mathematical context in which Chasles' work on the system of forces is inscribed. In the second section, his results are analysed and the thesis that they are a significant part of his global foundational programme is claimed.

The two main contributions given by Chasles to the study of the systems of forces are Chasles (1830c) and Chasles (1847). The latter paper is a specification and a clarification of the concepts already introduced in the former. Several references to the systems of forces also exist in the *Aperçu historique*. The best synthesis of the way in which Chasles contributed to explain the questions connected to the systems of forces is offered by Quetelet, who, while reviewing (Chasles, 1830c), wrote:

> When more forces urge a free solid body, they can be replaced in several different manners by other forces. The system of this new forces is defined to be equivalent to the system of the given forces. Two similar systems have mutual relations, most of which are continuously used in mechanics and are proved by analytical methods.
>
> In this writing, M. Chasles intended to prove these different relations and others, which are more general and new, in a purely rational manner, without resorting to analytical formulas.[1]

[1]Quetelet (1830, p. 246): "Quand plusieurs forces sollicitent un corps solide libre, on peut les remplacer d'une infinité de manières par d'autres forces; on dit quel le système de ces nouvelles forces est équivalent au système des forces proposées. Deux pareils systèmes ont entr'eux certaines relations, dont plusieurs sont d'un usage continuel dans la mécanique, et qu'on a coutume de démontrer par l'analyse.

P. Bussotti, *Chasles and the Projective Geometry*,
https://doi.org/10.1007/978-3-031-54266-4_4

This means: though some of the propositions proved by Chasles were new, most of them were already known. The most important novelty was the approach to the problem: for Chasles proved his theorems through purely geometrical methods without resorting to analysis. Furthermore, he framed them within a general picture. He invented, so to say, a new and more general environment in which these results could be read and interpreted. This environment is that of a geometrical foundation of mechanics, at least of its basic parts. From a conceptual point of view, Chasles' studies on the system of forces applied to a rigid body have to be seen in a line of continuity with those inherent to the geometrical movement of a rigid body. Chasles faced both the kinematical aspect (pure geometrical movement) and the causes (system of forces) which determine that state of movement of a rigid body. Therefore, though he worked on geometrical movement and on system of forces in the same period, the latter subject represents the second conceptual step towards the construction of a geometrical basis for mechanics, of which the former subject is the initial step.

To fully grasp the meaning of Chasles' work on forces' systems, a hint to the context in which they were conceived is important. In particular, it is worth recalling the results obtained by the French mathematicians between the end of the eighteenth and the beginning of the nineteenth century. As always while dealing with Chasles, the references to the German mathematicians must be prudently valuated because Chasles was not able to read German and, hence, he was not very well informed about the important progress made in Germany. Of course, there are some references to the German geometers, but Chasles' milieu is that of French (and, in part, Italian) mathematicians.

4.1 Systems of Forces: Chasles' Context

In Chasles (1830c), there is a strong connection between geometry and mechanics. In particular (as we will see), the link between the two disciplines is given by the relation between the areas of some geometrical figures and the representations of system of forces. Chasles mentioned some authors and their works as his reference points: Carnot's *Géométrie de Position* (Carnot, 1803); Poinsot's *Éléments de Statique* (Poinsot, 1860, first edition 1803); Poisson's *Traité de Méchanique* (Poisson, 1811. Arguably, the section concerning statics); Hachettes's *Éléments de géométrie a trois dimensions* (Hachette, 1817);[2] Binet's *Memoire sur la composition des forces et sur la composition des moments* (Binet, 1815); Giorgini's *Teoria*

M. Chasles s'est proposé dans cet écrit de démontrer ces diverses relations, et d'autres plus générales, ou nouvelles, d'une manière purement rationnelle, et sans le secours des formules analytiques".

[2]These references are given in Chasles (1830c, p. 97). With regard to Poinsot, Chasles referred to the fourth edition of his *Statique* (Ibid., p. 110).

analitica delle proiezioni (Giorgini, 1820);[3] Sturm's *Recherches analytiques sur les polygones rectilignes plans ou gauches, renfermant la solution de plusieurs questions proposées dans le présent recueil* (Sturm, 1824–1825);[4] Lhuilier's *Recherches polyèdrométriques* (Lhuilier, 1828).[5] Furthermore, Gergonne, Möbius and Giorgini are mentioned in a note on p. 112 because (as I will clarify) they reached results similar to Chasles'.

Carnot and Hachette are mentioned by Chasles for their geometrical achievements, in particular for the theorems on transversals regarding the relations between sides and angles of polygons as well as edges and solid angles of polyhedron and for the results on the projections of plane areas. Specifically, the third section of the analytical part of Hachette's treatise concerns the areas of figures projected onto the three coordinate orthogonal planes.[6] The reference to Sturm is important for a result concerning the centre of the mean distances and that to Lhuilier for a theorem on the areas and volumes with sign plus or minus. Therefore, these results are basically geometrical, though Sturm also referred to systems of forces. Hence, to offer a preliminary hint of the context which influenced Chasles, it is necessary to refer to Poinsot, Poisson, Binet and Giorgini. The most relevant results of the other authors mentioned by Chasles will be faced while dealing with his achievements, namely in Sect. 4.2.

The first 30 years of the nineteenth century represent an interesting period for the part of rational mechanics concerning the foundation of statics and dynamics and, hence, the equilibrium of the forces' systems as well as for the relations between geometry and mechanics. The problems of the composition and decomposition of forces dated back to the seventeenth century. Stevin discovered and used implicitly the parallelogram of forces. However, an extensive use of the forces' parallelogram is due to Varignon and to Newton's *Principia*.[7] Daniel Bernoulli, within statics, tried to offer a purely geometrical proof of this rule. During the eighteenth century, the study of the forces was extended and elaborated by several physicists. The most important were Euler and Lagrange. Euler, while facing the dynamics of a rigid body, introduced the concept of momentum of inertia of a body with respect to an

---

[3]Ibid., p. 102. Binet's work was published in 1815 but read at the Institut de France in 1814. Thus, Chasles referred to the year 1814.

[4]Ibid., p. 115.

[5]Ibid., p. 118.

[6]Hachette (1817, II, pp. 92–114).

[7]It is appropriate to point out that Newton introduced the decomposition of forces in the famous Corollary 1 to the laws of motion. However, as Westfall stresses, in the *Principia* two conceptions of force cohabit: 1) what afterwards was called "Newtonian force", namely $F = ma$, or $F = m\frac{dv}{dt}$, and 2) impulsive force, which in formulas can be indicated by $F = m\Delta v$ (Westfall, 1971). It is evident enough that in the mentioned corollary Newton is referring to an impulsive force. Therefore, he is decomposing velocities or quantities of motion, rather than Newtonian forces. This does not imply, obviously, that he did not think of the decomposition of Newtonian forces or of any other quantity we call "vector", but in the proposition assumed as evidence of such decomposition, he is developing a different operation.

axis and of momentum of a force with respect to an axis (Euler, 1752), being the momentum of a force with respect to a point well known.

The relations between geometry and dynamics became profound thanks to two fundamental theorems introduced by Laplace in his *Traité de Méchanique Céleste*: the first theorem concerns the *conservation of the mass centre's of a system* and the second one the *conservation of areas*. The law of areas' conservation regards the motion under the action of a centripetal force acting as the inverse square distance[8] and, as Caparrini points out, it is nothing but a scalar formulation of the principle of conservation of momentum (Caparrini, 2002, p. 156). It is, however, interesting the way in which Laplace formulated the principle of areas' conservation: if one considers a series of bodies subject only to their mutual attractions and projects the area described in an infinitesimal time *dt* by the radius vector joining two different positions of one of these bodies to the centre of mass of the system onto any two planes, the sum of these areas multiplied by the masses of the bodies is proportional to *dt*. Therefore, in a finite time, it is proportional to time. Thence, if, in the planes of the planetary orbits (be *n* the planets), in any given time, one draws an area equal to that of the orbit multiplied by the ratio between the mass and the planet's periodical time (let us call P_i this quantity), the sum of all these areas and their double product two by two and by the cosine of the planes' inclination is constant for equal given time intervals.[9] In symbols

$$\sum_{i=1}^{n} P_i^2 + 2 \sum_{j,l=1}^{n} P_j P_l \cos P_j P_l = \text{constant}$$

We will see that the form of this formula, which (Binet, 1815) clearly explains in the light of decompositions of forces and decomposition of moments, is important for Chasles' treatment of system of forces. However, before dealing with Binet's works, it is appropriate to recall those works by Poinsot, Poisson and Giorgini that are mentioned by Chasles. I will enter several details of the works produced by these authors because, in this case, it is fundamental to grasp the context in which Chasles inserted his discoveries.

4.1.1 Poinsot

The *Éléments de Statique* of Poinsot (first edition 1803) is the celebrated work in which the author introduced the concept of couple of forces. It is appropriate to

[8] In fact, the area law holds for every central force, though not acting according to the inverse square law.

[9] The principles of conservation of the mass centre and of conservation of areas are proved by Laplace in the first part, Book I, Tome I of his *Traité de Méchanique Céleste*, Chapter V, pp. 50–64 (Laplace, 1798–1825, I).

highlight that Poinsot's contribution has to be framed in the "rational re-organization" of mechanics. With this locution I mean the idea to offer a systematic, axiomatic, mathematical treatment of all parts of mechanics, and, in particular, with reference to this work, of statics. The very basic principles of mechanics, as that regarding the decomposition of forces, were used from a long time, but many physicists—among whom Poinsot—were not satisfied with the way in which they were presented. They searched demonstrations based upon few, clear—possibly geometrical—principles. The famous and admirable proof of the principle of virtual velocities given by Lagrange by means of a series of pulleys[10] was a sort of ideal reference point for the physicists who aspired to a rational organization of their science.[11] A meta-principle which, at that time, was of large use in mechanics represented a particular version of the principle of sufficient reason: it was the principle of symmetry. To give the most elementary example, let us consider the first axiom of Poinsot: two equal and opposite forces applied to a point are in equilibrium and two equal and opposite forces applied to the extremities of an inflexible bar are in equilibrium. As a justification for this axiom, Poinsot wrote: "Indeed, there is no reason for the movement to be produced towards one side rather than towards another side [. . .]".[12] Poinsot also applied this principle in the proofs of some theorems.

Poinsot's idea is to treat the composition and decomposition of forces and momenta relying upon the properties of parallel forces. For the first theorem he proved in the first section, entitled "Composition et décomposition des forces", is that any two parallel forces P and Q of the same sense applied to the extremities A and B of a rigid straight line have a resultant which is applied between A and B, which is parallel to P and Q and equal to their sum (ibid., pp. 18–19). The second theorem establishes the point of the segment AB in which the resultant has to be applied and proves that such a point C is that with respect to which the static momentum of the two forces is equal. The demonstration is interesting because Poinsot proved the theorem when the ratio of the two forces is rational and afterwards, demonstrated by an *ad absurdum* reasoning based on the first part of the proof that the theorem is true also when the modules of the forces have an irrational ratio. The proof of this theorem has strong similarities with those given by Archimedes in the propositions 6 and 7 in *On the equilibrium of planes*.

The first corollary of this theorem is remarkable and is a prelude to the introduction of the notion of couple of forces. Indeed, Poinsot considered (Fig. 4.1) three parallel forces P, Q and R which are in equilibrium on AB.

[10] Lagrange (1811, § 18, pp. 23–25).

[11] I do not address the question whether the idea to construct rationally the whole of mechanics is realizable. Mach, for example, thought it was not. I will only point out that, at the beginning of the nineteenth century, there were several slightly different approaches aiming at a rational re-organization of mechanics.

[12] 1803, p. 13): "Car il n'y a pas de raison pour que le mouvement naisse d'un côté plutôt que de l'autre [. . .]".

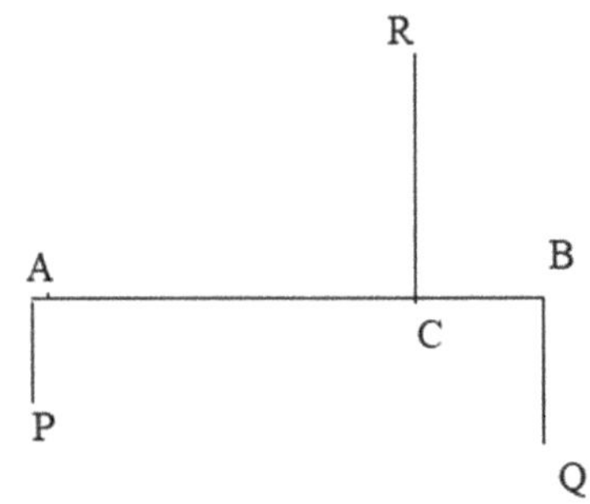

Fig. 4.1 Reproduction of the diagram used by Poinsot to prove the proposition expounded in the running text

Each of them, for example the force Q, considered in the opposite sense (i.e. the force $-Q$), can be interpreted as the resultant of P and R. Since P and Q urge in the same sense, $R = P + Q$ and $Q = R - P$. Thence, the resultant of two parallel forces acting in opposite sense is equal to their difference and urges towards the same sense as the bigger one. Given the two forces P and R, and the point C where their resultant is applied (the point called by Poinsot "the centre of parallel forces"), it is easy to prove that $R : Q = AB : AC$ (ibid., p. 24). If the intensity of P and R is the same, the resultant Q is null and its point of application goes to infinity. However, it is obvious that the two forces though having no resultant have an effect—though not a kinetic one—on the body on which they are applied.

Later on, Poinsot considered systems of parallel forces until reaching a theorem he defined "remarquable" (ibid., p. 30): be given a system of parallel forces applied to the points $A, B, C, D, \ldots$, which can be regarded as belonging to a straight line. If the action straight lines of these forces rotate at the same angle and in the same direction, while the remaining forces applied to $A, B, C, D, \ldots$, the point in which the resultant—if existing—is applied does not vary. This point is *the centre of the parallel forces* (ibid., p. 30). In conclusion, a system of parallel forces has always a resultant unless it is reducible to two equal and opposite forces.

Given these propositions on the parallel forces, Poinsot treated the composition and decomposition of any force as a consequence of the results obtained with regard to the parallel forces. The first two theorems (Théorème III, pp. 31–33 and IV, pp. 33–39) prove the parallelogram rule: the former establishes that the resultant of two forces P and Q, whose action lines AB and AC are incident, is directed along the diagonal of the parallelogram $ABCD$; the latter determines the intensity of the resultant as equal to such diagonal. I offer the basic elements of Theorems III's proof given by Poinsot because this is useful to clearly grasp the idea to construct the fundamental propositions of rational mechanics onto a mathematical-geometrical rigorous basis. Under this respect, Chasles' foundational programme can be seen as a part of this movement, though, as we have seen and as we will see, it went far beyond the aims of scholars who were essentially physicists and who, as a consequence, considered the mathematical foundation as a support to physics, rather than a reductionist idea.

Poinsot developed the following reasoning (Fig. 4.2):

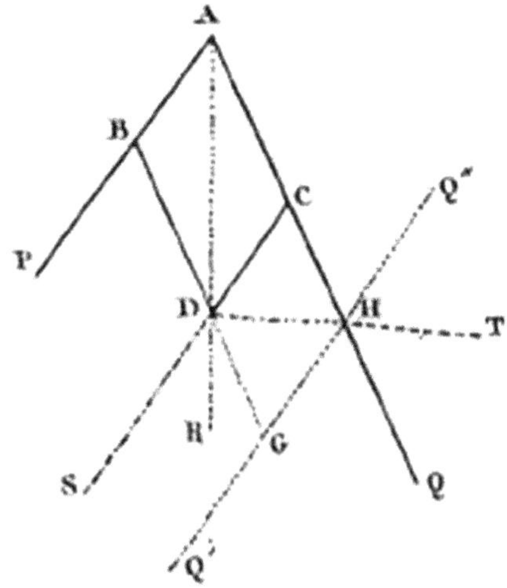

Fig. 4.2 The diagram used by Poinsot to prove the theorem expounded in the running text

After a brief reasoning to show that the resultant is applied to A and is coplanar with P and Q, he drew the parallelogram $ABCD$ and produced BD, so that $DG = DC$, so obtaining the rhombus $CDGH$. Let us apply the two opposite forces Q' and Q'' at G along GH such that, in intensity, $Q = Q' = Q''$. The resultant of the forces P, Q, Q', Q'' passes through D. For, being $Q = Q'$, the two parallel forces P and Q' are as the sides AB, AC—i.e. DC, DB—, or, because $DC = DG$, as DG and DB. Consequently, *because of the properties of the resultant of two parallel forces*, their resultant S is applied at D. Furthermore, since $Q = Q'$, their resultant T, if produced, divides the angle CHG of the rhombus $CDGH$ into two equal parts, so that it passes through D. Thence, the resultant of the forces S and T passes through D. Since the two forces Q' and Q'' have a null effect the resultant of the four forces P, Q, Q', Q'' coincides with that of P and Q. This implies that it passes through D and, therefore, through the diagonal AD.

Apart from the theorem on parallel forces, the only notions used in this proof are 1) the property that the resultant passes through A; 2) in the initial section of the first chapter entitled "Axiomes, Lemmes préliminaires, etc." Poinsot posed as axiom that the resultant of two equal forces bisects the angle between the forces (ibid., pp. 13–18).

After having treated the composition and decomposition of forces whose directions are cutting lines in space (ibid., pp. 39–43), Poinsot began the analysis of the couple of forces, which is relevant for Chasles' work.

Poinsot defined a couple of forces as the system (*ensemble*) of two equal and opposite parallel forces P, $-P$ not applied at a point (Fig. 4.3). The arm AB is the perpendicular between the directions of the two forces. This is exactly the case in which two parallel forces cannot be reduced to a sole resultant force. The momentum of the couple is defined, in intensity, as the product $P \times AB$. Therefore, Poinsot claimed that the couples are a cause of movement different from a single force and he added this consideration:

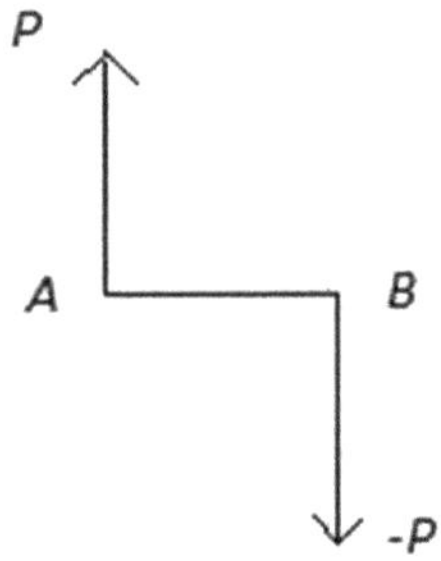

Fig. 4.3 A couple of forces P, $-P$ with arm AB

> To distinguish this new cause of movement, which is, in a sense, of a particular nature, we might call it *energy*. On the other hand, as we will see soon, the *energy* of a couple is measured by its *momentum*, so that it will be often possible to replace the second term to the first [. . .].[13]

Commentary: Poinsot's work is a significant step along the way which leads to the acquisition of the concept of vector. As it is manifest enough from this brief quotation and as will be seen in some details, Poinsot in his foundation of mechanics clearly distinguished, starting from a supposed state of rest, between a translational motion (or, in case of statics, the tendency to a translational motion) and a rotatory motion. He attributed the translational motion to a force and the rotatory motion to a couple. He understood that it was necessary to distinguish between two senses of rotation and, as I will clarify, he connected the momentum of a couple, to a straight line segment with the sign plus or minus. Therefore, he arrived very close to our idea to consider the momentum as a vector oriented according to the right-hand rule, but did not reach it exactly. In the specific case, it is clear that Poinsot fully guessed that a couple also has a "vector" character, which is something different from its intensity, and which should indicate the sense of rotation produced by the couple. However, the full identification of the vector character of a couple of forces was a far more difficult enterprise than the identification of the vector character of a single force. It is tempting to identify Poinsot's *momentum* with our momentum of a couple of forces and the *energy* with the absolute value of such a vector. However, this is, at least in part, misleading. Rather, it seems that *energy* is the general term indicating the capability to create a rotation, whereas *momentum* is the specific value of the energy when a single couple is considered. Poinsot regarded the momentum as the measure of energy. For, as we have seen, he wrote:

> [. . .] the action of a couple cannot be compared in any way to that of a single force. To distinguish this new cause of movement, which is of particular nature, the term *energy* will be used. As we will see later, the *energy* of a couple is measured by its *momentum*. Thus, It

[13] Poinsot (1803, p. 44): "Pour distinguer cette nouvelle cause de mouvement, qui est en quelque sorte d'une nature particulière, on pourrait l'appeler *énergie*. Au reste, comme on verra tout-à-l'heure que l'*énergie* d'un couple est mesurée par son *moment*, on pourra souvent substituer ce second mot an premier [. . .]". Italics in the text. On the concept of force in Poinsot, see Grattan Guinness (2014).

> will be often possible to replace the first term with the second one or to use sometimes the one for the other.[14]

It is interesting to reveal that Poinsot stressed the effect of the couples to be a rotation in a sense or in the opposite one, but that the idea of rotation is purely accessory to the couples because, as we read:

> [...] it is evident that the effect of any couple will be the body's rotation around the middle point of its lever's arm. It is easy to distinguish the couples' sense by distinguishing the couples which tend to produce a rotation in one sense from those which tend to produce a rotation in the opposite verse.... [However], the idea of rotation, which is purely accessory, will be useful only to create an image for our necessities.[15]

This quotation is important because it confirms patently Poinsot's aim to found statics on rational bases: it is necessary to explain clearly the causes of the rotational movements; for this aim a mathematical object was created—that of couple—with its essential associated quantities—momentum and energy. The properties of this object are developed mathematically, without taking into account that it was created to explain rotations or tendency to rotation; afterwards, the mathematical results—obtained independently of physics—are applied to physical situations. This is the typical rational attitude towards physics.

Coming back to the analysis of Poinsot's text, the first two important results (arts. 49 and 50 resp., pp. 45–52) are two lemmas which concern the most elementary cases in which a couple can be replaced by another couple.

In art. 49 Poinsot proved that a couple can be moved in its plane or from its plane to a parallel one and rotated arbitrarily in such a plane without any modification of its effect, provided that the new arm is attached to the first one (Fig. 4.4).

Lemma in art. 50 established that any couple $(P, -P)$ of arm AB can be replaced by another couple $(Q, -Q)$ of the same sense applied to the arm BC, different from AB, provided that $P : Q = BC : AB$.

As a "*Remarque*" to this Lemma, Poinsot formulated the important concept that, in order to compare the energy of different couples, it is possible to consider their momenta. If (ibid., art. 45, pp. 52–53) the couples act in the same or in parallel planes, they can be composed in a sole couple whose momentum is given by the algebraic sum of the various composing couples' momenta. The sign "+" or "−" depends on the sense of the couple.

After the lemmas, Poinsot proved a series of theorems. The second of them (ibid., arts. 55–63, pp. 53–61) is important especially for a remark added by Poinsot (ibid.,

---

[14] Poinsot (1803, pp. 43–44): "[...] l'effort d'un couple ne peut être comparé d'aucune manière à une simple force. Pour distinguer cette nouvelle cause de mouvement, qui est en quelque sorte d'une nature particulière, on pourrait l'appeler *énergie*. Au reste, comme on verra tout-à-l'heure que l'*énergie* d'un couple est mesurée par son *moment*, on pourra souvent substituer ce second mot an premier, ou les prendre quelquefois l'un pour l'autre". Italics in the text.

[15] Ibid., pp. 44–45: "[...] l'effet de chaque couple sera visiblement de faire tourner les corps autour du milieu de son bras de levier, et l'on distinguera facilement le sens des couples, en distinguant les couples qui tendent à faire tourner dans un sens, d'avec ceux qui tendent à faire tourner dans le sens contraire. [...] l'idée de rotation qui est purement accessoire, ne servira qu'à faire image au besoin".

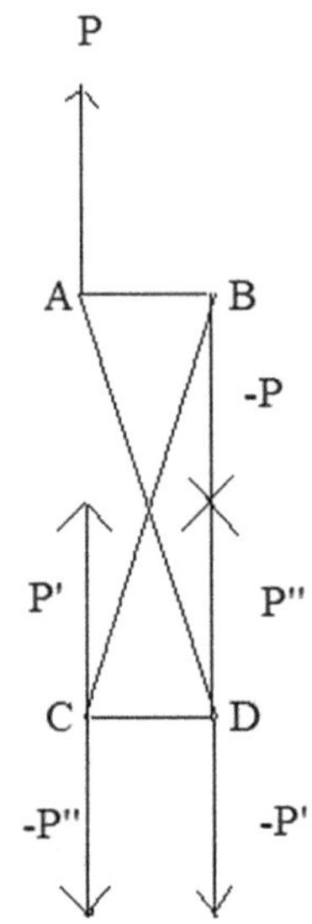

Fig. 4.4 A simplification of Poinsot's diagram used to prove Lemma in art. 49. To explain Poinsot's reasoning and way of expression, let us consider the easiest case: that in which the couple moves in its plane or in a parallel one, without any rotation of its arm. Be *AB* the arm of the couple $(P,-P)$. To the segment *CD*, parallel and equal to *AB* apply the two couples $(P',-P')$; $(P'',-P'')$ so that each single force has the same intensity as *P*. These two couples produce two equal and opposite rotations. Hence, their effect is null. The two forces *P* and P'' applied at *AD* give a resultant equal and opposite to that of the forces $-P$ and $-P''$ applied at *BC*. This means that the effect of the two couples $(P,-P)$ and $(P'',-P'')$ is null. Thus, only the couple $(P',-P')$ applied at *CD* remains and its effect is, consequently, the same as that of the couple $(P,-P)$. As Poinsot claimed the arm *CD* is attached (at least ideally) by *AD* and *CB* to the arm *AB*. The case in which the couples rotate is faced by Poinsot with similar devices

art. 59, p. 58). The theorem establishes that two couples in two planes cutting with a certain angle can be composed by a single couple. Furthermore, if the momenta of these couples are represented by two segments drawn with an angle equal to that of the two planes, so as to obtain a parallelogram, the momentum of the resulting couple is represented by the diagonal of such a parallelogram. The couples' plane will cut the angle of the two planes in which the composing couples lie as the diagonal of the parallelogram cuts the angle of the two adjacent sides (see Fig. 4.5 and its caption to clarify the meaning of Poinsot's theorem).

The fundamental concepts here expressed are two: 1) to any couple a segment is associable which represents the momentum of the couple. The inclination of this segment on the plane of the couple is arbitrary, but, once chosen a segment inclined of a certain angle α on the plane of the couple, the segment representing the other couple has to be inclined of α on the plane of the couple, and 2) these two segments are treatable exactly as if they were forces. Therefore, the couples have an algebraic treatment analogous to that of forces, provided that they are replaced by these segments which represent their momenta.

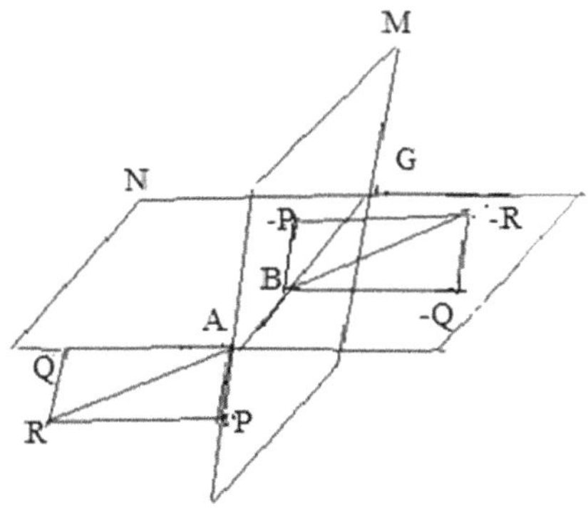

Fig. 4.5 Adaptation of Poinsot's diagram to prove the theorem enunciated in the running text. The two given couples (which are not drawn in the diagram) lie in the planes *AGM*, *AGN* cutting along *AG*. Change these couples into two equivalent with a common arm. This is possible on the basis of the two lemmas we have expounded, which proves the first part of the theorem. Let us now see the second and more important one. Wherever the first couple lies in the plane *AGM*, it is possible to pose it at right angle with the intersection line *AG* so that its arm *AB* lies on *AG*. With a similar operation the arm of the second couple can coincide with *AB*. The two forces *P* and *Q* (applied in *A*) of the two couples have a resultant *R*. Analogously $-R$ is the resultant of the two other forces of the couples $-P$ and $-Q$. Thus, we have a resultant couple (R,$-R$) applied at *AB*. Since these three couples have the same arm, their energies are respectively proportional to *P*, *Q*, *R* represented by *AP*, *AQ*, *AR*. Hence, the energy of the resultant couple *R* is given by *AR*. It is clear that the angles of the three lines *AP*, *AQ*, *AR* measure the angles of the three planes, so that the plane of the resultant couple divides the angles of the two other planes as the diagonal *AR* divides the angle of the two adjacent sides *AP*, *AQ*

What follows after the demonstration of this theorem is even more interesting (ibid., pp. 58–61). Poinsot claimed that the position of a couple can be determined by a straight line perpendicular to its plane rather than by the plane itself. For he introduced the concept of axis of a couple (*axe du couple*). He considered the parallelogram *ALGM* and, on the sides *AL*, *AM* and the diagonal *AG*, elevated three planes π, φ, ψ orthogonal to the plane of the parallelogram. The sides *AL* and *AM* can represent the momenta of two couples lying respectively in π and φ, while *AG*, which is the resultant of the two momenta, will represent the resultant couple lying in ψ. Now, through the point *A* and in the plane of the parallelogram, Poinsot drew three straight line segments AL', AG', AM', such that $AL' = AL$; $AG' = AG$; $AM' = AM$, respectively, perpendicular to *AL*, *AG*, *AM*. Obviously, they will also be perpendicular to π, φ, ψ. Each of these segments, which are perpendicular to the planes of the couples, is the *axe du couple* lying in the plane perpendicular to the segment.

A couple is, hence, represented by its axis and its energy. The length of the axis is a measure of the energy.

Therefore, the picture is rather complicated: there are two segments associated with the couples; one is the momentum, which can have any inclination on the plane of the couple; the other is the axis of the couple, which, one might comment, is the momentum when the angle momentum-plane of the couple is right. However, it is clear that Poinsot distinguished the concept of momentum from that of axis; I mean that he did not introduce the axis as a momentum which is perpendicular to the plane of the couple; he specified its genesis. In his mind, the axis is an object different from the momentum. Once again, one is tempted to identify the axis with the direction of our momentum. This assumption is—so to say—near to the truth, but it is not exactly the truth, because in Poinsot the two concepts (momentum and axis) continued to cohabit, though that of axis assumed primary importance. Furthermore, he would continue to speak of the sense of a couple as a condition to determine the sign of the axis, and he also arrived close to the idea of posing the axis either above or below the plane of the couple according to the sense of rotation.

The first chapter is closed by the general theorem of equilibrium: given any number of forces applied to a solid body, they can be reduced to a force and to a couple. Generally, the force and the couple lie in different planes (ibid., p. 65). Furthermore, he also proved that a system of forces applied to a body can be reduced to a sole resultant if and only if, after having reduced the system to a force and to a couple, the force is parallel to the plane of the couple (ibid., p. 69). Given these results, Poinsot proved that a system of forces is also reducible to two forces, whose directions are two skew straight lines (ibid., p. 71).

The other chapters of Poinsot's *Statique* are interesting, but they are less significant than the first one in reference to Chasles' works. Thence, I will not deal with them. Almost all of the Poinsot's results which are significant for Chasles are referred to his important essay entitled *Mémoire sur la composition des moments et des aires*, presented in 1804 at the Institut de France and published in 1806 in the *Journal de l'École Polytechnique* (Poinsot, 1806a).

One of the most important concepts introduced by Poinsot was that of "principal momentum" of a force with respect to a series of axes passing through a point.[16] This concept was fundamental for the development of statics and, in particular, to grasp Chasles' work on the system of forces. Thence, I will offer a detailed explanation of such notion.

Poinsot interpreted the notion of principal momentum and the consequent theorems in the light of the theory of couples.

Be a system of forces decomposed into a resultant force and a resultant couple, whose momentum is G.

Be L, M, N the components of G perpendicular to the three orthogonal axes x, y, z. Furthermore be λ, μ, ν the angles between the coordinate axes and a straight line perpendicular to the plane of the couple, which is the axis of the couple. The identities

[16] Poinsot (1806a) refers to the expression "moment maximum", but this is exactly the concept which, in the literature of that period, was called principal momentum. The locution *moment principal*, which becomes the usual one to indicate the concept introduced by Poinsot, is due to Poisson (1811, p. 114).

$$G^2 = L^2 + M^2 + N^2 \text{ and } \cos\lambda = \frac{L}{G}; \cos\mu = \frac{M}{G}; \cos\nu = \frac{N}{G}$$

hold. Therefore, the momenta of the couples are subject exactly to the same rules of composition and decomposition as the forces. Thence, the resultant momentum G can be mathematically treated as a force.

Indeed, the momenta L, M, N are respectively equal to the sum of the ordinary momenta[17] of forces with respect to the three axes x, y, z. Namely, they are equal to the sum of the products given by the projections of the force/s on the three coordinate planes multiplied by its/their distances from the axis perpendicular to any projecting plane.

Thus, it is possible to consider the resultant momentum G as if it were a force. If we search the component of G calculated on an axis a inclined of an angle θ with respect to the axis of the resultant couple (which is perpendicular to the plane of the couple), we obtain $G \cos \theta$.

If λ', μ', ν' are the angles between a and the coordinate axes, it is

$$\cos\theta = \cos\lambda\cos\lambda' + \cos\mu\cos\mu' + \cos\nu\cos\nu'$$

Consequently

$$G\cos\theta = L\cos\lambda' + M\cos\mu' + N\cos\nu'$$

This means that:

1. Among all the axes passing through the origin, the axis [b] of the resultant couple is that with respect to which the sum of the momenta is maximal.
2. The sum of the momenta is the same with respect to all the axes which form an equal angle [α] with the axis of the maximal momentum. These axes form a conic surface of angle [α] described around such axis.
3. The sum of the momenta is null with respect to all the axes which form a plane perpendicular to [b].[18]

This maximal momentum is the *principal momentum* of the given system of forces. Poinsot did not chose a priori the inclination that the momentum G of a couple must have on the plane of the couple itself. Now he is claiming that an only inclination exists which allows us to project the whole momentum on a single straight line so that the component on that straight line is equal in absolute value

[17] Poinsot spoke of "[. . .] somme des moments pris dans l'acception ordinaire pour rapport au trois axes, *x,y,z* [. . .]" (Poinsot, 1806a, p. 187).

[18] Ibid., p. 190: "1. Que de tous les axes qui passent par l'origine, l'axe de la couple résultant est celui par rapport auquel la somme des momens est la plus grande; 2. Que la somme des momens est la même par rapport à tous les axes qui font un même angle avec celui de plus grand moment, ou qui forment une surface conique décrite autour de lui sous cet angle; 3. Que la somme des momens est nulle par rapport à tous ceux qui font cet angle droit, ou qui forment un plan perpendiculaire à sa direction".

to the whole momentum and the components on any other straight line are equal to 0. This straight line is the axis of the couple. Therefore, Poinsot's maximal or principal momentum is the quantity that nowadays we call momentum of a couple of forces.

He claimed that Laplace reached the same results, but through different methods and concepts, i.e. without associating the axis with respect to which the sum of the momenta is maximal with the axis of the resultant couple (ibid., p. 12).

Poinsot clarified that so far some of the results he had obtained were already known—though using different concepts and methods—but that in the rest of his treatise, he was going to expound some completely new results.

So far, one might say that Poinsot determined a quantity intrinsically connected to any couple of forces. However, we have seen that, generally speaking, any system of forces can be decomposed into a resultant R and a resultant couple of principal momentum G. Suppose that we are studying the movement of a rigid body and we establish a coordinate system with origin in a point O of space. If we change the origin of our system, the value and the sense of the resultant will not change passing from a system of coordinates to the other, but the resultant couple will change in energy and its axis will also modify its inclination in space. At each position of the origin, thence, a different principal momentum will correspond. Poinsot posed the problem of determining the point assumed as the origin for which the value of principal momentum is minimum. He proved that this problem can be solved and also taught how to construct the axis with respect to which the principal momentum has its minimum. Poinsot called this axis "The central axis of the momenta" (ibid., p. 15). It has the feature to be parallel to the resultant force, and to have a well-determined distance from the resultant, as Poinsot proved (ibid., pp. 14–15). The origin of the system can be assumed in any point of such straight line for the principal momenta to have a minimum (ibid., p. 15).

This is exactly the property which is nowadays, too, one of the basic statements regarding the decompositions of systems of forces: for any system of forces, a straight line called central axis exists such that all and only its points, assumed as a centre, determine a system equivalent to that given composed of a force and a couple: the momentum with respect to such origins is parallel to the force. One might say that, among the infinite ways in which a system of forces can be reduced to a resultant and to a couple, the one referred to the central axis of the momenta is the optimal one: the one which allows the easiest calculation when a problem has to be solved. Therefore, Poinsot's results, besides being fundamental from a theoretical point of view, also have a remarkable value to solve problems connected to the movement of a rigid body.

Through his concepts, he was able to prove easily the Laplace's principles of gravity centre's movement conservation (ibid., pp. 20–22) and areas' conservation (ibid., pp. 22–23). I will focus on the latter principle because it is important to understand Chasles' work on the systems of forces. Poinsot explained the principle of areas' conservation and its connection with the momenta like this: be given a system of bodies and an arbitrary fixed point, which he called *focus*, a term often recurring in physics and mathematics at that time with slightly different meaning. If one draws the radius-vectors from this point to all the bodies of the system and projects all these straight lines onto a plane π, the momentum of any force urging a body of the system with respect to an axis perpendicular to π is proportional to the mass of the body by the area described by the projections of the radius-vectors on π. The principle of areas' conservation claims that the sum of the product of the bodies' masses by the areas described by the respective radius-vectors traced on π is proportional to time. Thence, for a given time interval, it is constant.

Therefore, if one does not take into account the mass (which is a constant eliminable through easy mathematical devices), the principle of areas' conservation can be replaced by the principle of momenta conservation. Thus, *the area and the momenta—hence, the couples—of the forces with respect to an axis are subject to the same mathematical treatment.* Poinsot clearly stated:

> This granted, since the areas drawn by the radius-vectors are nothing but the momenta of the forces, it is possible to apply to the compositions of areas all the results we have obtained for the composition of the momenta. [19]

This means that, among all the planes π passing through the focus, one exists for which the sum of the drawn areas has a maximum and, if L, M, N indicate the projection of the areas in π on three orthogonal coordinate planes, such maximal area will be $G^2 = L^2 + M^2 + N^2$. At the same time, as it is the case for the momenta, among the planes drawn through different points of space, there will be that for which the maximal projection has a minimum (ibid., p. 24). Poinsot was able to identify a series of interesting features of the planes for which the projected area has a maximum and to apply them to some problems concerning dynamics as well as the system of the world. However, as far as Chasles is concerned, this is enough.

Two brief comments seem to me necessary:

1) An immediate consequence of the concepts introduced by Poinsot is that a couple of forces can be assimilated to an area. The couple is represented by a segment whose intensity is that of its momentum and whose direction is that of its axis. That is, it is represented by what today we call the momentum of the couple. Consequently:
2) Poinsot's *Memoire* represents a fundamental step with regard to the association of a segment with an area. As we have seen, it is not yet possible to frame completely these results into the modern conception of a vector, but certainly they belong to the story of this concept's acquisition. The vectorial operations (scalar and vector products) are still missing, but the way towards vector algebra

[19] Ibid., p. 199: "Cela posé, puisque les aires tracées par les rayons vecteurs ne sont autre chose que les momens des forces, il s'ensuit qu'on peut appliquer à la composition des aires tout ce que nous avons dit de la composition des momens". I add: in fact, if we consider a segment as a force and we calculate its momentum with respect to a point O, it is exactly the double of the triangular area having as vertices O and the two extremes of the forces. There is, hence, a perfect homology between momenta of couples and areas.

is open. We will see that Chasles offered further contributions to the idea of associating segments to areas or volumes.[20]

4.1.2 Poisson

Poisson's approach to mechanics[21] in the text mentioned by Chasles, i.e. the *Traité de Méchanique*, 1811, is rather different from Poinsot's.[22] Poisson, when dealing with parallel forces had the purpose to show that their treatment can be made quite similar to that of a single force. Therefore, while Poinsot first stressed that a couple is a cause of movement different from a single force and afterwards tried to develop a unitary formalism for both the couples' momenta and the forces, Poisson tried to offer a unitary picture from the beginning, though, obviously, he was well aware that a Poinsot's couple produces a rotation. Poisson was more inclined than Poinsot to introduce elements drawn from algebra and mathematical analysis in the initial steps of his treatise, whereas Poinsot's approach is more geometrical.

At the beginning of the first book Poisson intended to offer a completely rigorous proof of the parallelogram-forces theorem. In the long demonstration (Poisson, 1811, pp. 11–17) he used concepts as expansion in Taylor series, differentials of high order, etc.: in sum, a mathematical apparatus which goes beyond geometrical constructions and elementary algebra. After this, he treated in any detail the equilibrium of system of forces both in plane and spatial configurations. He analysed the

[20] I share Caparrini's opinion: "Even if Poinsot's work had no immediate effect on the development of the mathematical theory of vectors, its importance for the history of vector calculus had been greatly undervalued. Most of the mathematicians of the nineteenth century who used vectorial methods in their works were deeply influenced by Poinsot's writings" (Caparrini, 2002, p. 159). A confirmation of this opinion is that in Crowe (1967) Poinsot is mentioned only once when Crowe quoted a passage by Tait. As a matter of fact, Poinsot's work is completely ignored by Crowe. Caparrini also claims that: "In substance, Poinsot offered a completely vectorial treatment of the rigid bodies' static". Caparrini (2006, p. 228): "In sostanza, Poinsot diede una trattazione completamente vettoriale della statica dei corpi rigidi". This opinion can be shared.

[21] Among the texts dedicated to Poisson's mechanics I mention, without any claim to be exhaustive, the seminal Arnold 1983–1984. This work is composed of ten papers appeared in the *Archive for History of Exact Sciences*. It offers a completely satisfying picture of Poisson's physics as well as of the historical context in which Poisson was educated and worked. In the bibliography, I will mention separately each of Arnold's articles because this is more comfortable for the reader. Important works on Poisson's science analysed in a historical perspective are also: Berger (2005) (there are many references to Poisson), Dupont (1963–1964); Metivier-Costabel-Dugac (eds.) (1981). A new and enriched edition of this work is Kosmann-Schwarzbach (2013); Neville Greaves (2013). Useful indications on Poisson are freely available at the site: https://www.math-info-paris.cnrs.fr/bibli/wp-content/uploads/2015/07/catalogue_exposition_Poisson_Berkeley1.pdf. This site reports the material of the exhibition: "Siméon-Denis-Poisson. Mathematics in the Service of Science". Exhibition at the Mathematics Statistics Library, University of California, Berkeley, 7 November–17 December 2014.

[22] In reference to Chasles, it is enough to deal with the first book—dedicated to statics—of Poisson's treatise.

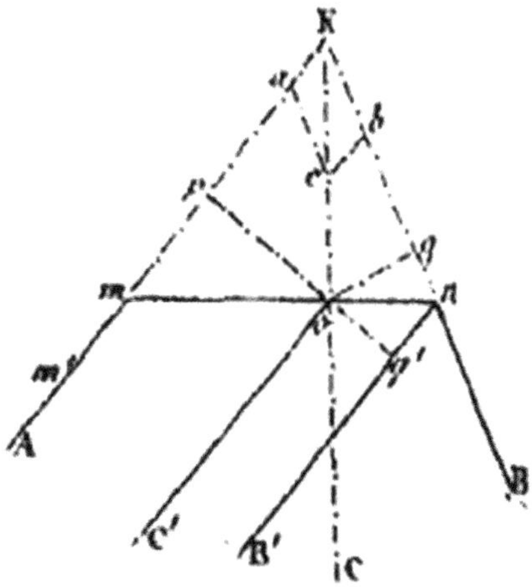

Fig. 4.6 The diagram used by Poisson in the reasoning expounded in the running text. (Poisson, 1811, table of diagrams posed at end of the volume)

case in which a material point is free or is constrained to move on given surfaces. In the entire first chapter, dedicated to the motion of a material point, there is no mention of parallel forces; they are introduced in the second chapter, concerning the equilibrium of a rigid body. This is an important point because Poisson followed a method opposite to Poinsot's: for Poisson proved the composition theorem for parallel forces starting from that of incident forces. It is worth referring to the basic elements of his reasoning (Fig. 4.6):

Poisson considered two forces P and Q applied at the extremities of a bar mn. Because of the principle according to which it is possible to move the application point of a force along its direction, one may apply two segments Ka and Kb proportional to P and Q at their intersection point K, to construct the resultant KC and to move it in the point O belonging to mn. By construction, it is $P : Q = aK : bK$ Hence $P : Q = \sin BKC : \sin AKC$. If, from O, the perpendiculars Op and Oq to the straight lines along which P and Q act are drawn, it is

$$Op = KO \sin AKC, \quad Oq = KO \sin BKC$$

so that $P : Q = Oq : Op$.

Therefore, he concluded, the two forces are inversely as the perpendiculars drawn on their direction from any point of the resultant (ibid., p. 37). Now there is the passage from incident to parallel forces because we read:

> This theorem is valid for any small amplitude of the components' angle AKB. It is also valid at limit, when this angle becomes null and the forces become parallel.[23]

[23] Ibid., p. 37: "Ce théorème ayant lieu, quelque petit que soit l'angle AKB des deux composantes, il s'ensuit qu'il subsiste encore à la limite, où l'angle devient nul et où les forces deviennent parallèles".

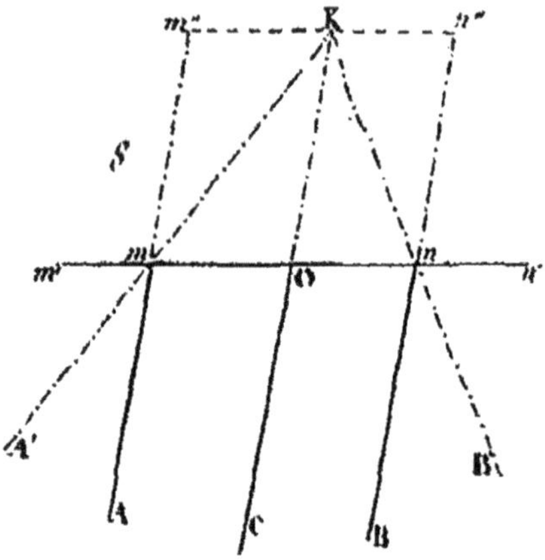

Fig. 4.7 The diagram used by Poisson for the composition of two parallel forces. The meaning of the symbols is clear. (Poisson, 1811, table of diagrams posed at end of the volume)

Therefore, contrary to Poinsot, the case of the resultant of parallel forces is conceptually reduced to that of incident forces. After this, Poisson developed all the classical theorems and formulas on parallel forces (ibid., pp. 37–43), in particular the formula according to which, if P and Q are two parallel forces, it is (Fig. 4.7)

$$nO = \frac{P}{Q-P} mn.$$

Thence, if the two applied forces are equal but have opposite senses, the resultant will have a null value and its point of application O will go to infinity.

As a commentary to confirm what the mathematical formulas claim, Poisson analysed the physical situation and stated that two equal and opposite parallel forces acting on two different straight lines have no resultant. If they had, there is no reason why it should be directed in the sense of one of the component forces rather than in the opposite sense. On the other hand, two forces of this kind are not reducible to a single force.

As in Poinsot, the principle of sufficient reason is applied, but the context is completely different: 1) the properties of the parallel forces are deduced from those of the incident forces and 2) there is no reference to Poinsot's concept of couple as a cause of movement different from "ordinary" forces.

To deal with the equilibrium of parallel forces without speaking of couples, Poisson introduced the concept of momentum of a force with respect to a plane π (*moment d'un force par rapport à un plan*, ibid., p. 49). This quantity is defined as the product of the force by the perpendicular drawn from its point of application to π. It is trivial to prove that the momentum of the resultant of an arbitrary number of parallel forces with respect to any given plane is equal to the sum of the momenta of these forces with respect to such a plane. Poisson added that the momentum of a force with respect to a plane has a sign: it is positive if the force and the ordinate of its application point have the same sign; it is negative otherwise (ibid., p. 49). After a brief series of reasoning, he was able to prove that for a system of parallel forces to be in equilibrium necessary and sufficient condition is that 1) the sum of the forces is equal to 0 and 2) the sum of their momenta with respect to two planes parallel to their direction is null (ibid., p. 54).

Commentary: the second condition is interesting if we compare it with Poinsot's condition according to which there is a rotation if and only if a couple is created. Let us, hence, consider a couple of forces (P,-P) and let us assume that the two coordinate planes xz and yz are parallel to the direction of the couple and that the forces are applied at the extremity of the couple's arm. These conditions do not limit the generality of the reasoning. Be x,x' the distances of the application points of P and $-P$ from the plane xz and be y,y' their distances from the plane yz. Poisson obtained the following system:

$$\begin{cases} Px - Px' = 0 \\ Py - Py' = 0 \end{cases}$$

It is evident that these equations cannot be satisfied together for any position of the couple with respect to the two planes.

The previous system derives for Poisson's general conditions of equilibrium with respect to two parallel planes (ibid., p. 53): he considered a set of parallel forces P,P', P'',... and obtained the following system of equations for the equilibrium of momenta with respect to two parallel planes: $\begin{cases} Px + P'x' + P''x'' + \ldots = 0 \\ Py + P'y' + P''y'' + \ldots = 0 \end{cases}$, of which the case of a couple of forces is a specific one.

While dealing with two parallel forces in a plane which are not reducible to one force, Poisson reached the same results as Poinsot, but he did not use explicitly the concept of couple, and he did not even mention Poinsot. For he stated that two equal opposite parallel forces acting on two different straight lines are not determined in intensity and direction because they can be replaced, in an infinity of manners, by two other parallel forces whose effect is equivalent, but which do not have either the same direction or the same intensity as the initial two (ibid., p. 56). This is, of course, true for the couples.

Poisson defined in the usual manner the momentum of a force with respect to a point and, comparing it with the momentum of a force with respect to a plane, he claimed that the latter is useful in the theory of parallel forces. Furthermore, he added this interesting consideration: the momenta with respect to a plane depend on the point of application of the force and not on its direction, whereas the momentum with respect to a point depends on the direction and is independent of the force's application point. Now, he continued in the following manner: the straight line along which the force acts and the perpendicular to such a straight line can always be considered positive quantities. Thence, whereas the momenta with respect to a plane can be positive or negative, those with respect to a point are always positive (ibid., p. 67).

Commentary: at first impression, it seems that Poisson considered the momentum with respect to a plane as a vector magnitude (or, more prudently, namely without superimposing more modern concepts to a mere description of Poisson's work, as a segment with a sign), while the momentum with respect to a point was regarded as a scalar quantity. However, this first impression has to be integrated because, while considering the momenta of a system of forces and that of the resultant with respect to a point, he claimed that the momentum of the resultant is equal to the sum of the force's momenta which tend to produce a rotation in the same sense as the resultant minus the sum of the forces' momenta tending to produce a rotation in the opposite sense (ibid., p. 71). In the equation in which two equivalent system of forces are analysed Poisson ascribed the sign "plus" to the momenta which tend to produce a rotation in a sense and the sign "minus" to those tending to produce the opposite rotation. Therefore, in this case, a sign is associated with the momentum with respect to a point according to the kind of rotation induced by the momentum. One might wonder whether this is a contradiction with what Poisson had claimed few pages before. It is a contradiction if we, in our modern perspective, consider the magnitudes as either vectors or scalars, but it is not in Poisson's perspective: for him the momentum with respect to a point is always positive, but it is possible to sum and subtract momenta considering the sense of rotation. The fact that the momentum produces or tends to produce a rotation does not mean that it is seen as a segment with an arrow; the sense is the sense of rotation, not the sense of the momentum. This association which is so natural for us was not for Poisson. The signs "+" or "-" indicate the sense of rotation induced by the momentum, but do not transform in themselves the momentum in a modern vector. Certainly, it is not a mere scalar quantity, but it seems to me an overreading to interpret it as a vector. In our perspective, it is an amphibious quantity, which, nonetheless, plays a perfectly comprehensible role in Poisson's reasoning.

The next step concerns the equilibrium of a system of n forces applied to a solid body or to a system of points rigidly constrained (ibid., pp. 75–80). If P^i indicates the forces, α^i, β^i, γ^i the angles of the direction of these forces with the three coordinate planes, and x^i, y^i, z^i the coordinates of the forces' application points, the three equations expressing the translational equilibrium (resultant equal to 0) are

$$\sum_{i=0}^{n} P^i \cos \alpha^i = 0; \sum_{i=0}^{n} P^i \cos b^i = 0; \quad \sum_{i=0}^{n} P^i \cos \gamma^i = 0$$

The three equations which express the rotational equilibrium are given by the forces' momenta with respect to the three coordinate planes. They are:

$$\sum_{i=1}^{n} P^i \left(y^i \cos \alpha^i - x^i \cos \beta^i\right) = 0; \sum_{i=1}^{n} P^i \left(y^i \cos \gamma^i - z^i \cos \beta^i\right)$$
$$= 0; \sum_{i=1}^{n} P^i \left(x^i \cos \gamma^i - z^i \cos \alpha^i\right) = 0$$

These are the six equations of equilibrium which existed, in a slightly different form, in Poinsot's, too.

A remarkable observation developed by Poisson is the following one: he stressed that it is commonly accepted as evident the fact that two skew forces do not have a resultant. However, if a proposition can be proved, it has to be, without referring to evidence. For Poisson highlighted that if two such forces had a resultant, it would be possible to determine a point in its direction so that the two given forces should be in equilibrium around this point. On the other hand, it is possible to draw a line through this point which cuts the direction of one of the two forces without belonging to the plane of the other force. In this manner, if one fixes this straight line the effect of the force cut by this line will be annihilated, but the other will produce a rotation around this line as an axis. Thence, the supposed equilibrium is impossible and, consequently, two skew forces do not have a resultant (ibid., p. 91).

This reasoning is a further and clear indication that also Poisson searched a rational foundation of mechanics. In this respect, he and Poinsot shared such idea, though they developed it in different manners. We will see that this was also Chasles' conception and that his foundational and reductionist programme was far more extreme than Poinsot's and Poisson's because it concerned a systematic reduction of the basic concepts of mechanics to basic concepts of geometry.

The next and final part of Poisson's *Mécanique* which is important for Chasles is the development of the momenta theory, to which Poisson dedicated the whole third chapter (ibid., pp. 99–118) of the first book of his *Traité*. This section is fundamental because Poisson clearly associated momenta and areas of certain figures and dedicated to this question a more extended treatment than Poinsot's. This is also one of the main aspects of Chasles' work on which he gave a more general view than that of scholars, as Poisson, who conceived such an association.

Poisson's investigation began by considering a force *AB* (see Fig. 4.8) and its momentum with respect to a point *C*, which, Poisson claimed, can be represented by the double area of the triangle *ABC*.

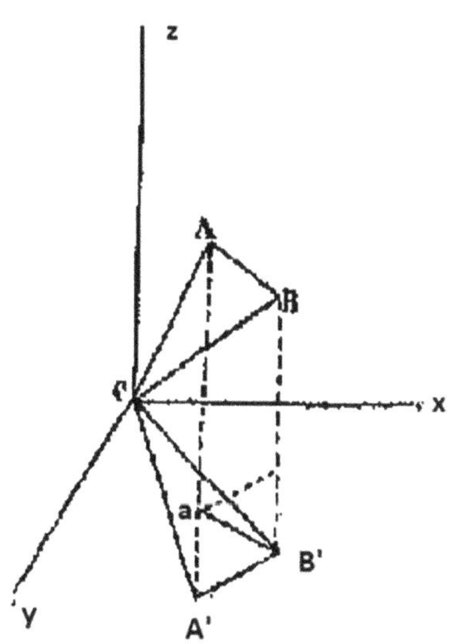

Fig. 4.8 The diagram used by Poisson to associate momenta with areas. (Poisson, 1811, table of diagrams posed at end of the volume)

If the triangle *ABC* is projected perpendicularly on any plane π passing through *C*, the projection will be the triangle $A'B'C$. Therefore, the projection of the momentum of *AB* is the momentum of $A'B'$, which is represented by the double area of the triangle $A'B'C$. The force *AB* can also be transported parallel to itself until *B* coincides with B' so that it is the segment aB'. Then, if one decomposes aB' in a component parallel and one perpendicular to π, the parallel component will coincide exactly with $A'B'$ (ibid., p. 100). Thence, if a force *P* is given and one intends to calculate its momenta *L,M,N* with respect to three planes passing through the momenta centrum, these momenta are the projections on the three planes of the double area of the triangle whose vertices are the extremes of the force and the momenta centrum. This proposition can be applied to the case in which the three planes are the three coordinate orthogonal planes and the momentum centre is the origin. Since it is

$$L = P(y\cos\alpha - x\cos\beta); M = P(x\cos\gamma - z\cos\alpha); N = P(z\cos\beta - y\cos\gamma)$$

it follows that the three quantities on the right member of any identity are the projections on the three coordinate planes of the double triangle whose vertex is the origin and whose side opposite to the origin is the segment representing the force *P* (ibid., p. 100–101). Poisson also observed that in order to calculate the inclination of a plane on the three coordinate planes, it is enough to calculate the inclination of the perpendicular to this plane passing through the origin.

Starting from here, a long conceptual way began which led Poisson to determine the important concept of *principal plane*. The steps towards this result are a series of propositions of which Poisson also offered detailed proofs. Several of these propositions were already known. Poisson's reasoning can be summarized like this: be given a system of planar areas $a, a^{'}, a^{''}, \ldots$ posed in different planes. Be also given an orthogonal reference system of coordinates. Project each area of the series on the three coordinate planes and indicate by $A, A^{'}, A^{''}$ the sum of these projections on them. Let us now consider another system of three perpendicular planes having their origin in the same point *m* as the first three. Be $B, B^{'}, B^{''}$ the sums of the areas' $a, a^{'}, a^{''}, \ldots$ projections on these three new perpendicular planes. If now (see Fig. 4.9) we indicate by *D,E,F* three lines drawn from *m* and respectively perpendicular to the first three planes and by $D^{'}, E^{'}, F^{'}$ three lines through *m* and perpendicular to the second three planes, nine angles are made up.

They are:

$$DmD' = \alpha; EmD' = \beta; FmD' = \gamma;$$

$$DmE' = \alpha'; EmE' = \beta'; FmE' = \gamma'$$

$$DmF' = \alpha''; EmF' = \beta''; FmF' = \gamma''$$

According to the theory of areas' projections, it holds:

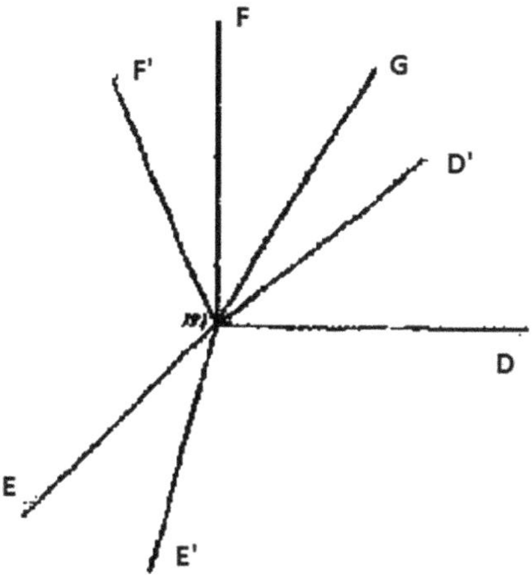

Fig. 4.9 Adapted from Poisson's 1811. I have changed the letters of the straight lines because Poisson used A, B, C and A', B', C' instead of D, E, F and D', E', F'. However, he used the capital initial letters also to indicate the projections of the areas and this could be a little confusing

$$\begin{aligned} B &= A\cos\alpha + A'\cos\beta + A''\cos\gamma \\ B' &= A\cos\alpha' + A'\cos\beta' + A''\cos\gamma' \\ B'' &= A\cos\alpha'' + A'\cos\beta'' + A''\cos\gamma'' \end{aligned} \tag{4.1}$$

and

$$B^2 + B'^2 + B''^2 = A^2 + A'^2 + A''^2.$$

This last equation shows that the sum of the projections is independent of the three orthogonal planes one considers (ibid., pp. 101–108).

Now Poisson developed the following interesting observation: from the last equation one derives

$$B = \sqrt{A^2 + A'^2 + A''^2 - B'^2 - B''^2}$$

This expression represents the sum of the areas projected on the first plane of the second system of three orthogonal coordinate planes. Of course, analogous formulas can be obtained for any of the quantities A, A', A'', B', B''.

Poisson considered the maximum that the radicand can assume. It is achieved when $B' = B'' = 0$. Thence, the expression $\sqrt{A^2 + A'^2 + A''^2}$ is the maximal sum of the projections on a plane of the areas $a, a', a'', \ldots$ (ibid., p. 109), which are posed in different planes.

Poisson claimed that this plane has important properties in mechanics that he will clarify in the prosecution of his treatise (ibid., p. 109).

The direction of this plane can easily be found taking into account the system (4.1), whose equations, in this specific case, are reduced to

$$A = B\cos\alpha; A' = B\cos\beta; A'' = B\cos\gamma$$

From which, since $B = \sqrt{A^2 + A'^2 + A''^2}$, the values of the cosines are immediately obtained (ibid., p. 109). In this way, a series of parallel planes has been individuated, but this implies no problem because Poisson proved that the sum of the areas' $a, a^{'}, a^{''}, \ldots$ projections has the same value for any plane having the same inclination on the principal plane and, therefore, also for planes parallel to that principal. Thence, it is enough to identify the direction of such a plane (ibid., pp. 110–111).

Now, Poisson applied all these results concerning the areas to the theory of momenta: for if one supposes that the areas $a, a^{'}, a^{''}, \ldots$ are triangles having a common vertex in the origin, they can be considered as half the momentum of a forces system $P, P^{'}, P^{''}, \ldots$ with respect to the origin. Every addend of the sum which makes up L will be the projection on the plane xy of the area representing the momentum of one of the forces composing the system. Analogously for M and N with respect to the planes xz and yz (ibid., p. 111).

Exactly as in the case of the areas, there is a plane among all those drawn through the centre of the momenta for which the sum of the forces' momenta, decomposed according to this plane, has a maximum. The value of such a sum has the form $\sqrt{L^2 + M^2 + N^2}$. This is easily comprehensible thinking that the momenta are represented by areas. Thence, in this case, the value L has the same role as the value A in the identity $B = \sqrt{A^2 + A'^2 + A''^2}$; the value M has the role of A' and N of A''. Therefore, if α, β, γ denote the angles that the perpendicular to this plane makes respectively with the axes z, y, x, one has:

$$\begin{aligned} \cos\alpha &= \frac{L}{\sqrt{L^2 + M^2 + N^2}} \\ \cos\beta &= \frac{M}{\sqrt{L^2 + M^2 + N^2}} \\ \cos\gamma &= \frac{N}{\sqrt{L^2 + M^2 + N^2}} \end{aligned} \tag{4.2}$$

For all the planes perpendicular to this one, the sum of the momenta is null. After a further brief consideration on the value of the momenta for planes inclined of a certain angle on the plane for which the sum of the momenta has a maximum, Poisson concluded:

> The composition of the momenta follows, hence, the same laws as the composition of forces. The biggest sum of the momenta and the perpendicular to its plane replace the resultant and its direction.[24]

[24] Ibid., p. 113: "La composition des momens suit donc les mêmes lois que celle des forces, la plus grande somme des momens et la perpendiculaire à son plan remplaçant la résultante et sa direction".

Poisson named *plan principal* the plane of the maximal sum and *moment principal* such a maximal sum (ibid., p. 114).

Commentary: the concepts introduced by Poisson are specular to Poinsot's. Both authors arrive at treating the momenta exactly as the couples; both the authors conceived the areas as objects formally equivalent to the momenta. The principal plane of Poisson coincides with the plane of the couple for Poinsot, and the principal momentum of Poisson is the maximal momentum of Poinsot. Poisson saw the results of the theory of momenta as an application of the theory of areas' projections (once the area is marked by a sign "plus" or "minus"); Poinsot, on the contrary, deduced some results on the projections of the areas from his theory of momenta. But, conceptually, the two itineraries are equivalent.

Through these concepts Poisson was easily able to express the equilibrium of a solid body: 1) if a body is free, a necessary and sufficient condition for it to be in equilibrium is that the principal momentum and the resultant are null; if the body is constrained at a point, it is sufficient that, considering the momenta with respect to that fixed point, the principal momentum is null; if the body is constrained at an axis, it is sufficient that the principal plane contains such axis and that the centre of momenta is in any point of the axis. For example, Poisson argued that, if the fixed axis is the z-axis, and if the principal plane contains it, the angle α of the system (4.2) will be right and, hence, $L = 0$ (ibid., pp. 114–115).

Finally, Poisson pointed out that, so far, he had supposed that the centre of momenta was in the origin of the coordinates, but it is possible to move such a centre everywhere. He wondered where such a point had to be posed for the principal momentum to assume a minimum value. Through a not difficult reasoning Poisson was able to determine such a minimal principal momentum when the forces have no resultant, namely the case in which they are reducible to a couple; though he did not use the term "couple", he spoke of "parallel and irreducible forces" (ibid., p. 116). Immediately afterwards he also solved the problem for the general situation in which both a resultant and a couple exist. In this case, the function which expresses the square of the principal momentum with respect to a point of variable coordinates (x_i, y_i, z_i) is a second-degree polynomial, whose minimum can easily be found by partial differentiation (ibid., pp. 116–117). The result of Poisson's reasoning (ibid., p. 117) is that the locus of the centres with respect to which the principal momentum is null is a straight line. If R indicates the resultant and X,Y,Z the coordinates of one of its extremes, supposing the other extreme in the origin, the square of the minimum principal momentum is (ibid., p. 118)

$$\frac{(NX + MY + LZ)^2}{R^2}.$$

4.1.3 Binet

As to Chasles' contributions to the study of force-systems, Binet plays a significant role since the formulas given in Binet (1815) for the resultant of a system of forces and for a system of momenta are those from which Chasles' speculations on this subject began.

First Binet considered a system of forces $P, P^{'}, P^{''}, \ldots$ applied at a point. Given a Cartesian orthogonal system of coordinates with axes x, y, z, Binet indicated by $\widehat{Px}, \widehat{P'x}, \widehat{P''x}, \ldots$ the angles of the forces' directions with the x-axis. Analogous symbols were used for the angle with the y and z-axes (Binet, 1815, p. 325).

Therefore, the sums of the components parallel to the three axes are:

$$P\cos\widehat{Px} + P'\cos\widehat{P'x} + P''\cos\widehat{P''x} + \ldots$$

$$P\cos\widehat{Py} + P'\cos\widehat{P'y} + P''\cos\widehat{P''y} + \ldots$$

$$P\cos\widehat{Pz} + P'\cos\widehat{P'z} + P''\cos\widehat{P''z} + \ldots$$

These expressions were designed by Binet respectively as:

$$\sum P\cos\widehat{Px}, \quad \sum P\cos\widehat{Py}, \quad \sum P\cos\widehat{Pz}$$

The square of the resultant R is hence:

$$R^2 = \left(\sum P\cos\widehat{Px}\right)^2 + \left(\sum P\cos\widehat{Py}\right)^2 + \left(\sum P\cos\widehat{Pz}\right)^2$$

Since thanks to well-known results, it is

$$\cos^2\widehat{Px} + \cos^2\widehat{Py} + \cos^2\widehat{Pz} = 1$$

$$\cos^2\widehat{P'x} + \cos^2\widehat{P'y} + \cos^2\widehat{P'z} = 1$$

$$\ldots$$

And

$$\cos\widehat{Px}\cos\widehat{P'x} + \cos\widehat{Py}\cos\widehat{P'y} + \cos\widehat{Pz}\cos\widehat{P'z} = \cos\widehat{PP'}$$

$$\cos\widehat{Px}\cos\widehat{P''x} + \cos\widehat{Py}\cos\widehat{P''y} + \cos\widehat{Pz}\cos\widehat{P''z} = \cos\widehat{PP''}$$

$$\ldots$$

The square of the resultant R assumes the form (ibid., p. 326).

$$R^2 = \sum P^2 + 2\sum PP' \cos \widehat{PP'}.$$

In our notation we would write, if the system is composed of n forces:

$$R^2 = \sum_{i=1}^{n} P_i^2 + \sum_{j,l=1}^{n} P_j P_l \cos \widehat{P_j P_l}$$

Binet pointed out that this theorem is important because it expresses an intrinsic property, namely that it depends only on the intensity of the forces and on the mutual angles between the forces composing the system and not on the angles between the action line of the forces and the coordinate axes (ibid., p. 326).

After this result, Binet faced the problem of the forces applied to a rigid body one point of which is fixed. The theory of the momenta of such forces is perfectly analogous, Binet claimed, to Poinsot's theory of couples (ibid., p. 328). Binet considered two not coplanar forces P, P' applied to a rigid body constrained to rotate around a fixed point F. If r and r' are the distances of the action lines of P and P' from F, their momenta with respect to F are $P_1 = Pr; P_1' = P'r'$.

Now, he stated, these momenta are respectively the same as those of two forces whose distances from F are equal to 1 and whose intensity is Pr and $P'r'$ (ibid., pp. 329–330). Therefore, they can be considered as forces which act tangentially to two circumferences whose radius is the unity, whose centre is F and whose planes are respectively those determined by F and by the action line of each of the two forces. The two circumferences imagined by Binet have two common points. Let us consider the action of the two momenta when applied to one of these points:[25] their action lines have an angle equal to that of the mutual inclination of the circumferences' planes. The resultant of the two momenta considered as forces will, thence, belong to their plane, which is tangent to the sphere of which the two unitary circumferences are great circles. Thus, exactly as it is the case of two "ordinary" forces, the resultant's squares of the two momenta are given by

$$P_1^2 + P_1'^2 + 2PP' \cos \widehat{PP'},$$

where $\cos \widehat{PP'}$ indicates the cosine of the angle between the direction of P and P' (ibid., pp. 330–331). With a further series of reasonings, Binet arrived at proving that, whatever the number of forces acting on a solid body constrained to rotate around a point is, a perfectly analogous law as that connoting the square of the forces' resultant holds for the momenta of the forces. This means that, if R_1 indicates such a resultant, it will be (ibid., p. 334):

[25] Binet had already argued that, given such circumference of radius 1, the effect of a force applied to one of its points does not change if the application point moves on the circumference (ibid., p. 329).

$$R_1^2 = \sum P_1^2 + 2\sum P_1 P_1' \cos \widehat{P_1 P_1'}$$

where the meaning of the symbols is clear and analogous to that relative to the forces.

Now Binet proved that the six equations given by Poinsot to express the equilibrium of a solid body can be replaced by the two equations (ibid., pp. 337):

$$R^2 = \sum P^2 + 2\sum PP' \cos \widehat{PP'} = 0 \tag{4.3}$$

$$R_1^2 = \sum P_1^2 + 2\sum P_1 P_1' \cos \widehat{P_1 P_1'} = 0 \tag{4.4}$$

relative, respectively, to the forces and to the momenta. The intensity of the resultant force can be expressed by Eq. (4.3). On the other hand, the effect of the couple arising from the translation of any force to the common point of the resultant is given by the product of this force by the distance of its direction from the point where it is moved. Thence, the three equations expressing the rotational component of the forces can be replaced by Eq. (4.4).

Binet summarized Poinsot's results in a clear and brief manner claiming that Poinsot proved that the effect of any system of forces on a solid body can be replaced by that of a force and of a couple. The force is the resultant of all the forces moved to a point of space, under the condition that the forces are moved remaining parallel to themselves. While moving m forces, m couples are produced. As Poinsot had proved, the effect of the couple generated by the translation of any force is measured by the product of the force by the distance of its direction from the point where the force is moved. The principles behind the couples' composition are the same as those typical of the momenta. Binet taught how to determine the direction of the resultant force and the resultant couple (ibid., pp. 337–338).

He continued his profound analysis claiming that, in Poinsot's theory, the effect of the couple deriving from the translation of every force F is measured by intensity of F by the distance from its direction and the point where this force has been transported.

Taking into account this observation and through a further series of other not difficult steps, Binet was able to prove that, between the angle of the resultant and of the resultant momentum, the following relation holds:

$$\sin \widehat{RR_1} = \left[\sum PP_1' \sin \widehat{PP_1'}\right] : RR_1$$

Afterwards Binet analysed the problem of the minimal principal momentum. For Poinsot had proved that, among all the couples deriving from the translation of all the forces composing the system to different points of space, there is a couple whose principal momentum has a minimum. This is the case when the plane of the couple is perpendicular to the resultant R. Furthermore, Poinsot had also proved that, naming $\mathcal{R}_1$ such a minimum value, it holds:

$$\mathcal{R}_1 = \left[\sum PP' \sin PP_1'\right] : R$$

To simplify, suppose the system is composed of two forces, but the situation is completely analogous for more forces: Binet replaced the momentum P_1 by its value Pr where r is the distance between the original direction of P and the point A where the forces have been transported. Through A draw a plane Δ parallel to the directions of the two forces in their original position. Given a system of forces, and considering the forces of the system two by two, different planes analogous to Δ will be determined. Binet was able to prove (ibid., pp. 339–340) that the quantity $\mathcal{R}_1$ can be indicated by the expression

$$\mathcal{R}_1 = \left[\sum PP'\delta' \sin PP'\right] : R$$

where the symbol δ' indicates the sum of the projections of the arms r and r' on a perpendicular to the plane Δ. Namely $\delta' = r \sin \widehat{r\Delta} + r' \sin \widehat{r'\Delta}$ (ibid., p. 340). For any pairs of forces belonging to the system, e.g., for the pair P, P'', you have a plane Δ' parallel to their directions and an expression $\delta'' = r \sin \widehat{r\Delta'} + r' \sin \widehat{r'\Delta'}$, and so on. Thus, Binet concluded that:

> If in space the forces $P, P', P'', \ldots$ are considered as represented by parts of their directions, the term $PP'\delta' \sin \widehat{PP'}$ will be the volume of the parallelepiped included between two planes parallel to Δ, one of which is drawn through P, the other through P' and included among four further planes, which are parallel two by two, and which are drawn through the straight lines joining the extremities of P to those of P'.[26]

Therefore, the minimum among the principal momenta of a system of forces is reduced to the sum of a series of parallelepipeds' volumes constructed in the manner described in the quotation.

The work by Binet is a significant step towards a precise rationalization of mechanics based on a general mathematical view of this discipline. He was actually able to organize several of the concepts introduced by Poinsot—who was one of the main sources of inspiration of Binet's paper—in a more perspicuous way, and he was also able to fully clarify the relation between the resultant of a system of forces and the resultant of the momenta produced by these forces when applied to a rigid body. The inventive work by Poinsot, where the author introduced and explained the different momenta (momentum, principal momentum, minimum of the principal momentum) related to the different choice of the way in which to decompose a system of forces in a resultant and a resultant couple, was followed by Binet's contribution. Here the material was organized in a more rational and perspicuous

[26] Ibid., pp. 340–341: "Si l'on considère dans l'espace les forces $P, P', P'', \ldots$comme représentées par des parties de leurs directions, le terme $PP'\delta' \sin \widehat{PP'}$ sera le volume du parallélipipède compris entre deux plans parallèles à Δ, l'un mené par P et l'autre par P', et sous quatre autres plans parallèles deux à deux, et conduits par les droites qui joindraient les extrémités de P à celles de P'".

way, though the ideas expressed were, in substance, already present in Poinsot. The reduction of the six Poinsot's equations for the equilibrium to two equations is a relevant step within this work of re-organization. Chasles will offer an even more general view.

Binet's contribution is a further step towards the link between geometrical quantities and physical quantities. Poinsot showed that the couples and the areas are subject to analogous projective rules and explicitly associated an area with the momenta of the forces. Binet proved that also a volume (that of a series of parallelepipeds) can be associated with the minimal principal momentum. Chasles would further analyse these relations between forces-momenta and areas-volumes. The theory of projections plays a fundamental role in the theoretical developments of all the three authors, Poinsot, Poisson and Binet. Hence, though the projective properties exploited by the three authors depend on classical and known features of the figures' projection, it is clear that modern projective geometry, which was developed around from the 1810s, might have offered a precise basis for mechanics. This was what Chasles did.

4.1.4 Giorgini

I examined in detail some aspects of Poinsot's, Poisson's and Binet's contributions because in these works there is a series of concepts which are useful to fully grasp the value of Chasles' results. Without referring to the difficult road which led these three scholars to enucleate such concepts, a thoughtful appraisal and comprehension of Chasles' work would be problematic. A reference to Giorgini's *Teoria analitica delle proiezioni* (1820), which was mentioned by Chasles, also seems appropriate to me.

Giorgini dedicated his essay to a systematic calculation of the value of the projections of segments and plane regions onto the two or, more often and more interestingly, three coordinate axes or planes. As the title indicates, his approach was analytical. What is interesting is that Giorgini identified a series of formulas, whose form is invariant according to the different problems dealt with. This form is that of Binet's Eqs. (4.3) and (4.4) and is the form to which Chasles referred. So, in the first section—named *Titolo I* by Giorgini—the author proved, among other theorems, that, given in the plane a system of straight line segments $a, a', a'', \ldots$, its maximal orthogonal projection on a single straight line has a value A subject to the equation:

$$A^2 = \sum a^2 + 2\sum aa' \cos \widehat{aa'} = A_x^2 + A_y^2 \tag{4.5}$$

After having analysed Binet's notation, Giorgini's is clear: it is only to remember that A_x, A_y indicate the projections on the axis x and y, respectively (Giorgini, 1820, p. 16).

Explanation: With regard to the explanation of formula (4.5), Giorgini supposed that an arbitrary reference frame (*x,y*) is assigned. Given a system of straight lines, the sum of the squares of orthogonal projections A_x^2 and A_y^2 on the *x*-axis and *y*-axis is also given. The sum $A_x^2 + A_y^2$ is an invariable quantity. However, as Giorgini claimed (ibid., p. 15), if the orthogonal coordinate axes vary, the values A_x, A_y will also vary. When one of the two values is 0, the other has its maximum. Therefore, the straight line A is that on which both the segments A_x, A_y of Eq. (4.5) are projected.[27] The direction of the straight line A can also be determined (ibid., p. 16), taking into account that

$$A_x = A\cos\widehat{Ax}; A_y = A\cos\widehat{Ay}.$$

This formula expresses the important theorem according to which the orthogonal projection of a system of straight lines on a straight line is given by the maximal orthogonal projection multiplied by the cosine of the angle between the line of maximal projection and the given line (ibid., pp. 15–16).

In *Titolo II*, the segments are in space and, hence, the coordinate axes are *x*, *y*, *z*. The formula is exactly the same as (4.5) (ibid., p. 30).

In *Titolo III*, entitled "Formulas concerning the projection of a flat area and of a system of flat areas onto three coordinate planes"[28] as Theorem 34 Giorgini proved that:

> The biggest orthogonal projection of a system of areas $m, m', m'', \ldots$ elevated to the square is equal to the function $\sum m^2 + \sum mm' \cos mm'$; and the projection of this system onto planes perpendicular to that of the biggest orthogonal projection is null.[29]

This means that the biggest orthogonal projection of a system of flat areas is subject to the same formal treatment as that valid for the orthogonal projection of a system of straight lines.

In our context, the further important *Titolo* is the sixth and last one because Giorgini applied the theory of the straight lines systems' and areas systems' projections to the composition of forces and momenta. As a consequence of the principle of the forces' parallelepiped and of that according to which the action of a force along a direction is given by the orthogonal projection of the force along such a direction, Giorgini illustrated this, so to say, principle of duality:

[27] To avoid confusion, one might indicate with A the value of the segment and with another letter the straight line of the axis including the segment A, but this is only a notational problem.

[28] Giorgini (1820, p. 31): "Formole relative alle projezioni sopra tre piani coordinati di una area piana e di un sistema di aree piane".

[29] Ibid., p. 93.: "La massima projezione ortogonale di un sistema di aree $m, m', m'', \ldots$ inalzata alla seconda potenza è uguale alla funzione $\sum m^2 + \sum mm' \cos mm'$; e le projezioni del sistema sopra piani perpendicolari a quello della massima projezione ortogonale sono nulle".

> [...] after having replaced the name 'straight line' with 'force', the name 'oblique projections' with 'components' and considering the orthogonal projections as the expression of the actions exerted by the forces along the straight lines of such orthogonal projections, so that the biggest orthogonal projection of the straight lines system represents a system of forces, it will be possible to name this biggest orthogonal projection as the maximal action of the forces' system.[30]

Hence, given a system of forces $a, a^{'}, a^{''}, \ldots$ applied at the same point in space and three orthogonal axes, the well-known conditions of equilibrium are

$$a_x + a_x' + a''_x + \ldots = A_x = 0; a_y + a_y' + a''_y + \ldots = A_y = 0; a_z + a_z' + a''_z + \ldots = A_z = 0.$$

Indicating with A the biggest projection of the system, the condition of equilibrium is, hence, expressible as:

$$A = \sqrt{A_x^2 + A_y^2 + A_z^2} = \sum a^2 + 2\sum aa' \cos \widehat{aa'} = 0$$

If the forces are not in equilibrium and R is their resultant, it is:

$$R = A = \sum a^2 + 2\sum aa' \cos \widehat{aa'}$$

Therefore, the expression with the two sums indicates the resultant of a system of forces (ibid., p. 59).

After having analysed the translational equilibrium, Giorgini studied the condition of the rotational equilibrium. If, given a system of forces $a, a^{'}, a^{''}, \ldots$, which are not reducible to a resultant, (x^i, y^i, z^i) indicate the distances between the application point of any force composing the system and the coordinate axes, the rotational equilibrium is represented by:

$$\begin{aligned} (ya_x - xa_y) + (y'a'_x - x'a'_y) + \ldots &= 0 \\ (xa_z - za_x) + (x'a'_z - z'a'_x) + \ldots &= 0 \\ (za_y - ya_z) + (z'a'_y - y'a'_z) + \ldots &= 0 \end{aligned} \tag{4.6}$$

Since, given a force AB, the origin of the system O and the perpendicular OP drawn from O to the straight line AB, the momentum of AB with respect to O is given by the product $OP \times AB$, the projection $OA^{'}B^{'}$ of the triangle OAB on the plane xy

[30] Ibid., p. 58: "[...] cambiato il nome di projezioni oblique in quello di componenti, e riguardando le projezioni ortogonali come l'espressione delle azioni esercitate dalle forze nel senso delle rette sopra le quali si sono formate le projezioni, in modo tale che le massima projezione ortogonale del sistema di rette rappresenti un sistema di forze, potrà essere chiamata la massima azione del sistema di forze."

will represent $\frac{1}{2}m_Z$, where m_Z is the component of the momentum calculated with respect to the origin in the plane xy. The area of the triangle $OA'B'$ is $ya_x - xa_y$ (ibid., Theorem 41, p. 47), so that $ya_x - xa_y = \frac{1}{2}m_z$. Analogous identities hold for all the expressions in the round brackets of system (4.6). This means that the system (4.6) is nothing but the system which indicates the annihilation of the components x,y,z of the momenta of all the forces composing the system, calculating the components with respect to the origin. Hence, the system (4.6) becomes (ibid., p. 61):

$$m_X + m'_X + m''_X + \ldots = M_x = 0$$
$$m_Y + m'_Y + m''_Y + \ldots = M_y = 0$$
$$m_Z + m'_Z + m''_Z + \ldots = M_z = 0$$

The sum of the momenta's components can be reduced to the sum of projection of flat areas. Thus, given the formula for the maximal projection of a system of flat areas on a plane, the formula of the rotational equilibrium is expressible as:

$$\sum m^2 + \sum mm' \cos \widehat{mm'} = 0$$

It is formally identical to the formula for the translational equilibrium. In the most general case, the resultant momentum M is expressed by $\sum m^2 + \sum mm' \cos \widehat{mm'} = 0$. Giorgini wrote this to be Binet's formula (ibid., p. 62). As a matter of fact, taking into account the relations highlighted by Poisson and by Giorgini himself between momenta and areas, what Giorgini found was an expression of the principal momentum of a system of forces. Indeed, as he wrote:

> From what previously proved, it follows that the principal momentum of a system of forces is represented in magnitude and direction by the biggest orthogonal projection of the areas which represent the forces' momenta [...].[31]

Giorgini faced then the problem already posed and solved by Poinsot, Poisson and Binet, i.e. to find the smallest principal momentum (ibid., p. 63). As he himself highlighted, he presented a method referable to Poisson (1811), pp. 115–118 (ibid., p. 65). He thus proved that a straight line—whose equation Giorgini wrote—can be determined such that the principal momentum has the same minimal value for all the points of such a line assumed as the centres of momenta. The value of this minimum momentum M' among the maxima is

[31] Ibid., p. 62: "Da quanto abbiamo superiormente dimostrato rilevasi che il momento principale del sistema di forze è rappresentato in grandezza e direzione dalla massima proiezione ortogonale dell'aree che rappresentano i momenti delle forze [...]".

$$M_{\prime} = \frac{A_x M_X + A_y M_Y + A_z M_Z}{A}$$

After having referred to these results ascribed by Giorgini to Poisson, he proved five new properties of the principal momenta of which, in relations to Chasles, only the first and easiest one has to be reported. This property is expressed by the equation (ibid., p. 65)

$$M_{\prime} = \frac{\sum am \sin \widehat{am}}{\sqrt{\sum a^2 + 2\sum aa' \cos \widehat{aa'}}}$$

With regard to the denominator, no explanation is necessary: it derives from the formula mentioned several times which indicates the biggest projection of a system of straight lines onto a straight line. The numerator expresses the theorem proved by Giorgini (ibid., p. 43) according to which the sum of the three products of any projection of a system of straight lines onto one of the three orthogonal axes by the projection of a system of flat areas onto the planes of the other two axes is equal to the sum of the products given by any straight line of the first system by any area of the second system by the sinus of the angle between such a straight line and the plane of the considered area. Since the forces are formally straight lines and the momenta are areas, Giorgini's formula is completely explained. He pointed out that this formula is particularly significant because it does not depend on the position of the three coordinate axes (ibid., p. 65).

Commentary:

1. Giorgini's work is important because the author organized a series of theorems concerning the segments' and areas' projections in a rational, perspicuous and ordered manner and through a clear analytical approach. Most of the results expounded by Giorgini were already known, but his re-organization has to be interpreted as a remarkable progress with respect to the previous state of the theory, which was still, at least in part, naïve. That of Giorgini is, then, a foundational work. Though he used an analytical approach, among the authors of the period under examination his way of thinking was close to Chasles'. As sources of inspiration for his work, Giorgini mentioned few authors: Hachette (ibid. pp. 28 and 29), in reference to his text *Traité des surfaces* (Hachette, 1813); a memoir by Monge (ibid., p. 53); Binet (1815) (ibid., p. 62); and Prony (1799) (ibid., p. 63). But the very reference point for Giorgini was Poisson (ibid., pp. 28, 38, 40, 65) who is mentioned for the third chapter of the section dedicated to static within his *Traité de Mécanique* (Poisson, 1811). This is the chapter I have analysed in detail. Giorgini gives the idea to interpret his work, at least in part, as a clarification, re-organization and extension of Poisson's ideas and results. Curiously, the author who was the inspirer of the idea to connect the momenta of the forces with the areas, namely Poinsot, is not mentioned by Giorgini. Perhaps, this omission has to be inserted within the picture drawn by

Grattan Guinness (2014) with regard to the scarce fortune of Poinsot's work in several scientific milieus.

2. Giorgini, following and developing Poinsot's and Poisson's train of thought, reached the results already obtained by the two French physicists through a complete analytical theory of projections: 1) the forces can be indicated by an oriented segment with a direction and a sense; 2) given a system of forces, its treatment can be reduced to that of a system of straight lines, of which Giorgini furnished explicitly the projective properties; 3) the momenta of a system of forces are associable with areas; 4) an oriented segment can be associated with the areas; and hence, 5) from a formal point of view, the system of forces and the system of momenta can be treated through the same mathematical technique. In clearer terms, both the forces and the momenta can be treated as oriented segments. Poinsot started from the momenta and showed that momenta and areas are passible of the same treatment; Giorgini started from the areas, but the results are the same. The important difference is that Giorgini framed these results within a complete mathematical theory. This told, it is noteworthy, however, that the association of a force with an oriented segment was considered unproblematic. The situation was far different with the momenta. In this case, it was necessary to associate a rotation and not a translation with an oriented segment. All the introduced concepts: momenta, principal momenta and minimum of the principal momenta, are a litmus paper that the system of notions used to deal with the rotations was not yet completely structured. These scholars perceived that the association of a rotation with a segment was appropriate but there were some aspects which had still to be completely grasped, and this association is not as elementary as that for translations. In a modern perspective, we know that these difficulties were absolutely justified: a force can be associated with a vector. But, in fact, quantities such as angular velocities, momenta, etc., in sum all those physical quantities which deal with rotations, are not vectors (more precisely: polar vectors) but pseudo- or axial vectors as any quantity deriving from the vector product of two vectors. This implies all the questions connected to the chirality of the reference system in use and to the sense of the vector product as well as to the different behaviour between vectors and pseudo-vectors with regard to the mirror reflection. For the vector product **c** of two vectors **a** and **b**, $\mathbf{c} = \mathbf{a} \times \mathbf{b}$, can be defined through the Levi-Civita symbol as $c_k = \varepsilon_{ijk} a^i b^j$ (the meaning of i,j,k is obvious), whereas this symbol is always missing while dealing with polar vectors. The association of an area with a vector enters into this reasoning because the signed area of a parallelogram is the vector product of its sides.

In a still more general and profound perspective, the axial vectors are anti-symmetric rank-2 tensors. In an m-dimensional space, these tensors have $\frac{m(m-1)}{2}$ coordinates, while the vectors have m coordinates.[32] When $m = 3$ the number of

[32] See Sonego (2016, pp. 356–357). The whole Appendix A (pp. 355–363) of such text is a useful guide to these questions. A satisfying bibliography is also referred to.

coordinates of such tensors and the vectors coincide. This is why we can consider the result of the vector product in three dimensions as a vector.

Now, let us focus on system (4.6). It is enough to concentrate on a single force and on its momentum with respect to the origin. Since the momenta are perfectly corresponding to the areas by which they are represented, the expressions $(ya_x - xa_y)$; $(xa_z - za_x)$; $(za_y - ya_z)$ are respectively the components on the axes z, y, x of the momentum of the force a with respect to the origin, where a_i indicates the orthogonal projections of the force on the three axes. While x,y,z are the components of the point where the perpendicular to the origin cuts the action line of the force. Therefore, if one indicates by d the distance between the origin and the straight line of the force and by a the force, these scholars had understood, in modern terms, that the momentum is the vector product between d and a. I mean that they had guessed the basic elements of this not easy concept. The concept of vector was still in its proto-history, not to speak of that of pseudo-vector. Therefore, these authors dealt with the analysed problems with a relatively poor mathematical apparatus—in fact, they were among the constructors of such an apparatus. Thence, the plurivoc approach with regard to the way to describe the rotations and to the way in which to associate them to an area and to a segment is an indication of the fact that the physical and the mathematical concepts used by these authors were clear. However, the mathematical apparatus was not yet completely suitable to express in a perspicuous and general manner the notions they had so ingeniously created.

4.2 The Main Contribution of Chasles on the Systems of Forces, 1830

The foundational character of Chasles (1830c) is conspicuous. This paper is divided into two parts; the first and more important one (ibid., pp. 92–112) concerns the proof that the whole theory of the forces, system of forces and momenta expounded by Poinsot, Poisson, Binet and Giorgini can be deduced from a single geometrical theorem. The references to these authors—especially to Poinsot and Giorgini—are numerous and regard important issues. Therefore, Chasles' contributions to the analysis of the system of forces and momenta are not as original as those concerning the movement of a rigid body, which Chasles began to develop in the same period. This notwithstanding, they are not less interesting from a conceptual and historical perspective because Chasles reduced to a single common root all the results concerning such a topic. This allowed him to clarify many ideas concerning the way in which an oriented segment can be associated to a rotation. As a matter of fact, he formulated the right-hand rule. In his studies on the rigid body's movement, Chasles connected several mechanical properties to some geometrical features of polygons and polyhedrons. In the case of the forces' systems, he also discovered many analogies between geometrical and mechanical concepts, so offering a new and unitary perspective to several results obtained by his predecessors, particularly by Carnot, on the theory of transversals and of polyhedron. A brief section on the mean distances is also inserted. These last two issues were developed in the second part of the paper (ibid., pp. 113–120).

Geometry offered, hence, the fundamental theorem. Afterwards the treatment has a dual evolution: Chasles proved that his theorems are valid for segments and forces; system of segments and system of forces; areas and couple of forces with their momenta. Thence, though the reference to modern projective geometry[33] is less evident than in the case of the rigid body's movement, Chasles' conceptual reference frame remains modern projective geometry.

Therefore, his works on the system of forces have to be interpreted as a further step towards the geometrization of physics and represent an important element in his foundational programme. Considering his idea that "the whole geometry is projective geometry"—to use an expression attributed to Cayley—and that the basis of rational mechanics can be reduced to geometry one might claim that, without risk of error, for Chasles the whole basis of science is projective geometry.

4.2.1 The Fundamental Theorem and Its Consequences

The theorem from which Chasles deduced all the properties concerning the systems of forces is this simple proposition:

> When two system of forces are given, if each force of the first system is multiplied by each force of the second system and by the cosine of the angle between the two forces, the sum of all these products will be the same as the sum of analogous products calculated while considering two other system of forces respectively equivalent to the two given.[34]

Chasles indicated by a, $a^{'}$, $a^{''}$, ... the forces of the first system and by b, $b^{'}$, $b^{''}$, ...those of the second one. Assumed that $\widehat{ab}$ indicates the angle between the directions of two any forces, one belonging to the first system and the other to the

[33] It is important to distinguish between the whole system of concepts developed in projective geometry starting from the beginning of the nineteenth century, i.e. theory of reciprocal polars, theory of duality, theory of the cross ratio and so on, which constitute what I call "modern projective geometry", from classical considerations on oblique and orthogonal projections regarding the relations between the projection of segments, system of segments, areas, system of areas. This doctrine, which includes the elementary metrical expressions that link a figure to its projection, had already been developed at the beginning of the nineteenth century and did not require a new theory. What was, in great part, new were the applications to rational mechanics. The authors who are Chasles' reference points for his studies on the system of forces already used these more classical and elementary results on projections.

[34] Chasles (1830c, p. 92): "Quand on a deux systèmes de forces, si on multiple chaque force du premier système par chaque force du second système, et per le cosinus de l'angle de ces deux forces, la somme de tous ces produits sera la même que la somme des produits semblablement faits à l'égard de deux autres systèmes de forces équivalens respectivement aux deux proposés".

second system, the quantity under examination was indicated by $\sum a.b\cos\widehat{ab}$.[35] To prove the theorem Chasles had the following brilliant idea: he argued that it is enough to demonstrate such property when one of the forces of the first system is replaced by its components with respect to an orthogonal reference frame. For, in the new system—because of the same reasoning—the theorem will be valid if any other force is replaced by its components. The passage from a system to an equivalent one is obtained by decomposing the forces of the original system and by recomposing them to obtain those of the new system. Thus, if Chasles were able to prove the proposition after having replaced one force with its components, then the whole theorem would have been proved. If a_x, a_y, a_z are the components of the force a, the system a, $a^{'}$, $a^{''}$, ... is replaced by a_x, a_y, a_z, $a^{'}$, $a^{''}$, ... This means that

$$a.b\cos\widehat{ab} = a_x \cos\widehat{a_x b} + a_y \cos\widehat{a_y b} + a_z \cos\widehat{a_z b}$$

This identity is true because, as Chasles pointed out (ibid., p. 93), it expresses the well-known condition according to which the orthogonal projection of the force a on the force b is equal to the sum of the orthogonal projections of the force a's components on the line of action of the force b. This proves the theorem.

Chasles clarified that the equation $\sum a.b\cos\widehat{ab}$=constant, which represents the proved theorem, is the only basis of his writing because it is subject to different geometrical interpretations (ibid., p. 93).

The first results concern the resultant of a system of forces applied to the same point. In this case, if A is the resultant of the forces belonging to the system a, $a^{'}$, $a^{''}$, ... and B is that of the forces belonging to the system b, $b^{'}$, $b^{''}$, ..., then the fundamental theorem claims that

$$A.B\cos\widehat{AB} = \sum a.b\cos\widehat{ab} \tag{4.7}$$

and, if the forces b, $b^{'}$, $b^{''}$, ... coincide with those of the a, $a^{'}$, $a^{''}$, ..., the theorem sounds (ibid., p. 94)

$$A^2 = \sum a^2 + 2\sum a.a^{\prime}\cos\widehat{aa^{\prime}} \tag{4.8}$$

Though Chasles did not specify the step from Eq. (4.1) to Eq. (4.2), it is clear that the first addend of the right side of (4.2) derives when a force is multiplied by itself and the other when two different forces are multiplied.

Commentary: each addend $a.b\cos\widehat{ab}$ of the sum $\sum a.b\cos\widehat{ab}$ is a signed scalar quantity. It depends on whether the angle between the directions of the two forces in $a.b\cos\widehat{ab}$ has to be considered acute or obtuse. This depends on the sense of the

[35] Nowadays, if m and n are the number of forces belonging respectively to the first and to the second system such quantity might be indicated like this: $\sum_{0<i\leq m, 0<j\leq n} \prod_{0<i\leq m, 0<j\leq n} a_i b_j \cos\widehat{a_i\ b_j}$.

forces. Thence the entire sum can be positive or negative. This granted, Chasles posed the following principle of duality:

> Be the forces represented in magnitude and direction by some straight lines. It will be possible to replace in the statement of theorem 1 the term *force* with the term *straight line*. In this case by *components* of a straight line you will intend the three projections on three any axes drawn from one of its points; by *systems of equivalent straight lines* two systems of straight lines, one of which will be formed by the decomposition and composition of the straight lines belonging to the other system, as if these straight lines were forces.[36]

What Chasles wrote is very interesting: the straight lines are a representation of the forces; since the fundamental geometrical theorem holds for the forces, it also holds for the object which represents the forces, i.e. the straight lines. Obviously, the inverse process would have been theoretically possible: to associate immediately a direction and a sense to a segment straight line and to derive by such object the duality straight lines-forces. But when Chasles wrote, the idea of seeing the force as an oriented segment was already present, while the idea to consider the object "oriented segment" in itself independently of the fact that it represents a force was not usual. As we have seen, the outline of such idea already existed in the works of other mathematicians. However, Chasles established the duality forces-oriented segments more clearly. This allowed him to consider the oriented segment as a new object in itself. Here Chasles is quite close to the concept of vector as abstract object. Towards the full acquisition of this concept, his capability to see the problems in their general form and his skill in understanding the general and common characters of different physical and mathematical situations were of great help. Undoubtedly his talent and inclination towards abstract geometry were fundamental.

Nowadays, the quantity $a.b \cos \widehat{ab}$ can obviously be considered as the scalar product between the two vectors a and b; the quantity $\sum a.b \cos \widehat{ab}$ is a sum of scalar products. As we will see, Chasles arrived close at conceiving several properties of the scalar products. Thence, though it is a mistake to claim that he reached the concept of scalar product, his work can also be interpreted as a step towards the not trivial acquisition of this concept.

The only result independent of the fundamental theorem introduced by Chasles is a lemma concerning the association between the projections of an area on three coordinate planes and a straight line segment. The lemma states: project a plane area onto any three coordinate planes yz, xz, xy. Consider a segment perpendicular to the plane of the area. Decompose it in three segments directed along three axes ox', oy', oz' respectively perpendicular to the three coordinate planes. The components of this segment will be to the projection of the plane area as the segment is to the area (ibid., p. 95).

[36] Ibid., pp. 94–95: "Les forces étant représentées en grandeur et en direction par des droites, on pourrait substituer dans l'énoncé du théorème 1 le mot *droite* au mot *force*. Alors on entendrait par *composantes* d'une droite ses projections sur trois axes quelconque menés par un de ses points ; et par *systèmes de droites équivalens*, deux systèmes de droites d'ont l'une serait formé par la décomposition et la composition des droites de l'autre système, comme si ces droites étaient des forces". Italics in the text.

The commentary added by Chasles to this lemma is very significant. For we read:

> In this manner, if the area π is represented by the segment straight line a, perpendicular to its plane, the projection of this area on three coordinate planes will be represented by the projections of the straight line a on three axes perpendicular respectively to these planes.[37]

Once again: Poinsot and Poisson had associated segments with flat areas and had arrived at conceiving that segments perpendicular to the plane of the areas are particularly suitable for such an association. But Chasles posed this proposition in general and clear terms, by means of a simple formulation and demonstration. He clearly associated the components of an area with the components of a segment. His view was more comprehensive and general than that of his contemporaries. Hence, though the idea to associate flat areas to segments is not Chasles' original, he developed such an idea in an almost modern and formal way.

Now Chasles connected this lemma with the fundamental theorem: be given a system of flat areas A posed in different planes. Consider a system of segments a perpendicular to the planes of the areas and proportional to the areas. Associate the two systems. Be the segments decomposed so as to make up an equivalent system b. In planes perpendicular to the straight lines of b, consider a system of areas which are to the segments of b in the same ratio as the areas of A are to the segments of a. A new system of areas B will be obtained. This new areas' system might have been obtained by projecting the areas of A in the same way as the segments of a have been decomposed. Chasles named *equivalent* the areas' systems A and B. The area corresponding to the resultant of all the straight lines is the *resultant area*. The projections onto three coordinate planes of the areas, which correspond to the projections of the straight lines perpendicular to the planes of the areas, are called *components* of the areas (ibid., p. 96). In this way Chasles' association between flat areas and segments straight lines perpendicular to such flat areas is complete, from a formal and a conceptual standpoint.

A further interesting annotation is Chasles' statement that the areas and the segments are always positive, but their components can be negative (ibid., p. 96).

Commentary: this means that the areas and the segments are subject to a calculus in which they can be considered as negative, thought, in themselves, the areas and the segments are positive. Thence, as a matter of fact, Chasles was arriving at conceiving the signs of a segment and of an area as a merely formal and conventional question, depending on the reciprocal position, orientation and sense as well as from the chosen system of coordinates. In this sense, too, he was getting close to important aspects of vector calculus.

Finally, Chasles pointed out that the source of inspiration of the previous lemma was Poinsot's decomposition of forces (ibid., p. 96). There is no doubt that he was referring to Poinsot's association between a couple and its axis developed in Poinsot (1803, pp. 58–61) and that I have reported. Nonetheless, though Chasles' lemma is

[37] Ibid., p. 95: "Ainsi, si l'aire π est représenté par la droite a perpendiculaire à son plan, les projections de cette aire sur trois plans coordonnés seront représentées par les projections de la droite a sur trois axes perpendiculaires à ces plans respectivement".

similar to Poinsot's, the general conclusions drawn by the former are not present in the latter. The picture drawn by Chasles is more general, formal and precise than Poinsot's ingenious reasoning. Chasles was perfectly aware of this feature connoting his work because he claimed that his fundamental theorem and the lemma allowed him to frame in a new unitary picture the beautiful theorems concerning the projections of the areas developed by Carnot in his *Géométrie de position*, by Hachette in his *Élémens de géométrie à trois dimensions* as well as by Poinsot in his *Statique* and by Poisson in his *Méchanique* (ibid., pp. 96–97). I have focused on Poinsot and Poisson because their works concern the treatment of forces and momenta, which is one of the subjects of Chasles' foundational programme in relation to physics. Let us now see how Chasles' approach permits to frame such a treatment within a general panorama connected to geometry.

The following step of Chasles' itinerary was the proof that the theorems obtained by Poinsot with regard to the couples and their momenta can be achieved within his own theory. The first significant theorem in this doctrine is the following one: be given two systems of flat areas $m, m', m'', \ldots$ and $n, n', n'', \ldots$. Multiply one of the areas belonging to the first system by an area of the second system and by the cosine of the angle between the two planes containing the two areas. The sum of all the products obtained in this way has a value which is constant when the two systems are replaced by two equivalent systems of areas.

This theorem is the same as the fundamental theorem if the term "force" or "straight line segment" is replaced by "area" and the angle between two straight lines is replaced by the angle between two planes. For the demonstration relies on the fundamental theorem and on the lemma which allows us to replace the areas with segments perpendicular to their planes and proportional to such areas (ibid., pp. 97–98). Therefore, two analogous corollaries as those of the fundamental theorems, which are expressed by Eqs. (4.7) and (4.8), hold:

1) If M is the resultant area of the first areas' system and N that of the second system's areas, it is

$$M.N\cos\widehat{MN}=\sum m.n\cos\widehat{mn}$$

2) If the areas of the second system are the same as those of the first system, it is:

$$M^2=\sum m^2+2\sum m.m'\cos\widehat{mm'} \tag{4.9}$$

A further corollary claims that, if the second system is composed only of an area, it is

$$M \cos \widehat{Mn} = \sum m \cos \widehat{mn} \tag{4.10}$$

which states that the sum of the orthogonal projections of a system of areas on any plane (in this case that of the area n) is equal to the orthogonal projection of the resultant area. It is easy to prove that this proposition also holds when the projections are not orthogonal (ibid., pp. 98–99). Now a consideration follows which connects Chasles' treatment directly with the problem of the couples and of the principal momentum of a system of forces. For, if the projections of the areas are orthogonal, the maximal projection of a system of areas on a single plane is obtained when this plane is that of the resultant. This follows from the fact that, in this case, the angle $\widehat{Mn} = 0$, so that the resultant of a system of flat areas is the biggest sum of the orthogonal projections of these areas on a plane (ibid., p. 99).

This set of results is now applied by Chasles to the theory of couples: since as Poinsot taught—and as we have seen—the couples are decomposable as the areas and are representable by straight lines which are perpendicular to their planes and proportional to their energy, it is possible to apply the theorem on the areas to the couples by replacing the term "area" with the term "couple", so obtaining:

> If two systems of couples are given and the product of any couple of the first system by any couple of the second system and by the cosine of the angle between the two planes is developed, the sum of all these products will maintain the same value when the two systems of couples are replaced by two respectively equivalent systems.[38]

Two corollaries, which are formally exactly the same as those expressed by Eqs. (4.9) and (4.10), are obtained by replacing the term "area" with "couple".

With regard to the momenta, Chasles pointed out that, given the momentum of a force with respect to a point, it has the same expression as the momentum of the couple made up of such force and of an equal force drawn in the opposite sense through the centre with respect to which the momentum is calculated.

The same argument valid for a force can be extended to systems of forces: consider a system of forces and their momenta with respect to a fixed point. They have the same expressions as the energies of the couples constituted by these forces and by equal forces drawn in the opposite sense with respect to the fixed point.

If the system of forces is replaced by an equivalent one and if you consider the momenta of these new forces with respect to the same point, they will have the same expression as the energies of the couples formed with these new areas. *The planes of the momenta are the same as those of the couples*, so that it will be possible to replace the couples of the previous theorem with the momenta of the forces, so to obtain the following theorem: given two system of forces A and B and the momenta with respect to a point O of the forces belonging to A as well as the momenta of the forces of B with respect to another point O', if each moment of the first system is

[38] Ibid., p. 99: "Si l'on a deux systèmes de couples, et qu'on fasse le produit de chaque couple du premier système par chaque couple de second système et par le cosinus de l'angle compris entre les plans des deux couples, la somme de tous ces produits conservera la même valeur quand on remplacera les deux systèmes par deux autres systèmes de couples, respectivement équivalens".

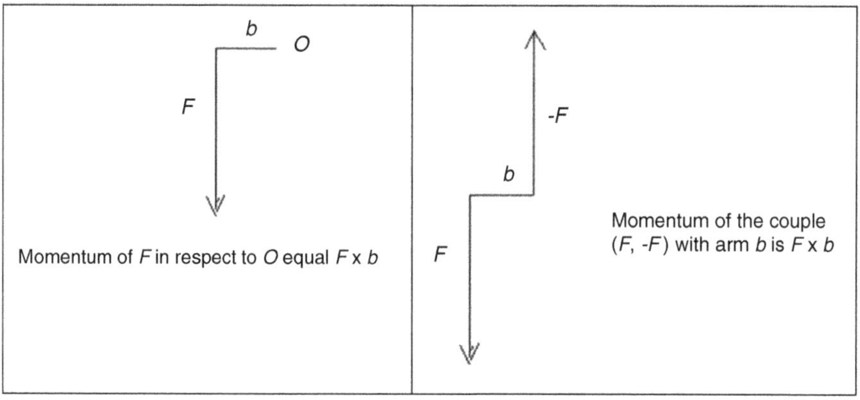

Fig. 4.10 The situation described by Chasles. The meaning of the symbols is clear. The momentum of F with respect to O is $F{\times}b$, as well as the momentum of the couple

multiplied by each of the second one and by the cosine of the angle between the planes of the two momenta, the sum of all these products will maintain the same value if two systems A' and B', respectively equivalent to A and B, replace A and B and their momenta with respect to O and O' are calculated (ibid., p. 100).

Commentary: I have written in italics the expression "the planes of the momenta are the same as those of the couples" because Chasles' assertion shows an interesting historical-conceptual aspect: Chasles is claiming that—for example in reference to Fig. 4.10—the plane of the momentum of the force F with respect to the point O is the same as the plane of the couple (F,-F). This shows that the momentum of a force with respect to a point had not been yet identified with a segment perpendicular to the plane of the force and of the arm. Poinsot had associated the momentum of a couple with a segment inclined on the plane of the couple. The direction of the couple's axis, i.e. the direction perpendicular to the couple's plane, was that chosen to pose the segment indicating the couple's momentum, but the momentum of a force with respect to a point was seen as an "object" (of the same kind as an area) lying on the plane itself of the force, not as a segment perpendicular to such plane. Chasles was well aware of the possibility to associate the momentum of a force with respect to a point with a segment perpendicular to the plane force-arm (see, e.g., ibid., p. 106). However, the object "momentum" in itself was seen as an area, an area to which a segment can be associated.

The same equations valid for the systems of equivalent forces also subsist for equivalent systems of momenta (ibid., pp. 100–101). For Chasles argued: be A the system of the momenta $m, m', m'', \ldots$ of a system of forces with respect to a point P. Be B the system of the momenta $n, n', n'', \ldots$ of a second system of forces calculated with respect to another fixed point P'. If A' and B' are two systems of forces respectively equivalent to A and B and you calculate their momenta $\mu, \mu', \mu'', \ldots$ and $\nu, \nu', \nu'', \ldots$ with respect to P and P', you obtain the formula

$$\sum mn \cos \widehat{mn} = \sum \mu\nu \cos \widehat{\mu\nu}$$

If B and B' are composed of a sole force, whose momentum with respect to P' is n, the previous formula is transformed into

$$\sum m \cos \widehat{mn} = \sum \mu \cos \widehat{\mu n}$$

Chasles claimed that the momentum m, being the product of two lines, represents an area. The quantity $m \cos \widehat{mn}$ is the orthogonal projection of this area on the plane of n (ibid., p. 101).

This is a strong confirmation that, in fact, the momentum of a force with respect to a point is imagined as an area rather than as a segment (the fact that a segment is associable with an area does not change this idea on the concept of a force's momentum).

Now Chasles considered the theorem—of which I have spoken in relation to Poinsot and Poisson—according to which any system of forces can be replaced by two forces F and F' one of which—be F—can be chosen so as to pass through the centre of momenta O. Obviously, its momentum will be null. Then, the momentum of F' will have a projection on any plane equal to the sum of the projections of the momenta of all the other forces. As Poinsot and Poisson had clarified—and as we have seen—if these projections are orthogonal, their sum has a maximum when the plane of projection is that of the momentum of the force F' (i.e. the plane individuated by F' and by O). This is the principal momentum of a system of forces. Thence, as Chasles claimed, the principal momentum of a system of forces with respect to a point is equal to the biggest sum of the orthogonal projections onto a plane of the momenta, calculated with respect to this point, of all the forces.

By means of his results, Chasles was able to calculate the value of the principal momentum of a forces' system: for, if in the previous theorem on the momenta of the forces belonging to two equivalent systems, all the forces of the first system are the same as those of the second system, the expression for the momenta of the two systems with respect to a point assumes the same form as Eq. (4.9), i.e. $\sum m^2 + 2\sum mm' \cos \widehat{mm'}$. Therefore, for the square of the principal momentum, an equation completely analogous to (4.9) holds, so that, if the principal momentum is indicated by M, one has:

$$M^2 = \sum m^2 + 2\sum m.m' \cos \widehat{mm'}$$

Chasles concluded this section of his paper claiming that this formula had been obtained in Binet (1815) through a completely different method and that in Giorgini (1820) such formula was reached by means of analytical procedures.

Furthermore, with Chasles' symbolism, the six equations of equilibrium for a solid body by Poinsot are reducible to the following two:

$$\sum a^2 + 2\sum a.a' \cos \widehat{aa'} = 0$$

$$\sum m^2 + 2\sum m.m' \cos \widehat{mm'} = 0$$

where $a, a', a'', \ldots$ are the forces applied to the rigid body and $m, m', m'', \ldots$ their momenta with respect to a point (ibid., p. 102).

Commentary: the sections so far explained—which include a little more than one-third of Chasles' paper—are particularly significant within his foundational programme because they are based only on a geometrical theorem applied to forces and a lemma concerning the projections of the areas. All the results achieved by the discoverers of the concepts that Chasles analysed (and above all by Poinsot, who is, so to say, the hero of these initial sections of Chasles' contribution) are expressed and, in great part, re-obtained developing reasonings based on the two initial statements, the fundamental theorem and the lemma. The steps by Chasles can be summarized like this: from a theorem concerning the geometry of forces, he passed to the geometry of segments; he associated the segments with the areas; he deduced properties of the areas formally equivalent to those connoting the segments; since the momentum of a force with respect to a point is, after all, an area, this allowed him to deduce the same properties for the system of momenta.

The problems dealt with by Chasles until this moment have involved straight lines and areas as well as the physical quantities he associated with such geometrical objects. He also proved propositions referred to volumes and associated physical quantities. In this case, in the products he considered a system of straight lines and a system of areas are involved. As in the previous case, all the propositions derive from the fundamental theorem and from the lemma.

In the first theorem (I indicate it by T1) of this kind he proved that, given a system of segments $a, a', a'', \ldots$, a system of flat areas $m, m', m'', \ldots$ and the product of each segment by each area and by the sinus of the angle between the segment and the plane of the area, the sum of the products is constant if the system of segments and the system of areas are replaced by two equivalent systems. The proof is relatively uncomplicated. For Chasles argued like this: be given the systems of segments $\alpha, \alpha', \alpha'', \ldots$ equivalent to $a, a', a'', \ldots$ and the system of areas $\mu, \mu', \mu'', \ldots$ equivalent to $m, m', m'', \ldots$. Be $b, b', b'', \ldots$ a further system of segments perpendicular to the planes of the first system of areas and proportional to such areas. It holds $m = kb$, $\sin\left(\widehat{a,m}\right) = \cos\left(\widehat{a,b}\right)$. Therefore:

$$\sum a.m \sin\left(\widehat{a,m}\right) = k\sum a.b \cos\left(\widehat{a,b}\right).$$

Be $\beta, \beta', \ldots$ a system of segments perpendicular and proportional to the areas' planes belonging to the second system so that it holds

$$\sum \alpha.\mu \sin\left(\widehat{\alpha,\mu}\right) = k \sum \alpha.\beta \cos\left(\widehat{\alpha,\beta}\right)$$

Since the two systems of areas are equivalent, the two systems of segments b, $b^{'}$, $b^{''}$, … and β, $\beta^{'}$, $\beta^{''}$, … are also equivalent. Furthermore, the two systems of segments a, $a^{'}$, $a^{''}$, … and α, $\alpha^{'}$, $\alpha^{''}$, … are equivalent for hypothesis. Thence, because of the fundamental theorem, one gets

$$\sum a.b \cos\left(\widehat{a,b}\right) = \sum \alpha.\beta \cos\left(\widehat{\alpha,\beta}\right)$$

Thus:

$$\sum a.m \sin\left(\widehat{a,m}\right) = \sum \alpha.\mu \sin\left(\widehat{\alpha,\mu}\right),$$

which is the proposition to prove (ibid., pp. 103–104).

If all the lines pass through a point, they will have a sole resultant A. Chasles also supposed that the planes of the areas also pass through that point and that M is their resultant area, then a corollary of this theorem follows which reads as

$$A.M \sin\left(\widehat{A,M}\right) = \sum a.m \sin\left(\widehat{a,m}\right)$$

Chasles pointed out that the quantity $a.m \sin\left(\widehat{a,m}\right)$ is three times the volume of pyramid having m as basis area and the extremity of the segment a as vertex. From this corollary a further interesting theorem follows: be given a system of segments through a point and a system of flat areas whose planes pass through such a point. If each area m is combined with each segment a and if the volume of the pyramid having m as basis and the extremity of a as vertex is considered, the sum of the volumes of all the so formed pyramids is equal to the volume of the pyramid whose vertex is the vertex of the resultant straight line and whose basis area is the resultant area. This is the maximal sum of all the orthogonal projections of the areas (as seen previously, ibid., p. 104).

Chasles applied these results to the systems of forces and reached a series of significant achievements by a uniform method and relying upon the two theorems I have already presented. He was constructing the basis for the initial sections of rational mechanics relying upon a few geometrical properties. This is one of the principal features of his vast foundational programme.

To expound Chasles' applications of his theorems to the system of forces, it is enough to recall that the forces are treatable as segments, the moments as areas and, since with any area a segment can be associated, the momenta can be represented by segments. The first result is immediately deducible by the dualities force-straight line, momentum-area through an application of T1. For Chasles considered two systems of forces A and A' and the momenta with respect to a point O of the forces belonging to A. If each of these momenta is multiplied by each force of A' and by the sinus between the direction of the force and the plane of the considered momentum, the sum of all these products is invariant when the two systems of forces are replaced by two respectively equivalent systems (ibid., p. 106).

For the next theorem, he represented the momenta m as segments. Given the second system of forces $b, b', b'', \ldots$ and the momenta $m, m', m'', \ldots$ of the first system with respect to a point, he considered the quantity of the previous theorem, i.e. $\sum b.m \sin(\widehat{b.m})$. Chasles deduced that its sign can be determined when the momenta are replaced by the straight lines perpendicular to their plane and proportional to their intensity and the cosine between b and such a straight line is calculated. However, he added this interesting consideration, which avoided resorting to the straight lines perpendicular to the momenta. As he wrote:

> Pose your eye at the extreme of the force b and direct the view towards the application point of this force. It is, then, easy to see in which sense the force, whose momentum m we have combined with the force b, tends to rotate. The sign plus or minus will be attributed to the term $b.m \sin \widehat{bm}$ depending on whether this straight line will tend to induce a rotation to the right or to the left. The same will be done for the other terms.[39]

This rule, which Chasles deduced by the position of the eye and, hence, relying on a visual appearance, is exactly the right-hand rule for the vector product of two vectors. Precisely for the vector product $b \times m$. Therefore, two conclusions can be drawn:

1) If the scholars who were Chasles' reference points, i.e. basically Poinsot and Poisson, arrived close to several properties of the concept of vector, Chasles arrived even closer. He apparently conceived the vector as an entity subject to some operations, though not yet as an element of an abstract vector space;
2) The fact that he used the eye's position is a further confirmation that a projective, visual background is the fundamental basis on which the whole of Chasles' idea is founded.

Chasles added further important results regarding the systems of equivalent forces, the most remarkable of which are connected to the signs of solid figure's volume. For, given two system of forces $A = a, a', a'', \ldots$ and $B = b, b', b'', \ldots$ and their momenta $M = m, m', m'', \ldots$; $N = n, n', n'', \ldots$ with respect to a point O, the two sums $\sum a.n \sin(\widehat{a, n})$ and $\sum b.m \sin(\widehat{b, m})$ will not change if the two systems of forces are replaced by equivalent systems (T2). Thence, the quantity

$$S = \sum a.n \sin(\widehat{a, n}) + \sum b.m \sin(\widehat{b, m}) \tag{4.11}$$

[39] Ibid., p. 106: "Car il est facile de voir qu'on peut, en plaçant l'œil è l'extrémité de chaque force b, et dirigeant la vue vers le point d'application de cette force, regarder dans quel sens la force dont on a combiné le moment m avec la force b, tend à tourner ; et donner au terme $b.m \sin \widehat{bm}$, le signe + ou le signe -, suivant que cette droite tendra à tourner à droite ou à gauche; et ainsi à l'égard des autres termes".

will not change. Now, Chasles argued (ibid., p. 107), if CD is a force of the system A and EF of the system B, these forces will contribute to the sum (4.11) for the amounts

$$2OCD \times EF \sin\left(\widehat{EF, OCD}\right); \quad 2OEF \times CD \sin\left(\widehat{CD, OEF}\right) \tag{4.12}$$

Let us analyse the first term: $2OCD$ indicates the product of CD by the distance between O and CD (it indicates the double of the area of the triangle OCD). It is, hence, the momentum of the force CD with respect to O. It corresponds to the letter m in the second addend of (4.11). The meaning of EF is clear; the angle (EF, OCD) is the angle between the direction of EF and the plane of the CD's momentum. Thence, the first term of (4.12) is an addend of the second sum in (4.11).

Chasles claimed and proved that the sum of the two terms in (4.12), divided by 6, is equal to the volume of the pyramid having its four vertices in C,D,E,F. Joining this consideration with the theorem T2, an immediate consequence is Chasles' Theorem VIII, which is the very cornerstone of this part of Chasles' paper:

> Given two systems of forces, if a tetrahedron is constructed on any force of the first system and on any force of the second system as opposite edges, the sum of the volumes of all these tetrahedra is constant if the two system of forces are respectively replaced by two equivalent system of forces.[40]

Chasles added some interesting considerations on the sign to ascribe to the volume of such tetrahedron. For he pointed out that the two addends $\sum a.n \sin\left(\widehat{a, n}\right) + \sum b.m \sin\left(\widehat{b, m}\right)$ which express the tetrahedron have the same sign: if, for example, the force b observed from the extremity of the force a produces a rotation rightwards, the force a seen from the extremity of b produces a rotation rightwards, as well. Thence, the two terms have the same sign.

From these considerations, it follows that to determine the sign of the volume of each tetrahedron, it is necessary to consider the sense of rotation of one of the two forces which form the tetrahedron with respect to the other force, when the view is directed in the opposite sense of this second force. Since the senses of rotation are two, to one sense the sing "+" is ascribed and to the other sense the sign "–" (ibid., pp. 108–109).

As it has always happened in the previous situations analysed by Chasles, he also considered the case in which the system of forces $b, b', b'', \ldots$ coincides with $a, a', a'', \ldots$. This consideration gave rise to an important corollary (C1). For each of the two systems of equivalent forces will be formed only by a single system of forces. Thus, Theorem 8 states that, given two systems of equivalent forces, the sum of the volumes of the tetrahedra constructed on the forces of the first system considered as opposite edges is equal to sum of the volumes of the tetrahedra constructed likewise

[40] Ibid., pp. 107–108: "Quand on a deux système de forces, si sur chaque force de premier système et chaque force du second système, comme arêtes opposées, on construit un tétraèdre, la somme des volumes de tous ces tétraèdres restera constant quand on substituera aux deux système des forces, deux autres systèmes qui leur soient équivalens respectivement".

on the forces belonging to the second system (ibid., p. 109). Therefore, taking into account that a system of forces applied to a solid body can be replaced by two forces, for any couple of two resultant forces, the tetrahedra constructed on these forces as opposite edges have the same volume (T3). Obviously, the volume is null if the two forces are coplanar. In this case, they either form a couple or have a sole resultant. Thence, the geometrical condition for two systems of forces applied to a solid body to have either a sole resultant or to coincide with a couple is that the sum of the volumes of the tetrahedron constructed on these forces, considered two by two and assumed as opposite edges, is null (ibid., p. 109). Finally, it is easy to deduce that when four forces are in equilibrium, the volume of the tetrahedron constructed on any two of them is equal to the volume of the tetrahedron constructed on the other two.

Commentary: this nucleus of theorems shows the meaning of Chasles' foundational programme because, at the end, the most important properties of the two forces' systems acting on a rigid body are reduced to geometrical elementary properties (the volume, in the specific case) of tetrahedra constructed considering two opposite edges as given by a force belonging to a system and a force belonging to the other system. The basic concepts of rational mechanics are connected to geometrical properties through the fundamental theorem and the following lemma, i.e. through two elementary propositions. However, the whole context presented by Chasles is anything but elementary: though not all the concepts derive from his own original ideas, the way in which he explained and used the association between areas and momenta, between areas and segments and, hence, between momenta and segments is far more precise than what the other mathematicians and physicists had done. The manner in which he conceived the right-hand rule is as clear as ingenious. The idea of considering the two terms $\sum a.n \sin\left(\widehat{a,n}\right) + \sum b.m \sin\left(\widehat{b,m}\right)$ to determine the sum of tetrahedra's volumes whose two opposite edges are made up by two forces as well as the connected argumentation is rather complicated and, in some parts, not completely clear. Thus, as we will see, it was replaced in Chasles (1847) by a perspicuous argument, for which only one of the two addenda has to be considered. He possibly realized that his arguments were not completely clear, and on pp. 111–112, he added an analytical proof of this proposition. Since the proof in Chasles (1847) is synthetic—which is more in line with the way in which he conceived this part of his studies—I will focus on it in Sect. 4.3.

An interesting commentary concerns the note with which Chasles closed the first part of his paper. For he pointed out that his Theorem, after his enunciation, had already been proved in Gergonne (1827–1828b) and in Möbius (1829). The situation is the following: Chasles had communicated the theorem to Gergonne without offering its proof, as Gergonne claimed (see anonym 1828, p. 187). In Gergonne (1827–1828b) a purely analytical proof is offered based on a rather long series of calculations. Gergonne himself added an interesting consideration: perhaps, he wrote, it might be possible to reach the proof of the theorem without resorting to complicate calculations, as those used by him, probably on the base of some considerations concerning the couples. However, this does not guarantee a priori the possibility to reach a proof, whereas the analytical method followed by him ensures, if the theorem is true, to arrive unequivocally at the demonstration, as he himself had done (Gergonne, 1827–1828b, p. 376). This is a useful indication of the way of thinking which connotes the supporters of the analytical methods!

On his part, Möbius offered an elegant and general demonstration based on some considerations concerning the momenta, specifically referring to his idea that, in space, the momenta of a force with respect to an axis can be represented by a tetrahedron (Möbius, 1829, pp. 170–180). Through this consideration, Möbius developed a series of further reasonings (ibid., pp. 181–184) which allowed him to prove Chasles' Theorem VIII.

Möbius developed a profound conception of the link between geometry and statics, which was fully expounded in his *Lehrbuch der Statik*, divided into two parts (Möbius, 1837). As he explicitly explained, the works of the French mathematicians, and above all, of Poinsot, were his fundamental source of inspiration. However, starting from these works, Möbius envisioned and implemented an entire theory which, under several respects, overcame the results of the other mathematicians. In his *Statik*, among other discoveries, he developed a complete theory of the systems of forces and studied the concept of momentum, till including the notion of principal momentum and of null plane (introduced by him) within a unitary theory. In the sections 71 and 72 of the fifth chapter, first part (ibid., pp. 118–122), Möbius analysed, once again, Chasles' Theorem VIII, a proposition which he, in the "Content" ("Inhalt", ibid., p. XV), defined "Marvellous proposition discovered by Chasles" ("Merkwürdiger von Chasles entdecketer Satz"). It seems, however, to exclude that Chasles (1847)—the paper written after Möbius' *Statik* and dedicated to the systems of forces, which I will analyse in the next subsection—was influenced by Möbius' work. Chasles mentioned Möbius' *Statik* in his *Rapport*. He expressed like this: "The excellent Treatise of Statics by Möbius, relies, specifically, upon the consideration of the couples". Chasles also recalled that Möbius (1829, 1837) addressed his theorem VIII (ibid., p. 59).[41]

Anyway, it is clear that the hint of a synthetic proof presented by Chasles in his paper under examination and expounded after Gergonne's and Möbius' demonstrations is not sufficiently elaborated. For sure, this is the reason why he re-proposed a synthetic and completely convincing proof in Chasles (1847).

The set of considerations concerning the volume of the tetrahedron are posed in a manner which may let us think that Chasles arrived close enough to our scalar triple product, but since this conception will emerge more clearly in Chasles (1847), I will consider it in Sect. 4.3.

[41] Chasles (1870, p. 14, Note 2): "L'excellent Traité de Statique de Möbius, notamment, repose sur la considération des couples". I avoid to enter the complex details of Möbius' concepts because my examination would be unavoidably lacunose. As to the literature, I restrict to mention Gray (1993). This is an excellent paper, which offers 1) a clear introduction to Möbius' barycentric calculus, 2) a description and explanation of the basic concepts introduced by Möbius' in his *Statik*, and 3) an examination of the link between geometry and statics in Möbius, also focusing on the role of duality and on Möbius' contribution to the understanding of the concept of vector.

Chasles' work aroused interest among physicists and mathematicians. The proofs given by Gergonne and Möbius of two of his theorems are significant under this respect. Quetelet's review (see Quetelet, 1830) of Chasles (1830c) is indicative as well. Levy in his paper *Mémoire sur quelques propriétés des systèmes de forces* (Levy, 1830) began his work from Chasles' Theorem VIII on the tetrahedron's volumes whose two opposite edges belong to two systems of equivalent forces. For Levy's paper begins quoting the enunciation of this theorem by Chasles (Levy, 1830, p. 261).

As a conclusive remark it is worth pointing out that, though many of Chasles' results expounded in this paper are not new, the way in which such results are conceived and presented is certainly new. They are clearly inserted within a general and foundational context. The context is that of the geometrical foundation of the basic mechanics' elements. Thence, Chasles (1830c) is an important part of his foundational programme. Once again, the early date of this paper shows that Chasles had in mind his programme from the almost initial steps of his scientific career, at least from the first steps of his career as a writer of scientific papers, which does not necessarily coincide with those of his career as a scholar. Thus, despite the fact that Chasles published almost nothing until the late 1820s, he never abandoned the study of geometry.[42] His foundational programme was developed and improved in the following years as it is testified by the fact that Chasles felt the need to offer more convincing proofs of some of his most significant statements as Theorem VIII of Chasles (1830c).

4.3 Chasles' Work on Forces After 1830

With regard to Chasles' foundational programme concerning mechanics and, in particular, the system of forces, the most important paper after Chasles (1830c) is Chasles (1847). However, some interesting considerations are also present in the *Aperçu*. Note XXXIV (Chasles, 1837a, pp. 408–416) is of particular importance. Nonetheless, since a section of this Note is dedicated to the system of forces in the context of a more general discourse on the philosophy of duality, I will face this Note in the sixth chapter. Thence, with respect to the *Aperçu*, I will only focus on a brief observation developed by Chasles in the memoir on duality. Though brief, this observation is significant because Chasles connected directly a result concerning the systems of forces with projective geometry and, specifically, with duality. He started from the well-known truth that a system of forces can be reduced to two forces, one of which, be F, is completely arbitrary, while the other one, be F', after

[42] On this aspect of Chasles' biography, see Bertrand (1892, pp. XLVII–L). This work by Bertrand is important for the reconstruction of Chasles' biography. It is an "Éloge historique" read at the Académie des Sciences on the 19th of December 1892. It was published in the *Mémoires de l'Académie des Sciences de L'Institut de France* in 1904. In the Bibliography, I will indicate it as Bertrand (1892). See also Riccardi (1881).

having chosen F, is determined both in intensity and in direction. It is necessary to take into account that if one point O of the segment indicating F is considered as the centre of the momenta, then the momentum of the whole system with respect to O coincides with the momentum of F' with respect to O. Therefore, Chasles claimed, the plane of the principal momentum of the forces of the system with respect to O passes through the force F' (ibid., p. 679). Here, a brief observation is maybe useful: Chasles is not considering now the momentum as an "oriented segment", but as a flat surface. This flat surface is equivalent to the double of the triangle whose vertices are O and the two points of the segment representing F'. Therefore, in Chasles' perspective the plane of this triangle is the plane of the principal momentum. Once again, the representation of the momentum as an oriented segment was not the only one: its representation as an area is still present. Now, Chasles continued, if F rotates around O, the force F' still belongs to the plane of the principal momentum of the forces' system with respect to O. Reciprocally, whatever the position of F and F' be in the plane of the principal momentum (in the first hypothesis F' belonged to such a plane), the other force will pass through the fixed point. Consequently, if one considers the points belonging to the plane of the principal momentum as those with respect to which the momenta of the forces' system are calculated, the planes to which the momenta belong pass through O. Therefore, if one extends this consideration from the points belonging to a plane to the points belonging to the whole space, it can be concluded that:

> *If in space a system of forces is considered and the planes of the principal momenta of these forces are taken into account, which are relative to all the points of a figure, this planes will envelop another figure which is* correlative *of the former*. That is: the planes relative to points belonging to a plane will pass through a point; the planes relative to points belonging to a straight line will pass through a second straight line, the planes relative to the points posed on a second degree surface will envelope another second degree surface, and so on.[43]

Commentary: this consideration developed by Chasles connects strictly such results with the basic assumptions of his memoir on duality and, hence, with projective geometry. Indeed, as already pointed out, the first theorem proved by Chasles in his memoir on duality (ibid., pp. 577) states that if a mobile plane is given in space and the parameters of the plane equation contain a point, which Chasles called *directeur*, then a) when the point moves on a plane, the mobile plane will rotate around a fixed point; b) when the point moves on a straight line, the mobile plane rotates around a second straight line; and c) when the point moves on a curved surface, the mobile plane rotates around another curved surface. If the first surface is of second degree, the second one also is (ibid., pp. 577–578).

[43] Ibid., p. 679: "*Si l'on conçoit dans l'espace un système de forces, et que l'on prenne les plans des momens principaux de ces forces, relatifs à tous les points d'une figure, ces plans envelopperont une seconde figure, qui sera* corrélative *de la première*. C'est-à-dire que les plans relatifs à des points situés sur un plan, passeront par un même point; les plans relatifs à des points situés en ligne droite, passeront par un même droite; les plans relatifs à des points situés sur un surface du second degré, envelopperont un autre surface du second degré, etc." Italics in the text.

Through the properties concerning the momenta of forces' systems, it is proved that if the point with respect to which the momenta are calculated moves on a plane (therefore, it can be considered a point *directeur*), the planes of the momenta relative to such movable point pass through another point (which is the point I have indicated with O). This given, the situation of Theorem 1 holds (Chasles, 1837a, p. 577). Therefore, merely through projective geometry, one deduces the other parts of the proposition on the forces' system, i.e.: the planes relative to points belonging to a straight line will pass through a second straight line, the planes relative to the points posed on a second-degree surface will envelope another second-degree surface and so on.

This shows how deep the interconnections between projective geometry and Chasles' considerations on forces' systems are. Projective geometry and a few propositions connected to trigonometry (which are inherent to a metric-projective way of thinking) represent the ground of Chasles' foundational programme for the basic parts of mechanics.

Almost all the theorems proved in Chasles (1847) are the same as those proved in the first part of Chasles (1830c). Nonetheless, there are substantial improvements in the later paper. For Chasles gave precise, brief and purely intrinsic proofs, i.e. independent of the subdivision of forces and momenta into their components in an orthogonal reference frame. There is only an explanation in which he resorted to the components. The language, which was redundant in Chasles (1830c), was made more perspicuous and concise. The only basis of the whole paper is the fundamental theorem. The lemma does not appear anymore. This means that he continued to reflect on the best manner of posing the basic propositions of his foundational programme.

Let us hence analyse the proof of the fundamental theorem. Be $a, a', a'', \ldots$ the forces of the first system, $b, b', b'', \ldots$ those of the second system. Be the meaning of the symbol $\sum ab\cos\widehat{(a,b)}$ the same as in Chasles (1830c). Be $A, A', A'', \ldots$ and $B, B', B'', \ldots$ the forces of two systems respectively equivalent to the first two. It is necessary to prove that the identity

$$\sum ab\cos\widehat{(a,b)} = \sum AB\cos\widehat{(A,B)} \tag{4.13}$$

holds.

The first member of (4.13) contains a series of addends of the form $ab\cos\widehat{(a,b)}$, $ab'\cos\widehat{(a,b')}$ and so on, whose sum does not change if the forces $b, b', b'', \ldots$ are replaced by the equivalent system $B, B', B'', \ldots$ because the term with the cosine expresses the product of the force a by the sum of the projections of the forces b, b', $b'', \ldots$ on the direction of a. This sum of projections does not change when the forces $b, b', b'', \ldots$ are replaced by the equivalent system $B, B', B'', \ldots$ (otherwise the two systems would not be equivalent). Applying this reasoning to all the forces a, a', a'', …, one has the identity

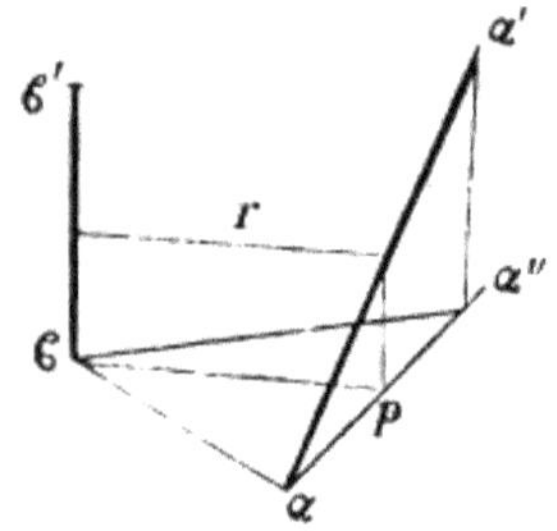

Fig. 4.11 The diagram used by Chasles to prove the theorem explained in the running text (Chasles, 1847, p. 219)

$$\sum ab\cos\left(\widehat{a,b}\right)=\sum aB\cos\left(\widehat{a,B}\right)$$

For the same reason, it will be

$$\sum aB\cos\left(\widehat{a,b}\right)=\sum AB\cos\left(\widehat{A,B}\right)$$

from which the fundamental theorem follows (ibid., p. 214).

Chasles pointed out once again that this theorem is the foundation (*fondement*) of several propositions concerning the systems of forces and momenta (ibid., p.214). It is not by chance that he himself used the term *fondement* to denote this part of his work.

To this theorem, the propositions concerning the systems of equivalent forces and equivalent momenta already proved in Chasles (1830c) follow (ibid., pp. 214–218).

A new important element added by Chasles with respect to his work published in 1830 is the rigorous proof of the proposition (now called Lemma) according to which the volume of the tetrahedron having as opposite edges the straight lines segment $a=\alpha\alpha^{'}$ and $b=\beta\beta^{'}$, whose distance r is given by $\frac{a.b.r.\sin\left(\widehat{ab}\right)}{6}$.

Chasles explained that (see Fig. 4.11) the volume of a tetrahedron does not change when an edge translates along its direction. Be drawn through α a plane π perpendicular to b, be $\alpha\alpha^{''}$ the projection of $\alpha\alpha^{'}$ on π and βp the perpendicular from β to $\alpha\alpha^{'}$. This perpendicular is equivalent to the distance r between a and b. The tetrahedron constructed on the two opposite edges $\beta\beta^{'}$ and $\alpha\alpha^{''}$ is equivalent to the tetrahedron having $\beta\beta^{'}$, $\alpha\alpha^{'}$ as opposite edges because the two tetrahedra share the vertices $\beta^{'}, \beta, \alpha$ and the fourth vertices $\alpha^{'}, \alpha^{''}$ belong to a parallel to the plane $\beta^{'}, \beta, \alpha$. Let us indicate by $S(\beta\alpha\alpha^{''})$ the surface of the triangles having these points as vertices. The volume of the tetrahedron $\beta\beta^{'}\alpha\alpha^{''}$ is given by

$$S(\beta\alpha\alpha'')\times\frac{1}{3}\beta\beta'=\frac{1}{2}\alpha\alpha''\beta p\times\frac{1}{3}\beta\beta'=\frac{1}{6}\alpha\alpha''.r.\beta\beta'=\frac{1}{6}\alpha\alpha'.\cos\widehat{\alpha'\alpha\alpha''}\times r.\beta\beta'=$$

$$\frac{1}{6}\alpha\alpha'.\beta\beta'.r\sin\left(\widehat{\alpha\alpha',\beta\beta'}\right).$$

A direct consequence of this geometrical lemma is a further lemma concerning mechanics, according to which the volume of the tetrahedron constructed on two forces F and F' is equal to the product of F by the momentum of F' with respect to F (or reciprocally) divided by 6. This follows directly from the previous lemma if, posed $F = \alpha\alpha^{'}$; $F^{'} = \beta\beta^{'}$, one takes into account that the momentum of F' with respect to F is exactly the (signed scalar) quantity $\beta\beta'.r\sin\left(\widehat{\alpha\alpha',\beta\beta'}\right)$.

Commentary: to claim that the momentum of F' with respect to F is a signed scalar quantity β is, perhaps, a little bit anachronistic since the momentum of a force with respect to an axis was still seen as a surface. The following quotation by Chasles testifies to the difficulty in passing from a conception in which the momentum is invariably identified with its "material" representation as a surface to the more abstract idea that the momentum is a physical quantity of which the surfaces are, in fact, only a "material" representation. As a matter of fact, Chasles claimed that the momentum is a surface, and never argued that it is represented by a surface. The idea of that period seems that the momentum with respect to a point and with respect to a straight line are two surfaces which can be represented by a segment, but, so to say, they are surfaces in themselves. My impression is that the distinction between the scalar character of the momentum with respect to an axis and the vector character of the momentum with respect to a point was not yet completely clear. Probably this depends on the fact that both of the momenta were seen as triangular areas, which induced these authors to think that they were objects of the same type. On the other hand, as we will see at the beginning of the next quotation, Chasles claimed that the momentum of the force $\alpha\alpha^{'}$ with respect to the force $\beta\beta^{'}$ is $\alpha\alpha^{''} \cdot \beta p$, i.e. a certain quantity which he immediately associated with the double area of the triangle $\alpha\beta\alpha^{''}$. Therefore, the situation is even more nuanced: Chasles and his contemporaries had probably the idea of two physical objects (the momentum of a force with respect to a point and with respect to a straight line) which are something different from their geometrical representation. However, since the only models which were available for them were those of classical (synthetic or analytic) geometry, after all, they considered these quantities as the geometrical objects to which they are directly associable, i.e. areas. Actually, they also imagined that such quantities can be associated with segments and, in some cases, with oriented segments. This indicates that they were in the phase of the passage from a representation by means of classical geometry to that of vector geometry. It is not easy to separate an ideal object (as a force's momentum) from its representation, i.e. to reason in terms of abstract structures independently of the representations of the object determined by these structures, as it was the case with the end-nineteenth century axiomatic. Furthermore, with regard to physics there is an additional question: it is necessary to apply the objects of the model to the physical reality and this can create further problems. All in all, Chasles and his contemporaries, such as Poinsot, Poisson and Möbius—seen from our point of view—were going in the direction of the momenta as abstract objects, and started to the road to vector calculus and to the concept of abstract structure (at least a geometrical abstract structure). From Chasles' point of view, his programme was the reduction of basic mechanics to geometry and in Chasles (1847) he developed his programme in a more perspicuous and general manner than in Chasles (1830c), though the fundamental ideas were unmodified.

As to the momenta Chasles wrote:

> The momentum of the force $\alpha\alpha'$ in respect to the force $\beta\beta'$ is $\alpha\alpha''.\beta p$, that is the double area of the triangle $\alpha\beta\alpha''$. This triangle is the projection of a triangle which will have the force $\alpha\alpha'$ as basis and a point of the force $\beta\beta'$ as a vertex. The double area of this triangle is the momentum of the force $\alpha\alpha'$ in respect to a point of $\beta\beta'$. It is then possible to claim that: *The momentum of a force in respect to a straight line is the projection of the force's momentum in respect to a point of the straight line, after having made the projection on a plane perpendicular to the straight line.*[44]

The theorem that, given two system of forces A and B, if a tetrahedron is constructed considering each force of A and of B as opposite edges, the sum of the volumes of all these tetrahedra will have the same value as an analogous sum whose addends are tetrahedra constructed with two systems of forces A' and B' respectively equivalent to A and B and its rigorous proof[45] allowed Chasles to demonstrate a series of further propositions (ibid., pp. 220–221). Among these propositions one also finds the theorems, (ibid., p. 222) of which I have already spoken, enunciated in Chasles (1830c, p. 109) and proved by Gergonne and Möbius.

The section in which Chasles introduced a new theorem with respect to Chasles (1830c) comprehends the last two pages (ibid., pp. 223–224, items 20, 21, 22) of Chasles (1847) and regards the action of two equivalent systems of forces on a body to which an infinitesimal movement is impressed. The final step of this brief subsection is the proof of the principle of virtual velocities in the context of Chasles' foundational programme. Since the next section of this book is dedicated to his contributions to the principle of virtual velocities, I postpone the considerations on such a principle.

In the course of this section, I have explained step by step the importance of Chasles' works on the systems of forces within his foundational programme. The dependence on fundamental geometrical theorems of his results concerning this part of mechanics has been made explicit. As Quetelet pointed out (Quetelet, 1830), though not all of Chasles' results were new, his approach was certainly innovative. He generalized and made it precise and perspicuous the geometrical approach to the basic concepts of mechanics, which, as we have seen, was already present in different forms in Poinsot, Poisson and Binet. Furthermore, Chasles used synthetic methods and, in this kind of studies, he almost never resorted to the coordinates. He used intrinsic methods, independent of the coordinates. He also gave important

---

[44] Chasles (1847, p. 220): "Le moment de la force $\alpha\alpha'$ par rapport à la force $\beta\beta'$ est $\alpha\alpha''.\beta p$; c'est le double e l'aire du triangle $\alpha\beta\alpha''$. Ce triangle est la projection d'un triangle qui aurait pour base la force $\alpha\alpha'$ et son sommet en un point de la force $\beta\beta'$. Le double de l'aire de ce triangle est le moment de la force $\alpha\alpha'$ par rapport à un point de la droite $\beta\beta'$. On peut donc dire que : *Le moment d'une force par rapport à une droite est la projection du moment de la force par rapport à un point de la droite, la projection étant faite sur un plan perpendiculaire à la droite*". Italics in the text.

[45] Here the complete proof is given, whereas in Chasles (1830c, pp. 107–108) it was given for granted.

contributions to the development of the concept which today we call "vector", though, for other respects, he never abandoned classical viewpoints, as that of the momentum as an area. For us, today, vector calculus clarifies unequivocally the association between an area and a vector through the vector product or between a volume and a scalar through the scalar triple product, but these concepts were not yet existing at that time; therefore, some of the uncertainties that can be detected in his studies are justifiable.

Chapter 5
The Principle of Virtual Velocities

Abstract From Lagrange's *Méchanique analytique* onwards, the principle of virtual velocities, or displacements, became a cornerstone of the whole mechanics. It was proposed with slightly different interpretations by several mathematicians between the end of the eighteenth and the first half of the nineteenth century. Chasles dedicated some reflections to this principle, although he did not develop a complex work like those on the displacement of a rigid body and on the system of forces. I prove that his contributions to the deduction and interpretation of the principle of virtual velocities belong to the development of his foundational programme. I argue for such thesis in the second section. The first section is, instead, dedicated to the contributions of the mathematicians who preceded Chasles. This is paramount to appreciate the homologies and the differences between his approach and those proposed by the other scholars. The Conclusion follows.

The most significant of Chasles' references to the principle of virtual velocities (in what follows sometimes abbreviated as PVV) are present in Chasles (1830c, 1843, 1847). Interesting considerations can also be found in the *Aperçu*. The tone of these references is different because in the first three works Chasles showed that the PVV is provable by means of the concepts he had introduced with regard to the movement of a rigid body and the system of forces, whereas in the *Aperçu* he connected the principle of virtual velocities to his philosophy of duality. Therefore, this section is divided into two subsections: in the first one, the elements of the PVV's history useful to grasp the meaning and significance of Chasles' contributions are analysed; in the second one, the technical contribution given by Chasles to the proof of the PVV will be examined. Though his considerations on the principle of virtual velocities are brief, they are significant to understand all the aspects of his foundational programme.

P. Bussotti, *Chasles and the Projective Geometry*,
https://doi.org/10.1007/978-3-031-54266-4_5

5.1 Proof of PVV: Chasles' Reference Authors

With regard to the history of the principle of virtual velocities, a certain amount of literature exists. I will refer to the work by Danilo Capecchi (see Capecchi, 2012) because it is a comprehensive text on this subject where the most important contributions and interpretations of the PVV are expounded, also with abundant references to the literature. Since Chasles mentioned no author in particular with regard to the PVV, I will add only those elements which I deem necessary for the reader to understand Chasles' novelties, referring, for the rest, to Capecchi's work.

5.1.1 Formulation of the PVV

Nowadays this principle is better known as virtual works principle or law. The term work in the modern meaning was introduced in mechanics by Coriolis[1] and progressively the locution "virtual works" became common, but, nonetheless, it is not universally used and even nowadays the locutions "virtual displacements" and "virtual velocities" are used to denote our principle.

It is possible to offer a completely formal definition of the principle of virtual works starting from the space of configurations. For a vector in $\mathbb{R}^{3N}$ (where N is the number of particles of a certain system) which is tangent to the configuration space of the system at a point X is called a virtual kinetic state or virtual displacement of the system in the configuration X at the time t.[2]

The PVV claims that

> *A necessary and sufficient condition for a mechanic system with smooth constraints to be equilibrated in a configuration C is that the virtual work of the active forces is not positive for any virtual displacement of the system developed from C. The virtual work is null for all the reversible displacements.*

The formal analysed definition of virtual displacement is not involved with the concept of infinitesimal motion because such notion is judged ambiguous, inaccurate, and—after all—meaningless by several physicists, especially by some modern experts in rational mechanics. In contrast to this more modern, formal and algebraizing tendency, the concept of virtual displacement has been traditionally associated with that of infinitesimal movement. For it is sufficient to recall that a

[1]Coriolis wrote, just before addressing the PVV: "Before dealing with the principle of virtual velocities, we also define what we call *virtual elementary work*, a name which, for reasons we will explain later, we replace to that of *virtual momentum*, used so far". Coriolis (1829), p. 11: "Définissons encore, avant d'en venir au principe des vitesses virtuelles, ce que nous appellerons *travail virtuel élémentaire* dénomination que, pour des motifs que nous expliquerons plus tard, nous substituerons à celle de *moment virtuel* qu'on a employée jusqu'à présent". Italics in the text.

[2]Capecchi (2012), pp. 15–26 and, for a more formal treatment, Sonego (2016), pp. 224–227. For this definition, see Sonego (2016, p. 226).

possible displacement is an infinitesimal ideal displacement in which the dependence on the constraints is taken into account as well as the dependence of the constraints on time (when the constraints are rheonomous), whereas a *virtual displacement*—the one in which we are interested—is an infinitesimal ideal displacement in which only the dependence on the constraints is considered without any reference to their possible dependence on time.

We will see that the contrast between the authors who relied on the concept of infinitesimal displacement and those who had a different opinion already existed at the beginning of the nineteenth century, though, obviously, the concepts were not formalized as they are nowadays.

In any case, the notion of virtual displacement is merely geometrical and has nothing to do with the actual motion of a system. This is the reason why it is useful in statics. The case in which the constraints do not depend on time is that in which the virtual displacements (or velocities) are tangent to the configuration space.

Given a system of n points, the virtual displacement of each point is, hence, under these conditions, parallel to the force applied at that point. Indicating by δL_v the virtual work, by δP_s the virtual displacement of the point P_s and by $\boldsymbol{F}_s$ the force applied to P_s, the PVV claims that the system is equilibrated if and only if

$$\delta L_v = \sum_{s=1}^{n} \boldsymbol{F}_s \cdot \delta P_s \leq 0$$

With regard to Chasles, it is enough to consider only the sign "equal".

5.1.2 A Hint to the Historical Context

If we consider the 60–70s of the nineteenth century as a final reference point (which is useful for Chasles' considerations and the spread of his ideas on PVV), for our history, one might speak of a proto-history of the PVV, which includes the long period between Hellenistic science and Johann Bernoulli's letter to Varignon in 1715; an ancient history of the principle which dates from 1715 to the first edition of Lagrange's *Méchanique analytique* (1788) and a modern history of the principle from 1788 onwards. The cultural milieu of Chasles is that of the French physicists and mathematicians, who worked under the inspiration of Lagrange's masterpiece. Therefore, I refer entirely to the first seven chapters of Capecchi (2012) for the contributions given before Bernoulli' letter to Varignon on 26th February 1715. This letter is so important in the history of the PVV that it is worth quoting it in full (Fig. 5.1):

> Conceive several different forces acting along different trends or directions to balance a point, line, surface, or body; conceive also to impress on the whole system of these forces a small motion either parallel to itself in any direction, or around a fixed point whatsoever: you will be glad to understand that with this motion each of these forces will advance or retire in

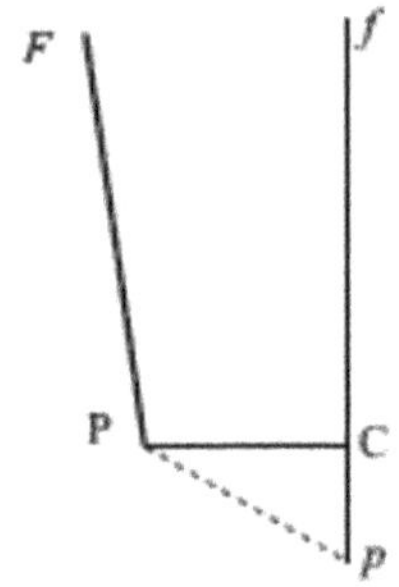

Fig. 5.1 Bernoulli's diagram to explain his concept of virtual velocity as reported in Capecchi (2012, p. 205)

> its direction, unless someone or more forces had their trends perpendicular to the direction of the small movement, in which case this force or these forces, neither advance nor retire anything. These advancements or retirements, which are what I call virtual velocities, are nothing but what each direction increases or decreases by the small movement. These increases or decreases are found by drawing a perpendicular to the end of the line of action of any force. This perpendicular will cut in the same line of action, displaced in a close position by the small motion, a small part that will measure the virtual velocity of this force. Take, for example, any point P in the system of forces that is in equilibrium, *F* one of those forces which push or pull the point P in the direction FP or PF; P*p* a small straight line that the point P describes because of the small motion, for which the trend FP takes the direction *f p*, which will be exactly parallel to FP if the small motion is made in all parts of the system along a given line, or will have, being prolonged, an infinitely small angle with FP if the small motion of the system is around a fixed point. So draw the perpendicular PC to *f p*, and you will have C*p* for the virtual velocity of the force *F*, so that C*p*×*F* is what I call energy. Note that C*p* is negative or positive relative with respect to the others: it is positive if the point P is pushed by the force *F*, and the angle FP*p* is obtuse and is negative if the angle FP*p* is acute, but otherwise, if the point P is pulled, C*p* will be negative when the angle FP*p* is obtuse, and positive when acute. All this being understood, I form this general proposition: *In any equilibrium of any forces in any way they are applied and following any directions, either they interact with each other indirectly or directly, the sum of the positive energies will be equal to the sum of the negative energies taken positively.*[3]

Bernoulli pointed out that the small movements that he called virtual velocities have to be assumed as parallel to the forces and that the small movements perpendicular to the forces have no effect in the progress of the forces along their direction (to express the concepts in Bernoulli's language). Bernoulli's principle of virtual velocities expresses only a necessary condition for the equilibrium. As Capecchi points out (ibid., p. 206), Bernoulli spoke of infinitesimal character of the virtual velocities, but he made no direct reference to the constraint conditions. Despite these limitations, Bernoulli formulated his principle in a form which is not substantially different from that to which also the mathematicians and physicists of the eighteenth and the first half of the nineteenth century referred to. There were, however, many problems which had to be clarified: 1) was Bernoulli's proposition a principle or could it be proved by means of more elementary propositions? Of course, it would

[3]Translation drawn from Capecchi (2012), p. 205. For the direct reference to Bernoulli's work, see Capecchi (2012), p. 205.

have been necessary to clarify the meaning of "elementary". In substance, was it a real principle or a theorem? 2) The ontological status of the infinitesimal displacements—equivalent to that of the infinitesimal velocities because the nascent virtual motion has to be considered (potentially) rectilinear and uniform—was regarded as problematic. Therefore, in the period following Bernoulli, some scientists began to wonder whether it were possible to avoid the resort to the infinitesimal displacements which were not clear from a physical standpoint and appeared close to the dangerous concept of actual infinitesimal; 3) with the progress of the research, the physicists understood that the conditions of the constraints had to be framed in the enunciation of PVV; 4) finally they proved that, if the constraints are smooth, the principle is also a sufficient condition for the equilibrium.

5.1.3 Lagrange

With some exceptions of which Capecchi speaks (ibid., chapter 9), there were no substantial improvements until the contributions given by Lagrange, which were fundamental. It is appropriate to recall that Varignon in his *Nouvelle méchanique* (Varignon, 1725) referred to Bernoulli's principle, but the letter of Bernoulli was not yet published when Lagrange dealt with the PVV (Capecchi, 2012, p. 250). Lagrange used the principle in his two memoirs on the libration of the Moon (Lagrange, 1763 and Lagrange, 1780, resp. See also Capecchi, 2012, pp. 242–252). However, the formulation given by Lagrange in his first edition of the *Méchanique analitique* is that which gave rise to a series of discussions about the PVV. Therefore, I will offer an idea of Lagranges' conception insofar as it is presented in the *Méchanique*. Lagrange formulated the principle like this:

> If any system of as many bodies or points one wishes, each solicited from any powers, is in equilibrium, and if this system is given an arbitrary small motion, under which each point passes along an infinitely small space, which will be its virtual velocity, the sum of powers, multiplied each by the space that the point where it is applied passes in the direction of that power, will always be zero, considering as positive the small spaces in the direction of power and as negative the spaces in the opposite direction.[4]

Lagrange did not assume the PVV as a pure postulate and tried to prove the statement. He assumed axiomatically the validity of the principle when applied to two forces acting, respectively, on the same point or on two points A and B of a

[4]Lagrange (1788), pp. 10–11: "Si un système quelconque de tant corps ou points que l'on veut, tirés chacun par des puissances quelconques, est en équilibre, et qu'on donne à ce système un petit mouvement quelconque, en vertu duquel chaque point parcoure un espace infiniment petit qui exprimera sa vitesse virtuelle; la somme des puissances, multipliées chacune par l'espace que le point où elle est appliquée, parcourt suivant la direction de cette même puissance sera toujours égale à zéro, en regardant comme positifs les petits espaces parcourus dans le sens des puissances, et comme négatifs les espaces parcourus dans un sens opposé". Translation drawn from Capecchi (2012), p. 253.

machine where the constraints are rigid and smooth. If P and Q denote the forces applied, respectively, on A and B and dP and dQ the virtual displacements (or velocities), Lagrange called the quantities of the form $P \cdot dP$ the *momentum* of the force P. Obviously this quantity is what from Coriolis onwards we call *virtual work* of a force. Therefore, Lagrange assumed that, in the case of equilibrium the equation

$$P \cdot dP + Q \cdot dQ = 0$$

holds. After that, he used the *principle of solidification* to prove the theorem when three forces are applied to three points. The principle of solidification was of large use in the mechanics of that period. It claims that if a system of particles (or points) is in equilibrium when the points are potentially movable, it is also in equilibrium if some of the points are fixed, namely the equilibrium cannot be disrupted by the addition of more connections between the particles (points).[5] Through this principle, Lagrange proved the PVV for three forces and through a reasoning by mathematical induction he reached the proof for any number of forces and points, so writing his famous equation for the PVV:[6]

$$P \cdot dP + Q \cdot dQ + R \cdot dr + \ldots = 0$$

In the fourth section of his masterpiece Lagrange considered explicitly the constraint equations $L = 0$; $M = 0$; $N = 0$; ... connected to a system of points whose coordinates are (x', y', z'); (x'', y'', z''); .... Lagrange took into account the differentials $dL = 0$; $dM = 0$; $dN = 0$; ... of the constraint equations and, by introducing the *Lagrange multipliers* λ, μ, ν, ..., he was able to write the PVV in the form

$$P \cdot dP + Q \cdot dQ + R \cdot dr + \ldots \lambda dL + \mu dM + \nu dN + \ldots = 0$$

where, now, the points P, Q, R,... have to be considered as if they were free (Lagrange, 1788, p. 46).

In the second edition (Lagrange, 1811) of his main work, which took the name of *Mécanique analytique*, Lagrange explicitly claimed that the PVV is anything but evident and that a proof based on more intuitive principles is necessary. He assumed the principle of pulleys. Lagrange's words are clear, for he wrote:

> As to the nature of the principle of virtual velocities, it is not so self-evident that it can be assumed as a primitive principle, but it can be considered as the expression of the general law of equilibrium, deduced by the two principles that we set out [that of the lever and

[5] For the definition of solidification principle, see, e.g., Targ (1979).

[6] Lagrange (1788), pp. 14–15). See also Capecchi (2012), pp. 255–256, where Lagrange's proof is summarised. A little imperfection in Lagrange's argument, which does not compromise its general validity, is also pointed out by Capecchi.

> composition of forces]. So in the proofs that are given of this principle it is always considered due to one of these, more or less directly. But, in Statics there is another general principle independent of the lever and the composition of forces, although the mechanicians will commonly refer it to them, which would seem to be the natural foundation of the principle of virtual velocities: you can call it *the principle of the pulleys*.[7]

Lagrange's proof covers both the necessary and the sufficient part of the PVV. As he claimed, his arguments do not depend on the principle of the lever and on that of forces decomposition. It is anyway appropriate to recall that to prove the necessary condition (if there is equilibrium, then the PVV holds) he also resorted to the principle of natural descent of the bodies, which is a very common experience, but is not deducible otherwise. For the sufficient part, Lagrange used the principle of sufficient reason (for details, see Capecchi, 2012, p. 262).

Lagrange offered a new proof of the PVV in his *Théorie des fonctions analytiques* (Lagrange, 1797, second edition 1813) and extended the considerations connected to the PVV to dynamics until he managed to include the PVV and the D'Alembert principle in a unique picture. Since the consideration of these topics is not necessary to understand Chasles' contribution to the PVV, I refer to Capecchi (2012), pp. 264–279.

Lagrange's work is the unavoidable reference point for any historical consideration concerning the PVV. After Lagrange, a debate on many aspects of PVV arose among French scientists. We have seen that Chasles' main sources in his woks on the movement of the rigid bodies and on the composition of forces, where he spoke of the PVV, were basically—among his contemporaries or almost contemporaries—Carnot, Poinsot, Poisson, Giorgini and, after 1840, Rodrigues. In the considerations dedicated by Chasles to the PVV, no author is mentioned, but it is presumable that he was influenced by the conceptions of these five scientists.[8] Therefore, I will try to summarize their ideas on the PVV. With regard to Carnot, I will basically refer to Capecchi (2012), pp. 282–299. As to Poisson and Poinsot I will refer directly to their

[7]Lagrange (1811), p. 23: "Quant à la nature du principe des vitesses virtuelles, il faut convenir qu'il n'est pas assez évident par lui-même pour pouvoir être érigé en principe primitif ; mais on peut le regarder comme l'expression générale des lois de l'équilibre, déduites des deux principes que nous venons d'exposer. Aussi dans les démonstrations qu'on a données de ce principe, on l'a toujours fait dépendre de ceux-ci, par des moyens plus ou moins directs. Mais il y a en Statique un autre principe général et indépendant du levier et de la composition des forces, quoique les mécaniciens l'y rapportent communément, lequel paraît être le fondement naturel du principe des vitesses virtuelles; on peut l'appeler le *principe des poulies*". Translation in Capecchi (2012), p. 260. Italics in the text. This proof is so well known that I do not refer to it. For the original demonstration, see Lagrange (1811), pp. 23–26. For a clear explanation, see Capecchi (2012), pp. 259–262. Mach's analysis and criticisms to Lagrange's demonstration (Mach, 1919, pp. 65–68. English translation of Mach 1883) are also well known. Capecchi (2012), pp. 263–264 argues that Mach's considerations are not completely convincing.

[8]Fourier gave three demonstrations of the PVV (Capecchi, 2012, pp. 321–328) and Laplace also gave one (ibid., pp. 332–334). However, Fourier is not mentioned in the works by Chasles in which he spoke of the rigid body's motion and of the forces' decomposition either he is mentioned in the *Aperçu*. It is to give for granted that Chasles knew Laplace's proof, but since it is included in Poisson's, I will refer directly to Poisson (1811), a work mentioned in Chasles (1830c), p. 97.

works, also taking into account some of Capecchi's considerations; insofar as Giorgini and Rodrigues are concerned, I will address directly their works because these authors are not mentioned by Capecchi. Chasles is not mentioned, either.

5.1.4 Carnot

Carnot had an empirical view as to the origin of the principles at the basis of mechanics and claimed that, despite their empirical origin, the fundamental laws of mechanics appear so evident and clear that they seem almost as logical as mathematical truths (Capecchi, 2012, p. 283). It seems that Carnot resorted to the dangerous and ambiguous criterion of the evidence. However, the meaning of what Carnot claimed is clear: the initial principles of mechanics are related to our most immediate experiences, the rest of this science is constructed mathematically.

A further consideration concerns the fact that Carnot had a mechanistic conception because—in his opinion—only impacts can determine the communication of motion; therefore, no action different from a direct interaction among bodies can be accepted in mechanics. His conception refuses, hence, any action at a distance and, though he used the term "force", it is rather unclear what he meant by this word. For sure, he was not referring to Newtonian forces because his paradigm was based on collisions rather than on the continuous transmission of forces (ibid., pp. 284–286). In the *Essay sur le machines en général* (Carnot, 1783. Consulted second edition 1786), Carnot established his two principles as follows:

> First law. The reaction is always equal and contrary to the action.
>
> Second law. When two hard bodies act each other, because of collision or pressure, i.e., because of their impenetrability, their relative velocity immediately after the mutual action is always zero.[9]

From his two principles applied to a system of n bodies connected by rods, Carnot deduced his first fundamental equation for mechanics. This principle is expressed by the following formula

$$\sum FV \cos q = 0,$$

[9]Carnot (1783). Consulted edition 1786, pp. 21–22: "Premiere loi: La réaction est toujours égale et contraire à l'action [. . .] Seconde loi : Lorsque deux corps durs agissent l'un sur l'autre, par choc ou pression, c'est-à-dire en vertu de leur impénétrabilité, leur vitesse relative, immédiatement après l'action réciproque, est toujours nulle". Translated in Capecchi (2012), p. 285. With regard to the basic concepts of Carnot's mechanics, it is at least necessary to mention the seminal (Gillispie & Pisano, 2014, second edition). In particular chapters 2, 3 and 4 offer a clear explanation of concepts such as geometrical motion, moment of momentum and moment of activity in reference to Carnot's *Essai sur le machines en general* as well as an examination of the following works by Carnot on the problem of equilibrium.

(in modern notation $\sum_{i=1}^{n} F_i V_i \cos q_i = 0$), where F indicates the force acting on a body of the system, V the velocity assumed by the point, and q the angle between the direction of the force and that of velocity.[10]

What is interesting from our point of view is that the first fundamental equation of Carnot's mechanics has the *form of a virtual works law*, though this equation was referred to a general movement, not only to a situation of equilibrium. Hence, the form is that of a virtual work law, but what we call "work" after Coriolis is true and not virtual in the case of a real motion.

Carnot gave his equation a slightly different form by separating the active forces from the constraint reactions and naming by V the velocity gained by the body as if it were free, by U the velocity "lost" because of the constraints, by Z the angle between U and V, and by m the mass of the body. The form of the fundamental equation became, hence:

$$\sum mVU \cos Z = 0$$

The second equation of mechanics is obtained by the first one taking into account the concept of geometrical motion.[11] This equation is very expressive, but it seems to me that it is also involved in concepts which are not completely clear. Carnot considered a generic geometrical motion u; he named U the velocity lost because of the constraints and z the angles between the motion and the lost velocity, so that his second fundamental equation gets the form

$$\sum mUu \cos z = 0$$

If the motion derived from the composition of the original motion u with the lost velocity U is virtual (hence, u too is virtual), then the second fundamental equation assumes the role of the principle of virtual velocities because, taking into account all the virtual motions derived by the composition of U and u, all the equations of equilibrium are obtained.[12]

A general observation has to be made: the concepts introduced by Carnot in the two fundamental equations as well as the very notion of virtual work (to use modern terms) imply the concept of scalar product. However, since neither the concept of

---

[10] Capecchi (2012), pp. 287–288.

[11] I have dealt with this concept in chapter 3 of this work. For the notion of geometrical motion in Carnot, see Capecchi (2012), pp. 289–290. Gillispie and Pisano claim that the notion of geometric motion prevented Carnot from using the problematic concept of infinitesimal displacement. They comment like this: "In effect, since in his theory *geometric motions* coincide with *velocities* and not with displacements this allowed Carnot to avoid, in the formulation of the *principle of virtual work* infinitesimal displacements, which could have produced some scientific embarrassment with respect to his assumptions" (Gillispie & Pisano, 2014, p. 376). Italics in the text.

[12] Capecchi (2012), p. 291.

vector nor that of vector operations were completely clear at that time, all these concepts can be interpreted as being part of an embryonic phase in the birth of vector conception. On the other hand, they create ambiguity because if the operation of scalar product is not clearly explained, it is not evident what the meaning of the product between two vector quantities is. It is intuitive, but it is not completely well founded from a logical point of view, even though it is clear from a physical standpoint.

If in the second fundamental equation the lost velocity U is replaced by the actual velocity V and the geometrical motion u is real and not virtual, being y the angle between motion and velocity, Carnot named the quantity muV cos y *moment of momentum* or *moment of the quantity of motion*. In a collision of a system of bodies the moment of momentum is conserved. This means that the following law holds:

$$\sum muV \cos y = \sum muW \cos x$$

The meaning of the symbols is clear.[13]

A corollary of the principle of conservation of the moment of quantity of motion introduces us to the idea Carnot had of the PVV, for he wrote:

> When a system of hard bodies changes its motion for imperceptible degrees, m is the mass of each body, V its velocity, p its moving force, R the angle between the direction of V and p, u the velocity which m would have if the system would take any geometrical motion, r the angle formed by u and p, y the angle formed by V and u, dt the element of time, it will hold any of two equations[14]
>
> $$\sum mVpdt \cos R - \sum mVdV = 0$$
> $$\sum mupdt \cos r - \sum mud(V \cos y) = 0$$

If one poses $Vdt = ds$, a differential form of the principle of conservation of living forces is obtained. The second one of the previous equations—apart from the constant m—can also be interpreted as the PVV. For the quantity udt cos r is the projection of the "imperceptible" motion of the body, whose mass is m, along the direction of the force p. Furthermore, if a body is in equilibrium $V = 0$, so that it holds

---

[13] Capecchi (2012), p. 291.

[14] Carnot (1783, consulted edition 1786), pp. 49–50: "Lorsqu'un système quelconques de corps durs change de mouvement par degrés insensibles; si pour un instant quelconque on appelle m la masse de chacun des corps, V la vitesse, p la force motrice, R l'angle compris entre les directions de V et p, u la vitesse qu'auroit m, si on faisoit prendre au système un mouvement quelconque géométrique, r l'angle formé par u et p, y l'angle formé par V et u, dt l'élément du temps, on aura deux équations

$\int mVpdt \cos R - \int mVdV = 0$

$\int mupdt \cos r - \int mud(V \cos y) = 0$". Translation in Capecchi (2012), p. 293.

$$\sum mupdt \cos r = 0,$$

namely PVV. Carnot claimed:

> Fundamental theorem. General principle of equilibrium and motion in machines.
> XXXIV. Whatever is the state of repose or of motion in which any given system of forces applied to a Machine, if it is given any geometric motion, without changing these forces in any respect, the sum of the products of each of them, by the velocity which the point at which they are applied will have in the first instant, estimated in the direction of this force, will be equal to zero.[15]

Therefore, the PVV is deduced as a corollary of the theorem of the conservation of the moment of quantity of motion: as a matter of fact as a corollary of the second fundamental equation stated by Carnot. He called the quantity *updt* cos *r* the "moment of activity" consumed by the force *p* during *dt*. This is the work. As Capecchi claims, Carnot was not the first to introduce this quantity, but he was the first to stress its fundamental role in mechanics.[16]

5.1.5 Poinsot

Poinsot is one of the most important reference points for Chasles, as we have already pointed out. Indeed, Chasles was one of the authors who most appreciated Poinsot's view of physics. We will see in Chapter 6 that Poinsot's ideas had a profound influence on Chasles' philosophy of duality. Poinsot's contributions to the comprehension of PVV are as profound as original. For sure, Chasles read and meditated these contributions.[17] Poinsot's basic idea is that the PVV is obscure and that it is mistakenly assumed to be the basis of the whole mechanics. No proof can make the comprehension and the logic behind PVV clear and perspicuous because its formulation implies a mysterious and obscure concept, that of infinitesimal displacement. Poinsot's criticism seems to me connected to the fact that, according to his opinion,

---

[15] Carnot (1783, consulted edition 1786), p. 68: "THÉORÈME FONDAMENTAL. Principe général de l'équilibre et du mouvement dans les Machines. XXXIV. Quel que soit l'état de repos ou de mouvement où se trouve un système quelconque de forces appliquées à une Machine, si on lui fait prendre tout-à-coup un mouvement quelconque géométrique, sans rien changer à ces forces, la somme des produits de chacune d'elles, par la vitesse qu'aura dans le premier instant le point où elle est appliquée, estimée dans le sens de cette force, sera égale à zéro". Translation in Capecchi (2012), p. 294.

[16] Capecchi (2012), p. 296. It is also interesting to highlight that Carnot thought that not the (Newtonian) force in itself is determinant for the movement, rather the work and that, furthermore, he used the term force in different meanings. Since this is not strictly connected to Chasles, I refer to Capecchi (2012), chapter on Carnot (pp. 281–297).

[17] Poinsot dealt with the PVV in several memoires (see Capecchi 335–351). His most significant and profound ideas were developed in *Sur la Théorie générale de l'équilibre et du mouvement des systèmes* (Poinsot, 1806b). Thence, I will analyse this work and, specifically, its later versions (see next note).

the infinitesimal displacements are not obtained as limits of finite displacements, but are introduced as a sort of actual infinitesimal movement, whose logical and physical status is, at best, ambiguous. For while dealing with Lagrange's proof expounded in the first edition of his *Méchanique analitique*, Poinsot wrote:

> But this research [by Lagrange] entails all the difficulties in which the principle itself is involved. This so general law, where the vague and strange ideas of the infinitesimal movements and of the equilibrium perturbation are involved, becomes even more obscure at a fine examination. And Lagrange's book offers nothing clear but the calculation procedure. It is evident that the clouds are not eliminated from the course of Mechanics, but, so to say, they gather at the origin itself of this science.[18]

This means that Poinsot did not search a proof of PVV based on other principles of mechanics (for example that of the lever or that of the composition and decomposition of forces, or both of them) in the attempt to find a satisfying way to apply the concept of infinitesimal motion. Rather, it was necessary to avoid any resort to such a concept and to found the theory of equilibrium on other principles which were free from such an ambiguous notion. In other terms: it was necessary to found the equilibrium theory on different bases and to show that the PVV, conveniently reformulated, is a corollary of such a theory (Poinsot, 1861, p. 264). This justifies why, as we will see, Poinsot resorted to principles which appear, under some respects, more complicated than PVV: they were perhaps more complicated, but they were free from the concept of infinitesimal displacement. Poinsot's perspective is thus different from that of the other authors I am analysing.

He based his considerations on five principles:

[18] Poinsot (1861), pp. 263–264: "Mais cette recherches ramena toutes les difficultés qu'on avait franchies par le principe même. Cette lois si générale, ou se mêlent des idées vagues et étrangères de mouvements infiniment petits et de perturbation d'équilibre, ne fit en quelque sorte que s'obscurcir à l'examen; et le livre de Lagrange n'offrant plus alors rien de clair que la marche des calculs, on vit bien que les nuages n'avaient paru levés sur le cours de la Mécanique que parce qu'ils étaient, pour ainsi dire, rassemblés à l'origine même de cette science". Poinsot published his memoire on the PVV *Sur la Théorie générale de l'équilibre et du mouvement des systèmes* in 1806 (see Poinsot, 1806b). In various editions of his *Éleméments de statique* this memoire was republished as an appendix to the text. It appears in the 1834 edition and is modified with respect to the original one. Obviously, the general ideas and the treatment of the problems is the same, but in the publications after 1806 Poinsot was completely clear in his criticism to the concept of infinitesimal movement. Whereas in Poinsot (1806b), though his ideas on this point are deducible from his treatment, they are not completely expressed, almost surely because of a form of respect towards Lagrange who was still alive. For example, the passage I have quoted does not exist in Poinsot (1806b). Furthermore, in Poinsot (1806b) there are two appendices, one concerning a necessary condition for the equilibrium of a system of points and another regarding the PVV, whereas in the following editions of his memoire inserted within his *Éleméments de statique*, a third note regarding a further property of the forces acting on a system in equilibrium is added. Beyond Poinsot (1806b), I have consulted the 10th (posthumous) edition of Poinsot's *Éleméments de statique*, where the memoire on the general theory of equilibrium is expounded as in the 1834 edition. Thence, while dealing with this memoire, I will indicate Poinsot (1861). This depends on the fact that Poinsot's thought is, starting from 1834, more clearly expressed than in Poinsot (1806b). Since Chasles' considerations on the PVV dates to the 40', the 1834 edition of Poinsot's *Éleméments de statique* was available to Chasles.

1) principle of solidification;
2) the forces which act on the points or bodies of a system in equilibrium can be decomposed along the lines joining two by two these points in pairs of equal and contrary forces (I indicate this principle as P2);
3) the principle of forces parallelogram;
4) the principle of perpendicularity according to which necessary condition for equilibrium is that any force is perpendicular to the surface or the curve which represents the constraints;[19]
5) the principle of separation: a system of points or bodies linked by more constraints can balance the sum of the forces which each constraint can balance.

Poinsot expressed his adherence to the first two principles as follows:

> One of the first elements of the general theory of equilibrium is this axiom: if some forces acting on a system of variable form are in equilibrium, the equilibrium will not be perturbed supposing that the system is made invariable in one fell swoop or, so to say, it is solidified.
> The conditions of equilibrium of the *solid* bodies must, hence, subsist in the equilibrium of all the possible systems. Because of this, it is possible to speak of *general properties* of equilibrium.
> It is possible to present these conditions in several ways, but it is appropriate to remember that they can be reduced to this sole condition: *the forces applied to the different points of a system can be decomposed, along the lines joining such points, in forces which are two by two equal and contrary.*[20]

To the formulation of P2 an interesting consideration follows, which is not a demonstration of the principle, because of course P2 is assumed, but which has to be interpreted as a proof of the correct manner to apply P2. For Poinsot spoke of "théorème" (ibid., p. 265). The reasoning developed by Poinsot is a sort of argument by mathematical induction based on geometry: the application is clear in the case of two forces. For the equilibrium, they have to be equal and contrary. If the forces are three and they are applied to the points A,B,C, it is necessary to imagine the points joined by three straight lines so that the applied forces, decomposed in a system of forces acting along AB, AC, BC, are such that those acting along each of these three lines are equal and contrary. If the number of the forces is $n > 3$, it is possible to consider three points A, B, C of application with the straight lines joining them. The points A, B, C represent a triangle (let us call it "the fundamental triangle"). For each of the other points D_i ($i = n - 3$) to which the forces are applied, it is possible to

[19] Capecchi (2012), p. 342 expounds the five Poinsot's principles.

[20] Poinsot (1861), pp. 264–265: "L'un des premiers éléments de la théorie générale de l'equilibre est cet axiome: que si des forces se font actuellement équilibre sur un système quelconque de figure variable, l'équilibre ne cessera point en supposant que le système soit rendu tout à coup invariable, ou vienne, pour ainsi dire, à se solidifier lui-même. Les conditions de l'équilibre des corps *solides* doivent donc se retrouver dans l'équilibre de tous les systèmes possibles, et c'est pour cela qu'on peut les nommer les *propriétés générales* de l'équilibre. On a présenté ces conditions de plusieurs manières ; mais on peut remarquer qu'elles se réduisent à cette condition unique, *que les forces appliquées aux divers points de système soient décomposable, suivant les droites qui les joignent, en forces deux à deux égales et contraires*". Italics in the text. For a hint to other physicists who applied the principle of solidification, see Capecchi (2012), p. 342.

consider the pyramid $ABCD_i$. It is now allowed to decompose a force acting on any vertex D_i of the pyramid into three forces acting along the edges AD_i, BD_i, CD_i. As for the equilibrium, it is necessary that the components of the forces acting on the vertices of the basis ABC in the direction AD_i, BD_i, CD_i are equal and contrary to the components of the forces acting in the same direction and applied to D_i. If, given the equilibrium, one deletes the equal and contrary forces applied to all the edges of the pyramids, only three forces applied to the vertices A,B,C will remain. But, since there is equilibrium, these three residual forces have to de decomposed along AB, AC, BC into forces, which are, in turn, equal and contrary (ibid., p. 265).

This is a proof of how to apply the principle. Though it might appear uselessly complicated, it has a great heuristic value because it allows us to understand Poinsot's way of reasoning, which was profoundly geometrical, although he was a physicist. This is a remarkable contact point with Chasles. Even though the latter was a geometer and Poinsot a physicist, both of them reasoned basically in geometrical terms. This means that the origin, the matrix of their thought was geometrical.

Poinsot continued highlighting that, if h is the number of points composing a system, the already shown decomposition determines $3h-6$ distances (Poinsot, 1861, p. 265). This depends on the fact that, given the fundamental triangle, three distances are determined (therefore $3h-6$ distances, with $h=3$), and for each point D_i, a pyramid having the fundamental triangle as basis is constructible, so that each point adds three further distances because only the connections with the vertices of the fundamental triangle are taken into account. On the other hand, if a Cartesian system of coordinates is introduced, each force can be decomposed into three forces along the directions of the coordinate axes, so that, given h points, there are $3h$ equations of the forces. Since the equilibrium is supposed, the forces acting along each of the $3h-6$ distances are equal and contrary and they form $3h-6$ (ibid., p. 266) unknowns. Hence six equations remain. These equations are the six equations which have to be fulfilled as a necessary condition for equilibrium. I have referred in detail to Poinsot's reasoning because it is emblematic of the geometrical way with which he addressed the foundational problems of mechanics. In this respect, his mentality was close to Chasles'.

After this important introductory section, Poinsot distinguished the case in which the points of the system are constrained by a sole equation or by several equations.

In the first case, he developed his geometrical reasoning based on the fundamental triangle: if four points are given; if m,n,p are the distances among the vertices of the fundamental triangle and q,r,s, the distances of the fourth point from the vertices of such triangle, the constraint equation can be written as

$$f(m,n,p,q,r,s)=L=\text{const} \tag{5.1}$$

The necessary condition for equilibrium is, hence, that the forces applied to the vertices of the pyramid are equal and contrary.

At this point, Poinsot applied the principles of solidification and of perpendicularity. If the distances q,r,s are solidified, Eq. (5.1) represents a surface which is function of m,n,p. The normal to this constraint surface along the directions m,n,p is, hence, given by the first derivative $f'(m), f'(n), f'(p)$. Thence, necessary condition for equilibrium is that the external forces are proportional to $f'(m), f'(n), f'(p)$.

If the distances m,q,r are solidified, then the forces have to be proportional to $f'(m), f'(q), f'(r)$ and so on. Thus, for equilibrium, it is necessary that the external forces applied at the edges of the pyramid are proportional to $f'(m), f'(n), f'(p), f'(q)$, $f'(r), f'(s)$.

Afterwards Poinsot began to analyse more specific conditions for equilibrium: he proved that, given a system composed of any number of bodies, the equilibrium needs that the applied forces are proportional to the first derivatives of the constraint conditions with respect to any distance. He argued once again by means of the principles of solidification and perpendicularity. He imagined that all the links of the system, four of them excluded, are solidified. He concluded, hence, that the situation is the same as when the system is composed of only four points. If four other links are not solidified, but all the remaining ones are, one can reason, once again, as in the case of four points. Separating the whole configuration into a series of pyramids by means of the expounded technique, for each pyramid the argument valid when only four points make up the system holds. In this way all the possible external forces compatible with the equilibrium are taken into account. Since the whole mechanic system is given by the juxtaposition of all the pyramids, the theorem is proved as it is proved for four bodies (Poinsot, 1861, pp. 269–270).

After this theory, in which he furnished the conditions of equilibrium in an intrinsic form, that is relying only upon the distances among the bodies of the system, while concretely calculating the expression of the forces, Poinsot used a Cartesian system of coordinates. It was rather easy for him to prove that, in a system of n points/bodies, where $P^{(i)}$ $(1 \leq i \leq n)$ is the force applied to the i-th point, given the equilibrium, the form of $P^{(i)}$ is

$$P^{(i)} = \sqrt{\left(\frac{dL}{dx^{(i)}}\right)^2 + \left(\frac{dL}{dy^{(i)}}\right)^2 + \left(\frac{dL}{dz^{(i)}}\right)^2}.$$

After having solved another apparent difficulty when the system is composed of more than four points (ibid., pp. 273–274), Poinsot concluded that, for any point $(x^{(i)}, y^{(i)}, z^{(i)})$, the components of the applied forces have to be proportional to

$$\frac{dL}{dx^{(i)}}; \frac{dL}{dy^{(i)}}; \frac{dL}{dz^{(i)}}.$$

As Capecchi points out, Lagrange had already obtained this result through a different method. For Lagrange the partial derivatives represented the directions of the constraint forces in an equilibrium situation, whereas for Poinsot, they are the directions of the external forces when the system is in equilibrium (Capecchi, 2012, p. 345).

Finally, Poinsot analysed the case in which there are several constraint equations:

$$f(m, n, p, q, r, \ldots) = L = 0$$
$$\varphi(m, n, p, q, r, \ldots) = M = 0$$
$$\psi(m, n, p, q, r, \ldots) = N = 0$$
$$\ldots$$

He proved easily that a necessary condition for equilibrium is that the forces are of the form

$$\lambda\sqrt{\left(\frac{dL}{dx^{(i)}}\right)^2 + \left(\frac{dL}{dy^{(i)}}\right)^2 + \left(\frac{dL}{dz^{(i)}}\right)^2},$$
$$\mu\sqrt{\left(\frac{dM}{dx^{(i)}}\right)^2 + \left(\frac{dM}{dy^{(i)}}\right)^2 + \left(\frac{dM}{dz^{(i)}}\right)^2},$$
$$\nu\sqrt{\left(\frac{dN}{dx^{(i)}}\right)^2 + \left(\frac{dN}{dy^{(i)}}\right)^2 + \left(\frac{dN}{dz^{(i)}}\right)^2},$$
$$\ldots$$

where the Greek letters indicate constants.

The problem now posed by Poinsot was whether these forces are the only ones which can mutually balance. In order to answer this question he applied the principle of separation and argued as follows (Poinsot, 1861, p. 280): suppose that all the equations, $L = 0$ excluded, are replaced by forces in equilibrium. The only remaining constraint equation will be $L = 0$ and the forces coherent with equilibrium will be proportional to $\sqrt{\left(\frac{dL}{dx^{(i)}}\right)^2 + \left(\frac{dL}{dy^{(i)}}\right)^2 + \left(\frac{dL}{dz^{(i)}}\right)^2}$. Analogously, it is possible to replace all the equations, excluded $M = 0$, with forces in equilibrium and to obtain that, in this case, the forces coherent with equilibrium will be proportional to $\sqrt{\left(\frac{dM}{dx^{(i)}}\right)^2 + \left(\frac{dM}{dy^{(i)}}\right)^2 + \left(\frac{dM}{dz^{(i)}}\right)^2}$ and so on. Thence, the principle of separation can be stated:

> Therefore, the forces which can balance in a system defined by more equations are nothing but the forces composed of those which will mutually balance separately, according to each equation.[21]

[21] Poinsot (1861), p. 280: "Ainsi, les forces qui peuvent se faire équilibre sur un système défini par plusieurs équations ne sont autre chose que les forces composées de celle qui s'y feraient équilibre séparément en vertu de chaque équation". In order to furtherly justify his principle, Poinsot added some very interesting, though not completely convincing, arguments, concerning the situation when more of one constraint equation is considered (particularly two or three equations). I do not have room to face this part of Poinsot's work, but I remark that it is worth of an attentive examination (see Poinsot, 1861, pp. 280–281).

At the end of his memoir Poinsot considered the components ($X^{(i)}, Y^{(i)}, Z^{(i)}$) of the forces applied at any point i of the system along the directions of the three coordinate axes, concluding (ibid., p. 283) that, for equilibrium, such components must fulfil the equations

$$X^{(i)} = \lambda\left(\frac{dL}{dx^{(i)}}\right) + \mu\left(\frac{dM}{dx^{(i)}}\right) + \ldots$$

$$Y^{(i)} = \lambda\left(\frac{dL}{dy^{(i)}}\right) + \mu\left(\frac{dM}{dy^{(i)}}\right) + \ldots$$

$$Z^{(i)} = \lambda\left(\frac{dL}{dz^{(i)}}\right) + \mu\left(\frac{dM}{dz^{(i)}}\right) + \ldots$$

In the *Conclusion de ce Mémoire* Poinsot highlighted that the whole of his statics was founded without any resort to PVV and that, rather, PVV can be deduced as a corollary from his theory. Since the deduction of PVV is explicit in the second appendix of Poinsot (1806b, the third in Poinsot, 1861), I will refer to such appendix.

Before expounding Poinsot's deduction of PVV as a corollary of his theory, it seems to me appropriate to highlight the novelty of his conception: he refused the concept of infinitesimal displacement because of its obscurity and indefiniteness. He also refused to interpret virtual velocity as an infinitesimal velocity associated with the infinitesimal displacement. The virtual velocity interpreted *à la* Poinsot is, to use modern terms, part of a general symbolism associated with the system of points/bodies under examination. Such a symbolism is that of the system's configurations. The basic concept is that of system, which cannot be assumed, according to Poinsot, in an intuitive and undefined manner: the system is equivalent to all the configurations compatible with the constraints. If the equilibrium of the system is required, the equation forces-velocities (I will clarify the meaning of this expression) has to be equated to 0. In this case the velocities are called "virtual". There is no reference to the infinitesimal nature of such velocities and there is no need of this reference. Poinsot did not address the problematic ontology of virtual velocities. He treated them in a formal manner as mathematical quantities within an equation which has to the equated to 0 for the equilibrium to be obtained. Thence, his conception of virtual velocities is deeply different from that which interprets the virtual velocities as something mysterious like "infinitesimal velocities that the body would assume in the initial instant of its motion but which it does not assume because there is equilibrium". It seems to me that Poinsot arrived close to conceive the space of configurations in almost modern terms and to consider the virtual velocities as belonging to the space tangent at the configuration space. Given these interpretations, the complicated principles he introduced as well as his complex geometrical itinerary to obtain the equilibrium equations can be read as the attempt to avoid the problematic concept of infinitesimal displacement. Therefore, he was developing an abstract mathematical structure within which to frame mechanics. Obviously, given the epoch, he was not able to express and prove the PVV directly in this abstract structure so to avoid both the concept of infinitesimal movement and his complicated argumentations presented in the text of his memoir.

In order to prove PVV, Poinsot considered the equations between the coordinates of the bodies

$$\begin{cases} f(x,y,z;x',y',z',\ldots)=0 \\ \varphi(x,y,z;x',y',z',\ldots)=0 \\ \ldots \end{cases} \tag{A}$$

Be given any velocity[22] compatible with the constraint conditions. If such impressed velocities are $\frac{dx}{dt}, \frac{dy}{dt}, \frac{dz}{dt}, \frac{dx'}{dt}, \ldots$, since they are compatible with the constraints, the following system (B) will hold, because of the theory Poinsot had expounded in his memoir

$$\begin{cases} f'(x)\dfrac{dx}{dt}+f'(y)\dfrac{dy}{dt}+f'(z)\dfrac{dz}{dt}+f'(x')\dfrac{dx'}{dt}+\ldots=0 \\ \varphi'(x)\dfrac{dx}{dt}+\varphi'(y)\dfrac{dy}{dt}+\varphi'(z)\dfrac{dz}{dt}+\varphi'(x')\dfrac{dx'}{dt}+\ldots=0 \\ \ldots \end{cases} \tag{B}$$

He multiplied now these equations by the constant coefficients λ, μ, ... whose nature is not necessary to specify, so obtaining the equation

$$\left\{\begin{matrix} [\lambda f'(x)+\mu\varphi'(x)+\ldots]\dfrac{dx}{dt} \\ +[\lambda f'(y)+\mu\varphi'(x)+\ldots]\dfrac{dy}{dt} \\ +[\lambda f'(z)+\mu\varphi'(z)+\ldots]\dfrac{dz}{dt} \\ +[\lambda f'(x')+\mu\varphi'(x')+\ldots]\dfrac{dx'}{dt} \\ +\ldots \end{matrix}\right\}=0 \tag{C}$$

He reminded the reader that in his memoir he had proved that the coefficients of $\frac{dx}{dt}, \frac{dy}{dt}, \frac{dz}{dt}, \frac{dx'}{dt}, \ldots$ are the general expressions of the forces which are in equilibrium in the system. If these forces are called X, Y, Z; $X'Y'$, Z'; ... it will be[23]

[22] Note that Poinsot did not enter the nature of such velocities. He claimed that the bodies of the system might have in fact ("actuellemenent") such velocities (Poinsot, 1861, p. 297).

[23] Curiously, in the following equation D, there is a typo in Poinsot (1861), p. 298 which does not exist in the original (Poinsot, 1806b), p. 238. For in Poinsot (1861) Eq. (D) is written $X\frac{dx}{dt}+Y\frac{dy}{dt}+Z\frac{dx'}{dt}+Y\frac{dy'}{dt}+\ldots=0$.

$$X\frac{dx}{dt}+Y\frac{dy}{dt}+Z\frac{dz}{dt}+X'\frac{dx'}{dt}+\ldots=0 \tag{D}$$

Instead of the components *X,Y,Z* multiplied by the respective velocities along their directions it is possible to consider the resultant *P* multiplied by the projection of the resultant velocity $\frac{ds}{dt}$ along its direction. Poinsot named $\frac{dp}{dt}$ such a projection, so obtaining (ibid., p. 298):

$$P\frac{dp}{dt}+P'\frac{dp'}{dt}+P''\frac{dp''}{dt}+\ldots$$

Thus, he concluded:

> That is, if the forces are in equilibrium in any system, the sum of their products by the velocities one wants to give the bodies, whatever being such velocities, provided that they are compatible with the constraints, will always be zero, by estimating these velocities along the directions of forces.[24]

Poinsot underlined that, if, instead of the velocities, whose nature is not necessary to specify, one formulates the principle resorting to the displacements as a measure of such velocities, the displacements must be infinitely small, thus resorting to the ambiguous and imprecise concept of infinitesimal displacement. Therefore, the final and correct formulation of PVV, derived as a corollary of the theorems expounded in his memoir, is expressed by Poinsot like this:

> When the different bodies of a system run any of the movements which do not violate in any way the link established between them, i.e. the system is continuously in one of those configurations allowed by the constraint equations, it can be sure that the forces that will be capable of being balanced in these configurations, when the system passes in them, are such that multiplied by the velocity of the bodies projected onto their directions, the sum of all these products is necessarily equal to zero.[25]

He claimed that in this manner, the PVV is not involved with the idea of infinitesimal displacements, an idea which introduces obscure elements (ibid., p. 299).

After a consideration on the sufficient conditions for equilibrium, Poinsot added that PVV has not to be interpreted as being referred to the configuration in an instant,

[24]Poinsot (1861), p. 298: "c'est-à-dire que, si des forces se font équilibre sur un système quelconque, la somme de leurs produits par les vitesses, quelles qu'elles soient, qu'on voudra imprimer aux corps, mais que leur liaison permet, sera toujours égale à zéro, en estimant ces vitesses suivant les directions des forces".

[25]Poinsot (1861), p. 298: "Lorsqu'on voit suivre aux différents corps d'une système des mouvements quelconques qui ne violent point le liaison établie entre eux, c'est-à-dire qui nous présentent continuellement le système dans des figures où les équations de condition subsistent, on peut être sûr que les forces qui seraient capables de se faire equilibre sur une de ces figures, dans le moment où le système y passe, sont telles, que, multipliées par les vitesses actuelles des corps projetées sur leurs directions, la somme de tous ces produits est nécessairement égale à zéro". Translation from Capecchi (2012), p. 349.

but to all the configurations coherent with the constraint conditions of the system. Thence, the PVV equation does not claim that one must take the forces so that it is satisfied, but that one has to choose the functions of the coordinates so that it is satisfied. Poinsot continued:

> Now, under the constraint conditions themselves, one knows that between the velocities that the bodies can simultaneously have, it must apply the linear equation (C), the coefficients of which are the derivatives of functions given with respect to the coordinates by which this velocity is estimated. The equation of moments says that the forces of equilibrium must be represented by the derivative of these functions, therefore, to prove it, it is necessary to show how these forces are actually equilibrated or it must look directly for what functions of the coordinates can represent the forces of equilibrium, as we did from the beginning.[26]

This is a litmus paper of the fact that Poinsot was close to conceive a general concept as that of space of configurations and to the idea that the principle was to refer to such general space. As a further confirmation of this idea, Poinsot claimed that every proof of PVV based on the functioning of a particular machine is a mere heuristic justification rather than an authentic proof, which can be obtained only in the context of a new conception of mechanics as the one he was trying to explain.

I have addressed almost all the details of Poinsot's reasoning because he was a particularly important author for Chasles who often referred to him. The two authors shared the idea that geometry should have been the basis of mechanics. It is not a coincidence that Poinsot concluded the appendix to his memoir on the equilibrium claiming that his intention had been that to reduce the PVV to the basic principles of statics and geometry. *Géométrie* (ibid., p. 301) is the word with which the appendix on the PVV ends. Chasles' philosophy, as we shall see, was influenced by Poinsot's views and results. Chasles' conception of PVV was partially different from Poinsot's, but both authors considered geometry as the foundation of mechanics. In the case of PVV, Chasles' idea of geometrical foundation did not exactly coincide with Poinsot's. Since a comparison between their ideas is paramount to fully grasp Chasles' approach, I offered all the details of Poinsot's proof.

5.1.6 Poisson

Poisson's approach to PVV is more traditional than Poinsot's. He did not criticize the concept of virtual movement as, so to say, potential infinitesimal movement and

[26]Poinsot (1861), p. 300: "Or, en vertu des conditions mêmes, on sait qu'il doit régner entre les vitesses simultanées que pourraient avoir les corps une équation linéaire identique (C), dont les coefficients sont les fonctions *primes* des fonctions données par rapport aux coordonnées suivant lesquelles on estime ces vitesses. L'équation des *moments* dit donc que les forces de l'équilibre doivent être représentées par ces fonctions primes; et, par conséquent, pour la démontrer, il fallait faire voir comment de telles forces se font effectivement équilibre, ou bien il fallait chercher directement quelles fonctions des coordonnées peuvent représenter les forces de l'Equilibre, comme nous l'avons fait d'abord". Translation from Capecchi (2012), pp. 350–351.

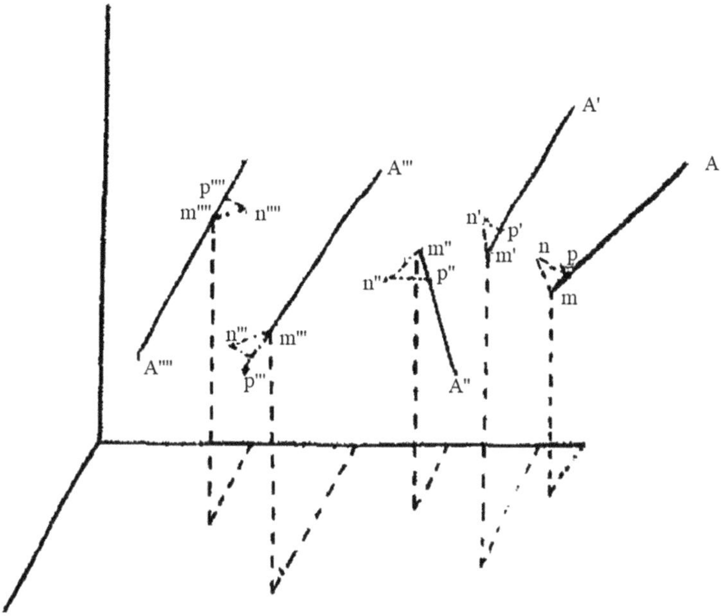

Fig. 5.2 The (adapted) diagram used by Poisson to describe PVV. All the figures in Poisson (1811) are in the planches at the end of the book

developed Laplace's proof completing and improving it. Chasles mentioned often the celebrate *Traité de Méchanique* (Poisson, 1811), where, in the seventh chapter of Book I, PVV is addressed. Thence, I will give a hint of Poisson's ideas, also to show the difference between his approach those of Poinsot and Chasles.

Poisson clearly stated that PVV originates from induction as a law which is always empirically verified in any system of balanced forces (ibid., p. 231).

He described PVV like this: $P, P', P'', \ldots$ are the given forces (see Fig. 5.2) whose directions are mA, $m'A'$, $m''A''$, …, applied to the points m, m', m'', …. These material points are tied by inextensible threads or other physical constraints, which oblige them to remain on a given curve or surface. Be given an *infinitesimal movement* ("mouvement infinement petit", ibid., pp. 231–232) to the whole system, so that m moves to n, m' moves to n', and so on. The segments mn, $m'n'$, $m''n''$, … are *infinitesimal* ("infiniment petits", ibid., p. 232). They are called the virtual velocities of the points m, m', m'', …, respectively. If each of these infinitesimal segments is projected along the direction of the force applied to the corresponding point, you obtain the component of the virtual velocity along the direction of the force. The sign "plus" will be attributed to those projections whose sense is the same as the force's, the sign "minus" if their senses are opposite. Indicating by p, p', p'', … these projections, PVV claims that, if the forces are in equilibrium, their sum multiplied by the respective virtual velocity's projection is equal to 0, namely the equation

$$Pp + P'p' + P''p'' + \ldots = 0$$

holds. Reciprocally, if the equation holds the forces are in equilibrium (ibid., p. 233).

Commentary: 1) Poisson used the concept of infinitesimal motion without any problem. This concept is, as a matter of fact, obscure: if by infinitesimal motion one intends a nascent motion, then the notion is, obviously, already present in Newton. But with the virtual displacement the situation is different because there is no displacement either infinitesimal or finite. Thence in the context of PVV, the infinitesimal motion should indicate a potential—not an actual—nascent motion, a very problematic concept, in fact; 2) Poisson connected immediately PVV to concrete physical situations where the constraints are explicit. In a sense, his approach is more concretely oriented than Poinsot's; 3) though the concept of vector did not yet exist, the way in which Poisson projected the virtual velocities on the forces (obtaining, in practice, the scalar product) confirms the idea I have already expressed: we are in the phase of the proto-history of vector calculus. There is no formalization, but several basic concepts exist.

After having offered two applications of PVV (ibid., pp. 233–238) Poisson proved easily the PVV when the system is composed of a sole point (ibid., pp. 238–244). In this simple case, the supposition that the virtual velocity is infinitesimal is not necessary (ibid., p. 240). Hence, I will not face this part of the proof.

Poisson clearly highlighted the importance of this principle: it allows us to obtain the equilibrium equations in any specific case without calculating the tensions of the constraints (ibid., p. 244). The PVV avoids the often difficult enterprise to work on the components (e.g., according to an orthonormal system of coordinates) of the forces applied to the system and to show that their sum is 0.

Let us look at Poisson's proof for systems composed of more than one point. Here the supposition of infinitesimal virtual displacements and velocities is necessary: given a number k of points, be $[m^i, m^j]$ the tension of the thread extended between the i-th and the j-th points $(0 \leq i, j \leq k)$. Be (m^i, m^j) the distance between the i-th and the j-th points. The letter $\delta^i_{'}$ indicates the variations of the distances relative to an infinitesimal displacement of the point $m^{'}$. Thus, e.g., if $i = 0$, the symbol will indicate the displacement of the point m, if $i = 1$ of the point m', ... Finally, the letter δ indicates the variation of the distance of two points as a consequence of their infinitesimal displacement (ibid., p. 245). Thence, e.g., as Poisson explained, if the point m has a virtual displacement to n and m' to n', it is:

$$\delta(m, m') = nn' - mm', \quad \delta_{'}(m, m') = nm' - mm', \quad \delta'_{'}(m', m) = mn' - mm'$$

Poisson's figure gives an idea of the situation (see. Fig 5.3):

Here, Poisson clarified a fundamental issue of his proof: the total variation δ can be decomposed into the sum of the partial variations $\delta_{'}$ and $\delta'_{'}$ only if the displacement is infinitesimal, for the distance $(m, m^{'})$ is a function f of m, $m^{'}$. If, given the infinitesimal movement from m to n and from $m^{'}$ to $n^{'}$, the powers of the

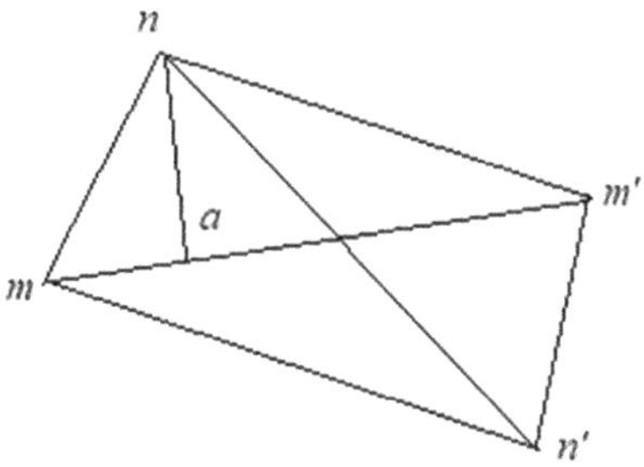

Fig. 5.3 Adaptation of Poisson's figure used to explain the previous equation

development of *f* bigger than the first one are neglected, the total displacement will be equal to the sum of the partial displacements, so that only for infinitesimal displacements the equation

$$\delta(m, m') = \delta_{'}(m, m') + \delta_{'}(m', m)$$

holds (ibid., pp. 245–246).

This told, for the equilibrium, it is necessary that the force P applied to m is equilibrated by the tensions $[mm']$, $[m, m'']$, $[m, m''']$, …, which can be considered separately. Poisson, hence, assumed a point n which is *infinitely close* ("infiniment voisin", ibid., p. 247) to m on the constraint surface. Be p, t, t', t'', … the projections of mn on the directions of the forces P, $[mm']$, $[m, m'']$, $[m, m''']$, …, respectively. Thence, since the theorem holds for one point m, it is possible to write the equation

$$Pp + [m, m']t + [m, m'']t' + \ldots \tag{5.2}$$

Since the line mn is *infinitesimal* ("infiniment petite", ibid., p. 247) and only because it is infinitesimal, given the projection ma of mn on mm' (see Fig. 5.3), it is easy to prove that $ma = mm' - nm'$. Therefore, it is

$$t = \delta_{'}(m, m'); t' = \delta_{'}(m, m''); t'' = \delta_{'}(m, m''), \ldots$$

This means that Eq. (5.2) can be written as

$$Pp + [m, m']\delta_{'}(m, m') + [m, m'']\delta_{'}(m, m'') + [m, m''']\delta_{'}(m, m''') + \ldots = 0.$$

Analogous equations hold for the points m', m'', …, namely:

$$P'p' + [m', m]\delta'_{'}(m', m) + [m', m'']\delta'_{'}(m', m'') + [m', m''']\delta'_{'}(m', m''') + \ldots = 0$$

$$P''p'' + [m'', m]\delta''_{'}(m'', m) + [m'', m']\delta''_{'}(m'', m') + [m'', m''']\delta''_{'}(m'', m''') + \ldots = 0,$$

$$\ldots$$

where $p', p'', p''', \ldots$ are the virtual velocities of the points $m', m'', m''', \ldots$ along the direction of the acting forces $P', P'', P''', \ldots$ (ibid., p. 248).

If all these equations are added, taking into account that $[m^i, m^j] = [m^j, m^i]$ and $(m^i, m^j) = (m^j, m^i)$, you get the equation

$$Pp + P'p' + P''p'' + \ldots + [m, m']\delta(m, m') + [m, m'']\delta(m, m'') + \ldots \tag{5.3}$$
$$+[m', m'']\delta(m', m'') + [m', m''']\delta(m', m''') +$$
$$\ldots + [m'', m''']\delta(m'', m''') + \ldots = 0.$$

Now, if, as in the case of equilibrium, one supposes that the points maintain the same respective distances, all the expressions $\delta(m^i, m^j)$ have value null so that the equation of the PVV

$$Pp + P'p' + P''p'' + \ldots = 0$$

is obtained.

Finally, this proof conceived for the systems of points is also valid for all the *infinitesimal movements* ("mouvements infiniment petits", ibid., p. 250) of a rigid body because the distances among the points of a rigid body are constant.

Poisson stressed that this proof is a development and improvement of that expounded by Laplace in the first book of his *Mécanique céleste* (ibid., p. 251).

Later on Poisson also offered an interesting proof of the sufficient condition (ibid., p. 252–257). I do not analyse this proof because the elements explained with regard to the necessary condition are enough to show clearly the difference between Poinsot's and Poisson's approaches. Capecchi points out that in the sufficient part of the proof a principle of dynamical character is used, whereas this was not the case with Poinsot's proof (Capecchi, 2012, p. 365).

Commentary: besides what stressed by Capecchi, it seems to me that there are at least two significant differences between Poinsot's and Poisson's approach to PVV:

1) Poinsot, for whom the concept of infinitesimal displacement in statics is an imprecise and obscure notion which has to be avoided, formulated a foundation of statics in which such concept was eliminated. The PVV, purged of any reference to infinitesimal movements or velocities, became a corollary of Poinsot's theory. Poisson did not criticize the concept of infinitesimal and virtual movement or distance. In contrast to this, it was one of the fundamental elements in his demonstration.
2) Poinsot's approach was *foundational*: he aspired to found statics on geometry and on few well-stated principles. Obviously, he also resorted to the basic concepts of mathematical analysis (as a matter of fact only that of derivative). Poisson's approach was not foundational, it was more traditional. He aspired to prove PVV within an already traced context, and did not at creating a new context. His methods were analytical. Geometry played a poor role in his demonstration. In

this respect, Chasles' approach is by far more similar to Poinsot's, though the former did not deny the validity of the concept of infinitesimal movement.

5.1.7 Giorgini

Giorgini was an important reference point for Chasles who, as already highlighted, often mentioned his works. We will see that Chasles' approach to the PVV resembles, under some respects, the one of Giorgini, who dedicated to the PVV the 12th chapter of his *Elementi di statica* (Giorgini, 1835, pp. 181–214).

As the other authors, Giorgini proved separately the necessary and the sufficient part of the principle. With regard to the necessary part, his idea was to prove the principle for a rigid body—which is the easiest case—and to extend his proof to systems of points and systems of rigid bodies. His demonstrative technique consisted in showing that the annihilation of the virtual works was equivalent to conditions of equilibrium which he had already stated in the previous sections of his book. Giorgini stressed the importance of PVV insofar as it allows us to express the equilibrium condition in a sole equation (ibid., p. 181). With regard to the sufficient part of the proof, it is analogous to Poisson's. I will concentrate on Giorgini's introductory considerations and on the necessary part of the proof because this is what is interesting in reference to Chasles.

To introduce PVV Giorgini started with an easy example: he considered an inflexible bar mOm', which can be right or curved (being O the centre of the bar, namely the point at which the bar is constrained) and two forces P, P' applied to m,m', under the condition that P,P',O are coplanar (Fig. 5.4).

Be $Op = p$; $Op' = p'$ the perpendiculars drawn from O to the direction of the two forces. Because of the equilibrium rules concerning the static momentum, which Giorgini had already analysed, the two forces will be in equilibrium if they tend to produce opposite rotations around O and if

$$Pp = P'p' \tag{5.4}$$

If, instead, the bar rotates, the points m and m' describe two circular arcs around O which have the same angular amplitude. Thence, they will be proportional to the radiuses Om and Om'. In the first instant of the motion, the directions of the points m and m' coincide with those of the tangents mA and $m'A'$, on which the infinitely small ("infinitamente piccoli", ibid., p. 182) arcs mn, $m'n'$ will lie. They are proportional to the radiuses Om and Om'. At the beginning of the movement, the velocities of the points m and m' are proportional to the two infinitesimal arcs mn and $m'n'$ and, hence, to the radiuses Om and Om'. Therefore, if along the tangent mA you choose the line $mA = Om$ to indicate the instantaneous velocity in m, then the straight line $m'A' = Om'$ along the tangent to $m'n'$ will indicate the instantaneous velocity of m' at the beginning of the motion. These velocities, which the points m and m' can assume at the beginning of the impressed motion without altering the

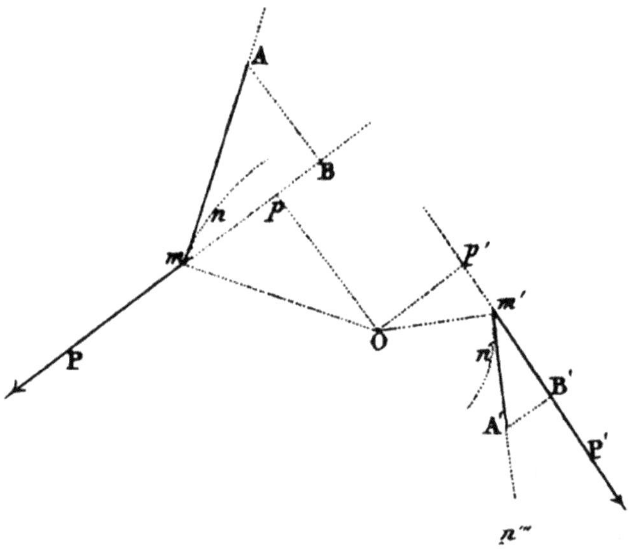

Fig. 5.4 The diagram drawn by Giorgini to prove the principle of virtual velocity in the case of a rigid bar. From Giorgini (1835). In the text by Giorgini the diagrams are in planches at the end of the book

conditions of the system, receive the name of virtual velocities ("velocità virtuali", ibid., p. 182).

A brief commentary is appropriate to show how problematic the concept of virtual velocity associated with an infinitesimal virtual motion is: Giorgini in the course of his explanation spoke of an impressed movement of rotation in the first instant of the motion (ibid., pp. 181–182). If this movement is interpreted as the motion considered in the instant in which it begins, such concept, though not completely transparent, is typical of classical mechanics and is the basis of the notion itself of instantaneous velocity. But the problem is that, if this movement is considered—as Giorgini did—in a static context, it is not a real movement because no motion exists. Rather, it is the movement that the body would assume if it rotated. It is only a potential instantaneous movement; in fact, a virtual infinitesimal movement; something which is both existing and not existing. These were, probably, the considerations which induced Poinsot to avoid the concept of infinitesimal displacement in reference to statics.

Giorgini continued claiming that the velocities represented by mA and $m'A$ can be decomposed according to the parallelogram rule. In particular, through very easy considerations concerning the orthogonal projections, the lines mB, $m'B'$ will be the virtual velocities of m and m' along the directions of the corresponding forces mP, $m'P'$ (ibid., p. 183). The two triangles mpO and ABm are equal as well as $m'p'O$ and $A'B'm'$ (the proof is very easy), so that

$$mB = Op = p; \quad m'B' = Op' = p'.$$

Replace these values in Eq. (5.4) connoting the equilibrium, you get

$$P \cdot mB = P' \cdot m'B',$$

where mB and $m'B'$ represent the virtual velocities of the points m and m'. If the virtual velocity is assumed as positive when its orthogonal projection on the direction of the force has the same sense of the force and negative in the opposite case, it will be

$$v = mB; \quad v' = -m'B'.$$

So that the equilibrium condition is expressed by the equation

$$Pv + P'v' = 0$$

which represents the PVV (ibid., p. 184). Therefore, in this simple case, the PVV expresses the same property which can be expressed otherwise taking into account the relations between the static momentum for two forces to be in equilibrium. The product Pv was called "*momentum of the force* with respect to the considered movement".[27] If one considers a system of points belonging to a rigid body or constrained in another manner, the PVV assumes the forms:

> For the equilibrium of any system of points, it is necessary that the momentum of the forces' system applied to these points relative to any considered movement compatible with the conditions of the points' system is null.[28]

Giorgini asserted that, in the elementary case of the bar, the momentum of a force considered in this new meaning (which is our concept of virtual work) is proportional to the momentum of the forces around the fixed point O, namely it is proportional to the static momentum. Thence, he continued, the new meaning of momentum has to be considered as an extension of the traditional one (ibid., pp. 186–187).

The relation between the virtual works and the momenta of a force with respect to an axis allowed Giorgini to prove the PVV for the rotational equilibrium of a rigid body. His interesting reasoning can be illustrated in this way (see Fig. 5.5):Let us consider a point m to which the force $F = mP$ is applied. Be Oz an axis to which the plane xOy containing the point m is perpendicular. Be Q the orthogonal projection of P on xOy. Let us join Q and m and draw the perpendicular Oq from O. According to

[27] Ibid., p. 186: "*momento della forza* respettiva relativo al movimento considerato". Italics in the text.

[28] Ibid, p. 186: "Per l'equilibrio di un qualsivoglia sistema di punti è necessario, che il momento del sistema di forze ad essi applicate, relativo a qualunque movimento compatibile colle condizioni che costituiscono il sistema di punti sia nullo".

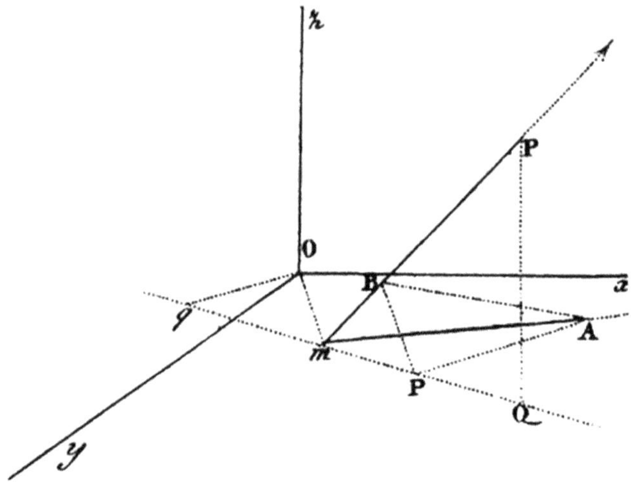

Fig. 5.5 The figure used by Giorgini to prove the principle of virtual velocity. From Giorgini (1835), planches of the figures at the end of the text

the definition of momentum of the force *mP* around the axis *Oz*, already provided by Giorgini (ibid., p. 44), the product $mQ \cdot Oq$ expresses such momentum. If the system of bodies (it is enough thinking of a rigid body), to which *m* belongs, begins to rotate around the axis *Oz*, all the points of the system will draw small arcs whose centre is in *Oz* and whose planes are perpendicular to *Oz*, while their radiuses are the perpendiculars from the considered point to the axis. The virtual velocities along these small arcs are, hence, proportional to the radiuses. As in the case of the bar, if one assumes a set of lines, respectively, equal to the radiuses on the tangent to these arcs and in the same direction of the rotation, these lines can be assumed to represent the virtual velocities of the corresponding points (ibid., p. 188). Thence, the segment *mA*—belonging to *xOy*—of the tangent at the arc described by *m* and equal to the arc's radius represents the virtual velocity of *m*. Thus, if the perpendicular *AB* from *A* to *mP*, which is the direction of the force *F*, is drawn, the straight line segment *mB* will be the virtual velocity of *m* along the direction of *F*. Ergo, the product

$$P \cdot mB = mP \cdot mB$$

is the momentum (in the new meaning) of this force relative to the movement of the solid body's rotation around *Oz*.

Giorgini's intention was to prove that the momentum $mP \cdot mB$ coincides with the momentum of *P* with respect to *Oz*, that is with the already presented expression $mQ \cdot Oq$. He observed that the ancient definition of momentum is less general (ibid., p. 188). Giorgini's proof is purely geometrical: be *AD* the perpendicular from *A* to *mQ*. Since, by construction, the plane *AmQ* is perpendicular to the plane *PmQ*, then *AD* will be also perpendicular to the plane *PmQ*. By construction *AB* is perpendicular to mP belonging to xOy. Consequently the straight line DB will be also perpendicular to mP. Therefore, the triangle mBD, which is right in B, is similar to mQP, which is right in Q, so that the proportion

$$mQ : mP = mB : nD \text{ (or the equation } mQ \cdot mD = mP \cdot mB)$$

holds. Analogously the triangles ADm and mqO, which are right in D and q, respectively, are equal, as it is easy to see. Thence, the identity $mD = Oq$ holds, so that the last equation gets the form

$$mQ \cdot Oq = mP \cdot mB$$

Giorgini drew this conclusion:

> The observed analogy between the two species of momenta proves that, in the condition of equilibrium expounded in n. 58 [equilibrium around a fixed axis of any system of forces], the now introduced momenta can replace the first species of momenta which had been there considered. Consequently, the enunciated principle of virtual velocities (n. 212) can be applied to the equilibrium of a solid body which can move around an axis. [...Therefore...] When a free solid body is in equilibrium, the momentum of the system of forces applied to the body, which is relative to any infinitesimal movement of mere rotation around an axis, is always null.[29]

This proves that the PVV is applicable to the rotational equilibrium of a rigid body. Giorgini was then able to prove easily that it also holds for translational equilibrium (ibid., pp. 190–191).

Relying on these results and on the composition of movements, he proved that the principle also holds necessarily for any system of points or rigid bodies in equilibrium (ibid., pp. 191–210). After that the sufficient condition is proved (ibid., pp. 210–212). Finally, he pointed out that the PVV can be applied only for the movements which satisfy the constraints conditions (ibid., pp. 212–214).

Commentary: the interesting issue in the presentation of the PVV offered by Giorgini is his clear explanation that the concept of (virtual) work (the momentum of a force in the new meaning, according to Giorgini) is a more general notion than the concepts of momentum of a force with respect to a point and with respect to an axis. "More general" means in this case that the new concept allows a unified treatment of a class of problems without the need of searching the polar momentum or the axial momentum of a force by specific constructions. Thus, it is not necessary to analyse any single case in order to realize if, for the equilibrium, the polar or the axial

[29] Ibid., pp. 189–190: "L'osservata analogia tralle due specie di momenti dimostra, che nella condizione di equilibrio del n. 58 alla prima specie di momenti ivi considerati possono essere sostituiti i momenti adesso introdotti e che in conseguenza l'enunciato principio delle velocità virtuali (n.212) si applica all'equilibrio di un corpo solido libero di muoversi soltanto intorno a un asse. [...Quindi...] Quando un corpo solido perfettamente libero è in equilibrio, il momento del sistema di forze ad esso applicate, relativo ad un qualsivoglia movimento infinitesimo di semplice rotazione attorno ad un asse qualunque, è sempre nullo".

momenta are required. It is enough to consider the force as a single object and the projection of the virtual velocity along the direction of the force. No operation, no decomposition of the force is needed. To use a locution dear to Mach: Giorgini fully realized and explained the economic value of the PVV. Of course, he was not the first; every physicist had understood such an economic value; but Giorgini's treatment is particularly illuminating. It is appropriate to remark that, though the concepts of polar momentum, axial momentum, and virtual work are used in a correct manner, their profound difference (namely, the vector character of the polar momentum and the scalar character of the other two quantities) was not yet felt as a fundamental element connoting these mechanical quantities. Implicitly these distinctions were clear, but the explicit distinction between vector and scalar quantities was not yet at the theoretical basis of physics.

With regard to the infinitesimal movements, Giorgini's ideas are not particularly original: he accepted the concept of infinitesimal displacement with all the ambiguities involved in it. He did not feel the need to separate the dynamical context in which the concept of infinitesimal movement makes sense from a static context in which, though the notion of infinitesimal virtual displacement permits to draw correct previsions, its physical meaning is ambiguous. In the context of statics, the idea of an infinitesimal movement has a pragmatic value, but, as to its theoretical meaning, there are conspicuous doubts which, probably, Poinsot was one of the few scientists to understand.

5.1.8 Rodrigues

The conclusions of Rodrigues (1840) are dedicated to the PVV applied to the equilibrium of a rigid body. The style of Rodrigues' considerations differs from Poisson's and Giorgini's because they are not inserted within handbooks but in a paper dedicated to advanced research. In this respect the style and the aims of his speculations on the PVV are more similar to Chasles'. A new detailed proof of PVV is not presented, rather the author developed some considerations on the way to interpret such principle as the basic element of statics. In spite of this more abstract character of Rodrigues' presentation, his approach is not far from Giorgini's: 1) Rodrigues connected the PVV to the conditions of equilibrium one obtains by using the concept of momentum of a force with respect to an axis; 2) he based the whole of his treatment on the idea of virtual movement as a (potential) infinitesimal movement. Taking into account that his essay dates to 1840, this shows how persistent such idea of infinitesimal movement was. It will also subsist in the following years. Rodrigues also proposed an interesting distinction between geometry, mechanics and statics, which presents some features that also recur in Chasles.

Indeed, according to Rodrigues, geometry considers the finite or infinitesimal displacements of rigid bodies insofar as they are produced by the *successive* action of causes or forces. Mechanics considers the consecutive displacements of the bodies with regard to the *simultaneous and prolonged* action of causes or forces. Statics is the first part of mechanics where infinitesimal or *virtual* movements are considered resulting from *simultaneous*, but not *continuous* actions of such causes or forces (Rodrigues, 1840, p. 434).

Therefore, in Rodrigues' view, geometry poses the basis for the correct decomposition of forces considered separately. The term "*successive*" seems to me referred to the operational way in which you can decompose geometrically a movement or a force. For example, when you decompose a force in two directions, it is indifferent whether you consider first the component in one direction or in the other direction. In a sense, Rodrigues is claiming the abstract and time-independent character of the geometrical operations with what nowadays we call vector magnitudes. These operations take place in a finite time, but the order of operations is indifferent. To be even more explicit: our gnoseological operation of decomposing a force takes place in a finite time, but it is the photograph of an abstract physical situation which is independent of time. Instead, mechanics considers the forces in their concrete action, when the interval of time tends to 0 and a continuous movement exists. Statics considers all the forces acting on a body in a certain instant, when the action of these forces does not produce any movement.

Rodrigues claimed that the forces applied to a solid body can act in two manners: 1) either they tend to induce a rotation around an axis 2) or they tend to move a point of the system so changing its coordinates.

With regard to case 1), consider the step from geometry to mechanics and analyse the situation when the time interval tends to 0. If the letters $\theta, \theta^{'}, \theta^{''}, \ldots$ denote the elementary or virtual rotations of a solid around certain axes, there is equilibrium if and only if the sum of the momenta of all the virtual rotations around any axis is null. Thence, if D indicates the distance between the axis and the instantaneous tangent to the rotation, while ν is the angle between such a tangent and the axis, the equilibrium condition is given by (ibid., p. 435)

$$\sum \theta D \sin \nu = 0 \tag{5.5}$$

This equation depends on the equilibrium of several successive or simultaneous infinitesimal displacements already analysed in Rodrigues (1840), pp. 419–421, where the author had proved that Eq. (5.5) includes, for infinitesimal movements, the six equations of equilibrium, three for the rotational equilibrium, and three for the translational equilibrium. This granted, it is easy for Rodrigues to prove that case 2) coincides, as a matter of fact, with case 1).

Therefore, considering n virtual displacements, whose rotations $\theta, \theta^{'}, \theta^{''}, \ldots$ are referred to a series of distances $D, D^{'}, D^{''}, \ldots$, being $\nu, \nu^{'}, \nu^{''}, \ldots$ the angles between the instantaneous tangent to the virtual rotation and the respective axes, Eq. (5.5) represents "[...] the algorithmic translation [traduction] of the equilibrium of a forces' system able to produce virtual or infinitely small translations [translations] proportional to the rotations $\theta, \theta^{'}, \theta^{''}, \ldots$. These forces are applied on the rotation axes positively or negatively according to the rotation's sign".[30] In sum: Eq. (5.5) expresses completely the equilibrium condition of a rigid body.

The reasoning by Rodrigues continued in the following manner: he imagined to replace the forces which produce virtual translations proportional to the rotations by finite magnitudes P, P', P'', ... proportional to, respectively, the impressed virtual translation. Thence, Eq. (5.5) is transformed into

$$\sum PD \sin \nu = 0.$$

He clarified that the term $PD \sin \nu$ expresses the static momentum of the force P with respect to the rotation's fixed axis ("[...] le *moment* statique de la force P, relative a l'axe fixe [...], ibid., p. 437). This is exactly our definition of the momentum of a force with respect to an axis.

This established, if the force P tends to produce the variations δp of the coordinate p, it will be $P\delta p = PD\theta \sin \nu$ and, hence, $\delta p = D\theta \sin \nu$.

Rodrigues added a brief physical explanation to this algebraic result: since this infinitesimal displacement—according to what Chasles and he himself had proved—is a rotation around an axis or two successive rotations around two axes, *virtually*, it can be considered either in its first rotation or in its unique rotation. Then, the infinitesimal arc described by the application point of the force P projected along the direction of P is equal to the variation of the coordinate p along which the force acts, which proves exactly that

$$\delta p = D\theta \sin \nu \text{ and } P\delta p = PD\theta \sin \nu.$$

Thus, the equilibrium equation

$$\sum PD \sin \nu = 0$$

becomes

$$\sum P\delta p = 0,$$

which is the PVV.

Before adding some interesting considerations on the nature of the constraints and on the application of the PVV to the mechanics of continua (ibid., pp. 439–440),

[30] Rodrigues (1840), pp. 436–437: "[...] la traduction algorithmique de l'équilibre d'un système de forces capables de produire des translations virtuelles ou infiniment petites proportionnelles aux rotations θ, θ', θ'',..., ces forces étant appliquées sur les axes de ces rotations, *positivement* ou *négativement*, selon le signe de ces rotations " Italics in the text. In modern notation Eq. (5.5) is written as $\sum_{i=1}^{n} \theta_i \cdot D_i \sin \nu_i = 0$.

which I will not address because they are not connected to Chasles' work, Rodrigues concluded the general part of his treatment, using these words:

> [The equation of virtual velocities] expresses that, given some forces in equilibrium in a solid system, if for any cause, this system is subject to an infinitesimal movement, the sum of the forces multiplied by the infinitely little spaces traversed by the points of this system, along the direction of the respective forces, has to be null and reciprocally. This is the enunciation of the principle of virtual velocities.
>
> This equation, algorithmically speaking, is far superior to the first one, though basically, it is not more general. However, it expresses as simply as possible the equilibrium law of all the systems in which the constraints conditions can be transcribed into linear equations among the variations of the coordinates of the different system's points.[31]

Rodrigues's treatment is interesting insofar as it connects his studies on the infinitesimal displacements of a rigid body to the PVV. In his work there is a clear link between the two questions: they are not treated separately, which is the case in most of the previous authors. This tract was also typical of Chasles' approach, who, in his *Rapport*, fully recognized the merit of Rodrigues because, as he wrote:

> Afterwards he [Rodrigues] analyses the case of infinitely small displacement and gives the analytical conditions for the equilibrium of successive infinitely small displacements. He compares these conditions to those connoting the equilibrium of a system of forces. This part of his Memoire seems to have been the principal purpose of Rodrigues, who had the aim to remark the separation between Geometry and Mechanics.[32]

On the other hand, Rodrigues was deeply involved in the concept of equilibrium as a condition physically identical to that of infinitesimal movement, with the several ambiguities this implies. Furthermore, let us analyse his distinction of the forces' action in two cases: 1) that in which the forces tend to produce a rotation of the solid body around an axis; 2) that in which they act simultaneously on the solid body tending, separately, to modify the coordinates of the system's diverse points. It seems to me that this distinction is superfluous exactly because of the unifying treatment of the solid body's equilibrium that he himself had offered (ibid., pp. 419–421).

---

[31]Rodrigues (1840), pp. 438–439: "[...] qui exprime que, des forces étant en équilibre dans un système solide, si, par une cause quelconque, ce système vient à être dérangé infiniment peu da sa situation actuelle, la somme des forces multipliées par les espaces infiniment petits parcourus par les points de ce système dans la direction de ces forces respectives devra être nulle, et réciproquement, ce qui est l'énoncé du principe des vitesses virtuelles.

Cette équation bien supérieure, algorithmiquement parlant, à la première, n'a pas au fond plus de généralité: mais elle exprime le plus simplement possible la loi de l'équilibre de tous les systèmes dans lesquels les conditions de liaison sont susceptible d'être traduites par des équations linéaires entre les variations des coordonnées des divers points du système".

[32]Chasles (1870), p. 159: "Puis il examine le cas du déplacement infiniment petit, et donne les conditions analytiques d'équilibre de plusieurs déplacements successifs infiniment petits, qu'il compare ensuite aux conditions d'équilibre d'un système de forces. Cette partie du Mémoire parait avoir été l'objet principal de Rodrigues, qui s'y propose de marquer le point qui sépare la Géométrie de la Mécanique".

5.2 Chasles' Approach to the Principle of Virtual Velocities

The PVV was addressed by Chasles at least on four occasions. In this section I will analyse two of them because in the *Aperçu historique* the reference to the PVV is within the context of his philosophical considerations on duality. Hence, this reference will be examined in the next chapter. Furthermore the reference in Chasles (1847), p. 224 is really only a brief note which adds nothing to the picture I am drawing.

The general features of Chasles' approach can be summarized like this: the PVV is a part of his foundational programme because he deduced this principle from his results concerning the infinitesimal movement of a rigid body and the systems of forces. The PVV is, thus, a further region of rational mechanics which can be reduced to a purely geometrical treatment. The idea to find the PVV on geometry is a common treat of Chasles' and Poinsot's approaches. However, there are also conspicuous differences between the two: 1) Chasles did not refuse the concept of infinitesimal movement. In contrast to this, such a concept was one of the bases of his approach. In this sense, his point of view is more traditional than Poinsot's; 2) on the other hand, the reduction of the principle to geometry is carried out more deeply in Chasles than in Poinsot. For the former, statics is exactly a section of geometry. In part, Chasles' approach resembles Rodrigues' because, as we have seen, Rodrigues also addressed the PVV relying upon his and Chasles' results on the infinitesimal movement of a rigid body. Finally, it is necessary to stress that Chasles' considerations on the PVV concern the case of a rigid body.

One finds the first reference to the PVV in Chasles (1830c), namely in the paper where the author offered his first proof of the famous theorem concerning the infinitesimal movement of a rigid body. The consideration on the PVV is extremely laconic and not very significant. For Chasles simply claimed that, since the infinitesimal movement of a rigid body is reducible to the movement of a screw in its nut and since for the screw the PVV is easily provable, then his theorem offers a rigorous demonstration of the PVV. He wrote:

> [...] when an infinitely small movement is impressed to a rigid body, in this body a straight line always exists which glides on itself while the entire body rotates around this line. Therefore, the movement of a body is nothing but that of a screw in its nut. From this, the principle of virtual velocities relative to a solid body solicited by any forces is deduced in the most rigorous manner. Indeed, since any virtual movement of this body is nothing but that which a screw can get in its nut, it is enough to prove this principle relatively to the screw. This presents no difficulty.[33]

[33]Chasles (1830i), p. 324: "[...] quand on imprime à un corps solide libre un mouvement quelconque infiniment petit, il existe toujours dans ce corps une certaine droite qui glisse sur elle-même pendant que les corps tourne autour cette droite; de sorte que le mouvement du corps n'est autre que celui d'une vis dans son écrou. De là, on conclut, de la manière la plus rigoureuse, le principe des vitesses virtuelles, relativement à un corps libre, sollicité par des forces quelconques, puisque tout mouvement virtuel du corps n'étant autre que celui qu'une vis peut prendre dans son écrou, il suffit de démontrer ce principe relativement à la vis; ce qui n'offre aucune difficulté".

Commentary: the identification between a virtual displacement and an infinitesimal displacement is conspicuous. Chasles is evidently considering a virtual displacement as a potential infinitesimal displacement which, from a mathematical point of view, is subject to the same treatment as that connoting the infinitesimal movement as a nascent movement, as the initial instantaneous phase of an actual movement. Today we call it "act of motion". He had proved his theorem for the infinitesimal movement, therefore—he thought—it can also be applied to the virtual movement. Since Chasles' theorem relies on a purely geometrical proof, then the principle of virtual velocity—at least insofar as a rigid body is concerned—which is founded on the concept of infinitesimal movement and which is the basis of statics, can also be reduced to geometry. Chasles did not pose the physical problem of distinguishing the virtual movement from the infinitesimal movement as a nascent motion. Thus, all the features ascribed by Chasles to the PVV are already present in this brief quotation dated to 1830. These ideas would assume a complete and clear form in his following papers.

The most important contributions given by Chasles to the comprehension of the PVV is Chasles (1843). The approach he adopted in this paper is paradigmatic of the way he conceived the PVV as well as the relations between statics and geometry. At the basis of his considerations is the analogy he posed between the rotations of a body around different axes and the system of forces. *This means the analogy between the theory of the infinitesimal displacement of a rigid body and the treatment of the system of forces.* As a matter of fact, it is appropriate to highlight that in 1843, as we have seen, Chasles had already developed several elements of his foundational programme for mechanics. New important insights would be added in Chasles (1847) with regard to the system of forces and in Chasles (1860–1861) with regard to the finite and infinitesimal displacement of a rigid body.

Chasles' considerations begin like this:

> When a body experiences an infinitely small displacement, which results from several simultaneous rotations around more fixed axes, if, on such axes, straight lines are drawn respectively proportional to these rotations and these lines are considered just as many forces which would solicit the body, the rectilinear element described by each point of the body under the action of these simultaneous rotations will be proportional to the principal momentum of the forces relative to this point.[34]

He replaced the infinitesimal rotations which generate the infinitesimal movement with segments belonging to the rotational axes and proportional to the rotations themselves. This granted, he considered these lines as forces acting on the body. The rectilinear element described by any point P of the body under the action of such forces, he claimed, is proportional to the principal momentum of the forces relative to this point, that is it is proportional to the momentum of such forces calculated with

[34]Chasles (1843), pp. 1429–1430: "Quand un corps éprouve un déplacement infiniment petit, résultant de plusieurs rotations simultanées autour de plusieurs axes, si l'on porte sur ces axes des lignes proportionnelles à ces rotations respectivement, et qu'on considère ces lignes comme autant de forces qui solliciteraient le corps, l'élément rectiligne décrit par chaque point du corps, en vertu de ce système de rotations simultanées, sera proportionnel au moment principal relatif à ce point".

respect to the axis passing through P for which the momentum has a maximum. Therefore, he continued, the properties concerning the rotations of a body around several axes and the properties of the rectilinear infinitesimal spaces described in consequence of such rotations by the body's points can be referred to the properties of a system of forces and to their momenta with respect to the points of the body. But, as highlighted in chapter 3, the displacement of a rigid body can be reduced to the rotation around two conjugate axes. Then, all the properties relative to two conjugated axes of rotations D and Δ are applicable to the systems of two forces which can replace a system of an arbitrary number of given forces (ibid., p. 1430). Furthermore, as in the theory of the rigid body's movement there is the axis of rotation X, in this case there is an axis with respect to which all the momenta are considered. Chasles stressed that this axis is exactly that which the illustrious author of the theory of couples (Poinsot) called the central axis of momenta ("l'axe central des moments", ibid., p. 1430). In this way a perfect dual correspondence between the theory of infinitesimal displacement of a rigid body and the system of forces is posed.

This clarified, Chasles began his analysis of the PVV. We read:

> The analogy between a system of forces which solicit a free solid body and the rotations which produce an infinitely small displacement of the body leads naturally to a demonstration of the principle of virtual velocities which shows how much the considerations of the movement and of the infinite in this principle correspond to purely static considerations.[35]

Commentary: as pointed out in the introduction to this section, according to Chasles the PVV is one of the stitches which show the inner connection between the theory of the rigid body's infinitesimal movement—in particular a rotational movement—and the theory of the system of forces. Both theories are, according to him, reducible to a geometric basis. The two theories allow us to prove the PVV, which is the fundamental proposition of the entire statics. Hence, statics, too, has a geometrical foundation. Chasles' assertion, according to which the movement and the infinite in PVV correspond to purely static considerations is interesting because it shows that he had the intention to distinguish clearly the domain of statics from that of dynamics. At all appearances, statics is considered by him as an appendix of geometry, while dynamics—where the action of the forces is actualized and the movement is real—implies the dimension of time which cannot be included within geometry. This is the reason why Chasles cared about the distinction statics-dynamics.

To prove the PVV for a rigid body, Chasles considered the forces $P, P', P'', \ldots$ which solicit a free rigid body and which are in equilibrium, while $Q, Q', Q'', \ldots$ are other any forces. He took into account all the tetrahedra whose opposite edges are,

[35] Chasles (1843), p. 1430: "L'analogie qui a lieu entre un système de forces sollicitant un corps solide libre et les rotations qui produisent un déplacement infiniment petit du corps, conduit naturellement à une démonstration du principe des vitesses virtuelles qui montre comment la considération du mouvement et de l'infini dans ce principe correspond à des considérations purement statiques".

respectively, the forces $P^{(i)}$, $Q^{(i)}$ and indicated the sum of their volumes with $\sum$ tétr(P, Q). This sum will hold the same value if each of the two systems of forces $P, P', P'', \ldots$ and $Q, Q', Q'', \ldots$ is, respectively, replaced by an equivalent system. The sum of tetrahedra's volumes is null because the system $P, P', P'', \ldots$ can be replaced by two equal and opposite forces F, F' as the forces $P, P', P'', \ldots$ are in equilibrium. Therefore, the sign of the volumes of the tetrahedra whose edge is the force F is opposite to the sign of the volumes of the tetrahedra whose edge is the force F'. This means that the sum of the volumes of all the tetrahedra is null. Reciprocally, if the sum is null, for any given system of forces $Q, Q', Q'', \ldots$, the forces $P, P', P'', \ldots$ are in equilibrium. Thus, Chasles summarized, the equilibrium condition of the forces $P, P', P'', \ldots$ is expressed by

$$\sum \text{tétr}(P, Q) = 0.$$

Since the volume of a tetrahedron can be obtained as the product of two opposite edges by their distance r by the sinus of their angle $(\widehat{P, Q})$ divided by 6, the previous expression gets the form

$$\sum P \cdot Qr \sin(\widehat{P, Q}) = 0.$$

Let us consider the force P. Chasles supposed that all the forces $Q, Q', Q'', \ldots$ are replaced by two forces, one of which is directed along P while the other one is a certain force q (ibid., p. 1431). The sum of the tetrahedra, one of whose edges is P, is then simply tétr(P, q), which can also be expressed in the form $P \cdot qr \sin(\widehat{P, q})$. The expression $qr \sin(\widehat{P, q})$, Chasles explained, is the projection on a plane perpendicular to the force P of the momentum of the force q with respect to a point of P. In modern terms: it is the momentum of the force q with respect to the axis P. From this moment Chasles' demonstration becomes extremely brachylogical. Let us follow it. He claimed:

> Thence, if you suppose that the forces $Q, Q', \ldots$ are in direction the rotation axes proportional to such forces, the momentum relative to a point of the force P will be equal to the rectilinear element which these rotations will make to describe to this point.[36]

Explanation: Chasles' approach resembles Rodrigues'. There are, however, significant differences. In particular, Rodrigues did not introduce the system of forces $Q, Q', \ldots$, but worked only with the given forces $P, P', P'', \ldots$. Nonetheless, the idea of proving the PVV starting from the momentum of the forces $Q, Q', \ldots$ with respect to the force P is close to Rodrigues' way of reasoning clarified in the previous section since in both cases the essential element is the concept of the momentum of a force with respect to an axis. The idea of describing the (virtual) rotational motion through

[36]Chasles (1843), p. 1431: "Si donc on suppose que le forces $Q, Q', \ldots$ soient en direction les axes de rotations proportionnelles à ces forces, le moment relatif à un point de la force P sera égale à l'élément rectiligne que ces rotations feront décrire à ce point".

a rectilinear element is also common to the two scholars. However, Rodrigues developed his argument by the explicit introduction of the concept of variation δp. Instead, Chasles worked at a more basic and elementary level, analysing the transcription of the general formula for equilibrium (granted the reduction of the forces $Q, Q^{'}, \ldots$ to two forces according to Chasles' procedure) $\sum P \cdot qr \sin\left(\widehat{P, q}\right) = 0$ when the forces $Q, Q^{'}, \ldots$ are considered the axes of rotation for the infinitesimal movement of the rigid body. This induced him to consider the momentum of the force q with respect to P and not, as in Rodrigues, the momentum of the force P with respect to an axis D. This operation allowed Chasles to develop a geometrical proof based on, so to say, concrete elements without the more abstract concept of variation.

First of all: Chasles supposed that the forces $Q, Q^{'}, \ldots$ were the axes of rotation of the infinitesimal movement of the rigid body. This is legitimate because the forces $Q, Q^{'}, \ldots$ are absolutely arbitrary. The term "proportionelles" is referred to the length of the segment to consider on the straight line indicating the direction of a force $Q^{(i)}$ and proportional to the intensity of the force itself.

Second step: Chasles regarded the forces $Q, Q^{'}, \ldots$ as axes of rotation. All their effects on the points belonging to the straight line of the forces P can be reduced to the effect of the force q. Therefore, the rotational movement—be it infinitesimal or actual—of the rigid body takes place perpendicularly to q. Since the arm r of the momentum of the force q with respect to P is the common perpendicular to q and P, the element p coincides exactly with the value qr. Actually, Chasles spoke of the momentum of the force q with respect to a point of P. He is referring to the application point of P. It seems to me necessary to add that Chasles' reasoning works because of the theorem according to which the axial momentum of the force q with respect to the force P does not change whatever the considered point of P be. Chasles is thus referring to the momentum of the force q with respect to the axis coinciding with the P-direction.

He continues in the following manner:

> Be p such a rectilinear element; the sum of the tetrahedra where the force P enters [namely: of which the force P is an edge] will be equal to $P.p\cos\left(\widehat{P, p}\right)$. For each of the other forces P $'$, etc. you will have a similar sum, so that the equilibrium equation will become
>
> $$\sum P.p\cos\left(\widehat{P, p}\right) = 0.$$
>
> which is the equation of virtual velocities.[37]

[37] Chasles (1843), p. 1431: "Soit p ces élément rectiligne; la somme des tétraèdres où entre la force P sera donc égale à $P.p\cos\left(\widehat{P, p}\right)$". Pour chacune des autres forces P', etc. on aura une somme semblable; de sorte que l'équation d'équilibre deviendra

$$\sum P.p\cos\left(\widehat{P, p}\right) = 0.$$

C'est l'équation des vitesses virtuelles".

Explanation: since p is perpendicular to q, it follows that $\sin\left(\widehat{P,q}\right) = \cos\left(\widehat{P,p}\right)$. The rest of Chasles' explanation is clear.

Chasles concluded:

> Therefore, in this principle of the virtual velocities, the rectilinear elements which are called *virtual velocities* express the principal momenta of another system of forces in reference to the application points of the proposed forces.[38]

Commentary: this consideration makes it explicit the link between the PVV, the geometrical movement of a rigid body, and the theory of the system of forces. For as we have seen at the beginning of this Sect. 5.2, suppose that a body experiences an infinitely small displacement derived from simultaneous rotations around more axes. If, on such axes, some straight lines—regarded as forces—are drawn proportional to the rotations, the rectilinear element described by each point of the body under the action of these simultaneous rotations is proportional to the principal momentum of the forces relative to this point. This is exactly the situation of the PVV. Therefore, the element p, which is the virtual velocity of the application point of the force P, represents the principal momentum of the forces Q, $Q^{'}$, $Q^{''}$, ... with respect to such application point. Obviously, things work in the same manner for all the forces of the system P, $P^{'}$, $P^{''}$, ... Hence, given that the concepts concerning the system of equivalent forces are formally analogous to those connoting the geometrical movement of a rigid body, which is based only on geometrical considerations, this implies that the PVV is also susceptible of a geometrical foundation within Chasles' foundational programme.

Chasles' research on the equilibrium principle for a rigid body continued offering a different formulation of the PVV. He explained that, given a system of forces P, $P^{'}$, $P^{''}$, ...in equilibrium applied to a rigid body which experiences an infinitesimal (virtual) movement, this body experiences a rotation around to each of the forces. Considered, i.e., the force P, this rotation is inversely as the projection of the trajectory of a point belonging to the direction of P.

I add a simple example: if P were the principal axis of rotation, the infinitesimal movement of a point of P would be a mere translation along the straight line P corresponding to itself in the movement. Therefore, if θ, $\theta^{'}$, $\theta^{''}$, ... indicate the rotations around, respectively, P, $P^{'}$, $P^{''}$, ..., the equation of virtual velocities will get the form $\sum \frac{P}{\theta} = 0$, so that Chasles concluded:

> When more forces applied to a free solid body are in equilibrium, if an infinitely small movement is given to the body, through which it experiences a rotation around each force, the sum of these forces divided respectively by these rotations, is null. Reciprocally, if this sum is null for any infinitesimal movement, the forces will be in equilibrium.

---

[38] Chasles (1843), p. 1431: "Ainsi, dans ce principe des vitesses virtuelles, les éléments rectilignes qu'on appelle les *vitesses virtuelles* expriment les moments principaux d'un autre système des forces par rapport aux points d'application des forces proposées". Italics in the text.

> In this manner, the equilibrium of a system of forces is expressible considering the body's rotations around these forces, analogously as considering the rectilinear elements described by the points of these forces.[39]

Chasles expressed the PVV both in the classical translational form and in a rotational form. In both cases the reference to the geometrical treatment and concepts is explicit so that this principle fits into his geometrical foundational programme.

5.3 Conclusion

In comparison with Poinsot, Chasles accepted the concept of infinitesimal movement as a fictitious, virtual movement, by means of which the conditions of equilibrium can be studied. His resort to the geometrical motion and to the properties of the system of forces—which are formally analogous to those of the geometrical motion—is developed more profoundly than it is in Poinsot. In this respect, Chasles had a more purist attitude than Poinsot because his reduction of statics to geometry is pushed to the extreme limit. Whereas it can be said that Poinsot's attitude is more purist with regard to the problem of the infinitesimal motion as a virtual motion, a conception which he refused.

The considerations of the momenta with respect to an axis of rotation as well as part of the geometrical treatment represent a view shared by both Giorgini and Chasles.

Finally, the introduction of the geometrical motion is a feature which Chasles' treatment shares with Rodrigues'.

However, in no one of the other authors, who highlighted the geometrical aspects of the PVV, both in the meaning of this principle and in its demonstration, the foundational aspect connected to geometry is so pronounced as it is in Chasles. Poisson's treatment is, so to say, mixed because both geometrical and analytical aspects subsist in his explanation and proof of the PVV.

The inclusion of the PVV within his foundational programme is a further confirmation of how broad Chasles' view was. It was so wide that philosophy also belongs to it, as the next chapter will clarify.

---

[39] Chasles (1843), pp. 1431–1432: "Quand plusieurs forces qui sont appliquées à un corps solide libre se font équilibre, si l'on donne au corps un mouvement infiniment petit, par suite duquel il éprouvera une rotation autour de chacune des forces, la somme de ces forces divisées par ces rotations, respectivement, est nulle; et réciproquement, si cette somme est nulle quel que soit les mouvement infiniment petit du corps, les forces se feront équilibre.

Ainsi l'équilibre d'un système de forces s'exprime par la considération des rotations du corps autour de ces forces, de même que par le considération des éléments rectilignes décrits par des points de ces forces".

Chapter 6
Chasles' Philosophy of Duality

Abstract In this chapter, I present the thesis that Chasles' foundational programme is not restricted to geometry and physics, but it also concerns philosophy in its epistemological and ontological aspects. Dualisms became the creed of Chasles' philosophy. The basis of this idea derived from the principle of duality, born in projective geometry and extended by Chasles to several aspects of mechanics and philosophy. The chapter is divided into five sections. In the first one, I offer a reading of how Chasles interpreted duality within projective geometry. In this context, the principle of continuity, which Chasles reformulated as principle of contingent relations, plays a prominent role. In the second section, the decisive role of duality in geometry and physics is discussed. The third section regards the extension of this principle to philosophy and the interpretations offered by Chasles in this regard. As Steiner was something like an alter ego for Chasles with regard to projective geometry, Poinsot was an analogous figure with respect to physics and to the philosophy of duality. Therefore, in the fourth section a comparison between Poinsot's and Chasles' ideas on duality is developed. Final considerations on Chasles' concept of duality close the chapter.

Chasles developed the idea that duality is the fundamental metalaw of the entire universe. I refer to it as to a metalaw and not to a law because duality does not prescribe that a single determined phenomenon or state of things will develop in a certain manner, but that if a certain phenomenon evolves in a certain manner than another phenomenon exists which is the dual of the former and which, changed the contingent situations connoting the two phenomena, evolves as the former from a structural point of view. The two dual phenomena have the same structure. Hence, known one of them, the other is also known. Furthermore, duality does not concern a single region of the being, but all of it; be it an ideal being (as in mathematics) or a real being (as in physics). As soon as we are able to look at the phenomena in their general terms, we discover the existence of duality, which is, hence, even more general than other metalaws as the cause–effect principle, which only concerns the physical universe. Moreover, duality is not merely a gnoseological principle, as, e.g., the principle of sufficient reason. It is an ontological meta-principle because it claims that, given a phenomenon, the dual one exists. If we compare duality with the

P. Bussotti, *Chasles and the Projective Geometry*,
https://doi.org/10.1007/978-3-031-54266-4_6

principle of sufficient reason, we discover that there are other differences besides the fact that the former is more broadly applicable and indicates an ontological property of the universe. As a matter of fact, the latter is a negative principle since it claims only indirectly the existence of a phenomenon insofar as it establishes the opposite one to be impossible because some conditions of symmetry would be violated, while duality positively determines the existence of a state of things A because the dual state A' exists. A and A' are not causally connected, but structurally. Thus, the philosophy of duality is a real view on the world, a scientific *Weltanschauung*.

References to philosophy of duality exist in several of Chasles' works, but the main text is the *Aperçu*, where he dedicated a long note to this question (Note XXXIV, Chasles, 1837a, pp. 408–416) and where many interesting considerations also exist along with the whole essay. Chasles was perfectly aware that his philosophical proposal was still in an initial, if not embryonic, phase. However, he was able to support his ideas with some noteworthy examples and to show the contexts which should confirm the existence of the universal duality metalaw. From the genetic standpoint, the duality law of projective geometry was the source of inspiration which induced Chasles to think duality to be a universal property of phenomena. It was not restricted to projective geometry, whose law of duality appeared, thence, as a particular declination of the general metalaw. Therefore, this confirms that projective geometry was the real root of the whole of Chasles' speculation, also with regard to those parts of his work which went beyond projective geometry itself. Thus, with regard to Chasles' philosophy of duality, it seems to me appropriate to consider four subjects: 1) the principle of the contingent relations in reference to duality. This is a very important principle, which Poncelet called principle of continuity and which he used extensively in his works. It is strictly connected with the introduction and use of the imaginary in geometry; 2) the principle of duality as a common root of the subjects I have dealt with in this book, namely, projective geometry, the theory of the rigid body movement, the theory of the systems of equivalent forces; 3) the extension of duality beyond geometry and beyond subjects which are directly connected to geometry. In particular: the examination of phenomena which might justify the idea of a general duality metalaw; 4) a comparison with Poinsot's view of duality.

To each of these subjects a subsection will be dedicated.

6.1 The Principle of Contingent Relations and Duality

Chasles' ideas on the principle of contingent relations (in what follows PCR) are significant because they are connected with two topics: the former is a methodological question and concerns the relations between analytic and synthetic or pure or rational geometry, the second problem regards the legitimacy of the PCR as a demonstrative means after Cauchy's criticisms. For Cauchy (1820–1821) ascribes a heuristic value to Poncelet's principle of continuity, but not a demonstrative one. Cauchy, jointly with Arago and Poisson, was asked by the Académie to review

(Poncelet, 1820). He wrote a report in which many aspects of Poncelet' work are positively valued, but the principle of continuity is criticized as a means of proof.

We will see that duality has a deep link with Chasles' examination of PCR. His analysis begins with a valuation of Monge's descriptive geometry, which offered a systematic means to transform the properties of a spatial figure in those of a flat figure. Before Monge all the operations involving a plane projection of a three-dimensional figure as the stone cutting, the perspective, the theory of fortifications, the gnomonics were developed without any precise and rigorous criteria. With his descriptive geometry Monge solved most of these problems, but the importance of Monge went far beyond his results: he restored dignity to rational geometry because descriptive geometry is the graphical translation of properties concerning pure geometry. Thence, Monge's results were a fundamental stimulus to begin the construction of the rational doctrine on which—from a conceptual standpoint—descriptive geometry is based, namely projective geometry. Monge's research was a fundamental step for pure geometry to regain its importance within the context of mathematical and physical sciences. Chasles clearly stated:

> In this way, Geometry can regain more easily its generality and intuitive evidence on mechanics and on physical sciences.[1]

Chasles also stressed that Monge's epigones were able to prove several and new propositions of plane geometry through his technique. As underlined in the Introduction, this consists in transforming the three-dimensional figures into flat figures by projecting them onto two orthogonal planes which, after the projections, are supposed to rotate one on the other. The set of these two projections can be used to prove plane theorems (ibid., pp. 191–193). Chasles called this technique *Transmutation of the figures* ("Transmutation des figures", ibid., p. 194) and offered several examples in which it is usable. It seems to me appropriate to refer at last to the first example presented by Chasles: suppose you have to find the intersection point of three planes. This will lie at the intersections of the three straight lines along which the three planes cut two by two. The projections of these three straight lines on one of the two projection planes pass through the same point. Thence, the following theorem of plane geometry can be deduced: in a plane, be given two triangles whose sides concur two by two in three points belonging to a straight line L. Through any point of the plane, draw three straight lines to the vertices of the first triangle. Produce them until cutting L in three points. If these points are joined, respectively, to the three vertices of the second triangle through three straight lines, these lines will concur in the same point (ibid. p. 192, Fig. 6.1).

As it is evident, a corollary of this proposition is Desargues' theorem.

Other geometrical techniques of transforming the figures exist: Chasles mentioned stereographic projection and perspective. However, he wondered: do principles exist which can reduce the different techniques of geometrical transformation to

[1] Chasles (1837a), p. 190: "La Géométrie devint ainsi un état de repandre plus aisément sa généralité et son evidence intuitive sur la méchanique et sur les sciences physico-mathématiques".

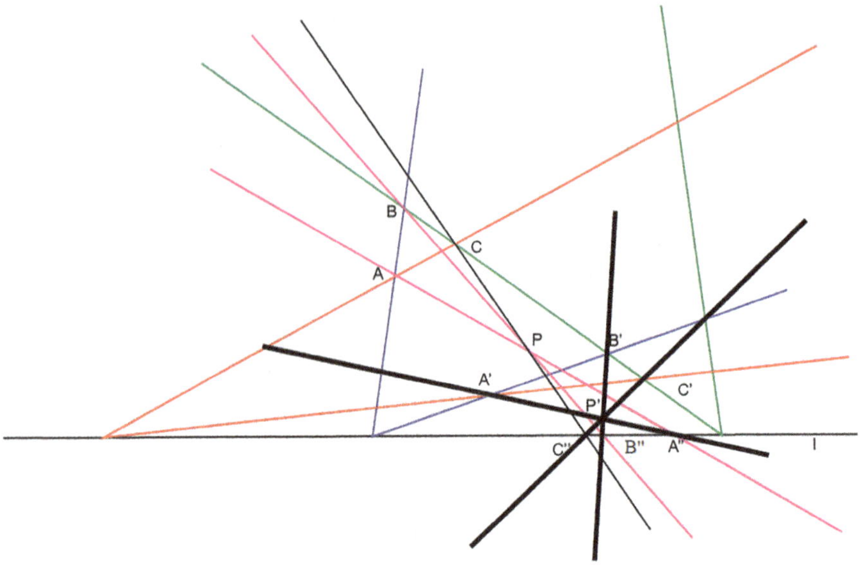

Fig. 6.1 Reconstruction of the figure described by Chasles: consider the triangles *ABC* and *A'B'C'*. The three couples of homologous sides be (*AB*, *A′B′*); (*AC*, *A′C′*); (*BC*, *B′C′*). They meet on the straight line *l*. Consider the point *P* and draw *PA*, *PB*, *PC*. They cut *l* in *A″*, *B″*, *C″* resp. Join *A″* with *A′*; *B″* with *B′*; *C″* with *C′*. The straight lines *A′A″*, *B′B″*, *C′C″* mutually cut in the point *P′*, as Chasles' theorem states

a common root? If this were the case, pure geometry could rely on few principles and obtain the same generality and perspicuity as analytical geometry. Furthermore, it could solve a series of problems which cannot be transformed in the ordinary system of Cartesian coordinates. The answer is affirmative: these principles are duality and homography. Chasles claimed:

> As we will show, several methods which are reducible to two general principles of extension (that is *duality* and *homography* of the figures) that we will describe in this essay, are the methods of transformations we are looking for.
>
> This kind of methods, whose utility seems to us well established, deserves to be cultivated. If we are not wrong, the geometers who are going to meditate on this subject will appreciate even more the philosophical importance of the method of transmutation we have tried to draw from Monge's descriptive geometry.[2]

[2]Chasles (1837a), p. 196: "Diverses méthodes qui se rattachent, comme nous le ferons voir, aux deux principes généraux de l'étendu, la *dualité* et l'*homographie* des figures, que nous démontrons dans ces écrit, sont de telles méthodes de transformation.

Ces sortes de méthodes, dont l'utilité nous parait bien constatée, méritent d'être cultivées; et, si nous ne nous abusons, les géometres qui voudront méditer sur cet objet apprécieront davantage l'importance philosophique de la méthode de transmutation que nous avons cherché à faire ressortir des principes de la Géométrie descriptive de Monge". Italics in the text.

Thus, Chasles considered duality and homography as the two philosophical methods of geometry. This means that they are the two methods which can give pure geometry all the necessary generality and perspicuity.

But there is something more: besides transmutation there is a further method in Monge's geometry, which has a more subtle connection with duality and that extends the concept of duality beyond the technical meaning in which this term has been used in projective geometry. Chasles described such a method in the following terms:

> It consists of this: let us consider a figure of which some general properties have to be proved in general circumstances of construction, where the presence of certain points, planes or lines, which, in other circumstances are imaginary, facilitate the proof. Apply the so proved theorem to the case of the configuration in which these points, planes and lines would be imaginary, namely consider it to be true in all the general circumstances to which it is referred.[3]

Chasles explained how this principle worked by means of a simple example drawn from a theorem proved by Monge: if some cones circumscribed to a second-degree surface have their vertices on a straight line *l*, then the planes of their contact curves with the surface belong to a straight line *r*. To demonstrate this proposition Monge drew two tangent planes from *l* to the surface. The contact curves of the cones will pass through the two contact points of the tangent planes. All the planes of such contact curve will pass, hence, through *r*. Therefore, the theorem is proved if it is possible to draw two tangent planes from *l* to the surface. However, Monge claimed that the theorem is also true even if the two tangent planes through *l* cannot be drawn, i.e., it is true for any position of *l* (Chasles, 1837a, p. 198).

Chasles expounded this principle claiming that it is based on an *invariant*, which depends on the constructive conditions. If these do not change, the principle can also be applied when some elements of the construction become imaginary. Given the invariance of the constructive conditions, it can happen that *contingently* some elements are real or imaginary, as the tangent planes drawn from *l* in the previous example. This is why Chasles named this method the *principle of contingent relations*. It is the same principle that Poncelet called principle of continuity. Chasles was aware of how problematic it was to justify a priori the legitimacy of this principle as a demonstrative means. I will explore his argumentations. However, it is first of all appropriate to highlight the relations he saw between PCR and duality. Chasles wrote:

> In the future it might be possible to found the principle of contingent relations on some metaphysical principle of the figurative extension connected to the idea of homogeneity.

[3]Chasles (1837a), pp. 197–198: "qu'elle consiste à considerer la figure, sur laquelle on a à démontrer quelque propriété générale, dans des circumstances de construction générale, où la présence de certains points, de certains plans ou de certaines lignes, qui dans d'autres circonstances seraient imaginaires, facilite la démontration. Ensuite, on applique le théorème qu'on a ainsi démontré aux cas de la figure où ces points, ces plans st ces droites seraient imaginaires; c'est-à-dire, qu'on le regarde comme vrai dans toutes les circonstances de constructions générales que peut présenter la figure à laquelle il se rapporte".

> Sometimes this is occurred in natural sciences, particularly in those of organised bodies. The PCR seems already to belong to some general principle of duality, as that which seems inherent to such bodies where two kinds of elements are recognised: permanent elements and variable elements; fixity and movement.[4]

Thus, the PCR fits into the more general picture of the duality principle. To what kind of duality is referred the PRC? There are two immediate answers:

1) Duality permanent/variable. The constructive conditions of the geometrical problems to which PCR is applicable are permanent. In the previous example drawn from Monge, the permanent condition is the construction of the tangent planes through the line *l*. The variable or contingent condition is that such planes are either real or imaginary. In this way Chasles considered that PCR expresses a fundamental dual difference which holds for every mathematical problem: that between invariant aspects and variable aspects of the problem. Therefore, this crucial kind of duality is involved while dealing with the PCR.
2) Duality fixity/movement. It is connected to the former one, but is not exactly identifiable with it. I give an example: Suppose we have in the plane a circumference and an external point. A rotating straight line centred on the point can be drawn. The fixed elements are the point and the circumference, the movable element is the straight line. During the movement it assumes all the possible positions with respect to the circumference: it can be external, tangent or internal. The solutions of the system straight line-circumference will be, hence, respectively imaginary, real and coincident, real and distinct. The permanent elements, associated with fixity are the circumference and the point. The variable elements, associated with the movement, are the positions of the line with respect to the circumference and the kind of solutions of the system. The permanent is associated with fixity and variable to movement. This means that the language of the movement can be used to express the distinction permanent/variable, but, from a conceptual point of view the duality permanence/variation is more general than fixity/movement. For the first duality refers to a more abstract conceptual scheme, whereas the second one is the transcription of such a scheme into physical terms, which can also be applied to mathematical situations. According to my interpretation of Chasles' thought, the dual scheme permanent/variable is also applicable, for example, to a living being, where what is permanent—at least temporarily—is the organism. What is variable are the relations among its parts, independently of the existence of a local motion, that is a movement. It seems to me that, while referring to "organized bodies", Chasles was thinking precisely of this extended use of the principle of duality. Therefore: the principle of duality is the general

[4]Chasles (1837a), p. 204: "Ce prinicipe de relations contingentes sera peut-être basé un jour su quelque principe métaphysique de l'étendue figurée, tenant à des idées d'homogénéité, telles que celles qu'on apportées quelquefois dans les sciences naturelles, particulièrement dans celles des corps organisés; il semble appartenir déjà à quelque principe général de dualité, tel que celui que paraissent présenter ces mêmes corps où l'on a à reconnaitre deux genres d'éléments, eléménts permanents, éléments variables; fixité et mouvement".

archetype according to which the being is organized. It can be called a meta-principle. The PCR is an aspect of this archetype. In reference to the PCR, the duality permanent/variable is the general scheme of which the duality fixity/movement is an application.

However, these are not the only considerations which Chasles connected to the PCR as a product of the principle of duality. As a matter of fact, the nature of imaginary in geometry can also be explained taking into account the PCR. I have already outlined the problem of the imaginary in geometry, but Chasles dedicated to it some other and more interesting considerations which allow us to perfectly understand his view. The duality in this case is real/imaginary. How could we interpret it? Chasles claimed that this duality is a manifestation of the contingent aspect of the duality contingent/permanent. He claimed:

> It seems to us that the doctrine of *contingent* relations can offer a further advantage: it can give a satisfying explanation of the term *imaginary*, now used in pure Geometry, where it expresses a rational *being* without existence. Nonetheless, it is possible to ascribe it some properties, that we use momentarily as an auxiliary means and to which you apply the same reasoning as those applied to a real and palpable object. This idea of imaginary, which might appear at a first glance obscure and paradoxical, acquires so, in the theory of contingent relations, a clear, precise and legitimate meaning. Under this respect, our distinction between intrinsic and permanent properties of the figures and their momentaneous and contingent properties might be of some utility.[5]

First of all, it is appropriate to highlight that Chasles, just like the majority of the French geometers of his epoch (we are in 1837), did not have an axiomatic idea of imaginary quantities, such that, if their properties are not self-contradictory or do not contradict other known and accepted properties, they have the same mathematical existence as the real quantities. It is not necessary to wait for Hilbert and his abstract axiomatic conception to reach this idea of the imaginary: in 1832, Gauss published the *Theoria residuorum biquadraticrum. Commentatio secunda* (Gauss, 1832). Even though his analysis was restricted to the complex integers, Gauss offered a satisfying axiomatization of the properties connoting the complex numbers. He overcame the difficulties and the inaccuracies typical of the brilliant but naïve use of the complex numbers in the eighteenth century mathematics, as it was typical, for example, of Euler.[6] Gauss did not refer to any kind of existence of the imaginary

[5] Chasles (1837a), p. 207: "La doctrine des relations *contingentes* nous semble pouvoir offrir encore une avantage; c'est de donner une explication satisfaisante du mot *imaginaire*, employé maintenant en Géométrie pure, où il exprime un *être* de raison sans existence, mais auquel on peut cependant supposer certaines propriétés, dont on se sert momentanément comme d'auxiliaires, et auquel on applique les mêmes raisonnements qu'à un objet réel et palpable. Cette idée d'imaginaire, qui parait au premier abord obscure et paradoxale, prend donc, dans la théorie des relations contingentes, un sens clair, précis et légitime. Sous ce rapport, la distinction que nous avons faite entre les propriétés intrinsèques et permanentes des figures, et leur propriétés fugitives et contingentes, paraitra peut-être de quelque utilité" Italics in the text.

[6] A synthetic but good picture to guess how problematic the interpretation of the imaginary quantities was at least until the half of the nineteenth century is presented by Chasles himself in his *Rapport* (Chasles, 1870, Chapter I, section XIX entitled *Nouvelle doctrine des imaginaires*,

quantities beyond their correct use within the mathematical context. He was not referring to an intuitive, absolute, noetic world of which mathematics is a transcription and in which the real quantities exist and those imaginary do not. Gauss identified mathematical existence, at least implicitly, with a coherent system of definitions (D) and axioms (A). If an object satisfies (D) and (A), then it has a mathematical existence.

Instead, it seems that Chasles believed in the existence of a noetic, absolute realm where the real quantities "exist", whereas he possibly interpreted the imaginary quantities as a mere mathematical means. These do not have the same level of existence as the real quantities, but have, nonetheless, their form of subsistence. They are the conceptual and linguistic tools through which the mathematicians can examine the behaviour of the real quantities. It seems to me that the following quotation illustrates Chasles' idea clearly:

> As a matter of fact, it is possible to consider the expression *imaginary* only as indicating a condition of the figure such that some parts, which will be real in another condition of the figure, ceased to exist. For it is possible to get an idea of an imaginary object only by representing, at the same time, an object of space in a condition of real existence. Thence, the idea of *imaginary* would be meaningless, if it were not always associated with the actual idea of the real existence of the object itself to which it is applied. These are, thus, the relations and the properties that we have called *contingent*. They offer the key of the *imaginary* in Geometry.[7]

In order to clarify the way in which the PCR is usable within the duality real/imaginary, it is useful to consider Chasles' example. First of all, he explained the general concept: given a figure with imaginary parts, it is always possible to conceive another figure in which such parts become real. However, you are not allowed to operate on the first figure considering as real those parts which are imaginary. For example, if the expression of a point on a straight line is imaginary, this point will be imaginary, as well, and constructing this point as if it were real would be a serious mistake (Chasles, 1837a, p. 369). The point constructed in this manner does not belong to the figure, and not even to the problem proposed. If we treat such a point as a real one and construct real entities basing on it as on a real one, we make a serious error. That point is an instrument which might be useful to deduce properties of real objects, but it cannot be treated as a real entity. Chasles' example clarifies his way of reasoning: given a hyperbola and any pair of conjugate diameters, the directions of such diameters are real, but the length of one of them is

pp. 60–63) where the opinions and interpretations concerning this subject expressed by several important mathematicians are referred to.

[7]Chasles (1837a), p. 368: "En effet, on ne peut regarder l'expression *imaginaire* que comme indiquant seulement un état d'une figure dans lequel certaines parties, qui seraient réelles dans un autre état de la figure, on cessé d'exister. Car on ne peut se faire l'idée d'un objet imaginaire qu'en se représentant en même temps un objet de l'espace, dans un état d'existence réelle; de sorte que l'idée d'*imaginaire* serait vide de sens, si elle n'était toujours accompagnée de l'idée actuelle d'une existence réelle du même objet auquel on l'applique. Ce sont donc les relations et propriétés que nous avons appellées *contingentes*, qui donnent le clef des *imaginaires* en Géométrie". Italics in the text.

imaginary.[8] However, the square of this distance between two imaginary elements is a real number. Hence, all the properties of the conjugate diameters of an ellipse in which only their squares are involved can be applied to the hyperbola too. But the properties in which the lengths of the diameters, instead of those of their square, are used cannot be applied to hyperbola. For the length of the imaginary axis (or diameter) of the hyperbola is imaginary. The lines and the points obtained considering such a distance to be real do not belong to hyperbola, but to another figure (ibid., p. 369). This figure, in a general sense, though not in the strict projective meaning, can be considered dual of hyperbola. It is the figure one constructs considering as real the parts which in the first figure are imaginary. As an example, Chasles considered an equilateral hyperbola and the circle whose diameter is the hyperbola's real axis. The square of all the chords of the circle perpendicular to this axis is real. If the feet of the chord on the axis is internal to the circle, the chord is real too. Otherwise, it is imaginary. But supposing that such a chord has a real length, you will obtain a point of the equilateral hyperbola. Thus, the imaginary chords of the circle, when their lengths are considered real, produce a real chord of the equilateral hyperbola and vice versa. In this sense, the two figures are dual.

With regard to the legitimacy of the PCR, Chasles recognized that an *a priori* justification is not possible. Nonetheless, an *a posteriori* justification is indeed possible. It is exactly based on the way in which a certain figure is constructed and on the difference between permanent and contingent parts. To apply the PCR it is, namely, necessary to distinguish the properties which are permanent and which subsist independently of being the contingent parts real or imaginary (Chasles, 1837a, p. 200). The PCR is, thence, usable when the transition of contingent parts from being real to being imaginary does not change the constructive conditions of the permanent parts. These geometrical considerations have an analytical counterpart. The PCR cannot be used if, given the general constructive conditions, its application would impose to change any element different from the signs of the variables' coefficients, for example the signs of the exponents. In this case PCR cannot be applied. Chasles wrote:

> Transcribe the general constructive circumstances of which we have spoken into Analytical Terms. If you will find that it is necessary to change something more than the signs of the variable quantities' coefficients (for example, the signs of the exponents of these quantities), then you will have to refrain from applying this principle.[9]

---

[8] Ibid., p. 369. Today we prefer to claim that one of the two conjugate diameters cuts the hyperbola in two complex points, but Chasles' concept is, anyway, clear. It is, perhaps, appropriate to point out that, for the geometers of Chasles' epoch (and also for those living until the end of the nineteenth century), the distance between two points of coordinates, for example, $(0, 3i)$, $(0, 2i)$ was not a real number, but the complex number $\sqrt{(3i)^2 - (2i)^2} = \sqrt{-5} = i\sqrt{5}$.

[9] Chasles (1837a), pp. 200-201: "Ainsi, par exemple, on devra se garder d'appliquer ce principe aux questions dans lesquelles, si l'on voulait faire entrer dans l'Analyse les circonstances générales de construction dont nous avons parlé, ou trouverait qu'il a aurait à changer autre chose que les signes des coefficients des quantités variables; par exemple les signes des exposants de ces quantités".

Told in a different manner, as Chasles himself claimed (ibid., p. 201): the PCR cannot be used when the problem transcribed into analytical terms needs the use of definite integrals, whereas its use is legitimate when the geometrical questions need only the resort to finite analysis.

These argumentations developed by Chasles make the use of the PCR plausible and give an idea of the cases in which its utilization is legitimate. However, they do not offer a rigorous justification of the principle, and Chasles himself did not think of offering such a justification. Instead, he aspired to give a relatively precise idea of the use and of the heuristic behind the principle. In this respect, his considerations are useful and illuminating.

Chasles (ibid., Note XXIV, pp. 357–359) clarified that he preferred to use the expression "principle of contingent relations" rather than the traditional one "principle of continuity" (which he ascribed, in its general formulation, to Leibniz) because in the latter denomination the use of infinity and of the movement is implicit. This is useful in order to grasp the heuristic and the meaning of the principle. However, it is possible to avoid the explicit resort to the language of the continuous movement and, hence, to infinity, by introducing the two notions of permanent and contingent relations. They are purely geometrical as they are connected to the constructive conditions of a problem or a theorem and do not need the use of infinitary concepts. Thus, Chasles preferred to use this denomination of the principle, which is more general.

Finally: each principle becomes significant when its application is clarified. In the interesting and long note XXV of the *Aperçu* (ibid., pp. 359–368), Chasles explained how to find the three axes of an ellipsoid given three conjugate diameters by using the PCR. At the beginning of the Note he solved the "plane version" of this problem, namely: given two conjugate diameters of an ellipse, determine their axes in direction and length. I will only focus on Chasles' solution concerning the problem of the ellipse for the axes' direction because it is sufficient to give a clear idea of all the elements connoting the application of the PCR.

Chasles argued that: if two conjugate diameters of a hyperbola rather than those of an ellipse are given, it is easy to construct the hyperbola's axes (Fig. 6.2). For given the direction of the two diameters—which are two real lines—and their length (suppose a to be the length of the real diameter and b that of the imaginary one), draw a parallel p to the diameter b from the extremity A of the diameter a. The straight line p is tangent to the hyperbola at A. If, with A as middle point, two opposite segments, each one equal to b are drawn on p, the extremities of the two segments belong to the hyperbola's asymptotes. It is, thence, possible to draw the two asymptotes. If you bisect their angle and its supplementary, you obtain the direction of the axes (ibid., pp. 359–360).

If one thinks of constructing the axes of an ellipse with the same method as that used for the hyperbola, then one has to avoid the use of the contingent elements existing in the previous construction, that is the elements which connote the hyperbola, but not the other conics. These elements are clearly the asymptotes. Thence, the expounded construction must be used as a basis to develop another construction where the asymptotes do not appear anymore.

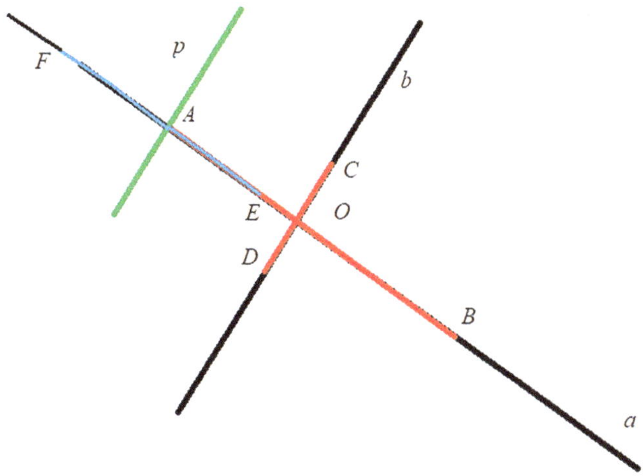

Fig. 6.2 Reconstruction of the diagram described by Chasles: a=real conjugate diameter; b=imaginary conjugate diameter; AB=length of the real diameter; CD=length of the imaginary diameter; p=parallel from A to b. It is tangent to the hyperbola; AF=AE=CD. The points F and E belong to the hyperbola's asymptotes

In order to develop such construction, Chasles considered the two points R and S where p cuts the asymptotes as the two foci of a conic c passing through the centre O of the hyperbola. Since the two axes of the hyperbola bisect the angle between the two straight lines OR and OS and its supplementary, they are, respectively, the tangent and the normal to the conic c. Thence, the conic c of which the centre of the hyperbola is a point is tangent to one of the hyperbola's axes (ibid., p. 360).

This conic for whose construction the asymptotes of the hyperbola have been used can, nonetheless, be constructed without resorting to the asymptotes. For the axes of c are known because they are the tangent and the normal to the hyperbola at the point A. The eccentricity of this conic, being given b, is also easily determinable. Thus, Chasles could argue:

> If the tangent and the normal at a point A of a hyperbola are considered as the principal axes of a conic passing through the centre of the hyperbola and having its eccentricity directed along the normal equal to the conjugate diameter of that passing through the point A, this conic is necessarily tangent to one of the two principal axes of the hyperbola.[10]

[10] Ibid., p. 360: "Si l'on regarde la tangente et la normale en un point A d'une hyperbole, comme les axes principaux d'une conique qui passerait par le centre de l'hyperbole, et qui aurait son excentricité, dirigée suivant la normale, égale au diamètre conjugué de celui qui aboutit au point A, cette conique sera nécessairement tangente à l'un des deux axes principaux de l'hyperbole". In the context of this theorem, Chasles supposed that a conic had four foci, two of them being real and two imaginary, and two eccentricities, one real, directed along the tangent and the other imaginary, directed along the normal. This distinction is useless for my aims and the eccentricity mentioned in the quotation can be considered as the magnitude which today we call precisely eccentricity.

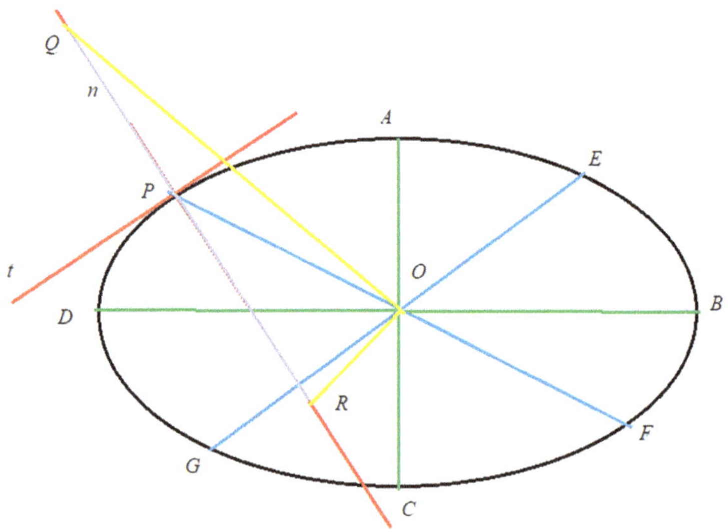

Fig. 6.3 Reconstruction of the figure described by Chasles: *n*=normal to the ellipse at point *P*; *t*=tangent to the ellipse at *P*; *BD*=ellipse's major axis; *AC*=ellipse's minor axis; *PF*= diameter containing the point *P*. *EG*=diameter conjugate at *PF* (parallel to the tangent at *P*). *QP*=*PR*=*OE*. *OR*, *OQ*=the two straight lines equally inclined on the ellipse's principal axes

Chasles highlighted that the property expressed by this theorem is independent of the hyperbola's asymptotes which are exploited to determine it. All the parts of the figure used in the theorem also exist in the ellipse, ergo it is possible to argue that:

> Consider the tangent and the normal at a point of an ellipse as the principal axes of a conic [I name it *C*] passing through the centre of the ellipse. Suppose that *C*'s eccentricity along the normal [we can simply claim "the eccentricity"] to be equal to the ellipse's diameter conjugate to that containing the point considered on the ellipse. This conic will be tangent to one of the ellipse's principal axes.[11]

If the vector radiuses from the two foci of *C* to the centre of the ellipse are drawn, they form equal angles with the principal axis of the ellipse to which *C* is tangent. Therefore, Chasles concluded (Fig. 6.3):

> Consider the normal to an ellipse at a point. From both sides of this point take two segments equal to the half-diameter conjugate to that containing the point. Furthermore, from the extremities of these segments draw two straight lines to the ellipse's centre. These lines are equally inclined on the principal axes of the ellipse.[12]

[11] Ibid., p. 361: "Si l'on regarde la tangente et la normale, en un point d'une ellipse, comme les axes principaux d'une conique qui passerait par le centre de l'ellipse, et qui aurait son excentricité, prise sur la normale, égale au diamètre conjugué de celui qui aboutit au point pris sur l'ellipse, cette conique sera tangente à l'un des deux axes principaux de l'ellipse".

[12] Ibid., p. 361: "Si, sur la normale en un point d'un ellipse, on prend, de part et d'autre de ce point, deux segments égaux au demi-diamétre de celui qui aboutit à ce point, et que, des extrémités de ces

This solves the problem of finding the direction of the ellipse's axes.

After having clarified the way in which Chasles conceived the use of the PCR, some final considerations on PCR/duality are appropriate: the way of reasoning, the philosophy, behind the PCR is profoundly involved with duality in an extended meaning: the conic C_1 whose axes are the normal and the tangent to the conic C and the conic C itself of which you have to find the axes' direction can be interpreted as two dual entities where the duality is given by the substitution normal/tangent with axis/axis. However, there is also a more profound meaning which connects PCR to duality: from a methodological point of view, these two principles are means used in a certain phase of a proof in order to demonstrate a property of a given configuration for which a direct argumentation would be difficult. However, at the end of the demonstrative process, the use of the PCR, as well as the use of duality disappear and one obtains a proof which, from a mathematical point of view, does not bring any traces of its origin due to PCR or to duality. In our example, the PCR is used to obtain the construction of a hyperbola and it is a fundamental element in the proof. However, thanks to further properties of the hyperbola, it is possible to obtain a proof in which the contingent entities of the hyperbola (the asymptotes) disappear and this second proof can be applied to the ellipse. Analogously: if you have to prove a theorem, for example in projective space, for configurations of planes and you are not able to find a direct proof, you can try to analyse the dual configuration of points. If you are able to find a proof for such a configuration, you also obtain a proof for the former, making the replacements typical of spatial projective geometry. At this point, you have obtained a proof by duality, but the demonstration for your configuration appears as a usual self-contained proof without any apparent trace of its dual origin. In this respect, PCR and duality show a further profound link. Obviously, it would be a mistake to exaggerate the analogies because there are also significant differences: the duality law is universally accepted and it is possible to offer a rigorous analytical proof for it, whereas a rigorous formulation of the PCR (or principle of continuity) seems problematic and the principle has always raised doubts among the mathematicians who cared about the rigorous foundation of mathematics (Cauchy was one of them). It was often considered as a mere heuristic means rather than a demonstrative procedure.[13] These differences notwithstanding, the PCR fits into the general philosophy of duality conceived by Chasles to frame science within a philosophical unitary picture.

segments, on tire deux droites au centre de l'ellipse, ces droites seront également inclinées sur l'un des deux axes principaux de l'ellipse".

[13] On the other hand, it seems to me appropriate to remember that—as stressed in the first chapter—Chasles himself argued that, once given a proof by duality, it would be desirable to find a direct demonstration of a certain statement.

6.2 Duality as a Common Root of Projective Geometry, Movement of a Rigid Body, Systems of Forces

I have dealt with Chasles' view on the concept of duality within projective geometry in the first section of this book. Hence, I will now refer to his general ideas on duality and to the links Chasles found between duality in projective geometry and duality in the theory of rigid body's movement and in the theory of systems of forces. This will be a further element to grasp the nature of his foundational programme based on the main concepts of projective geometry.

The most important issue to point out is the following: the origin of the duality principle in projective geometry derives from the theory of reciprocal polars. This was the first theory which was proper of modern projective geometry and was the milestone on which, in the first twenty years of the nineteenth century, projective geometry was developed as an abstract and new branch of mathematics. Poncelet's *Traité* can be interpreted as the reference text in which the entire development of the previous researches was systematized with the addition of new significant elements. It is not a coincidence that Poncelet himself, after the *Traité* wrote a series of fundamental memoirs on the theory of reciprocal polars (Poncelet, 1827, 1828b, 1829a, 1829b). Chasles refused to continue seeing the principle of duality as a sort of mere extension of the theory of reciprocal polars and to interpret duality only in the light of the concept of pole/polar. He reversed the relation between principle of duality and theory of reciprocal polars: the former is not, so to say, only a formalization of the latter. It is its real conceptual basis. As it is often the case, the historical-chronological order is the opposite with respect to the conceptual order: the principle of duality is the general law of which the theory of reciprocal polars is only an application, though an important one. According to ideas he had developed at least starting from the end of the 20s, in the *Aperçu* Chasles expressed his convictions on the generality of the duality law, as a true foundational principle, which must not be confused with its applications. He wrote unambiguously:

> The theory of reciprocal polars transforms a figure into one of a different kind (in which the planes and the points correspond respectively to points and to planes of the given figure). The properties of the figures are transformed into properties of new figures, which establishes a permanent *duality* of the forms and of the properties of the extended figures. As to such theory, we have already announced (*Annales de Mathématiques*, tom. XVIII, p. 270) that it is not at all the only method to obtain this scope: several other methods exist, which highlight this *duality* and whose application is easy.[14]

[14]Chasles (1837a), p. 224: "Quant à la théorie des polaires réciproques, qui sert à transformer les figures en d'autres figures de genre différent (dans lesquelles les plans et les points correspondent respectivement à des points et à des plans des figures proposées), et à convertir les propriétés de cette figures en propriétés des figure nouvelles, ce qui établit une *dualité* permanente des formes et des propriétés de l'étendu figurée, nous avons déjà annoncé (*Annales de Mathématiques*, tom. XVIII, p. 270), que cette théorie n'est point une méthode unique pour ces fins: il en existe plusieurs autres, qui mettent en évidence cette *dualité*, et qui sont d'un usage facile dans leurs applications". Italics in the text.

First of all, he proposed, as an example, the extension of the concept of pole/polar beyond the rectilinear spatial configurations and used the properties of the reciprocal polars theory in reference to well-known properties of the supplementary figures on a spherical surface. After that he proved that this kind of polarity can be considered as a specific case of a more general dual transformation. Secondly and more importantly, he referred to the general concept of reciprocity of which the theory of reciprocal polars is an application.

With regard to the first issue, Chasles considered the example of the sphere and its geometry: in the sphere each figure has its supplementary, in which arcs of great circle correspond to the points of the first figure. Nowadays the two figures are called exactly mutually polar.[15] The example of two polar spherical triangles is the easiest of two polar figures on the spherical surface. Therefore, the theory of reciprocal polars can be extended and applied to the supplementary figures on the spherical surface. If on a sphere two supplementary figures are given and, from the centre of the sphere, the two figures are projected onto a plane, two perspective figures will be obtained of which one is the transformed of the other. This offers a clear example of duality (Fig. 6.4).

After that, Chasles proved that the configuration obtained projecting the two polar spherical figures onto a plane can be achieved also without any reference to the spherical figures, working directly in the plane. The proof of this proposition implies a planar transformation which can also be extended to three-dimensional figures. Chasles thus obtained a new general principle of spatial transformation which reads as follows:

> Be given a figure in space. From any fixed point draw the radiuses to all the points of this figure. On these radiuses (or on their prolongation beyond the fixed point), draw lines respectively proportional to them. Through their extremities, draw the planes perpendicular to the radiuses. All these planes will envelope a second surface which will be the TRANSFORMED of the given one, as it is considered according to the DUALITY principle. That is, to the planes of the given figure, points will correspond in the new figure. When these planes will pass through a point, these points will belong to a plane.[16]

---

[15] I recall the meaning of polar figures on a sphere. I consider the example of triangles (Fig. 6.4). On a sphere of centre O, be given two distinct and not diametrically opposed points A and B. The line passing through O and orthogonal to the plane OAB meets the sphere at two points called poles of the plane OAB. For a triangle ABC drawn on a sphere, call C' the pole of the plane OAB located on the same hemisphere as C. Be the points A' and B' constructed in the same way. The triangle $A'B'C'$ is called the polar triangle of triangle ABC. By construction, the great circles $C'B'$ and $C'A'$ intersect the great circle AB at right angles. The same happens for the two great circles $B'A'$ and $B'C'$, for the great circle AC, etc. The sides of the polar triangle are therefore each perpendicular to two sides of the original triangle. The transformation that associates a triangle with its polar triangle is involutory, i.e. the polar triangle of triangle $A'B'C'$ is triangle ABC.

[16] Chasles (1837a), p. 225: "Étant donnée une figure dans l'espace; que, d'un point fixe, pris arbitrairement, on mène à tous les points de cette figure des rayons, et que sur ces rayons (ou bien sur leurs prolongements au-delà du point fixe), on porte des lignes qui leur soient respectivement proportionnelles; que, par les extrémités de ces lignes, ou mène des plans perpendiculaires aux rayons; tout ces plans envelopperont une seconde figure qui sera la TRANSFORMÉE de la proposée, comme on l'entend dans le principe de DUALITÉ. C'est-à-dire

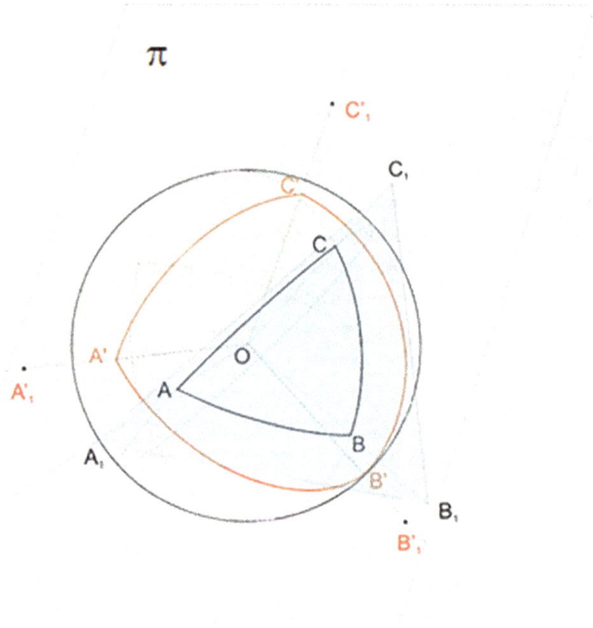

Fig. 6.4 Reconstruction of the situation described by Chasles: the spherical triangles ABC and $A'B'C'$ are polar figures on a spherical surface. From the centre O of the sphere, project them onto the plane π. Their projection gives rise resp. to two triangles $A_1B_1C_1$ and $A_1'B_1'C_1'$ with the features described by Chasles

Furthermore, if these lines are taken in the direction of the radiuses, the planes drawn through their extremities can be regarded as the planes polar to the points of the given figure in reference to a sphere having its centre in the fixed point (ibid., p. 226). Therefore, Chasles concluded:

> Our way of transformation includes the theory of the *reciprocal polars* considered in the sphere. It is more general because in the theory of polars the planes corresponding to the points of a given figure are always drawn between these points and the centre of the sphere, while, in our procedure these planes can be drawn beyond the fixed point which represents this centre.[17]

qu'aux plans dans la figure proposée, correspondront des points dans la nouvelle figure, et quand ces plans passeront par un même point, ces points seront sur un même plan". Capitals in the text. It is necessary to highlight that here "proportional" means "inversely proportional". This transformation is the transformation by reciprocal vector rediuses.

[17]Ibid., p. 226: "Notre mode de transformation comprend donc celui de la théorie des *polaires réciproques* considérées dans la sphère; et il est plus général que celui-ci, en ce que, dans la théorie des polaires, les plans correspondant aux points d'une figure proposée sont toujours menés entre ces points et le centre de la sphère, tandis que, dans notre mode de transformations, ces plans peuvent être menés au-delà du point fixe qui représente ce centre". Italics in the text.

This clarifies Chasles' view: duality is a general means of transforming figures which is far more extended than the transformations by reciprocal polars. The transformation just described is an extension of the theory of polars, which indicates that polarity is a specific dual transformation. It is possible to connote the dual reciprocity in an absolutely general form. This is what Chasles did in the FTD (fundamental theorem of duality) (ibid., pp. 577–578), which opens his memoir on duality and which I have analysed in the first chapter of this book. This theorem, according to Chasles, includes all the dual transformations inherent to projective geometry, whose basis is that to a point of the given figure a plane corresponds in the derived figure and reciprocally to any point of the derived figure a plane in the original figure corresponds (ibid., p. 228). He expressed clearly his ideas according to which all the dual kinds of transformations are reducible to such a theorem and to its consequences, which he developed in his crowned memoir on duality. He wrote:

> But all these methods, as that of *deformation*, of which we spoke above, can be replaced by a sole and unique principle. It is more general and extended than each of them. This principle which constitutes a complete doctrine of the figures' *transformation* has its source in a single theorem of Geometry, which seems to us to be the first reason of this property inherent to the forms of extension. I mean *duality*, on which the learned geometers have already written, but without discovering—despite the philosophical views they introduced in this part of Geometry—its primordial principle, independent of any particular doctrine.[18]

Chasles, thence, pointed out that the different opinions of the geometers—and, in my view, he was referring to Poncelet and Gergonne[19]—were centred on particular aspects concerning the nature of duality, but, before him, no mathematician grasped either the real essence of such a general principle or the fact that it is based on a sole

[18]Ibid., p. 228: "Mais toutes ces méthodes peuvent être remplacées, comme celle de *déformation*, dont nous avons parlé ci-dessus, par un seul et unique principe, plus général et plus étendu que chacune d'elles. Ce principe, qui constitue une doctrine complète de *transformation* des figures, prend sa source dans un seul théorème de Géométrie, qui nous parait être la raison première de cette propriété inhérente aux formes de l'étendu, la *dualité*, sur laquelle de savants géomètres ont déjà écrit, mais sans remonter, malgré les vues très-philosophiques qu'ils ont apportées dans cette partie de la Géométrie, à son principe primordial, indépendant de toute doctrine particulière". Italics in the text. I remind the reader that Chasles associated the term "deformation" to homology and the term "transformation" to duality with the idea that a homology deforms a figure, but does not change its nature because (in space) it transforms points into points, straight lines into straight lines, and planes into planes, while a duality changes the nature of the figure because it transforms points into planes, straight lines into straight lines, and planes into points.

[19]Among the numerous works where Poncelet expressed his conception of duality I refer to: Poncelet (1822, 1828b, 1829a, 1829b, 1857). As to the works dedicated by Gergonne to duality, I mention: Gergonne (1825–1826, 1826–1827a, 1826–1827b, 1827–1828a, 1828–1829, 1847). The secondary literature on duality, in the different concepts held by Gergonne and Poncelet of this notion as well as on the priority polemics between the two mathematicians—a polemic in which Plücker was also involved—is abundant. As a matter of fact, almost every work dealing with projective geometry in the period under examination dedicates some considerations to this issue. I prefer to mention Lorenat (2015a) as reference-indication, because the author presents an abundant and updated literature on this subject, which is completely satisfactory for any reader who intends to explore this theme beyond Chasles' contributions and conception. In the running text, I will analyse Gergonne's view of duality to the extent that it is useful to spread light on Chasles'.

theorem. The reason is that they were still tied to the idea that duality was, in its essence, reducible to the polar reciprocity.[20] Chasles also explained that such a transformation is specific because it is an involutory transformation, which is not the case if one considers a general reciprocity based on the duality principle. With regard to duality as a general method, Chasles used the same dichotomous couple permanent/contingent he had already developed for the properties of specific configurations. He claimed that the involutory character of the polarities is a contingent property of the reciprocities, not a permanent one. As we read:

> The theory of polars has been so far the only means used to transform the figures. Thence, you might believe that the figures due their concordance, or reciprocity of forms, of which we have already spoken, to the identity of construction which takes place in this theory of polars.[21] This would be a serious mistake. This identity of construction is an accidental, a particular property typical of the figures which produce the theory of polars and which also subsists in other kinds of transformations; but it is not this property that determines the *duality* of extension. As a matter of fact, it does not subsist in diverse other ways of transformation, and particularly, in the one, as we will show, which includes all the others as corollaries, or particular cases. Thence, we will not do any use of this identity of construction and we will exclude it from the explanation of our doctrine of transformation, as an extraneous datum, which occurs only in particular and accidental circumstances.[22]

This extended use of the couple permanent/contingent is a further confirmation of the idea that Chasles was developing a whole foundational programme based on few concepts, of which duality is one of the most significant. His entire construction is consistent: the idea of dichotomy characterizes Chasles' thought; it is a prototypical way of thinking that he applied on several circumstances and occasions. The dichotomy permanent/contingent is one of the most important within this context.

It should also be recalled that he highlighted the existence of a sole relation through which the dual transformations are expressible because it is an invariant of any projective transformation (be it a homography or a reciprocity). This relation is the cross ratio, or anharmonic ratio, as Chasles named it, of four collinear points (or planes) or of four straight lines belonging to a pencil (ibid., p. 255). In any dual transformation the cross ratio is conserved.

[20]This was particularly true for Poncelet, not for Gergonne, as we will see.

[21]Chasles is referring to the fact that a polarity is an involutory transformation.

[22]Chasles (1837a), p.229: "La théorie de polaires ayant été jusqu'à ce jour le seul moyen employé pour la transformation des figures, on pourrait croire qu'elles doivent leur concordance, ou réciprocité de formes, dont nous parlions tout à l'heure, à l'identité de construction qui a lieu dans cette théorie des polaires. Ce serait un erreur grave. Cette identité de construction est une propriété accidentelle, particulière aux figures que produit la théorie des polaires, et qui se présente aussi dans d'autres modes de transformation; mais ce n'est point elle qui donne lieu à la *dualité* de l'étendue; et, en effet, elle n'existe point dans diverses modes de transformation, et notamment dans celui qui, comme nous le ferons voir, comprend tous les autres comme corollaires, ou cas particuliers. Aussi nous ne ferons aucun usage de cette identité de construction, et nous l'écarterons de l'exposition de notre doctrine de transformation, comme y étant étrangère, et ne s'y rencontrant que par circonstance particulière et accidentelle". Italics in the text.

In order to further highlight the foundational character of Chasles' work, it is worth pointing out that he also thought of the invention of a specific terminology to connote the reciprocal figures. For he claimed that the name *correlative* is a too generic term, used on several occasions and with different meanings, and that the term duality is also generic. Thence, to determine a dual projective transformation he proposed to replace the term "*dualité*" with "*diphanie*", the term "*figures duales*" with "*figures diphaniques*" (ibid., p. 255). As far as I know, however, Chasles himself never used this terminology. Notwithstanding this, the only fact that he had thought of such a terminology is an indication of his foundational approach. He also added that the principle of duality gives an advantage to pure geometry on analytical geometry because the classical Cartesian coordinates are not suitable to express the theorems in which duality is involved (ibid., pp. 256–257). Of course, this depends on the fact that in classical Cartesian coordinates the objects at infinity of projective geometry cannot be expressed. This does not mean that geometry has to renounce to the use of algebra, but that the system of classical Cartesian coordinates has to be replaced by systems of coordinates suitable to express the projective properties. Chasles explicitly claimed that the principle of duality leads naturally to a new system of analytic geometry based upon a new system of coordinates. Thus, such a principle is so strong and general as to also include analytic geometry, which, hence, becomes an application of the basic geometrical notion of duality (ibid., pp, 257–258).

As well known, in the period 1827–1830 Möbius introduced the barycentric coordinates and Plücker the system of coordinates which hold his name. It is, however, to point out that Chasles too, in 1830, conceived two systems of projective coordinates. In a contribution appeared in the *Correspondance mathématique* of Quetelet (Chasles, 1830b) he wrote to have read that Plücker would have proposed a coordinate system suitable to be applied to projective geometry. Chasles highlighted that he was thinking of a system of projective coordinates for a while now. His basic idea is the following one: given in space three fixed points A, B, C (the points should not be collinear), draw three parallel axes a, b, c through them. A plane α cuts a, b, c resp. in three points A', B', C'. The distances AA', BB', $CC' = (x,y,z)$ are the coordinates of A', B', C'. A plane is, thus, represented by three coordinates. Thence, an equation of the form $F(x,y,z) = 0$ represents a surface enveloped by an infinity of planes. If $F(x,y,z)$ is a first-degree polynomial, the equation $F(x,y,z) = 0$ represents all the planes passing through a point, so that it is the equation of a point given as a bundle of planes. Chasles also offered an example in which his system of coordinates is applied. In the same contribution he also envisioned another system of coordinates: given three coordinate planes, the coordinates (x',y',z') of a point are not obtained drawing three parallel lines to the three axes x,y,z, and considering their projections on the planes (yz, xz, xy) resp., but considering the straight line joining the point to three fixed points of the axes. Thus, be Ox, Oy, Oz the three coordinate axes; A,B,C their fixed points. Suppose you look for the coordinates of a point m. Draw the three planes (m, A, B); (m, A, C); (m, B, C). They saw the three coordinate axes in three points a,b,c. The ratios $\frac{Oa}{Aa}$; $\frac{Ob}{Bb}$; $\frac{Oc}{Cc}$ are the coordinates of m (Chasles,

1830b, p. 84). This is an evidence that Chasles neglected no aspect of geometry connected to his foundational programme.[23]

The grandiose view envisioned by Chasles can be summarized by the following quotation, where he claimed that, probably, the principles of duality and homography are the expression of an even more general law of extension which was still unknown. A law—or some laws—which should have the same role as the PVV for statics or of Newton's law for the celestial phenomena. Chasles wrote:

> After the considerations we will develop on the nature and use of the two principles of duality and homography, it will be possible to think that, in the science of extension, some primordial, very great and fruitful laws should exist, as in Analysis the infinitesimal calculus, which has summarized and perfected all the methods of the quadratures and of the maxima, as in mechanics the principle of virtual velocities, from which Lagrange deduced all the others, and in celestial phenomena the great law of Newton. You might think—I claim—that the two easy theorems of Geometry, from which the two principles of duality and homography derive, are those that better approximate—given the actual situation of Geometry—these great general laws, which are still unknown.[24]

It is difficult to guess of which general law or laws of extension Chasles was thinking. One might refer to the development of algebraic geometry to which Chasles himself gave fundamental contributions; but this branch of mathematics can be interpreted as an extension of the pure geometry created in the first 30 years of the nineteenth century rather than as a new foundational part of geometry. However, under a plurality of aspects, algebraic geometry is more general than projective geometry. Perhaps something more satisfying might be the analysis situs or topology, whose modern development coincides with the second part of Chasles' life (though Chasles did not contribute to topology). Here new basic concepts and new more general transformations than the projective ones are introduced and studied. Without any doubts the bicontinuous and biunivocal transformations opened the door to a more profound level of the science of extension. Thence, it can be assumed that Chasles would have been satisfied by topology's foundational aspect. However, the answer to the posed question is uncertain, though intriguing; what is sure is that, in the course of his scientific career, Chasles regarded duality as a milestone through which projective geometry could be considered as the basis of other scientific subjects, in particular of kinematics and statics.

---

[23] For a succinct history and conceptualization of the coordinate systems, see, e.g., Coolidge (1940, Book II, Chapters II and III, pp. 141–179).

[24] Ibid., pp. 266–267: "Après les considérations que nous venons de développer, sur la nature et la destination des deux principes de dualité et d'homographie, on pensera peut-être que s'il doit exister, dans la science de l'étendue, quelques lois primordiales vraiment grandes et fécondes, comme en Analyse le calcul infinitésimal, qui a résumé et perfectionné toutes les méthodes de quadratures et de maxima, comme en mécanique le principe des vitesses virtuelles, d'où Lagrange a tiré tous les autres, comme dans les phénomènes célestes la grande loi de Newton; on pensera peut-être, dis-je, que les deux simples théorèmes de Géométrie, d'où dérivent les deux principes de dualité et d'homographie, sont de ceux qui approchent le plus, dans l'état actuel de la Géométrie, de ces grandes lois générales qui nous sont encore inconnues".

I have explained the relations between the theory of reciprocity and those of the movement of a rigid body and of the systems of forces, respectively, in Chaps. 3 and 4. Therefore, I will add only some details to clarify further Chasles' ideas.

With regard to the infinitesimal movement of a rigid body *B*, in Sect. 3.2 we have seen the deep connection between the general theory of reciprocity and the theory of *B*'s infinitesimal movement. It is also appropriate to point out that Chasles claimed and expounded the reason why the infinitely small movement can be explained by resorting to the theory of polar reciprocity (which—as told—is a specific case of duality). He wrote:

> In the kind of transformation by an infinitely small movement, there is identity of construction, as in the theory of polars: that is, the planes normal to the trajectories of the points of the first figure envelop a second figure. If this figure had been constructed and had been subject to the same movement as the former, the planes normal to its trajectories would have enveloped such first figure.[25]

In the *Rapport*, while commenting in retrospect his own results obtained in the *Aperçu*, Chasles, with regard to the theory of duality, recalled that in the infinitesimal displacement of a figure the planes normal to the trajectories of its points envelop a correlative figure. He also recalled that the considerations on the system of forces led to the same conclusions because, given a system of forces in space, considered the planes of the principal momenta of the forces relative to any point of the figure, these planes envelop a correlative figure.[26] Chasles also summarized his results concerning the relations he had discovered between the generatrices of a one-sheeted hyperboloid and the infinitesimal movement of a rigid body: given a point P in a plane π and drawn a series of straight lines belonging to π through P, so that each line meets two generatrices of the hyperboloid belonging to the same system, these generatrices are called *associated*. The main properties of the associated generatrices are that any plane cuts each couple of generatrices in two points and the chords joining these points converge to the same point. Reciprocally if from any point of space straight lines are traced which lie on two associated generatrices, all these straight lines belong to a plane passing through the point. As we have seen this point is the *focus of the plane*. Chasles then spoke of the conjugate straight lines D and Δ and pointed out that:

> We do not enter into further details concerning the indication of the properties of these systems of conjugate straight lines, which will also subsist in a very different question, that of the infinitesimal movement of a rigid body.[27]

---

[25] Ibid., pp. 229-230: "Dans le mode de transformation par voie de mouvement infiniment petit, il y a identité de construction, comme dans la théorie des polaires: c'est-à-dire quel les normaux aux trajectoires des points d'une première figure enveloppent une seconde figure telle, que si elle eût été construite, et qu'elle eût éprouvé le même mouvement que la première, les plans normaux à ses trajectoires envelopperaient la première figure".

[26] Chasles (1870), p. 85. For the theorem on the system of forces, see Chasles (1837a), p. 679.

[27] Chasles (1870), p. 111: "Nous nous arrêtons dans cette indication des propriétés de ces systèmes de droites conjuguées, qui vont se retrouver ci-après dans une question fort différente, celle du déplacement infiniment petit d'un corps solide".

A further section of the Rapport devoted to *Propriétés géométriques du movement infiniment petit d'un corps solide dans l'espace* (ibid., pp. 114–116) is dedicated to Chasles' discoveries concerning this subject explained in Chasles (1843), which I have expounded in Sect. 3.4.2. It is here appropriate only to stress that, as the properties of the infinitesimal movement can be obtained starting from those of the one-sheeted hyperboloid, the opposite is also true, that is those of the one-sheeted hyperboloid can be deduced from those of the infinitesimal movement. Chasles wrote: "From there [the theory of the infinitesimal movements of a rigid body] several properties of the generatrices of the hyperboloid derive" (ibid., p. 116).

This is a clear litmus paper that shows that duality in an extended meaning was the basis of Chasles' foundational programme: there is the duality focus of a plane-plane; the duality between the two associated straight lines and, extending the concepts from the objects to the theories, there is a perfect duality between theory of the hyperboloid and theory of the infinitesimal movements of a rigid body. This kind of dualities highlights Chasles' idea that the formal structure connoting the relations between objects of a theory and between theories is more significant than the objects and the theories considered as single theoretical objects. Insofar as duality shows the perfect reciprocity between points and planes, in a sense, from a projective—not from a metrical—perspective, points and planes are equivalent objects; insofar as the theory of the infinitesimal movements and that of the hyperboloid are reciprocal, they can be considered as a sole theory. Therefore, thanks to his reflexions on duality, it seems to me that Chasles grasped, at least *in nuce*, several significant issues: 1) rather than the objects, the structures are important; 2) the concept of equivalence is relative: within a metric theory the planes and the points are different objects; in a broader theory, they can be regarded as equivalent objects; 3) the concept of object can be extended to the theories and they themselves can be considered as subject of study. I am perfectly aware that my interpretation risks to offer a too modernized view of Chasles' thought. In particular: I am not exactly claiming that he was a predecessor of modern axiomatic, which was developed at the end of the nineteenth century, and not even of the modern epistemological speculation on the nature of the theories developed in the second half of the twentieth century. But it is certainly interesting to speculate on the possible ideas of an author which might go beyond the conceptual panorama connoting his time. It seems to me that Chasles had several ideas which, opportunely modified and changed, are in harmony with diverse modern conceptions.

Furthermore, Chasles' speculations on the rigid body movement and on the systems of forces as well as the interest with which he followed the works connected, also in a broad meaning, with these theories, are a clear evidence of the profound links that he saw between the basic sections of mechanics and geometry and also of his tendency to reduce mechanics to its geometrical root as much as possible. It is not a coincidence that in the *Rapport* numerous sections are dedicated to the role that geometry plays in physics. Only to give some examples: Chasles dedicated a subsection (ibid., pp. 314–342) to the *Mémoire sur les propriétés géométriques du mouvement le plus general d'un corps solide* (1865) by Résal, where the author reached new properties connoting the general movement of a rigid body by considering two infinitesimal consecutive movements. One of Résal's starting points was exactly Chasles' work. Transon's and Bresse's contributions expounded significant results concerning the velocities and the accelerations of the diverse points belonging to a rigid body, a subject which Chasles had not faced. In the *Rapport* there is also a satisfying description of Haton's memoir in 1867 where Haton studied some methods of transformation connoting geometry and mathematical physics (ibid., pp. 343–344). Chasles did not restrict to the relation geometry-mechanics. He also dealt with the relations between the undulatory theory of light and geometry. In a long section of the *Rapport*, the 17th section of the first chapter (ibid., pp. 47–54) which is entitled to Fresnel, Chasles did not restrict to the results obtained by Fresnel, but offered a picture, though a synthetic one, of the relations between geometry and the undulatory theory of light between the early nineteenth century and the 1860s. At the beginning of this section he explained that:

> The questions concerning mathematical physics are often related to considerations on curved surfaces and lead, thence, to some results which enter the domain of geometry.[28]

After having synthetized the results of Fresnel, Ampère, Cauchy, Herschel, Hamilton, Lloyd and MacCullagh, Chasles pointed out that Plücker—who did not know Hamilton's and Lloyd's studies—in 1839 achieved important results concerning the general form of the luminous waves through the concepts of projective geometry. Chasles wrote:

> Some considerations of analytical geometry, based on the theory of reciprocal polars, led this eminent geometer not only to the proof of already known theorems, but also to several new properties, which establish deep relations between the wave's surface and the ellipsoid necessary to its construction.
>
> Therefore, one sees that: *The reciprocal polar of the waves' surface in reference to a concentric sphere is a new waves' surface*.[29]

On the other hand, as it is the case with the properties of the one-sheeted hyperboloid which are deducible from those of the infinitesimal movements, the results concerning the waves theory and obtained through geometry are, in turn, usable to highlight further geometrical properties. In the *Aperçu*, Chasles had already stressed the connections between the undulatory theory of light, optics and geometry: in a long note to the observation that Fresnel—to explain the polarization of light—replaced Huygens' ellipsoidal waves with a fourth-degree surface, Chasles referred to numerous theorems and results on the theory of light obtained by several authors and wrote:

---

[28] Ibid., p. 47: "Les questions de Physique mathématique donnent souvent lieu à la considération de surfaces courbes et conduisent ainsi à des résultats qui rentrent dans le domaine de la Géométrie".

[29] Ibid., p. 52: "Des considérations de Géométrie analytique, fondées sur la théorie des polaires réciproques, conduisirent l'éminent géomètre non-seulement à la démonstration des théorèmes déjà connus, mais à diverses propriétés nouvelles, qui établissent des relations intimes entre la surface de l'onde et l'ellipsoïde qui sert à sa construction.

Ainsi il vit que: *La polaire réciproque de la surface des ondes par rapport à une sphère concentrique est une nouvelle surface des ondes*". Italics in the text.

> From this theorem [of Fresnel], the beautiful laws of polarization recently discovered by illustrious physicists, and particularly those of Biot and of Dr. Brewster, offer immediately geometrical properties of the ellipsoid and, in general of the second degree surfaces.
>
> Thence, these optical phenomena, which have already spread such a vivid clearness on any subject pertinent to the intime structure of the crystallized bodies, can offer the same help in the study of rational Geometry.[30]

With regard to another subject concerning optics, in which Chasles was directly interested and to which he had given contributions, namely the problem of the caustics, he stressed the profound links with projective geometry. In the *Aperçu*, he wrote:

> [...] and the new *Theory of caustics*, through which M. Quetelet reduces to some principles of elementary Geometry this important and difficult part of optics, for which all the means offered by the Analysis would have not been sufficient.
>
> This theory which, at a first glance, would appear extraneous to the methods [of projective geometry] of which we have already spoken, might, thence, be connected, under certain respects, with such methods and receive a useful help from them. The singular reapproaching developed by Mr. Quetelet between the theory of caustics and that of stereographic projections represent a first evidence; we will have the opportunity to offer other examples.[31]

A long note (ibid., pp. 220–221) to the mentioned assertion makes it clear that, while referring to other examples showing the role of projective geometry within physics, Chasles was referring to other examples drawn from optics and from the infinitesimal rotations of the rigid body with their correlated properties.

In an interesting section of the *Rapport* entitled *Transformation par rayons vecteurs réciproques* (Chasles, 1870, pp. 140–146), Chasles, besides highlighting the numerous uses of this transformation within geometry, referred once again to Quetelet's works on caustics. He pointed out that Quetelet called "inverse" a figure obtained from another by applying the transformation through reciprocal radius vectors and mentioned two propositions in which Quetelet in a paper published in 1827 proved that: 1) the polar of a curve c with respect to a circle has the secondary

---

[30] Chasles (1837a), p. 105 note: "D'après ce théorème, les belles lois de polarisation, découvertes dans ce derniers temps par d'illustre physiciens, et particulièrement celles de M. Biot et du docteur Brewster, donnent immédiatement des propriétés géométriques de l'ellipsoïde, et en général des surfaces du second degré.

Ainsi ces phénomènes optiques, qui ont déjà jeté une si vive clarté sur tout ce qui tient à la structure intime des corps cristallisés, peuvent apporter les mèmes secours dans l'étude de la Géométrie rationnelle".

[31] Chasles (1837a), p. 220: "[. . .] et la nouvelle *Théorie des caustiques*, par laquelle M. Quetelet a réduit, à quelques principes de Géométrie élémentaire, cette partie importante et difficile de l'optique, à laquelle ne pouvaient suffire toutes les ressources de l'Analyse.

Ces théories, qui semblent au premier abord étrangères aux méthodes dont nous venins de parler, pourraient pourtant s'y rattacher sous certains rapports, et en recevoir d'utiles secours. Les singuliers rapprochements que M. Quetelet a faits entre sa théorie des caustiques et celle des projections stéréographiques, en sont une première preuve; nous aurons occasion ailleurs d'en donner d'autres". Italics in the text.

caustic of c as inverse; 2) the polar of any flat curve, after two successive stereographic projections, becomes similar to its inverse (Chasles, 1870, p. 141).

These evidences are sufficient to confirm my thesis: Chasles, throughout his entire career, developed a foundational programme which relied on the idea that the basic concepts of projective geometry represent the milestone for many sections of science, in particular of physics. Besides his own contributions, the meticulous attention with which he read and interpreted the works of the other mathematicians and physicists are a clear litmus paper of his ideas. As already told and as it is evident from the broad use of the doctrine of projections within physics, many authors resorted to modern geometry (either in its "pure" or analytical form) within a physical context, but Chasles was the only one who developed a wide foundational programme as the next section will furtherly clarify.

6.3 Duality as a Universal Law

The extension of duality law beyond the context of projective geometry is the subject of an interesting Note Chasles wrote in the *Aperçu*, note XXXIV (Chasles, 1837a, pp. 408–416). Before dealing with the conceptions developed by Chasles, it is appropriate to recall that the two authors who most influenced Chasles with regard to duality, namely Poncelet and Gergonne,[32] had two different views on this principle.

There is no doubt that the results obtained by Poncelet in his theory of reciprocal polars are the first achieved by using the duality principle. They make up a complete theory and precede Gergonne's reflexion on duality by a few years. Therefore, from a mathematical point of view, they are Chasles' source of inspiration. However, Poncelet's conception of duality was rather narrow: he was convinced that duality coincided with polar reciprocity, that its importance should not be overestimated and accused Gergonne's conceptions to be too philosophical. Poncelet explicitly claimed that Gergonne developed his conception of duality "d'une manière très-philosophiques" (Poncelet, 1826–1827, p. 265). Significantly in the rest of his paper, Poncelet refers only to his own results on polarity while dealing with duality.

In contrast to Poncelet, Gergonne had a broader view of duality, though his mathematical results were less innovative and precise than Poncelet's. Furthermore, Gergonne fell into the error of considering the dual of a curve of order n a curve of class n and thought that no metric property could be treated through graphical methods, which Poncelet and Chasles proved not to be the case. These weaknesses notwithstanding, Chasles' philosophical conception of duality is closer to Gergonne's.

[32] Möbius, Plücker, and Steiner were also fundamental for their conceptions of duality. But, Chasles' ideas were forged during his education and the French mathematicians were those who had a preponderant influence on him.

The first paper in which Gergonne used duality was Gergonne (1824–1825), a very interesting text: the author considers some theorems on polyhedrons proved by Legendre and points out that this latter regarded such propositions as not connected truths. In contrast to this, these theorems can be divided into couples where one member of the couple can be obtained automatically from the other simply by replacing the term "face" with the term "vertex" and the term "side" with "edge". The truth of this assertion can be deduced considering the reciprocal polar of a polyhedron with respect to a quadric (Gergonne, 1824–1825, pp. 157–158). Thus, by means of Euler theorem $F+V = E+2$ and the principle of duality, Gergonne gave extremely simple and unitary demonstrations of many of the theorems Legendre had proved otherwise and was able to deduce new propositions. In particular, through a relation, which is indicated by the number (4) in Gergonne's text, he was able to deduce all Legendre's theorems. This relation is obtained through an appropriate algebraic manipulation of Euler's formula (ibid., p. 160).

For clarity, I give three examples of the kind of propositions proved by Gergonne, which are demonstrated through simple and appropriate algebraic manipulations of Euler's identity and through duality. As theorem 2, he proved that in every polyhedron the number of faces having an odd number of sides is even. Dually, the number of vertices where an odd number of edges join is even (ibid., p. 160). As theorem 6, Gergonne demonstrated that if a polyhedron has no trihedral or tetrahedral vertex, then it has twelve pentahedral vertices. Dually: if a polyhedron has no triangular or quadrangular face, then it has at least twelve pentagonal faces (ibid., p. 161). As theorem 14: if a polyhedron with pentagonal faces has only trihedral vertices, the number of these vertices is 20. Dually: if a polyhedron with pentahedral vertices has only triangular faces, then these faces are 20 (ibid., p. 162).

The first paper explicitly dedicated to duality where long methodological observations appear is Gergonne (1825–1826). The author distinguished between metric and graphical properties. In order to study the first ones, calculations are necessary, whereas for the second they are not needed, though to fully achieve a graphical geometry without calculations, it is necessary to pass from the plane to space. It is, thence, paramount to wonder whether the distinction between geometry of plane and geometry of space is intrinsic to the subject or if it depends on our, so to say, metrical and, hence, not sufficiently profound view (ibid., pp. 209–210). In the plane the duality point-straight line exists, while in space point-plane. Gergonne used explicitly the locution "cette sorte de dualité" (ibid., p. 210). As examples of dual propositions, he refers to some results of Coriolis, to the theorems on polygons and polyhedrons he himself had proved in the previously mentioned paper, to the theory of reciprocal polars and to several properties of spherical geometry (ibid., p. 211).

After that, he expounded, by means of the double column notation, 15 propositions which represent the fundamental initial truths of his theory. They regard both plane and spatial geometry. To give an idea, the first proposition states that, in space, two distinct points determine a straight line and dually two distinct planes also determine a straight line (ibid., p. 212). The last proposition claims that, given two polyhedrons A and B, such that the number of faces of A is the same as the number of vertices of B, the polyhedron A is defined *circumscribed* to B if the B's vertices belong to the planes of A's faces. Dually: if the numbers of the A's vertices are the same as B's faces, the polyhedron A is defined *inscribed* to B if the planes of B's faces contain the vertices of A (ibid., p. 216). Here Gergonne inserts an interesting epistemological observation on the language of science which shows that he was going to find a theoretical doctrine of mathematical duality of which the known examples are only the first truths: "[. . .] only when science has reached a high degree of maturity, one can hope to appropriately develop its language".[33] He was constructing the language of duality because the obtained results show duality to be a fundamental doctrine of geometry needing a specific language, which is composed not only of words, but also of new visual objects such as the double column notation.

Thus, Gergonne's project was to develop the doctrine of duality beyond the polar reciprocity. Chasles shared this idea. In the rest of his memoir, Gergonne demonstrated Desargues theorem as well as its dual version concerning trihedrons instead of triangles, and the converse proposition with its dual (ibid. pp. 217–219). Some theorems which can be seen as corollaries of Desargues' theorem and his converse proposition follow. Through duality, it is also possible to obtain theorems holding for a spherical surface as soon as the term "straight line" is changed with "great circle". A series of theorems on polyhedrons and on skew polygons follow, where the incidence properties of the faces and the edges are analysed and their dual nature is highlighted (ibid., pp. 223–230) and reduced, ultimately, to Desargues' theorems, its dual and its reciprocal proposition. For example, the first theorem proved by Gergonne claims that, given two tetrahedra such that the straight lines joining their corresponding vertices cut is a point, then the straight lines delimiting their faces are coplanar. Dually, if the terms "point" and "plane" are interchanged, while the term "straight line" is preserved, one has the proposition that, given two tetrahedra such that the straight lines delimiting their faces are coplanar, then the straight lines determining their corresponding vertices mutually cut in a point (ibid., p. 223).

Gergonne concluded his paper claiming an important thesis: duality is not a property of a particular transformation, but a general principle of extension which is universally valid for all the graphical properties:

> We believe that what we said is sufficient to avoid any dispute on two points of mathematical philosophy: 1) a remarkable part of geometry exists in which the theorems mutually correspond exactly two by two. This is also true for the reasoning necessary to prove them, which depends on the nature itself of extension; 2) this part of geometry, which will have a large extension if the curved lines and surfaces will be also included, can be completely developed independently from calculations and knowledge of any metric property of the considered quantities.[34]

---

[33]Gergonne (1825–1826), p. 216: "[. . .] ce n'est seulement que lorsqu'une science est déjà parvenue en un assez haut degré de maturité qu'on peut espérer d'en bien faire la langue".

[34]Gergonne (1825–1826), pp. 230–231: "Nous croyons en avoir dit suffisamment pour mettre hors de toute contestation ces deux points de philosophie mathématique, savoir I.° qu'il est une partie assez notable de la géométrie dans laquelle les théorèmes se correspondent exactement deux à deux,

In the numerous papers Gergonne dedicated to duality in connection with his polemic against Poncelet (Gergonne, 1826–1827a, 1826–1827b, 1827–1828a, 1828–1829, 1847), he confirmed the opinions expressed in the text I have analysed: duality is a general law of extension which goes far beyond Poncelet's polar reciprocity, it holds for graphical properties, it shows the profound properties of space. It is appropriate to add that in Gergonne (1826–1827a) the author developed a series of epistemological reflections which are a clear evidence of the foundational value that he attributed to duality: in modern mathematics, it is useful to trace a few essential truths, so that all the propositions are mutually linked. While dealing with non-metrical properties, all the propositions are double, which implies a small number of basic principles. This is shown by Poncelet's polar theory, by the theorems of Pascal and Brianchon, obtained without any calculation, on which many important properties of projective geometry are based (Gergonne, 1826–1827a, pp. 214–229). The relations of graphical geometry, based on few principles, among which duality is the most important, have a remarkable heuristic value and make the proofs of many theorems superfluous (ibid. p. 233).

Enriques (1907–1910, p. 83) points out that the first idea to reduce some metric properties to graphical ones was due to Poncelet and was developed by Chasles. Kötter (pp. 162–165) claims that Poncelet discovered duality through the theory of reciprocal polars, that he gave a precise mathematical form to his theory and that he also showed that some metric properties are passible of a dual treatment. Gergonne, in an initial phase of his speculation on duality, was stimulated by Poncelet's reciprocal polars, but he regarded duality as a fundamental law of nature (it is better to say of extension, in my opinion, because as far as I know, Gergonne refers to duality as a general law of extension, not of nature) and had a more profound view than Poncelet. Kötter also stresses the role of Möbius and Plücker in the birth of duality (ibid., pp. 166–171). Fano (1907–1910, pp. 231–233) states that Poncelet first understood the importance of duality, but that Gergonne offered fundamental generalizations. Schönflies (1907–1910, pp. 398–401) expresses an idea close to Fano's: Poncelet was fundamental for the theory of reciprocal polars, but Gergonne achieved a more general conception of duality. Lorenat (2015a, pp. 119–162) dedicates a chapter ("Polemics in public: controversies around methods, priority and principles in geometry", pp. 119–162) to the problem of duality in the polemic Gergonne-Poncelet touching all the most important issues. To testify Gergonne's philosophical interests, Gerini (2016, p. 6) recalls that Gergonne himself inaugurated (arguably in 1817) at the Faculty of Science in Montpellier, what can be called one of the first courses in epistemology. It was named "Philosophie des Sciences". The author also points out that John Stewart Mill attended such course in 1820.

ainsi que les raisonnemens qu'il faut faire pour les établir, et cela en vertu de la nature même de l'étendue ; 2.° que cette partie de la géométrie, qui prendrait une très-grande étendue, si l'on voulait y comprendre les lignes et les surfaces courbes, peut être complètement développée indépendamment du calcul et de la connaissance d'aucune des propriétés métriques des grandeurs que l'on considère."

The conceptions inherited by Chasles with regard to the principle of duality are, hence, interesting: there was a great mathematician, Poncelet, who had developed a working theory of duality, namely the theory of reciprocal polars. He thought that the principle of duality, in substance, coincided with such a theory. There was a geometer, Gergonne, who, as a mathematician, was inferior to Poncelet, but who had interesting methodological and philosophical ideas on duality: he thought it was a fundamental law of extension and should be the basic principle of geometry. Chasles was as able as Poncelet as a mathematician, and with regard to duality, his ideas were even broader than Gergonne's: duality is not only a law of extension. It also concerns the physical world and is applicable beyond geometry. Thus, Chasles developed the property of duality in mathematics beyond Poncelet's reciprocal polars and the properties of duality in philosophy beyond Gergonne's ideas duality to be the basis of extension. It is, according to Chasles, the fundamental law of the whole nature.

After having traced this picture, it makes sense to come back to Chasles' conception of duality.

At the beginning of Note XXXIV, he identified the conceptual change that duality law induced in geometrical thought: he faced the subject in a broad and, to use his expression, philosophical manner. He claimed that, with respect to traditional geometry, duality has made it evident the possibility to consider not only the point as "element", but also the plane (while dealing with spatial geometry) or the straight line (while dealing with plane geometry). For he wrote that duality highlights the existence of

> [...] a constant relation which connects all the geometrical truths two by two: therefore, so to say, two kinds of *Geometry* exist. These two Geometries are distinguishable for a circumstance which is very important to be remarked: in the former the *point* is the unity and, so to speak, the *element*, or the *monad* which you use to form the other parts of extension. This is the basis of ancient Geometry and of analytical Geometry.
>
> In the latter Geometry the *straight line* or the *plane*—depending on whether you are operating in a plane or in space—are considered as the *primitive being*, or the *unity*, necessary to make up all the other parts of extension.[35]

Chasles stressed that Gergonne and Poncelet had the merit to highlight the value of duality law within projective geometry, but he claimed for him himself the merit to have understood the general value of duality as a law (or better, metalaw) of nature. For already in Note V of the *Aperçu* Chasles had remarked that geometry has traditionally be considered the science of the measure, whereas modern geometry

---

[35]Chasles (1837a), p. 408: "[...] une relation constante qui lie deux à deux toutes les vérités géométriques; ce qui fait pour ainsi dire deux genres de Géométrie. Ces deux *Géométries* se distinguent par une circostance qu'il est très-important de remarquer: dans la première le *point* est l'unité, et pour ainsi dire l'*élément*, ou la *monade* dont on se sert pour former les autres parties de l'étendue; c'est là la base de la philosophie de la Géométrie ancienne et de la Géométrie analytique.

Dans la seconde Géométrie, on regarde le *droite*, ou le *plan*, suivant qu'on opère sur un plan ou dans l'espace, comme l'*être primitif*, ou l'*unité* qui doit servir à former toutes les autres parties de l'étendu". Italics in the text.

has to be regarded as the more general science of extension in which the relations of form and position of the figures are taken into account (ibid., pp. 288–289). The ancient geometry was based on a specific philosophical conviction: that monistic, whose origin Chasles ascribed to the Pythagoreans. It seems rather natural that he was thinking of Pythagoreans' idea that the point was the monad representing the ontological basic element on which any mathematical speculation had also to be grounded. Chasles was evidently thinking that such an idea was as a sort of imprinting which permeated geometry and science, even including the theories of authors who, for other aspects, did not share the Pythagoreans' conceptions. In contrast to this, modern projective geometry highlights the importance of duality. According to Chasles, a deep reflection has to lead us towards the idea that duality is a general natural law. He wrote:

> The numerous dualisms detectable in the natural phenomena as well as in the different parts of human knowledges tend, on the contrary, to let us to suppose that a constant duality or double unity is the true principle of nature.[36]

This quotation is noteworthy: Chasles is referring to duality both as an ontological criterium intrinsic to nature as well as a gnoseological basis to develop human knowledge. It seems that he is interpreting geometry as a natural science. To be more precise: from the standpoint of gnoseology and of history of human knowledge geometry, and specifically projective geometry, is the basic discipline because a precise idea of duality has been reached only through geometry and because the concepts of projective geometry are the basis of several notions belonging to kinematics and to statics. However, from an ontological point of view, the duality of projective geometry is the first emergency grasped by human mind of the general natural duality metalaw, if, as Chasles seems to accept, the structure of our knowledge is a mirror of the structure of being. An important question, whose answer seems to me rather complicated and conjectural has to be asked: what relations did Chasles think to exist between physical and geometrical space? As far as I know, there is no explicit answer to this question. We have seen that in his thought there are several elements which, at the end of the nineteenth century, connoted the abstract concept of hypothetical deductive system, but this is not enough to claim that Chasles considered projective geometry as a mere formal system independent of its relations with the physical space. He understood that a more abstract approach to axioms and transformations had to be posed, but, once established (in Chasles' case implicitly, or at least, in part implicitly) the formal apparatus necessary to have a perspicuous geometrical system, does this system, if fulfilled with the appropriate objects, correspond to the physical space? Though no explicit answer exists in Chasles' works, the answer seems to be in the affirmative, which is deducible from the continuous connections Chasles established or tried to establish between geometry and natural science as well as between geometry and technique.

[36] Chasles (1837a), p. 290: "Les dualismes nombreux qui se remarquent dans les phénomènes naturels, comme dans les différentes parties des connaissances humaines, tendent, au contraire, à nous faire supposer qu'une *dualité* constante, ou double unité, est le vrai principe de nature". Italics in the text.

The already mentioned Note XXXIV of the *Aperçu* is a litmus paper of this situation. For after the incipit I have described, and before dealing with duality in natural sciences, Chasles considered an example drawn from technique, which can be connected to geometry. He took into account what he called the arts of construction ("les arts de construction", ibid., p. 409) and, in particular the work of a turner, who can operate in two manners: either he maintains at rest the work he has to construct and makes to move the lathe, or, vice versa—which is what in general happens—he maintains at rest the lathe and makes to move the work (ibid., p. 409). Chasles claimed that we are in front of a dual mode of construction, which relies on a general dual geometry, that is an aspect of geometrical duality not immediately connected with duality in projective geometry. He wrote:

> Here we have in the arts a well pronounced and constant duality of description.
>
> It is known that each of these constructions depends, in every circumstance, on geometrical principles; thence, in the theories concerning the two ways of construction a constant duality exists.[37]

The geometrical principle which is the basis of the duality described for the work of the turner is the one which allows us a double description of a curve through a stiletto. In a broad sense, this question belongs to the problem of finding mechanical means suitable to describe curves, of which the ellipsographs or the parabolographs are examples. This geometrical principle is founded on an extremely easy law of duality ("une loi de dualité extrémement simple", ibid., p. 409) Chasles described like this:

> When a plane figure moves in its plane, one of its points describes a curve;
>
> The movement of this figure is determined by constant relations which have to take place between the figure itself and some fixed points or lines drawn in its plane;
>
> These points and these lines considered together will form a second figure which remains at rest while the former moving;
>
> Now consider the first figure in one of its positions and suppose it to be at rest; afterwards make to move the second figure so that it always maintains the same conditions of position in respect to the first figure;
>
> On the mobile plane of the second figure, a fixed stiletto located in a point which describes the first figure will draw a curve co-moving with such a plane. This curve will be identically the same (apart from the position) as that previously drawn by the point describing the first figure when it was in movement.[38]

[37] Ibid., p. 409: "Voilà donc, dans les arts, une dualité de description, bien prononcée et constante.

On sait que chacune de ces constructions repose, dans chaque circonstance, sur des principes géométriques: il existera donc aussi, dans les deux théories relatives à ces deux modes de construction, une dualité constante".

[38] Ibid., pp. 409–410: "Quand une figure plane est en mouvement dans son plan, l'un des ses points décrit une courbe;

Le mouvement de cette figure est déterminé par des relations constantes, qui doivent avoir lieu entre elle et des points ou des lignes fixes tracées dans son plan;

Ces points et ces lignes forment, par leur ensemble, une seconde figure, qui reste fixe pendant le mouvement de la première:

Since Chasles aspired to offer a general and precise formulation of his idea, the linguistic aspect of his description appears rather odd and complicated, but the meaning is clear: if a certain mobile figure moves on another fixed figure, a curve is described; under appropriate conditions, if the role of the mobile and the fixed figures is inverted, the curve described is the same.

I refer here only to the first example given by Chasles because it is sufficient to fully clarify his view: it is well known that, if two vertices of a triangle move on two fixed straight lines *r* and *t*, the third vertex describes an ellipse. Here, the mobile figure is the triangle and the fixed one is given by *r* and *t*. The duality claimed by Chasles establishes that: be the triangle maintained at rest and move the two straight lines always passing through two vertices of the triangle. Then, when two sides of an angle whose form is invariable slide on two fixed points, a fixed stiletto located in any point draws an ellipse on the mobile plane to which the angle belongs (ibid., p. 410).

Here the duality is between the mobile and the fixed figures, which change their role in the two constructions of an ellipse and between points and straight lines, because in the first figure the points are mobile while the straight lines are at rest; the opposite situation takes place in the second figure.

Chasles also gave the example of dual constructions concerning the hypocycloid and the conchoid of Nicomedes. He clarified that his principle is also applicable when a curve is regarded as an envelope of tangents with the only consideration that, in this case, an instrument different from a stiletto has to be used. The same theory can also be applied to the three-dimensional figures. Thence, Chasles concluded:

> Here a duality of *doctrines* is shown. It concerns the double mechanical description of the bodies, which is well pronounced and which is based on a single and unique theorem, as it is the case for the properties of extension.[39]

Thus: Chasles was inspired by duality in projective geometry and extended the dual principle beyond projective geometry. This last quotation is interesting: he saw a profound homology between the just mentioned theorem in which the duality for the construction of the figures is stated and the FTD. In the former there is a figure which remains at rest and a mobile one. The role of these figures can be mutually changed. In the FTD there is a mobile figure and a fixed figure. In the three cases Chasles addressed, the mobile figures are: 1) point moving on a plane-mobile plane; 2) point moving on a straight line–mobile plane; 3) point moving on a curved

---

Que l'on considére maintenant la première figure dans une de ses positions, et qu'on la suppose fixe, puis, qu'on fasse mouvoir la seconde figure, de manière qu'elle se trouve toujours dans les mèmes conditions de position par rapport à la première figure;

Un stylet fixe, placé au point décrivant de la première figure, tracera, sur le plan mobile de la seconde figure, une courbe mobile avec ce plan, et qui sera identiquement la même (sauf la position) que celle qu'aura tracée d'abord le point décrivant de la première figure, quand celle-ci était en mouvement".

[39] Ibid., p. 411: "Ainsi voilà une *dualité* de doctrines, concernant la double description mécanique des corps, qui est bien prononcée, et qui repose, comme celle des propriétés de l'étendue, sur un seul et unique théorème". Italics in the text.

surface–mobile plane. The fixed figures are, respectively, given by: 1′) fixed point; 2′) fixed straight line; 3′) fixed curved surface. The pairs 1)–1′); 2)–2′); 3)–3′) express the principle of duality in projective geometry which Chasles considered identical, from a formal standpoint, to the principle of duality used for the construction of the figures described in Note XXXIV.

It seems to me appropriate to point out that Chasles used a language based on the movement of the figures. In part such a language was common to the geometry of that period, but in part this also indicates that Chasles looked at the geometrical situations in a dynamical manner; the geometrical figures were for him as ideal objects whose behaviour was like that of the physical objects, but without material properties such as mass. Therefore, his conception is extremely multifaced: from a certain point of view, he considered the geometrical properties in a rather abstract manner. This shows that his doctrine shares some aspects with abstract axiomatic; from another standpoint, the geometrical figures were, for him, as objects "living" in a proper universe. That is: the axioms determine a structure which can be fulfilled with diverse objects. Thence, if, from an absolutely abstract point of view what is important is the structure, it is true that the same structure can be instantiated with different objects. The object does not disappear in the structure. Objects and structures are referred to two different ontological levels: the most general, prototypical and profound level is that of the structures; under this level, the objectual one exists insofar as different sets of objects can satisfy the laws of a structure and the most general and fundamental of these laws is the duality law.

A further evidence of this conception is shown by Chasles' idea that duality law might also be valid for the material world so that the science dealing with it—namely, physics—has also to take into account duality.

With regard to the universal duality existing in the universe, Chasles' ideas play an important role in the general structure of his foundational programme, but they are less defined and precise than his speculations concerning duality in geometry. Probably this is an unavoidable problem when one tries to extend a principle from a field in which it is well circumscribed and defined to a more general—in Chasles' case even to a universal—level. Chasles was aware that his conceptions were hypothetical and conjectural. Nonetheless, he appeared convinced of their truth. Before entering the details, it is appropriate to refer to the Chasles' general view. In the universe two kinds of movements exist: translational movements and rotational movements. All the others can be considered a sum of these two. The celestial bodies are an example of this situation: they rotate around their axes and they translate in space (obviously Chasles is not claiming that their movement is rectilinear, but only that they also have a local motion, ibid., p. 411). Modern mechanics, based on the inertia principle, posed a clear distinction between the two movements. Chasles interpreted such a distinction considering the bodies' natural and elementary movement to be the rectilinear (and uniform) one. In this respect there is no duality between translations and rotations: the rotations need a force, the rectilinear uniform translations do not (ibid., pp. 411–412). However, this granted, is there an ontological level at which translations and rotations are perfectly symmetrical, or, dual, so that a more profound theory discovering more thorough universal laws than the ones known can be constructed? This is the problem posed by Chasles. For he asked the question:

> However, is it possible to conceive a mathematical theory where the two inseparable movements of the celestial bodies of the Universe would play the same role? Then, the principle which should unify these two theories and which should be necessary to pass from the one to the other—as the theorem on which we have based the geometrical *duality* of the resting extension, and that was necessary to connect the two ways of bodies' mechanical description—, this principle, I claim, might spread a clear light on the principles of natural philosophy.
>
> Is it be possible to foresee where the consequences of such *duality* principle would halt? After having connected two by two all the natural phenomena and the mathematical laws governing them, could this principle ascend to the causes themselves of the phenomena?[40]

Therefore, if such a universal duality law, which should be the basis both of the translational and rotational movements, existed, it would be a metalaw, a meta-principle. To use Chasles' terms, this would be a principle which might spread light on the other principles of natural philosophy. This principle might, furthermore, offer the true causal link behind the mathematical laws which are the bases of natural phenomena.

In favour of his idea, he argued that the most general law of nature is Newton's universal gravitation. If the universal duality existed, a dual law of universal gravitation should also exist. Through this law all the phenomena explained by Newton's should also be explained. However, if Newton's law were the only able to explain the celestial phenomena, this would not exclude the existence of the universal duality law: it might be the case that Newton's law be correlative of itself. In this case, this self-duality would be the decisive proof in favour of the fact that Newton's law is the very milestone of the universe (ibid., p. 412).

This told, Chasles clarified that the discovery of the dual universal law has to start considering the infinitesimal movements because, as to the finite movements, the existence of centrifugal forces establishes an irreducible difference between rotational and translational movements (ibid., p. 412). It seems to me that he was thinking of Newton's bucket experiment and of the inertial character of the centrifugal forces existing only in the rotational movements. Thence, the dual doctrine whose final aim should be to prove the existence of duality in the universe (and the dual or self-dual character of the law of universal gravitation) must have its basis in the study of the infinitesimal movements.

---

[40]Ibid., p. 412: "Mais ne peut-on pas supposer, maintenant, que les deux mouvements inséparables des corps de l'Univers doivent donner lieu à des théories mathématiques, dans lesquelles ces deux mouvements joueraient identiquement le même rôle? Et alors, le principe qui unirait ces deux théories, qui servirait à passer de l'une à l'autre, comme le théorème sur lequel nous avons basé la *dualité* géométrique de l'étendue en repos, et celui qui nous a servi à lier entre eux les deux modes de description mécanique des corps, ce principe, dis-je, pourrait jeter jour sur les principes de la philosophie naturelle.

Peut-on prévoir même où s'arrêteraient les conséquences d'un tel principe de *dualité*? Après avoir lié deux à deux tous les phénomènes de la nature, et les lois mathématiques qui les gouvernent, ce principe ne remonterait-il point aux causes mêmes de ces phénomènes?". Italics in the text.

The first manifestation of the universal duality is deducible from Lagrange's works: for in the first edition of his *Mécanique Analytique* Lagrange taught how to decompose a rotational movement around a straight line passing through a fixed point O into three rotational movements around three orthogonal axes passing through O. Lagrange's formulas are similar to those by which a rectilinear movement is decomposed into its three orthogonal components.[41] In the second edition of his masterpiece, Lagrange also explained how to construct the three axes (ibid., pp. 412–413). Therefore, in analogy with the rectilinear movements of a point, it is possible to decompose the rotations around different axes passing through a point. Here the duality is rectilinear movement of a point—rotational movement around one axis.

The analogy between the decomposition of translations and rotations is, in fact, even deeper because it can be extended to rotations taking place around axes which do not cut in a point. After having recalled his own results on the infinitesimal rotation of a rigid body, Chasles remarked that if several rotations around different axes are given, it is possible to adopt a technique similar to that used by Lagrange to decompose a single rotation. This technique is, in substance, that which Rodrigues and Chasles considered when dealing with the PVV: along each axis of rotation draw a segment line proportional to the rotational movement. Consider all these segments as a system of forces acting on the body. These forces can be composed in two forces and their two action lines be regarded as the two rotation axes which replace the original ones. The rotations along these two axes are, thence, represented by the length of the segments proportional to the intensity of the two forces. In this way the infinitesimal movement of rotation is reduced to the consideration of two forces (ibid., p. 413). Here the duality is rotation around two axes—representation of two forces. The opposite process would also be possible, that is to interpret two forces as derived from a series of rotational movements which were reduced to the movements around two axes.

But Chasles' enquire continued: he supposed that each rotation of a body around an axis is associated with a rotating plane passing through the axis, so that, given a body rotating around different axes, a system of planes will be associated with the system of rotations. This is perfectly analogous, Chasles claimed, to consider the forces producing the rectilinear movement of a body as applied to a point of the body which lies in the direction of the movement or of the soliciting force (ibid., p. 414). Here you have a double duality 1) plane-point; 2) rotation-translation.

In the rotation each plane associated with the rotational movement will rotate around an axis belonging to the plane itself. This movement was called by Chasles *effective rotation* ("*rotation effective*", ibid., p. 414). Whereas the partial rotation of the body around the axis belonging to the plane was named *impressed rotation*

[41] Lagranges's formulas which express the decomposition of a rotational movement into three components are explained in Lagrange (1788, pp. 337–389) in the first two chapters of the sixth section entitled "Sur la rotation des corps". In the first chapter, Lagrange proposes the general formulas for the rotational movements; in the second chapter, he focuses specifically on the problem of the rigid body.

("*rotation imprimée*", ibid., p. 414) to the plane. The impressed rotation does not depend only on the effective rotation because it also depends on the rotations of the other planes, so that it can be considered as a component of the effective rotation. Thence, Chasles continued his reasoning by claiming that the effective rotation of a plane is the result of the combination of the impressed rotation with the rotations impressed to the other rotating planes. Therefore, given the analogy between the treatment of translations and rotations and relying on the results I have explained with regard to the movement of the rigid body and to the decompositions of forces, Chasles claimed:

> When a solid body is subject to several simultaneous rotations around different axes; if, through these axes, you conceive to draw some planes in the body, these planes will experience effective movements on themselves;
>
> If you make the product of the effective rotation of each plane by its impressed rotation and by the cosine of the angle between the [two] axes of these two rotations, the sum of these products will be a constant quantity, whatever the planes drawn through the axes of rotation are;
>
> This quantity will be equal to the sum of the squares of the impressed rotations, plus the double product two by two of these rotations by the cosine of the angle between the axes.[42]

Now: if a body subject to some rotations is in equilibrium, and if an infinitely small movement is impressed to the body, the planes through the axes of rotation will experience rotations on themselves. Chasles named them *virtual rotations* of these planes. On these bases he claimed that a principle of virtual rotations analogous to that of virtual velocities exists. He wrote:

> It will be possible to express the equilibrium condition of a body through an equation which will offer a *principle of virtual rotations* analogous to the principle of *virtual velocities*. This is the principle:
>
> Consider different planes of a solid body subject to some rotations around different axes contained in the planes themselves. Give an infinitesimal movement to the body and, for each plane, carry the product of its impressed rotation by its effective rotation and by the cosine of the angle between the two axes of rotation. Then I say that necessary and sufficient condition for these rotations to be in equilibrium is that the sum of all these products is equal to 0.[43]

---

[42] Ibid., p. 414: "Quand un corps solide est soumis à plusieurs rotations simultanées autour de divers axes; si par ces axes, on conçoit menés des plans dans le corps, ces plans éprouveront des mouvements effectifs sur eux-mèmes;

Si l'on fait le produit de la rotation effective de chaque plan, par sa rotation imprimée, et par le cosinus de l'angle que font entre eux les axes de ces deux rotations, la somme de ces produits sera une quantité constante, quels que soient les plans menés par les axes de rotation:

Cette quantité sera égale à la somme des carrés des rotations imprimées, plus le double de la somme des produits deux à deux de ces rotations par le cosinus de l'angle que comprennent leur axes".

[43] Ibid., p. 414: "La condition d'équilibre du corps pourra s'exprimer par une équation qui nous offrira un *principe des rotations virtuelles*, analogue au principe des *vitesses virtuelles*. Voici ce principe:

Quand différents plans d'un corps solide sont soumis à des rotations autour de différents axes contenus dans ces plans; pour que ces rotations se fassent équilibre, il faut que si l'on donne au corps un mouvement infiniment petit quelconque, et qu'on fasse, pour chaque plan, le produit de sa

Therefore, Chasles concluded that, at least with regard to statics and infinitesimal movements, there is a perfect duality between the traditional approach to mechanics based on the pair *infinitesimal translation-point* and the new dual approach based on the pair *infinitesimal rotation–plane*. He also asserted that his new dual doctrine of the infinitesimal movements should be included in kinematics, the new branch of rational mechanics named by Ampère and to which Chasles himself had given fundamental contributions (ibid., p. 415).

As a commentary, one might add that the idea of a possible duality between infinitesimal rotations and infinitesimal translations is confirmed by the fact that the sum of infinitesimal rotations is commutative, exactly as the sum of translations, whereas the sum of finite rotations is not commutative. Therefore, while dealing with finite motions, rotations cannot be considered dual of translations because: 1) from a geometrical-kinematical point of view, their sum is not commutative; 2) from a dynamical standpoint, the existence of centrifugal forces poses an irreducible difference between rotations and translations, while, at infinitesimal level, the duality devised by Chasles could not be excluded.

Chasles' conception of duality derives from the duality of projective geometry point-plane. This is evident from the way in which Chasles developed his ideas regarding the infinitesimal rotational movements based on the concept that either the point or the plane can be regarded as the element of extension.

He was aware that what he called the *double Dynamique*—a locution so clear that no commentary is necessary—was in an embryonic phase and that important branches of physics still escaped it. He mentioned celestial mechanics where the traditional approach based on the Cartesian coordinates—through which no duality is expressible—seem to be the most suitable to face astronomy and physical astronomy. This notwithstanding, it seems—though he did not express explicitly this idea—that he judged this situation as a temporary phase in the history of physics, which might be overcome by a successive phase in which the whole mechanics could be formulated in dual terms. In this context, it is comprehensible that he ascribed a particular importance to Poinsot's theory of couples. For in this theory he saw the seeds and the traces of the possible new global double dynamics. He was explicit:

> The theory of couples, which we now mention, seems to us a doctrine completely conformal to the ideas of correlation which we are developing. [. . .]. For, in effect, the couples play the same role as the single forces. The latter appear to produce the movement of translation, and the couples the movement of rotation. The ones and the others are subject to the same mathematical laws of composition and decomposition. Thence, we can consider this elegant theory of the couples as an eminently brilliant conception, which is indispensable as an introduction to a complete theory of the double Dynamics, of which we have spoken.[44]

rotation imprimée par sa rotation effective, et par le cosinus de l'angle que font entre eux les axes de ces deux rotations, il faut, dis-je, et il suffit que la somme de tous ces produits soit égale à zéro". Italics in the text.

[44] Ibid., p. 415: "La théorie des couples, que nous venons de citer, nous parait une doctrine tout à fait conforme aux idées de corrélation que nous venons de développer. [. . .]. Partout, en effet, les couples jouent le même rôle que les simples forces; celles-ci semblent destinées au mouvement de

Probably in the attempt to prove that his ideas were not completely isolated, Chasles mentioned a passage of Comte's *Cours de philosophie positive* where the latter wrote that Poinsot's couples are the fundamental element to make the movement of rotation as natural as that of translation, because the couple is the natural element of the movement of rotation as the force is the natural element of the movement of translation (ibid., p. 416).

It seems to me that Comte's assertions are too generic to be interpreted as a proof in favour of Chasles' dual universal conception. There is no doubt that Chasles was deeply influenced by Poinsot's theory of couples. It induced profound reflections on the links between rotations and translations because, in fact, Poinsot created or discovered (according to the different conceptions) an object which is to rotations as a force is to translations. Chasles had the idea that Poinsot's concepts could be the first discovered notion of a dual and general mechanics.

6.4 Chasles and the Dual Aspects of Poinsot's *Théorie nouvelle de la rotation des corps*

Note XXXIV is closed by a brief observation concerning the work by Poinsot *Théorie nouvelle de la rotation des corps*,[45] a text published after that Chasles had already composed Note XXXIV, as he himself remarked (ibid., p. 416). In his book Poinsot introduced a series of considerations concerning the rotational motions which were interpreted by Chasles as an important step in favour of the existence of the dual dynamics. Chasles' reference to Poinsot's work is appropriate because the *Théorie nouvelle de la rotation des corps* encompasses numerous aspects which show how many ideas were shared by Chasles and Poinsot, though the latter was always reluctant to mention the works of other authors. We have already seen the existence of this *communaté a penser*, of which the *Théorie nouvelle* represents a further and significant step.

Both of them considered geometry as the basic doctrine for the foundation of science. In Poinsot (1834, 1851) the author claimed explicitly that his basic idea was to offer a treatment of the rotational motion of a rigid body (be it free, constrained to rotate around and axis or constrained to rotate around a point) which allowed the reader to follow instant by instant, as in a series of photograms, the instantaneous movement of the body as well as the continuous movement during a finite time. His aim was to present a picture of the rigid body's rotational movement decomposed into its geometrical components. These components are influenced by the forces and

translation, et les couples au mouvement de rotation; les unes et les autres sont soumis aux mêmes lois mathématiques de composition et de décomposition. Nous pouvons donc regarder cette élégante théorie des couples comme une conception éminemment heureuse, et qui était indispensable, comme introduction à une théorie complète de la double Dynamique, dont nous avons parlé".

[45]Poinsot (1834). Consulted edition 1851. I indicate this text as Poinsot (1834, 1851).

the couples of forces acting on the body. Poinsot offered a dynamical treatment, not a merely kinematical one. However, the true originality of this treatment was its geometrical description based on the two curves he called *polhodie* and *herpolhodie*.[46] Poinsot was clear in explaining that Euler's and Lagrange's contributions had solved most of the problems concerning the rigid body's rotation, but the geometrical visualization of such a rotation was something no one had appropriately studied and expounded. The two great mathematicians offered the calculations to solve these problems, but they provided no explanation either of the causes determining such a movement or of the way to visualize the movement. Whereas Poinsot had intention to offer both of them. In an interesting *Réflexion générale* inserted in his work he wrote:

> Therefore, we are led through the only reasoning to a clear idea which the geometers are not anymore obliged to draw from the analysis' formulas. This is a new example which shows the advantage of this new [geometrical] method: it is simple and natural in order to consider the things in themselves. This allows us to visualise the whole evolution of the movement in the course of the reasoning. For, if one is content—as it is usually the case—with translating the problems into equations and with managing them by means of the calculus' transformations to find the solution, he will find often that this solution is even more hidden in these analytical symbols than the nature of the proposed question itself. Therefore, the art which guides us to the discovery is not in the calculus, but in this cared consideration of things. Here our spirit tries—first of all—to obtain a general idea [of phenomena]. To this aim he tries, through the analysis in a proper sense, to decompose the problems into simpler ones and to see eventually the problems as if they were composed of these simple things of which our spirit has a complete knowledge. This does not mean that things are, in fact, composed in this manner, but this is the only way through which we can get an idea of them and, hence, to know them.[47]

In a comparison between Chasles' and Poinsot's thought this long quotation is remarkable. Both of the authors shared the idea that geometry should be the basis of science, but the way in which they declined such an idea is, in part, different: Chasles was convinced that geometry was the ontological basis of the world, not only the epistemic foundation of our knowledge. According to Chasles—we have seen—duality is inscribed in the universe and belongs to our system of knowledge insofar

[46]Poinsot (1834, 1851), p. 94. The English translation for the names of the two curves are polhode and herpolhode, respectively.

[47]Poinsot (1834, 1851), p. 78: "Nous voilà donc conduits par le seul raisonnement à une idée claire quel es géomètres n'ont pu tirer des formules de l'analyse. C'est un nuovel exemple qui montre l'avantage de cette méthode simple et naturelle de considerér les choses en elles-mêmes, et sens les perdre de vue dans le cours du raisonnement. Car, si l'on se contente, comme, on le fait d'ordinaire, de traduire les problèmes en équations, et qu'on s'en rapporte ensuite aux transformations du calcul pour mettre au jour las solution qu'on a en vue, on trouvera lé plus souvent que cette solution est encore plus cachée dans ces symboles analytiques, qu'elle ne l'était dans la nature même de la question proposée. Ce n'est donc point dans le calcul que réside cet art qui nous fait découvrir; mais dans cette considération attentive des choses, où l'esprit cherche avant tout à s'en faire une idée, en essayant, par l'analyse proprement dite, de les décomposer en d'autres plus simples, afin de les revoir ensuite comme si elles étaient formées par la réunion de ces choses simples dont il a une pleine connaissance. Ce n'est pas que les choses soient composées de cette manière, mais c'est notre seule manière de les voir, de nous en faire un idée, et partant de les connaitre".

as it is a mirror of the being. Whereas Poinsot did not refer to the universe in itself. His way of reasoning seems close to Kant's: we can say nothing as to things in themselves; but, as to our knowledge, geometry is the most satisfactory means through which we can create an image of phenomena suitable for our spirit, so that phenomena themselves appear under our control and comprehension. Calculus is an instrument, a fundamental instrument—which Poinsot himself will develop after the geometrical descriptions of the rigid body's motion—but, in itself, it does not allow us to obtain a profound understanding of phenomena.

For Chasles, geometry is a universe in itself, in this respect he can be considered a Platonic, whereas, according to Poinsot, geometry is a product of our mind.

In reference to the use of analysis—in the modern meaning—the two authors also share the basic idea: the analytical calculations are only an instrument, but the comprehension of the mathematical and physical facts must not rely exclusively on them. There is a more profound level that geometry and, so to say, a qualitative study of mathematics and physics can reach to disclose the truth in a clear manner. This level is precluded to the mere calculations. We will see that Chasles' works on the attraction of ellipsoid are as qualitative as those by Poinsot on the rigid body rotation.

While developing the order of ideas which led Poinsot to the concepts of polhode and herpolhode and to the famous visualization of the phenomenon of the equinoxes precession through the so-called Poinsot cones[48]—which are an application of the general theory of the polhodes and herpolhodes—he created several concepts. If they are interpreted in the light of the notion of duality, they seem to substantiate some of Chasles' ideas with regard to the central role of duality itself. Poinsot underlined the duality between the classical concepts used to describe the translational motions and those he employed to describe the rotational one, but he did not draw so general ideas on the role of duality as Chasles did. However, it is comprehensible that the latter used Poinsot's work to claim his thesis. It is, thence, significant for my aim to refer to the most important ideas expressed by Poinsot in the *Théorié nouvelle* because they might be interpreted as a support to Chasles' conception of duality.

First of all Poinsot, referring to conceptions already explained in his *Statics*, observed that the couples of forces whose axes pass through a point as well as the rotations themselves can be composed in the same manner as the forces acting on a point. It is enough to consider the momentum of the couple, which can be

[48] In the next pages, I will offer a description of Poinsot's cones as far as it can be useful to spread further light on Chasles' and Poinsot's conception of duality. Poinsot (1853) is specifically dedicated to this question. On Poinsot's cones, there is a vast literature, I only mention: Benettin (2005), in particular pp. 49–62; Bhat and Crasta (2018) (improved version 2019); De Champ (1867); De Iaco Veris (2018), in particular section 7.9 entitled "Unsymmetrical Body Not Subject to External Moments (Geometric Solution)", pp. 1016–1022; Hecht (2011); Teodorescu (2007) (original Romanian edition 2002), in particular pp. 319–321. All these works do not have a historical approach, but explain the movements of Poinsot cones either in a manuals perspective or in an advanced research perspective. The explanations offered by De Iaco Veris seem to me particularly clear and perspicuous.

represented as a segment of the couple's axis proportional to the angular velocity produced by the couple on a uniform body, e.g. a homogeneous sphere, to which it is applied. Therefore, the couples can be composed by treating their momenta as forces and the rotations can be composed by treating their angular velocities as segments proportional to the momenta. Poinsot offered a geometrical proof of this theorem (Poinsot, 1834, 1851, pp. 5–8). He stressed that angular velocity—independently of its causes—of a rotating body can be treated as an oriented segment for which he established the right-hand rule (ibid., p. 5).

From Chasles' point of view this can be interpreted as the existence of a duality between the composition of the translations and the rotations around a series of axes passing through a point as well as a duality between the causes of the translations (forces) and the causes of the rotations (couples).

After dealing with compositions of rotations around parallel axes, Poinsot introduced an important concept, that of couple of rotations: two rotations $p, -p$ having the same angular velocity in modulus, but opposite in sense, generate a couple of rotations, an object which, he claimed, cannot be reduced to a single rotation, as it produces a *sui generis* rotation, namely a translation (ibid., p. 10).

Here we have a perfect duality, too: a couple of equal and opposite translations along parallel lines produces a rotation, analogously a couple of rotations produces a translation.

Poinsot then addressed the problem of the rotation around axes located in any position in space and proved Chasles' theorem without naming Chasles (ibid., pp. 13–14).

However, he wrote something which probably made Chasles happier than a direct quotation of his work. Poinsot wrote:

> You observe the prefect symmetry of these compositions of rotations and of forces. They are identical, at all. For if one, at the beginning, had given the name of *force* to the cause able to produce a rotation around an axis, he would have found, for these new forces, a Statics completely similar [to ours]. With the only difference that, in this new Statics, the simple forces (considered as moved to the gravity centre of the body) would correspond to our couples in the usual Statics, and the couples would correspond to our simple forces.[49]

Poinsot added that for our spirit it is more comfortable to consider the translational motions as the basic ones, but this is not in the order of nature, only in our mental order. Therefore, dynamics can be treated in two different manners. They only depend on giving the name of force to the cause of a translation or of a rotation. Poinsot claimed that this is an important philosophical question, which, I add, is strictly connected to the preference ascribed to the treatment of phenomena in inertial or non-inertial reference frames.

[49] Poinsot (1834, 1851), p. 14: "On voit la parfaite symétrie de cette composition des rotations et de celle des forces: elles sont presque identiques ; car, si l'on avait primitivement donné le nom de *force* à la cause capable de faire tourner sur un axe, on aurait eu pour ces nouvelles forces une Statique toute semblable. Seulement, dans cette-ci, les simples forces (toujours considérées comme transportées au centre de gravité du corps) auraient répondu à nos couples dans la Statique ordinaire, et les couples auraient répondu à nos couples forces". Italics in the text.

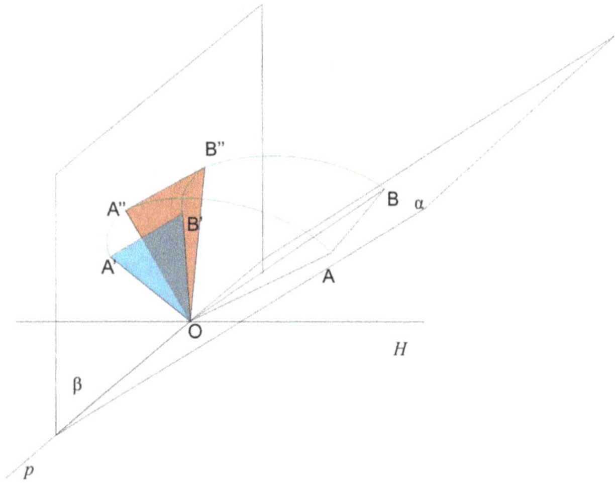

Fig. 6.5 Diagram representing the situation described by Poinsot

Poinsot's short quotation is easily interpretable as a real hymn to duality: there is a perfect symmetry between translations–their causes and rotations–their causes. The one is not the foundation for the other. It depends on the point of view you decide to adopt. It is, thence, not surprising that Chasles mentioned Poinsot's work as an indication that his ideas on duality were correct.

However, duality plays a fundamental role in the entire construction devised by Poinsot: the true aim of Poinsot was to describe the motion of a rigid body rotating around a point because he claimed that, while the rotation of a body around an axis is clear, that of a body rotating around a point is not (ibid., p. 15). Therefore, he reduced the latter to the former: Poinsot considered the motion of the body in an instant. If the body M rotates around the point O, instantaneously such a motion can be considered a rotation around an axis passing through O. In an instant, the axis' direction does not change. Consider two points A, B belonging to M, then the triangle OAB (which lies in the plane α) will be transformed into $OA'B'$ (which lies in the plane β). This infinitesimal rotation can be composed as follows: 1) a rotation p around the intersection line OS of the planes containing the two triangles, through which the triangle OAB is moved into the triangle $OA''B''$ belonging to the plane of $OA'B'$; 2) a rotation q around the perpendicular OH to this plane by which A'' is moved in A' and B'' in B' (Fig. 6.5).

But these rotations can be composed into a single rotation θ through an axis OI. This axis is named by Poinsot instantaneous axis ("axe instantané", ibid., p. 16). Through the same reasoning it will be possible to determine the instantaneous axis of rotation in the following instant ("instant suivant", ibid., p. 15).

In this manner the rotation around a point is reduced to the rotation around a series of instantaneous axes.

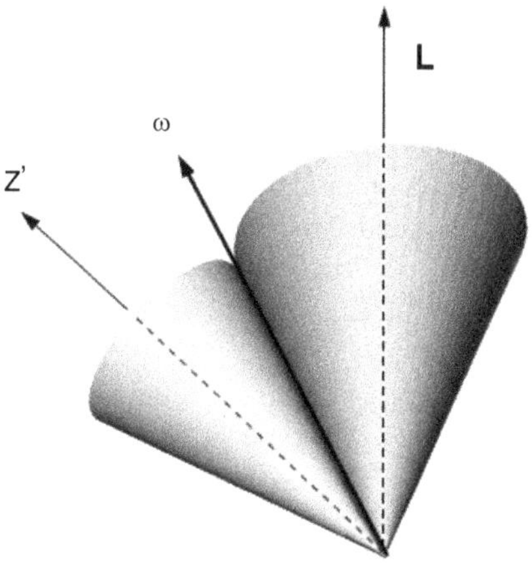

Fig. 6.6 An image which shows the Poinsot cones. This diagram can be useful for the reader to follow the reasoning expounded in the running text

Poinsot made his visual-geometrical description even deeper and clear: since the instantaneous axis of rotation passes always through the point O, it describes a fixed cone in absolute space and a mobile cone in the rotating body (Fig. 6.6).

Poinsot imagined a fixed sphere with its centre in O cutting the two conical surfaces in two curves which can be considered as the bases of the two cones. The curve σ traced on the surface of the fixed cone is, in turn, fixed, while the curve s drawn on the mobile cone is fixed in the body, but mobile jointly with the body, in absolute space (ibid., p. 17). Thence, after a brief and intuitable series of reasoning Poinsot reached the conclusion that the continuous motion of a rigid body around a point can be explained in the following way:

> In any manner a body moves rotating around a fixed point, this movement cannot be nothing but that of a cone, whose vertex is in this point, and which rolls without slipping on the surface of another fixed cone having the same vertex.[50]

Poinsot pointed out that in the single instant the instantaneous axis of rotation is fixed with respect both to the body and to the absolute space, while in an interval of time it is mobile in the absolute space (ibid., p. 19).

After having introduced the concept of angular velocity, Poinsot was able to easily prove the fundamental theorem according to which if the two curves s and σ and the angular velocity θ around the instantaneous axis OI are given, then the body's movement is completely determined (ibid., p. 20).

Commentary

1. The first commentary concerns the use made by Poinsot of the concept of instant: we have seen that he was very critical of the concept of virtual infinitesimal motion as used in the principle of virtual velocity. It is necessary to claim that

[50]Poinsot (1834, 1851), p. 18: "De quelque manière qu'on corps se meuve en tournant autour d'un point fixe, ce mouvement ne peut être autre chose que celui d'un certain cône, dont le sommet est en ce point, et qui roule actuellement, sans glisser, sur la surface d'un autre cône fixe de même sommet".

Poinsot's notion of instant—rigorously speaking—is not less mysterious than that of virtual infinitesimal motion. For he spoke of an axis which is at rest in an instant and which moves in a successive instant, as if something like an infinitesimal actual entity—the instant—might exist, as if the temporal continuous were composed in act of instants. Without referring to modern conceptions of space and time, which cannot be ascribed to scientists who lived in the nineteenth century, Poinsot's view is, strictly speaking, wrong. However, we might interpret his idea as an approximation to reality: if the interval of time is small enough, nothing relevant can happen and the state of rest of the instantaneous axis is an acceptable approximation of reality. Therefore "instant" would not mean "zero time", but "a time sufficiently small for nothing to happen". In a visual description of the movement as that devised by Poinsot this is surely acceptable. Whereas the fiction of the virtual infinitesimal movement is not acceptable—or, at least, it is not an approximation of reality, but a formal, an abstract concept superimposed to reality—because, as a matter of fact, no movement exists. Thence, the instant is a model of reality which is an approximation of reality itself, while the virtual infinitesimal motion is not an approximation of reality.

2. Duality plays also here a fundamental role. For the continuous motion of a body is described by means of two dual figures: the two cones, one fixed in absolute space and the other mobile in absolute space and fixed at the rotating body. The two curves σ and s are also two dual entities used by Poinsot to visualize the continuous motion of a rigid body. It seems to me that Poinsot's idea is perfectly in line with Chasles' description of the geometrical infinitesimal motion of a rigid body: indeed, as we have seen, Chasles proved that such a motion can be determined by means of two rotations around the two conjugated straight lines D and Δ. In the same manner, the geometrical properties of the continuous motion can be determined by the two curves σ and s. If the development of the motion in time is also required, then it is necessary to add the knowledge of angular velocity.

Thence, duality—interpreted in a broad sense and not only restricted to the duality principle of projective geometry—is a feature connoting the intimate structure of Poinsot's work. It is, therefore, comprehensible that Chasles mentioned the *Théorié nouvelle* as a writing in which some elements of the general duality law devised by Chasles could be found. To be more precise, the dualities expounded in Poinsot's work are a manifestation of such a general law. Not only does a perfect symmetry between the treatment of translations and the treatment of rotations exist, but also within the theory of rotations a symmetry between the cone drawn in absolute space and that drawn in the body subsists.

In the following analysis offered by Poinsot, duality continues to be a possible interpretative key of his work, a key which was clearly used by Chasles. Thence, I will focus on the dual aspects of Poinsot (1834, 1851), only mentioning his general train of thought for the reader's convenience. I will not go into details, unless they are directly connected to our reasoning on duality.

Poinsot was able to describe the precession of the equinoxes as the rolling of the mobile cone S on the fixed cone Σ (ibid., p. 24) and he could prove Chasles' theorem in a manner different from Chasles' (ibid., pp. 25–26) because he used the notion of angular velocity whereas Chasles' proof was purely geometrical. In this respect, Chasles' attitude was more purist than Poinsot's because the latter, though offering a visual and geometrical treatment, used the concept of angular velocity, whereas Chasles resorted only to geometrical concepts.

After the kinematical treatment Poinsot began the dynamical one, focusing on forces and their decomposition. With regard to the rotational movement with angular velocity θ of a rigid body around an axis, he considered the mass element (he named *molecule*) dm in an infinitesimal interval of time, which he posed to be unitary. Thence, the expression of the force is $F = \theta r dm$, where r indicates the distance of the molecule from the axis. Consider the axis of rotation as z-axis and decompose the force along the three perpendicular coordinate axes. If x,y,z are the coordinates of the molecule in this reference frame, whose origin O can be posed in any point of the rotation axis, then F can be decomposed into the three components parallel to the axes. Transport these three components parallel to themselves and apply them to O. Under these hypotheses, it is:

$$\begin{cases} X = \theta y dm \\ Y = -\theta x dm \\ Z = 0 \end{cases}$$

where X, Y, Z represent the components of the force along the three coordinate axes. If L, M, N indicate the three couples of forces producing the rotation around the three coordinate axes, their momenta, according to the general theory of momenta will be, respectively, $(Yz - Zy)$, $(Zx - Xz)$, $(Xy - Yx)$. Hence:

$$\begin{cases} L = \theta x z dm \\ M = -\theta y z dm \\ N = \theta(y^2 + x^2) dm \end{cases}$$

By integrating these quantities in dm you get all the forces and couples determining the motion of the body (ibid., pp. 30–32).

Some corollaries concerning the systems of forces and couples when the rotation axis passes through particular positions, e.g. the gravity centre of the body, follow. After these corollaries Poinsot inserted one of the most remarkable sections of his entire book. It is entitled "Des forces centrifuges qui naissent de la rotation" (ibid., pp. 35–41): at the beginning, Poinsot considered an instantaneous force $F = \theta r dm$ tangent to the molecule dm. If the body rotates around an axis, a centripetal force $-dm\theta^2 r dt$ is produced (r is the distance of the molecule from the axis). Though he did not use this expression, it is clear that he judged this analysis of forces sufficient if the problem is considered in what nowadays we call an inertial reference frame.

However, Poinsot added the consideration that the only real force is F. The centripetal force intervenes only to produce the rotation around the axis (ibid., p. 37). Then, if you consider the instantaneous situation as experienced by the molecule dm, it feels three actions: 1) the push of F; 2) a force $-dm\theta^2 rdt$ which tends to move dm towards the axis, 3) a force $+dm\theta^2 rdt$ which tends to move dm radially outside from the instantaneous centre of rotation.[51] He named $+dm\theta^2 rdt$ the "centrifugal force". An important part of Poinsot's treatment is developed basing on centrifugal forces, thence, we would say, on a non-inertial reference frame. He proved that all the centrifugal forces acting on the molecules of the rigid body can be transported in the instantaneous centre of rotation O and they can be decomposed in a resultant force π and in a resultant couple χ, whose expressions he obtained easily.

Poinsot stressed that the principal axes of inertia of a body can be obtained in two different manners: either by searching when the actual forces θrdm are reducible to a single couple N perpendicular to the axis or by searching the axis around which the body has to rotate for all the centrifugal forces $\theta^2 rdm$ which derive from the rotation to be mutually in equilibrium. A similar analysis can be developed when the body is constrained to rotate around a fixed point (ibid., pp. 39–40).

An interesting problem addressed by Poinsot, which is not only important in theoretical physics, but also for practical applications to engineering and architecture is the following one: if a body is mobile around an axis, when it is urged by a couple of forces N perpendicular to the axis, it is required to determine: a) its angular velocity, b) the percussion to which the axis is subject in the first instant after the application of the couple; c) the continuous pressure it has to bear because of the centrifugal forces produced by the rotation (ibid., pp. 42–45). The treatment is developed resorting to centrifugal forces.

Furthermore, be P the force and K the couple which, combined with N, produce a spontaneous (ibid., p. 42) rotation in the body. This means a rotation which causes no pressure on the rotational axis. Be G the couple resulting from K and N. If the axis is fixed and can only rotate in its position, during the movement, P and G are conserved within the body. However, if the axis becomes free, the general principles of dynamics prove that, though P and G are not conserved within the body, they are in absolute space (ibid., p. 45), or, as we would say, in an inertial reference frame.

[51] Poinsot wrote: "Then it is possible to imagine that in the instant dt the first two forces are used to make the molecule freely rotate along an arc of circle θrdt, while the third force $+dm\theta^2 rdt$ makes to move this molecule far from the centre". Ibid., p. 37: "Mais alors on peut imaginer que pendant l'instant dt, les deux premières forces sont employées à faire tourner librement la molécule par un arc de cercle θrdt, tandis que la troisième force $+dm\theta^2 rdt$ tire cette même molécule dans le sens où elle tend à l'éloigner du centre". It is to point out that the notion of inertial reference frame was explicitly introduced by Lange relying upon the works by Neumann (see Lange 1885a, 1885b, 1885c, 1886, 1902; Neumann 1870). In principle, in Newton's work the distinction between inertial and not inertial reference frame is implicit, but evident. Nonetheless, the full acquisition of the concept of inertial reference frame was not an easy one and required a long time. In the next pages, I will stress that Poinsot's ideas on the reference frames were not completely clear, though his treatment led to no physical mistake.

These are the principles known, respectively, as *conservation of the forces* and *conservation of the couples* (ibid., p. 45). Given this analysis developed by Poinsot himself, it is clear that the centrifugal forces have nothing to do with the conservation of the forces and of the couples because such forces do not exist in the absolute space and Poinsot wrote explicitly that such forces cannot modify the resultant and the couple of the system (ibid., p. 46). This notwithstanding, he added the following consideration:

> However, since these centrifugal forces are not in mutual equilibrium within the body, the conservation of the force R and of the couple G is rather obscure because in any instant dt a force πdt and a couple χdt are born, which are not null.[52]

This claimed, Poinsot proved directly that the resultant and the couple deriving from the centrifugal forces have no effect on the conservation of forces and couples in absolute space.

From our point of view this proof by Poinsot is a non-sense because it is simply obvious, as Poinsot himself highlighted, that in absolute space (or from a modern point of view, in an inertial reference frame) no centrifugal force exists. Thence, this is a litmus paper that Poinsot did not distinguish the inertial from the non-inertial reference frames in a completely clear manner, though in his treatment of centrifugal forces there is no mistake. The idea to resort to a direct proof shows a clear goal of Poinsot, namely, to offer a treatment of physics which was as most visual and direct as possible.

With regard to duality, obviously the treatment of the problems in terms of centripetal or of centrifugal forces has nothing to do either with duality in projective geometry or with a duality intrinsic in nature. Thence, this part of Poinsot's works is probably less suitable for Chasles to insert it directly within his dual vision. Nonetheless, if I may use this locution, there is a dual flavour in Poinsot's treatment because, at least from a gnoseological and a methodological point of view, the distinction between a treatment in terms of centripetal or centrifugal forces can be interpreted as a sort of duality, in a very broad sense of this term. Thence, probably Chasles interpreted this part of Poinsot's work as inserted within a general picture in agreement with his ideas on duality.

Next Poinsot faced the theory of the momenta of inertia.[53] The momentum of inertia of a particle dm with respect to an axis, being r the distance between the particle and the axis, is defined as r^2dm. The momentum of inertia of the whole body is, hence, $\int r^2dm$. Given a Cartesian system of coordinates x,y,z, Poinsot was easily able to prove that the momentum of inertia of the body with respect to an axis inclined with the angles α, β, γ on the three coordinate axes is (ibid., pp. 50–51)

---

[52] Ibid., pp. 46–47: "Cependant, comme ces forces centrifuges ne se font point équilibre entre elles sur le corps, il reste ici une certaine obscurité sur la conservation de la force R et du couple G, puisqu'il naît à chaque instant dt une force πdt et un couple χdt qui ne sont point nuls d'eux mêmes".

[53] The concept of moment of inertia is due to Euler (1752). He treated analytically the problems connected to the inertia of the rigid body. A complete and geometrical treatment is due to Poinsot.

$$\int r^2 dm = \cos^2\alpha \int (y^2 + z^2)dm + \cos^2\beta \int (x^2 + z^2)dm$$
$$+ \cos^2\gamma \int (x^2 + y^2)dm$$
$$-2\left(\cos\alpha\cos\beta \int xydm + \cos\alpha\cos\gamma \int xzdm + \cos\beta\cos\gamma \int yzdm\right).$$

He, thence, identified the quantities which nowadays, apart from dm, constitute the tensor of inertia:

$$\int (y^2 + z^2)dm, \quad \int (y^2 + z^2)dm, \quad \int (x^2 + z^2)dm$$
$$\int xydm, \quad \int xzdm, \quad \int yzdm.$$

He named these quantities, respectively, as A, B, C, l, m, n.

Poinsot's two next steps were: 1) suppose the value H for the momentum of inertia of the body be given. He was able to determine the axes with respect to which the value of the momentum of inertia is exactly H, being the angles α, β, γ variable. The set of these axes is given by the generatrices of a cone centred in the origin. Though Poinsot did not use this name, we might call it "cone of inertia" (ibid., p. 52); 2) he was also able to find the principal axes of inertia, namely those axes for which the value of the integrals l,m,n is 0 (ibid., pp. 53–64).

The latter is a very important problem. Poinsot was able to identify these axes as the three perpendicular diameters of the second order conic surface whose vertex is in the origin and whose generatrices are the straight lines around which the body has the same momentum of inertia (ibid., p. 53). He then analysed the variation of the momentum of inertia of the body with respect to the three principal axes determining the extremal properties of such axes (ibid., pp. 56–59).

Poinsot drew a general methodological conclusion which is important insofar as duality is concerned. As he claimed:

> Therefore, when you have to consider only the momenta of inertia, you can always abstract from the form of the body. Rather, you can always suppose that this body is reduced to a more regular one as an ellipsoid or a simple rectangular parallelogram, having the same principal momenta of inertia as the irregular body. [...]. Through this consideration, the problem of the rotation of a body is clarified by replacing a simpler and more easily conceivable figure, as in the movement of translation you reduce the body to a single point which is the gravity centre.[54]

[54]Ibid., pp. 59–60: "Donc quand on ne considère que les moments d'inertie, on peut toujours faire abstraction de la figure du corps, ou plutôt on peut toujours supposer réduite à celle de quelque corps plus régulier tel qu'on ellipsoïde, ou même un simple parallélipipède rectangle, qui aurait les mêmes moments principaux d'inertie. [...] Par cette considération, on éclairait le problème de la rotation

As a regular figure whose axes are the principal axes of inertia, Poinsot chose an ellipsoid, which he called "central ellipsoid". Nowadays it is named "ellipsoid of inertia".

What is important to highlight in my perspective is the perfect duality that Poinsot established between the translations and the rotations insofar as their dynamics is concerned. If one would like to use a proportion, one might claim that, according to Poinsot

Translations : gravity centre = rotations : central ellipsoid.

This section of Poinsot's work seems to me easily interpretable in terms of duality. When Chasles read it, he very likely found consistent indications in favour of a general duality law of nature. Thence, this confirms that Chasles' reference to Poinsot's *Théorié nouvelle* was appropriate to claim his thesis on the existence of a general dual law.

Poinsot closed the first part of his work by proving what nowadays is known as Steiner theorem (ibid., pp. 60–61) and determining all the straight lines with respect to which the moment of inertia of a rigid body is the same. The generatrices of the cone of inertia are a subset of the set of all such lines (ibid., pp. 61–64).

In the second (out of three) part of his work, Poinsot solved the problem of the rotation of free bodies. He introduced a series of fundamental concepts: the first is that of "arm of inertia": given an axis with respect to which the momentum of inertia of a body is mK^2, the quantity K is defined by Poinsot "arm of inertia". We read:

> The arm of inertia is nothing but the side of the mean square among the squares of the distances of all the equal molecules of the body from the considered axis.[55]

Now, Poinsot continued: if, given a system of coordinates x, y, z and being α, β, γ the three arms of inertia of the body with respect to the three coordinate axes, for the problem concerning the rotational movement of the free body, it is enough to know the centre of rotation O, the direction of the three axes and the three arms of inertia with respect to such axes (ibid., p. 66). Then, it is sufficient to construct a symmetrical figure with O as centre and x, y, z as orthogonal axes of symmetry. Thus, such a figure can be the following ellipsoid

$$\alpha^2 x^2 + \beta^2 y^2 + \gamma^2 z^2 = R^4$$

being R a constant. The ellipsoid can be written in its canonical form

des corps, en substituant une figure plus simple et plus facile à concevoir, comme dans le mouvement de translation on réduit le corps à un seul point qui est le centre de gravité".

[55] Ibid., p. 66: "[...] le bras de l'inertie, n'est autre chose que le côte du carré moyen entre les carrés des distances de toutes les molécules égales du corps à l'axe que l'on considère".

$$\frac{x^2}{a^2}+\frac{y^2}{b^2}+\frac{z^2}{c^2}=1$$

with $a=\frac{R^2}{\alpha};b=\frac{R^2}{\beta};c=\frac{R^2}{\gamma}$.

Poinsot was able to prove that for such an ellipsoid, given any diameter, the body's momentum of inertia with respect to this diameter is reciprocal to the diameter's square. Because of this remarkable property he named this figure *central ellipsoid* ("*ellipsoïde central*", ibid., p. 67). By means of the concept of central ellipsoid Poinsot was able to describe the movement of a free rigid body starting from the first instant in which the movement begins and extending his examination to the continuous movement. He determined the way in which the angular velocity varies during the movement and offered minute descriptions of the couples and of the forces involved in the movement, referring them both to absolute space and to the point of view of the ellipsoid. He remarked once again (ibid., p. 72) the analogy between the central ellipsoid and the centre of gravity. He concluded the first chapter of this second part with a "General reflexion" ("Réflexion générale", ibid., pp. 78–81) in which he clearly claimed that the great results obtained by Euler and Lagrange on the rotation of a rigid body, though irreprehensible from an analytical point of view, lacked any geometrical evidence. So that one sees that the calculations work, but he does not understand why they work. His aim was exactly to offer a geometrical and visual representation of the phenomena.

The second chapter is fundamental for our examination of duality: in the first chapter, Poinsot proved that if G is the couple producing the rotation of the central ellipsoid, the plane of such a couple is always at a constant distance h from the centre O of the ellipsoid (ibid., p. 76). But the centre of the ellipsoid is unmovable in absolute space and the plane of the couple is always parallel to itself. Therefore, this plane which touches the central ellipsoid in the instantaneous centre of rotation[56] is the same plane in absolute space. Then, the movement of the ellipsoid is such that the ellipsoid is always in contact with a plane in absolute space. The ellipsoid rotates at any instant on the vector radius from the centre to the contact point. Thence, it rolls without slipping on the fixed plane (ibid., p. 76). Now the problems posed by Poinsot in the second chapters are: 1) what curves does the instantaneous pole of rotation describe on the surface of the central ellipsoid? 2) What curve on the plane fixed in absolute space?

With regard to the first question, Poinsot developed the following reasoning: he indicated the curve with s. Obviously, in general, it is a skew curve. It can be found thinking of the series of points in which an ellipsoid of axes a, b, c is touched by a plane that is always at a distance h from the ellipsoid's centre. To give an idea of the form of such a curve Poinsot used the expression "elliptical wheel" ("roué elliptique", ibid., p. 82). Given the equation of the ellipsoid

[56] The instantaneous pole of rotation is defined by Poinsot as the intersection between the instantaneous axis of rotation and the surface of the central ellipsoid (ibid., p. 70).

$$\frac{x^2}{a^2}+\frac{y^2}{b^2}+\frac{z^2}{c^2}=1,$$

the distance of the centre from the tangent plane is expressed by

$$1:\sqrt{\frac{x^2}{a^2}+\frac{y^2}{b^2}+\frac{z^2}{c^2}},$$

where the letters x,y,z indicate now the coordinates of the contact point. Since this expression has to be equal to h, you get the equation

$$\frac{x^2}{a^2}+\frac{y^2}{b^2}+\frac{z^2}{c^2}=\frac{1}{h^2}.$$

If you intersect this equation with that of the ellipsoid, you obtain the system of equations

$$\begin{aligned}\frac{b^2-a^2}{b^4}y^2+\frac{c^2-a^2}{c^4}z^2&=\frac{h^2-a^2}{h^2}\\ \frac{a^2-b^2}{a^4}x^2+\frac{c^2-b^2}{c^4}z^2&=\frac{h^2-b^2}{h^2}\\ \frac{a^2-c^2}{a^4}x^2+\frac{b^2-c^2}{b^4}y^2&=\frac{h^2-c^2}{h^2}.\end{aligned}$$

These equations express the projection of the curve on the three coordinate planes.

Supposing $a>b>c$, the length of the line h will always be included between that of a and c. From the previous equations it is possible to see that the curve is projected in a whole ellipse on one of the two planes perpendicular to the plane ac, in an arch of ellipse on the other plane and in an arch of hyperbola on the plane perpendicular to the mean axis b. If h=b the curve is plane and it is an ellipse (ibid., p. 83). The radius vector of s is

$$u=\sqrt{x^2+y^2+z^2}.$$

With regard to the curve σ described by the instantaneous pole of rotation on the fixed plane, Poinsot developed this brilliant reasoning (ibid., p. 84): the curve s—whose vector radius is indicated by u—can be considered as the basis of a conic surface whose centre is the centre O of the ellipsoid. During the movement this cone rotates around its generatrix OI, where I represents the instantaneous pole of rotation. Therefore, the curve s, while rotating the generatrix of the cone, draws the plane curve σ described by the instantaneous pole of rotation in absolute space. This means that the infinitesimal arcs ds of the curve s are equal to the infinitesimal arcs $d\sigma$ of the

Fig. 6.7 The central ellipsoid: *OA* is the radius vector of the polhode; *OB* is the distance between the fixed plane and the centre of the ellipsoid; *AB* is the radius vector of the herpolode. Retrieved from Frosali (2014), Third part, p. 8

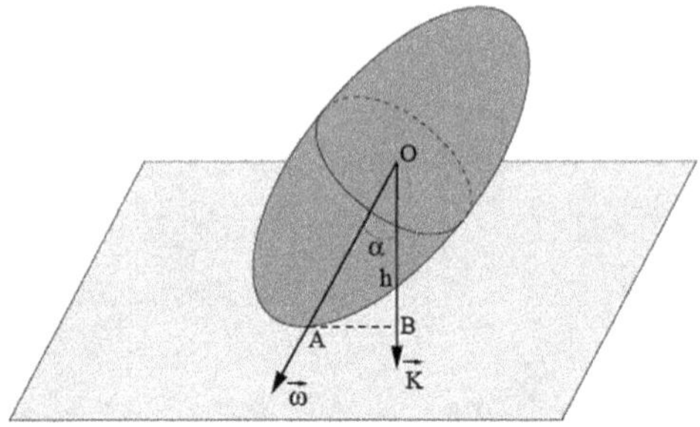

curve σ. Assuming the foot *P* of the perpendicular drawn from the centre *O* of the ellipsoid onto the fixed plane as the origin from which the radius vector ν of σ is calculated, such a radius vector ν is the continuous projection of u onto the fixed plane. Thus, Poinsot concluded his intuitive description like this:

> Thence, you see that the curve σ described by the instantaneous pole in absolute space is a plane curve regularly undulating around a fixed centre; that is, a curve formed by a series of equal and regular waves whose vertices are equidistant and which serpentines infinitely between two concentric circles of which it touches the one or the other circumference alternatively.[57]

Poinsot studied the different shapes and properties of these curves as well their consequences on the movement of the central ellipsoid (ibid., pp. 86–93). Since these curves are strictly related to the instantaneous pole of rotation, Poinsot named them "Polhodie", the curve s and "Herpolhodie" the curve σ (Figs. 6.7 and 6.8).

The rest of Poinsot's book offers a very detailed analytical description of the motion of a rigid body relying upon the properties of the polhode and of the herpolhode. Thus, after the geometrical, visual and intuitive description, Poinsot also transcribed into equations all the concepts which he had developed geometrically.

With regard to Chasles' conception of duality, the work of Poinsot and the specific way in which he reached the concepts of polhode and herpolhode are important, because, as we have outlined, the description of the dynamics of the rigid body in terms of these two curves perfectly fits with the geometrical treatment offered by Chasles in terms of the two conjugate straight lines *D* and Δ. The two geometrical objects s-σ and D-Δ are perfectly compatible with a dual vision of reality which goes beyond the duality connoting projective geometry. Chasles possibly thought that, as he himself had proved, the kinematics of the rigid body can be described in terms of dual entities and, as Poinsot had proved, the dynamics of the rigid body can also be described through dual entities. This might be interpreted as a

---

[57] Ibid., p. 85: "On voit donc que la courbe σ décrite par le pôle instantané dans l'espace absolu, est une curve plane régulièrement ondulée autour d'un même centre ; c'est-à-dire une courbe formée par une suite d'ondes égales et régulières, dont les sommets son équidistants, et qui serpente à l'infini entre deux cercles concentriques dont elle va toucher alternativement l'une et l'autre circonférence".

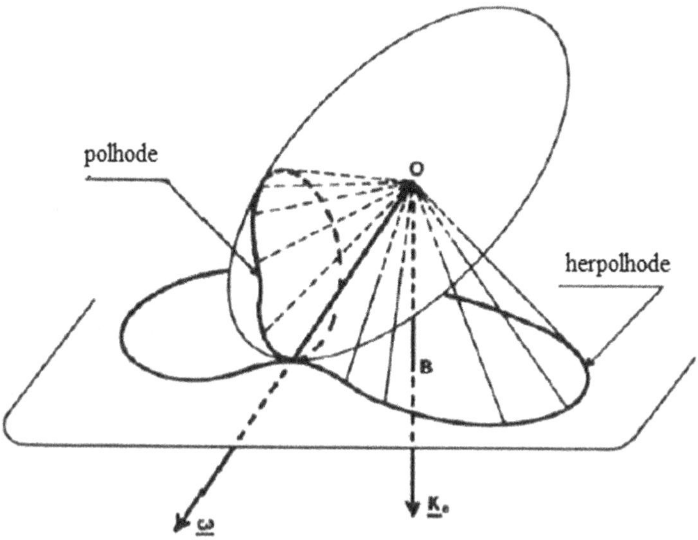

Fig. 6.8 A representation of the polhode and of the herpolode with the respective cones of which the two curves are the bases. Retrieved from Frosali (2014), Third part, p. 9

good indication that a general dual law exists in the universe of which the duality used to explain the movement of a rigid body is one of the numerous applications. Furthermore, Chasles very likely felt completely comfortable while reading Poinsot's theory of the polhode and herpolhode because he was the first author who, at the beginning of the 1830s, had offered a satisfying treatment of the spherical conics[58] and, after all, the polhode can be interpreted as, so to say, an ellipsoidal conic, because the surface cut by the cone is an ellipsoid and not a sphere. Thence, Poinsot's was an order of mathematical and physical ideas which, in great part, fit with Chasles', though the former never formulated an explicit general theory of duality.

Chasles' philosophy of duality as well as the connections between his and Poinsot's thought were not at the centre of the mathematical and scientifical debate. As a matter of fact, the mathematicians and scientists who considered these ideas were not numerous. However, there is at least a mathematician who was particularly interested in Chasles' general ideas and who also pointed out their links with Poinsot's *Théorie nouvelle*. This mathematician was Ernest de Jonquières. The work where he addressed this subject is a text that I have already analysed in Chap. 3 of this book: *Mélanges de Géométrie pure* (Jonquières, 1856). In the first chapter (ibid., pp. 1–54) dedicated to the geometrical infinitesimal motion of the rigid body, apart from Monge and Poncelet, who are mentioned once (Monge on p.1

[58] On the spherical conics, see the two fundamental memoires Chasles (1830h) and Chasles (1831). These two works were translated into English in 1841 (see Chasles, 1841). A later work on the spherical conics is Chasles (1860b).

and Poncelet on p. 11), only Chasles and Poinsot are quoted. In the first two sections dedicated, respectively, to the graphical and metric properties of the infinitesimal movement of a rigid body only Chasles is mentioned, but in the third section, entitled "Analogy between the rotations of a body around different axes and the system of forces" (ibid., pp. 42–54), there are continuous cross references to Chasles-Poinsot as well as a reference to the philosophy of duality.

Jonquières began the third section proving a theorem we have already seen in Poinsot (1834, 1851) and in note XXXIV of Chasles' *Aperçu*: when a body experiences an infinitesimal movement deriving from several rotations around diverse axes, if straight lines proportional to these rotations are drawn on the axes and if these lines are considered as forces solicitating the body, the rectilinear element described by each point of the body will be proportional to the principal momentum of the forces with respect to this point (Jonquières, 1856, p. 42). After having explained the proof of the theorem, Jonquières continued like this:

> Therefore, all the properties concerning the rotations of a body around diverse straight lines, and those concerning the rectilinear spaces described by the points of the body, will produce corresponding properties of a forces' system. These properties regard the forces themselves and their momenta in reference to different points of space and reciprocally.
>
> Following the consequences of this analogy, referring both to Chasles Memoire [Chasles, 1843] and to the methods used by Poinsot in his *Théorie nouvelle de la rotation de corps* (first part, first chapter), you arrive easily to see that "a perfect symmetry exists between the composition of the rotations and that of the forces".[59]

Jonquières considered the property according to which both in the case in which the infinitesimal movement of a rigid body can be attributed to forces or to rotations (namely, to couples of forces) it can be reduced to the rotation of a screw in its nut. He deemed it as a remarkable evidence which testifies the philosophical plausibility of Chasles' idea according to which a universal law of duality exists. It seems to me appropriate to quote these arguments developed by Jonquières because he clearly saw the contiguity of Chasles' and Poinsot's thought and because he understood and, to all appearances, shared Chasles' conviction regarding the existence of a dual universal law which goes beyond the duality connoting projective geometry. As Jonquières wrote:

> This is a thing [the duality rotation-translation] which is surely very remarkable from a philosophical point of view; let us add that this perfect analogy of which we will give further confirmations, does not exist only in the doctrine [but also in nature]. For, if the most general

[59] Jonquières (1856), pp. 43–44: "Il suit de là que toutes les propriétés relatives aux rotations d'un corps autour de diverses droites, et aux espaces rectilignes décrits par les points du corps, donnent lieu à autant de propriétés d'un système de forces, relatives à ces forces elles-mêmes et à leurs moments par rapport aux différents points de l'espace; et réciproquement.

En suivant les conséquences de cette analogie, soit à l'aide du Mémoire de M. Chasles, soit en se reportant aux méthodes employées par M. Poinsot dans *Théorie nouvelle de la rotation de corps* (Ier partie, chapitre Ier) on arriverait facilement à voir: «qu'il existe une parfaite symétrie entre la composition des rotations et celles des forces»". The words included between « » are quoted from Poinsot (1834, 1851), p. 14. Literally Poinsot wrote: "on voit la parfaite symétrie entre la composition des rotations et celles des forces".

movement of a solid body can be attributed only to simple forces applied to it, it can also be attributed in the same way to pure rotations around different axes. Indeed, the body's movement is always reduced ultimately to that of a screw which moves in its nut, whatever the idea we have of the first cause of such a movement be. This remarkable fact is proved by Poinsot in an elegant manner in his beautiful mentioned work (art. 36, 1st chapter).

The science of forces offers, thence, an impressive example which embraces the whole of geometry and of which you find the general proof in the memoire which follows the *Aperçu historique*.

On the other hand, the particular duality which we here highlight has been already pointed out in the Note 34 of the *Aperçu historique* by M. Chasles who considers "*the universal dualism* as the great law of nature and as reigning in all the parts of the human being's knowledge".[60]

In the third paragraph of this quotation Jonquières states that duality in projective geometry has been proved by Chasles in the memoir following the *Aperçu*. He is clearly referring to crowned Chasles' memoir on duality. It is significant that, as to duality, Jonquières did not quote either Poncelet or Gergonne. This is an evidence which shows that, according to his opinion, Chasles was the first author who gave a precise, scientific statement and a profound mathematical justification to the projective duality law. Whereas, before him, the law of duality had been applied and discussed without guessing all the nuances of this concept. With regard to the universal law of duality devised by Chasles, Jonquières was convinced of its existence. It is not a coincidence that, while dealing with a set of subjects connected to this general duality, he mentioned jointly Chasles and Poinsot. I remark that Jonquières mentioned Poinsot's proof of Chasles' theorem rather than the demonstration given by Chasles' himself.

Insofar as the PVV is concerned, Jonquières based his treatment (ibid., pp. 51–54) on Chasles' ideas. In particular, he connected the PVV to the system of forces and to the principle of virtual rotation relying completely on Chasles' results, with particular reference to Note 34 of the *Aperçu*.

Thence, Jonquières' treatment of the infinitesimal rigid body movement as well as his general considerations show that: 1) Chasles was correct in identifying a deep similarity—though as I have tried to clarify not an identity—between his and

[60]Ibid., pp. 44–45: "C'est assurément là une chose très-remarquable au point de vue philosophique; ajoutons que cette analogie parfaite, dont nous verrons plus loin de nouvelles confirmations, n'existe pas seulement dans la doctrine. Car, si le mouvement le plus général d'un corps solide peut être attribué uniquement à de simples forces qui lui sont appliquées, il peut l'être aussi et identiquement à de pures rotations autour de divers axes, puisque le mouvement du corps se réduit toujours en dernière analyse, et quelle que soit l'idée qu'on se fait de la cause première du mouvement, à celui d'une vis qui se meut dans son écrou, fait remarquable qui se trouve également démontré par M. Poinsot, dans le bel ouvrage que j'ai déjà cité (art. 36 du chap. Ier). La science des forces offre donc, à son tour, un exemple frappant de cette *loi de dualité* qui embrasse toute la géométrie, et dont on trouve la démonstration générale et complète dans le Mémoire qui fait suite a l'*Aperçu historique*.

Au reste, la dualité particulière que nous remarquons ici avait déjà été signalée dans le Note 34 de *Aperçu historique* [...] par M. Chasles, qui regarde « *le dualisme universel* comme étant le grande loi de la nature, et comme régnant dans toutes les parties des connaissances de l'esprit humain »". Italics in the text. The words included between « » are quoted from Chasles (1837a), p. 409.

Poinsot's conception; 2) though not many scientists were interested in the physic-philosophical speculations of Chasles on duality, someone was.

6.5 Final Considerations on Chasles' Conception of Duality

The idea that a duality is the basis of being and, hence, of human knowledge is not original by Chasles. The history of philosophy offers so many and well-known examples that it is useless to mention them here. However, the case of Chasles is so remarkable that it deserves attention because the genesis of his general conception of duality is based on mathematics, and, to be more specific, on projective geometry. There is not an a priori idea, either philosophical or theological, which induced Chasles to think of duality. As a matter of fact, there is neither an a priori idea on how the law of the physical universe might be structured. Rather: 1) Chasles was impressed by duality in projective geometry; 2) he understood that through concepts deriving from projective geometry the kinematics of a rigid body could be clearly explained; 3) he realized that the systems of forces as well are subject to a geometrical treatment involving duality; 4) the principle of virtual velocities for a rigid body can also enter within this picture of duality; 5) more in general: rotations and translations of a rigid body are subject to a dual treatment; 6) Poinsot's work reinforced his ideas; 7) Chasles supposed that duality might be extended at least to the whole mechanics until including Newton's theory of gravity. Therefore, from a genetic point of view, his ideas on duality are born within a mathematical context and are extended to a mechanical one where the kinematics and the dynamics of a rigid body offers the empirical evidence in favour of Chasles' thesis. This purely mathematical genesis connotes Chasles' ideas.

It is appropriate to remark that this conception does not derive from an incompleteness in physics because there is no physical problem which Chasles' idea of duality might solve. This is also true for the foundational issues, for example connected to the action at a distance of Newtonian forces. Chasles' view has a philosophical basis: its genesis relies upon his conviction that a superior dual law exists which had, until that moment, escaped the researches of the physicists and of the mathematicians.

There are many differences, but under some respects Chasles' way of proceeding has a certain similarity with Kepler's in *Mysterium Cosmographicum* and in *Harmonice Mundi*.[61] Let us consider, for example, the *Mysterium*: Kepler created the theory according to which the regular Platonic polyhedrons are interposed

[61] With regard to *Mysterium Cosmographicum*, see Kepler (1596). An excellent English translation with a clear notes apparatus is Kepler (1981). A French translation, with a huge amount of useful considerations, is Kepler (1984). As to the *Harmonice*, see Kepler (1619). For an English translation, see Kepler (1997). On these two works, there is a vast literature. I restrict to mention two very classical and excellent contributions. On *Mysterium*, see Field (1988) and on *Harmonice*, see Stephenson (1994). Among my works on Kepler, jointly with Raffaele Pisano and Brunello Lotti,

among the planetary spheres. This theory has many aspects. Duality is a remarkable one because Kepler proposed a dual division of the platonic polyhedron in primary and secondary, and, which is by far more profound and significant from a mathematical point of view, he guessed the duality of the regular polyhedrons' group of rotations.[62] Though duality is part of some aspects of Kepler's argument, its basis is not duality, but rather a tripartition of the universe in regions which correspond to the three people of the Holy Trinity.

In Kepler, elements deriving from religion, from a Platonic-Pythagorean philosophical background, from his Copernican convictions and from his mathematical knowledge converge in order to create the theory expounded in *Mysterium*. Hence, Kepler's inspiration is far from being based only on mathematics. Therefore, the differences with Chasles—given also that the latter lived more than two centuries after Kepler—are obviously enormous.

However, there is a common root: in both cases, the personal need felt by the author to determine a more general and unitary principle which is both ontological and epistemological is the stimulus which gave rise to the theory. Kepler's arguments expounded in the section of *Mysterium* connected to the theory of Platonic polyhedral solve no problem of Copernican theory[63] and were also judged rather bizarre by many Copernicans. The case of Galilei is emblematic: Kepler sent Galileo a copy of his *Mysterium*. Galileo, probably after having read only the initial pages of Kepler's work, where the author expresses his Copernican convictions, answered on 4 August 1597 with an enthusiastic letter, where he wrote:

> [. . .] I promise you that I will read your book with serenity of spirit because I am sure to find there many beautiful things. Actually, I will do that the more willingly the more for many years I have accepted Copernicus's doctrine. Starting from these principles I was able to discovery the causes of many natural effects which remain inexplicable on the basis of the current theories.[64]

Galileo sent no other letter to Kepler on the *Mysterium*, likely because he did not share the whole of the theory explained in that book. When Galileo read the entire text, he possibly realized that the only aspect he shared with Kepler was Copernicanism.

I have dedicated two publications to the themes here outlined. See Pisano and Bussotti (2012) and Bussotti and Lotti (2022, chapter 3).

[62] On this subject, see Kepler (1596, KGW, I, pp. 29–30). For the interpretation of Kepler's passage, see Pisano and Bussotti (2012), p. 128; Field (1988), p. 55; Stephenson (1987), p. 80.

[63] In contrast to this, a long series of considerations in favour of the Copernican system were developed by Kepler in the initial chapters of *Mysterium* and in the course of the whole book. They show that such a system is far more plausible than the Ptolemaic one.

[64] Galilei, EN X, p. 68: "[. . .] pollicebor me aequo animo librum tuum perlecturum esse, cum certus sim me pulcherrima in ipso esse reperturum. Id autem eo libentius faciam, quod in Copernici sententiam multis abhinc annis venerim, ac ex tali positione multorum enim naturalium effectuum caussae sint a me adinventae, quae dubio procul per comunem hypothesim inexplicabiles sunt".

Both in Kepler and Chasles the principle is not something vague, but it is expressible in precise mathematical forms: Kepler's principle of order has a manifestation in the archetypical geometrical and musical structure of the universe. In Chasles, such a principle has a concrete manifestation in the duality to which the movement of a rigid body is reducible.

In both cases, we are in presence of ideas deriving from the desire of the two authors to determine a more basic foundation of science with respect to the then given one. One might say that an epistemic conception of order led both Kepler and Chasles to formulate their ideas, which, for the rest, are very different.

Another interesting example that suggests a possible comparison with Chasles is that of Wolfgang Pauli. This is directly connected to duality. As is well known, Pauli adhered to the so-called Copenhagen interpretation of quantum mechanics. Therefore, he accepted the duality wave/corpuscle and the statistical aspects of quantum mechanics as a datum which could not be explained through a more classical theory as, e.g., the pilot wave theory. In other terms, the duality wave/corpuscle and the statistic aspects of nature cannot be reduced to a more profound theory showing that they are only a sign of our ignorance, as is the case with thermodynamic with respect to classical mechanics. These aspects are intrinsic in nature, according to Pauli. However, the latter had some problems with the epistemic status of the duality wave/corpuscle. He thought that such a strange behaviour of the elementary components of the universe had to be the manifestation of a more general duality which went beyond the borders of physics traced at that time. He thought that such a duality was universal and also included the mental and the psychological aspects. His deep interest in Jung's theory of collective archetypes depends on the dual-opposite aspects that every archetype of the collective unconscious assumes (the *coincidentia oppositorum*) and on his analysis of his own dreams. The interpretation of the debate Kepler-Fludd, jointly with several ideas on the relations between Jung's archetypes theory and quantum mechanics, is an evidence that Pauli believed in a more profound law which should include both the physical and mental-psychological aspects. This theory, of which quantum mechanics and Jung's theory of archetypes are only a partial manifestation, should not be something vague or mysterious, it should have been developed into precise mathematical terms.[65] Pauli's idea of duality is connected to his conception of symmetry. Therefore, when, at the

[65] Among Jung's works where the dual character of unconscious is proclaimed and where connected themes are developed, I mention (in the bibliography I refer to the original date of publication, but to the title of the English translation): Jung (1911/1912, 1918, 1921, 1928, 1934, 1936, 1939, 1951). With regard to Pauli, his most remarkable contributions on the possible links between some concepts of Jung's analytical psychology—the most important of which is that of collective archetypes—and notions of physics are Pauli (1948, 1950, 1952, 1954) (in this case too, I will refer to the original date of publication, but to the title of the English translation). The epistolary Pauli-Jung is also important to understand Pauli's position with respect to the problem of duality and, more generally, on the relations physics-psychology. For the English translation, see Pauli and Jung (2014). As regards the abundant literature on the connections between Pauli and Jung, I mention: Atmanspacher et al. (1995), Bussotti (2007), Gieser (2005), Halpern (2020), Laurikainen (1988), Lindorff (2004), Miller (2010), Tagliagambe and Malinconico (2011, 2018).

beginning of the 1950s, the not-conservation of parity in the weak interactions was discovered, Pauli—probably thinking that a more profound duality and symmetry should exist—proved the conservation of CPT (charge, parity, time). Probably he considered that such a symmetry corresponds to the more profound level of physics he was looking for.[66]

The case of Pauli is interesting in a comparison with Chasles: Pauli developed his conception of a more general dual law due to his partial dissatisfaction with the interpretation of the duality wave/particle. His general ideas on symmetry were useful for him to obtain important scientific results, independently of whether such ideas of symmetry are correct.

Chasles was dissatisfied with the status of geometry and was convinced that projective geometry might offer a more profound foundation to several mathematical concepts. In principle, he was not dissatisfied with the status of physics either with its interpretation, but argued that a general duality law not yet discovered should exist. The dualities that he and other scientists (particularly Poinsot) had discovered were particular instantiations of such more general law. It is not easy to establish how much this conviction led him to new discoveries beyond geometry, but likely it played a certain role in some of them. Particularly, it seems to me that his most important and thorough contribution to the difficult problem of determining the attraction of an ellipsoid towards objects external to the ellipsoid's surface is influenced by his conception of duality, as we will see in the next chapter.

A final comment on Chasles' philosophy of duality: it seems to me appropriate to point out that—as any duality conception—the one developed by Chasles is linked to a certain idea of symmetry. In our case such a symmetry was discovered first in geometry and only afterwards Chasles supposed that it existed in the universe. The symmetry in the treatment of the rotations and of the system of forces is an example. Obviously, as we have seen, the study of the projective transformations is a subject which can easily induce a conviction in the existence of duality and symmetry.

Chasles' ideas on projective geometry and his attempt to reduce metrical properties within a projective context were carried out in the second part of the nineteenth century (though the merits of Chasles were not always recognized) as, e.g., the development of the Cayley-Klein metric shows. His conception of a general dual law was, instead, not successful in the form proposed by Chasles, if one excludes a few scientists, as Jonquières. On the other hand, this conception was problematic insofar as mechanics is concerned because it was based on a restricted set or results, almost all of which related to the mechanic of the rigid body. Chasles' ideas on how to extend this supposed duality to a domain different from the study of the rigid body's movements were rather vague and did not find a precise and convincing formulation either in Chasles or in successive scientists. It is, however, appropriate to recall that many branches of modern mathematics and physics depend on duality in a way not far from Chasles' ideas. Only to give a significant example: given a vector space V on a field K, a concept which is fundamental to offer a rigorous and intrinsic

[66]The first proofs of the CPT invariance were offered by Lüders (1954) and Pauli (1955).

definition of tensor is that of dual space V^* —defined as the set (which is easy to prove to be a vector space)—of all the linear functionals $f: V \to K$. As known, the elements of V are called vectors, those of V^* covectors or 1-forms and a tensor T is a multilinear application

$$T : \underset{h}{V \times V \times V} \times \underset{k}{V * \times V * V *} \to K.$$

The vector space V and the space of its linear functionals are dual in a sense similar to that in which two theorems of projective geometry are dual. For given a property of V, it is possible to deduce a corresponding property of V^*, though in a less easy and direct manner with respect to the way in which duality law can be applied in projective geometry. Therefore, V and V^* have the same structure, as, in projective geometry, the space of points and the space of planes have the same structure. Thus, duality in many branches of modern mathematics and physics resembles Chasles' ideas.

A further reflexion is necessary: to claim a general principle of duality does not necessarily mean to have a philosophical conception assimilable to the dualism because this implies the existence of two fundamental principles as, to give an example, the two Manichaean principles of the good and the bad. Whereas the duality, at least if we would like to give a philosophical interpretation of mathematics' development, implies that two different sets of mathematical objects are regulated by the same formal laws. The principle is exactly only one and it is the principle according to which we are able to transcribe, through a mere mechanical terminological change, the properties of a set of objects to another set. This is the duality law, which is a principle of symmetry. Therefore, the principle is the duality, and it is a single structural principle, not two principles. According to Chasles such a principle is valid in the universe as well.[67]

Ergo, in spite of the fact that the philosophical section of Chasles' thought is incomplete and, in fact, only outlined, it is very interesting because, beyond its value as a historical document of his conception, which is not restricted to geometry, it induces us to reflect on many questions:

1) how numerous and variegated are the manners in which a philosophical conviction can arise. In this case, Chasles was working on duality in projective geometry; he was impressed by the power of this property and though that it was a universal property. This is not surprising: in Descartes' time, technique was having an important development. Descartes constructed his cosmos as a machine. Nowadays it is the time of artificial intelligence and some scientists think that the cosmos is a sort of immense computer. It is rather common that a scientist is influenced by the most important tendencies of his epoch and by the

[67] Though not being directly connected with the aims of my book, it would be interesting to understand whether Pauli believed in the existence of a single dual principle *à la* Chasles or if he believed in a very dualism, namely in the existence of two principles.

subject he is studying taking into account these tendencies. Generalization is typical of science. Even when this concept is not well defined, it is still a real mark of distinction of the creative scientific activity;

2) the problem of duality is an important part of the history of modern and contemporary mathematics, as well as of significant sections of contemporary physics. Chasles has a decisive role in the construction of the general idea of mathematical duality;
3) the philosophical problem of the relations between a principle of duality as that existing in several parts of mathematics and a very dualistic philosophy at which bases there is not a single principle (though being a principle of duality), but two principles, is a noteworthy conceptual issue.

Chapter 7
Chasles and the Ellipsoid Attraction

Abstract This chapter is divided into seven sections. The problem of the ellipsoid attraction was a subject to which many of the greatest physicists and mathematicians devoted their research, starting from Newton's *Principia* onwards. Therefore, when Chasles had begun to deal with this topic the literature was already abundant. In the first section, I recall the main steps concerning the history of this subject because they are fundamental to understand the novelty of Chasles' approach. The other sections are dedicated to his works, methods and results. Chasles campaigned for 6 years, from 1837 to 1842, to get to the bottom of the matter. His aim was methodological: he was going to determine the ellipsoid attraction by means of a synthetical approach and succeeded. The projective properties of the conics and of the transformations between conics, in particular polarity, played a crucial role in his research. Therefore, the attraction of ellipsoid is a further result that can be included within Chasles' foundational programme.

Chasles dedicated six memoirs to the problem of the attraction of an ellipsoid towards an external point: Chasles (1837e; 1837f; 1837, 1846; 1838; 1840; 1842), the brief note (Chasles, 1839) and several considerations in other of his works, especially in the *Report* (Chasles, 1870). The problem of the ellipsoid attraction was studied from Newton's time. Maclaurin gave a fundamental contribution and several important mathematicians such as Laplace, Legendre, Ivory, Gauss, Poisson and Jacobi faced the problem under different perspectives. However, the basic question, i.e., the way in which an ellipsoid attracts an external point (without, hence, dealing with subjects as the equilibrium configuration of a fluid mass) was, in substance, completely solved by Laplace and Legendre. Therefore, it is to wonder why Chasles dedicated so many efforts to face a problem which had already been solved. The answer concerns the methodologies of solution. With the exclusion of the oldest works, namely those of Newton and Maclaurin and in part the ones by Ivory and Gauss, all the other solutions were basically analytical, full of calculations which solved the problem, but rarely made the solution perspicuous. It is not a coincidence that, e.g., Laplace and Poisson who also offered analytical solutions criticized the obscurity of Legendre's procedure. Chasles, instead, considered this problem in a different manner: starting from his first work on this subject, he had the intention to

P. Bussotti, *Chasles and the Projective Geometry*,
https://doi.org/10.1007/978-3-031-54266-4_7

individuate a core of geometrical properties which represented—jointly with the inverse square law—the real conceptual essence of the problem. Obviously, he also used the analytical technique—this is unavoidable—but his new idea was the determination of the geometrical core at the basis of the problem. This is one of the reasons why his works on the attraction of the ellipsoid might appear, for some aspects, repetitive. In each of them Chasles extended the geometrical core of his inquiry and began the analytical treatment only when pure geometry was not sufficient anymore. The geometrical aspect was pushed further on at any new memoir published by Chasles, until reaching its definitive form in the most important work he dedicated to this subject, i.e., Chasles (1837, 1846). It is necessary to point out that Chasles (1837, 1846) is a memoir which was presented at the Académie royale des Sciences in 1837, but was published only in 1846. I will insert this memoir at the end of the essays Chasles published in 1837. However, since this contribution appeared 9 years after the presentation at the Académie, he had the possibility to reconsider its content. Thus, it has to be regarded as the expression of a mature phase in Chasles' studies on the ellipsoid attraction.

This is the general line along which Chasles' works on the ellipsoid have to be interpreted. It is obvious that each of them has its specificities: for example, Chasles (1838, 1842) connect the issue of the ellipsoid attraction with other branches of physics, in particular with the theories of electricity and heat. Chasles (1842) extends some properties valid for the attraction of the ellipsoid to closed bodies of any form, and so on. Thence, each contribution deals with slightly different aspects of the attraction problem, but the general red line which connects them is the progressive extension of synthetic geometric methods, till reaching in Chasles (1837, 1846) a projective foundation of the ellipsoid attraction problem.

I will consider many mathematical details of each contribution because, without a mathematical analysis, it is absolutely impossible to trace the evolution of Chasles' thought in a convincing manner. I dedicate a long section to this problem since it has been at the centre of the physical investigations for a long time. It is an important topic as it concerns the attraction of the celestial bodies, the distribution of charge on a conductor and the isothermal surfaces. Thence, it ties many sections of physics and I deem significant that Chasles was able to find a geometrical foundation for them.

The works of Chasles on the theory of attraction were well known in the nineteenth century and at the beginning of the twentieth century. To give just some examples: Catalan in 1841 wrote a thesis on the ellipsoid attraction where Chasles' researches play a prominent role. Sturm in 1842 commented the contribution by Chasles on the general theory of attraction, which was read on the 11th February 1839 at the Académie des Sciences, though it was published in 1842 (Chasles, 1842). In his contribution Sturm offered a more elementary way to deduce the two fundamental theorems proved by Chasles in his memoir. Liouville in 1842 wrote an addiction to this memoir by Chasles. Bartholomew Price in his treatise of infinitesimal calculus (Price, 1856) dedicated several pages (pp. 288–294) to Chasles' theory of the attraction of a thin ellipsoidal shell as explained in Chasles (1838, 1840). Chelini in his researches on the ellipsoid mentions Chasles, jointly with Legendre, Laplace Ivory, Gauss, Rodrigues and Poisson among the most important scholars who faced the problem of the ellipsoid attraction. An entire section is dedicated to Chasles (Chelini, 1862a, pp. 38–41). Specifically, Chelini reported and commented with admiration Chasles' proof of the theorem that the potentials of two confocal elliptic shells with respect to an external point are proportional to their masses (ibid., p. 40). In the monumental Schell (1870), the author reserves 12 pages to Chasles' synthetical method (pp. 701–712). Here he offers a clear, though rather synthetic, picture of the way in which Chasles developed his global theory of the ellipsoid's attraction, focusing in particular on Chasles (1838). Schell rightly identifies a line of continuity between Newton's-Maclaurin's and Chasles' approaches because all the three authors gave a geometrical setting to their theories (Schell, 1870, pp. 701–702). Todhunter (1873), a very classical and complete text on this topic, mentions Chasles several times, even if no chapter is devoted to him. Darboux in 1905 pointed out that, within Chasles' defence of pure geometry, his synthetic solution inherent to the ellipsoid attraction is also included. Gray (1907) gave a valuable judgement of Chasles' work. Other examples might be added, but those mentioned are sufficient in order to understand that Chasles' contributions were well known. Throughout the twentieth century the situation changed. References to Chasles' work on the theory of ellipsoid attraction became more sporadic, or, at least, they were inserted in very brief notes or observations where only Chasles' works were mentioned, but there were few or no explanations on his methods, only recalling that they were synthetic. This is the case both with essays of historical character and with contributions on advanced physics. Probably one of the reasons of this situation is precisely that Chasles' geometrical procedures were considered as difficult and obsolete, so that they became secondary in the panorama of the research on gravitational problems. For example, Chandrasekhar (1969) never mentions Chasles, even in the historical introduction. The historians of mathematics and physics did not dedicate much care to this aspect of Chasles' production in recent times. It should be highlighted that Michel in 2020 wrote a valuable paper on this subject, but his principal aim was to determine the value of generality in Chasles' geometry rather than to discover a very foundational programme. Thence, under many respects, Chasles' work on the ellipsoid attraction has to be rediscovered. This is what I will try to do by offering a precise interpretative line: Chasles' contributions on the ellipsoid attraction belong to his foundational programme.

After this introduction, I will offer some historical notes on the way in which the attraction problem was addressed before Chasles. These notes will not be complete, their only aim is to give the general panorama and to clarify the environment in which Chasles developed his thought. For more technical details I refer to Schell (1870) and, above all, Todhunter (1873).

After that the subsections dedicated to Chasles' contributions will follow and eventually the conclusive remarks.

7.1 A Hint to the History of the Ellipsoid Attraction Until the 30s' of the Nineteenth Century

The aim of this subsection is not to trace a history of this topic, but only to offer those elements which might be useful to understand the role of Chasles' works within such a history. Basically, I will refer to the historical preface with which Chasles opened his memoir and to Todhunter (1873). Where necessary, some direct references to the works of the authors will also be presented.

The problem of the ellipsoid attraction depends on three variables: 1) the position of the point attracted by the ellipsoid with respect to the ellipsoid, i.e., if it is external, it lies on the surface or it is internal; 2) the form of the ellipsoid, namely if it is a revolution ellipsoid or if it has three different axes; 3) the homogeneity or inhomogeneity (and in this last case the kind of inhomogeneity) of the ellipsoid.

In the history of the ellipsoid attraction, different authors treated the connections among these three items in different ways.

Newton posed the bases of the problem when he, in the 12th section of his *Principia*, investigated the attraction of the spherical bodies and when in the 13th section he explored that of not spherical bodies. In any case his hypothesis is that the attractive force of any particle is inversely proportional to the square of the distance. The 70th proposition is the celebrated theorem according to which, given a spherical shell an internal point perceives no attractive action, while (Proposition 71) an external point is attracted towards the centre of the spherical shell with a force which in inversely proportional to the square of the distance from the point to the centre of the shell. Propositions 72–75 concern the attraction of full homogeneous spheres, whereas the 76 regards not homogeneous spheres whose density varies according to a specific law. In the following propositions Newton also analysed the case in which the forces are proportional to the distance (elastic forces). In Proposition 79 he determined the attraction of a zone of an infinitely thin spherical shell on a point at the centre of the shell. Propositions 83 and 84 (the last one of the 12th section) show the possibility of calculating the attraction of a homogenous spherical segment on a particle posed on the axis of the segment. The 13th section is the one which is more interesting in our history because Newton examined the attraction of not spherical bodies. Proposition 91 is fundamental. It is that mentioned by Chasles (1837, 1846, p. 2). Newton required to determine the attraction of a corpuscle posed on the axis of a revolution solid in the condition that the attractive force satisfies the inverse square law. With regard to the attraction of an ellipsoid of revolution, Newton determined the ratio between the attraction of the ellipsoid on a point along the ellipsoid's axis and the attraction of a sphere whose diameter is equal to the rotation axis of the ellipsoid (prop. XCI, corollary 2). He also proved that if the corpuscle is located within the ellipsoid along its revolution axis, the attraction will be proportional to the distance of the corpuscle from the centre. Furthermore, a shell composed of two concentric and homothetic ellipsoidal surfaces exerts no action on an internal point (Prop. XCI, corollary 3). As Todhunter pointed out, Newton's propositions can be extended to points belonging to the axes of ellipsoids which are not of revolution (Todhunter, I, 1873, p. 8).[1] The magnificent methods used by Newton are those typical of his infinitesimal geometry. That is, given a theorem to prove or a problem to solve, Newton develops his reasoning through the canons of synthetic geometry. The infinitesimal quantities are introduced only in the final part of the reasoning itself in order to obtain the instantaneous magnitudes. Newton's method is, thus, profoundly geometrical.

The next author mentioned by Chasles who gave fundamental contributions to the solution of the problem we are dealing with is Maclaurin. He addressed this question in Maclaurin (1740),[2] but in a more precise and convincing way in Maclaurin (1742), the famous *A Treatise on fluxions*, in two volumes. The problem of the ellipsoid attraction is addressed in the second volume. In this work, the author proved some interesting propositions on the attraction of infinitesimal cones and of frustum of cones (Todhunter, pp. 133–134). The first article directly connected to the attraction of the homogeneous ellipsoid is the n. 630 where Maclaurin proved, with a method slightly different from Newton's, the proposition according to which a particle within an ellipsoidal shell composed of two concentric and homothetic ellipsoids of revolution is urged by no force. As Todhunter (ibid., p. 134) points out, a series of Maclaurin important theorems are those at the articles nn. 631–635 where he decomposed the attraction of an ellipsoid of revolution on a particle into two components, one perpendicular to the axis and the other parallel to the axis. Then, the first component varies as the distance of the particle from the axis and the second component as the distance from the equator plane. Maclaurin's proof is a noteworthy geometrical and creative demonstration. As a matter of fact, he decomposed the ellipsoid into particular infinitesimal pyramidal solids and, to reach his conclusion, he developed a reasoning involving such solids (Maclaurin, 1742, II, pp. 523–527). After these propositions, a series of theorems follows which

[1] As to Newton's theory of the ellipsoid attraction, the fundamental statement is the Corollary 2 of Proposition XCI. Newton's demonstration of this corollary is extremely brachylogic. Almost all the necessary details are missing. The first text which offers a complete proof of Newton's corollary is the Geneva Edition of Newton's *Principia*. It was published between 1739 (first book) and 1742 (third book). This edition reports Newton's Latin text. The three editors Le Seur, Jacquier (the only two mentioned in the frontispiece) and Calandrini added a series of notes which are more extended than Newton's text itself. In these notes, all of the Newton's proofs are clearly and detailed explained. The Geneva Edition was republished in Glasgow in 1822 in an improved form because some mistakes of the first publication were emended. Thus, I will refer to the Glasgow edition with the indication Newton (1739–1742, 1822). The reader will find all the explanations on Newton's theory of ellipsoid on pp. 396–401 (notes 542–545 (z)) of such edition. Raffaele Pisano and I dedicated several works to the Geneva Edition. See, e.g., Bussotti and Pisano (2014a, 2014b) and Pisano and Bussotti (2016). A paper on this subject is also Guicciardini (2015). Recently Pisano and I published a paper specifically on Newton's theory of ellipsoid attraction and on the connected explanation given by the Geneva Edition (Pisano & Bussotti, 2022). After the writing of this paper, but before its publication, a student of mine at the Department of Mathematics, University of Udine, wrote and defended a valuable degree dissertation on the ellipsoid attraction in Newton and Maclaurin (Floriduz, 2021).

[2] This is the famous memoir of Maclaurin on the tides. It was published in 1741, though composed in 1740. Therefore, in the References, I indicate it as Maclaurin (1741).

are dedicated to the equilibrium figures of the fluids, where Maclaurin first introduced the concept of level surface (Todhunter, I, p. 136), but this research, though close to the problem of the ellipsoid attraction, is not exactly identifiable with it. Thence, I will not address the details. In the art. 642 (Maclaurin, 1742, II, pp. 532–533) Maclaurin proposed to calculate the attraction of an ellipsoid of revolution at the pole and at the equator. In the articles 644–647 he solved the problem. It is worth referring to what Todhunter claimed with regard to Maclaurin's results:

> Maclaurin then in his Articles 644. . .647 investigates accurate expressions for the attraction of any ellipsoid of revolution on a particle at the pole or at the equator. The investigations are conducted in the manner of the time by representing the attractions by areas of certain curves, and finding the areas by the method of fluents. The results agree with those obtained by analysis, and presented in modern works on Statics. Maclaurin's processes are remarkable specimens of ingenuity, considering the date of their publication [. . .]. (Todhunter I, p. 138).

Given the use Maclaurin made of geometry, it is not a coincidence that Chasles considered him as a source of inspiration.

The articles 648–652 (Maclaurin, 1742, II, pp. 539–542) are those fundamental for our history. For in these articles Maclaurin proved a series of propositions which, in their totality, are known as Maclaurin theorem on the attraction of the ellipsoid of revolution. Todhunter (1873, p. 139) summarizes Maclaurin's results as follows: given two confocal ellipses, rotate both of them around their major axes. Ergo, two ellipsoids of revolution are generated. The attraction of the two ellipsoids (which can be called confocal ellipsoids of revolution) on a particle external to both and placed on the prolongation of the revolution axis or in the equatorial plane is as the volume of the ellipsoids.

Chasles expressed Maclaurin theorem in these terms:

> The attractions which two ellipsoids of revolution, described with the same foci, exert on the same point, external to their surfaces, and posed on the axis of revolution or in the equatorial plane, are as the masses of the two ellipsoids.[3]

Since the ellipsoid is supposedly homogeneous, speaking of volumes or of masses is equivalent. In the article 653 Maclaurin argued that his theorem is valid for any couple of ellipsoids provided that the attracted point is on the prolongation of any of the three principal axes (Maclaurin, 1742, II, p. 543; Chasles, 1837, 1846, p. 3). His demonstrations are based on a geometrical reasoning[4] (in which analytical elements as the fluxions and the quadratures are involved. This is unavoidable) coherent with Newton's style. Before Chasles, he was the last author to frame the problem of the ellipsoid attraction within a geometrical context.

---

[3]Chasles (1837, 1846), p. 3: "Les attractions que deux ellipsoïdes de révolution, décrits des mêmes foyers, exercent sur un même point, extérieur à leur surface et situé sur l'axe de révolution ou dans le plan de l'équateur, sont entre elles comme les masses de deux ellipsoïdes".

[4]Todhunter (1873, pp. 140–145) offers clear explanations of Maclaurin's proofs. For a complete treatment of Maclaurin's proofs on the ellipsoid attraction, see Floriduz (2021).

After Maclaurin's work, Chasles (ibid., p. 3) claimed that three problems had still to be solved:

1) For the ellipsoid of revolution, the case in which the attracted point is posed in any place of a meridian plane.
2) For the ellipsoid with three different axes, the case where the attracted point is external or on the surface of the ellipsoid.
3) Generalization to any point of space of the theorem on the ratio of the attractions of two confocal ellipsoids on a point on the prolongation of a principal axis. In the generalization, not only the intensity, but also the direction of the attractions has to be determined. The final idea should be to reduce the general question of the ellipsoid attraction on an external point to the attraction of a point on the surface.

In our history an important author is Clairault: the most remarkable contribution by Clairault concerns the form of the Earth. His main work is Clairault (1743). In his research, he investigated the problem of the attraction of an ellipsoid since the form of the Earth is approximately ellipsoidal. In a memoir written in 1737, but published in 1741 (Clairault, 1737, 1741), he was able to determine an approximate expression for the attraction of an oblate ellipsoid[5] on any point of its surface (Clairault, 1737, 1741, pp. 19–25; Todhunter, 1873, I, p. 85). In Clairault (1743) this proof is not presented because in the meantime Maclaurin had found the exact solution, so that Clairault reproduced Maclaurin's proof (Clairault, 1743, p. 157), whereas he developed a proper method to calculate the attraction exerted by an oblate heterogeneous ellipsoid on the points of its surface (ibid., pp. 233–243; Todhunter, 1873, I, p. 85). A further memoir in which Clairault addressed the problem of the ellipsoid attraction is Clairault (1738). Here the author determined the attraction of an ellipsoid of revolution which is composed of similar shells of variable density. The density varies according to a well-determined law and exactly it varies as $fr^p + gr^q$, where f, g,p are constants and r is the variable indicating the polar half-axis of the shell. He supplied an approximate solution to this problem. In Clairault (1743) he abandoned the hypothesis that the strata are similar and indicated the density with a general function (Todhunter, 1873, I, p. 87). In this work, besides the results of which I have already spoken, Clairault obtained other significant achievements: 1) he found the approximate value of the attraction on an external point of an almost circular oblate ellipsoid (Todhunter, 1873, I, pp. 206–208, where Clairault's method is clarified); 2) in the third chapter of his work, he discussed the variation of gravity at the surface of an ellipsoid of revolution composed of strata whose density and ellipticity are variable (whereas, as seen, in Clairault (1738) the strata were similar). He was able to prove that the diminution of gravity from the poles to the equator varies as the square of the cosine of the latitude (Todhunter, 1873, I, p. 215; art. 330). Relying upon the theorems proved so far, Clairault was able to obtain other significant

[5]Given an ellipse of axis a and b with $a>b$, the oblate ellipsoid is the solid of revolution obtained by the rotation of the ellipse around its minor axis b, the prolate ellipsoid is the solid of revolution obtained by the ellipse's rotation around its major axis a.

results: if C is the centre of an almost spherical ellipsoid of revolution and M an external particle, cut MC the ellipsoid in n; then, the attraction of the solid on M is almost the same as that of an ellipsoid of revolution of equal volume whose rotational axis is Nn. Through this proposition he was able to find the famous theorem which also nowadays is called "Clairault theorem": to determine the ratio between the difference of gravity at the pole P and gravity at the equator E as well as the gravity at the equator for an almost spherical oblate ellipsoid of rotation (as the Earth is approximately). This ratio is

$$\frac{P-E}{E}=\frac{5}{2}j-\varepsilon_1,$$

where j is the ratio between the centrifugal acceleration and gravity acceleration at the equator and ε_1 is the ellipticity of the oblate ellipsoid along a meridian.[6] Clairault's hypothesis is that the strata are ellipsoidal of revolution around a common axis and almost spherical. Each stratum is homogeneous, but the densities vary from stratum to stratum according to a given law; 3) given a shell of constant density and bounded by two concentric and homothetic oblate ellipsoids, Clairault was also able to determine the component of the attraction perpendicular to the radius vector joining a point to the centre of the shell (Todhunter, 1873, I, pp. 222–223).

Therefore, the results obtained by Clairault are crucial. Notwithstanding this, Chasles dedicated only few lines to Clairault in his historical introduction in Chasles (1837, 1846) and, in substance, ascribed to him few improvements with respect to Maclaurin. He almost reproached Clairault because, though having developed personal analytical methods, he did not apply them to the problem of the ellipsoid attraction (Chasles, 1837, 1846, pp. 3–4). Probably the reason of Chasles' attitude depends on the fact that he was more interested in tracing a general theory of the ellipsoid attraction rather than to analyse how things work in many particular cases, namely for ellipsoids of specific forms. On the contrary, this was exactly the method followed by Clairault because his principal purpose was to study the form of the Earth, which is (almost) an ellipsoid of revolution of specific form.

Chasles, instead, emphasized the importance of D'Alembert's works. He actually wrote that D'Alembert first addressed the question of the ellipsoid attraction in general terms and interpreted D'Alembert as the real successor of Maclaurin because, through geometrical considerations, he was able to extend to any ellipsoid the propositions for the internal or superficial points discovered by Maclaurin for the ellipsoids of revolution. However, in the general case he was not able to integrate the formula because it is expressed by an elliptical function. This problem he found in the general solution induced D'Alembert to think that Maclaurin theorem was not true. However, after a deeper analysis he found three different proofs of such a theorem, though he developed no further generalizations of this proposition (Chasles, 1837, 1846, pp. 4–5).

[6] For a complete and clear proof of Clairault theorem, see Todhunter (1873, I, pp. 218–221).

In order to achieve a more detailed picture of D'Alembert's work on the ellipsoid attraction, the contribution (D'Alembert, 1754, 1756) is essential. The second section of this book is entitled *De la figure de la Terre considérés physiquements* and the second chapter of this section *De l'attraction d'un sphéroïde sur les corpuscles placés à sa surface; et de la figure qui en résulte pour ce sphéroïde.* Here, he addressed the problem to determine the attraction on a point of the surface of an ellipsoid of revolution. To determine such an attraction, the fundamental step developed by D'Alembert was to find the tangential attraction at the point. In particular, he determined the tangential attraction towards the pole. This is a laborious analytical work described in detail by Todhunter (1873, I, pp. 287–290) and the final formula itself is rather complicated. All the cases of attraction towards internal, superficial or external points with respect to an ellipsoid are restricted to ellipsoids which, though not of revolution, have a shape close to that of a sphere. It is worth giving an idea of how D'Alembert tried to determine the general attraction of an ellipsoid (the problem of which Chasles spoke in his historical introduction). The work in which D'Alembert developed his method is *Sur l'attraction des sphéroides elliptiques*, which is his 53rd memoir (D'Alembert, 1780, 7, pp. 102–207. Remarks to such memoir, pp. 208–233). D'Alembert decomposed the ellipsoid into slices, but he always reached an elliptical integral for any attempted decomposition. Todhunter (1873, I, p. 403) explains that D'Alembert's ideas can be so summarized: 1) suppose that the attracted particle P is at the end of the half-axis c and that a is another half-axis of the ellipsoid. Draw a plane passing through P and through the tangent at P which is parallel to the axis $2a$. Let the plane rotate around the tangent so to cut the ellipsoid into wedge-shaped slices (Fig. 7.1); 2) it is also possible to develop the same construction considering the axis $2b$; 3) a further decomposition is possible supposing that a rotating plane passes through the axis $2c$. This plane cuts the ellipsoid into wedge-shaped slices. This decomposition is similar to that used by Maclaurin; 4) finally, the ellipsoid can be cut by a plane which is always at right angles to the axis $2c$ (Todhunter, 1873, I, p. 403). In this way it is possible to construct the integral which offers the attraction of the ellipsoid on P, but this is an elliptical integral. However, given this geometrical construction, it is comprehensible that Chasles included D'Alembert among those mathematicians who addressed the problem of the ellipsoid attraction geometrically rather than analytically (Chasles, 1837, 1846, p. 5).

Another important author in this history is Lagrange, who dedicated two memoirs to the ellipsoid attraction: Lagrange (1773, 1775; 1775, 1777).[7] Chasles expressed a general opinion on this part of Lagrange's work: Lagrange tried to obtain the attraction of a general ellipsoid on the external points through purely analytical methods, but, though having clarified many aspects of the problem, his results did

[7] The first memoir of Lagrange to which I refer appeared in *Noveaux Mémoires de l'Académie de Berlin* for the year 1773, which was published in 1775. Thence, I refer to it as Lagrange (1773, 1775). It is entitled "Sur l'attraction des sphéroïdes elliptiques". The second memoir is an addition to the first one; it appeared in the same journal for the year 1775, which was published in 1777. Thus, I refer to it as Lagrange (1775, 1777).

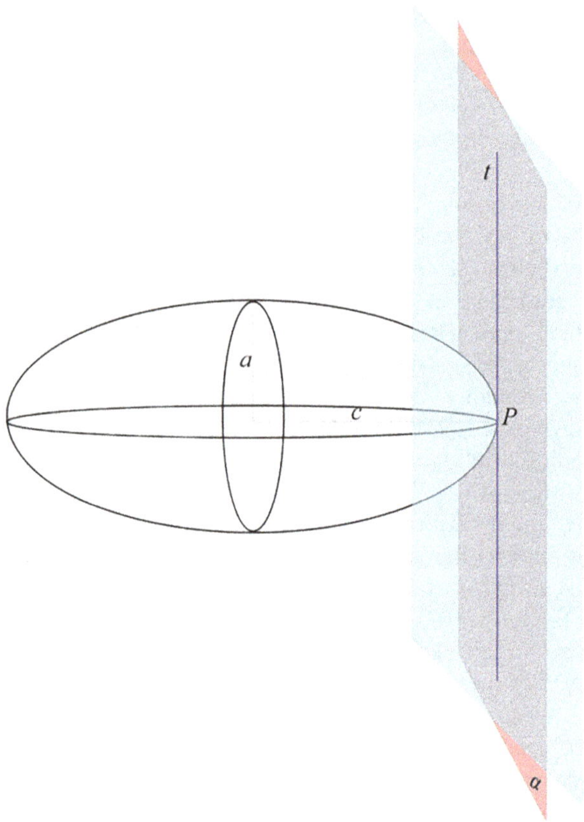

Fig. 7.1 Reconstruction of the figure described by D'Alembert in item 1): c is the half-axis of the ellipsoid at which extremity the attracted particle P is located. As original position of the plane α, I have chosen the one indicated in red. The plane rotates around the tangent t to the ellipsoid at P which is parallel to the half-axis a. The planes' position after the rotation is indicated in blue

not go beyond those obtained by Maclaurin and D'Alembert (Chasles, 1837, 1846, p. 5). In particular, Chasles stressed that in Lagrange (1773, 1775) he achieved the results which D'Alembert obtained, with a different mathematical technique as to the attraction of an ellipsoid with three different axes on internal or superficial points. Lagrange imagined the ellipsoid composed of little cones with their vertices at the attracted point, a technique which then became standard and reduced the question to calculate a double integral whose first integration presents no difficulty. For the external points, which was the real problem, Lagrange was able to use his technique only for ellipsoids of revolution and for points along the axis. D'Alembert showed that such a method is also applicable to points at the equator, but these results had already been obtained otherwise by Maclaurin (ibid., p. 5). A very interesting observation made by Lagrange in this memoir is that the general treatment of the attraction problems using rectangular coordinates is difficult. This observation led him to the study of transformation formulas between a Cartesian system of coordinates and other systems, which is useful for the analysis of triple integrals. In this context, he offered the formula $dxdydz = r^2 \sin \varphi d\theta d\varphi dr$. Though his results on the transformation formulas are not completely satisfactory, they are fundamental

(Todhunter, 1873, I, p. 453). The most relevant achievements by Lagrange with regard to the attraction of a general ellipsoid on internal points can be expressed as follows: if we consider the point which lies on a radius drawn from the ellipsoid's centre, the attraction varies as the distance from the centre; the attraction of a shell composed of two homothetic and concentric ellipsoidal surfaces on an internal point is null; the component of the attraction parallel to an axis varies as the perpendicular distance from the plane which contains the other axis (ibid., p. 454). With regard to the attraction of an ellipsoid with three different axes on an external point situated on the prolongation of an axis, Lagrange pointed out the difficulty to calculate the integral which expresses the attraction. He doubted that it was calculable (ibid., p. 455).

Lagrange (1773, 1775) is an important memoir because the author extended Maclaurin's theorem to an ellipsoid with three different axes (Todhunter, 1873 I, pp. 456–458).

Thence, as argued by Chases, D'Alembert and Lagrange, in a geometrical and analytical way respectively, applied to general ellipsoids what Maclaurin had proved for ellipsoids of revolution on the internal or superficial points. As to the external points, they only gave new demonstrations of Maclaurin's theorem (Chasles, 1837, 1846, pp. 5–6).

Chasles summarized the situation. He individuated two problems which had still to be solved while dealing with an external point: 1) to calculate the attraction for a point posed on the meridian plane of an ellipsoid of revolution; 2) to generalize the theorems to the confocal ellipsoids with three different axes (ibid., p. 6).

At this point of the story, two protagonists come on stage: Legendre and Laplace. In substance they solved the problem of the ellipsoid attraction in a definitive manner, but their methods raise some doubts: in fact, the researches on the ellipsoid attraction continued after Legendre's and Laplace's works so that more convincing and easier methods of proof could be found.

Legendre 1782 (published 1785; I will indicate as Legendre, 1782, 1785) proved that Maclaurin theorem on the attractions exerted on an external point by confocal ellipsoids of revolution holds for any position of the attracted point and he supposed that the theorem of the attraction of an ellipsoid with three different axes might be subject to the same generalization. This generalization was obtained in Legendre (1788). As Chasles pointed out, this result was obtained overcoming enormous analytical difficulties for Legendre to reach the integral formula expressing the theorem (Chasles, 1837, 1846, p. 6). The procedure is awkward and difficult to follow.

In the meantime, Laplace in 1783 presented a direct proof of this theorem (this memoir was published in the year 1786, though the issue of the journal is 1783. I indicate as Laplace, 1783, 1786). Laplace arrived at the demonstration deducing the formula which expresses the attraction for the external points from that of the attraction for the superficial points (ibid., p. 6). However, Chasles argued, this demonstration which is based on the series does not appear completely strict and satisfactory, though based on profound analytical considerations introduced by Laplace (ibid., p. 6).

Therefore, Legendre and Laplace solved the problem of the ellipsoid attraction through two different procedures: Laplace proved that Maclaurin theorem can be generalized, whereas Legendre, regardless of Maclaurin theorem, reached directly a formula expressing the attraction exerted on any external point. Chasles pointed out that both solutions leave something to be desired because Laplace's is founded on the use of series; Legendre's needs long and painful calculations. Because of this, their solutions, though complete, did not offer the final answer to the ellipsoid's attraction problem (ibid., p. 7).[8]

As a matter of fact, Chasles summarized Laplace's and Legendre's researches on the ellipsoid attraction, but the situation is rather faceted because the two authors dedicated several memoirs to this problem which are inserted, at least for Laplace, within the general problem of finding the form of equilibrium of a fluid mass as well as of determining precisely the form of the Earth. There are various phases in Laplace's thought. The main difference among them is that in the initial works he did not use the concept of potential, while later on he did (Todhunter, 1873, II, p. 1). I try to give just an idea of Laplace's procedures, referring to Todhunter II for a complete treatment. In an appendix at Laplace (1772, 1776), entitled *De l'Équilibre de spheroïdes homogènes* pp. 536–554 (this memoir was published in the year 1776, though the issue of the journal is 1772, second part. I indicate as Laplace (1772, 1776), Laplace proved that the variation of the gravity law at the surface of a solid (in substance an ellipsoid), whatever the form of equilibrium be, is the same as that for an ellipsoid of revolution. Laplace used the polar expression $r^2 dr \sin \theta d\theta d\varphi$ for the mass element and the decomposition of attraction in a component along the radius vector and in a component perpendicular to the radius vector. He approximated some expressions through series. The proof is complicated, but correct, though Laplace did not justify all his steps in a perspicuous manner (Todhunter, 1873, II, pp. 5–12). A further memoir is Laplace 1782, 1785 (this memoir was published in the year 1785, though the issue of the journal is 1782. I indicate as Laplace, 1782, 1785). Laplace 1783, 1786 (this memoir was published in the year 1786, though the issue of the journal is 1783. I indicate as Laplace, 1783, 1786) analyses the case in which gravity varies as the n-th power of the distance. As Todhunter stressed, these initial memoirs are related to D'Alembert research (ibid., p. 18).

The treatise (Laplace, 1784) is a significant work insofar as Laplace introduced the concept of potential—already used by Legendre—though not yet that of Laplacian (ibid., p. 30). Through the notion of potential V and using the differential quotients of V Laplace proved the generalization of Maclaurin theorem, namely: given two confocal ellipsoids, their attractions on an external point will be as their volumes, i.e. they will have the same direction and their intensity will be as the volumes of the two ellipsoids. In modern terms: the potentials of two confocal ellipsoids on an external point are as their volumes (Todhunter II, p. 31; ibid., I,

[8]Chasles expressed the same opinion on the works by Legendre and Laplace concerning the ellipsoid attraction in Chasles (1870), p. 104.

p. 139). In the *Méchanique Céleste* an improved version of this proof is presented in the third book, § 5 and § 6. In the fifth section of Laplace (1784), the problem of the attraction of the ellipsoid towards an internal point is faced. This result is also referred to in the second chapter, third book of the *Méchanique Céleste.* Laplace reduced the attraction of points lying on straight lines parallel to an ellipsoid's axis to a definite integral, but he was convinced that it could not be integrated elementarily. In fact, he was right because it is an elliptical integral. He also proved that a particle within an elliptical shell whose surfaces are similar is in equilibrium (ibid. II, p. 32).

The memoir (Laplace, 1782, 1785) is significant. It is indeed worth pointing out that here Laplace introduced the equation that today holds his name. He presented first the equation in polar coordinates, whereas the Cartesian form is expounded in the book 11, § 11 of the *Méchanique Céleste*. He applied the concept of potential to find the attraction of ellipsoids of rotations which are almost spherical (ibid., pp. 56–60).

Laplace 1787, 1789 (this memoir was published in the year 1789, though the issue of the journal is 1787. I indicate as Laplace, 1787, 1789) is also an important contribution because, while studying the nature and the behaviour of Saturn ring, he applied the concept of potential and Laplace equation valid for the points external to the ring. Through these instruments, Laplace reached several noteworthy conclusions for which we refer to Todhunter (1873, II, pp. 65–69).

In the *Méchanique Céleste* the author applied systematically the concept of potential and his equation to calculate the attractions of particular bodies. In the first tome, second book of his masterpiece, first of all he applied his equation to a spherical shell, expressed its potential in spherical coordinates and reached all the results obtained by Newton both for internal and external points, but through his analytical method based on the potential (Todhunter, 1873, II, pp. 176–183). However, Todhunter reveals that a part of Laplace's reasoning is unsatisfactory because he changed the form of V passing from the hollow part of the shell to the external points without justification, in an arbitrary manner. This depends on the fact that Laplace equation does not hold if the attracted particle belongs to the body itself. As known, Poisson replaced Laplace equation with his own (ibid., p. 177). Despite this arbitrariness, Laplace's work was of fundamental importance because he taught how to use the concept of potential within the theory of attraction and, in general, within celestial mechanics. The second tome, third book of the *Méchanique Céleste* is crucial because in the first chapter a complete account of a homogeneous ellipsoid's attraction both towards internal or external points is presented: Laplace proved that, if the origin of the coordinate system is the attracted particle and A, B, C denote the components of the attractions parallel to the three orthogonal axes, these values are expressible as:

$$A = \iiint \sin\theta \cos\theta dr d\theta d\phi$$

$$B = \iiint \sin^2\theta \cos\phi dr d\theta d\varphi$$

$$C = \iiint \sin^2\theta \sin\varphi dr d\theta d\varphi$$

being r, θ, φ polar spherical coordinates.[9] The limits of integration have to include the whole attracting mass.

Laplace obtained that, if M is the mass of the ellipsoid, $a,b,c,$ its axes; f,g,l the coordinates of the attracted point and $\lambda^2 = \frac{b^2-a^2}{a^2}$; $\lambda'^2 = \frac{c^2-a^2}{a^2}$, the expressions for A, B,C can be written as

$$A = \frac{3fM}{a^2}L, \quad B = \frac{3gM}{a^2}\frac{d\lambda L}{d\lambda}, \quad C = \frac{3hM}{a^2}\frac{d\lambda' L}{d\lambda'},$$

where $L = \int_0^1 \frac{x^2 dx}{\sqrt{(1+\lambda^2x^2)(1+\lambda'^2x^2)}}$ and this is the integral which cannot be calculated elementarily (ibid., pp. 184–185).

Through what just proved Laplace was able to offer a slightly improved demonstration of the generalized Maclaurin theorem he had already proved in Laplace (1784). In the first chapter of the third book he was also able to express the attraction of an ellipsoid on an external or superficial point through a single integral (ibid., p. 188). The second chapter of the third book is the one in which Laplace presented the development in series relative to the formulas of the ellipsoid attraction, where what nowadays are called Laplace functions and Laplace coefficients are introduced. For this problem I refer to Todhunter (1873, II, pp. 188–192). Laplace also considered the case in which the ellipsoid is not homogeneous and the inhomogeneity varies according to certain laws. Furthermore, he obtained several other interesting results with regard to the attraction of the bodies. For example, while dealing with the Saturn ring, Laplace determined the attraction of an infinite cylinder on an external particle (ibid., p. 210, ff.), always resorting to the concept of potential. Therefore, Laplace's work on the attraction of the ellipsoid was enormous and fundametal. A great part of the following works dealing with this general problem attemped to offer an easier demonstration of Maclaurin extended theorem than that given by Laplace. Thence, Laplace should be considered as important as Maclaurin in this history. Before analysing Ivory's work that is a fundamental reference point for Chasles, I will present some aspects of Legendre's approach.

At the beginning of his memoir Legendre (1782, 1785) (one of the two Legendre's memoirs to which also Chasles referred) the author presented a very brief summary on the state of the art insofar as the ellipsoid attraction was concerned, claiming that the generalization of Maclaurin theorem was still missing. The first result he obtained was to prove that, given two confocal ellipsoids with three

[9] Instead of the Greek letters θ and φ for the angles, Laplace used resp. p and q (Laplace, 1799, p. 6).

different axes, the attraction that they exert on an external point situated on the prolongation of one of the axes is as the mass of the ellipsoid.

This is a partial extension of Maclaurin theorem. Legendre considered the ellipsoid $\frac{x^2}{a^2}+\frac{y^2}{b^2}+\frac{z^2}{c^2}=1$ of mass M and searched the attraction it exerts on a point lying on the x axis (for the other axes, the reasoning is exactly the same) at a distance h from the origin. He wrote the mass element as r^2 cos $\phi d\phi d\psi dr$, where r is the distance of any point of the ellipsoid from the attracted point, ϕ is the angle between r and its projection on the plane (x,z) and ψ the angle between this projection and the x-axis. It is easy to prove that the attraction along the x-axis is

$$\cos^2\phi \cos\psi d\phi d\psi dr$$

The integration in dr was calculated between the limits r_1 and $r_2$where these expressions indicate the limiting radii vectors drawn from the attracted particle to the ellipsoid in the direction assigned by the angles ϕ and ψ. After a certain number of calculations on the differential forms and an analysis to establish the integration limits for ϕ and ψ, Legendre arrived at determining the attraction for a point along the x-axis as

$$\frac{3Mh}{\left(a^2-c^2\right)\sqrt{\left(h^2-a^2+b^2\right)}}\int_0^{\frac{\pi}{2}}\sqrt{\left(\frac{h^2-a^2+b^2+\left(c^2-b^2\right)\sin^2\theta}{h^2-a^2+b^2+\left(a-b^2\right)\sin^2\theta}\right)}$$

$$-\sqrt{\left(\frac{h^2-a^2+c^2}{h\cos\theta d\theta}\right.}$$

where it is possible to assume $\sin\phi=\frac{b\sin\theta}{\sqrt{h^2-a^2+b^2}}$, varying θ in the interval $\left(-\frac{\pi}{2},\frac{\pi}{2}\right)$. Since for the confocal ellipsoids $a^2-b^2; a^2-c^2; b^2-c^2$ are constant, the attraction is proportional to the mass of the ellipsoid (Todhunter, 1873, II, pp. 21–23). This is a very important result and, in substance, the first really significant improvement after Maclaurin theorem. Legendre claimed that Maclaurin theorem is valid whatever the position of the attracted particle is, but, as we have seen, Laplace was the first to prove this fundamental theorem. In the attempt to demonstrate the general theorem, Legendre was, anyway, able to prove a significant proposition: if the attraction of a revolution solid is known for every external point lying on the prolongation of the axis, it is also known for every external point. This proof is obtained through development in series. In this context, Legendre introduced two fundamental notions: 1) the coefficients which are nowadays known as Laplace coefficients (ibid., pp. 23–25); 2) the function we now call *potential* and which indicated the sum of the elements of a body divided by their distances from a fixed point.[10] Using

[10]It is worth highlighting that Legendre ascribed the merit for the invention of the potential to Laplace because he wrote to have directly deduced such a concept from a private communication by

the concept of potential and Laplace coefficients, Legendre (and Laplace) obtained formulas valid until the third order, as Poisson proved in 1826 (ibid., p. 27). Todhunter expresses this opinion on Legendre's memoir: "In conclusion we may affirm that no single memoir in the history of our subject can rival this in interest and importance" (ibid., p. 28).

Legendre (1788) is the second memoir to which Chasles referred. It is entitled *Mémoire sur les Intégrales Doubles*. In the introductory section, the author explained that his aim was to find a means which might lead to a valuation of the formula on the attraction of any ellipsoid so as to obtain the proof of the extended Maclaurin theorem: if two ellipsoids (he used the terms elliptical spheroid, but was referring to a general ellipsoid, not to one of revolution) have their principal sections described with the same foci, the attractions they exert on the same point have the same direction and are as the masses. Legendre looked for a proof of this proposition which did not depend on Maclaurin theorem, as it was the case with Laplace's proof (Todhunter, 1873, II, p. 74). Legendre recognized Laplace's crucial contributions to the study of the double integrals (ibid., p. 75). The memoir is divided into four parts: in the former, Legendre offers a theory of the transformations of double and multiple integrals which is similar to Lagrange's and which is not completely rigorous, but it is not much used in the rest of the memoir (ibid., pp. 75–76). The second section provides the general formulas for the attraction of an ellipsoid on an external point, which produces a double integral. He used polar coordinates. The third section concerns the case in which the attracted particle lies in one of the principal planes of the ellipsoid. As Todhunter points out, one of the two integrations can be made, in this case, through the ordinary methods, without adopting new procedures of cutting the ellipsoids into elements (ibid., p. 76). With regard to this section of Legendre's memoir Todhunter confirms what Chasles had already claimed: the proof is sound, but very tortuous and elliptical, so that the reader has to reconstruct several missing passages in a complicated context (ibid., p. 76–77). The fourth section concerns, indeed, the general problem of the attraction of an ellipsoid on an external point. As Todhunter points out, "the process is very laborious" (ibid., p. 77), but it is, at least interesting, to refer to the way in which Legendre reasoned to develop the integral which supplies the attraction. He constructed, according to a determined law, a series of conical surfaces having their vertices in the attracted point. According to such law, it makes sense to define one of these cones as the most external one. Such a cone is assumed as cone touching the ellipsoid. Legendre's integration consists in determining the attraction parallel to an axis by the portion of ellipsoid included between two indefinitely close conical surfaces of the series. The cones of the series depend on a parameter ω which is zero for the tangent cone and has its maximum when the cone degenerates into a straight line (ibid., p. 77).

We will see that the idea to use the cones in order to calculate the ellipsoid's attraction was not typical only of Legendre. Poisson and Chasles also resorted to it,

---

Laplace (Todhunter, 1873, II, p. 25). The first who indicated this function with the name potential was Georg Green in 1828; Gauss used the term in 1840 (ibid., p. 26).

though through methods different from Legendre's. Todhunter's commentary expresses the difficulty and tortuosity of Legendre's procedure, while also emphasizing its success, as he wrote:

> Now the remarkable fact is that Legendre succeeds in obtaining an expression *free from the integral sign* which represents the resolved attraction of one of these portions of a conical shell: and when we look at the very laborious process by which the result is obtained, we may safely pronounce it is one of the most extraordinary mathematical facts ever performed (ibid., p.77. Italics in the text).

After the laborious work to which we have alluded, Legendre reached the following conclusion insofar as the attraction of a homogeneous ellipsoid of mass M on an external point is concerned: be a,b,c, the half-axes of the ellipsoid, and f,g,h the coordinates parallel to the three axes of the external point for which the ellipsoid's attraction is calculated. The attraction parallel to the a half-axis is

$$\frac{3Mf}{k}\int_0^1 \frac{x^2dx}{\sqrt{\{k^2+(b^2-a^2)x^2\}}\sqrt{\{k^2+(c^2-a^2)x^2\}}},$$

where k denotes the greatest root of the equation (ibid., p. 82)

$$1=\frac{f^2}{k^2}+\frac{g^2}{k^2+b^2-a^2}+\frac{h^2}{k^2+c^2-a^2}.$$

As Todhunter stresses, this means that the attraction depends only on the mass M and on b^2-a^2 and c^2-a^2. But this is exactly Laplace's theorem, namely given two confocal ellipsoids the attractions they exert on the same point, external to both, are in the same direction and are proportional to the masses. The difference is that Legendre, though through a complicated method, deduced this generalization of Maclaurin theorem without referring to Maclaurin theorem itself, as, instead, Laplace did (ibid., p. 82).

Thus, with Laplace and Legendre the problem of the ellipsoid's attraction is solved, at least as far as the basic statements are concerned, namely the attraction of a homogeneous ellipsoid on any point of space. The elliptical integral providing the attraction was found and the generalized Maclaurin theorem was proved. This problem is the first one of a series of further and more difficult questions connected with the forms of equilibrium of a fluid mass, with the forms of the celestial bodies and so on, but, as to Chasles, this is the significant aspect. However, as Chasles himself highlighted, the methods by Legendre and Laplace were complicated and, in part, unsatisfactory.[11]

---

[11] An interesting paper on the contributions given by Legendre and Laplace to the theory of the ellipsoid's attraction is Pecot (1993). The author focuses on a controversy between the two mathematicians, which was, in substance, connected to a priority problem.

A fundamental contribution is Ivory (1809) since he obtained the same results as Legendre and Laplace through an easy reasoning, a part of which was also used by Chasles in his memoirs on the ellipsoid's attraction. Chasles commented Ivory's works like this:

> In 1809, M. Ivory discovered a beautiful and easy solution, which is founded on a curious property of the ellipsoids described with the same foci. This property establishes a simple relation between the attraction of an ellipsoid on an external point and the attraction of another confocal ellipsoid on an internal point. Through this property you immediately derive the formula for the attraction on the external points (consequently, Maclaurin theorem) from the known formula for the attraction of the internal points.[12]

Ivory clarified at the beginning of his memoir that his results were not new from a physical point of view and that his method was inspired by Laplace. However, both Legendre's and Laplace's proofs of the generalized Maclaurin theorem were long and complicated, whereas he offered a simpler one, which has to be regarded as a progress in such a difficult issue. Ivory explicitly wrote:

> LE GENDRE has given a direct demonstration of the theorem of LA PLACE [what I have called generalised Maclaurin theorem], by integrating the fluxional expressions of the attractive forces; a work of no small difficulty, and which is not accomplished without complicated calculations. In the *Mecanique Celeste*, the subject of attractions of ellipsoids is treated by LA PLACE after the method first given by himself in the Memoirs of the Academy of Sciences, founded on the theory of series and partial fluxions. It was in the study of LA PLACE'S work, that the method I am about to deliver, was suggested; and it will not be altogether unworthy of the notice of the Royal Society, if it contribute to simplify a branch of physical astronomy of great difficulty, and which has so much engaged the attention of the most eminent mathematicians (Ivory, 1809, p. 347. Capitals and italics in the text).

Ivory developed his reasoning when the point attracted by the ellipsoid is everywhere, namely internal, on the surface or external to the ellipsoid. After a brief and general introduction on the way to decompose the attractive forces according to three perpendicular directions (ibid., pp. 348–351), he applied his method to determine the attraction on an internal point. In this occasion, as Todhunter (1873, II, p. 223–224) points out, there is the frequent use of the process of transformations of variables within a double integral. However, since the real novelty of Ivory's memoir is the possibility to treat easily the attraction of an ellipsoid on an external point, I will focus on this aspect of his work, also taking into account that such an aspect is the one mentioned and used by Chasles.

Ivory theorem reads:

> If two ellipsoids of the same homogeneous matter have the same excentricities, and their principal sections in the same planes; the attractions which one of the ellipsoids exerts upon a

[12] Chasles (1837, 1846), p. 7: "En 1809, M. Ivory en trouva une belle et facile solution, fondée sur une curieuse propriété des ellipsoides décrits des mêmes foyers, qui établit une relation simple entre l'attraction d'un ellipsoïde décrit des mêmes foyers, sur un point intérieur; relation au moyen de laquelle on conclut immédiatement, de la formule connue pour l'attraction sur des points intérieurs, l'attraction sur des points extérieurs, et, par suite, le théorème de Maclaurin".

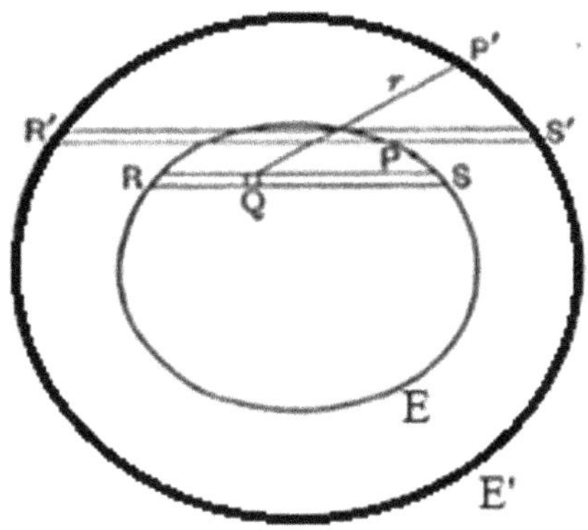

Fig. 7.2 The diagram used by Ramsey to explain Ivory's work. Retrieved from Ramsey (1940, p. 172)

> point in the surface of the other, perpendicularly to the planes of the principal sections, will be to the attractions which the second ellipsoid exerts upon the corresponding point in the surface of the first, perpendicularly to the same planes, in the direct proportion of the surface, or areas, of the principal sections to which the attractions are perpendicular (Ivory, 1809, p. 355).

I follow here the excellent description of Ivory's proof offered by Ramsey (1940), text which simplifies some passages developed by Ivory, without modifying Ivory's method and procedure:

Consider (Fig. 7.2) two confocal ellipsoids E and E': Be $E = \frac{x^2}{a^2} + \frac{y^2}{b^2} + \frac{z^2}{c^2} = 1$ the internal one and $E' = \frac{x^2}{a'^2} + \frac{y^2}{b'^2} + \frac{z^2}{c'^2} = 1$ the external and be $P \equiv (x, y, z)$; $P' \equiv (x', y', z')$ two points belonging respectively to E and E'. Ivory calls *corresponding* the two points P and P' if

$$\frac{x}{a} = \frac{x'}{a'}; \frac{y}{b} = \frac{y'}{b'}; \frac{z}{c} = \frac{z'}{c'} \tag{A}$$

(Ramsey, 1940, p. 172; Ivory, 1809, p. 355). Assume that the two points of the figures are corresponding. Consider the elementary strip RS of E parallel to the x-axis and of cross-section $dydz$. Let its corresponding element be $R'S'$ of E' parallel to the x-axis and whose cross-section is $dy'dz'$. Then it holds

$$\frac{dydz}{dy'dz'} = \frac{bc}{b'c'}.$$

If $f'(r)$ is the law of the force at the distance r, while ρ is the density of both ellipsoids and Q indicates the variable position of the element $dxdydz$ within the strip RS, the component parallel to the x-axis of the attraction exerted by the strip RS on P' is

$$
\begin{aligned}
&-\rho dydz \int f'(r)\cos P'QSdx = -\rho dydz \int f'(r)\left(-\frac{dr}{dx}\right)dx \\
&= \rho dydz\{\, f(P'S) - f(P'R)\}.
\end{aligned}
$$

With the same reasoning it is possible to prove that the attraction exerted on P by the strip $R'S'$ to be

$$
\rho dy'dz'\{\, f(PS') - f(PR')\}
$$

It is easy to deduce from Eq. (A) that for any couple of corresponding points it holds $PS^{'} = P^{'}S$; $PR^{'} = P^{'}R$ (Ramsey, 1940, p. 173). Therefore, the ratio of the x-component of the RS attraction on P' and the x-component of the $R'S$' attraction on P will be

$$
\frac{dydz}{dy'dz'} = \frac{bc}{b'c'}. \tag{B}
$$

If now, all the strips like RS and $R'S'$ of the two ellipsoids are considered, you obtain

$$
X : X' = bc : b'c', \tag{C}
$$

where X is the x-component of the attraction of the first ellipsoid on P' and X' the x-component of the attraction of the second ellipsoid on P.

This is exactly Ivory theorem which, as Ramsey (ibid. p. 173) points out, is valid for any law of attraction. Furthermore, the attraction of E' on P is known because P is internal at E' and its value is $X^{'} = -A^{'}\rho x$, where x is the x-coordinate of P and A' is the same function of a',b',c' as A is of a,b,c, this is depending on the fact that the two ellipsoids are confocal.[13] Thence, it is

[13]The expression denoted by Ramsey with A, B, C, A', B' C' is more easily explained referring directly to Ivory's work. We have: $A = \int \frac{2xdydz}{(x+y+z)^{\frac{3}{2}}}$; $B = \int \frac{2ydxdz}{(x+y+z)^{\frac{3}{2}}}$; $C = \int \frac{2zdxdz}{(x+y+z)^{\frac{3}{2}}}$; $A' = \frac{k}{h}\int \frac{2x'dy'dz'}{(x\prime+y\prime+z\prime)^{\frac{3}{2}}}$; $B' = \frac{k'}{h'}\int \frac{2y'dx'dz'}{(x\prime+y\prime+z\prime)^{\frac{3}{2}}}$; $C' = \frac{k''}{h''}\int \frac{2z'dx'dy'}{(x+y+z)^{\frac{3}{2}}}$, where the letters with the apex are referred to the external ellipsoid, those without to the internal ellipsoid, the letters k^{i} indicate the axes of the internal ellipsoid and the letters h^{i} indicate the axes of the external ellipsoid (Ivory, 1809, pp. 362–363). Through a theorem he had already proved (ibid., pp. 350–356), Ivory was able to demonstrate a relation among such magnitudes. Specifically: $A = \frac{kk'k''}{hh'h''}\int \frac{2x'dy'dz'}{(x\prime+y\prime+z\prime)^{\frac{3}{2}}}$; $B = \frac{kk'k''}{hh'h''}\int \frac{2y'dx'dz'}{(x\prime+y\prime+z\prime)^{\frac{3}{2}}}$; $C = \frac{kk'k''}{hh'h''}\int \frac{2z'dx'dy'}{(x\prime+y\prime+z\prime)^{\frac{3}{2}}}$.

$$X = -\frac{bc}{b'c'}A'\rho x = -\frac{abc}{b'c'}A'\rho\frac{x}{a} = -\frac{abc}{a'b'c'}A'\rho x'$$

Analogous relations hold for the y and z components of the attraction, so that the attraction exerted by the ellipsoid E on the external point P' can be expressed, in components, like this:

$$\left.\begin{aligned} X &= -\frac{abc}{a'b'c'}A'\rho x' \\ Y &= -\frac{abc}{a'b'c'}B'\rho y' \\ Z &= -\frac{abc}{a'b'c'}C'\rho z' \end{aligned}\right\},$$

in the last equation write $C'\rho z'$ with the correct form of the apices being a', b', c' the half-axes of E', confocal with E and passing through P' and A', B', C' are the same functions of a', b', c' as A, B, C are of a, b, c (ibid., p. 173).

A direct corollary of this system of equations is the generalized Maclaurin theorem: the attractions of two confocal ellipsoids on an external point are proportional to their masses. For given the point P' and the attractions of two different confocal ellipsoids, one of half-axes a, b, c, and density ρ, the other of half-axes a_1, b_1, c_1 and density ρ_1, the components x, y, z of the attractions will be as $\rho abc : \rho_1 a_1 b_1 c_1$, so that they are proportional to the masses and have the same directions (ibid., p. 174).

Thus, through the brilliant geometrical idea of two corresponding points, Ivory was able to offer a relatively easy demonstration of a theorem which tested the capabilities of two mathematicians as Laplace and Legendre who supplied far more complicated proofs than Ivory's. This author was one of Chasles' sources of inspiration since a simple geometrical reasoning allowed Ivory to overcome the difficulties in which Legendre's and Laplace's analytical methods were involved, in spite of the fact that these methods were successful. A remarkable aspect of Ivory's reasoning is that he was able to reconduct the problem of the attraction of an ellipsoid towards an external point to that towards an internal point, whose solution was known. His other brilliant idea was, given the point P external to the ellipsoid E, whose attraction on P has to be calculated, to consider the confocal ellipsoid E' passing through P. These two brilliant geometrical ideas were the bases of Ivory's success.

In 1813, Gauss published a fundamental and celebrated memoir on the ellipsoid attraction. He developed a new approach, through which both the cases of internal and external points could be addressed. This allowed him to prove the generalized Maclaurin theorem. Rodrigues (1815), through an analytical method similar to Gauss', but based on a different decomposition of the ellipsoid, obtained the same results as Gauss (Chasles, 1837, 1846, pp. 7–8). Gauss' and Rodrigues' methods were more general than Ivory's. This notwithstanding, Chasles commented on the way in which the French mathematicians considered the solutions of Ivory, Gauss and Rodrigues:

> But the elegant theorem of M. Ivory which, joined to Lagrange's analysis for the case of the internal points, offers such an easy and brief solution to this problem, attracted so much the attention of the geometers that the beautiful memoir by M. Gauss and the remarkable solution by M. Rodrigues where, implicitly, the consideration of an infinitely thin shell included between two similar ellipsoids exists, remained, so to say, not applied [...] so that it was Lagrange's method with the theorem of M. Ivory as a complement which most geometers used in their works. Thanks to this solution, for a long time the question of the ellipsoid's attraction was considered closed and completed.[14]

I offer an outline of Gauss' method, which is absolutely general since, in principle, it holds for the attraction of any body and for any law of attraction which is a function of the distance between the attracted point and an element of the attracting body. Obviously, in order to reach a specific solution it is necessary to make the nature of the attracting body and the law of attraction explicit. It is difficult to claim whether Gauss' approach is geometrical or analytical because, as it is often the case with many of his works, Gauss was able to use both geometrical and analytical properties in a brilliant way, thus resolving the problem of the attraction for external, internal and superficial points in a uniform manner. Therefore, I think that his approach goes beyond the distinction analytical-synthetical: he offered an elegant solution which does not involve complicated calculations and which exploits only some elementary geometrical and analytical considerations. I will concentrate only on the results that are useful to solve the problem of the ellipsoid attraction.

Gauss used two decompositions of the attracting body: as to the first one, he considered a finite body bordered by a continuous surface or by more continuous surfaces in the case in which the body has one or more cavities (Gauss, 1813, p. 3). He considered the surface divided into elements ds. Be P a point belonging to ds whose coordinates in an orthogonal reference frame are x,y,z. Be PX, PY and PZ straight lines parallel to the axes and directed as the axes and be PQ the normal at the surface directed outwards. Be $M \equiv (a, b, c)$ the attracted point and pose $PM=r$. Gauss indicated the angles made by the straight line PM with, respectively, PX, PY and PZ by MX, MY, MZ and the angles between PQ and PX, PY, PZ, PM, respectively. through QX, QY, QZ, QM (ibid., pp. 3–4). He developed the following geometrical construction: consider a plane α perpendicular to the x-axis, $x=\alpha$ so that α is less than the smallest value of x on the surface of the body. He projected (Fig. 7.3) the body on the plane α and considered an element $d\Sigma$ of the projection and

[14]Chasles (1837, 1846), p. 8: "Mais l'élégant théorème de M. Ivory, qui, joint à l'analyse de Lagrange pour le cas des points intérieurs, complétait une solution facile et briève de la question, fixa tellement l'attention des géomètres, que le beau mémoire de M. Gauss, et la solution remarquable aussi de M. Rodrigues, où se trouvait, implicitement, la considération d'une couche infiniment mince comprise entre deux ellipsoïdes semblables, reştèrent, pour aussi dire, inaperçus. [...] aussi ce fut la méthode de Lagrange, avec la théorème de M. Ivory comme complément, que la plupart des géomètres adoptèrent dans leurs ouvrages; et cette solution fit regarder, pendant longtemps, la question de l'attraction des ellipsoïdes comme tout à fait close et complète".

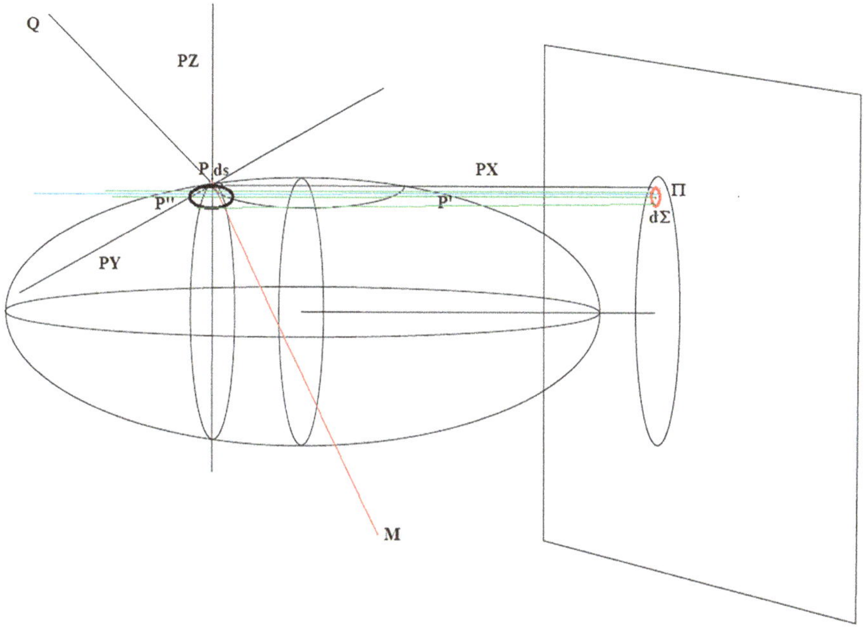

Fig. 7.3 Reconstruction of the diagram described by Gauss: the ellipsoid is the attracting body. *ds* is an element of the body's surface. The point *P* belongs to *ds*. *PX, PY, PZ* are the straight lines from *P* parallel to the three orthogonal axes which represent the principal lines of a Cartesian system of coordinates (not drawn). *PQ* is the normal to the surface. *M* is the attracted point. The ellipse on the plane is the ellipsoid's projection. $d\Sigma$ is an element of the ellipse. I have chosen $d\Sigma$ so that it is the projection of *ds* (this is not necessary). Π is a point internal to $d\Sigma$. From Π the blue straight line parallel to x is drawn so that it cuts the ellipsoid in P' and P''. From the board of $d\Sigma$ straight lines (in green) perpendicular to the plane are drawn. *ds* is, thus, the cylindrical projection of $d\Sigma$ on the body

one of its points Π from which he drew the perpendicular to the plane, which cuts the body in an even multiplicity of points $P^{'}$, $P^{''}$, $P^{'\,'\,'}$, ... If from the board of the element $d\Sigma$ straight lines perpendicular to the plane are drawn, they will form a cylindrical surface which will cut the body in the elements $ds^{'}$, $ds^{''}$, $ds^{'\,'\,'}$, ... Since $d\Sigma$ is the projection on the plane of the elements $ds^{'}$, $ds^{''}$, $ds^{'\,'\,'}$, ..., it will hold $d\Sigma = \pm ds^{'} \cos QX^{'} = \pm ds^{''} \cos QX^{''}$, ... where the sign "+" has to be assumed if the angle is acute, the sign "-" if it is obtuse. It is easy to see that the QX^{2i-1} are obtuse and the QX^{2i} are acute. Therefore, it will be

$$d\Sigma = -ds' \cos QX' = +ds'' \cos QX'' = -ds''' \cos QX''', \ldots$$

Since the number of elements is even,

$$- ds' \cos QX' + ds'' \cos QX'' - ds''' \cos QX''', + \ldots = 0.$$

So that Gauss can deduce his first theorem which will be used in the problem of the ellipsoid attraction:

$$\int ds \cos XQ$$

extended to the whole body is equal to 0. More generally, it also follows that

$$\int T \cos QX + U \cos QY + Z \cos QZ = 0.$$

if T,U,V represent, respectively, rational functions of the sole y,z; x,z; x,y (ibid., pp. 4–5).

Through a brilliant and simple reasoning based on the explained division in infinitesimal elements, Gauss imagined that the infinitesimal cylinder is full of matter of uniform density. He was able to separate the attraction of the part of the cylinder out of the body and within the body. Through an integration, he then reached this important theorem (his Third Theorem, ibid., p. 6): the attraction of a body on a point M, in the direction parallel and opposite to the x-axis is obtained by the formula

$$- \int Frds \cos QX \tag{D}$$

The symbol Fr can be explained like this: Gauss imagined the cylinder divided into infinitesimal strips parallel to the bases of the cylinder. Be a point whose x-coordinate is ξ, while η, ζ are the other coordinates (ibid., pp. 4–5). The volume of the cylinder included between the bases of which one coordinate is ξ will be $d\Sigma d\xi$. Be ϱ the distance between M and the point (ξ, η, ζ) and $f\varrho$ the attraction law. Then this slide of cylinder will attract the point M with a force equivalent to $d\Sigma d\xi f\varrho$. Gauss then proved that the attraction element on M along the x-axis is given by $-f\rho d\rho d\Sigma$, while the attraction of the whole cylinder until the point (ξ, η, ζ) will be $\int f\varrho d\varrho$ varying ρ between a point of the basis Σ with fixed coordinates and the point (ξ, η, ζ). Gauss denoted such an integral with $F\rho$. If r', r'', r''', … indicate the distances between a point of Σ and points equivalent at (ξ, η, ζ), but belonging to the surface of the body, then the attraction of the parts of the infinitesimal cylinder included within the body will be given by

$$(Fr' - Fr'' + Fr''' - \ldots)d\Sigma = - Fr'ds' \cos QX' - Fr''ds'' \cos QX'' - Fr''' \cos QX''' - \ldots$$

Applying the same reasoning to all the elements $d\Sigma$, the formula (D) follows.

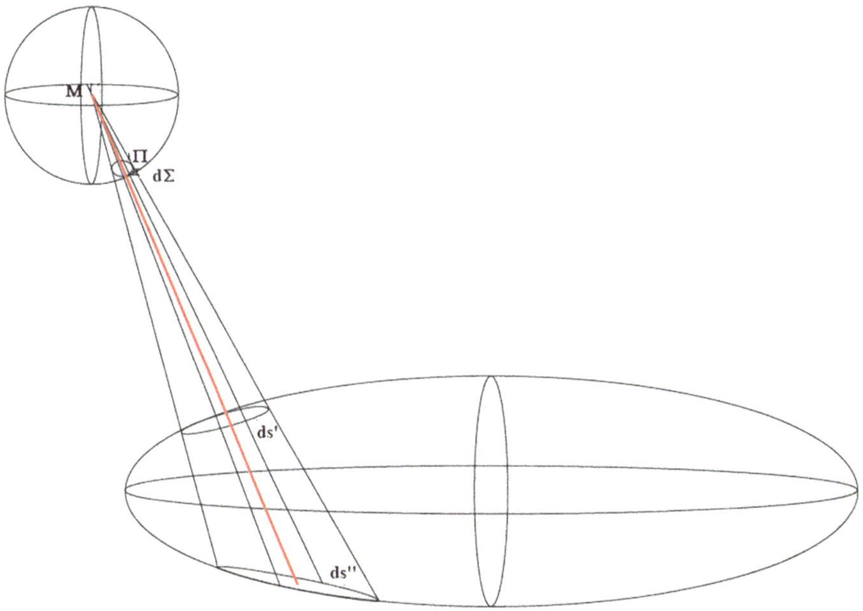

Fig. 7.4 Reconstruction of the diagram which represents the steps of Gauss' reasoning based on his second decomposition of the attracting body in infinitesimal elements

This theorem can, obviously, be applied to the ellipsoid, but in order to reach a more specific and expressive form for the attraction Gauss resorted to a further division of the body in infinitesimal elements: he considered a spherical surface of unitary radius whose centre is the attracted point M. Divide the spherical surface into infinitesimal elements $d\Sigma$ and be Π a point of $d\Sigma$. Draw the straight line $M\Pi$ and be $P^{'}, P^{''}, P^{'''}, \ldots$ the points where such a line cuts the surface of the attracting body (ibid., p. 7). Gauss distinguished the three cases in which M is external, internal to the body, or belonging to its surface. For my aim, it is possible to consider M external to the body. Now Gauss considered the set of the straight lines drawn from M to the border of the element $d\Sigma$. These straight lines are the generatrices of a cone. On the surface of the body they form the elements $ds^{'}, ds^{''}, ds^{'''}, \ldots$ to which the points $P^{'}$, $P^{''}, P^{'''}, \ldots$, respectively, belong (ibid., p. 7, see Fig. 7.4).

One of the most important results obtained by Gauss through this new decomposition is the Theorem 4 according to which $\int \frac{ds \cos MQ}{r^2}$ extended to the whole surface of the body is equal to 0, to -2π or to -4π depending on the external, on-surface or internal position of the point M with respect to the body's surface.

To reach the law of attraction he developed a reasoning similar to that used to explain the attraction when the element is cylindrical. For given the distance ρ from M, Gauss supposed the force be a function $f\rho$ of such a distance and the density of the body be uniform. The cone of basis $d\Sigma$ is supposed full of matter. He imagined to divide the cone through spherical surfaces centred in M, infinitely close one to the other and such that the radius of the biggest of these surfaces is ρ. The element of the cone relative to the division of the sphere whose radius is ρ will be $\varrho^2 d\varrho d\Sigma$ and the force exerted by this element on M will be $\varrho^2 f\rho d\varrho d\Sigma$. If $\int \varrho^2 f\varrho d\varrho$ is expressed by $\phi\varrho$, then $d\Sigma(\Phi\varrho - \Phi 0)$ indicates the attraction exerted on M by the cone's part included between the vertex M and the distance ρ. In general, $d\Sigma(\Phi\varrho' - \Phi\rho)$ indicates the attraction of the cone's part included between ρ' and ρ. Therefore, if $r', r'', r''', \ldots$ indicate the distances between M and the intersection of the cone's elements with the body, so that these distances also identify an element of the body, the attraction of the body's part included within the cone towards M in the direction $M\Pi$ is

$$d\Sigma(-\Phi r' + \Phi r'' - \Phi r''' + \ldots) = -\frac{ds'\Phi r' \cos MQ'}{r'^2} - \frac{ds''\Phi r'' \cos MQ''}{r''^2} - \frac{ds'''\Phi r''' \cos MQ'''}{r'''^2} - \ldots$$

If this expression is multiplied by cosMX, one will get the attraction exerted by the part of the cone included within the body in the direction of the x-axis and opposite sense. If you integrate on the whole body, you will get the global attraction along the x-axis, which is (Theorem 6, ibid., pp. 10–12)

$$-\int \frac{ds\Phi r \cos MQ \cos MX}{r^2}. \tag{E}$$

Gauss also addressed the question for the cases when M is internal to the body or lies on its surface through a reasoning similar to that expounded for the case in which M is external.

Now I will concentrate on the way in which he applied his technique to the attraction of a homogeneous ellipsoid in the case in which M is external to the ellipsoid.

Be given the ellipsoid

$$\frac{x^2}{A^2} + \frac{y^2}{B^2} + \frac{z^2}{C^2} = 1.$$

To solve the problem, he introduced two variables p and q such that

$$x = A\cos p; y = B\cos p \sin q; z = C \sin p \sin q.$$

After a series of long, but not difficult transformations, Gauss was able to determine the body's element as

$$ds = dp \cdot dq \cdot ABC \cdot \psi \cdot \sin p$$

where $\psi = \left(\frac{x^2}{A^4} + \frac{y^2}{B^4} + \frac{z^2}{C^4}\right)^{\frac{1}{2}}$ (ibid., p. 18).

Applying this value of the element at the formula (D), which expresses Theorem 3, the attraction X parallel to the x-axis will be

$$X = \iint dp \cdot dq \frac{BCx \sin p}{rA} = \iint dp \cdot dq \frac{BC \cos p \sin p}{r}$$

If the attraction X is posed equal to $ABC\xi$, it will be, hence,

$$\xi = \iint \frac{dp \cdot dq \cdot \cos p \sin p}{Ar} \tag{F}$$

Applying the formula (E) (Theorem 6), you will get:

$$\xi = -\iint \frac{dp \cdot dq \cdot \sin p}{r^3}(a-x)\left(\frac{(a-x)x}{A^2} + \frac{(b-y)y}{B^2} + \frac{(c-z)z}{C^2}\right). \tag{G}$$

Because of the Theorem 4, it will be (ibid., p. 19)

$$\iint \frac{dp \cdot dq \cdot \sin p}{r^3}\left(\frac{(a-x)x}{A^2} + \frac{(b-y)y}{B^2} + \frac{(c-z)z}{C^2}\right) = 0.$$

Gauss considered the values A,B,C as three variables α, β, γ so that $\alpha^2 - \beta^2$ and $\alpha^2 - \gamma^2$ are constant. As a matter of fact, he took into account a series of ellipsoids which are confocal with that whose axes are A,B,C. In particular, he focused on the situation in which the axes of one of such ellipsoids have an infinitesimal difference with respect to A,B,C.

From Eq. (F) it follows that

$$\alpha\xi = \iint \frac{dp \cdot dq \cdot \cos p \sin p}{r}.$$

Considering the variation δ of this quantity, it will be

$$\alpha\delta\xi + x\delta\alpha = \iint \frac{dp \cdot dq \cdot \cos p \sin p \delta r}{r^2}$$

where $r^2 = (a - x)^2 + (b - y)^2 + (c - z)^2$, so that (ibid., p. 19).

$$r\delta r = -(a-x)\delta x - (b-y)\delta y - (c-z)\delta z$$

Through a series of calculations Gauss was able to prove that

$$r\delta r = -\alpha\delta\alpha\left(\frac{(a-x)x}{\alpha^2}+\frac{(b-y)y}{\beta^2}+\frac{(c-z)z}{\gamma^2}\right).$$

Thence

$$\alpha\delta\xi+\xi\delta\alpha=\delta\alpha\iint\frac{dp\cdot dq\cdot x\sin p}{r^3}\left(\frac{(a-x)x}{\alpha^2}+\frac{(b-y)y}{\beta^2}+\frac{(c-z)z}{\gamma^2}\right). \tag{H}$$

Now Gauss multiplied Eq. (G) by $\delta\alpha$ and subtracted the result from Eq. (H). After having replaced A,B,C through α, β, γ, he obtained (ibid., p. 20)

$$\alpha\delta\xi=\delta\alpha\iint\frac{dp\cdot dq\cdot a\sin p}{r^3}\left(\frac{(a-x)x}{\alpha^2}+\frac{(b-y)y}{\beta^2}+\frac{(c-z)z}{\gamma^2}\right).$$

However, for an external point the right part of this equation is equal to 0, so that $\delta\xi=0$. This means that ξ is constant. Let us recall that Gauss had posed the attraction equal to $ABC\xi$. This means that he regarded the attraction as divided into a constant part ABC which is proportional to the volume of the ellipsoid and, thence, to its mass, and a potentially variable part ξ. The whole of his resoning has shown that, in fact, ξ is constant too. Furthermore, the argumentation has been developed by taking into account a series of confocal ellipsoids. Thus, the general conclusion of Gauss' procedure is that, given two confocal ellipsoids, their attraction on a point external to both is proportional to their mass. This is the generalized Maclaurin theorem. Through an analogous technique Gauss was able to solve the problem for internal points and for points posed on the surface of the ellipsoid.

I have dedicated many specifications to Gauss' method because of its generality and simplicity in comparison with those used by Legendre and Laplace to prove the generalized Maclaurin theorem. Gauss added an interesting *Additamentum* (ibid., pp. 23–24) to his memoir where he explained that Laplace communicated to him the existence of Ivory's (1809) memoir. Gauss recognized the great elegance of Ivory's technique through which the attraction for an external point had been reduced to that for an internal point. However, he remarked that in treating the attraction for the internal points Ivory resorted to infinite series which are not always convergent and that, despite some similarities, his own method and Ivory's are substantially different.

The next and last fundamental author in the history briefly narrated by Chasles is Poisson. Among the various memoirs that Poisson dedicated to our subject, Poisson (1835a) is particularly important and is a reference point for Chasles (1837, 1846). In this memoir Poisson used the same decomposition as Lagrange: he resorted to infinitely thin pyramids having their vertices at the attracted point (supposed external). In this way Poisson obtained a double integral, which cannot be integrated elementarily (whereas this is possible for internal points). It is, however, possible to transform the double integral in a simple one. To this aim, Poisson first transformed it into a triple integral. In doing so, he was led to decompose the ellipsoid in another manner, that is in infinitely thin shells included between two concentric, similar and similarly posed ellipsoidal surfaces (Chasles, 1837, 1846, pp. 9–10). Through two integrations Poisson was able to express the attraction of the shell, but to obtain such integrations, he had to change the reference frame and to assume as coordinate axes the three principal axes of the cone circumscribing the external surface of the shell and having its vertex in the attracted point (ibid., p. 10). We will see that this change of reference frame is crucial for Chasles, because it is the one that he used in the fundamental memoir written in 1837, 1846 I am presenting and commenting. Therefore, Poisson was the first who obtained an expression for the attraction of the infinitely thin shell. When this expression is integrated on the whole ellipsoid, you get the elliptical integral which expresses the ellipsoid's attraction on an external point. I will not consider the details of Poisson's analysis because many of them will represent the starting points of Chasles' work and I will recall those details while dealing with the analysed works by Chasles.

In these elements of history on the ellipsoid attraction I have had no claim to be exhaustive. Rather I have pointed out the works of the authors mentioned by Chasles and who, hence, represented a source of inspiration for his works on this subject.

7.2 *Mémoire sur l'attraction des ellipsoides*, 1837e

This memoir is not a foundational one. I analyse some results achieved here by Chasles because, after the development of his synthetic theory of the ellipsoid attraction (Chasles, 1837, 1846), such results will be included within it. Thence, they are useful for the reader to fully grasp the width of Chasles' view and achievements.

The starting point of this work is a famous result obtained by Poisson (1835a) which concerns the attraction exerted on an external point S by an infinitely thin ellipsoidal shell included between two s.s.p. and concentric ellipsoidal surfaces. Poisson was able to determine the intensity and the direction of the attraction. In particular, he proved that the direction is along the axis of the cone whose vertex is S and which is circumscribed to the external surface of the shell (I call this theorem T1). The attraction of a whole ellipsoid was thus obtained by an integration.[15]

In this memoir Chasles developed, so to say, an interesting mathematical exercise because he inverted the order of the deductions. For he supposed that the general formula expressing the attraction of a whole ellipsoid is known and from this formula he deduced several properties concerning the attraction of an ellipsoidal

[15] For the moment I do not distinguish between the case in which the ellipsoid is homogeneous and those in which its density varies according to a function of the distance from the ellipsoid's centre because insofar as my considerations on Chasles' work are concerned, this is not, at the moment, essential. I suppose the ellipsoid to be homogeneous.

shell as well as other features of physical phenomena treatable with the same techniques as those used to study the ellipsoid's attraction. Chasles wrote explicitly:

> My aim is to show that these different results and other ones which are related to them, can be easily deduced from the ordinary formulas for attraction of a homogeneous ellipsoid on an external point independently of the method through which such formulas have been obtained.[16]

Established an orthogonal Cartesian reference frame with its origin in the centre of the ellipsoid E and assumed as axes the principal axes A,B,C of the ellipsoid, the known formula for the attraction along the x-axis is

$$4\pi\rho x' \frac{ABC}{A_1} \int_0^1 \frac{v^2 du}{\sqrt{A_1^2 + v^2 (B^2 - A^2)} \sqrt{A_1^2 + u^2 (C^2 - A^2)}}$$

where (x', y', z') are the coordinates of the attracted point S and A_1, B_1, C_1 are the axes of the ellipsoid passing through S and whose principal sections are confocal with those of E (ibid., p. 245).

Through the change of variable $v = \frac{A_1}{A} u$, the previous integral was transformed into

$$4\pi\rho x' BC \int_0^{\frac{A}{A_1}} \frac{u^2 du}{\sqrt{A^2 + u^2 (B^2 - A^2)} \sqrt{A^2 + u^2 (C^2 - A^2)}}$$

which allowed Chasles, after a series of passages, to arrive at proving that the attraction exerted by an infinitely thin shell included between two concentric and homothetic surfaces on a point S external to the shell is directed along the normal at this point to the ellipsoid through S whose principal sections are confocal with the external surface of the shell (ibid., pp. 250–251, I call this theorem T2).

He obtained these results through considerations connected to the equation of the ellipsoid and to trigonometry (besides, obviously, those connected with mathematical analysis). Thence, they are not tied to pure geometry, and for this reason I will not consider them.

However, he reached the proof of Poisson's theorem T1 through T2 thanks to a result he had obtained in pure projective geometry. For he wrote:

> It is known from a theorem of geometry I have proved some years ago that when some second degree surfaces have their principal sections described with the same foci, if any point of space is assumed as the common vertex of cones, each of them circumscribed to one

[16]Chasles (1837e), pp. 244–245: "Je me propose de faire voir que ces divers résultats, et quelques autre qui s'y rapportent, peuvent être déduits assez facilement des formules ordinaires pour l'attraction d'un ellipsoïde homogène sur un point extérieur, sans qu'on ait besoin de connaitre la méthode qu'on a suivie pour arriver à ces formules".

surface, all these cones have their principal axes which are normals to three surfaces confocal with the given ones, passing through the vertex of the cone.[17]

As I will show in Sect. 7.4, this is one of the fundamental theorems used by Chasles to construct his synthetic solution for the attraction of the ellipsoid. For through this theorem, as Chasles pointed out (ibid., p. 251), T1 follows immediately as a consequence of T2.

After having obtained the formula for the intensity[18] and the attraction's direction of an infinitely thin ellipsoidal shell on an external point, Chasles stressed that his results can also be extended to domains of physics which go beyond mechanics. When in Chasles (1837, 1846) he developed a complete theory of the ellipsoid attraction through a geometrical method, these results were reinterpreted: for Chasles, they became an evident litmus paper of the soundness of his foundational programme. Therefore, it makes sense to illustrate these results as soon as Chasles referred to them, though this happened in his work dating from 1837 which is not a foundational one, but almost a mathematical-physical divertissement.

The solution concerning the attraction of an infinitely thin shell of an ellipsoid—Chasles claimed (ibid., pp. 253–254)—derives from the inverse square law. This same law, with the only difference that the force can also be repulsive, holds in the theory of electricity. In particular, Chasles considered the electricity distribution on the surface of an ellipsoidal conductor. As he explained, when a body is electrified, within the conductor the equilibrium is established and the electric fluid in excess is drawn at the surface which is in equilibrium because of the pressure of the surrounding air. The electric charges are, hence, distributed on an infinitesimal shell contiguous at the surface. The features of the charge distribution are the following:

1. The whole charge distribution is included between two similar and concentric surfaces where the external surface is that of the ellipsoid.
2. There is no action on any point internal to the ellipsoid.
3. The repulsive action exerted by a point of the external surface on the ambient air is normal to the surface, which produces the equilibrium state of the shell.
4. The repulsive force is proportional in any point at the thickness of the shell.[19]

[17] Chasles (1837e), p. 251: "On sait, par un théorème de géométrie que j'ai donné il y a quelques années, que quand plusieurs surfaces du second degré ont leurs sections principales décrites des mêmes foyers, si un point quelconque de l'espace est pris pour le sommet commun d'autant de cônes circonscrits à ces surfaces, tous ces cônes ont les mêmes axes principaux, qui sont les normales aux trois surfaces décrites des mêmes foyers que les proposées, et passant par le sommet commun des cônes".

[18] I will deal with any detail of the formula providing the intensity of an ellipsoidal shell attraction in Sect. 7.3. In this context, it is enough to remember that such an attraction is proportional to the mass of the shell.

[19] Chasles (1837e, p. 253) recalls that this result was obtained by Poisson in his memoir *Sur la distribution de l'électricité à la surface des corps conducteurs* (Poisson, 1812–1813). This ponderous work by Poisson is divided into two parts. The first one was read at the Académie on 9 May and 3 August 1812. The second part was read on 6 September 1813. Both of them were published in the 12th tome of the *Mémoires de la classe des Sciences mathématiques et physiques de l'Institut*

Poisson—Chasles highlighted—had pointed out that his theorem on the attraction of an infinitely thin ellipsoidal shell can be extended to the electricity theory. Furthermore, it is useful to determine the direction that a little body posed close to the surface of the electrified ellipsoid assumes while approaching the ellipsoid's surface due to the attractive electric force.

Chasles added that, according to what proved in the theory of an ellipsoid's shell attraction, this direction is exactly that of the instantaneous axis of the cone circumscribed to the ellipsoid and whose vertex is the attracted point. This direction coincides with that of the normal at a second ellipsoid drawn through the point and whose principal sections are described with the same foci as those of the attracting ellipsoid. Such consideration—Chasles continued—is also useful to grasp other phenomena. For example (ibid., p. 254), in order to identify the surface, lying on which little bodies under the attractive or repulsive attraction of the ellipsoid are in equilibrium without sliding wherever they are posed on the surface. He is referring to what nowadays we call equipotential surfaces: if a body moves on the same equipotential surface, the field exerts no work on it. As to the line along which a little body moves under the action of the ellipsoidal shell, it is possible to prove that it is a curve formed by a series of infinitesimal lines which are the consecutive normals to a series of ellipsoids confocal with the external surface of the ellipsoidal shell. It is a fourth-degree skew line determined by the intersection of the one-sheeted and the two-sheeted hyperboloids confocal with the external surface of the shell. On each of these hyperboloids these are curvature lines (ibid., p. 254).

Besides remarking the beautiful language with which Chasles described the continuous motion of the little body separating it in its instantaneous acts of motion, it is worth recalling that all these results would automatically become part of the geometrical attraction of the ellipsoid when Chasles developed it in (Chasles, 1837, 1846). I do not go into further details of this memoir dating 1837 because, though interesting, they are not necessary to trace Chasles' foundational programme.

7.3 *Mémoire sur l'attraction d'une gouche ellipsoidale infiniment mince*, 1837f

Although the fundamental memoir by Chasles is, insofar as my aims are concerned, Chasles (1837, 1846), in Chasles (1837f) there are several geometrical results applied to the study of the attraction of an infinitesimal ellipsoidal shell which are interesting and which will be presented in a systematic and more mature form in Chasles (1837, 1846), where they are founded on some achievements of polar theory (as we will see). Therefore, in this subsection, I will not give the details of such

Impérial de France (in the Napoleonic period the *Institut de France* assumed the denomination *Imperial*). Chasles indicated this memoire referring to the year 1811 because this is the indication which appears in the frontispiece, but I prefer to indicate it as Poisson (1812–1813).

results since I will refer to them in Sect. 7.4. Instead, it seems to me more interesting to focus on some homologies pointed out by Chasles between the theories of electricity and health and that of the ellipsoid attraction. Chasles had already highlighted some of them in the previous essay, but here he is more explicit and exhaustive and the most relevant sections of this memoir concern the concept which nowadays is called *potential* and that of *level surfaces*. Thence, when he founded the theory of ellipsoid attraction in a geometrical manner, the results shown in this memoir were also included in his foundational geometrical programme, which, hence, embraced a not negligible part of the physics known at that time.

As to the specific problem concerning the direction of the attraction of an ellipsoidal infinitesimal shell on an external point S, Chasles developed the same reasoning (Chasles, 1837f, pp. 266–271) I will expound in Sect. 7.4.2. He thus reached the conclusion that the attraction exerted on an external point S by an infinitely thin shell included between two ellipsoidal, concentric and homothetic surfaces is directed along the normal from S to the ellipsoid which can be drawn through S such that its principal sections are confocal with the external surface of the shell.

As to the value of this attraction, Chasles resorted to the concept of *potential*. In this case, too, we will see in Sect. 7.4.2 that his argumentation will be different; but the reasoning on potential is remarkable since it is connected to geometrical properties of the second-degree surfaces and it holds for any vector field with a potential. Surely Chasles in his seminal memoir 1837, 1846 chose to present his theory in a different manner because the new presentation was connected to the graphic properties of the second-degree surfaces he expounded in the first section of his memoir. Whereas, in Chasles (1837f), starting from some very basic features of the level surfaces and through the concept of potential, he obtained, basically through geometrical methods, advanced properties of the level surfaces. His arguments testify to the continuous intention to reach a unified and broad set of ideas of which the theory of ellipsoid attraction is the basic element, but which can be extended to other fields of physics, too. At the end, he would envisage the geometrical road. His presentation concerning the properties of the level surfaces can be interpreted as an attempt to reach a unitary view of many physical problems, though not, properly speaking, a step of his foundational programme. Thence, I will present the basic elements of this train of thoughts without considering all the details.

Chasles argued that, if

$$\frac{x^2}{a^2}+\frac{y^2}{b^2}+\frac{z^2}{c^2}=1$$

is the equation of the external surface of the ellipsoidal shell, the equation of the level surfaces can be expressed like this:

$$\frac{x^2}{a_1^2} + \frac{y^2}{a_1^2 + b^2 - a^2} + \frac{z^2}{a_1^2 + c^2 - a^2} = 1$$

being a_1 a variable parameter which determines each surface.

If X, Y, Z are the components of the body's attraction on a point of coordinates (x,y, z), the differential equation of the level surface through this point is

$$Xdx + Ydy + Zdz = 0.$$

Now Chasles introduced the concept of potential: for he wrote that the components of the attraction of a body are the differential coefficients of a function V which—he remarked—is the sum of the body's molecules, divided by the distances from the point[20] (x,y,z), so that

$$\frac{dV}{dx} = X; \quad \frac{dV}{dy} = Y; \quad \frac{dV}{dz} = Z. \tag{7.1}$$

Therefore, this follows

$$\frac{dV}{dx}dx + \frac{dV}{dy}dy + \frac{dV}{dz}dz = 0. \tag{7.2}$$

Hence, $dV = 0$ and V is constant. This means that for any point of the same level surface the function V holds the same value. Thus, this value is a function of the sole parameter a_1, which varies from a level surface to another one. If the value of V were known, the attractions X, Y, Z would also be known. Chasles added that the function V, when the attracted points are external to the attracting body, fulfils (ibid., p. 272) the equation

$$\frac{d^2V}{dx^2} + \frac{d^2V}{dy^2} + \frac{d^2V}{dz^2} = 0 \tag{7.3}$$

Finally, he claimed that, as shown by Lamé while dealing with the subject of isothermal surfaces, this equation can be integrated if the level surface is known (ibid., pp. 272–273).[21]

[20] It seems to me appropriate to mention Chasles' original expression. Ibid., p. 272: "Mais on sait que les composantes de l'attraction d'un corps sont les coefficiens différentiels d'une même fonction V qui est la somme des molécules du corps, divisées respectivement par leurs distances au point (x,y,z)".

[21] For the introduction and use of Eq. (7.3) Chasles referred to Laplace's *Méchanique celeste*, book 2, p. 137 and to Poisson's *Mémoire sur la théorie du Magnétisme en mouvement*. This important memoire was read at the Académie on the 10th July 1826, but was published in the *Mémoires de l'Académie des Sciences* in the issue 1823. This is simply due to the delay in the publication of the

Commentary: considering V to be the gravitational potential and G_x, G_y, G_z the components of the gravitational field in a given reference frame, in modern notation Eq. (7.1) are written as

$$\frac{\partial V}{\partial x}=G_x;\ \frac{\partial V}{\partial y}=G_y;\ \frac{\partial V}{\partial z}=G_z$$

For well-known reasons, we prefer to place a sign "minus" before the partial derivatives. These equations show that the field is the gradient of the potential. In modern notation Eq. (7.2) is written like this

$$\frac{\partial V}{\partial x}dx+\frac{\partial V}{\partial y}dy+\frac{\partial V}{\partial z}dz=0$$

which indicates that, on a level surface, the total potential difference is null.

Equation (7.3) shows that in an empty space $\nabla^2 V=0$, which represents Laplace's equation for gravitational fields.

Chasles' conception of potential seems to me rather interesting. With regard to this function, he wrote, as we have seen: "une même fonction V qui est la somme des molécules du corps divisées respectivement par leurs distances au point (x,y,z)". Therefore, the potential is conceived in an almost modern manner. For we write the gravitational potential of a mass m as $V=-\gamma\frac{m}{r}$, whereas Chasles would have written $V=\frac{m}{r}$.

The concept of potential is not Chasles' original. It is due to Legendre and Laplace, as outlined in Sect. 7.1. They used this notion in their studies on ellipsoid attraction, but the way in which Chasles exploited the concept of potential to deduce the attraction of an ellipsoidal thin shell allowed him to obtain an easier and more direct solution. It is also significant that, after having introduced the geometrical elements of his theory (see next subsection), the analytical elements used in Chasles (1837f) relied on a solid geometrical basis.

Chasles continued considering the equation

$$\frac{x^2}{a_1^2}+\frac{y^2}{a_1^2+b^2-a^2}+\frac{z^2}{a_1^2+c^2-a^2}=1$$

as that which expresses the level surface passing through S. By means of a series of analytical steps (ibid., pp. 273–276) in which the properties of the function potential and of this level surfaces are used, he reached the conclusion that the attraction A of an infinitely thin ellipsoidal shell can be expressed as

issues, which appeared in 1827. Thence, I indicate this memoire as Poisson (1827). Chasles specified that Lamé proved that this equation is integrable (Lamé, 1837).

$$A = 2\frac{Cp}{a_1 b_1 c_1}$$

where $b_1^2 = a_1^2 + b^2 - a^2; c_1^2 = a_1^2 + c^2 - a^2; p = \frac{1}{\sqrt{\frac{x^2}{a_1^4} + \frac{y^2}{b_1^4} + \frac{z^2}{c_1^4}}}$. It remains to determine the constant C.

These results were obtained through a series of reasonings analogous (in substance) to those I will expound in detail in Sect. 7.4.

Chasles' final achievement for the formula of attraction was:

$$A = 4\pi\rho bcda\frac{p}{a_1 b_1 c_1}, \tag{7.4}$$

where a is the major axis of the ellipsoidal shell's external surface, da is the thickness of the shell in the a-direction, b,c are the two other axes of the shell's external surface and ρ is its density.

Chasles claimed that, given the attraction of an ellipsoidal infinitesimal shell, the attraction of the whole ellipsoid on an external point can be deduced considering the ellipsoid as composed of infinitesimal shells, each of them included between two homothetic surfaces. It is therefore necessary to calculate an integral of a variable which can be posed equal to $u = \frac{a}{a_1}$. Then, Chasles claimed, the calculation is exactly the same as that he had expounded in the previous memoir, but developed in inverse order (ibid., p. 279). However—and this is significant in my perspective—he announced his intention to offer a completely geometrical foundation to the problem of the ellipsoid attraction. For he wrote:

> Therefore, the solution we have given for the attraction of an infinitely thin ellipsoidal shell is, in fact, a complete solution to the problem of the attraction of a heterogeneous ellipsoid. I will provide a further solution to this problem which will rely only on simple geometrical considerations.[22]

Chasles pointed out the novelty of his approach because he stressed that Laplace, Poisson and Pontécoulant[23] had used Eq. (7.3) to calculate V in a spherical shell or in

[22] Chasles 1837f, p. 280: "Ainsi, la solution que nous venons de donner pour l'attraction d'un couche infiniment mince, est, dans le fait, une solution complète du problème de l'attraction d'un ellipsoïde hétérogène. Je donnerai ailleurs une autre solution de ce problème, qui ne reposera absolument que sur de simples considérations de géométrie". Chasles is referring to a heterogeneous ellipsoid because this is the most general case. Obviously, in this case, the density ρ itself is a variable function and not a constant. In (Chasles 1837, 1846), he will show that, when ρ varies according to a particular law in function of the distance between the ellipsoid's centre and its external surface, the resulting function can be integrated by quadratures, whereas when the ellipsoid is homogeneous, it produces an elliptical integral. I will give only a hint to this problem while addressing (Chasles, 1837, 1846).

[23] With regard to Laplace, Chasles referred to the *Mécanique céleste*, book 2, p. 138. As to Poisson he mentioned Poisson (1826). Chasles also recalled the contribution given by Pontécoulant in his *Systéme du Monde* (Pontécoulant, 1829–1834, volume 2, p. 329).

a sphere, but that the application of such result to the case of the ellipsoid is his own original (ibid., p. 280).

Chasles continued connecting his results to geometry. He stressed that the property according to which the level surfaces of an ellipsoidal shell are ellipsoids having their principal section described with the same foci is a new property of these surfaces and added:

> It is possible to express [such a property] in geometry saying that the sum of the molecules of an infinitely thin ellipsoidal shell, reciprocally divided by their distances from a point external to the shell, has a constant value for all the points belonging to an ellipsoid described with the same foci as the external surface of the shell.[24]

This assertion is equivalent to stressing that such surfaces are equipotential.

Here Chasles added a commentary, which is useful to grasp his mentality: he claimed that the confocal second-degree surfaces, already analysed by Dupin, Binet and Lamé, seem particularly suitable to study natural phenomena. Beyond the intrinsic geometrical interest behind these surfaces, Chasles developed a series of inquiries on them (he mentioned pp. 384–399 of the *Aperçu*) also with the intention to apply their properties to natural phenomena. This will happen in Chasles (1837, 1846), but already in 1837f the idea of using (in a fundamental and foundational manner) geometry as the basis for physics existed, though Chasles (1837f) is basically an analytical essay.

In the second section of his memoir (ibid., pp. 281–291) Chasles studied the attraction of an infinitely thin ellipsoidal shell on different points of space. This section has some interesting elements because he offered some geometrical and dynamical specifications connoting the attraction of an ellipsoidal shell. First of all, he remarked that the formula $A = 4\pi\rho bcda\frac{p}{a_1b_1c_1}$ shows that, considering the point S belonging to a level surface, the attraction is proportional to the distance of the plane tangent at the ellipsoid in S from the centre of the shell because p indicates exactly such a distance. Therefore, given a level surface, the maximal attraction takes place at the extremity of the major axis of the ellipsoid and the minimal attraction at the extremity of the smallest axis. However, the attraction can assume a more expressive form because, told Ω the area determined in the ellipsoid whose major axis is a_1 by the diametral plane parallel to the plane tangent in S, it is $p\Omega = \pi a_1b_1c_1$, so that

$$A = 4\pi^2\rho bcda\frac{1}{\Omega}.$$

Chasles claimed that this expression is simpler than the previous one. For it shows that the attractions exerted by an ellipsoidal shell on different points of space are

[24] Ibid., p. 280: "On peut l'exprimer, en géométrie, en disant que la somme des molécules d'une couche ellipsoïdale infiniment mince, divisées respectivement par leurs distances à un point pris au dehors de la couche, a une valeur constante pour tous les points situés sur un même ellipsoïde décrit des mêmes foyers que la surface externe de la couche".

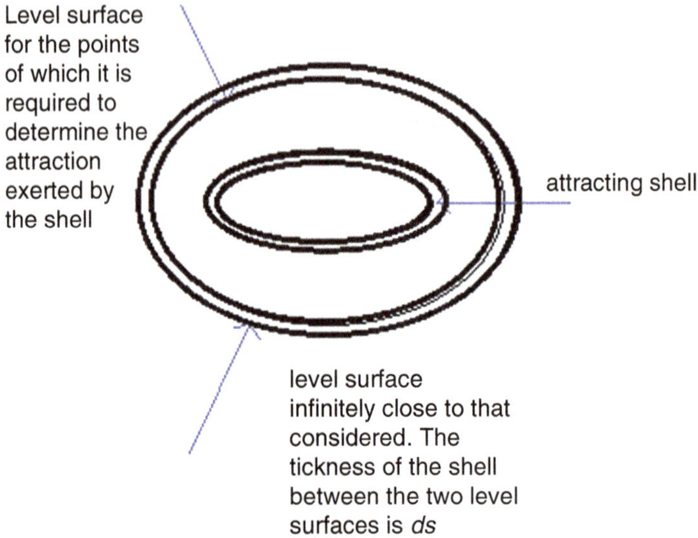

Fig. 7.5 My outlined diagram in which the basic elements of Chasles' reasoning reported in the running text are represented

inversely as the areas drawn in the sections of the ellipsoid having major axis a_1 passing through *S* by the diametral plane parallel to the plane tangent to this ellipsoid at this point.

As a commentary, it is worth pointing out that this new formula for the attraction appears to Chasles more expressive since it connects directly such an attraction with a tangible geometrical quantity as an area. This is a further confirmation of the profound geometrical approach behind the whole work by Chasles on the ellipsoid attraction. This approach reached its highest point in Chasles (1837, 1846).

After a brief series of calculations Chasles was able to determine the attractions exerted by a shell *L* on the different points of a level surface *NS* reaching the conclusion that such attractions are inversely as the thickness of the shell included between *NS* and the infinitely close level surface (see Fig. 7.5). The formula obtained by Chasles is (ibid., pp. 282–283)

$$A = 4\pi\rho bcda\frac{1}{b_1c_1}\frac{da_1}{ds}$$

where *b* and *c* are the mean and the small axes of the external surface of *L*, *da* is the differential of the major axis (that is the difference between the major axis of the external and the internal surface of *L*), b_1, c_1 are the mean and the small axes of the internal level surface *NS*, da_1 is the differential of the major axis of the level shell (namely, the difference between the major axis of the external and the internal surface of the shell determined by the two level surfaces), *ds* is the thickness of this shell.

After a series of further analytical reasonings Chasles reached the following result: be given two infinitely thin ellipsoidal shells. Be the first included within the surfaces of two concentric and homothetic ellipsoids, and the second one between the surfaces of two ellipsoids described with the same foci as the external surface of the first shell. Be also given the linear elements measuring the thickness of the second shell, the attractions exerted by the former shell on these elements are equal (ibid., p. 284). Therefore, given the form of the level surfaces, the attraction exerted by the attracting shell on the linear elements measuring the thicknesses of the shell included between two infinitely close level surfaces is equal (ibid., p. 284).

This is a significant result achieved by Chasles as it clarifies the way in which the different points belonging to a level surface are attracted by an ellipsoidal shell with the given characteristics.

Chasles' next step consisted in analysing the relation between the attractions exerted by an infinitesimal ellipsoidal shell on different level surfaces. He used the important concept of "corresponding points", a concept introduced by Ivory (see section 6.1., pp. 351–354) to prove his theorem.

Chasles considered two level surfaces indicated, respectively, by (a_1), (a_1'). If $d\omega$ is an infinitesimal portion of (a_1), according to formula expressed by Eq. (7.4), it is

$$Ad\omega = 4\pi\rho bcda \frac{pd\omega}{a_1 b_1 c_1}.$$

Considering (a_1') the formula will be

$$Ad\omega' = 4\pi\rho bcda \frac{p'd\omega'}{a_1'b_1'c_1'},$$

so that

$$\frac{Ad\omega}{A'd\omega'} = \frac{pd\omega}{a_1 b_1 c_1} : \frac{p'd\omega'}{a_1'b_1'c_1'}.$$

It is possible to create a biunivocal correspondence between each point (x,y,z) of (a_1) and a point $\left(x\frac{a_1'}{a_1}, y\frac{b_1'}{b_1}, z\frac{c_1'}{c_1}\right)$ of (a_1'). Chasles stressed that these are the points called by Ivory "corresponding" (ibid., pp. 285–286). After some brief passages Chasles proved that

$$A'd\omega' = Ad\omega \tag{7.5}$$

Namely: an attracting shell exerts the same attraction on two corresponding superficial elements belonging to two any ellipsoids (a_1) and (a_1') confocal with the external surface of the shell. Now Chasles introduced a result he had proved in the memoir on homography postponed to his *Aperçu* (Chasles, 1837a, p. 814): be given two corresponding elements on two ellipsoids fulfilling the conditions of this

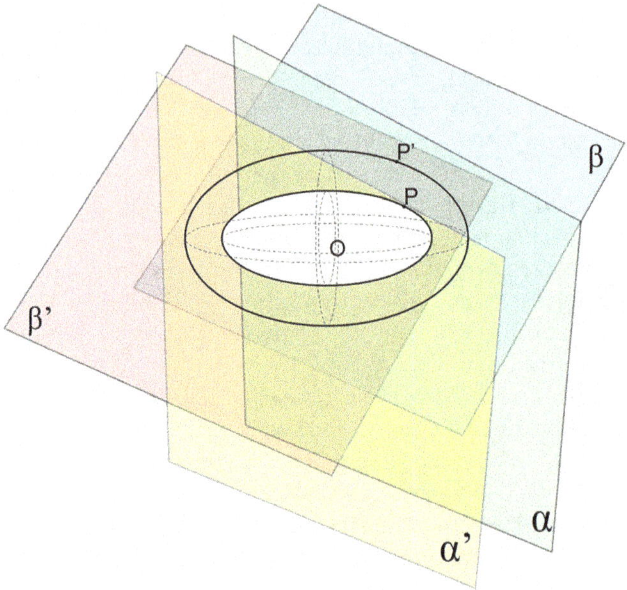

Fig. 7.6 Diagram representing the figure described by Chasles. P and P' are the two corresponding points in the sense of Ivory (Chasles indicates them by m and m'). The plane α is tangent to the internal ellipsoid at P; the plane α' is that parallel to α and passing through the common centre of the two ellipsoids O. The plane β is tangent to the external ellipsoid in P'; the plane β' is that parallel to β and passing through the common centre of the two ellipsoids

theorem. Their surfaces are as the areas of the sections drawn in the two ellipsoids by two diametral planes parallel to the planes of the two elements, namely to the planes tangent to the two ellipsoids from two corresponding points m,m' belonging respect. to $d\omega$, $d\omega'$ (Fig. 7.6).

It is therefore possible to conclude that:

> [...] the attractions exerted by the attracting shell on two corresponding points m,m' belonging to two any ellipsoids (a_1) and (a_1') are inversely as the areas of the sections drawn in these two ellipsoids by the diametral planes parallel resp. to their tangent planes in the points m,m'.[25]

Chasles' final result concerning the attraction exerted by an ellipsoidal shell whose external surface *ES* is confocal with the level surfaces—and, hence, is a level surface itself—regards the case in which the attracted point belongs to *ES*. Then, if $d\sigma$ is the element of *ES* corresponding to $d\omega$, the attraction exerted by the

[25] Ibid., p. 287: "... les attractions que la couche attirante exerce sur deux point m,m' situés sur deux ellipsoïdes quelconques (a_1), (a_1') et CORRESPONDANS entre eux, sont en raison inverse des aires des sections faites dans ces deux ellipsoïdes par les plans diamétraux parallèles, respectivement à leurs plans tangents aux points m,m'". Capitals in the text.

shell on $d\sigma$ will be $4\pi\rho d\varepsilon d\sigma$, being $d\varepsilon$ the thickness of the shell in the point where $d\sigma$ is posed, so to have

$$Ad\omega = 4\pi\rho d\varepsilon d\sigma.$$

Chasles highlighted that $d\varepsilon d\sigma$ is the volume of the shell having $d\sigma$ as basis. Thence, $d\varepsilon d\sigma = dv$, from which Chasles concluded that the action exerted by an attracting shell on a surface element of a level surface (a_1) is proportional to the portion of the shell's volume determined by a surface element corresponding to the first one (ibid., p. 287).

Four interesting pages follow where he offered a more general perspective on these theorems which he had obtained in relation to the attraction exerted on the different points of a level surface by an ellipsoidal shell. As always, this more general and wide view derives from geometry, which is a further and unequivocal litmus paper of Chasles' foundational intentions. For Chasles explicitly claimed:

> These different theorems can get a simpler form, which is more interesting, if the notion of corresponding points—which holds for any two ellipsoids—is replaced by another one, which is equally characterizing, but is typical of two confocal ellipsoids.
>
> The property consists in the fact that these two corresponding points—on two confocal ellipsoids—belong to the same intersection line of two hyperboloids (one of which is one sheeted and the other two sheeted) confocal with the ellipsoids[26] (Fig. 7.7).

Chasles, after having proved this property (ibid., pp. 288–289), was able to conclude that, given two confocal ellipsoids, the intersection curve of two hyperboloids (one of which is one-sheeted and the other two-sheeted) confocal with the two ellipsoids cuts the two ellipsoids in two corresponding points. It is possible to prove that such a curve is skew and of the fourth order. Obviously, the curve is orthogonal to the ellipsoids' surfaces. Chasles claimed that, in the light of this new property, the notion of corresponding points can be applied to level surfaces relative to any attracting shell.

Chasles now developed the following geometrical reasoning: he considered any curve on the external surface of an ellipsoidal shell. Through the points of the curve, it is possible to draw a one-sheeted and a two-sheeted hyperboloid confocal with the ellipsoid. As seen, the two hyperboloids mutually cut along a skew line. The set of all these lines make up a surface having the form of a canal curve whose base is on the external surface of the shell. All the intersection-curves of the canal surface passing through the ellipsoids (a_1), (a_1'), (a_2''), ... are composed of corresponding points in the sense of Ivory, and the portion of surfaces drawn on the ellipsoids will

---

[26] Ibid., pp. 287–288: "Ces divers théorèmes peuvent prendre des énoncés plus facile et qui offriront plus d'intérêt, si l'on remplace la notion de points *correspondans*, qui appartient à deux ellipsoïdes quelconques, par une autre, également caractéristique, mais particulière aux ellipsoïdes décrits des mêmes foyers.

Cette propriété consiste en ce que deux points *correspondans*, sur deux ellipsoïdes décrits des mêmes foyers, sont sur une même ligne d'intersection de deux hyperboloïdes à une et à deux nappes décrits des mêmes foyers que les ellipsoïdes".

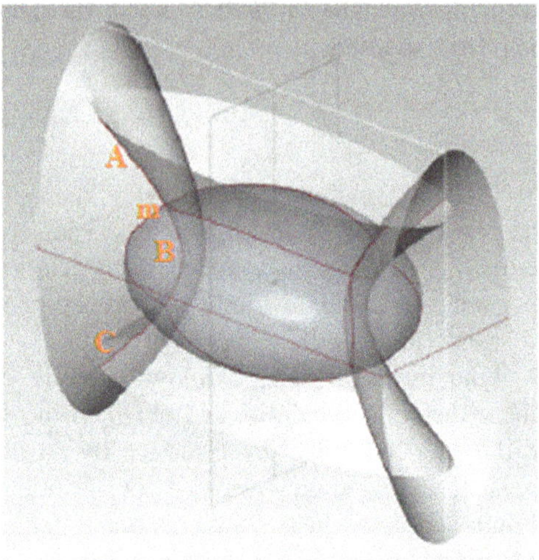

Fig. 7.7 In this figure which represents three confocal quadrics, an ellipsoid, a two-sheeted hyperboloid and a one-sheeted hyperboloid, one of the lines of which Chasles spoke is the one I have indicated by ABC. Instead, m is the point to which the point m' will correspond. When the points m, m' vary on two any ellipsoids of the array, a line as ABC will exist which represents an intersection line of a one-sheeted hyperboloid and a two-sheeted hyperboloid of the array

also be corresponding. Therefore, as Chasles concluded, the theorem expressed by Eq. (7.5) can now assume this form:

> If you consider a canal whose width is infinitely small, and whose curvilinear wedges are the trajectories orthogonal to the ellipsoids (a_1), ... and if, in any canal, several sections perpendicular to its wedges are drawn, the attractions of the attracting shell on these sections will be normal to them and will be reciprocally equal. Their common value will be proportional to the volume intercepted in the attracting shell by the canal produced until the internal surface of the shell.[27]

Commentary: as already pointed out, this purely geometrical section of Chasles (1837f) is clear evidence that he was trying to introduce geometrical elements within the theory of the ellipsoid attraction in the attempt to supply a solid geometrical foundation to the whole theory. This happened in the memoir which he read at the Académie in 1837, but which was published 9 years later (Chasles, 1837, 1846): However, in Chasles (1837f) he was developing a series of considerations which made the use of geometry within this theory wider and wider, with the intention to find it through geometrical concepts. One might claim that the different ways developed in Chasles (1837f) are the "general proofs", something like a sort of conceptual experiments, for the seminal memoir written in 1846. Not all these conceptual experiments and points of views converged in Chasles (1837, 1846), but the road to follow was already clear in his mind.

[27] Ibid., pp. 290–291: "Si l'on conçoit un canal d'une ouverture infiniment petite, dont les arêtes curvilignes soient ls trajectoires orthogonales aux ellipsoïdes (a_1)... et si l'on fait, dans ce canal, diverses sections perpendiculaires à ses arêtes, les attractions que la couche attirante exercera sur ces sections, leur seront normal respectivement et seront égales entre elles ; leur valeur commune sera proportionnelle au volume intercepté dans la couche attirante par le canal prolongé jusqu'à la surface interne de la couche".

In the third section of his work (pp. 291–297) Chasles determined the gravitational potential of an ellipsoid. I do not deal with the details of his demonstrative method, but I refer to the result, which is not new,[28] though Chasles' procedures are slightly different from those used by his predecessors and contemporaries: given an ellipsoid of half-axes a,b,c, where a is the major half-axis, and density ρ, the potential is given by the elliptical integral of first type (ibid., p. 297)

$$V = 2\pi\rho\ abc\ da \int_0^1 \frac{du}{\sqrt{a^2 + u^2(b^2 - a^2)}\sqrt{a^2 + u^2(c^2 - a^2)}}, \tag{7.6}$$

which can also be written as

$$V = \frac{2\pi\rho abcda}{\sqrt{a^2 - b^2}\ \sqrt{a^2 - c^2}} \int_0^1 \frac{du}{\sqrt{u^2 + \frac{a^2}{b^2 - a^2}}\sqrt{u^2 + \frac{a^2}{c^2 - a^2}}}. \tag{7.7}$$

I will focus on the similarities highlighted by Chasles between the attraction of the ellipsoid and other sections of physics because this is useful to fully grasp the width and profoundness of his ideas.

As a matter of fact, the fourth section (ibid., pp. 297–303) is dedicated to the analogies between the properties of the attraction of an ellipsoidal shell and the heat law of a body in thermal equilibrium. Chasles considered the case of a homogeneous body in which the surfaces are maintained at a constant temperature which varies from surface to surface. When such a body reaches the thermal equilibrium, the temperature is a function V of the coordinates which fulfils:

a) the law 3)

$$\frac{d^2V}{dx^2} + \frac{d^2V}{dy^2} + \frac{d^2V}{dz^2} = 0$$

and

b) the property that in the body, isothermal surfaces exist such that, for their points, $dV = 0$ so that V is a constant (ibid., p. 297). Thence, Chasles continued, this function satisfies law 3) and is the function of a parameter which determines the isothermal surfaces.[29] The function V, determined by Eq. (7.6)

[28] See Laplace, *Méchanique celeste*, Tome I, Book II, § 11, pp. 135–138 (Laplace, 1798–1825).

[29] I have slightly changed Chasles' notation because his original might create a little ambiguity.

$$V = 2\pi\rho\, abc\, da \int \frac{da_1}{\sqrt{a^2 + u^2(b^2 - a^2)}\sqrt{a^2 + u^2(c^2 - a^2)}}$$

fulfils, from a formal point of view, exactly the conditions a) and b). This is the function we call "potential" and that Chasles indicated as representing "the sum of the molecules of the ellipsoidal shell, respectively, divided by their distances from a point external to the shell on a level surface" (ibid., p. 298).

The general function

$$V' = C \int \frac{da_1}{\sqrt{a^2 + u^2(b^2 - a^2)}\sqrt{a^2 + u^2(c^2 - a^2)}} + C' \tag{7.8}$$

also satisfies conditions a) and b). This is, hence, the equation expressing the temperature of a solid envelope included between two level surfaces. The identity in the forms of the expressions 6) and 8) implies, as Chasles remarked (ibid., p. 298), that the isothermal surfaces are the level surfaces relative to the attraction of an infinitely thin ellipsoidal shell. He, thus, observed, with his typical accentuation of the geometrical aspects of any proof:

> From this, a synthetic proof of the beautiful results obtained analytically by Lamé in his Memoir on the isothermal surfaces of second degree follows. But the consideration of the attraction of the ellipsoidal shell leads us to further results which seems to us to hold some interest.[30]

These results concern a further link between heat theory and the theory of the ellipsoid attraction as well as the concept of heat flow.

To reach his first achievement on the heat flow, Chasles reasoned like this: being $b_1 = \sqrt{a_1^2 + b^2 - a^2}; c_1 = \sqrt{a_1^2 + c^2 - a^2}$, the integral (7.8) becomes

$$V' = C \int \frac{da_1}{b_1 c_1}$$

Given two infinitely close isothermal surfaces, indicated by (a_1), $(a_1 + da_1)$, it will be

$$dV' = C \frac{da_1}{b_1 c_1}$$

[30]Ibid., pp. 298–299: "On a de la sorte une démonstration synthétique des beaux résultats obtenus analytiquement par M. Lamé, dans son Mémoire sur le surfaces isothermes du deuxième degré. Mais la considération de l'attraction de la couche ellipsoïdale va nous conduire à divers autre résultats qui nous paraissent offrir quelque intérêt".

Be ds the shell's thickness, S a point belonging to (a_1), $d\omega$, the area element to which S belongs. From the heat theory, it follows that the heat quantity, indicated by ΔQ, which, in a unitary interval of time, traverses $d\omega$, is given by[31]

$$\Delta Q = K\frac{dV'}{ds}d\omega = KC\frac{d\omega}{b_1c_1},$$

where K is the thermal conductivity.

Since in the paragraph 13 Chasles has proved that $\frac{da_1}{ds} = \frac{p}{a_1}$, finally it is

$$\Delta Q = KC\frac{pd\omega}{a_1b_1c_1}.$$

As $Ad\omega = 4\pi\rho bcda\frac{pd\omega}{a_1b_1c_1}$, it will be

$$Ad\omega = \frac{4\pi\rho bcda}{KC}\Delta Q,$$

that means

$$\frac{Ad\omega}{\Delta Q} = C''$$

This formula is noteworthy because it holds for any point of each isothermal surface, not only for (a_1) (ibid., p. 301) since $d\omega$ is not connected to a specific surface; it is a generic area element. Thence Chasles could conclude:

> Consider an envelop determined by two confocal ellipsoidal surfaces, subject to constant sources of heat and cold. When it has reached its equilibrium insofar as its internal temperature is concerned, the heat quantity which, in a unitary time, traverses a surface element located in any place within the envelop on one of its isothermal surfaces, is proportional to the attraction exerted on this element by an infinitely thin shell whose external surface is the internal surface of the envelop and whose internal surface is a further ellipsoidal surface concentric and homothetic to the former.[32]

[31] With regard to the equations concerning thermal conductivity Chasles referred to: 1) the celebrated memoire by Fourier, *Théorie du mouvement de la chaleur dans les corps solides* appeared in the issue 1819–1820 of the *Mémoires de l'Académie des Sciences,* p. 207. This issue was published in 1824. I indicate this memoir as Fourier (1819–1820); 2) Poisson (1835b, p. 98).

[32] Ibid., pp. 301–302: "Quand une enveloppe terminée par deux surfaces ellipsoïdales décrites des mêmes foyers, et soumises à des sources constantes de chaleur et de froid ; est parvenue à son équilibre de température intérieure, la quantité de chaleur qui traverse, dans l'unité de temps, un élément superficiel pris en un lieu quelconque de l'intérieur de l'enveloppe sur une de ses surfaces d'égale température, est proportionnelle à l'attraction qu'exercerait sur cet élément une couche infiniment mince, terminée extérieurement par la surface interne de l'enveloppe et intérieurement par une seconde surface ellipsoïdale concentrique et homothétique à cette première".

Since the flow depends on the normal at the considered surface, Chasles easily concluded that:

> The heat quantity which traverses, in a unitary interval of time, a surface element posed in any manner within a body is proportional to the attraction exerted on this element, in the direction of the normal by the attracting shell.[33]

A further and final result connected to geometry is the following one: let us recall that in the formula $\Delta Q = KC\frac{pd\omega}{a_1b_1c_1}$, the term $p = \frac{1}{\sqrt{\frac{x^2}{a_1^4}+\frac{y^2}{b_1^4}+\frac{z^2}{c_1^4}}}$ indicates the distance between the plane tangent at the ellipsoid in $S \equiv (x, y, z)$ and the centre of the ellipsoid, so that Chasles concluded:

> The heat flux in different points of an isothermal surface is proportional, as intensity, to the distances of the planes tangent to the surface at these points from the centre of the envelop.[34]

In the fifth and final section of his memoir (ibid., pp. 304–316) Chasles extended the results he obtained for an ellipsoidal infinitely thin shell to shells whose form is not ellipsoidal. He proved that, with the appropriate modifications, several of such results are extensible to the attraction of a shell of any form. I will not consider most of these results because they are not very significant for my aims. Rather, I will concentrate on a series of geometrical considerations which Chasles applied to the attraction theory in connection with the heat theory, where also references to the electricity theory are involved. This is a further evidence that shows the broadness of Chasles' interests in physics and in his attempt to extend the use of geometry within physics as much as possible.

After having obtained some specific results on the attractive shells of any form and on the isothermal surfaces of a solid envelope, the problem posed by Chasles is the following: *given a solid envelope subject to constant heat sources, and in temperature equilibrium, determine the attracting shell corresponding to the solid envelope in which the heat movement takes place* (ibid., p. 309), *namely which is the attracting shell whose level surfaces are the isothermal surfaces of the given envelope*. In general—Chasles highlighted—this is a very difficult question, but some basic ideas can be supplied. His considerations arise from what he had proved for the ellipsoidal case where the envelope is determined by two confocal ellipsoidal surfaces and the attractive infinitely thin shell is included between two ellipsoidal surfaces which are homothetic and concentric, under the condition that the external surface of the shell coincides with the internal surface of the envelope. If this shell is considered as that most external of an ellipsoid, the electricity theory foresees that this shell contains the electric fluid which accumulates at the surface of the ellipsoid

[33] Ibid., p. 302: "La quantité de chaleur qui traverse, dans l'unité de temps, un élément superficiel placé d'une manière quelconque dans l'intérieur du corps, est proportionnelle à l'attraction exercée sur cet élément, dans le sens de sa normale, par la couche attirante".

[34] Ibid., p. 303: "Les flux de chaleur en différens points d'une surface isotherme ont leurs intensités proportionnelles aux distances des plans tangens à la surface en ces points, au centre de l'enveloppe".

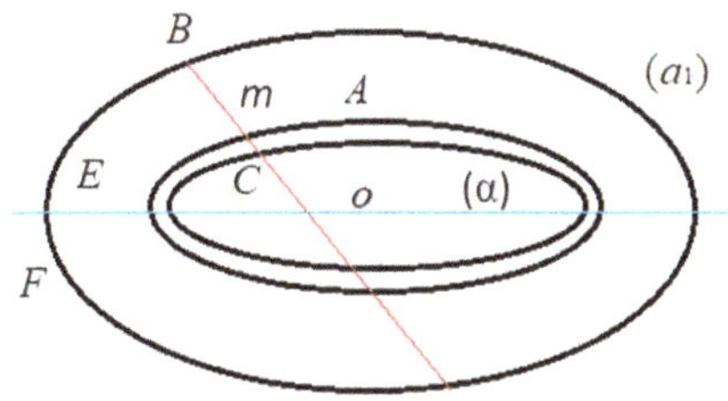

Fig. 7.8 A hint of the configuration described by Chasles. Legenda: A=the medial surface; (a_1) =the external surface of the infinitesimal envelope; (α)=the internal surface of the infinitesimal shell; BmC=the normal at m to A; $Bm=ds$; $Cm=\delta\varepsilon$; FED=major axis of the three surfaces; $EF=da$; $ED=\delta\alpha$

(ibid., p. 310). In the light of this truth, Chasles developed the following heuristic-geometrical hypothesis:

> It is, hence, possible to suppose and to put forward a hypothesis: for an envelop of different form, the attracting body will be the infinitely thin shell which forms the electric fluid spread on the internal surface of the envelop, if it is considered as the surface of a conductive body. This electric shell is probably the same as that which also forms the accumulated heat on a surface devoid of emissive power.[35]

However, Chasles did not restrict his considerations to this heuristic idea. Continuing to reason geometrically, he had a further idea:[36] the basic concept is always the same, namely the attempt to determine features of the second-degree surfaces—and in particular of the ellipsoids—which can reasonably be extended to other geometrical surfaces. The configuration Chasles is analysing consists of three surfaces A, (a_1) and (α) (see Fig. 7.8): the medial surface A is, at the same time, the internal surface of the ellipsoidal envelope of which the isothermal surface (a_1) is the external surface and the external surface of the attractive shell whose internal surface is (α). Chasles pointed out that the two ellipsoidal surfaces (a_1) and (α) hold different characteristics because (a_1) has its principal section confocal with those of A, while (α) is concentric and homothetic to A (ibid., p. 310). Despite the differences between the two surfaces, they share an important common feature: α

For the segments formed by (a_1) and (α) on each normal at A form a constant product. More specifically: be drawn the normal at A from the point m; be ds the segment of normal between (a_1) and A; be $\delta\varepsilon$ the segment of normal between A and (α), then the identity

[35] Ibid., p. 310: "On peut donc supposer, et essayer cette hypothèse, que, pour une enveloppe d'une autre forme, le corps attirant sera la couche infiniment mince que formerait le fluide électrique répandu sur la surface interne de l'enveloppe, si on la considérait comme la surface d'un corps conducteur. Cette couche d'électricité est probablement la même que celle que forme aussi la chaleur accumulée sur une surface dépourvue de pouvoir émissif".

[36] He wrote explicitly: "Voici un autre aperçu fondé sur des considérations géométriques". (Ibid., p. 310).

$$\delta\varepsilon \cdot ds = da \cdot \delta\alpha \tag{7.9}$$

holds.[37] This implies that, to form the surface (α) by means of (a_1), it is necessary to draw, on the normals in different points of the surface A, segments which are inversely as the shell included between A and (a_1). The extremities of these segments will form (α). After having developed this reasoning, Chasles claimed that this geometrical relation between (α) and (a_1) is a property worth being known in itself. Furthermore, it can also be useful to supply a first idea on the general relation which has to take place between the level surface infinitely close to the external surface of an attracting infinitely thin shell and the internal surface of this shell itself (ibid., p. 311).

Commentary: with regard to this last assertion, it is rather difficult to understand what Chasles meant exactly. In particular, it is difficult to argue if he thought that the construction by means of which the surface (α) is constructible through (a_1) is valid for all or, at least, for a wide part of geometrical surfaces (which is false), or if he thought that for other classes of surfaces something equivalent to Eq. (7.9) might hold. Hence, the construction of (α) by means of (a_1) would be, anyway, possible, though not in a such easy manner as it is the case for ellipsoidal shells and envelopes.

This clarified the importance which Chasles ascribed to geometry should be highlighted. By means of some geometrical considerations—supported by algebraic and analytical methods—he determined the relations between attractive ellipsoidal shells composed of concentric and homothetic surfaces and isothermal surfaces. In substance, the isothermal surfaces coincide with the level surfaces of a gravitational field produced by an ellipsoid. This allowed him to offer an important contribution to a field of research in which physicists and mathematicians such as Laplace, Ivory, Poisson and Lamé—to mention just a few—had already offered some crucial insights. Chasles' geometrical approach was, in a sense, an exception since the other authors resorted to analytical methods, or to geometrical methods when they could offer an easier treatment of the problems. Whereas, for Chasles, the resort to geometrical methods was not only a question of simplicity—which, of course, plays an important role—but a foundational issue. For him, the resort to geometry was not only instrumental, it was essential, insofar as geometrical concepts and constructions are ontologically the bases of our universe, and, hence, had to be posed at the ground of our knowledge.

Furthermore, the last considerations on the surfaces A, (a_1), (α) highlight that, for Chasles, geometry is not only the discipline which—in 1837f—offers several elements to solve problems within dynamics and which, in Chasles (1837, 1846), will become its essential foundational element. Geometry is also the discipline which can furnish heuristic means and reasoning to reach some truths difficult to achieve by analytical instruments. Geometry, with its capability to offer generalizations, can represent an immense source of inspiration for new ideas on dynamics which might be right or wrong and which, of course, have to be verified.

[37] The proof of this identity is in Chasles (1838, p. 307).

All these aspects confirm that Chasles (1837f) is an important step towards the idea that geometry can play a decisive role within dynamics. This role will become foundational in Chasles (1837, 1846).

In the last section of Chasles (1837f, ibid., pp. 311–316), other significant results were obtained, but what expounded so far is enough for my aim.

7.4 *Mémoire sur l'attraction de l'ellipsoïdes. Solution synthétique*

This is by far the most important memoir written by Chasles on the ellipsoid attraction because, after a historical introduction, Chasles inserted a very long section (Chasles, 1837, 1846, pp. 17–41) which is the longest and the most interesting one of the whole treatise. Here he clearly expounded the geometrical theorems which are the bases of his theory. Specifically, he proved that such theorems express projective properties connected to the theory of reciprocal polars. Thence, coherent with his entire foundational programme, projective geometry will be also the basis of his theory of ellipsoid attraction. After this section, the one dedicated to the proof of Maclaurin theorem through the synthetic apparatus explained in the previous section follows (ibid., pp. 41–55). From a foundational point of view, these are the two most relevant part of Chasles' essay. Thus, I will consider numerous mathematical details explaining them and also adding my comments. The third section (ibid., pp. 55–62) concerns the attraction of an ellipsoidal shell on an external point, and the fourth one (ibid., pp. 62–76) the attraction of the whole ellipsoid on an external point. Though with less details, I will examine these sections with care, as well. Finally, I will focus on the geometrical construction of a coefficient (ibid., pp. 76–79) because it shows that Chasles intended to use geometry as much as possible in any phase of his complex architecture.

7.4.1 *The Geometrical Results Expounded in the* Mémoire

In the purely geometrical section of his *Mémoire* Chasles explained a series of results which he used to offer a synthetical solution to the problem of the ellipsoid attraction. As already outlined, these results are deeply connected to the doctrine of polarity, that is with the then most important and wide section of projective geometry. This is clear evidence of the foundational character of Chasles (1837, 1846). In what follows I will explain the main steps of his reasoning in detail, also presenting a series of figures which do not exist in Chasles' text and which can give the reader an idea of the geometric situations he described. I will divide this section into subsections, each dedicated to the most significant results proved in this second part of the *Mémoire*.

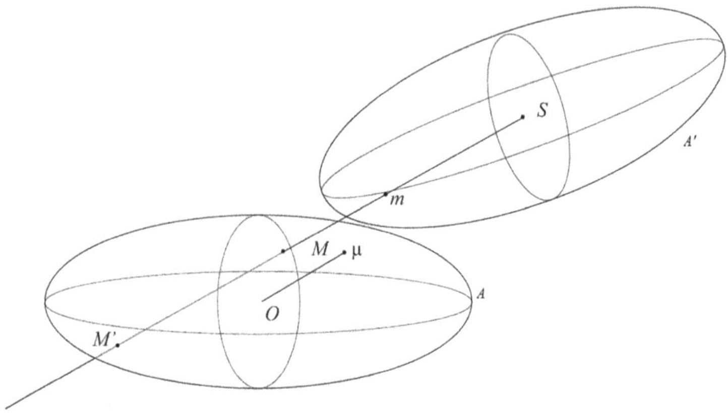

Fig. 7.9 Reconstruction of the figure described by Chasles

A) The first important theorem.

The first proposition (I call it T1) widely used by Chasles along his work is the following one, which he proved at the beginning of his research (Fig. 7.9):

> Rotate a transversal around a fixed point S and cut a second degree surface A in two points M, M'. Be the half-diameter of the surface parallel to the transversal $O\mu$. If on this straight line a segment Sm proportional to $\frac{O\mu^2}{MM'}$ is considered, the geometrical locus of the point m is a second degree surface A' whose centre is S.[38]

There are several steps through which Chasles arrived at proving this theorem. The first one is the following proposition for whose proof I refer directly to Chasles (1837, 1846) because in the demonstration classical properties of the second-degree surfaces and, more in general, of the algebraic surfaces are used which are not directly connected to polarity.

The conditions of the theorem are the following ones: be U and V two s.s.p. second-degree surfaces (as a matter of fact, two ellipsoids) with similarity ratio equal to λ, and centres, respectively, in two different points G and S. Be H, H' the points where the straight line SG cuts U, then it holds:

1. If a transversal rotates around the point S and cuts the first surface U in two points Π, Π' and the second one V in a point π (Fig. 7.10), the constant relation

[38] Chasles (1837, 1846), p. 23: "Si autour d'un point fixe S on fait tourner une transversale qui rencontre une surface du second degré A en deux points M, M', et que, $O\mu$ étant le demi-diamètre de cette surface mené parallèlement à la transversale, on prenne sur celle droite un segment Sm proportionnel au rapport $\frac{O\mu^2}{MM'}$, le point m aura pour lieu géométrique une surface du second degré A', ayant son centre en S".

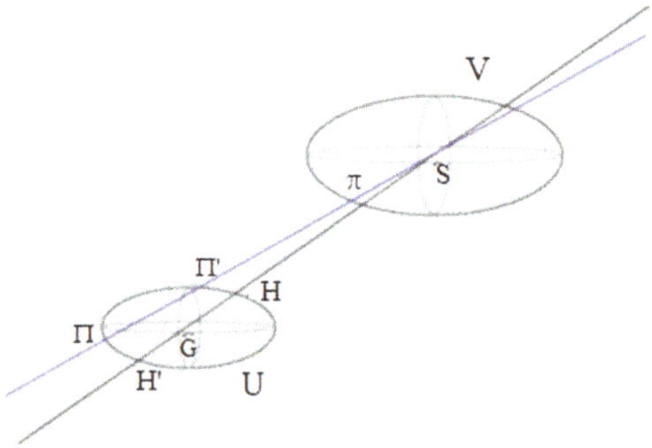

Fig. 7.10 Reconstruction of the figure described by Chasles

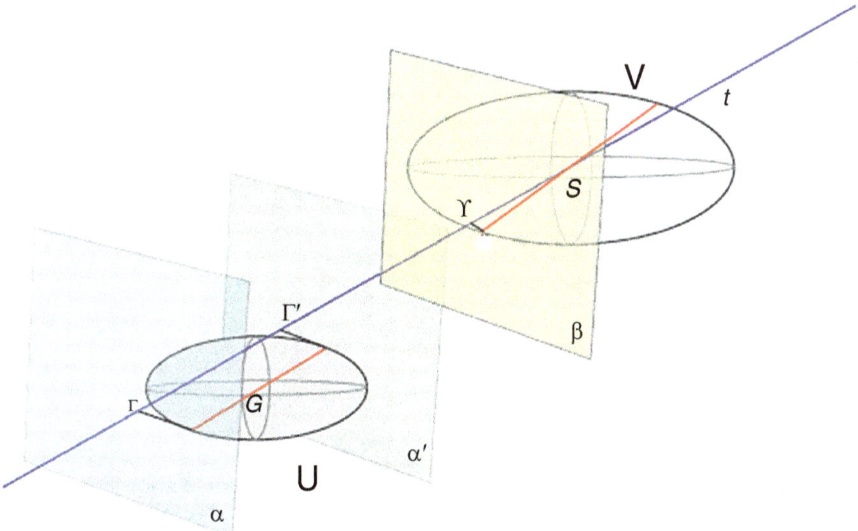

Fig. 7.11 Reconstruction of the figure described by Chasles

$$S\Pi \cdot S\Pi' = \lambda^2 \left(\frac{SG^2}{GH^2} - 1 \right) \cdot S\pi^2 \tag{7.10}$$

holds (Fig. 7.10).

2. If two planes tangent to the former surface and a plane tangent to the latter are drawn and if all of the three are perpendicular to the transversal, which they cut in the three points Γ, Γ, γ, respectively, it will hold (see Fig. 7.11).

$$S\Gamma - S\Gamma' = 2\lambda \cdot S\gamma. \tag{7.11}$$

After this introductory theorem, the section concerning polarity begins: Chasles considered the polar transformation of U and V with respect to a sphere concentric

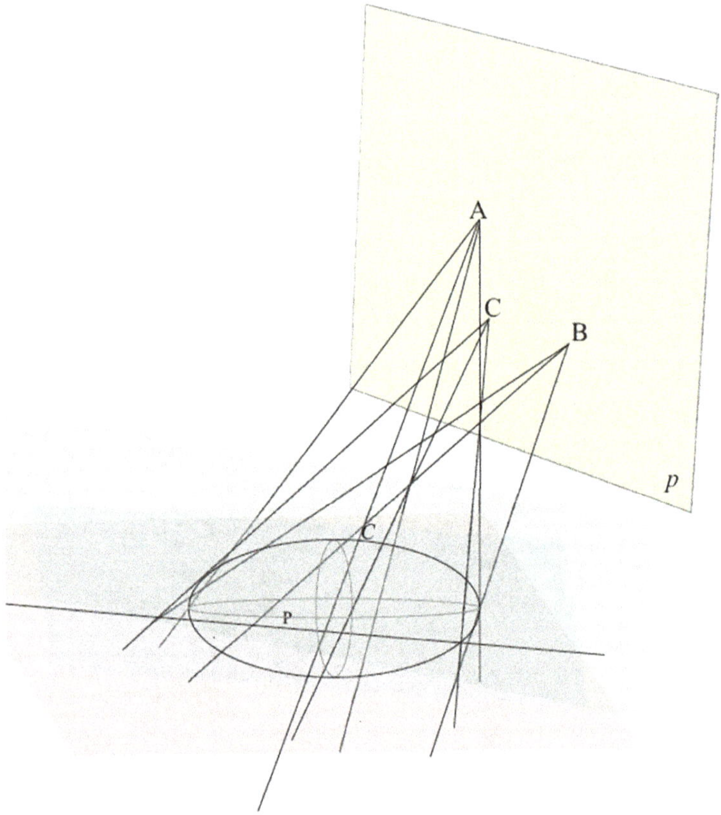

Fig. 7.12 Reconstruction of the figure described by Chasles. Be p the plane of which the focus with respect to the second-degree surface C (in this case an ellipsoid) is searched. I have chosen three points A,B,C on p. They are assumed as vertices of three cones which are tangent to the ellipsoid. They cut the ellipsoid in three conical sections whose planes are indicated by the colours pink, clear blue and blue. P, which is the point where the three planes cut, is the pole of p. In this case P is within the ellipsoid since p is external to the ellipsoid

with V. Two other surfaces U' and V' will be the transformed of U and V. To the points Π and Π' of U, two planes tangent to U' will correspond. This is in the nature of any polar transformation, but the particular position of these planes depends on the fact that the polarity is developed according to a sphere concentric with V.

To explain the situation clearly, Chasles added a note where in less than one page he expounded the main features of polarity. It is appropriate to illustrate this explanation (ibid., pp. 18–19). He clarified that, given a second-degree surface C, the pole of a plane p with respect to C is the point P through which the planes of the contact-curves of C and of the cones circumscribed at C and having their vertices on p pass (Fig. 7.12). It is possible to prove that there is a single point with these features. The plane p is the polar plane of P with respect to C. If p cuts the surface, its pole is the vertex of the cone circumscribed at C along the intersection curve.

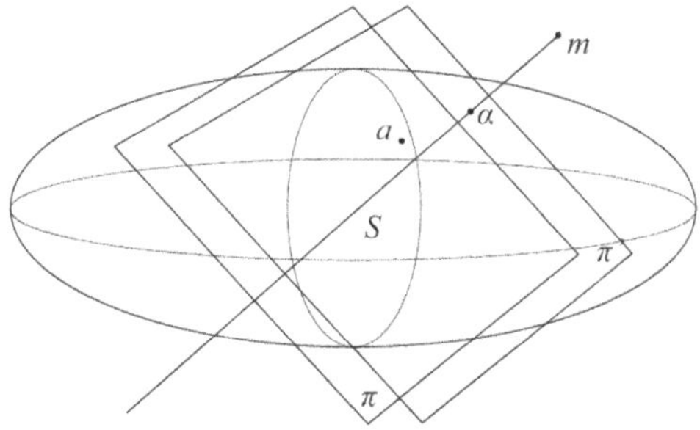

Fig. 7.13 Reconstruction of the figure described by Chasles

If now you consider the polar transformed U' of the surface U, it will be given as the envelope of the polar planes of U, according to the correspondence points of U-planes tangent to U' (the converse is also true). Therefore, given two points A, B of U and the straight line AB, the straight line in which the two polar planes of A and B with respect to C mutually intersect corresponds to AB.[39] Chasles recalled that it is possible to construct directly the polar plane of a point m with respect to the quadric C with centre S. For it is possible to prove that it is sufficient to draw the half-diameter $S\alpha$ passing through m, and to determine, on this straight line, the point a such that $Sa = \frac{S\alpha^2}{Sm}$ and to draw from a the plane parallel to the plane conjugate[40] to $S\alpha$ (Fig. 7.13).

Thence, if the surface C with centre S is a sphere, this plane is perpendicular to the radius passing through m and its distance from S is inversely as the distance of m from S.

This clarified, the planes corresponding in the polarity to the points Π and Π' belonging to U are tangent to U'. They are perpendicular to $S\Pi$ and their distances from S are inversely as the distances $S\Pi$ and $S\Pi'$ of the points S from the two tangent planes. Analogously to the point π of V a plane tangent to V' corresponds whose distance Sp from S is inversely as $S\pi$. If the radius of the sphere C is equal to 1, the three relations are written as:

$$SP = \frac{1}{S\Pi}, SP' = \frac{1}{S\Pi'}, Sp = \frac{1}{S\pi}$$

If you replace these values in Eq. (7.10), you get

$$\frac{1}{SP} \cdot \frac{1}{SP'} = \lambda^2 \left(\frac{SG^2}{GH^2} - 1\right) \cdot \frac{1}{Sp^2} \qquad (7.12)$$

[39] Chasles offered here a general explanation valid for any algebraic surface. I restrict this analysis to the second-degree surfaces because they are the only ones important in our context.

[40] I recall that, given a quadric Q and a diameter d of Q, the plane conjugate to d is that passing through the centre of Q and cutting the system of chords parallel to d.

Now, Chasles continued (ibid., pp. 19–20), the surface U can be considered an envelope of planes and its polar surface U' with respect to C as a surface of points. Thence, to the two planes tangent to U and perpendicular to $S\Pi$, the two points M, M' determined by the intersection of this straight with U', correspond. The segments SM and SM' are inversely as the distances $S\Gamma$, $S\Gamma'$ of S from the two tangent planes. The same reasoning holds for V: to the plane tangent at V in π, the point m belonging to V' corresponds. This is the point where $S\pi$ cuts V'. Clearly Sm is inversely as $S\gamma$, so that Eq. (7.11) is transformed into

$$\frac{1}{SM} - \frac{1}{SM'} = 2\lambda \cdot \frac{1}{Sm} \tag{7.13}$$

The next step of Chasles' argument consists in replacing in Eq. (7.12) the ratio $\frac{SG}{GH}$ which belongs to the configuration U-V with its corresponding in the configuration U'-V'.

To reach this result, he remarked that to the centre G of U the polar plane of S with respect to U' corresponds. This is the case because: 1) to G, centre of U, the plane at infinity corresponds in a polarity with respect to U; 2) to the plane at infinity, considering the polarity with respect to C, the centre S of C corresponds; 3) in a polarity with respect to U', to the point S, the polar plane of S with respect to U' corresponds. But since U' is the polar transformed figure of U, this means that in the polarity with respect to C, to the point G the polar plane of S with respect to U' corresponds. To the two points H and H', two tangent planes to the surface U' correspond. Since the polarity is developed with respect to a sphere, these three planes are perpendicular to the straight line SG. They cut the straight line SG in three points g, h, h' such that

$$Sg = \frac{1}{SG}; Sh = \frac{1}{SH}; Sh' = \frac{1}{SH'}$$

With easy calculations Chasles reached the result

$$\frac{SG}{GH} = \frac{Sh}{hg}.$$

The two planes tangent to U' are parallel to the polar plane of S. This means that they touch the surface at the extremities of the diameter through S. Be D one of these extremities and G_1 the point where the diameter cuts the plane polar of S. Consider the triangle ShD cut by the straight line gG, parallel to its basis hD. It is $\frac{Sh}{hg} = \frac{SD}{DG_1}$, so that $\frac{SG}{GH} = \frac{SD}{DG_1}$. If O is the centre of U', it is possible to prove through easy calculations that $\frac{SD}{DG_1} = \frac{SO}{OD}$, so that, finally, it is

$$\frac{SG}{GH} = \frac{SO}{OD}.$$

Thence, Eq. (7.12) is transformed into

$$\frac{1}{SP} \cdot \frac{1}{SP'} = \lambda^2 \left(\frac{SO^2}{OD^2} - 1 \right) \frac{1}{Sp^2}. \tag{7.14}$$

Thus, Chasles can conclude (ibid., p. 21) that for the two transformed surfaces U', V', the two following equations hold:

$$\frac{2\lambda}{Sm} = \frac{1}{SM} - \frac{1}{SM'}$$

$$Sp^2 = \lambda^2 \left(\frac{SO^2}{OD^2} - 1 \right) SP \cdot SP'.$$

Therefore, from these two equations, the two following propositions are deduced:

Given a quadric A, if a transversal a rotating around a fixed point S cuts the quadric in M and M' and the point m on this straight line is assumed such that

$$\frac{2\lambda}{Sm} = \frac{1}{SM} - \frac{1}{SM'} \tag{7.15}$$

being λ a constant, it holds:

1°) The point m belongs to a second-degree surface A' whose centre is in S;
2°) If SP and SP' are the segments determined on a by the two planes tangent to A and perpendicular to a and if Sp is the segment determined by a plane tangent to A' and parallel to the other two, it holds (ibid., pp. 21–22).

$$Sp^2 = \lambda^2 \left(\frac{SO^2}{OD^2} - 1 \right) SP \cdot SP'. \tag{7.16}$$

Now, Chasles' first aim has been achieved because Eq. (7.15) will enable us to construct the points belonging to the surface A', whereas Eq. (7.16) its tangent planes.

Through easy algebraic transformations, Eq. (7.15) can be written as

$$Sm = 2\lambda \left(\frac{SO^2}{OD^2} - 1 \right) \frac{O\mu^2}{MM'} \tag{7.17}$$

being $O\mu$ the half-diameter of A parallel to SM.

From this the theorem T1 mentioned at the beginning of this section immediately follows:[41] "If a transversal rotating around a fixed point S cuts a second-degree surface A in two points M, M', and the half-diameter of the surface parallel to the transversal is $O\mu$, if on this straight line a segment Sm proportional to $\frac{O\mu^2}{MM'}$ is considered, the geometrical locus of the point m is a second-degree surface A' whose centre is S".

Commentary: It is appropriate to point out that the surfaces now indicated by Chasles with A and A' are those that in the previous treatment had been named U' and V'. It seems to me that the change in the denomination is due to a precise reason: the whole argumentation which led Chasles to prove the previous theorem is based on the properties of the polarities. On the other hand, in the statement of the theorem there is no reference to such a doctrine. It has been used as an instrument to reach a property of the second-degree surfaces which will be fundamental in the course of the entire Chasles' essay. However, as far as the surfaces A and A' are concerned, no reference to polarity exists. Therefore, Chasles named the surfaces A and A' because what was important was the indication that they have correlative properties and not the fact that these properties have been obtained considering such surfaces as the polar transformed of two other surfaces U and V which are s.s.p. Now A and A' are two any surfaces fulfilling the conditions of the theorem, independently of their origin. In any case, the doctrine of polarity is the methodological and demonstrative basis of the fundamental theorem posed by Chasles as initial proposition for his theory of the ellipsoid attraction. For this theorem concerns metric–projective properties of the second-degree surfaces, which are obtained through a polarity with respect to a specific sphere, exactly as other metric properties had been obtained—as we have seen—by means of a polarity with respect to a paraboloid. This is a clear indication that he considered projective geometry as the basic element of a vast part of science. Obviously, given this basis, the specific elements of any single problem or sector of science have then to be considered and added to the properties obtained through projective geometry. We will see that, as far as this was possible, Chasles tried to extend the use of geometry in the problem of the ellipsoid attraction. Furthermore, in the course of his *Mémoire*, he proved several important properties relying on the features of the polar transformations.

[41] For what we will see in the part C) of this subsection it is appropriate to recall that, if Sa, Sb, Sc are the half-axes of the surface A and Sm any other half-diameter, making with them the angles θ, φ, ψ resp., the identity 1) $\frac{1}{SM^2} = \frac{cos^2\theta}{Sa^2} + \frac{cos^2\varphi}{Sb^2} + \frac{cos^2\psi}{Sc^2}$ holds. Furthmore, if AA', BB', CC' are the segments intercepted on the surface A on Sa, Sb, Sc and $O\alpha$, $O\beta$, $O\gamma$ are the half-diameters of A parallel to Sa, Sb, Sc, it is possible to prove that the following formula holds: 2) $\frac{MM'^2}{O\mu^2} = \frac{AA'^2}{O\alpha^2} cos^2\theta + \frac{BB'^2}{O\beta^2} cos^2\varphi + \frac{CC'^2}{O\gamma^2} cos^2\psi$ (ibid., pp. 23–24). Here Chasles makes a certain linguistic abuse because he uses the letters A, A' to indicate both two surfaces and the two extremes of a segment, but I prefer to remain faithful to this notation—which, anyway, does not imply any conceptual ambiguity—rather than to introduce new letters.

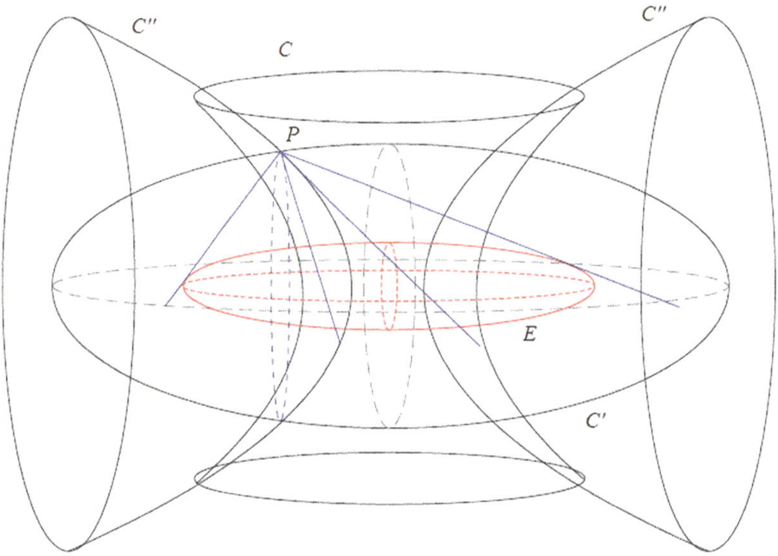

Fig. 7.14 Reconstruction of the figure described by Chasles. C, C', C'' are three surfaces described with the same foci of the ellipsoid E. They mutually cut in the point P from which the cone tangent to E is drawn. I have not drawn the three normals to C, C', C'' because the diagram would have been unclear

B) The second step of Chasles' geometrical argument

While addressing the problem of the attraction exerted by an ellipsoid on an external point, some properties of the second-degree surfaces which have their principal sections described with the same foci are fundamental. Thence, the geometrical demonstrative iter by Chasles continues connecting the just proved theorem with the properties of second-degree surfaces whose principal sections are confocal conics, used to study the ellipsoid attraction.

The second fundamental theorem proved by Chasles states that:

> The principal axes of any cone circumscribed to a second-degree surface are the normals to three surfaces passing through the vertex of the cone, so that these surfaces have their principal sections described with the same foci as the given surface.[42] (Fig. 7.14).

The first step to prove this theorem (I call it T2) is based on Eq. (7.17), which expresses T1: the value $\frac{O\mu^2}{MM'}$, to which the diameter Sm of A' is proportional, varies according to the positions of the points M and M' on A. Chasles wondered for what positions of such points this ratio is constant. The answer is that such positions of Sm are determined by the intersection of the surface A' with a concentric sphere.

[42] Chasles (1837, 1846), p. 29: "Les axes principaux d'un cône quelconque circonscrit à une surface du second degré sont les normales aux trois surfaces qu'on peut faire passer par la sommet du cône, de manière qu'elles aient leurs sections principales décrites des mêmes foyers que celles de la surface proposé".

Therefore, the set of the straight lines which are the positions of Sm for a given value of such segment are the generatrices of a second-degree cones. All the cones generated by the different values of Sm have, hence, the same principal axes.[43] To determine these axes when the point S is external to A Chasles argued that if the transversal SM (or Sm) is tangent to A at M, then $MM' = 0$. Thence Sm is infinite, which proves that A' is, in this case, a hyperboloid whose asymptotic cone is the cone circumscribed to A and having its vertex in S. This means that the three principal diameters of A' have the same directions as the principal axes of the cone circumscribed to A (ibid., p. 25).

Chasles claimed that, when S is inside A, the previous reasoning is not applicable because the cone becomes imaginary. However: 1) while dealing with the specific problem of the ellipsoid attraction, the difficulty concerns the attraction of an external point. In this case the expounded argument is satisfying; 2) Chasles also offered a more general reasoning in which the case of the internal points is included, as well. But for his aim, the more specific reasoning highlights important properties of the cone he used in the course of his *Memoir.*

Chasles' next step consists in proving that, when two second-degree surfaces A and B have their principal sections described with the same foci, if two planes tangent to these surfaces are drawn so as to be mutually parallel, then the difference of the distances' squares of these planes from the centre of the surface is constant (ibid., p. 25). The reasoning is purely geometrical. It is developed like this: be p and p' the considered distances; a,b,c the half-axes of A and a',b',c' those of B. Be α, β, γ the cosinuses of the angles between p and a,b,c, respectively, (obviously they are the same as those between p' and a',b',c'). Chasles proved that the following identity holds

$$p^2 - p'^2 = \left(a^2 - a'^2\right)\left(\alpha^2 + \beta^2 + \gamma^2\right). \qquad (7.18)$$

Since both the factors on the right side are constant and since $\alpha^2 + \beta^2 + \gamma^2 = 1$, the difference $p^2 - p'^2$ is equal to $a^2 - a'^2$.

Chasles also proved the reciprocal proposition: if two concentric quadrics are such that the difference of the distances' squares of their common centre from two any parallel planes which are tangent to the two surfaces is constant, the principal sections of the surfaces are confocal (ibid., pp. 26–27).[44]

---

[43] I recall that a second-degree cone has three principal axes. The principal axes of the surface are defined as the three straight lines which determine, two by two, the three planes with respect to which the cone has an orthogonal symmetry. They are called the principal planes of the cone. Consider, for example, an elliptic cone. Be a the principal axis internal to the cone. The other two principal axes are external. To find them, it is sufficient to cut the cone with a plane perpendicular to a. It intersects the cone in an ellipse. The two axes of the ellipse are parallel to the two other principal axes of the cone. In the degenerate case in which the directrix of the cone is a circumference, the cone has only the internal axis.

[44] The proofs of these two theorems resort to known elements of the synthetical theory of second-degree surfaces and are rather clear.

Now Chasles considered the surfaces A' and B' which are the polar transformed in the polarity analysed in the previous subsection. Referring to Eq. (7.16), be given the surfaces A and B whose principal sections are described with the same foci. Draw the tangent planes to the two surfaces perpendicularly to a transversal from a fixed point S. Be P, P' and Q,Q', the points where such planes cut the transversal. Take two straight line segments Sp and Sq such that

$$Sp^2 = \nu \cdot SP \cdot SP', \qquad Sq^2 = \nu \cdot SQ \cdot SQ'$$

being ν a constant. Finally, from the extremities p and q of these two segments draw two perpendicular planes to the transversal. Under these conditions, the two planes envelop two concentric second-degree surfaces A and B' whose centre is S. Chasles, showing the difference $Sp^2 - Sq^2$ to be constant, proved that the principal sections of A' and B' are confocal (ibid., pp. 27–28). Therefore, the three axes of the A' and B' have the same directions.

According to the considerations on the cones circumscribed to a second-degree surface explained in the previous page, the axes of A' are, in direction, the three principal axes of the cone circumscribed to A whose vertex is S and the axes of B' are, in direction, the three principal axes of the cone circumscribed to B whose vertex is S. Thence, these cones have the same principal axes (ibid., pp. 28–29). Thus, Chasles can conclude that, given two second-degree surfaces whose principal sections are confocal, if a point of space is considered the common vertex of two cones circumscribed to the surfaces, these two cones have the same principal axes (ibid., p. 29).

Now Chasles developed the following brilliant and purely geometrical reasoning: if the common vertex of one of the two cones circumscribed, respectively, to A and to B—for example to B—belongs to B itself, it will be transformed into its tangent plane and one of the principal axes of the cone will be normal to B. The other two principal axes of the cone circumscribing A will be the normals to two other surfaces passing through its vertex S, so that their principal sections will be confocal with the principal sections of A. Thence, finally T2 is deduced:

> The principal axes of any cone circumscribed to a second degree surface are the normals to three surfaces passing through the vertex of the cone, so that these surfaces have their principal sections described with the same foci as the given surface.[45]

Therefore, the three axes of the surface A' are directed along the normals of three surfaces passing through S and having their principal sections confocal with those of A. This result does not depend on the position of S with respect to A. Hence, it is possible to determine the direction of the axes of A' also in the case in which S belongs to or is inside A.

[45] Chasles (1837, 1846), p. 29: "Les axes principaux d'un cône quelconque circonscrit à une surface du second degré sont les normales aux trois surfaces qu'on peut faire passer par le sommet du cône, de manière qu'elles aient leurs sections principales décrites des mêmes foyers que celles de la surface proposée".

Commentary: while developing a foundational programme, an essential element is the interconnection among the different sections of the programme itself. For every single problem and every branch of mathematics need specific concepts and methods, but the programme is the more profound the more its concepts and methods include different branches of mathematics. Insofar as the ellipsoid attraction is concerned, some concepts of projective geometry—in particular, that of polarity—represent the foundation of Chasles' work. The whole reasoning expounded here relies on a specific polar transformation, which confirms Chasles' basic idea: only modern geometry (namely, projective geometry) can allow us to face the question of ellipsoid attraction with a simplicity and a generality which escape the analytical methods. Furthermore, the developments described by Chasles permit to frame this difficult dynamical problem within his foundational programme, thus showing that a common root to the whole of exact sciences—or at least to a great part of them—exists.

Two deeply interconnected, but not identical, aspects have to be remarked:

1) The first one is the methodological aspect. The situation is clear: Chasles repeated more than once in his *Memoir* that he was going to show that the synthetic method offers the same results as that analytical, but with a more consistent and unitary picture. Here there is no ambiguity as to the term "synthesis-synthetic". From the first page of his work Chasles used this term in a specific way: to work on the general properties of the configurations he is using without resorting, in an initial phase, to the equations of the objects he is dealing with. The fact that he did not draw the diagrams depends—beyond the technical difficulties to show some configurations and their impossibility when some elements become imaginary—on his idea that the diagram might appear linked to the specific drawn figure, whereas the synthetic method has to be as general as possible. Hence, a single figure, which might seem a didactical support, is, in fact, a strong deviation from the true spirit of the synthetic method. Thence, no figure (or very few) exists in Chasles' (and not only in Chasles') works. Therefore, the reader might be surprised to observe that, after the initial theorems, Chasles resorted to the classical equation which, in analytical geometry, indicates the figures. Is he not coherent? The answer is that he is perfectly coherent: after having framed the problems within the synthetic picture, the use of equations is a helpful and probably indispensable means to specify the elements which have been constructed through synthesis. However, in Chasles' perspective, it makes sense to introduce them when the synthetic picture has been clarified without resorting to coordinate systems. In his view, analytical geometry is an *ancilla* of synthetic geometry, but a very useful *ancilla*. It is not a coincidence that he was one of the inventors of the projective coordinate systems, which, anyway, are not used in the *Mémoire*.
2) The structural and ontological aspects are connected with the methodological one. However, they cannot be identified with the latter. The basic "objects" of the abstract universe and, hence, of science, are the projective transformations, not exactly the geometrical figures in themselves. The geometrical figures are, so to

say, the undifferentiated, the passive objects on which the projective transformations act. If one prefers to use some terms that are referable to Greek philosophy, one can say that the figures are the *chora*, the projective transformations are the *nous*, and that the geometer is the *demiourgos*. Out of metaphor, Chasles is claiming that mathematics is, before than a science of objects, a science of structures. The basic structures are, according to him, the projective transformations. This told, I am not claiming that his concept of transformation as a structure can be identified, or is, so to say, a precursor, of the concept of group of transformation *à la* Klein, since in order to reach such a concept it is necessary to further develop the abstract view of geometry and of the whole of mathematics. But Chasles was going, in part, towards this direction. I say "in part" because, in any case, he thought that the foundations of geometry and of science were to be found in a section of geometry itself, not, possibly, in a new and more comprehensive abstract discipline.

Finally, I point out once again the unitarity character of Chasles' thought. When he proved T2, he first demonstrated this theorem when S is a point external to A and afterwards for any position of S. In the first case he stressed that he was using contingent properties of the configuration insofar as he supposed that the whole reasoning was based on a contingent position of the point S. These properties, he claimed (ibid., p. 25 note), are also useful to guess what happens when S is inside A and the cone becomes imaginary. Nonetheless, they are contingent because they separate the particular positions of S with respect to A. Furthermore, as Chasles knew well, in the specific case of the ellipsoid attraction only the position in which S is external to A is interesting because when S is internal or on A, other far easier properties than those to which Chasles resorted are sufficient. While, in the second proof of this theorem, derived from other considerations, permanent properties are used and, hence, it is not necessary to speak of the position of S with respect to A either of cones which can be real, degenerated into a real plane, or imaginary.

C) The third and final step of Chasles' geometrical introduction.

The final problem Chasles solved in his complex and profound introduction to the geometrical theory of the ellipsoid attraction is the following one: to find expressions for the ratios $\frac{O\alpha^2}{AA'}, \frac{O\beta^2}{BB'}, \frac{O\gamma^2}{CC'}$ which are independent of any reference to the surface A'. The reasoning is framed like this: the plane tangent at A' drawn from the extremity a of the half-diameter Sa is perpendicular to Sa. Be P, P' the points where the planes tangent to A and perpendicular to Sa cut Sa.

Applying Eq. (7.16), you get

$$Sa^2 = \lambda^2\left(\frac{SO^2}{OD^2} - 1\right)SP \cdot SP'.$$

Indicating with O' the point where the plane ζ drawn from the centre of A and perpendicular to Sa cuts Sa, this equation is transformed into

$$Sa^2 = \lambda^2 \left(\frac{SO^2}{OD^2} - 1 \right) \left(SO'^2 - O'P^2 \right).$$

If through S a plane parallel to ζ is drawn, it will be tangent to one of the surfaces Si passing through S and whose principal sections are confocal with A.[46] Thence, if the major half-axis of the surface A is called a, and that of Si is called a_1, the following result is deduced by applying Eq. (7.18):

$$SO'^s - O'P^2 = a_1^2 - a^2,$$

so that

$$Sa^2 = \lambda^2 \left(\frac{SO^2}{OD^2} - 1 \right) \left(a_1^2 - a^2 \right)$$

If the same theorem is applied to the two other surfaces Si, you will obtain the values of the three axes of A' as a function of the axes of A and of the major axes of the surfaces Si, i.e.,

$$Sb^2 = \lambda^2 \left(\frac{SO^2}{OD^2} - 1 \right) \left(a_2^2 - a^2 \right)$$

$$Sc^2 = \lambda^2 \left(\frac{SO^2}{OD^2} - 1 \right) \left(a_3^2 - a^2 \right).$$

Through another passage (ibid., p. 31) Chasles achieved the expressions for $\frac{O\alpha^2}{AA'}, \frac{O\beta^2}{BB'}, \frac{O\gamma^2}{CC'}$ in function of the four axes a, a_1, a_2, a_3, so eventually reaching the equation

$$\frac{MM\prime^2}{O\mu^4} = 4 \left(\frac{SO^2}{OD^2} - 1 \right) \left(\frac{cos^2\theta}{a_1^2 - a^2} + \frac{cos^2\varphi}{a_2^2 - a^2} + \frac{cos^2\psi}{a_3^2 - a^2} \right).$$

After this result, through a series of not difficult algebraic transformations (ibid., pp. 31–33), he was able to express the values of each half-diameter a_i as a function of the three axes a,b,c of A and of the coordinates of the point S. This was useful to reach a result which will be fundamental for the synthetic theory of the ellipsoid attraction. Such result is expressed by the formula (ibid., p. 33):

[46] I highlight that such conclusion follows applying T2. The surface of which Chasles speaks is one of those I have indicated by *S1, S2, S3*.

$$\left(a_1^2-a^2\right)\left(a_2^2-a^2\right)\left(a_3^2-a^2\right)=a^2b^2c^2\left(\frac{SO^2}{OD^2}-1\right). \tag{7.19}$$

The last theorem to which it is necessary to refer has a complicated enunciation, but its proof is based on a series of not difficult algebraic transformations (ibid., pp. 33–35). It states that:

Be given two ellipsoids whose principal sections are confocal and whose major half-axes are resp. a, a'. Consider three second-degree surfaces of major half-axes a_1, a_2, a_3, which pass through a point S and which are confocal with the two ellipsoids. Draw two transversals l, m through S, of which l is arbitrary, whereas m is such that the angles φ, ψ and φ', ψ' made by l and m resp. with the normals in S to second and third surface fulfil the following relations

$$\frac{\cos\varphi}{\cos\varphi'}=\frac{\sqrt{a_2^2-a^2}}{\sqrt{a_2^2-a'^2}},\qquad \frac{\cos\psi}{\cos\psi'}=\frac{\sqrt{a_3^2-a^2}}{\sqrt{a_3^2-a'^2}}.$$

Be E, F and E', F' the points where l and m cut the two ellipsoids resp. and Oe, Oe' their two half-diameters parallel to l and m. Finally be D, D' the points where the straight line SO drawn from S to the common centre of the two ellipsoids cut their surfaces. Under all these conditions, the two ratios $\frac{Oe^2}{EF}$, $\frac{Oe'^2}{E'F'}$ will be in a constant ratio for any pair of transversals l, m. Specifically, it will be:[47]

$$\frac{Oe^2}{EF}:\frac{Oe'^2}{E'F'}=\frac{\sqrt{a_1^2-a^2}}{\sqrt{\frac{SO^2}{OD^2}-1}}:\frac{\sqrt{a_1^2-a'^2}}{\sqrt{\frac{SO^2}{OD'^2}-1}}.$$

This expression, through Eq. (7.19), can be transformed into another one which Chasles used while dealing with the ellipsoid attraction. Such expression is (ibid., p. 35):

[47] Chasles (1837, 1846), pp. 33–34: "Étant donnés deux ellipsoïdes dont les sections principales sont décrites des mêmes foyers, et dont les demi-axes majeurs sont a and a' ; et étant pris arbitrairement dans l'espace un point S; Que l'on conçoive les très surfaces du second degré que l'on peut faire passer par ce point, de manière qu'elles aient les mêmes foyers que les deux ellipsoïdes ; et soient a_1, a_2, a_3 les demi-axes majeurs de ces trois surfaces; Que par le point S on mène deux transversales, la première dans une direction arbitraire, et la seconde dans une direction telle, que les angles φ, ψ et φ', ψ' que ces deux droites, respectivement, feront avec les normales en S à la seconde et a là troisième surface, aient entre leurs cosinus les relations $\frac{\cos\varphi}{\cos\varphi'}=\frac{\sqrt{a_2^2-a^2}}{\sqrt{a_2^2-a'^2}}$, $\frac{\cos\psi}{\cos\psi'}=\frac{\sqrt{a_3^2-a^2}}{\sqrt{a_3^2-a'^2}}$; Soient E, F et E', F' les points où ces deux transversales rencontrent, respectivement, les deux ellipsoïdes proposés, et soient Oe, Oe' les demi-diamètres de ces ellipsoïdes, parallèles aux deux transversales; soient enfin D, D' les points où la droite SO menée du point S au centre commun des deux ellipsoïdes rencontre leurs surfaces. Les deux rapports $\frac{Oe^2}{EF}$, $\frac{Oe'^2}{E'F'}$ seront entre eux dans une raison constante, quelles que soient les deux transversales; Cette raison sera $\frac{Oe^2}{EF}:\frac{Oe'^2}{E'F'}=\frac{\sqrt{a_1^2-a^2}}{\sqrt{\frac{SO^2}{OD^2}-1}}:\frac{\sqrt{a_1^2-a'^2}}{\sqrt{\frac{SO^2}{OD'^2}-1}}$".

$$\frac{Oe^2}{EF} : \frac{Oe'^2}{E'F'} = \frac{abc}{a'b'c'} : \frac{\sqrt{a_2^2 - a^2}\sqrt{a_3^2 - a^2}}{\sqrt{a_2^2 - a'^2}\sqrt{a_3^2 - a'^2}}. \tag{7.20}$$

I will provide the reader with a table in which I summarize the most important results obtained by Chasles in the purely geometrical parts of his *Mémoire* and their interconnections.

---

1) U, V= two second-degree surfaces which are similar and similarly posed. U', V'=polar transformed of U and V with respect to a sphere concentric with V.
2) Be Π, Π' two points of U. G=centre of U, S=centre of V. The points S, Π, Π' are collinear, π is the point where $S\Pi$ cuts V.
3) In the polarity, to Π, Π' two tangent planes to U' correspond. They are perpendicular to $S\Pi$.
4) After several passages Chasles proves the two fundamental equations referred to U', V':
$\frac{2\lambda}{Sm} = \frac{1}{SM} - \frac{1}{SM'}$
$Sp^2 = \lambda^2\left(\frac{SO^2}{OD^2} - 1\right) SP \cdot SP'$.
where λ is a constant; m is the point of V' corresponding to the tangent plane of V which is perpendicular to $S\pi$; M, M' are the points of U' which correspond to the planes tangent to U and perpendicular to $S\Pi$ (i.e. $S\pi$); O is the centre of U'; D=one of the extremities where one of the two planes tangent to U' (which are parallel to the plane polar to S) touch U'; $SP = \frac{1}{S\Pi}$; $SP' = \frac{1}{S\Pi'}$; $Sp = \frac{1}{S\pi}$.
5) Now Chasles considers two surfaces A, A'. If S is any point of space and a transversal a rotating around S cuts A in M,M' and A'; if m is a point belonging to the locus such that
$\frac{2\lambda}{Sm} = \frac{1}{SM} - \frac{1}{SM'}$
Then: 1°) The point m belongs to a second-degree surface A' whose centre is S.
2°) If SP and SP' are the segments determined on a by the two planes tangent to A and perpendicular to a and if Sp is the segment determined by a plane tangent to A' and parallel to the other two, it holds.
$Sp^2 = \lambda^2\left(\frac{SO^2}{OD^2} - 1\right) SP \cdot SP'$.
Note: A,A' are the surfaces that, beforehand, Chasles had called U',V'.
6) Being $O\mu$ the half-diameter of A parallel to SM, Chasles requires that the ratio $\frac{O\mu^2}{MM'}$ gets a constant value and he finds that the transversals Sm for which this is the case are the generatrices of a second-degree cone. If you assume different values of such a ratio, you obtain different cones, but they have the property to share the same principal axes.
7) The three axes of A' are, in direction, the three principal axes of every cone circumscribed to A.
Note: Chasles proved this property supposing that S is external to A. He claimed that he would have reached a general proof which does not depend on the position of S with respect to A.
8) When two second-degree surfaces A and B have their principal sections described with the same foci, if two planes tangent to these surfaces are drawn which are mutually parallel, the difference of the distances' squares of these planes from the centre of the surface is constant and vice versa.
Note: A and B are the surfaces Chasles had called U and V.
9) Chasles poses a condition to prove that A' and B' are confocal from which you deduce that A' and B' are exactly the polar transformed of A and B, namely the "old" U' and V'.
10) A', B' have their axes directed along the same straight lines (deduction from 9).
11) Given two second-degree surfaces whose principal sections are confocal, if a point of space is considered the common vertex of two cones circumscribed to the surfaces, these two cones have the same principal axes.
12) The principal axes of any cone circumscribed to a second-degree surface are the normals to three surfaces $S1$, $S2$, $S3$ passing through the vertex of the cone, so that these surfaces have their

---

(continued)

principal sections described with the same foci as the given surface. Therefore, the three axes of the surface A' are directed along the normals of three surfaces passing through S and having their principal sections confocal with those of A.

13) So far Chasles' reasoning has been developed supposing that S is external to A. Through what proved in 12) and through the equations in 5), he proved that any proposition is true also if S is inside A or belongs to A.
14) Determination of the ratio $\frac{MM'}{O\mu^2}$ independently of any element of the surface A' and relying only upon elements of A, *S1, S2, S3*.
15) Determination of such ratio only relying upon elements of A.
16) Deduction of the fundamental equation $\left(a_1^2 - a^2\right)\left(a_2^2 - a^2\right)\left(a_3^2 - a^2\right) = a^2b^2c^2\left(\frac{SO^2}{OD^2} - 1\right)$, where a_i are the major half-axes of the surfaces *Si*.
17) Deduction of the last theorem mentioned in the running text.

Commentary: the propositions I have now expounded represent the geometrical structure on which Chasles edified his results on the ellipsoid attraction. From Newton's and Maclaurin's epoch, no mathematician had used geometry is such a broad manner as Chasles did in order to deal with dynamics subjects. Hence, in what follows I will specify in which part of Chasles' dynamical reasoning the results obtained in the geometrical section of his *Mémoire* are exploited to face the problem of the ellipsoid attraction. This is necessary to fully grasp what it means that geometry, and, in particular, projective geometry is the basis of his whole science. I will not, hence, consider all the details of Chasles' argumentation. Rather, I will focus on his use of geometry within it.

7.4.2 Chasles' Proof of Maclaurin Theorem

In the second section of his *Mémoire* Chasles proved the GMT. For he proved a theorem which is a generalization GMT. It states that:

> Be given two ellipsoidal shells of any thickness, included between two concentric ellipsoidal surfaces, which are s.s.p.. Be the external as well as the internal surfaces confocal.
>
> Be the density in any point of the two shells proportional to the same power of the distance of this point from the centre of the shell divided by the half-diameter of the external surface on which the point lies. Then:
>
> The attraction exerted by the two shells on a point external to their surfaces will have the same directions and will be mutually as the masses of the two shells.[48]

[48] Chasles (1837, 1846), pp. 54–55: "Si l'on a deux couches d'une épaisseur quelconque, comprises chacune entre deux surfaces d'ellipsoïdes concentriques, semblables et semblablement placés, et si les surfaces externes des deux couches sont décrites des mêmes foyers, ainsi que leurs surfaces internes; La densité, en chaque point de chacune des deux couches, étant proportionnelle à une même puissance de la distance de ce point au centre de la couche, divisée par le demi-diamètre de sa surface externe, sur lequel de point est situé. Les attractions que les deux couches exerceront sur un même point situé au dehors de leurs surfaces, auront la même direction et seront entre elles comme les masses des deux couches".

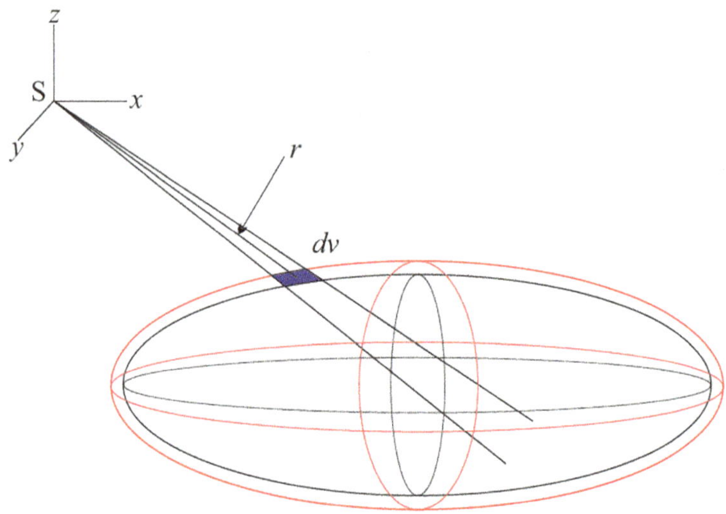

Fig. 7.15 Diagram reproducing the situation described by Chasles. The diagram is referred to the shell L1 which is included between the ellipsoid drawn in red and that drawn in blue

Therefore, there are two shells, namely *A*,*A*' the internal surfaces of the two shells and *B*,*B*', those external. *A*,*B* are s.s.p.; A',B' are s.s.p. *A*, A' are confocal. *B*, B' are confocal.

As Chasles pointed out (ibid., p. 55), if the internal surfaces of both shells are reduced to a point GMT is obtained.

The argumentations developed by Chasles have many facets and nuances and are connected with geometry, different reference frames and treatment of differential equations. For it is obvious that, while facing dynamics, the differential equations are unavoidable as soon as the difficulty-level of the addressed problems progressively increases. Chasles gave a geometrical rather than an analytical support to the necessary differential equations.

In what follows my main point will be to stress the passages of Chasles' reasoning where the properties expounded in the first section of the *Mémoire* and, hence, tied to projective geometry via polarity are used. Therefore, I will concentrate on such properties, but, obviously, an outline of the whole of Chasles' reasoning has to be referred to.

The argumentation begins considering two ellipsoidal shells L1 and L2 in an ellipsoid. The surfaces which delimit each shell are s.s.p. The external surfaces of the two shells have the principal sections confocal.

The first operation carried out by Chasles consists in determining the attraction exerted by a shell's (e.g. shell L1, see Fig. 7.15) element of volume on an external point *S*. The initial step is to establish an element of volume *dv* for any shell. Chasles constructed an infinitesimal cone with its vertex in *S* cutting the shell. If ρ is the shell's density and *r* the distance of *dv* from *S*, the attraction exerted by *dv* on *S* is

$$\rho \frac{dv}{r^2}.$$

If a Cartesian orthogonal system of coordinates is centred in S with axes SA, SB and SC and being θ, φ, ψ the angles between r and the three axes, respectively, the attraction along the axes will be

$$\rho \frac{dv}{r^2} \cos\theta, \quad \rho \frac{dv}{r^2} \cos\varphi, \rho \frac{dv}{r^2} \cos\psi,$$

Passing to spherical coordinates and naming ω the angle between the planes (r, SA) and (SB,SC), the three attractions can be written as

$$\rho \frac{dv}{r^2} \cos\theta, \quad \rho \frac{dv}{r^2} \cos\theta \cos\omega, \rho \frac{dv}{r^2} \cos\theta \sin\omega.$$

This granted, it is easy to calculate dv, whose expression will be

$$dv = r^2 dr \sin\theta d\theta d\omega.$$

According to Newton's law, the attractions exerted on S by dv and by their components along SA, SB, SC are, respectively:

$$\varrho dr \sin\theta d\theta d\omega$$

$$\varrho dr \sin\theta \cos\theta d\theta d\omega$$

$$\varrho dr \sin^2\theta \cos\omega d\theta d\omega$$

$$\varrho dr \sin^2\theta \sin\omega d\theta d\omega$$

Through these operations, Chasles found the element of attraction, which is necessary for this problem. Now, there is the first direct use of geometry, though not of the considerations connected to polarity which are expounded in the previous section. For being dr the portion of r included between the two surfaces delimiting the shell, Chasles claimed that to calculate its value analytical geometry can be used, but he avoided the tedious and long calculation based on the analytical treatment through the following geometrical consideration (ibid., pp. 43–45). Notations: E, F=points where r cuts the external surface of L1, Oe=half-diameter of such surface parallel to r; D,D'=the two points where the straight line joining the centre O and S cuts the surface. Because of a fundamental property of the ellipsoid, it is

$$\frac{SE \cdot SF}{Oe^2} = \frac{SD \cdot SD'}{OD^2}.$$

If G indicates the middle point of EF, this equation, after a simple passage, can be written as:

$$SG^2 - GE^2 = \frac{Oe^2}{OD^2}\left(SO^2 - OD^2\right).$$

Now Chasles considered the internal surface of L1. Since it is concentric and s.s.p. with respect to that external, the ratio $\frac{Oe}{OD}$ is constant (ibid., p. 44). Thence, the only variables of the previous equation are GE, OD. Therefore, by differentiating, you get:

$$GE \cdot dGE = \frac{Oe^2}{OD^2} OD \cdot dOD.$$

Since OD is the half-diameter of the external surface and $Od - dOD$ that of the internal one, their ratio is the same as that of two any homologous half-diameters a and a-da, being the surfaces similar, so that it also holds $\frac{dOD}{OD} = \frac{da}{a}$. After some brief and easy passages, you get

$$dr = 2\frac{Oe^2}{EF}\frac{da}{a},$$

so that the attraction exerted by dv and their components are, respectively,

$$2\varrho\frac{Oe^2}{EF}\frac{da}{a}\sin\theta\cos\theta d\theta d\omega$$

$$2\varrho\frac{Oe^2}{EF}\frac{da}{a}\sin^2\theta\cos\omega d\theta d\omega$$

$$2\varrho\frac{Oe^2}{EF}\frac{da}{a}\sin^2\theta\sin\omega d\theta d\omega.$$

Considering L2, and indicating with an accent the elements of L2 which correspond to L1, you will obtain the same relations. Hence, if one considers the ratio of the components along SA of the attractions exerted by the volume elements of the two shells, one obtains (ibid., p. 46)

$$\frac{\varrho\frac{Oe^2}{EF}\frac{da}{a}\sin\theta\cos\theta d\theta d\omega}{\varrho'\frac{Oe'^2}{E'F'}\frac{da'}{a'}\sin\theta'\cos\theta' d\theta d\omega'}. \tag{7.21}$$

Chasles continued as follows:

> The major half-axes of the two external surfaces of the two shells are a, a'. For hypothesis, these two surfaces have their principal sections described with the same foci. Consider the three surfaces passing through the point S, so that their sections are described with the same

foci as those of the two proposed surfaces. Be a_1, a_2, a_3 their major half axes. Assume their normals at S as the three perpendicular axes SA, SB, SC, of which, so far, we had not yet determined the direction.[49]

He argued that there is no loose in the generality of the whole argument supposing that the radius vector from S to dv has any direction, but that the radius S-dv' has a direction such that the cosine of the angles that this direction makes with SB and SC has a constant ratio with the cosine of the angles that the former radius makes with SB and SC.

Chasles is legitimate to rely on such an assumption because his argument will prove that the final result is independent of the values of these angles, which are, hence, used in the proof, but which disappear in the final result.

Thence, he supposed that such ratios were, respectively,

$$\sqrt{\frac{a_2^2 - a'^2}{a_2^2 - a^2}} \text{ and } \sqrt{\frac{a_3^2 - a'^2}{a_3^2 - a^2}}.$$

The following identities hold:

$$\frac{\cos(r', SB)}{\cos(r, SB)} = \frac{\cos\varphi'}{\cos\varphi} = \frac{\sin\theta'\cos\omega'}{\sin\theta\cos\omega} = \sqrt{\frac{a_2^2 - a'^2}{a_2^2 - a^2}}.$$

Therefore, as Chasles claimed, the two identities (the second one is deduced exactly as the one shown) follow:

$$\sin\theta\cos\omega = \sin\theta'\cos\omega' \cdot \sqrt{\frac{a_2^2 - a^2}{a_2^2 - a'^2}}$$

$$\sin\theta\sin\omega = \sin\theta'\sin\omega' \cdot \sqrt{\frac{a_3^2 - a^2}{a_3^2 - a'^2}}$$

After a brief calculation Chasles reached the equation

[49] Chasles (1837, 1846), p. 46: "Les demi-axes majeurs des surfaces externes des deux couches sont a, a'. Ces deux surfaces ont, par hypothèse, leur sections principales décrites des mêmes foyers. Que l'on conçoive les très surfaces que l'on peut faire passer par le point S, de manière que leurs sections principales soient décrites des mêmes foyers que celles des deux surfaces proposées; soient a_1, a_2, a_3 les demi-axes majeurs de ces trois surfaces, et prenons leurs normales au point S pour les trois axes rectangulaires SA, SB, SC dont jusqu'ici nous n'avions pas déterminé la direction".

$$\sin^2\theta = \frac{\cos^2\omega'}{\cos^2\omega}\sin^2\theta' \frac{a_3^2 - a^2}{a_2^2 - a'^2}.$$

By differentiating this equation with respect to θ, θ' and after further calculations (ibid., p. 47) he arrived at the differential equation

$$\frac{\sin\theta\cos\theta d\theta d\omega}{\sin\theta'\cos\theta' d\theta' d\omega'} = \frac{\sqrt{\left(a_2^2 - a^2\right)\left(a_3^2 - a^2\right)}}{\sqrt{a_2^2 - a'^2)\left(a_3^2 - a'^2\right)}}$$

Replacing these values in Eq. (7.21), the ratio of the volume-elements' attractions of the two shells gets the form (ibid., p. 48)

$$\frac{\varrho\frac{Oe^2}{EF}\frac{da}{a}\sqrt{\left(a_2^2 - a^2\right)\left(a_3^2 - a^2\right)}}{\varrho'\frac{Oe'^2}{E'F'}\frac{da'}{a'}\sqrt{a_2^2 - a'^2)\left(a_3^2 - a'^2\right)}}$$

Now there is an essential step in which a result proved in the first section of the *Mémoire* is exploited. For through Eq. (7.20), which states that

$$\frac{\frac{Oe^2}{EF}\cdot\sqrt{\left(a_2^2 - a^2\right)\left(a_3^2 - a^2\right)}}{\frac{Oe'^2}{E'F'}\cdot\sqrt{a_2^2 - a'^2)\left(a_3^2 - a'^2\right)}} = \frac{abc}{a'b'c'},$$

the previous expression of the ratio between the attractions of the volume elements can be written as

$$\frac{\rho abc}{\rho' a'b'c'}$$

Through another analytical reasoning (ibid., pp. 48–49), Chasles was able to prove that such a ratio is the same as that between the masses of the two shells. For indicating with dV, dV' the volumes of the two shells he proved that the previous ratio is equivalent to

$$\frac{\rho dV}{\rho' dV'}$$

This result shows that the ratio of the component along *SA* of the two shells' attractions is constant and equivalent to the ratio of the shells' masses. The same argument is applicable to the components of the attractions along the axes *SB, SC*.

Therefore, the first fundamental conclusion obtained by Chasles through his geometrical methods is the following one:

> Given two infinitely thin ellipsoidal shells, both of them included between similar and concentric surfaces of an ellipsoid, if the external surfaces of these two shells have their principal sections described with the same foci, the attractions exerted by these two shells on a point of space external to their surfaces will have the same direction and will be as the ratio of the masses of the two shells. These shells are supposed to be homogeneous, but of any density.[50]

Now Chasles is ready to prove MT and GMT. For be given two ellipsoids whose principal sections are confocal and whose axes are, respectively, A, B, C; A', $B'C'$. Both ellipsoids can be considered as composed of infinitely thin shells, each included between two ellipsoidal surfaces which are similar to those of the ellipsoid to which the shells belong. Thence, told L1 and L2 one shell of the first ellipsoid and one of the second, respectively, the axes of the external surface of L1 will be nA, nB, nC; those of L2 will be $n'A'$, $n'B'$, $n'C'$. If $n=n'$, L1 and L2 are told correspondent shells.

For what follows it should be pointed out that each ellipsoid can be thought of as composed of correspondent shells.

This clarified, Chasles (ibid., pp. 50–51) proved, through a reasoning based on elementary properties of the ellipsoid, that both the external and the internal surfaces of two correspondent shells are ellipsoidal surfaces whose principal sections are confocal. Then, the attractions exerted by two correspondent shells on an external point will have the same direction and their intensity along any straight line passing through the attracted point will be as

$$\frac{\rho nB \cdot nC \cdot dnA}{\rho' nB' \cdot nC' \cdot dnA'} = \frac{\rho BCAn^2 dn}{\rho' B'C'A'n^2 dn} = \frac{\rho BCA}{\rho' B'C'A'}. \tag{7.22}$$

If the densities do not vary, such a ratio is constant. This means that the sum of the components of the attractions of all the ellipsoidal shells have this same ratio. Thence, MT follows:

> The attractions exerted by two homogeneous ellipsoids whose principal sections are confocal on a point external to their surfaces have the same direction and are mutually as the masses of the two ellipsoids.[51]

Chasles pointed out that this proposition is known as the Maclaurin theorem, although Maclaurin proved it only when S is on the prolongation of one of the ellipsoid's axes (ibid., p. 52).

---

[50] Chasles (1837, 1846), p. 49: "Étant données deux couches ellipsoïdales, infiniment minces, comprises chacune entre deux surfaces d'ellipsoïdes semblables et concentriques; Si les surfaces externes de ces deux couches ont leurs sections principales décrites des mêmes foyers, les attractions que ces couches exerceront sur un même point de l'espace situé au dehors de leurs surfaces externes, auront la même direction, et seront entre elles dans le rapport des masses des deux couches; Ces couches étant supposées homogènes, mais de densité quelconque, l'une et l'autre".

[51] Chasles (1837, 1846), p. 52: "Les attractions des deux ellipsoïdes homogènes dont les sections principales sont décrites des mêmes foyers, exercent sur un même point situé au dehors de leurs surfaces, ont la même direction et sont entre elles comme les masses des deux ellipsoïdes".

The proof of the GMT is a simple derivation of the reasoning I have expounded. The generalization consists in the fact that it is not necessary that the two ellipsoids are homogeneous: their densities can vary from point to point, provided that it is constant in an infinitely thin shell and that it varies from shell to shell proportionally to a rational power of the major axis of the external surface.[52]

Commentary: I have focused, in particular, on those passages of Chasles' reasoning where the synthetic properties proved in the first section of his *Mémoire* are used. They are employed in fundamental passages. Thence, since a basic element of these synthetic properties is the polarity analysed in the previous subsection, such a projective transformation plays a decisive role in Chasles' whole argument.

This shows that projective geometry is the basis of Chasles' foundational programme in two senses: 1) its properties and transformations enter directly in the proofs of theorems used in branches of science apparently far from projective geometry. The example of the ellipsoid attraction is emblematic; 2) the geometrical metric properties—in this specific case, those of the ellipsoid—which Chasles used in his synthetic approach rely ultimately—as any metric property, according to Chasles—upon a projective foundation, as we have seen in the first section of this book.

The term "synthetic" has to be referred to the foundational approach Chasles used in his *Mémoire*, exactly because the initial theorems, namely those propositions which represent the basic elements of the entire reasoning, are proved through synthetic geometry. This does not mean that Chasles did not use analytical concepts and procedures. He did in two senses: 1) he used analytical geometry because he resorted to the equation of the ellipsoids; 2) he used mathematical analysis insofar as he employed differential equations. This implies no contradiction with his synthetic approach: I have already pointed out the way in which he employed analytical geometry. As to mathematical analysis, its use is unavoidable while dealing with infinitesimal concepts, which are necessary to solve the dynamical problems. Newton himself, who in the *Principia* used his synthetic methods, at the end of any reasoning, when the instantaneous quantities were necessary, developed a passage to the limit, so creating his infinitesimal geometry. It is true that no differential equation exists in the *Principia*, but the final stage of any Newton's proof based on a passage at the limit of certain quantities implies a differential equation. Newton's method is, so to say, even more geometrical than Chasles' who, anyway, used explicitly differential equations. But, in Chasles, too, the basic argument is founded on synthetic properties of the analysed configurations. Thence, there is a red line Newton-Maclaurin-Chasles. For these authors geometry is the basis of dynamics, too.

In what follows I will address the further developments of Chasles' *Mémoire*. In particular, I will focus on the sections of his argumentation where the synthetic elements play a significant role.

[52] For such a proof, which does not involve elements connected with the foundational problems I am dealing with, I refer directly to Chasles (1837, 1846, pp. 52–55).

7.4.3 The Attraction Exerted by an Ellipsoidal Shell on an External Point

So far, through his synthetic methods, Chasles has determined ratios of attractions. The next step concerns the determination of an infinitely thin shell on an external point. Chasles' argumentation is divided into two parts: in the former, he determined the attraction exerted by an infinitesimal shell on a point *S* posed on the external surface of the shell. In the latter he applied the achieved results to determine the attraction in the case in which the point is external.

We have seen that the expression of a shell's attraction towards a point *S* can be written as $\rho dr \cdot \sin\theta \cdot d\theta \cdot d\omega$, where the meaning of the symbols is that I have already clarified. The problem consists in giving a specification of dr as a function of known elements. Suppose that *S* belongs to the external surface of the shell (Fig. 7.16). Be *SA* the normal to the surface and *SIE* the generatrix of the infinitesimal cone through which the element of volume is determined. Since the cone is infinitesimal, Chasles is legitimate to consider the point *I* as the locus in which the element of volume is concentrated (ibid., p. 56) and to regard *SI* as dr. Since the infinitesimal triangle *SAI* is rectangle, it is

$$SI = \frac{SA}{\cos \widehat{ISA}} = \frac{SA}{\cos\theta}.$$

Hence, the expression of the volume element gets the form

$$\rho SA \frac{\sin\theta d\theta d\omega}{\cos\theta}.$$

Since the attraction of the element in *E* has exactly the same value as that in *I*, the global attraction exerted by the two elements is

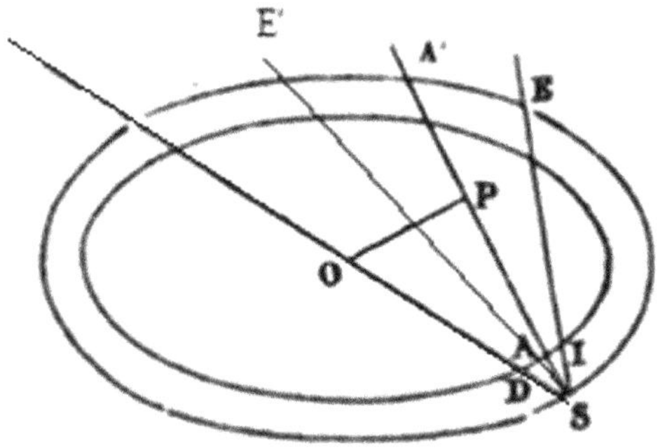

Fig. 7.16 The diagram used by Chasles to calculate the attraction of a shell on a point located on the external surface of the shell itself. Retrieved from Chasles (1837, 1846, p. 56)

$$2\rho SA\frac{\sin\theta d\theta d\omega}{\cos\theta}.$$

If in the plane *EAS* a transversal specular to *SIE* with respect to *SAA'* is considered and if the angle $A'SE'$ is indicated by θ', the angle ω is unchanged. It is, hence (ibid., p. 57):

$$2\rho SA\frac{\sin\theta' d\theta' d\omega}{\cos\theta'} = 2\rho SA\frac{\sin\theta d\theta d\omega}{\cos\theta}.$$

The resultant of these two attractions is directed along *SA* because the transversals *SE* and *SE'* have the same inclination with respect to *SA*. Since this reasoning can be developed for any transversal passing through *S*, Chasles concludes:

> The attraction exerted on any point [*S*] of the surface by of an infinitely thin shell included between two concentric, s.s.p. ellipsoidal surfaces is directed along the normal of this surface at this point.[53]

The whole attraction of the shell on *S* can be calculated considering *SA* as the element *dr*, so that the double integral of $2\rho SA\,\sin\theta d\theta d\omega$ has to be calculated for $0 \leq \theta \leq \frac{\pi}{2}$; $0 \leq \omega \leq 2\pi$. Through an easy reasoning Chasles (ibid., p. 58) reached the result

$$4\pi\rho SA.$$

Therefore, the attractions exerted by an ellipsoidal infinitesimal shell on the points of its external surface are proportional to the thickness of the shell in the considered point (ibid., p. 58).

Chasles found another expression for such attraction (ibid., pp. 58–59): being *O* the centre of the ellipsoid and *OP* perpendicular to *SA*, the infinitesimal triangle *SDA* is similar to that finite *SOP* (Fig. 7.17). Thence, applying this property and considering that the two surfaces of the shells are s.s.p., the attraction gets the form

$$4\pi\rho\frac{da}{a}SP.$$

So that, given a straight line through *S*, the attraction of an infinitely thin shell on a point of its external surface is proportional to the segment determined on the normal at this point by the diametral plane perpendicular to such a straight line (ibid., p. 59). This implies that this attraction has its maximum when the point *S* belongs to the major axis of the ellipsoid.

[53]Chasles (1837, 1846), p. 57: “L’attraction qu’une couche infiniment mince, comprise entre deux ellipsoïdes semblables, concentriques et semblablement placés, exerce sur un point quelconque situé sur sa surface externe, est dirigée suivant la normale à cette surface en ce point”.

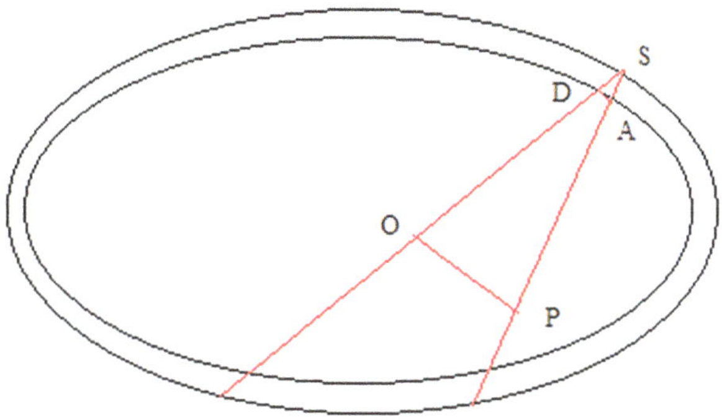

Fig. 7.17 Diagram used by Chasles to develop the last part of the reasoning concerning the attraction of an ellipsoidal shell on a point belonging to the shell's surface. With respect to Fig. 7.13, I have drawn only the elements necessary in this part of Chasles' argumentation

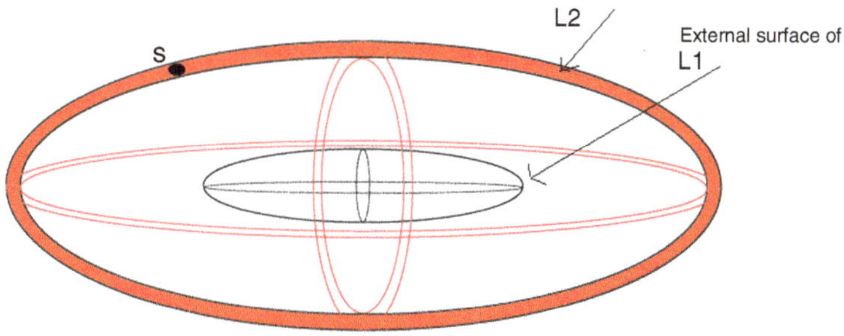

Fig. 7.18 Reconstruction of the figure described by Chasles

This proved, Chasles was able to determine the attraction of an ellipsoidal infinitesimal shell L1 on an external point, reasoning like this: be given L1 and the external point *S*. Through *S* it is possible to draw a shell L2 whose limiting surfaces have their principal sections described with the same foci as the external surface of L1 (Fig. 7.18). According to MT the attractions of L1 and L2 have the same directions and are as the masses of the two shells. Thence, as to the direction this conclusion is drawn:

> The attraction of an infinitely thin ellipsoidal shell on an external point is directed along the normal to the ellipsoid passing through this point such that its principal sections have the same foci as those of the external surface of the shell.[54]

But since this normal, as Chasles had proved in the first geometrical part of his *Mémoire*, coincides with the axis of the cone circumscribed to the external surface of the shell, he could conclude:

[54] Chasles (1837, 1846), p. 59: "L'attraction qu'une couche ellipsoïdale infiniment mince exerce sur un point extérieur est dirigée suivant la normale à l'ellipsoïde mené par ce point, de manière que ses sections principale aient les mêmes foyers que celles de la surface externe de la couche".

> The attraction of an infinitely thin ellipsoidal shell on an external point is directed along the axis of the cone whose vertex is in this point and which is circumscribed to the shell.[55]

He remarked that his synthetic proof allowed him to reach the same result as that obtained by Poisson through an analytical demonstration. Chasles wrote explicitly:

> This theorem is that at which M. Poisson arrived in the course of his analytical solution of the problem of the external points' attraction.[56]

It is now easy to calculate the value of the attraction since the ratio of two attractions is known. For if a_1, b_1, c_1 are the principal half-axes of the ellipsoid through S, which is the external surface of L2, and ρ_1 is the density of L2, then the attraction of L2 on S is

$$4\pi\rho_1\frac{da_1}{a_1}SP.$$

The masses of L1 and L2 are, respectively, $4\pi\rho bc \cdot da$; $4\pi\rho_1 b_1 c_1 \cdot da_1$. Since the attractions are as the masses the attraction exerted by L1 on S is (ibid., p. 61)

$$4\pi\rho\frac{abc}{a_1 b_1 c_1}SP\frac{da}{a}.$$

Chasles concluded this section with two interesting considerations, the former is a methodological statement which shows clearly the foundational character of his *Mémoire*, whereas the latter also involves questions concerning content.

As to the first consideration, it is enough to refer to Chasles' words because they are clearer than any comment. As he wrote:

> We could have directly determined, in magnitude and direction, the attraction of the shell on an external point, without before calculating the attraction on a point situated on the external surface and without resorting to the theorem on the attractions of two shells whose principal sections are described with the same foci. But the method we have followed has offered the advantage to use only simple geometrical considerations, without any calculation, which is the purpose at which we aimed. We will give the other manner [the analytical one] to solve the question at the end of this memoir.[57]

---

[55] Ibid., pp. 59–60: "L'attraction qu'une couche ellipsoïdale infiniment mince exerce sur un point pris au dehors de sa surface externe est dirigée suivant l'axe du cône qui a son sommet en ce point et qui est circonscrit à la couche".

[56] Ibid., p. 60: "Ce théorème est celui auquel est parvenu M. Poisson dans le cours de sa solution analytique du problème de l'attraction sur des points extérieurs".

[57] Ibid., p. 61: "Nous aurions pu déterminer directement, en grandeur et en direction, l'attraction de la couche sur un point extérieur, sans calculer préalablement l'attraction sur un point situé à la surface externe, et sans nous servir du théorème sur les attractions de deux couches dont les sections principales sont décrites des mêmes foyers; mais la marche que nous avons suivie nous a offert l'avantage de ne faire usage que de simples considérations de géométrie, sans aucun calcul, et c'est là le but que nous nous sommes proposé. Nous donnerons dans un dernier paragraphe, à la fin de ce mémoire, l'autre manière de résoudre la question".

With regard to the issue connected to content: in the previous subsection, we have seen that Chasles, also relying on the achievements of other scholars, had shown that the results concerning the attraction exerted by an ellipsoidal shell on an external point are also useful in the theory of electricity and in the heat theory. For example, as Chasles remarked, they offer the expression of the temperature in any point of a solid envelope included between two ellipsoidal surfaces whose principal sections are described with the same foci, when these surfaces are subject to constant sources of heat. Chasles thus concluded (ibid., p. 62) that it is now possible to have synthetic proofs of these properties.

Commentary: The section of Chasles' *Mémoire* analysed above testifies to my thesis according to which Chasles had the intention to develop an entire foundational programme for geometry and for relevant part of physics whose common basis is represented by the concepts, methods and transformations of projective geometry. In this subsection, Chasles proved that in order to determine the attraction exerted by an ellipsoid's shell on an external point it is possible to resort to graphical concepts. Furthermore, since important sections of the electricity and heat theories are treatable in the same manner as the attraction of an ellipsoid on an external point, such graphic concepts can also represent the basis for these branches of physics which go beyond dynamics.

To complete the panorama described so far, it is necessary to develop one last step: to pass from the attraction of an ellipsoidal shell on an external point to the attraction of an entire ellipsoid. This is what Chasles carried out in the next section of his work. After having analysed such a section it will be possible to understand until what extent the foundational programme by Chasles was successful.

7.4.4 *Further Results of the* Mémoire

The other results obtained by Chasles in his *Mémoire* concern:

1) The attraction of an ellipsoid on an external point.
2) The attraction of a finite ellipsoidal shell on an ellipsoid.
3) Case of a heterogeneous ellipsoid whose density varies according to certain laws, so to obtain an integral which can be calculated elementarily.
4) Geometrical construction of a coefficient necessary in the researches in 1), 2), 3).

I will focus on the points 1) and 4) because 1) is, so to say, the aim for which the whole *Mémoire* was written and 4) is significant as a further indication of the powers of geometrical methods.

To understand the way in which Chasles obtained his result concerning the attraction of an ellipsoid on an external point, it is necessary to clarify which were the figures he considered: 1) the ellipsoid E whose attraction has to be calculated. Be A, B, C its axes; 2) an ellipsoidal shell concentric with the ellipsoid. The axes of its external surface are indicated by a,b,c; 3) a first auxiliary ellipsoid passing through the point S, whose principal sections are confocal with those of the external surface

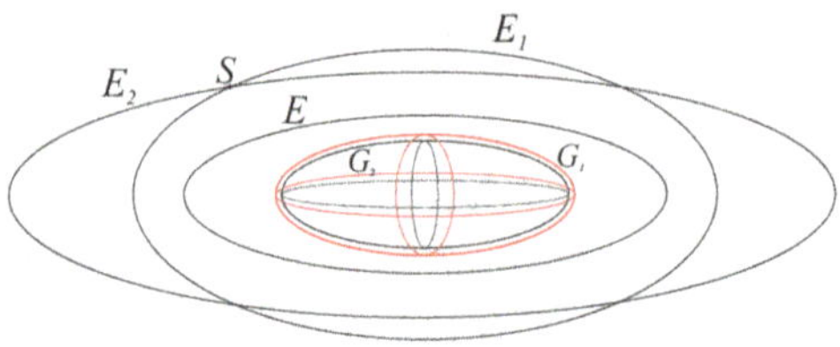

Fig. 7.19 Reconstruction of the figure described by Chasles. G_1 and G_2 represent, respectively, the external and the internal surface of the shell. The auxiliary ellipsoid E_1 has its principal sections described with the same foci as E. The auxiliary ellipsoid E_2 has its principal sections described with the same foci as G_1

of the shell. Its axes are indicated by a_1, b_1, c_1; 4) a second auxiliary ellipsoid passing through S, whose principal sections are confocal with those of E and whose axes are indicated by A_1, B_1, C_1 (Fig. 7.19).

As axes of the reference frame Chasles assumed the principal axes of E and as origin its centre O. Be S the external point of coordinates (x,y,z). As we have seen, its attraction on S is

$$4\pi\rho\frac{abc}{a_1b_1c_1}SP\frac{da}{a}$$

where the symbols have the same meaning as in the previous subsection and the attraction is directed along the normal in S to the first auxiliary ellipsoid.

Through an easy reasoning (ibid., p. 63) it is possible to determine the component of the attraction along each of the coordinate axes. Let us consider the x-axis. If e indicates the angle between SP and the x-axis, Chasles was able to determine the relation

$$\cos e = \frac{SP \cdot x}{a_1^2},$$

so that the attraction assumes the form

$$4\pi\rho\frac{abc}{a_1b_1c_1}\frac{SP^2}{a_1^2}x\frac{da}{a} \tag{7.23}$$

Therefore, the global attraction of the ellipsoid will be given by

$$\int_{a=0}^{a=A}\frac{abc}{a_1b_1c_1}\frac{SP^2}{a_1^2}\frac{da}{a}.$$

This integral is not very expressive because it seems that there is a proliferation of variables, but—Chasles recalled (ibid., p. 63)—as a matter of fact, the only variable is a, thence, it is needed to express the other variables as a function of a.

This can be obtained by thinking that the external surfaces of the shell and E are s. s.p., so that

$$\text{a)}\ \ b = a\frac{B}{A}, \qquad \text{b)}\quad c = a\frac{C}{A}$$

and that the shell and the first auxiliary ellipsoid have the principal sections described with the same foci, so that

$$\text{c)}\quad b_1^2 = a_1^2 + b^2 - a^2; \quad \text{d)}\ \ c_1^2 = a_1^2 + c^2 - a^2.$$

The problem becomes now determining the value of a_1 as a function of a. This can be done by thinking that the first auxiliary ellipsoid passes through S and, hence, its equation is $\frac{x^2}{a_1^2} + \frac{y^2}{b_1^2} + \frac{z^2}{c_1^2} = 1$. Replaced for the three axes the values in c), d) and those in a) and b), a second-degree equation in a, a_1, will be obtained, so that a_1 will be expressed as a function of a.

Finally, through a further brief reasoning Chasles proved (ibid., pp. 64–65) that

$$\text{e)}\ \ SP^2 = \frac{1}{\frac{x^2}{a_1^4} + \frac{y^2}{\left[a_1^2 + a^2\left(\frac{B^2}{A^2}\right) - 1\right]^2} + \frac{z^2}{\left[a_1^2 + a^2\left(\frac{C^2}{A^2}\right) - 1\right]^2}}.$$

After another algebraic passage, and having posed $u = \frac{a}{a_1}$, the relation between a and a_1 can be written as

$$u^2x^2 + \frac{y^2}{\frac{1}{u^2} + \left(\frac{B^2}{A^2} - 1\right)} + \frac{z^2}{\frac{1}{u^2} + \left(\frac{C^2}{A^2} - 1\right)} = a^2.$$

Differentiating this expression in u and a, you get

$$\left\{\frac{u^4}{a^4}x^2 + \frac{y^2}{\left[\frac{a^2}{u^2} + a^2\left(\frac{B^2}{A^2} - 1\right)\right]^2} + \frac{z^2}{\left[\frac{a^2}{u^2} + a^2\left(\frac{C^2}{A^2} - 1\right)\right]^2}\right\}\frac{a^3}{u^3}du = da.$$

According to e), it is possible to prove that the quantity within curly brackets is $\frac{1}{SP^2}$, so to obtain

$$\frac{a^2}{u^2}\frac{1}{SP^2}\frac{du}{u} = \frac{da}{a}$$

and replacing u with its value $\frac{a}{a_1}$, you get

$$\frac{SP^2}{a_1^2}\frac{da}{a}=\frac{du}{u}$$

Replacing this value in Eq. (7.4) which expresses the elementary attraction, Chasles (ibid., p. 66) obtained, for the expression to integrate

$$4\pi\rho x\frac{abc}{a_1b_1c_1}\frac{du}{u}=4\pi\rho x\frac{bc}{b_1c_1}du.$$

It is only a question of elementary algebra to write the value $\frac{bc}{b_1c_1}$ in function of A, B,C (which, I remember, are the axes of the ellipsoid whose attraction has to be calculated), so achieving, for the whole attraction, the integral

$$4\pi\rho xBC\int\frac{u^2du}{\sqrt{A^2+u^2\left(B^2-A^2\right)}\sqrt{A^2+u^2\left(C^2-A^2\right)}}.$$

The last step consists in determining the limits of integration for the variable u. But this is an easy enterprise, because, as we have seen, when such limits are expressed in the variable a, the integral has to be calculated between 0 and A. Since $u=\frac{a}{a_1}$, when a=0, u=0 and when a=A, $a_1=A_1$, the expression which offers the attraction of an ellipsoid along the x-axis on an external point is given by (ibid., p. 67).

$$4\pi\rho xBC\int_0^{\frac{A}{A_1}}\frac{u^2du}{\sqrt{A^2+u^2\left(B^2-A^2\right)}\sqrt{A^2+u^2\left(C^2-A^2\right)}}.$$

The integration cannot be developed with elementary methods. Analogous expressions hold for the attraction along the y and the z-axes.

A final consideration regards the value of A_1, which can be expressed as a function of A,B,C (ibid., p. 66) through the equation

$$\frac{x^2}{A_1^2}+\frac{y^2}{A_1^2+\left(B^2-A^2\right)}+\frac{z^2}{A_1^2+\left(C^2-A^2\right)}=1.$$

I will dedicate the next subsection to the geometrical solution offered by Chasles to this equation and to his explanations conceived to determine which, among the solutions of such equation, is that representing A_1.

Commentary: in this section, where Chasles found the attraction of an ellipsoid on an external point, the properties of two s.s.p. ellipsoids and of two ellipsoids whose principal sections are described with the same foci play a fundamental role. Thence, the direct use of geometry is important in this circumstance, too. However, the most remarkable utilization of geometry is indirect, insofar as it has been used to prove that the attraction of an ellipsoidal shell is $4\pi\rho\frac{abc}{a_1b_1c_1}SP\frac{da}{a}$. Most of the elements through which Chasles constructed this part of the proof are algebraic and analytical (as it is unavoidable), given that an integral has to be calculated and the calculation has to be reduced to a sole variable. Nonetheless, the analytical procedures used by Chasles rely on the geometrical bases of which I have spoken more than once. So that, Chasles is *completely legitimate to consider his proof synthetical, based on elements deriving from the then modern geometry, namely, projective geometry.*

A further significant element in favour of this thesis is represented by the geometrical demonstration, given by Chasles, concerning the solution of the equation through which the value of A_1 can be deduced. This is a clear indication of Chasles' idea to use geometry in the widest possible manner also in questions which might be faced with purely algebraic or analytical methods. Thence, his reductionist programme has many facets: the most relevant is that projective geometry offers the foundation for the entire geometry and for important branches of physics, but there is also a general methodological issue: geometry is used by Chasles in the widest possible way.

7.4.5 *Geometrical Determination of the Value of* A_1

The last step of Chasles' geometrical theory of the ellipsoid's attraction is the determination of the value of A_1 in function of known elements, which is what he realized in the fifth section of his *Mémoire*. This section is strictly connected with the first one in which the geometrical properties of the second-degree surfaces useful in the problem of the ellipsoid's attraction were proved by Chasles in a purely synthetic manner. Thence, this fifth section is the perfect final step for the whole synthetic itinerary conceived and developed by Chasles. At the beginning of the section he wrote:

> In this memoire I have used only simple geometrical considerations, as I had announced. But the formulas for the coefficient A_1, which is not given explicitly and which depends on a third degree equation, have still to be determined. You may require to complete this geometrical solution of the problem of the ellipsoid attraction and to construct the solution of this equation through a graphical construction.
>
> The properties of the second degree surfaces proved in the first section will offer an easy solution to this problem.[58]

Therefore, Chasles' statement is clear: he will use graphical constructions to find the solutions of the equation expressing A_1.

---

[58] Chasles (1837, 1846), p. 76: "Je n'ai fait usage dans ce mémoire que de simples considérations géométriques, ainsi que je l'avais annoncé; mais il reste dans les formules un coefficient A, qui n'est pas donné explicitement et qui dépend d'une équation du troisième degré. On peut demander, pour compléter cette solution géométrique de problème de l'attraction des ellipsoïdes, d'effectuer par une construction graphique la résolution de cette équation. Les propriétés des surfaces du second degré démontrées dans le premier paragraphe, procurent une solution facile de cette question".

He considered the ellipsoid E whose attraction on the external point S has to be studied and whose axes are A,B,C and the auxiliary ellipsoid E' whose principal sections are confocal with those of E and whose principal axes are A_1, B_1, C_1. He considered two parallel planes tangent, respectively, to E and E'. Named p, p_1 their distances from the centre of E and E', according to a result obtained in section 7.4.1. (see my Eq. 7.18), it holds

$$p^2 - p_1^2 = A^2 - A_1^2 \rightarrow A_1^2 = A^2 - \left(p^2 - p_1^2\right).$$

The elements A and p are known, the problem is, hence, transformed into the possibility to draw a tangent plane to E', so to determine p_1. However, the point S is given and, through one of the theorems proved in the first section and that Chasles has most widely utilized, the normal to E' in S is the principal axis of the cone circumscribed to E, so that the problem is transformed into the following one:

> Given a second-degree surface and considered a point of space as the vertex of a cone circumscribed to the surface, it is required to determine the principal axis of this cone.[59]

Therefore, Chasles recalled that, given a set of quadrics whose principal sections are confocal, all their circumscribed cones passing through a point have the same principal axes. Since, among these surfaces, there are two for which an axis of theirs is null, in the considered set there are two conics. In this case, the circumscribed cones are those whose bases are the two conics. Chasles continued (p. 77) by claiming that, given the equation of E

$$\frac{x^2}{A^2} + \frac{y^2}{B^2} + \frac{z^2}{C^2} = 1, \quad A > B > C$$

the two conics are the ellipse in the plane x,y whose equation is

$$\frac{x^2}{A^2 - C^2} + \frac{y^2}{B^2 - C^2} = 1$$

and the hyperbola in the plane x,z whose equation is (for more mathematical details, see the table)

$$\frac{x^2}{A^2 - B^2} - \frac{y^2}{B^2 - C^2} = 1.$$

[59] Chasles (1837, 1846), p. 77: "Une surface de second degré étant donnée, et un point pris dans l'espace étant regardé comme le sommet d'un cône circonscrit à la surface, déterminer les axes principaux de ce cône".

I give here some brief mathematical explanations for the equations of the two conics which Chasles obtained.
A system of confocal centred quadrics (by "confocal" I mean that they have the principal sections described with the same foci, to use Chasles' language) is formed by ellipsoids, one-sheeted hyperboloids and two-sheeted hyperboloids. This system can be described by one equation with a real parameter d:
$\frac{x^2}{a^2+d}+\frac{y^2}{b^2+d}+\frac{z^2}{c^2+d}=1,\ a>b>c>0$
If $d > -c^2$, the quadric is an ellipsoid.
If $-b^2 < d < -c^2$, the quadric is a one-sheeted hyperboloid;
If $-a^2 < d < -b^2$, the quadric is a two-sheeted hyperboloid.
If $d = -c^2$, you get the ellipse $\frac{x^2}{a^2-c^2}+\frac{y^2}{b^2-c^2}=1$ in the plane $z=0$.
If $d = -b^2$, you get the hyperbola $\frac{x^2}{a^2-b^2}+\frac{z^2}{c^2-b^2}=1 \to \frac{x^2}{a^2-b^2}-\frac{z^2}{b^2-c^2}=1$ in the plane $y=0$.
These are exactly the two conics of which Chasles spoke, whose equations, hence, do not depend on the value of the parameter d.
If $d = -a^2$, you get the equation of the plane $x=0$.

Since the two cones whose vertex is S and whose bases are the two conics that have the same axes as the cone circumscribed to the ellipsoid, the problem is furtherly reduced to determine the system of principal axes shared by the two cones (ibid., p. 78). Chasles named these conics "*conique focales* de l'ellipsoide" ("focal conics of the ellipsoid", ibid., p. 78).

The following part of Chasles' reasoning is the most evident litmus paper of the prominent role of projective geometry within the whole of his ellipsoid's attraction theory. For it is based on a series of results directly deriving from the theory of polars. The results are the following (ibid., p. 78):

1) Any transversal plane cuts the two cones whose bases are the focal conics of the ellipsoid along two conics and the three common axes of the two cones in three points. Each of these points has as polar with respect to the two conics the straight line joining the other two points.
2) In the plane of the two conics only a system of three points exists fulfilling this property. These points are the intersecting points of the diagonals and of the opposite sides of the complete quadrilateral having its vertices in the intersection points of the two conics (ibid., p. 78).

The determination of these points is enough to solve the problem of finding the normal drawn from S to the auxiliary ellipsoid. For Chasles took into account the two focal conics of the ellipsoid and considered one of the two—he chose the hyperbola—as the basis of the cone with vertex in S. He then constructed the conic determined by the intersection of the cone with the plane of the ellipse. This conic cuts the focal ellipse of the ellipsoid in four points which can be considered as the vertices of a quadrilateral. The intersection point of the two diagonals of this quadrilateral belongs to the required normal, which, hence, is determined.

For the plane tangent at S to the auxiliary ellipsoid is determined and, hence, the distance p_1 of this plane from the centre will be known. It is enough to draw a tangent plane to the ellipsoid E which is parallel to that determined and tangent to the auxiliary ellipsoid E'. Thence, named p its distance from the centre, the solution of the equation

$$A_1^2 = A^2 - \left(p^2 - p_1^2\right)$$

can be obtained. Hence, the problem is solved.

This concludes my analysis of Chasles' memoir, where the foundation of an important physical theory, which is also connected with electricity and heat theory, was offered by relying upon projective concepts. Thence, through his successful results Chasles proved that his geometrical foundational programme was nothing utopistic, but a precise mathematical and philosophical idea through which he reached the following results: 1) deduction of several fundamental metrical properties from a graphic background; 2) foundation of the theory of the rigid body's movement 3) global explanation of the theory of the systems of equivalent forces; 4) new interpretation of the principle of virtual velocities; 5) geometrical theory of the ellipsoid's attraction and connected problems. Thence, for important branches of geometry and sections of the then known physics, Chasles' foundational programme provided the mathematicians and the physicists with new stimulating, general and working ideas.

At the end of his memoir (Chasles, 1837, 1846, pp. 86–87), the author added a brief consideration which was written immediately before the publication, thence, in 1845–1846: he wrote that some time after the presentation of this memoir at the Académie, he offered a simpler solution of the attraction of ellipsoids in Chasles (1838). In a further memoir he proposed general theorems on the bodies' attraction where the infinitely thin shells have the property to exert no action on the internal points and that, as to the external points, have properties similar to the generalized Maclaurin theorem. These theorems concern not only the problem of gravitational attraction, but also the behaviour of the electricity in a conductor (Chasles, 1842). Let us now see the most interesting of the developments obtained by Chasles.

7.5 The Period 1838–1840

In the period 1838–1840 Chasles published three contributions on the attraction of the ellipsoid. The first (Chasles, 1838) entitled *Nouvelle solution du problème de l'attraction d'un ellipsoïde hétérogène sur un point extérieur* presents a series of results connected to the relations between ellipsoidal shells given by two concentric and homothetic ellipsoids and ellipsoidal shells given by ellipsoidal surfaces whose principal sections are confocal. We have already seen several of these relations while dealing with Chasles (1837f) and the complete treatment offered in Chasles (1837, 1846). This memoir published in 1838 is the one in reference to which Chasles claimed that his results are presented in a simpler form. I will not face this memoir because I should repeat a series of reasonings, most of which had already been expounded in Chasles (1837f) and others—in a more complete way—in Chasles (1837, 1846). It is only worth pointing out that the basis of Chasles' train of thoughts is the geometrical proposition, according to which if two ellipsoids have their principal sections described with the same foci and if S and m are two arbitrary points of the ellipsoid and S',m' the two corresponding points in the meaning of Ivory, then the two segments Sm' and $S'm$ are equal (Chasles, 1838, p. 903). This is a further confirmation of the geometrical itinerary towards the solution of the ellipsoid attraction problem developed by Chasles.

The second writing of this period (Chasles, 1839) is a very brief contribution where the author proved no theorem, but enunciated some interesting propositions worth being presented and commented on.

The other paper (Chasles, 1840) is long and important enough. Thence, I will refer to it insofar as some elements useful to clarify Chasles' foundational programme exist.

Let us begin with Chasles (1839). This brief paper refers to two general theorems on the attraction and on the heat theory. Both theorems regard bodies of any form and can, hence, be interpreted as a prosecution of Chasles' order of ideas expressed at the end of Chasles (1838). The first theorem reads as follows: given any body limited by a close surface, consider a level surface S—in respect of the attraction—of the body and suppose that such surface includes a homogeneous infinitely thin shell whose thickness, in different points, is inversely as the distances of these points from the infinitely close level surface S, then: 1) the shell exerts no action on any point within its internal wall; 2) the attraction exerted by the shell on an external point has the same direction as the attraction exerted by the body itself on such point. These attractions are as the attracting masses (Chasles, 1839, p. 209).

Clearly these theorems are the extension to closed bodies of any form of theorems valid for the ellipsoidal shells. Chasles pointed out that this theorem has a useful application: for when a level surface is known, it is possible to reduce the calculation of the body's attraction to that of an infinitely thin shell. The given body itself can be an infinitesimal shell. This is the case while dealing with an infinitesimal shell included between two homothetic and concentric ellipsoids, so to obtain the attraction of the shell on an external point and, by integrating, the attraction of the whole ellipsoid composed of homogeneous shells of variable densities (in particular cases of density variation, an integration through quadrature is possible). Chasles added now an observation which confirms furtherly my thesis concerning his foundational programme based on geometry. We read:

> So that this problem, considered from a general point of view, is devoided of the great difficulties which it presents when it is addressed with considerations tied to the specific, particular form of the body. This case seems to offer a new example of the advantages of the *generalization* in geometry in order to simplify the theories, which, in this way, hold an intuitive clearness.[60]

[60]Chasles (1839), p. 210: "De sorte que ce problème, envisagé ainsi d'un point de vue général, se dépouille des grandes difficultés qu'il avait présentées quand on l'attaquait par des considérations restreintes et toutes spéciales à la forme particulière du corps. Cet cas parait offrir un nouvel

Here there is a further role for geometry: its capability of generalization. Its prospective, synthetic view can be useful to have some ideas which, with specific methods tied to the particular form of a figure or a configuration—though translated into analytical terms—could not be obtained. As at the end of Chasles (1837f), geometry is seen as a source of new ideas and as a fundamental heuristic means.

The second theorem concerns the heat theory, a subject already addressed in Chasles (1837f). It states that: be given a homogeneous envelope included between two closed surfaces subject to two heat sources. Suppose that the body reached the temperature in which it is in thermic equilibrium. Within the body, consider its internal isothermal surfaces. Consider also the heat quantity which, in a unitary time, traverses one of these surfaces under the supposition that such heat quantity forms, on this surface, an infinitely thin shell whose attractive power satisfies the natural law,[61] then the shell fulfils three properties: 1) it exerts no action on a point within the shell's internal surface (this property is analogous to the first one of the previous theorem); 2) the attraction exerted on an external point is the same, in direction and intensity, whatever the considered isothermal surface is; 3) The attraction exerted by this shell on a surface element of an isothermal surface external to the shell is directed along the normal to this element. Its intensity is equal to the heat quantity which flows from this element during the unitary time by 4π (ibid., p. 210). Chasles claimed that this theorem solves easily a problem "which had undoubtedly presented some difficulties to the analysis" (ibid., p. 210): as well known, an electric fluid distributed on the surface of a conductor forms a thin shell which exerts no action on the internal points. If the body is an ellipsoid, the shell is limited by a second ellipsoidal surface similar to the first one. But more in general, Chasles continued: can a given surface belong to a sole infinitely thin shell which forms the electric fluid that satisfies the property to exert no action on an internal point? For example, if one surface of the shell is an ellipsoid, is it necessary that the second one is an ellipsoid too? (Ibid., p. 211). The previous theorem gives a negative answer to this question because the two surfaces limiting the homogeneous envelope have only to be closed, there is no hypothesis on their form. Therefore, as Chasles concluded:

> A given surface can always cover an infinity of infinitely thin shells which satisfy the property to exert no action on any internal point.[62]

Therefore, Chasles showed once again that a geometrical approach can offer solutions and does indeed solve problems when an analytical one might present some difficulties. Thence, though brief, Chasles (1839) is a paper whose value should not be underestimated if one wants to understand Chasles' conception.

---

exemple des avantages de la *généralisation* en géométrie, pour simplifier les théories et y répandre une clarté intuitive".

[61] With the expression "loi naturelle", Chasles was referring likely to Newton's law of the inverse square.

[62] Ibid., p. 211: "Une surface donnée peut toujours recouvrir une infinité de couches infiniment mince jouissant de la propriété de n'exercer aucune action sur un point intérieur quelconque".

Chasles (1840) is an important paper because in it the author gave a more solid and general foundation to the synthetic methods he had used in Chasles (1838). The difference between Chasles (1840) and Chasles (1837, 1846) consists in the fact that in the former memoir he used pure geometry to find the theory of ellipsoid attraction. However, he did not offer the whole geometrical apparatus necessary to give a complete foundation to the geometrical properties of the second-degree surfaces used in this theory because this apparatus had already been supplied in Chasles (1837, 1846). Thence, consistent with the development line I am presenting, Chasles (1840) represents a further step towards the geometrical foundation of the ellipsoid attraction theory.

At the beginning of this memoir Chasles clearly stated that his intention was to furtherly develop his research on the ellipsoid attraction through "la méthode synthétique" (Chasles, 1840, p. 465) of Maclaurin and to generalize such a method beyond what he had already done in Chasles (1838). The different analytical methods used by D'Alembert, Lagrange, Legendre and Laplace offer the solution, but they do not allow a comprehensive understanding of the attraction phenomenon. They do not reveal its true nature (ibid., pp. 465–466). Chasles added here a noteworthy passage which I quote because it expresses his point of view in the clearest manner:

> The procedure I have followed consists in a comparison—molecule by molecule—of the attractions of two ellipsoids considered in Maclaurin theorem. In a sense, this shows the origin and the first cause of this peculiar and celebrate theorem. This solution is easy in itself, but it requires the knowledge of several new properties of the second-degree surfaces. The proof of these properties entails geometrical considerations which cannot be achieved without some difficulties and which might induce to think this solution to be hardly suitable for a practical use, at least until the geometrical methods, neglected from one century, have regained strength. In this proposed solution, I avoid these considerations by comparing, first of all, the attractions of two Maclaurin's ellipsoids. In this manner, a sole property of these surfaces is sufficient. This is a well-known proposition. It is that on which the beautiful Ivory theorem is based.[63]

What Chasles meant is clear: a geometrical foundation of the ellipsoid attraction which is satisfactory and complete from a logical point of view needs several difficult (projective) properties of the second-degree surfaces. This is exactly what he did in 1837 and published in Chasles (1837, 1846). However, since modern

[63] Ibid., p. 466: "La marche que j'ai suivie consiste à comparer de molécule à molécule les attractions des deux ellipsoïdes considérés dans la théorème de Maclaurin; ce qui montre, en quelque sorte l'origine et la cause première de ce singulier et célèbre théorème. Cette solution est simple en elle-même ; mais elle nécessite la connaissance de plusieurs propriétés nouvelles des surfaces de second degré ; et la démonstration de celle-ci entraine dans des considérations géométriques qui se sont pas sans quelque difficulté, et qui peuvent faire regarder cette solution comme peu susceptible d'un usage pratique, du moins jusqu'à ce que les méthodes géométriques, fort négligées depuis un siècle, aient repris faveur. Dans la solution actuelle, j'évite ces considérations, en comparant, tout d'abord, de couche à couche, les attractions des deux ellipsoïdes de Maclaurin. De cette manière, une seule propriété de ces surfaces me suffit, et cette proposition n'est pas inconnue: c'est précisément celle sur laquelle repose le beau théorème de M. Ivory, sur l'attraction des ellipsoïdes".

projective geometry was not universally well known among mathematicians and physicists, he also offered a further synthetic solution. It is less complete and foundational than the complete one developed in Chasles (1837, 1846), but it is easier, simpler and more intuitive than the solutions obtained through analytical methods. Such a solution does not reach the profound truth of the question (as it is the case for Chasles, 1837, 1846), but, is more expressive than those purely analytical. This is the clear line of thought proposed by Chasles.

His idea was to calculate the attraction on an external point exerted by an infinitely thin ellipsoidal shell included between two similar and concentric surfaces. The ellipsoid is considered as composed of these shells whose density either varies according to specific laws or is constant. Therefore, an integral will be eventually necessary (ibid., p. 467).

The geometrical proposition exploited by Chasles is the same used in Chasles (1838): if two ellipsoids have their principal sections described with the same foci, given the points S,m belonging to the former, and given on the latter the two Ivory-correspondent points S',m', it is $Sm' = S'm$.

Chasles (ibid., p. 467) considered two homothetic and concentric ellipsoids A, B

$$\frac{x^2}{a^2}+\frac{y^2}{b^2}+\frac{z^2}{c^2}=1, \quad \frac{x^2}{a^2}+\frac{y^2}{b^2}+\frac{z^2}{c^2}=n^2. \tag{7.24}$$

Given any point of space $m = (x, y, z)$, let be $m' = (x', y', z')$ its corresponding point so that the three identities hold:

$$\frac{x}{x'}=\frac{a}{a'};\ \frac{y}{y'}=\frac{b}{b'};\ \frac{z}{z'}=\frac{c}{c'},$$

being a', b', c' three arbitrary coefficients. Under these conditions two homothetic and concentric ellipsoidal surfaces A', B' composed of the m'-points will correspond to A and B. A first property of these two pairs of surfaces is that any part of the volume included between two of them (A' and B') is to the part of volume included between the first two surfaces in the constant ratio $\frac{a'b'c'}{abc}$ because the relations between the coordinates of two corresponding points supply the equation

$$dx'dy'dz' = \frac{a'b'c'}{abc}dxdydz$$

Since in the problem of the ellipsoid attraction, the surfaces whose principal sections are described with the same foci are important, Chasles chose the three values a', b', c' in a particular manner and specifically such that

$$a^2 - b^2 = a'^2 - b'^2$$

$$a^2 - c^2 = a'^2 - c'^2$$

These posed, the surfaces *A*, *A'* as well as *B*, *B'* are surfaces whose principal sections are described with the same foci (ibid., p. 468). Now Chasles added a further assumption: be the surfaces *A* and *B* as well as *A'* and *B'* infinitely close, so that they envelop an infinitesimal shell. He supposed the value *n* in Eq. (7.24) to be infinitely smaller than 1. Under this supposition it is easy to prove that the thickness of the two shells calculated along the same principal axis are as the half-diameters of the external surfaces directed along such axis. This means that, if *da* and *da'* indicate the mentioned thickness, it is

$$\frac{da}{da'} = \frac{a}{a'}.$$

Chasles considered two fixed points *S* and *S'* belonging, respectively, to *A* and *A'*. If *m,m'* are two other points belonging to those surfaces and *dv*, *dv'* the volume elements of the two shells at the points *m,m'*, it is easy to prove that

$$\frac{dv}{mS'} : \frac{dv'}{m'S} = \frac{abc}{a'b'c'}.$$

Considering all the molecules of the two shells, you obtain the equation

$$\sum \frac{dv}{mS'} : \sum \frac{dv'}{m'S} = \frac{abc}{a'b'c'}.$$

This is the ratio of the volumes of the two shells. Chasles remarked:

> This equation expresses a geometrical property of two shells which—*alone*—can offer the entire solution of the ellipsoid's attraction.[64]

It seems to me that no commentary is necessary: Chasles identified a geometrical property which, alone, is able to supply a solution to ellipsoid's attraction theory. It is paramount to point out that by "geometrical property" Chasles meant a property which is the basis for the entire theory; for sure, he did not mean that the use of calculus and algebra can be avoided. This is impossible. The aspects which differentiate Chasles' treatment from those offered by the other contemporaries of his epoch are:

1) A purely synthetical geometric property is the initial theorem as well as the conceptual core of the entire theory.
2) The calculations can be simplified by using such geometrical property and, more generally, geometrical means.

[64] Ibid., p. 469: "Cette équation exprime une propriété géométrique de deux couches, qui va nous suffire, *seule*, pour résoudre toute la question de l'attraction des ellipsoïdes". Italics in the text.

This does not mean that it is either possible or convenient to avoid the use of reference frames and—hence—of equations of curves and the use of derivatives and integrals. The conceptual basis of the problem is reduced to a geometrical property, but the problem is not purely geometrical, hence, it is impossible to avoid other mathematical means.

The geometrical situation is this: we have two shells C, C', the external shell C is composed of two homothetic and concentric surfaces A, B; the internal shell C' is composed of two homothetic and concentric surfaces A',B' corresponding, respectively, to the first two. The surfaces A, A' are those external to the two shells; B,B' those internal. The surfaces A,A' have the principal sections described with the same foci; they are concentric, but, obviously, not homothetic. S' is a fixed point on A'; m a mobile point on A; S is a fixed point on A; m' a mobile point on A'.

Chasles claimed that, if the position (x, y, z) of the point m is the variable quantity, the differential coefficients of the function $\sum \frac{dv}{mS'}$ are the components of the attraction of the shell C on the point S'. He proved this assertion in a long note (ibid., pp. 470–471) starting from Newton's law, according to which the attraction the molecule dv exerts on S' is $\frac{dv}{mS'^2}$. From here, by means of infinitesimal geometrical reasoning, really worthy of a Leibniz or a Newton, Chasles proved his assertion.

Given the positions of the two shells C, C', the point S' lies within the internal wall of C. Thence, because of a well-known theorem proved by Newton, C will exert no action on S', so that the differential coefficients of $\sum \frac{dv}{mS'}$ are null. Hence, such a sum is constant for any position of S' within C. Therefore, the value of $\sum \frac{dv'}{m'S}$ is also constant for any position of S on A. Ergo, Chasles can conclude that, given an infinitely thin shell included between two homothetic and concentric ellipsoidal surfaces, the sum of the shell's molecules divided by their distances from a point external to the surface [the point S in our case] is constant for any position of this point on an ellipsoid whose principal sections are described with the same foci as those of the external surface of the shell (ibid., p. 471). The value of the sum is

$$\sum \frac{dv'}{m'S} = \frac{a'b'c'}{abc} \sum \frac{dv}{mS'}.$$

If C'' is a shell having the same properties as C', namely included between two homothetic and concentric surfaces A'', B'' described, respectively, with the same foci as A and B, it is easy to prove that the following relation holds (ibid., p. 472).

$$\frac{\sum \frac{dv'}{m'S}}{\sum \frac{dv''}{m''S}} = \frac{a'b'c'}{a''b''c''}, \tag{7.25}$$

where the meaning of the symbols is clear. Thence, given two ellipsoidal infinitely thin shells with the described features, the sums of the molecules divided by their distances from the same point external to the two shells, these two sums are as the volumes of the two shells (ibid., p. 472).

It is thus possible to determine the *direction* of the two attractions, the *ratio of their intensities*, and the *absolute value of such attractions* (ibid., p. 473).

With regard to the direction, since $\sum \frac{dv'}{m'S}$ is constant for any position of S on A, the surface A is normal to the direction of the attraction exerted by the shell C' on S. This depends on the fact that, passing from S to an infinitely close point s, the expression $\frac{d\sum \frac{dv'}{m'S}}{Ss}$ is the component of the attraction exerted by C' on S along the direction of the element Ss. If this element belongs to A, it is $\sum \frac{dv'}{m'S}$ constant and its derivative is null, so that the attraction along Ss will be null. This is the case for all the directions of the element if it belongs to A. Consequently, the attraction is directed along the normal at A. Thence, Chasles can conclude that the attraction exerted on an external point S by an infinitely thin shell included between two homothetic and concentric ellipsoids is directed along the normal to the ellipsoid drawn through S so that its principal sections have the same foci as those of the external surface of the shell (ibid., p. 473). In an interesting note (ibid., pp. 473–474) he claimed that Poisson obtained another expression for the direction of a shell's attraction which—as we have seen—coincides with the axis of the cone circumscribed to the shell and having its vertex in the attracted point.[65] Steiner (1834) also presented a simple geometrical proof based on Newton's theorem according to which a shell exerts no action on a point within its wall. Chasles (1837f, p. 269) also offers a demonstration. However, the proof here expounded has the advantage of making the level surfaces relative to the attraction of a shell known. These surfaces had not yet been introduced in the theory of attraction. The results achieved in this section of physics can be extended to the theory of electricity and heat, given their analogy with the theory of attraction.

The same theorem might be expressed also stating that the level surfaces relative to the attraction of a shell are ellipsoids whose principal sections are described with the same foci as those of the external surface of the shell, which is itself a level surface (ibid., p. 474).

With regard to the ratio of the intensity between the attractions exerted by two ellipsoidal shells, Eq. (7.25), which can be written as

$$\frac{d\sum \frac{dv'}{m'S}}{dx} : \frac{d\sum \frac{dv''}{m''S}}{dx} = \frac{a'b'c'}{a''b''c''},$$

offers the answer, where the x-axis can assume any direction. This equation claims that two shells with the described characteristics exert on the same external point two attractions having the same direction and whose intensities are as the masses of the two shells (ibid., p. 474).

As to the problem of determining the attraction of a single shell, Laplace had already supplied the answer (ibid., pp. 475–476), but, Chasles continued, it can be obtained directly without resorting to Laplace theorem. The reasoning here presented by Chasles is rather easy:

[65] Chasles refers to the results in Poisson (1835a).

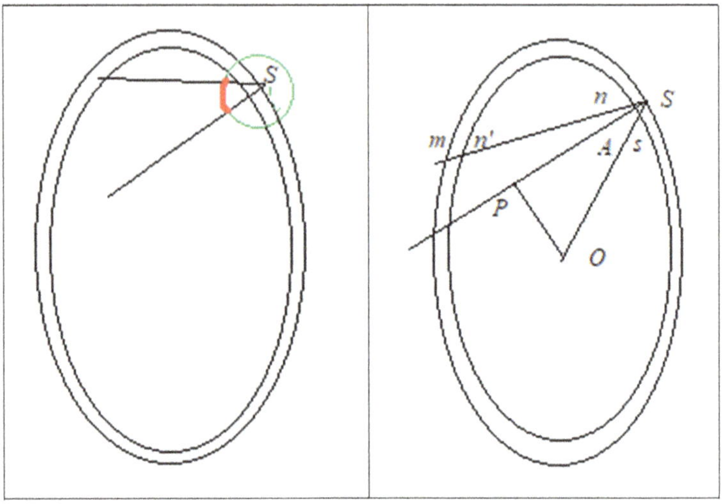

Fig. 7.20 Reconstruction of the figures described by Chasles to determine the attraction of a single ellipsoidal shell

Be (left part of Fig. 7.20) the point *S*, which is located on the external surface of the shell, the vertex of a cone. Be also *S* the centre of a sphere whose radius is equal to 1 and σ a superficial element of the sphere which is infinitesimal with respect to ε that is the thickness of the shell. If the radius of the sphere is *r* (green sphere, Fig. 7.20 on the left), the surface of its element intercepted in the cone will be $r^2\sigma$ (in red, Fig. 7.20 on the left). This means that the portion of the shell included in the cone can be decomposed into infinitesimal elements of volume whose value is $r^2\sigma dr$. Each element is responsible of the force $\rho\sigma dr$, being ρ the density of the shell. The total force can be obtained integrating on dr. Be (right part of Fig. 7.20) m,n,n' the points where the generatrices of the cone cut the external and the internal walls of the shell. The integral has to be calculated between the points *S* and *n* and between n' and *m*. The result of this integration will be $\rho\sigma(Sn + mn') = 2\rho Sn \cdot \sigma$, as in the ellipsoid $mn' = Sn$.

Be *SPA* the normal to the shell at *S*, where *A* is the point in which this normal cuts the internal surface. Given that the infinitesimal triangle *nSA* can be considered rectangle, it is

$$SA = Sn \cdot \cos \widehat{nSA}.$$

Replacing *Sn* in the expression of the attraction one gets

$$2\rho \frac{SA \cdot \sigma}{\cos \widehat{nSA}}.$$

The component of the attraction along the normal is, hence, $2\rho SA \cdot \sigma$.

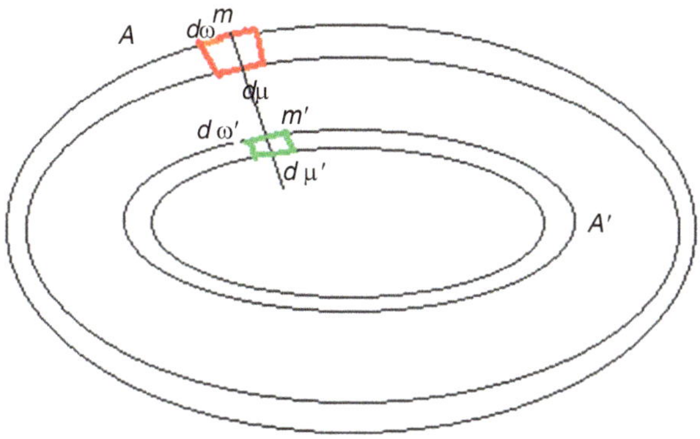

Fig. 7.21 Reconstruction of a figure showing Chasles' corresponding elements. The black straight line represents the normal at *A* and *A′* through *m* and *m′*. The figures represent ellipsoids, but the construction is valid for any shape because, in this case, the shape is not a significant element. It is sufficient that the shells are closed. The internal unnamed surfaces are the internal surfaces of the two shells

The area σ, indicated in red in the left part of Fig. 7.21 can vary between 0 and a hemisphere. This integral has the value 2π, hence the global attraction is

$$4\pi\rho \cdot SA$$

being *SA* the thickness of the shell. This is exactly Laplace formula (ibid., pp. 477–478).

Chasles also gave a more expressive form for the previous formula and in the last part of his work (ibid., pp. 479–484) calculated the attraction for the entire ellipsoid too, but for my aims this is enough.

I have considered several details of Chasles (1840) because this paper represents an important step towards the geometrization of this part of dynamics. Chasles' purpose here was to offer a brief way to determine the attraction of an ellipsoidal shell relying upon synthetic geometrical properties, and he was successful. Therefore, Chasles (1840) perfectly fits to his idea of conceiving the geometrization of science. This notwithstanding, the foundation proposed here is based on the fact that the attraction of an ellipsoidal shell can be calculated starting from a geometrical theorem, and the projective geometrical solution had already been presented in 1837, though published only in 1846 (Chasles, 1837, 1846). Thence, the words by Chasles that I have reported at the beginning of my analysis of Chasles, 1840 leave no doubt that, in 1840, he continued to rely on the whole geometrical apparatus expounded in Chasles (1837, 1846), by developing and specifying it. To be more precise: insofar as geometry was concerned, such apparatus had already been constructed by Chasles and the other mathematicians who had developed projective geometry. It was necessary to apply it at the problem of the ellipsoid attraction.

7.6 *Théorèmes généraux sur l'attraction des corps*, 1842

This paper is significant because Chasles, with his typical geometrical way of thinking, deduced some general properties of the attraction for bodies and shells of any form and claimed that the features of the ellipsoid's attractions could be deduced from these more general theorems. Not all the aspects of the attraction problem can be faced in general terms without specifying the kind of attracting body. We will see until which level Chasles was able to generalize. This paper is based on few geometrical theorems, though a broad analytical apparatus is unavoidable. Thence, it can be considered as a part of Chasles' geometrical foundational programme, even if the kind of theorems proposed here are slightly different from those we have seen and despite the abundant use of analytical technics. Given the interest of this paper, I will focus on many details.

Chasles's (1842) principal purpose is to prove the two theorems he had enunciated in the brief note (Chasles, 1839). However, there are many other interesting observations in this memoir.

The first section (Chasles, 1842, pp. 18–23) is a remarkable historical and methodological introduction to the memoir: Chasles began his exposition by pointing out that, from when it was discovered, the inverse square law resulted to be also valid in electrostatic and magnetic theory. Thence, it has assumed an absolutely central role within physics. Nonetheless, with the exclusion of the attraction exerted by body of well-specified shapes, the scientific community is far from having obtained a general theory of the attraction (or analogously of electrostatic) valid for all the bodies. This depends on the enormous difficulties to treat the differential equations deriving from the specific addressed problem (ibid., p. 18).

As a matter of fact, this memoir offers a unified treatment of the aspects shared by gravity and electrostatic theory.

Chasles stated that the only two solid results are those due to Laplace.

The first is the so-called Laplace equation for a point external to the attracting body, which, indicated by V the gravitational potential, is, in modern notation, $\nabla^2 V = 0$. As Poisson proved, it becomes $\nabla^2 V = -4\pi\rho$ (ρ being the density of the body) for a point internal to the body (ibid., p. 19).

The second result concerns the property of an infinitesimal shell to exert no attraction on internal points and to exert, on a point of its external surface, an attraction directed along the normal drawn from the point to the surface and whose intensity is equal to the shell's thickness in the point by 4π.

These properties also concern electric and magnetic shells.

Chasles highlighted the novelty of his approach: he considered the level surfaces. This choice is appropriate because the attraction exerted by a body on each of its level surfaces is directed along the normal from the point belonging to the surface to the body (ibid., p. 19). Chasles vindicated this merit, recalling his memoir (Chasles, 1838), where he introduced the equipotential surfaces for a gravitational and electric field deriving the idea from the level surfaces of a temperature field (ibid., p 19). Now his intention was to prove that the concept of equipotential surface is also useful to prove the two theorems he had expounded in Chasles (1839). Restricting to gravity theory, Chasles was going to prove the two following propositions that are valid for any closed shell constructed on the level surfaces of any body:

1. Each of these shells exerts no action in the space included within its internal surface. Consequently, it has the form of an electric shell in equilibrium.
2. The attraction each shell exerts on an external point has the same direction as the attraction exerted on this point by the whole body. The intensities of the two attractions are as the shell's mass is to the body's mass. Thence from the shell's attraction it is possible to deduce the attraction of the body and vice versa. He added that, if the shell has an appropriate density, its attraction can be the same as that of the entire body (ibid., p. 20).

These theorems, Chasles claimed, also solve the problems connected to the distribution of charge in a conductor which has been studied for a long time by the scientists. With regard to ellipsoidal shells included between two concentric and homothetic surfaces, the situation was well known and Coulomb developed an experiment which confirmed the validity of the two theorems for these specific ellipsoidal shells. Chasles also recalled that the theoretical calculations by Poisson confirmed the Coulomb's experiments, so that for ellipsoidal shells, several results had been achieved. Thence, Poisson can be considered as the real inventor of the mathematical theory of electricity, but a general theory extended to bodies of any shape was still lacking (ibid., p. 21).

Chasles claimed that the researches developed with analytical methods are satisfactory from a physical point of view because—he seemed to think—they include the most common forms for the body attraction and for the distribution of electricity in a conductor. However, from a geometrical point of view they are unsatisfactory because they are not general. Therefore, once again, geometry represents the basis for a generalization: in this case, the generalization of the inquire from shells of ellipsoidal form to shells of any form. It is clear that, during the treatment, an advanced analytical apparatus is necessary. Nonetheless, the basic ideas behind the extension of Laplace's theorems are geometrical. Thence, once again, geometry plays a decisive role: in this case through its capability of generalization. This is a further facet of Chasles' foundational programme. He clearly claimed:

> But, though the researches of the geometers of which we have spoken might appear satisfactory from a physical standpoint, nevertheless, from a geometrical point of view, they include only surfaces of a very particular and restricted form. It is necessary, of course, to want to know the equilibrium shells for surfaces of more general form.[66]

In contrast to this, his theorem is general because:

[66]Ibid., p. 21: "Mais, quelque satisfaisantes que puissent paraître, sous le point de vue physique, les recherches des géomètres dont nous venons de parler, cependant, sous le point de vue géométrique, elles ne comprennent que des surfaces d'une forme particulière et très-restreinte. On doit naturellement désirer de connaître les couches d'équilibre pour des surfaces d'une forme plus générale".

> Our theorem solves the question for an infinity of surfaces, for all the level surfaces relative to the attraction of a body or a system of bodies. If you consider any of these surfaces, provided that they envelop the body or the system of bodies, the theorem offers the possibility to assign the shell on which the electricity will be distributed.[67]

As in Chasles (1838), he considered the potential V (whose paternity he ascribed to Laplace) as the function which represents the sum of the molecules of a body divided by their distance from any point. The level surface can be constructed because it fulfils the equation V = constant (ibid., p. 21). Chasles posed this problem: given a level surface as external surface of a shell, how is it possible to construct the internal surface of the shell so that the shell satisfies the previous condition 2? (Ibid., p. 22). The determination of such shell is the main subject of this memoir. A shell of this type, Chasles claimed, is that which forms the electric fluid in a conductor. The heat theory—developed by Lamé—deals with similar shells whose external surfaces are the isothermal surfaces. Chasles also mentioned Gauss' work in 1840 on the general attraction of the bodies as well as some contact points between Gauss' works and his own, in particular insofar as the theorems which are mentioned in Chasles (1839) and proved in this memoir are concerned.[68]

It is worth following Chasles' reasoning because it is connected both with geometrical and analytical considerations as well as with what today we call vector analysis.

Chasles considered a body, each point of which attracts with Newton's law. Then, level surfaces which are external, internal and partially external, partially internal to the body exist, unless the body itself is a level surface. In this case only internal and external surfaces exist. Since the problems of the gravitational as well as the electric fields are interesting for points eternal to the body, Chasles took into account the external level surfaces (ibid., p. 23).

Be A a level surface, and be considered its normal in a point m. Indicate dn the portion of such normal external to A and included between A and an infinitely close level surface. Be K a constant coefficient which will be considered a second-order infinitesimal. Now, along the normal, take within A a segment whose length is $\frac{K}{dn}$. While varying m, its extremity describes a further surface. This is the surface internal to A, which, together with A, forms the required infinitesimal shell, whose variable thickness is $\frac{K}{dn}$ (ibid., p. 23). Chasles proved that this shell plays an important role in gravity and electricity theory.

If $d\omega$ and $d\mu$ indicate, respectively, the surface element and the volume element of A around m, it will be

[67] Ibid., p. 21: "Notre théorème résout la question pour une infinité de surfaces, pour toutes les surfaces de niveau relatives à l'attraction d'un corps ou d'un système de corps. Qu'on prenne arbitrairement l'une de ces surfaces, pourvu qu'elle enveloppe le corps ou le système de corps, le théorème donne le moyen d'assigner la couche suivant laquelle s'y distribuera l'électricité".

[68] The seminal work in which Gauss explained the complete mathematical foundation of the potential theory is Gauss (1840). Chasles is referring to such work.

$$d\mu = \frac{Kd\omega}{dn}.$$

If A' is a further level surface and m' lies on the orthogonal trajectory of A', it holds

$$d\mu' = \frac{K'd\omega'}{dn'},$$

where the meaning of the symbols is obvious (ibid., pp. 23–24).

Two points are defined as "corresponding" by Chasles if they belong to two level surfaces and to the same line orthogonal to both surfaces. Thence, m and m' are two corresponding points (Fig. 7.21). Corresponding surface elements are those included in an infinitesimal canal whose edges are the orthogonal trajectories to the level surfaces. Finally, the corresponding volume elements are those included in the little canal.

Therefore, the body's attraction on m is directed along the normal dn at A. Passing from A to an infinitely close level surface the attraction of the body on m will be $\frac{dV}{dn}$, while the attraction of the element $d\omega$ will be $\frac{dV}{dn}d\omega$. The same reasoning holds for the shell whose external surface is A'. As we have seen in Chasles (1838, p. 308), he had proved that the attraction exerted by the body on these elements is equal. Thence, it is (ibid., p. 24)

$$\frac{dV}{dn}d\omega = \frac{dV'}{dn'}d\omega'. \tag{7.26}$$

This means that, passing from the elements to an entire level surface, the sum of the body's attraction on a level surface is constant independently of the surface. Chasles' intention is to prove that its value is $4\pi M$, being M the mass of the body, namely it holds

$$\iint \frac{dV}{dn}d\omega = 4\pi M.$$

Commentary: As a matter of fact, this result is equivalent to Gauss' flow theorem applied at the gravitational field because the field is the gradient of the potential. Thence, transcribed into modern terms and using the concept of scalar product, what Chasles is claiming is that the global flow Φ across a level surface, expressible by $\int \vec{G} \cdot d\omega$, is equal to $4\pi M$, which, apart from the sign minus and the presence of Newton's constant in the second member of the equation (which are merely formal aspects), is exactly the result obtained by Gauss. This might be considered, or at least, interpreted, as a result of vector analysis.

I would like also to point out the way in which Chasles introduced the constant K. He defined it to be a constant which is a second-order infinitesimal ("un coefficient constant qui sera un infiniment petit de deuxième ordre", ibid., p. 23). As we have already stressed in the commentary to section 3.4.4., this is a strange expression because the reader might think that Chasles believed in the existence of actually infinitesimal quantities included in the structure of the *continuum*, which would be something extremely problematic because the structure of the continuum itself would be substantially modified. As in the case of my previous commentary, I am inclined to think that Chasles did not have the intention of introducing new and problematic mathematical entities. Rather the coefficient K can be interpreted as a value which remains infinitesimal with respect to the infinitesimal dn. In other terms, from an intuitive point of view: it is a constant which can be considered very small also with respect to the segment dn which is a potential infinitesimal of first order with respect to a finite length. The role of K would be, so to say, an instrumental role, an element of the symbolism useful to develop the calculations. This explained, it is indubitable that this entity does not have a completely clear ontological status, though its mathematical usefulness is clear. This proves that, in the epoch I am considering, the language and also some concepts of mathematical analysis had not yet reached a completely perspicuous status. It should also be added that mathematicians, as Chasles and as Newton himself, who had a strong geometrical inclination might have had a tendency to ontologize—though, probably, only from a linguistic point of view—the entities of which they were speaking because, in spite of the fact that they did not believe in the real existence of such objects, their geometrical imaginations guided them to feign such entities as geometrical objects which are very little. They had a sort of intuition of these objects, but I tend to exclude that they believed in their real mathematical existence.

I will present most of Chasles' proof because it offers a brilliant reasoning: the function V is the sum of terms of the form $\frac{dM}{r}$, where M is a mass-element of the body and r its distance from m. Therefore, the term $\frac{dV}{dn}d\omega$ is composed of a series of terms of the form $dM\frac{d\frac{1}{r}}{dn}d\omega$. Thus, $\int\frac{dV}{dn}d\omega$ is composed of a series of terms of the form $dM\frac{d\frac{1}{r}}{dn}d\omega$ of which every term produces the integral $dM\int\frac{d\frac{1}{r}}{dn}d\omega$. It has to be extended to the whole surface A, so that the final integral equation will be

$$\iint\frac{dV}{dn}d\omega=\iiint dM\iint\frac{d\frac{1}{r}}{dn}d\omega$$

After having developed this clever argument, Chasles worked on the left member of the equation to prove, applying Gauss theorem, that its value is $4\pi M$ (ibid., p. 25). In words, the result of the theorem is so expressed by Chasles:

> The sum of the numerical values of the attractions a body exerts on the surface elements of one of its level surfaces, when this surface embraces the body everywhere, is equal to the mass body by 4π.[69]

[69] Ibid., pp. 25–26: "La somme des valeurs numériques des attractions qu'un corps exerce sur les éléments superficiels d'un de ses surfaces de niveau, quand cette surface entoure le corps de toutes parts, est égale à la masse du corps multipliée par 4π".

He added that this theorem has a general validity because it is applicable to the attraction of a body on any surface—not necessarily, hence, a level surface—provided that this surface envelops the body everywhere. Thence, when a body is enveloped everywhere by a closed surface, the sum of the body's attraction on the surface elements, calculated along the normals to these elements, is equal to the body's mass by 4π. The proof is the same as the previous one (ibid., p. 26 n.). Since dV is constant insofar as it is referred to an equipotential surface, it is easy to reach the conclusion that

$$dV = \frac{4\pi MK}{\mu}.$$

Hence, dV is a function of the shell's volume constructed so that the level surface is the external part of the shell (ibid., p. 26).

Now Chasles, reworking Eq. (7.26), was able to draw a first important conclusion which connects the element volume of a shell with the whole volume of a shell. For this equation can be written as

$$\frac{Kd\omega}{dn} : \frac{K'd\omega'}{dn'} = \frac{KdV}{K'dV'}.$$

Since the second member is constant for any level surface, the first also is and it expresses the ratio of two corresponding volume elements relative to the surfaces A and A'. As this ratio is constant, it is evident that it is the ratio of the entire volumes themselves (ibid., p. 27), so that it holds

$$\frac{d\mu}{\mu} = \frac{d\mu'}{\mu'}. \tag{7.27}$$

Thence Chasles concluded:

> [...] if you conceive a canal orthogonal to all the level surfaces, the volumes intercepted by this canal in two shells will be as the volumes of the two shells.[70]

He pointed out that Eq. (7.27) represents a *geometrical property* and that such a property led him to discover several features of the function V applied to the bodies and to the shells as well as to theorems concerning the bodies attraction (ibid., p. 27).

Thence, once again, a geometrical property (which, certainly, has been obtained also using analytical means) is the basis for several developments of mechanics, electric theory and heat theory.

Now Chasles continued his refined reasoning by combining geometrical and analytical considerations: from a geometrical point of view, he considered two shells

[70] Ibid., p. 27: "[...]si l'on conçoit un canal quelconque orthogonal à toutes les surfaces de niveau, les volumes qu'il interceptera dans deux couches seront entre eux dans le rapport des volumes des deux couches".

on A and A' with the further supposition that the two surfaces are infinitely close. Being m a point of A, while S is a point external to the two shells, the distance m-S was indicated by Chasles through ρ. He considered the expression $\frac{d\mu}{\rho} : \mu$. The variation $\delta\left(\frac{d\mu}{\rho} : \mu\right)$ of this expression while passing from volume $d\mu$ to the corresponding infinitely close volume $d\mu'$ depends only on ρ because $\frac{d\mu}{\mu}$ is constant. Hence, after easy passages Chasles obtained (ibid., p. 28)

$$\delta\left(\frac{d\mu}{\rho} : \mu\right) = -\frac{d\mu}{\mu} : \frac{\delta\rho}{\rho^2}.$$

If i denotes the angle $m'mS$ between the straight line $mS = \rho$ and $mm' = dn$, it will be

$$-\delta\rho = dn \cos i.$$

This depends on the fact that the directional derivative is maximal along dn, being dn perpendicular to the two surfaces. The sign minus depends on the opposite orientation of mm' and mS.

After a brief calculation Chasles (ibid., p. 27) obtained the variational equation

$$\delta\left(\frac{d\mu}{\rho} : \mu\right) = \frac{K}{\mu}\frac{d\omega \cos i}{\rho^2}.$$

Extending this expression to all the elements of the surface, you get the integral equation (ibid., p. 28)

$$\delta\iint\left(\frac{d\mu}{\rho} : \mu\right) = \frac{K}{\mu}\iint\frac{d\omega \cos i}{\rho^2}. \tag{7.28}$$

From this equation, Chasles deduced a series of important results for shells of any form. Let us see some of them.

In the case in which S is external to the two shells, it is known from the flux theorem that the integral in the right member of Eq. (7.28) is null, hence it is also the expression on the left, which implies (ibid., p. 28) that

$$\iint\left(\frac{d\mu}{\rho} : \mu\right) = \text{constant}$$

The expression $\int\frac{d\mu}{\rho}$ indicates the sum of the molecules composing the shell divided by their respective distances from S. Chasles indicated this integral by v, so that v' will be an analogous sum relative to another shell. Since the quantity v/μ is constant for any shell, he arrived at the formula, which represented one of the aims of his paper:

$$\frac{v}{\mu} = \frac{v'}{\mu'}. \tag{7.29}$$

Namely:

> The sum of a shell's molecules divided by their respective distances from an external point is to the volume of the shell in a constant ratio, whatever the shell is.
>
> We can conclude consequently that the shells have all the same external level surfaces.[71]

Through a similar reasoning applied at the case in which the point S is internal to the shell Chasles arrived at establishing the important identity, which involves the potential:

$$\frac{v}{\mu} = \frac{V}{M}, \tag{7.30}$$

where M indicates the entire mass of the body. In words, Chasles expressed his result like this:

> The sum of a shell's molecules divided by their respective distances from an internal point [that is v] is to the mass of the shell [μ], as the sum of the molecules of the body divided by their respective distances from a point of the external surface of the shell [which is an equipotential surface] is to the mass of the entire body.
>
> It follows an important property and feature of our shell, namely that the sum of the molecules of a shell divided by their respective distances from an internal point is constant, whatever this point is.[72]

Chasles obtained a further interesting result through this brilliant reasoning: suppose that in Eq. (7.29) the shell μ' envelops μ and that the external point S is infinitely close to μ'. Since the function v' varies with continuity, the theorem still holds when S lies on μ'. The situation is the same if in Eq. (7.30) the point S internal at the external surface of μ can approach such a surface indefinitely. Thence, considering two points S and S_1, of which S is very close to the surface and S_1 lies on the surface, the difference in the value of v will be the smaller the nearer the point S is at the surface. But since the value of v is constant for any internal position of the point, it is possible to conclude—given the continuity of the function at stake—that it is also the same for the point S_1. Thence the following important conclusion follows:

---

[71] Ibid., p. 28: "La somme des molécules d'une couche, divisées par leurs distances respectives à un point extérieur, est au volume de la couche, dans un rapport constant, quelle que soit la couche.

Nous pouvons conclure tout de suite de là que: Les couches ont toutes les mêmes surfaces de niveau extérieures".

[72] Ibid., p. 29: "La somme des molécules d'une couche, divisées par leurs distances respectives à un point pris dans son intérieur, est à la masse de la couche, comme la somme des molécules du corps, divisées par leurs distances respectives à un point de la surface externe de la couche, est à la masse du corps.

Il résulte de là une propriété importante et caractéristique de nos couches, savoir, que: La somme des molécules d'une couche, divisées par leurs distances respectives à un point pris dans son intérieur, a une valeur constante, quel que soit ce point".

> The sum of the molecules of a shell divided by their respective distances from any point of the external surface of the shell is constant and equal to the sum of these molecules divided by their respective distances from a point within the shell.
>
> The first part of this theorem shows that the external surface of a shell is a level surface relative to the attraction of this shell.[73]

Through a further brief reasoning, Chasles was able to prove that Eq. (7.30) is also valid in the case of an external point S, so that he stated:

> The sum of the molecules of a shell divided by their respective distances from an external point is to the sum of the molecules of the body divided by their distances from the same point as the mass of the shell is to the mass of the body.[74]

This is an important result, which actually fulfils the initial scope of Chasles' memoir insofar as his purpose was to find a relation between the potential and the mass of a shell and the potential and the mass of the whole body for the attraction towards an external point.

Such a result implies immediately that what is valid for ellipsoidal shells whose external surface is a level surface of the ellipsoid is also valid for shells of any shape. That is: the attractions exerted by a body and by a shell on a point external to the shell have the same direction and are, in magnitude, as the mass of the body is to the mass of the shell. This theorem is the successful conclusion of Chasles' long conceptual iter.

Before offering my comments, I would like to conclude with this long and perspicuous explanation given by Chasles:

> We have always spoken of attraction in the enunciation of our theorems, but it is evident that our statements also suit while considering bodies with a repulsive power which acts according to the same inverse square law. It is possible to consider what we have always called attracting body as a set of different masses, some of which endowed with attractive power and other with repulsive power. The proof of the different theorems is the same. The only condition to observe is that the level surfaces are closed and that they envelop all the masses. They will be closed if they have no sheet at infinity. This is always the case when the masses have the same sign, that is when all of them are either repulsive or attractive. For, in this case, in the equation V=constant all the terms have the same sign. Consequently, for the surface to have points at infinity, necessarily the second member has to be equal to zero. This does not happen when the masses have the same sign, but it is possible if they have different signs.

[73] Ibid., p. 30: "La somme des molécules d'une couche, divisées par leurs distances respectives à un point quelconque de la surface externe de la couche, est constante et égale à la somme de ces molécules divisées par leurs distances respectives à un point pris dans l'intérieur de la couche.

La première partie de ce théorème montre que: La surface externe d'une couche est une surface de niveau relative à l'attraction de cette couche".

[74] Ibid., p. 31: "La somme des molécules d'une couche, divisées par leurs distances respectives à un point extérieur, est à la somme des molécules du corps, divisées par leurs distances respectives au même point, comme la masse de la couche est à la masse du corps".

> It, thence, follows that the previous general theorems lead to the solution of the attraction problem of the ellipsoids, at which I had arrived through considerations based on specific properties of these bodies.[75]

Commentary: This memoir deserves a particular role within Chasles' production on the problem of attraction: first of all, it is not dedicated to the attraction of an ellipsoid, though it draws inspiration from the results obtained by Laplace, Poisson, Chasles himself on the attraction of the ellipsoid and from Lamé's work on heat theory. Indeed, Chasles obtained several theorems on the general attraction of shells which are level surfaces of bodies having any form and from the attraction of these shells he deduced some theorems on the attractions of the bodies themselves. Gauss had proved many of Chasles' theorems before him. However the results of the latter are original since he had already mentioned two of them in Chasles (1839) and because his method and Gauss' are different. The other difference between this memoir and the others which he dedicated to the argument of a body's attraction is that in Chasles (1842) the use of geometry is less wide than in the other memoirs. Therefore, one might ask why I have dedicated such a long treatment to this work. The reason is, in any case, connected to the use of geometry because, as always, the basic and essential idea that guided Chasles to his discoveries is a geometrical one. As a matter of fact, the idea of considering an infinitesimal segment of the normal dn external to the surface A and the infinitesimal segment K/dn internal to A (through which the second surface of the shell is constructed) is typically geometrical. The construction of the shell is, thence, developed by a pure geometrical reasoning. To be more precise: this happens through a reasoning of infinitesimal geometry, so to say, in Newton's style. Therefore, though the development of the proofs relies most on considerations concerning differential equations and integrals, the basic idea is geometrical. Furthermore, this idea allowed Chasles to generalize properties valid for the attraction of an ellipsoid to closed bodies of any shape. Thence, this paper is noteworthy as it shows the generalizing power of geometry when this discipline is used with Chasles' mastery. Thus, it is correct to claim that geometry is also the basis of this memoir, which, therefore, is an element of Chasles' foundational programme,

---

[75] Ibid., p. 33: "Nous avons toujours parlé d'*attraction* dans l'énoncé de nos théorèmes, mais il est évident qu'ils conviennent au cas où l'on considérerait des corps doués du pouvoir *répulsif*, suivant la même loi du rapport inverse du carré des distances. On peut même considérer ce que nous avons toujours appelé le *corps attirant*, comme un assemblage de diverses masses douées les unes du pouvoir attractif, et d'autres du pouvoir répulsif. La démonstration des divers théorèmes reste la même; la seule condition à observer, c'est que les surfaces de niveau soient fermées et qu'elles enveloppent toutes les masses. Elles seront fermées si elles n'ont pas de nappes à l'infini. Cela a toujours lieu quand toutes les masses sont de même signe, c'est-à-dire toutes attractives ou toutes répulsives; car alors, dans l'équation V= *constante*, tous les termes sont de même signe, et conséquemment, pour que la surface eût des points à l'infini, il faudrait que le second membre fût égal à zéro: ce qui n'a pas lieu quand les masses sont de même signe, mais ce qui peut avoir lieu quand elles sont de signes différents.

On reconnaîtra aisément que les théorèmes généraux qui précèdent conduisent à la solution du problème de l'attraction des ellipsoïdes, à la quelle je suis parvenu par d'autres considérations fondées sur les propriétés particulières de ces corps". Italics in the text.

even though in a manner different from the others. It is a very rich and global programme constructed as a tree, which has its roots (projective geometry), its trunk (the metric properties proved through specific graphic constructions), its principal branches (theory of rigid body, theory of the system of forces, theory of the ellipsoid's attraction) and its secondary branches (principle of virtual velocities, and the results achieved in Chasles, 1842). The ground on which this tree grows is represented by the philosophy of duality.

7.7 Final Comments

The importance ascribed by Chasles to the problem of the ellipsoid attraction within his geometrical foundational programme is conspicuous if you analyse the *Rapport* (Chasles, 1870), a late work in which Chasles showed that his mind did not change in the course of the years with regard to his foundational idea. For he dedicated many pages to the scientists and mathematicians who faced this problem and valuated his own work important insofar as it offered a geometrical solution to the question. The best way to offer a commentary of Chasles' work is, first of all, to propose the commentaries which he himself left.

The first reference to the ellipsoid problem in Chasles (1870) is to the doctoral thesis of Rodrigues (1815) where he treated the question of the ellipsoid in a manner substantially analogous to Gauss'—though, Chasles commented, probably Rodrigues did not know Gauss' work (Chasles, 1870, p. 38). However, he first introduced a new element: the consideration of an infinitesimal ellipsoidal shell included between two homothetic and concentric ellipsoids and of the element of volume belonging to such a shell. This way of decomposition will be fundamental for Poisson (1835a), where the power of this way of decomposition is fully explored and realized beyond Rodrigues' treatment.

The first important commentary of Chasles on his own work concerns Note XXXI of the *Aperçu* and an *Addition* to the same work (see, respectively, Chasles, 1837a, pp. 384–399 and p. 556), because in such a note he proposed a series of theorems concerning the second-degree surfaces. Among these theorems, one also finds those I have presented in Sect. 7.4.1 and that Chasles used for his geometrical approach to the problem of the ellipsoid attraction. He wrote explicitly:

> The theorems included in the note XXXI have the necessary generality, apart from two of them. While indicating these two theorems for the researches of the geometers, the author [Chasles himself] adds that their generalization would permit undoubtedly to remove the difficulties which the problem of the attraction of the ellipsoids on the external points presents. As a matter of fact, in a simple note at the end of the historical part of this work (p. 556) there is a proposition which, because of its generality, satisfies to the double

> remarked *desideratum*. Thus, the generalization of these theorems allowed to treat the problems of the ellipsoid attraction with new and very fruitful considerations.[76]

The second chapter of Chasles (1870) is almost entirely dedicated to comment the discoveries connected with the *Aperçu*. Among these investigations, he dedicated a particular care to those related to the ellipsoid attraction. While commenting the application of synthetic methods to this problem he wrote:

> As a matter of fact, the feature of the speculations of pure geometry is that they can be naturally extended and offer, with a continuous concatenation, unexpected resources. Thus, the question of the attraction is seen by the author [Chasles himself] under many viewpoints, which will produce several different Memoirs, and will be extended to the general problem of the attraction of a body of any shape.[77]

After this consideration, he commented on all his memoirs on the ellipsoid attraction, starting from Chasles (1837e). He pointed out that through a single change of variable[78] of which no one had thought before, he was able to deduce from the formulas for the ellipsoid attraction the attraction of an infinitely thin ellipsoidal shell included between two homothetic and concentric ellipsoidal surfaces. Such an attraction results perpendicular to the ellipsoid through the attracted point and confocal to the external surface of the shell. This consideration allowed him to develop a series of reasonings on the level surfaces relative to the attraction of a shell and of a heterogeneous ellipsoid. Poinsot considered these results important. He wrote:

> First of all, as to the *direct* and complete analytical solution of the problem of the homogeneous ellipsoids' attraction, it is necessary to point out that M. *Legendre* first reached this result. He arrived at the formulas of the quadrature which express the components of the attraction of an ellipsoid on any external point. From these formulas—through some very easy transformations—it is possible to find everything found afterwards on this subject, as M. Chasles has shown in the last issue *Journal de l'École Polytechnique*. Actually, M. Legendre arrived at his formulas be means of very long and complicated calculations. While, I arrived *directly* at this formula, namely without resorting to Maclaurin theorem, which has been recognised by M. Chasles in the historical part of his Memoir [. . .].[79]

---

[76]Chasles (1870), p. 95: "Les théorèmes contenus dans cette Note XXXI ont toute la généralité désirable, à l'exception de deux. En signalant ces deux-là aux recherches des géomètres, l'auteur ajoute que leur généralisation permettrait sans doute de lever les difficultés que présente le problème de l'attraction des ellipsoïdes sue les points extérieurs. Effectivement, dans une simple note, à la fin de la partie historique de l'ouvrage (p. 556), se trouve une proposition qui, par sa généralité, satisfait au double *desideratum* signalé. Et depuis, ces théorèmes généralisés ont permis de traiter le problème de l'attraction des ellipsoïdes par des considérations nouvelles et très-fécondes".

[77]Ibid., pp. 101–102: "C'est effectivement le propre des spéculations de pure Géométrie de s'étendre naturellement et d'offrir, par un enchaînement continu, des ressources inattendues. Aussi la question de l'attraction se présenta-t-elle à l'auteur sous plusieurs points de vue, qui donnèrent lieu à divers Mémoires, et s'étendirent même au problème général de l'attraction d'un corps de forme quelconque".

[78]Chasles is referring to the change $v = \frac{A_1}{A} u$.

[79]Poinsot (1838), p. 869–70: "Et d'abord, en ce qui regarde la solution analytique *directe* et complète du problème de l'attraction des ellipsoïdes homogènes, il faut bien remarquer que

And furthermore:

> It is known that the consideration of a shell derives naturally for the formulas relative to the homogeneous ellipsoid, as the author himself [Poisson] recognises in his Note and as M. Chasles had already remarked in the cahier de l'École Polytechnique I have quoted above.[80]

Thus, Poinsot fully recognized the relevance to Chasles' first work on the ellipsoid attraction.

While commenting on his own work (Chasles, 1837f), he pointed out once again the geometrical aspect of his discoveries because he wrote:

> The author [Chasles himself] determines directly the attraction's expression of the infinitely thin shell by means of simple geometrical considerations; while in the previous memoir the formulas for such attraction had been deduced from those known relative to the attraction of an ellipsoid. He is led to a new property of the *corresponding* points which, in his memoir on the attraction of the homogeneous ellipsoids, Ivory had considered on two confocal ellipsoids. Through these properties he was able to reduce the calculation of the attraction on external points to the case of internal points. This property is: *the locus of a series of corresponding points on the homofocal ellipsoids is a trajectory orthogonal to their surface.*[81]

Chasles added that this last theorem can be extended to the confocal paraboloids and hyperboloids and that the properties of the corresponding points pave the way to generalize the theorems of the ellipsoids to the attraction of bodies of any shape. Furthermore, he stressed that in Chasles (1837f) analogies between the properties of the attraction of an ellipsoidal shell and the heat laws for a body in temperature equilibrium are shown as well as the features of the level surfaces relative to the attraction of a body (ibid., p. 103).

M. *Legendre* est le premier qui y soit arrivé. Il est parvenu aux formules de quadrature qui expriment les composantes de l'attraction d'un ellipsoïde sur un point quelconque extérieur et de ces formules, par quelques transformations très simples, on peut tirer tout ce qui a été trouvé depuis sur cette matière, comme M. Chasles l'a fait voir dans le dernier numéro du *Journal de l'École Polytechnique*. M. Legendre, il est vrai, n'est parvenu à ses formules que par des calculs très longs et très compliqués; mais il y est parvenu *directement*, je veux dire sans employer le théorème de Maclaurin c'est ce que M. Chasles a reconnu dans la partie historique de son Mémoire [. . .]". Italics in the text.

[80] Ibid., p. 871: "On sait d'ailleurs que la considération d'une couche pouvait s'offrir naturellement par les formules relatives à l'ellipsoïde homogène, comme l'auteur lui-même en convient dans sa Note, et comme M. Chasles l'avait déjà remarqué dans le cahier de l'École Polytechnique que j'ai cité plus haut".

[81] Chasles (1870), pp. 102–103: "L'auteur détermine directement, par de simples considérations de Géométrie, l'expression de l'attraction de la couche infiniment mince, qu'il avait déduite, dans le Mémoire précédent, des formules connues de l'attraction d'un ellipsoïde. Il est conduit à une propriété nouvelle des points *correspondants* que, dans son Mémoire sur l'attraction des ellipsoïdes homogènes, Ivory avait considérés sur deux ellipsoïdes homofocaux, et par lesquels il ramenait immédiatement le calcul de l'attraction sue des points extérieurs au cas des points intérieurs. Cette propriété consiste en ce que: *Le lieu d'une série de points correspondants sur des ellipsoïdes homofocaux est une trajectoire orthogonale à leur surface*".

The way in which he commented on his own works is, hence, a clear evidence of the importance he ascribed to his foundational geometrical programme.

Afterwards Chasles traced a very synthetic history of the problem of the ellipsoid attraction until Poisson (1835a). In an interesting note (Chasles, 1870, p. 105, note 3) he referred to Poisson's opinion according to which only the analysis (*analyse* is the term used by Poisson) can solve difficult problems regarding physics, whereas the synthesis (*synthèse* is Poisson's term) is not suitable. Poisson claimed that, with the important exception of Newton and Maclaurin, all the most modern results in physics and, specifically, those concerning the ellipsoid attraction, have been obtained by analytical methods. As a matter of fact, Poisson stated that, for modern physics, only analytical methods can offer general solutions. In contrast to this, despite the results achieved through analytical methods, Chasles asserted that a direct and rigorous proof of Maclaurin theorem (in fact, the generalized Maclaurin theorem) was still missing. Chasles' memoir presented at the Académie Royale des Sciences in 1837 and published in 1846 (Chasles, 1837, 1846) fulfilled this gap. He pointed out that he was able to obtain his results thanks to some geometrical theorems proved in the *Aperçu*. He also referred to the already mentioned report written by Poinsot (1838) on his work and wrote:

> [This work] was the subject of a report by M. Poinsot, where you find some considerations on analysis and synthesis suitable to inspire faith in the resources of pure geometry. It has been too neglected from two centuries in favour of the improvement of the new calculus by Leibniz and Newton, to which the geometers, with natural emulation, have consecrated all their efforts. But the words themselves of the illustrious reporter find here their natural place because they certify a progress of geometry and, above all, because they show the need of a simultaneous development of this part of mathematics and of the analysis in a proper sense.[82]

As a matter of fact, Poinsot claimed that Chasles' remarkable memoir offers a new example of the elegance and clearness of geometry to solve the most difficult and obscure questions. Though the modern analytical methods by Lagrange, Laplace and Legendre leave nothing to be desired, they cannot be considered as a proof of the superiority of analysis if compared to the ancient methods. For the clear and brilliant method used by Chasles permitted a briefer and more elegant solution than the analytical procedures. After all, neither analysis nor synthesis can be neglected, but they have to be used together as the most complete instrument of the human spirit (Poinsot, 1838; in Chasles, 1870, pp. 106–107).

---

[82] Ibid., p. 106: "Il a été le sujet d'un rapport de M. Poinsot, où se trouvent des considérations sur l'analyse et la synthèse, bien propres à inspirer confiance aux jeunes géomètres dans les ressources de la pure Géométrie, trop négligée depuis près de deux siècles en faveur de perfectionnement des nouveaux calculs de Leibnitz et de Newton, auquel les géomètres, avec un émulation naturelle, ont consacré tous leurs efforts. Mais les paroles mêmes de l'illustre rapporteur trouvent ici leur place naturelle, parce qu'elles constatent un progrès de la Géométrie, et surtout parce qu'elles montrent la nécessité d'une culture simultanée de cette partie des Mathématiques et de l'analyse proprement dite".

Therefore, Poinsot ascribed an extreme relevance to Chasles (1837, 1846), for he saw in this memoir, the text which affirmed the importance of pure geometry—in the form of projective geometry—within the context of modern mathematics. This is a further confirmation that Poinsot and Chasles shared the idea that geometry was a fundamental branch of mathematics also for application to physics. Chasles went beyond Poinsot because he constructed a true foundational programme.

Chasles also analysed his other memoirs on the attraction of the ellipsoid, but what referred to here is enough to fully grasp his ideas on this question. Obviously, he did not restrict to value the importance of geometry within his own works, but also in reference to those of other scholars. For example, while dealing with Bonnet's memoir on the orthogonal isothermal surfaces (Bonnet, 1845a) and on the theory of elastic bodies (Bonnet, 1845b) he, sharing and referring to a judgement expressed by Lamé, pointed out the meaning of geometry within Bonnet's work and the use of infinitesimals in issues connected to geometrical considerations (Chasles, 1870, p. 204–205).

Chapter 8
Conclusion

In mathematics and science, the concept of foundation is used with two slightly different meanings.

1) A discipline is founded when a set of axioms, definitions and procedures is offered which clarify the logic and the methods typical of such discipline. For instance, Euclid founded geometry because he first offered an axiomatization of such branch of mathematics, clarified what properties were deducible by resorting only to the first four axioms (*Elements* I, props. 1–28) and introduced the concept of proportion (book V) through which the most general transformation of his geometry—the similarities—can be studied (book VI). He also extended his notions to space geometry (books XI-XIII). This does not mean that Euclid invented geometry. This part of mathematics had already a long history before Euclid,[1] but until his work no mathematician had felt the need to specify the logical and conceptual bases of geometry. Thus, Euclid's is a foundational work.

If we think of mathematical analysis, the picture is analogous. The great mathematicians of the seventeenth and eighteenth centuries created calculus and its main concepts, but the logical bases of calculus were not sound until the work of mathematicians such as Bolzano, Cauchy and Weierstrass who lived in the nineteenth century and clarified fundamental concepts as those of continuity of a function, limit and derivative in a completely satisfying way.

This implies that the foundational aspect arises after that a branch of mathematics has already reached a vast set of results and the mathematicians begin to research whether these results are well founded and are logically consistent. From a chronological standpoint, the foundational analysis is, thus, always successive to the

[1] I am not claiming that Euclid was not an original geometer. He was. Likely the theory of parallels is due to Euclid. However, in the context considered in this Conclusion, Euclid's foundational contribution is the most relevant one.

P. Bussotti, *Chasles and the Projective Geometry*,
https://doi.org/10.1007/978-3-031-54266-4_8

creative and inventive phase. Federigo Enriques expressed clearly the necessity to distinguish the inventive phase from the foundational one and the idea that the former precedes the latter:

> Archimedes, following his rigid logical criterion as well as the prevalent scientific opinion in the academic milieu of Alexandria, thinks that the true author of a theorem is the one who first supplied a (true), namely an impeccable, demonstration rather than the one who first arrived at the theorem, through a more or less precise method. Nowadays, many mathematicians, who are not interested in history, reach the same conclusion as Archimedes. With the best intentions, they would deprive the founders of infinitesimal calculus of their discoveries in favour of the critical thinkers such as Cauchy, Weierstrass or Dini who have made calculus rigorous two centuries later.[2]

There are also some cases in which the foundational aspect is simultaneous with the creative one. Newton founded physics in a rigorous manner, through definitions and axioms or laws of motion. At the same time, he was one of the most creative physicists ever existed. Nonetheless these are exceptions. Generally speaking, history teaches us that the foundational activity begins after that the creative activity has produced an abundant mess of results.

2) The outlined picture implies that each branch of mathematics has a proper, a local foundation. Classical geometry had, mechanics had, calculus had. Therefore, as pointed out in Sect. 2.4, it is inappropriate to think that the foundational problems were dealt with only in the period 1870–1910, the one known as the epoch of the debate on the foundations of mathematics. However, it is true that such period had a peculiarity: the logical foundation of concepts used in every branch of mathematics was looked for. Weierstrass, Cantor and Dedekind defined the notion of real number in three different ways. However, all the three reduced the real numbers to particular infinite classes of rational numbers and, from an extensional standpoint, all the three definitions identify the usual real numbers. The entity "real number" is reduced to the more primitive entity "infinite class of rational numbers". In that period, the notion of actual mathematical infinity was thoroughly unravelled by Cantor who created set theory and posed the notion of set at the basis of the whole mathematics. The actually infinite objects, that caused so many paradoxes and doubts before Cantor, were framed within a precise mathematical theory, which was furtherly specified and clarified when it was axiomatized by Zermelo in 1908. Cantor's discussions on the concept of mathematical existence, developed in Cantor (1883), inspired Hilbert's interpretation of this so crucial notion. Frege's conception was even more extreme: he

[2]Enriques (1924–1927), I, I, p. 22: "Archimede, sia per proprio rigido criterio logico, sia per ossequio all'opinione scientifica dominante nell'ambiente accademico alessandrino, ritiene che il vero autore d'un teorema sia, non colui che per primo vi è giunto con un procedimento più o meno rigoroso, ma colui che ne ha fornito una (vera) dimostrazione, cioè una dimostrazione impeccabile. Molti matematici contemporanei sono dello stesso avviso, e – non preoccupati della storia – giungerebbero, colle migliori intenzioni del mondo, a spogliare per esempio i fondatori del calcolo infinitesimale, delle loro scoperte, a favore dei critici, come Cauchy e Weierstrass o Dini, che due secoli più tardi vi hanno portato il rigore delle dimostrazioni".

tried to offer a logical foundation to the notion of natural number and to the method of mathematical induction. Hilbert, also resorting to Pasch's work, developed the modern axiomatical method and the modern concept of mathematical existence. Mathematics became a science of structures rather than of objects. At the end of the nineteenth century, paradoxes as those of Burali Forti, Cantor and Russell shook the bases of the mathematical building. Thus, a new foundational season began. It was characterized by the complete axiomatization of the theories and by the attempt to prove the coherence of mathematics itself, which means, ultimately, of arithmetic. This extremely synthetic picture of a very complex and articulated movement on which much specialized literature exists is sufficient to realize that in the period 1870–1910 the whole foundation of mathematics was at stake. It was not question of single branches, but of the entire discipline.

What was Chasles' foundational idea? From a certain standpoint, it was the idea of a local foundation: the discipline from which his speculation started was geometry. He was convinced that the basic concepts of projective geometry, as that of cross ratio, duality and projective transformation, might found the whole of geometry, even including metric properties. So that one might say that, according to Chasles, the entire geometry is projective geometry.

At a first impression, it seems that Chasles was not interested in a global foundation of mathematics, which included concepts also belonging to analysis or number theory. However, to a more careful reflection, the issue is not so easy: we have seen that he attempted to revitalize Newton's infinitesimal geometry. Therefore, it seems to me plausible that he also had in mind to construct a geometry within which the basic concepts of calculus were offered an interpretation and a foundation different from the merely analytical one. This kind of speculation is not profoundly developed by Chasles, but the way in which he structured his work on the ellipsoid attraction makes this reading plausible. Thus, the foundation of calculus should also be included within a geometrical picture. If affine and metric geometry as well as mathematical analysis rely upon concepts reducible to projective geometry, this discipline should have been considered the ground for most of mathematics known in Chasles' epoch, at least insofar as the period 1830–1850 is concerned. However, this is only a part of Chasles' programme.

As a matter of fact, he intended to prove that the basic notions of projective geometry are also at the basis of the theory of top, the problem of forces' and couples' composition, the principle of virtual velocities and of the ellipsoid attraction. Since several aspects of the heat and electricity theories can be treated with the same means used for the ellipsoid attraction, this implies that a remarkable sections of physics could be considered—broadly speaking—as applications of projective geometry.

To this picture, further elements have to be added. They concern methodological issues: as pointed out in the course of this book, Chasles did not share the idea that analytical and algebraic methods ought to be completely replaced by purely geometrical procedures. With regard to Chasles' methodological conception, it is interesting to point out that illustrious mathematicians and historians of mathematics had different opinions: Morris Kline, for example, wrote: "Though he defended pure geometry Chasles thought analytically but presented his proofs and results geometrically. This approach is called the "mixed method" and was used later by others". (Kline, 1972, p. 850). Eric Straume in a contribution, which is interesting though full of typos and not much cared from a bibliographical standpoint, includes Chasles, jointly with Möbius, Plücker, Cayley and Salmon among the geometers who preferred analytical methods, while Carnot, Poncelet, Steiner, von Staudt and Cremona would be synthesists (Straume, 2014, p. 54). In the following pages, he adds that Chasles' position is more nuanced than that of the other geometers who supported analytical methods. However, he is convinced that Chasles' basic inspiration is analytical. Traditionally Chasles is included among the synthesists. Basically, I share this position. In the course of this book, I tried to show in which meaning Chasles was a synthesist. Nonetheless, it is true that he did not despise analytical methods. Chasles in the *Aperçu* is clear: the ultimate foundation of projective geometry is synthetic, but when analytical methods allow a remarkable simplification, they can be used. On the other hand, if he was not a synthesist, why did he try to find synthetically such difficult doctrines as that of the ellipsoid attraction? Anyway, Chasles was aware of the advantages deriving from algebra, analytic geometry and calculus. His idea was: analytical methods are useful instruments. However, they do not grasp the essence, the profound nature of mathematical concepts and physical phenomena. This is a prerogative of the geometrical method. Chasles expressed this opinion more than once and his numerous works in pure geometry are a confirmation. This is the reason which lies behind his effort to prove that the problem of ellipsoid attraction can be achieved through a geometrical and genetic method which reveals the true essence of the problem. Hence, in his view, the analytical methods might be replaced by those geometrical, or better, might be founded on them. This is true in principle. In fact, the analytical procedures can be more convenient and shorter, even though in several circumstances this is not the case. It is obvious that, given the difficulty of the problems faced by Chasles, it makes no sense to claim that "geometrical" means "completely lacking of any equation or algebraic and analytical symbolization". In his essays concerning pure geometry and written between the end of the 20s and the beginning of the 30s, the use of equations is almost completely missing. However, when Chasles developed the concept of cross ratio and applied geometrical methods to physical problems, the resort to equations, algebraic symbolization and analytical concepts became unavoidable. The terms "synthetic", "purely geometric" methods mean, in this case, methods that offer the basic concepts and theorems to construct a theory, which, after this foundational statement, is developed also resorting to the entire available set of analytical knowledges.

Notwithstanding this, Chasles intended to highlight that such algebraic and analytical objects and procedures rely upon a geometrical basis. Since the basis of geometry is the doctrine of projections, this implies such a disciple, with its most elementary concepts and methods, to be the basis of vast sections of mathematics and physics.

Besides these considerations, it is also appropriate to recall that Chasles is one of the authors who understood that a deeper way to conceive geometry required the step from a geometry of objects to a geometry of transformations. Thus, he studied profoundly the projective transformations, specifically reciprocities and homographies, and framed the most used transformation of the then projective geometry—polarity—within the general panorama of reciprocities and, hence, of duality.

Therefore, he had a broad view of the foundation of mathematics and science. His conception was not local. It was global. He intended to show the dependence of vast parts of mathematics and of science on few concepts connoting projective geometry. He did not wonder what a real number or a natural number is, either he developed a speculation on actual infinity. This depends on the fact that, when Chasles' foundational production was more intense, namely between the end of the 20s and the end of the 40s, such questions were not yet felt as problematic from the mathematical milieu. However, Chasles' view of foundation was a global one, if we think that he also tried to construct a philosophy relying upon this view: the philosophy of duality. Thus, he did not only overcome the domain of mathematics, including physics within his programme. He also went beyond exact sciences, including philosophy in such a programme.

Some comments are necessary: Chasles was not the only mathematician who understood the importance of the cross ratio: we have seen that Möbius and Steiner also did. Furthermore, Möbius had probably a more profound view than Chasles' with regard to the importance of the abstract notion of transformation in geometry. In connection with the relevance ascribed to the synthetic methods, it is sure that Chasles inherited it from Poncelet and shared such a view with many other mathematicians, among whom Steiner himself. Poncelet and Steiner were more intransigent than him about the primacy to ascribe to the synthetic methods with respect to those analytical. As to the principle of virtual velocity, Rodrigues developed an order of ideas not far from Chasles'. The concept that the system of forces and the momenta of the couples ought to be treated geometrically was shared by Chasles and Poinsot, though it was developed in different manners and surely Poinsot preceded Chasles. The geometrical speculations on the movements of a rigid body and the geometric theory of the ellipsoid attraction were, instead, absolutely original contribution by Chasles. Thus, he shared different sections of his foundational programme with other mathematicians. Nonetheless, the idea to connect all these ideas in an entire programme which included vast and important sections of mathematics, physics and, in part, philosophy is an absolutely original proposal envisioned and developed by Chasles along his mathematical and scientifical career. His view on the foundations of mathematics was broader than the view of any other mathematician of his epoch. This is the reason why he deserves a special place within European mathematics among the late 20s and the late 40s of the nineteenth century. This does not mean that other foundational contributions were less important than those given by Chasles. Only to give some examples: the foundation given by Gauss to number theory in the *Disquisitiones Arithmeticae* and to differential geometry in the *Disquisitiones generales super superficies curvas* are mathematical contributions whose quality overcomes anything Chasles ever wrote. However, they are contributions to local foundations, not to a global foundation of mathematics. Probably this also depends on Gauss' idiosyncrasy to face methodological problems and questions having a philosophical flavour. Cauchy's *Course d'Analyse* is a further example of local foundation. Chasles was the only one to conceive a global foundation of mathematics.

A further question which is appropriate to address is the attempt to understand how much Chasles' foundational programme was successful. Here, the answer is necessarily diversified: his idea that metric geometry can be interpreted as a part of projective geometry is the successful first step of a mathematical programme which finds its crowing in the Cayley's and Klein's discovery of the projective metric (Cayley, 1859; Klein, 1871, 1873). From a conceptual point of view, the two mentioned works of these mathematicians can be considered as the conclusion of an itinerary begun by Chasles, even though, from a historical perspective, probably Cayley and Klein were not directly influenced by Chasles. With regard to the central importance ascribed by Chasles to the notion of cross ratio and, particularly, to that of harmonic ratio, Staudt's proof that a harmonic group can be obtained with purely geometrical means can be interpreted as a further step in the realization of Chasles' programme: the notion of cross ratio can be reduced to purely graphical concepts, which was not yet the case in Möbius', Steiner's and Chasles' work. Thus, it seems to me that Chasles' programme for geometry, or better, a completion of it carried out by Staudt, and by Cayley and Klein was successful.

In respect of the idea to reduce the movement of a rigid body to the transformations of projective geometry, it had some followers, as we have seen in the third chapter. Some of these followers, as Jonquières were enthusiastic about Chasles' conceptions and also shared the possibility to construct the philosophy of duality. However, most mathematicians and physicists preferred to give an analytical form to their results concerning the movement of a top. Thus, Chasles' theorem on the screw motion became a classical result in kinematics of the rigid body, but his geometrical methods were, generally, not followed by the community of mathematicians and physicists, with some remarkable exceptions. This is also true with regard to Chasles' ideas on the systems of forces and systems of momenta.

Similar considerations can be developed insofar as the attraction of the ellipsoid is concerned. Chasles' work was known and appreciated because it gave a rational basis to such a difficult problem, which the scientists were discussing from many decades. On the other hand, Ivory, Gauss and Poisson, after the works by Legendre and Laplace, reached results comparable with those obtained by Chasles. Their methods, that, *lato sensu*, can be defined as analytical, became the standard procedures to furtherly develop the problems connected to the ellipsoid attraction (see, e.g., Chandrasekhar, 1969). There were mathematical physicists as Chelini, who tried to develop a geometrical approach to the problem of the ellipsoid attraction, but most scientists followed the analytical one.

A further important aspect, which is strictly connected to the former, regards the relation between synthetic and analytical methods in geometry. As a matter of fact, the idea conceived by Poncelet and shared by Chasles and other geometers that the true nature of the geometrical figures and transformations can be obtained only through synthetical methods, while those analytical-algebraic are a way to make easier the procedures, never disappeared in the nineteenth century. There were illustrious mathematicians who claimed such idea until the end of the nineteenth-beginning of the twentieth century. I offer some examples of their ideas because they are indicative.

Giuseppe Veronese in his *Fondamenti di geometria* wrote:

> Thence, according to this method [that analytical] the distance between two points is a number. In fact, the distance can be represented by a number, but the analytical method does not say us what the distance is because, from a geometrical point of view, it is not a number. Analogously, the straight line, the plane and the spaces with three, etc., n dimensions are not, geometrically, the equations or the auxiliary analytical forms which represent them.[3]

Veronese continued his examination of the comparison between analytical and synthetical methods in geometry writing:

> [. . .] Geometry cannot be restricted to know that, e.g., a given surface exists. It also wants to know the laws through which it is possible to construct such surface.[4]

Apart from the reference to the n-dimensional space, these words might have been written by Chasles! Veronese is claiming that geometry has a proper ontology which is not grasped by the analytical methods, but only by those constructive-synthetical. The analytical procedures are useful instruments to simplify the calculations and to make the solution of a problem briefer, but the approach appropriate to geometry is the synthetic one, an approach which Veronese developed beyond Euclidean geometry. For he edified the non-Archimedean geometry through the synthetic method. Veronese's position is a clear echo of that claimed many years before the publication of his book by mathematicians such as Poncelet, Chasles and Steiner.

Henri Poincaré developed an order of ideas not far from Chasles' and Veronese's. The context in which Poincaré proposed his epistemological view is more advanced because he wrote his philosophical/methodological works after that analysis situs had replaced projective geometry as the most general geometrical doctrine, after the paradoxes of set theory and the development of Russell's advanced logistic approach and of Hilbert's abstract axiomatic. However, the conceptual core of Poincaré's observations has the same meaning as Chasles' and Veronese's. All of them think: 1) there is a world of geometrical objects and structures. They are connected by internal laws holding their proper nature. This nature cannot be revealed either through an algebraic approach (use of equations) or by a logical analytical procedure which

---

[3]Veronese (1891), p. XXI: "Così secondo questo metodo [analitico] la distanza di due punti è un numero: ma se la distanza è rappresentabile con un numero, il metodo analitico non dice che cosa sia la distanza, perché essa non è geometricamente un numero; come la retta, il piano, gli spazi a tre ecc. n dimensioni non sono geometricamente le equazioni o le forme analitiche ausiliarie che li rappresentano".

[4]Veronese (1891), p. XXIII: "[. . .] la geometria non può contentarsi di sapere ad es. Che esiste una data superficie, ma vuole conoscere le leggi della costruzione della superficie stessa".

consists in reducing the different parts of a demonstration to more and more elementary steps. A geometrical concept of demonstration has to be intuited as a whole, otherwise their meaning gets lost. One can say that this is the profound idea behind the notion of what a "synthetic method" is. These considerations, if referred to the epoch in which Chasles developed his foundational programme, mean that the treatment with equations, that is unavoidable in advanced questions, must rely upon a conceptual basis supplied by notions expressed through synthetic geometry. While, referring this order of ideas to Poincaré's epoch, this means that the logical and analytic separation of a proof in its elements, the development of a theory of demonstration and the examination of the logic behind geometry make sense only if the general meaning of such operations is grasped before their division into single steps. Therefore, in Chasles', Veronese's and Poincaré's view the nature of geometry is synthetic, rather than logical or analytical, though algebra, analysis and logic are fundamental either to simplify the treatment of the problems or to develop physics and advanced geometry, or to grasp the fine-tuning of the logic behind the functioning of demonstrations.

Poincaré wrote explicitly:

> [Through logic and analysis] we see how the questions can be answered, we no longer see how and why they are put. This shows us that logic is not enough; that the science of demonstration is not all science and that intuition must retain its role as complement, I was about to say, as counter-poise or as antidote of logic.[5]

And continued by claiming that:

> The logician cuts up, so to speak, each demonstration into a very great number of elementary operations; when we have examined these operations one after the other and ascertained that each is correct, are we to think we have grasped the real meaning of the demonstration? Shall we have understood it even when, by an effort of memory, we have become able to repeat this proof by reproducing all these elementary operations in just the order in which the inventor had arranged them? Evidently not; we shall not yet possess the entire reality; that I know not what which makes the unity of the demonstration will completely elude us.[6]

The way of thinking applied by Poincaré to the relation between mathematics considered as reducible to logic and mathematics regarded as a discipline whose meaning goes beyond logic is exactly the same as that which led Chasles to claim

[5]Poincaré (1958), p. 21. First French edition 1905. I consulted the 1908 edition. Poincaré (1908), p. 25: "[...] on voit comment les questions peuvent se résoudre, on ne voit plus comment et pourquoi elles se posent. Cela nous montre que la logique ne suffit pas; que la Science de la démonstration n'est pas la Science tout entière et que l'intuition doit conserver son rôle comme complément, j'allais dire comme contrepoids ou comme contrepoison de la logique".

[6]Poincaré (1958), pp. 21–22. Poincaré (1908), p. 26: "Le logicien décompose pour ainsi dire chaque démonstration en un très grand nombre d'opérations élémentaires ; quand on aura examiné ces opérations les unes après les autres e qu'on aura constaté que chacune d'elles est correcte, croira-t-on avoir compris le véritable sens de la démonstration? L'aura-t-on-compris même quand, par un effort de mémoire, on sera devenu capable de répéter cette démonstration en reproduisant toutes ces opérations élémentaires dans l'ordre même ou les avait rangées l'inventeur? Évidemment non, nous ne posséderons pas encore la réalité tout entière, ce je ne sais quoi qui fait l'unité de la démonstration nous échappera complétement".

that the analytical methods have a synthetic foundation. In both cases, these two mathematicians thought that mathematics, and specifically, geometry has a meaning in itself which is not reducible to a series of analytical operations. Obviously, the epoch was different, so that the manner in which Chasles and Poincaré expressed their—so to say—"synthetic spirit" was different, but the mentality was similar, in spite of the fact that Poincaré was not a Platonist, while Chasles probably was.

The analogies between the thought of these two great mathematicians also concern further aspects: both of them were interested in determining the most general geometry where quantitative concepts, as the metric ones, are not included among the basic notions, but are obtained by specifying the nature of other concepts, which, in their general formulation, are not metric. Given the epoch in which Chasles wrote his foundational contributions, he identified with projective geometry such a general doctrine. Poincaré claimed that projective geometry is still a specific doctrine, though being more general than metric geometry, because notions such as that of straight line, conic, etc. are still separated.[7] The true absolutely general branch of geometry is analysis situs, the most profound doctrine of the bi-continuous transformations (Poincaré, 1908, pp. 65–69; 1958, pp. 40–41).

In *Science et méthode* Poincaré wrote:

> The problems of Geometry of Position [analysis situs] would perhaps not have presented themselves if only the language of analysis had been used. Or rather I am wrong, for they would certainly have presented themselves, since their solution is necessary for a host of questions of analysis, but they would have presented themselves isolated, one after the other, and without our being able to perceive their common link.[8]

If we change the expression "analysis situs" with projective geometry, these words by Poincaré could have been pronounced by Chasles: "the problems of

[7]Beyond what highlighted by Poincare as to the generality of analysis situs in comparison with projective geometry, there are specific problems for a complete synthetical foundation of projective geometry itself. Coolidge at pp. 226–227 of his valuable (Coolidge, 1934) wrote: "The two great difficulties have now been mentioned—a sound theory of complex points and something definite to replace the theory of general equations" (Coolidge, 1934, p. 226). The possibility to replace the theory of general equations is tied to that of offering a synthetic and solid framework to the projective doctrine of the curves and surfaces of degree greater than 2. Coolidge referred to the profound research by von Staudt and Cremona in the attempt to overcome such difficulties, but also stressed their weak or, at least, problematic, aspects. He claimed that the most ambitious and complete results were carried out in Kötter (1887). However, Coolidge himself proposed a different synthetic approach developing von Staudt's conception of polar (Coolidge, 1915), because in 1915 he ignored that Thieme (1879) had developed the same train of thought (Coolidge, 1934, p. 22). However, the author concluded: "And yet I now doubt whether it would be worthwhile to follow such a lead for the algebraic approach is so much easier and more satisfactory. It is for this reason that the latest advances in projective geometry, before it finally flickered out, lay in a closer examination of the fundamental postulates" (Coolidge, 1934, p. 227).

[8]Poincaré (2003), p. 44. First French edition 1908. I consulted the 1920 edition. Poincaré (1920), p. 40: "Les problèmes de l'*Analysis Situs* ne se seraient peut-être pas posés si on n'avait parlé que le langage analytique ; ou plutôt, je me trompe, ils se seraient posés certainement, puisque leur solution est nécessaire à une foule de questions d'analyse ; mais ils se seraient posés isolément, les uns après les autres, et sans qu'on puisse apercevoir leur lien commun".

Projective Geometry would perhaps not have presented themselves if only the language of analysis had been used".

Thus: the train of thought conceived by mathematicians as Poncelet, Chasles and Steiner was influential until the end of the nineteenth century, as the examples of such great geometers as Veronese, Poincaré and Enriques show.[9] Hence, this part of Chasles' foundational programme was not lost, though it had a substantial development due, in a first phase, to the research of Von Staudt (late 40s–50s) and later of Cayley and Klein (late 50s–middle 70s), and, in a second phase, to the fact that projective geometry, from the end of the nineteenth century, was progressively replaced by topology as the most general science of abstract space. The synthetic spirit of the great projective geometers of the first half of the nineteenth century was embodied in the way in which Poincaré developed *analysis situs* and the Italian school of algebraic geometry conceived algebraic geometry itself.

Other parts of Chasles' foundational programme were less successful: the geometrical treatment of rigid body's movement and of the system of forces was followed by few scholars, even if one might claim that the development of vector calculus formalized and algebraized some of the concepts envisioned by Chasles and Poinsot.

With regard to advanced problems of rational mechanics, such that the ellipsoid attraction Chasles was an exception already in his epoch because most physicists used analytical methods and continued to use them also after the publication of Chasles' memoirs on this subject.

However, it should be pointed out that the debate between analytical and synthetic methods continues to subsist, though in different forms. I would like to recall, only as examples, the magnificent works by Tristan Needham *Visual Complex Analysis* (Needham, 1997) and *Visual Differential Geometry and Forms: A Mathematical Drama in Five Acts* (Needham, 2021). In the first work, the author founds complex analysis on a series of geometrical reasonings, so that the reader can understand the common geometrical root behind complex derivation, Möbius transformation, movement in the complex plane, models of non-Euclidean geometries and complex integration. In particular, the author makes it clear the geometrical meaning of the analytical functions. This geometrical theme permeates his whole treatment. In the second text, also relying upon results already presented in the former, Needham treated differential geometry using formulae the least possible. Of course, this does not mean that he does not use formulae, which is simply impossible. Needham is conscious of the new foundational character of his work and speaks of "Newtonian Genesis" of his ideas (Needham, 1997, pp. viii–ix). He explicitly writes:

> The basic philosophy of this book is that while it often takes more imagination to find a picture than to do a calculation, the picture will always reward you by bringing you nearer to the Truth (Needham, 1997, p. 222).

---

[9]With regard to Enriques and to a comparison between his way of conceiving geometry and Poincaré's, see Bussotti (2006b).

Needham shows interesting forms of duality through geometry (see, e.g., the Kasner-Arnold Theorem, ibid., pp. 246–247), develops a geometrical proof of the Argument Principle (ibid., pp. 3244–3245) and offers a geometrical treatment of complex integration and Cauchy theorem (ibid., chapter 9, pp. 427–449). *Mutatis mutandis*, this book might have been written by Chasles! Therefore, although a minority, the preference for a synthetic approach is still alive in modern mathematics.

I would like to conclude this book with a consideration: the work of an author is the more stimulating, the more it is connected to scientific problems born after its publication, also considering fields of research far from that to which such a work belongs. Chasles was not interested in the theory of perception and the birth of neurosciences dates to a later period if compared to the works of Chasles I have analysed. However, his distinction between graphical and metric properties and the attempt to reduce the latter to the former makes me think of the following question: do regions of our brain exist which are responsible for the creation of our graphical notions and other regions which are responsible for the creation of the metric notions? As far as I know, neurosciences, which are nowadays so important and advanced, do not have an answer to this question, which connects two mathematical disciplines directly to the structure of our brain. Therefore, if this were the case, the axioms of such branches of geometry would lose their absolute formal character, but would represent two different mental processes inscribed in our physiological-anatomical structure. Both Enriques and Poincaré, also relying upon the studies of experimental psychology carried out in the second half of the nineteenth century, developed profound considerations with regard to the way through which we achieve the notion of distance: it cannot be given by mere visual perception, without any intervention of the muscles which allow us to change our perspective.[10] However, they explained how the perception of distance arouses, I am posing a problem concerning the structure of our brain.

[10]See Poincaré (1917) (first edition 1902), Chapter IV: "L'Espace et la Géométrie", pp. 68–91. English translation of the first edition, Poincaré (1905). Poincaré (1908), Chapter III: "La Notion d'Espace", pp. 59–95; Chapter IV; "l'Espace et ses Trois Dimensions", pp. 96–136. Enriques (1901). Enriques (1906), Chapter IV: "La Geometria", pp. 261–347, where Enriques also criticized Poincaré's conventionalism, he called "the new nominalism". English translation (Enriques, 1914). See also Bussotti (2006b, pp. 63–79).

References

Acerbi, F. 2011. Perché una dimostrazione geometrica greca è generale. In G. Micheli, F. Repellini (Eds.), *La scienza antica e la sua tradizione. IV Seminario di studi (Gargnano, 13-15 ottobre 2008)*, pp. 25-80. Milano-Fontevivo: Cisalpino Istituto Editoriale Universitario.

Altmann, S. 1989. Hamilton, Rodrigues and the Quaternion Scandal. *Mathematics Magazine*, 62, 5: 291-308.

Altmann, S. 2007. Olinde Rodrigues, mathematician and social reformer. *Gazeta de Matemática*, 152: 40-48.

Altmann, S. L., & Ortiz, E. L. (Eds.). 2005. *Mathematics and Social Utopias in France. Olinde Rodrigues and His Times*. American Mathematical Society-London Mathematical Society.

Atmanspacher, H. - Primas, H. - Wertenschlag-Birkhäuser, E. (Eds.). 1995. *Der Pauli - Jung Dialog Und seine Bedeutung für die moderne Wissenschaft*. Berlin-Heidelberg: Springer.

Ampère, A.M. 1834. *Essai sur la philosophie des sciences, ou exposition analytique d'une classification naturelle de toutes les connaissances humaines*. Paris: Bachelier, Imprimeur-Libraire pour les Sciences.

Andersen, K. 1991. Desargues' method of perspective: its mathematical content, its connection to other perspective methods and its relation to Desargues' ideas on projective geometry. *Centaurus* 34 (1): 44-91.

Anglade, M. – Briend, J.Y. 2017. La notion d'involution dans le 'Brouillon Project' de Girard Desargues. *Archive for History of Exact Sciences*, 71 (6): 543-588.

Anglade, M. – Briend, J.Y. 2019. Le diamètre et la traversale: dans l'atelier de Girard Desargues. *Archive for History of Exact Sciences*, 73 (4): 385-426.

Anonymous (un Abonné). 1822–1823. Géométrie élémentaire. Sur la construction du cercle tangent à trois cercles donnés. *Annales de mathématiques pures et appliquées* 13: 199–200.

Anonymous. 1828. Review of *Annales de Mathématiques pures et appliquées*; par. M. Gergonne. Tome XVIII, n° 12, juin 1828. *Bulletin des sciences mathématiques*, X:185-187.

Arago, F. 1853. *Biographie de Gaspard Monge, ancien membre de l'Académie des Sciences*. Paris: Firmin Didot (read at the Academy of Science on 11 May 1846).

Arana, A. - Mancosu, P. 2012. On the relationship between plane and solid geometry. *Review of Symbolic Logic*, 5 (2): 294-353 (2012)

Arnold, D.H. 1983a. The mécanique physique of Siméon Denis Poisson: the evolution and isolation in France of his approach to physical theory (1800-1840). I. Physics in France after the revolution. *Archive for History of Exact Sciences*, 28, 3: 243-266.

Arnold, D.H. 1983b. The mécanique physique of Siméon Denis Poisson: the evolution and isolation in France of his approach to physical theory (1800-1840). II. The Laplacian program, *Archive for History of Exact Sciences*, 28, 3: 267-287.

P. Bussotti, *Chasles and the Projective Geometry*,
https://doi.org/10.1007/978-3-031-54266-4

Arnold, D.H. 1983c. The mécanique physique of Siméon Denis Poisson: the evolution and isolation in France of his approach to physical theory (1800-1840). III. Poisson: mathematician or physicist? *Archive for History of Exact Sciences*, 28, 4: 189-197.

Arnold, D.H. 1984d. The mécanique physique of Siméon Denis Poisson: the evolution and isolation in France of his approach to physical theory (1800-1840). IV. Disquiet with respect to Fourier's treatment of heat. *Archive for History of Exact Sciences*, 28, 4: 299-320.

Arnold, D.H. 1983e. The mécanique physique of Siméon Denis Poisson: the evolution and isolation in France of his approach to physical theory (1800-1840). V. Fresnel and the circular screen. *Archive for History of Exact Sciences*, 28, 4: 321-342.

Arnold, D.H. 1983f. The mécanique physique of Siméon Denis Poisson: the evolution and isolation in France of his approach to physical theory (1800-1840). VI. Elasticity: the crystallization of Poisson's views on the nature of matter. *Archive for History of Exact Sciences*, 28, 4: 343-367.

Arnold, D.H. 1983g. The mécanique physique of Siméon Denis Poisson: the evolution and isolation in France of his approach to physical theory (1800-1840). VII. Mécanique physique. *Archive for History of Exact Sciences*, 29, 1: 37-51.

Arnold, D.H. 1983h. The mécanique physique of Siméon Denis Poisson: the evolution and isolation in France of his approach to physical theory (1800-1840). VIII. Applications of the mécanique physique. *Archive for History of Exact Sciences*, 29, 1: 53-72.

Arnold, D.H. 1983i. The mécanique physique of Siméon Denis Poisson: the evolution and isolation in France of his approach to physical theory (1800-1840). IX. Poisson's closing synthesis: traité de physique mathématique. *Archive for History of Exact Sciences*, 29, 1: 73-94.

Arnold, D.H. 1984. The mécanique physique of Siméon Denis Poisson : the evolution and isolation in France of his approach to physical theory (1800-1840). X. Some perspective on Poisson's contributions to the emergence of mathematical physics. *Archive for History of Exact Sciences*, 29, 4: 287-307.

Atmaspacher, H. – Primas, H. – Wertenschlag-Birkhäuser, E. (Eds.). 1995. *Der Pauli-Jung-Dialog und seine Bedeutung für die moderne Wissenschaft*. Berlin-Heidelberg: Springer.

Atzeni, F.M. 2014–2015. I movimenti rigidi da Euler al Programma di Erlangen. PhD Thesis. Università degli Studi di Cagliari. Dipartimento di Matematica e Informatica. Dottorato di Ricerca in Matematica e Calcolo scientifico. Ciclo XXVIII.

Baldus, R. 1923. Zur Steinerschen Definition der Projektivität. *Mathematische Annalen* 90 (1–2): 86–102.

Barbin, É. 2019. Monge's Descriptive Geometry: His Lessons and the Teachings Given by Lacroix and Hachett. In É. Barbin-M. Menghini-K. Volkert (Eds.), *Descriptive Geometry, The Spread of a Polytechnic Art. The Legacy of Gaspard Monge*, pp. 3-18. Cham: Springer.

Belhoste, B. 1989. Les origines de l'École Polytechnique. Des anciennes écoles d'ingénieurs à l'École centrale des Travaux publics. *Histoire de l'éducation*, 42: 13-53.

Belhoste, B. 1994. De l'École des ponts et chaussées à l'École centrale des travaux publics. Nouveaux documents sur la fondation de l'École polytechnique. *Bulletin de la société des amis de la bibliothèque de l'École polytechnique*, 11: 1-69.

Belhoste, B. 1998. De l'École polytechnique a Sàratoff, les premiers travaux géométriques de Poncelet. *Bulletin de la Sabix*, Édition électronique: 1-23.

Belhoste, B. 2003. *La formation d'une technocatie. L'École polytechnique et ses élèves de la Révolution au Second Empire*. Paris: Belin.

Bellavitis, G. 1860. *Sposizione dei nuovi metodi di geometria analitica*. Venezia: Segreteria dell'Istituto Veneto.

Benettin, G. 2005. Physical Applications of Nekhoroshev Theorem and Exponential Estimates. In A. Giorgilli (Ed.), *Hamiltonian Dynamics Theory and Applications Lectures given at the C.I.M. E.-E.M.S. Summer School held in Cetraro, Italy, July 1-10, 1999* (pp. 1–76). Berlin-Heidelberg-New York: Springer.

Berger, M. 2005. *Cinq siècles de mathématiques en France*. Paris: Association pour la diffusion de la pensée française, 2005.

Bertrand, M.J. 1892. *Éloge historique de Michel Chasles, membre de l'Académie des Science*. Lu dans la séance de l'Académie des Sciences du lundi 19 décembre 1892. Published in *Mémoires de l'Académie des Sciences de L'Institut de France*, 1904, 47: XXXIX-LII.

Bhat, S. P., & Crasta, N. 2018. Rolling Cones, Closed Attitude Trajectories, and Attitude Reconstruction. *The Journal of the Astronautical Sciences, 65*, 261–290.

Binet, J. P. M. 1815. Memoire sur la composition des forces et sur la composition des momens. *Journal de l'école Polytechnique, X, 17*, 321–348.

Biosemat-Martagon, L. (Ed.). 2010. *Eléments d'une Biographie de l'Espace Projectif.* Nancy: Presses Universitaires de Nancy.

Blåsjö, V. 2009. Jakob Steiner's Systematische Entwickelung: The culmination of classical geometry. *The Mathematical Intelligencer* 31 (1): 21–29.

Bonnet, P.O. 1845a. Mémoire sur la théorie des surfaces isothermes orthogonales. *Journal de l'École Polytechnique*, 18: 141-164.

Bonnet, P.O. 1845b. Mémoire sur la théorie des corps élastiques. *Journal de l'École Polytechnique*, 18: 171-191.

Borgato, M.T. 2006. Il fusionismo e i fondamenti della geometria. In L. Giacardi (Ed.), *Da Casati a Gentile, momenti di storia dell'insegnamento secondario della matematica in Italia*, pp. 127-157. Lugano: Lumières Internationales.

Bottazzini, U. 1990. *Il flauto di Hilbert. Storia della matematica moderna e contemporanea.* Torino: UTET.

Bresse, J. 1853. Mémoire sur un théorème nouveau concernant les mouvements plans et l'application de la cinématique à la détermination des rayons de courbure. *Journal de l'École Impériale Polytechnique*, 20: 89-115.

Breton, P. 1838. Application d'un principe de Mécanique rationnelle à la résolution de quelques Problèmes de Géométrie. *Journal de Mathématiques pures et appliquées*, 1, 3: 488 - 494

Brianchon, C.J. 1817. *Mémoire sur les lignes du second ordre; faisant suite aux recherches publiées dans les journaux de l'École Royale Polytechnique.* Paris: Bachelier.

Brisse, C. 1870. Mémoire sur le déplacement des figures. *Journal de Mathématiques pures et appliquées*, 2, 15: 281-314.

Brisse, C. 1874. Sur le déplacement fini quelconque d'une figure de forme invariable. *Journal de Mathématiques pures et appliquées*, 2, 19: 221-264.

Brisse, C. 1875. Sur le déplacement fini quelconque d'une figure de forme invariable (suite et fin). *Journal de Mathématiques pures et appliquées*, 3, 1: 141-180.

Bussotti, P. 1997. *Giuseppe Veronese e i fondamenti della matematica*. Pisa. ETS.

Bussotti, P. 2006a, *From Fermat to Gauss. Indefinite descent and methods of reduction in number theory*. Augsburg: Rauner.

Bussotti, P. 2006b. *"Un mediocre lettore". Le letture e le idee di Federigo Enriques*. Lugano: Agorà Publishing – Lumières Internationales.

Bussotti, P. 2007. Il concetto di archetipo in Jung: origine, significato e rapporto con le atre scienze. *Sentieri. Itinerari di psicopatologia – psicosomatica – psichiatria*, VII: 81-101.

Bussotti, P. 2015. *The Complex Itinerary of Leibniz's Planetary Theory: Physical Convictions, Metaphysical Principles and Keplerian Inspiration*. Basel: Springer-Birkhäuser Verlag.

Bussotti, P. 2019. Michel Chasles' foundational programme for geometry until the publication of his *Aperçu historique*. *Archive for History of Exact Sciences*, 73, 3: 261-308.

Bussotti, P. 2020. Review of Anglade, M. – Briend J. Y. 2019. Le diamètre et la traversale: dans l'atelier de Girard Desargues. Zbl 1442.01004. *Zentralbatt für Mathematik*.

Bussotti, P., & Lotti, B. 2022. *Cosmology in the Early Modern Age. A Web of Ideas*. Cham: Springer.

Bussotti, P., & Pisano, R. 2014a. On the Jesuit Edition of Newton's *Principia*. Science and Advanced Researches in the Western Civilization. *Advances in Historical Studies, 3*(1), 33–55.

Bussotti, P., & Pisano, R. 2014b. Newton's *Philosophiae Naturalis Principia Mathematica* "Jesuit" Edition: The Tenor of a Huge Work. *Rendiconti Lincei: Matematica e Applicazioni, 25*(4), 413–444.

Cantor, G. 1883. *Grundlagen einer allgemeinen Mannigfaltigkeitslehre*. In E. Zermelo (Ed.), *Gesammelte Abhandlungen mathematischen und philosophischen Inhalts: Mit erläuternden Anmerkungen sowie mit Ergänzungen aus dem Briefwechsel Cantor-Dedekind* (pp. 165–209). Berlin: Springer.

Cantù, P. 1999. *Giuseppe Veronese e i fondamenti della geometria*. Milano: Unicopli.

Caparrini, S. 2002. The Discovery of the Vector Representation of Moments and Angular Velocity. *Archive for History of Exact Sciences*, 56: 151-181.

Caparrini, S. 2006. Le origini del calcolo vettoriale nella geometria e nella meccanica. *Bollettino della Unione Matematica Italiana. Sezione A. La Matematica nella Società e nella Cultura*, 9-2: 227-230.

Capecchi, D. 2012. *History of Virtual Work Laws*. Milan-Heidelberg-New York-Dordrecht-London: Birkhäuser.

Carnot, L. 1783. *Essai sur les machines en général*. Second edition 1786. Dijon: Defay.

Carnot, L. 1801. *De la correlation des figures de géométrie*. Paris: Duprat.

Carnot, L. 1803. *Géométrie de position*. Paris: Duprat.

Carnot, L. 1806. *Mémoire sur la relation qui existe entre les distances respectives de cinq points quelconques pris dans l'espace suivi d'un essai sur la théorie des transversales*. Un appendice leur est adjoint: *Digression sur la nature des quantités dites négatives*. Paris: Courcier.

Catalan, E. 1841. *Attraction d'un ellipsoïde homogène*. Thèse de mécanique présentée a la Faculté des Sciences de Paris, 1841. Paris: Bachelier.

Catastini, L. 2004. Il giardino di Desargues. *Bollettino della Unione Matematica Italiana*, S. 8, vol 7-A, *La Matematica nella Società e nella Cultura*, 2: 321-435.

Catastini, L. – Ghione, F. 2005. Nella mente di Desargues tra involuzioni e geometria dinamica. *Bollettino della Unione Matematica Italiana*, S. 8, vol 8-A, *La Matematica nella Società e nella Cultura*, 1: 123-147.

Catton, P., & Montelle, C. 2012. To Diagram, to Demonstrate: To Do, to Dee, and to Judge in Greek Geometry. *Philosophia Mathematica, 20*(1), 25–57.

Cauchy, A. 1820–1821. Géométrie des courbes. Rapport à l'académie royale des sciences. *Annales de Mathématiques pures et appliquées*, 11: 69-83

Cayley, A. 1859. A sixth memoir on quantics. In *The collected mathematical papers of Arthur Cayley*, II, 1889, 561–592. Cambridge: Cambridge University Press.

Ceccarelli, M. 2000. Screw axis defined by Giulio Mozzi in 1763 and early studies on helicoidal motion. *Mechanism and Machine Theory*, 35, 6: 761-770.

Ceccarelli, M. 2007. Giulio Mozzi (1730-1813). In M. Ceccarelli (Ed.), *Distinguished Figures in Mechanism and Machine Science. Their Contributions and Legacy* (Vol. Part 1, pp. 279–293). Dordrecht: Springer.

Chandrasekhar, S. 1969. *Ellipsoidal Figures of Equilibrium*. New Haven and London: Yale University Press.

Charbonneau, L. 1993. Les mathématiciens au pouvoir: la Révolution Française. *Plot*, 64/65: 33-43.

Chelini, D. 1862a. Della legge onde un ellissoide eterogeneo propaga la sua attrazione da punto a punto. *Memorie dell'Accademia delle Scienza dell'Istituto di Bologna* 2,1: 3-52.

Chelini, D. 1862b. Dei moti geometrici e loro leggi nello spostamento di una figura di forma invariabile. *Memorie dell'Accademia delle Scienza dell'Istituto di Bologna* 2,1: 361-430.

Chemla, K. 1998. Lazare Carnot et la généralité en géométrie. Variations sur le théorème dit de Menelaus. *Revue d'histoire des mathématiques*, 4: 163-190.

Chemla, K. 2016. The value of generality in Michel Chasles's historiography of geometry. In *The Oxford handbook of generality in mathematics and the sciences*, ed. K. Chemla, R. Chorlay, and D. Rabouin, 47–89. Oxford: Oxford University Press.

Clairault A. 1737, 1741. Investigationes aliquot, ex quibus probetur Terrae figuram secundum Leges attractionis in ratione inversa quadrati distantiarum maxime ad Ellipsin accedere debere. *Philosophical Transactions*, 445:19-25.

Clairault, A. 1738. An Inquire concerning the Figure of such Planets as revolve about an Axis, supposing the Density continually to vary, from the Centre towards the Surface. *Philosophical Transactions*, 449:277-306.
Clairault, A. 1743. *Théorie de la Figure de la Terre, tirées des Principes de l'Hydrostatique*. Paris: David Fils.
Clifford, W. K. 1878. *Elements of dynamics. An introduction to the study of motion and rest in solid and fluid bodies* (vol I. Kinematics). London: Macmillan.
Coolidge, J. L. 1900. A Purely Geometric Representation of all Points in the Projective Plane. *Transactions of the American Mathematical Society, 1*(2), 182–192.
Coolidge, J.L. 1915. The meaning of Plücker's equations for a real curve, *Rendiconti del Circolo Matematico di Palermo*, XL: 211-216.
Coolidge, J.L. 1934. The rise and fall of projective geometry. *The American Mathematical Monthly* 41(4): 217–228.
Coolidge, J.L. 1940. *A history of geometrical methods*. Oxford: Oxford University Press
Coriolis, G.G. 1829. *De Calcul de l'Effet des Machines*. Paris: Carilian-Goeury.
Crowe, M.J. 1967. *A History of Vector Analysis. The Evolution of the Idea of a Vectorial System*. New York: Dover.
Culmann, K. 1866. *Die graphische Statik*. Zürich: Meyer & Zeller.
Dahan-Dalmedico, A. 1986. Un texte de philosophie mathématique de Gergonne: Mémoire inédit déposé à l'Académie de Bordeaux. *Revue d'histoire des sciences*, 39, 2: 97-126.
D'Alembert, J.1749. *Recherches sur la précession des équinoxes et sur la nutation de l'axe de la Terre dans le système newtonien*. Paris: David l'aîné.
D'Alembert, J. 1754–1756. *Recherches sur différens points importans du système du monde*. In three parts. First and second parts 1754, third part 1756. Paris: David.
D'Alembert, J. 1780. Sur l'attraction des sphéroïdes elliptiques (plus remarques). In *Opuscules mathématiques ou Mémoires sur différens sujets de géométrie, de méchanique, d'optique, d'astronomie*. Tome 7, LIII Mémoire: 102-233. Paris: Claude-Antoine Jombert.
Dandelin, J.P. 1827. Propriétés projectives des courbes du second degré. *Correspondance mathématique et physique, publiée par A. Quetelet* III: 9–12.
Dandelin, J.P., and J.D. Gergonne. 1825–1826. Géométrie pure. Usages de la projection stéréographique en géométrie. *Annales de mathématiques pures et appliquées* 16: 322–327.
Darboux, G. 1905. A survey of the development of geometric methods. *Bulletin of the American Mathematical Society*, 10 (11): 517-543.
Daston, L.J. 1986. The Physicalist Tradition in Early Nineteenth Century French Geometry. *Studies in History and Philosophy of Science*, 17, 3: 269-295.
De Champ, B. 1867. Sur les forces centrifuges mises en usage par Poinsot dans sa théorie de la rotation des corps. *Nouvelles annales de mathématiques*, 2, 6: 362-366.
De Iaco Veris, A. 2018. *Practical Astrodynamics*. Cham: Springer.
Delcourt, J. 2011. Analyse et géométrie, histoire des courbes gauches de Clairaut à Darboux. *Archive for History of Exact Sciences, 65*, 229–293.
De La Hire, P. 1685. *Sectiones Conicae*. Paris: Michallet.
De La Hire, P. 1698. *La Gnomonique ou méthode universelles, pour tracer des horloges solaires ou cadrans sur toutes sort*. Paris: Thomas Moette.
Del Centina, A. 2016a. On Kepler's system of conics in *Astronomiae pars optica*. *Archive for History of Exact Sciences*, 70, 6: 567-589.
Del Centina, A. 2016b. Poncelet's porism. A long story of renewed rediscovery, I. *Archive for History of Exact Sciences*, 70, 1: 1-122.
Desargues, G. 1639. *Brouillon project d'une atteinte aux événemens des rencontres d'un cone avec un plan. Paris*.
Dhombres, J. 2018. *L'École polytechnique et ses historiens*. Open Edition Journals. FMSH - Fondation Maison des sciences de l'homme. http://journals.openedition.org/bibnum/1154
Dimmel, J.K. – Herbst P.G. 2015. The Semiotic Structure of Geometry Diagram: How Textbooks Diagrams Convey Meaning. *Journal for Research in Mathematics Education*, 46, 2: 147-195.

Dupin, C. 1819. *Essai historique sur les services et les travaux scientifiques de Gaspard Monge*. Paris: Bachelier.

Dupont, P. 1963–1964. Esame storico-critico del contributo di d'Alembert, Eulero, Poisson, Poncelet ed altri al concetto dell'asse istantaneo di rotazione nei moti rigidi con un punto fisso. *Atti dell'Accademia delle Scienze di Torino. I. Classe di Scienze fisiche, naturali*, 98.

Enriques, F. 1898, second edition 1904. *Lezioni di geometria proiettiva*. Bologna: Zanichelli.

Enriques, F. 1901. Sulla spiegazione psicologica dei postulati della geometria. *Rivista di filosofia*, 4 (3): 171-195.

Enriques, F. 1906. *Problemi della Scienza*. Bologna: Zanichelli.

Enriques, F. 1907–1910. Prinzipien der Geometrie. In *Encyclopädie der mathematischen Wissenschaften mit Einschluss ihre Anwendung*. Dritter Band. Ester Teil, erste Häfte, pp. 1-129. Leipzig: Teubner.

Enriques, F. 1914. *Problems of science*. Chicago-London: The Open Court Publishing Company.

Enriques, F. 1922, reprint 2003. *Insegnamento dinamico con scritti di Franco Ghione e Mauro Moretti*. La Spezia: Agorà, 2003.

Enriques, F. 1924–1927. L'evoluzione delle idee geometriche nel pensiero greco. In F. Enriques (Ed.), *Questioni riguardanti le matematiche elementari*, part I, tome I: 1-40. Bologna. Zanichelli. Anastatic reprint 1983.

Esquisabel, O.M. – Raffo Quintana, F. 2021. Fiction, possibility and impossibility: three kinds of mathematical fictions in Leibniz's work. *Archive for History of Exact Sciences,* 75 (6): 1-35.

Euler, L. 1748. *Introductio in analysin infinitorum*, second volume. Lausanne: Apud Marcum-Michaelem Bousquet et Socios. In *Opera Omnia*, series 1, vol. 9.

Euler, L. 1752. Découverte d'un nouveau principe de mécanique. *Memoires de l'Academie Royal des Sciences*, 6, 185-217. In *Opera Omnia*, series 2, vol 5 : 81-108.

Euler, L. 1765a. Recherches sur la connoissance mécanique des corps. *Mémoires de l'académie des sciences de Berlin*, 14: 131-153. In *Opera Omnia*, series 2, vol 8: 178-199.

Euler, L. 1765b. Du mouvement de rotation des corps solides autour d'un axe variable. *Mémoires de l'académie des sciences de Berlin*, 14: 154-193. In *Opera Omnia*, series 2, vol. 8 : 200-235.

Euler, L. 1767. Du mouvement d'un corps solide quelconque lorsq'il tourne autour d'un axe mobile. *Memoires de l'academie des sciences de Berlin* 16: 176-227. In *Opera Omnia*, series 2, vol 8: 313-356.

Fano, G. 1907–1910. Gegensatz von synthetischer und analytischer Geometrie in seiner historischen Entwicklung im XIX. Jahrhundert. In *Encyclopädie der mathematischen Wissenschaften, mit Einschluss ihre Anwendung*. Dritter Band. Ester Teil, erste Häfte, pp. 221-288. Leipzig: Teubner.

Fano, G. 1915. Exposé parallèle du développement de la géométrie synthétique et de la géométrie analytique pendant le XIXe siècle. In *Encyclopédie des Sciences Mathématiques Pures et Appliquées*, Tome II (premier volume), pp. 185-259. Paris: Gabay.

Fermat, P. 1891–1922. *Oeuvres de Fermat,* four volumes *plus Supplément aux tomes I-IV*. Paris: Gauthier-Villars.

Ferreirós, J. 2016. *Mathematical Knowledge. An Interplay of Practices*. Princeton and Oxford: Princeton University Press.

Field, J.V. 1987. Linear perspective and the projective geometry of Girard Desargues. *Nuncius* 2 (2): 3-40.

Field, J.V. 1988. Kepler's geometrical cosmology. Chicago: The University of Chicago Press.

Field, J.V.-Gray, J. 1987. *The geometrical work of Girard Desargues*. New York-Berlin: Springer.

Floriduz M. 2021. Gravità: il problema dell'attrazione dei corpi sferici ed ellissoidali in Newton e Maclaurin. Degree dissertation. Department of Mathematics, University of Udine. Supervisor Paolo Bussotti.

Fourcy, A. 1828. *Histoire de l'École Polytechnique*. Paris: Chez l'auteur, a l'École Polytechnique.

Fourier, C. 1819–1820. Théorie du mouvement de la chaleur dans les corps solides. *Mémoires de l'Académie Royale des Sciences de l'Institut de France*, IV (1824): 185-525.

Frosali, G. 2014. *Complementi di meccanica razionale. III parte: dinamica dei corpi rigidi.* Firenze: Università degli Studi di Firenze.

Galilei, G. 1890–1909. Edizione Nazionale delle Opere. Direttore A. Favaro. 20 vols. Firenze: Barbera. Abbreviated as EN.

Garza, E.P.- Quintanilla, M.E.P. 2011. Benjamin Olinde Rodrigues, matemático y filántropo, y su influencia en la Física Mexicana. *Revista mexicana de física E*, 57 : 109-113.

Gaultier, L. 1813. Mémoire sur les moyens généraux de construire graphiquement les cercles déterminés par trois conditions, et les sphères déterminés par quatre conditions. *Journal de l'École polytechnique* 16: 124–214.

Gauss, C. F. 1813. Theoria attractionis corporum sphaeroidicorum ellipticorum homogeneorum methodo nova tractata. *Commentationes Societatis Regiae Scientiarum Gottingensis recentiores, 2*, 1–24. In *Werke*, vol. V: 1-22. Göttingen: Gesellschaft der Naturwissenschaften zu Göttingen, 1867.

Gauss, C.F. 1832. Theoria residuorum biquadraticorum. Commentatio secunda, Societati Regiae tradita 1831. *Commentationes Societatis Regiae Scientiarum Gottingensis recentiores*, 7. In *Werke*, vol. II: 93-178. Göttingen: Gesellschaft der Naturwissenschaften zu Göttingen, 1876.

Gauss, C.F. 1840. *Allgemeine Lehrsätze in Beziehung auf die im verkehrten Verhältnisse des Quadrats der Entfernung wirkenden Anziehungs und Abstossungs-Kräfte*. Leipzig: Weidmann. In *Werke*, vol. V: 195-242. Göttingen: Gesellschaft der Naturwissenschaften zu Göttingen, 1877.

Gergonne, J. D. 1812–1813. Géométrie analytiques. Théorie analytique des pôles des lignes et des surfaces du second ordre. *Annales de mathématiques pures et appliquées, 3*, 293-302.

Gergonne, J.D. 1817. De l'Analyse et de la Synthèse dans les sciences mathématiques. *Annales de Mathématiques pures et appliquées*, VII: 345-373.

Gergonne, J. D. 1824–1825. Géométrie élémentaire. Recherche de quelques-unes des lois générales qui régissent les polyèdres. *Annales de mathématiques pures et appliquées, 15*, 157–164.

Gergonne, J.D. 1825–1826. Philosophie Mathématique. Considérations philosophiques sur les élémens de la science de l'etendue. *Annales de mathématiques pures et appliquées*, 15: 209-231.

Gergonne, J.D. 1826–1827a. Géométrie de situation. Recherches sur quelques lois générales qui régissent les lignes et surfaces algébriques de tous ordres. *Annales de Mathématiques pures et appliquées,* 17: 214-252.

Gergonne, J.D. 1826–1827b. Réflexion sur le précédent article. Propriétés générales de l'etendue. *Annales de Mathématiques pures et appliquées*, 17: 272-276.

Gergonne, J.D. 1827–1828a. Polémique mathématique. Réclamation de M. le capitaine Poncelet (extraite du bulletin universel des annonces et nouvelles scientifiques); avec des notes. *Annales de Mathématiques pures et appliquées,* 18: 125.

Gergonne, J.D. 1827–1828b. Démonstration d'un théorème de M. Chasles. *Annales de Mathématiques pures et appliquées,* 18: 372-377.

Gergonne, J.D. 1828–1829. Géométrie de situation. Double théorème de géométrie à trois dimensions. Annales de mathématiques pures et appliquées, 19: 114-119.

Gergonne, J.D. 1847. Note sur le principe de dualité en géométrie. *Académie des Sciences de Montpellier. Mémoires de la de la Section des Sciences*, 1: 61-64.

Gerini, C. 2000. Les "Annales" de Gergonne: apport scientifique et épistémologique dans l'histoire des mathématiques. PhD dissertation, UFR Lettres. Université Aix-Marseille I. Université de Provance, Presses Universitaires du Septentrion Villeneuve d'Asq, 2003.

Gerini, C. 2016. Joseph-Diez Gergonne (1771-1859), Professeur et Recteur d'Académie à Montpellier. Éditeur du premier grand journal de l'histoire des mathématiques et de leur enseignement: les Annales de Gergonne (1810-1831). Https://hal.archives-ouvertes.fr.

Gieser, S. 2005. *The Innermost Kernel Depth Psychology and Quantum Physics. Wolfgang Pauli's Dialogue with C.G. Jung*. Berlin-Heidelberg: Springer.

Gilbert, M.P. 1861. *Recherches Sur Les Propriétés Géométriques Des Mouvements Plans*. Bruxelles: Hayez. Imprimeur de l'Académie Royale.

Gillispie, C.G. 1971. *Lazare Carnot savant*. Princeton: Princeton University Press.

Gillispie, C., & Pisano, R. 2014. *Lazare and Sadi Carnot. A Scientific and Filial Relationship* (2nd ed.). Dordrecht-Heidelberg-New York-London: Springer.

Gillispie, C., – Youschkevitch, P. 1979. *Lazare Carnot savant et sa contribution a la théorie de l'infini mathématique avec trois mémoires inédits de Carnot*. Paris: Vrin.

Giorgini, G. 1820. *Teoria analitica delle proiezioni*. Lucca: Tipografia Ducale di Francesco Bertini. Reproduced in *Atti dell'Accademia Lucchese di Scienze Lettere ed Arti*, I (1821): 29–96.

Giorgini, G. 1835. *Elementi di statica*. Firenze: Pezzati.

Giorgini, G. 1836. Intorno alle proprietà geometriche dei movimenti di un sistema di punti di forma invariabile. Memoria del Professore Cavaliere Gaetano Giorgini ricevuta adì 15 Marzo 1830. *Memorie di Matematica e Fisica della Società Italiana delle Scienze*, XXI: 1-54.

Glas E. 1986. On the dynamics of mathematical change in the case of Monge and the French Revolution. *Studies in History and Philosophy of Science*, 17, 3: 249-268.

Graf, J.H. 1897. *Der Mathematiker Jakob Steiner von Utzenstorf*. Bern: K. J. Wyss.

Grattan-Guinness, I. 1990. *Convolutions in French Mathematics, 1800-1840*. Basel: Springer.

Grattan-Guinness, I. 2005. The *Ecole Polytechnique*, 1794-1850: Differences over Educational Purpose and Teaching Practice. *The American Mathematical Monthly*, 112, 3: 233-250.

Grattan-Guinness, I. 2014. From anomaly to fundament: Louis Poinsot's theories of the couple in mechanics. *Historia Mathematica*, 41: 82-102.

Gray, A. 1907. The attraction of ellipsoidal shells and of solids ellipsoids at external and internal points, with some historical notes. *The London, Edinburgh and Dublin philosophical magazine and journal of science*, 6th series, April 1907: 385-413.

Gray, J. 1980. Olinde Rodrigues' paper of 1840 on Transformation Groups. *Archive for History of Exact Sciences, 21*, 375–385.

Gray, J. 1993. Möbius' geometrical mechanics. In Fauvel, J – Flood, R. – Wilson, R. (Eds., 1993), *Möbius and his band. Mathematics and Astronomy in Nineteenth-century Germany*, pp. 78-103. Oxford-New York-Tokyo: Oxford University Press.

Gray, J. 2005. 1822 Jean Victor Poncelet, *Traité des propriétés projectives des figures*. In I. Grattan-Guinness (Ed.), *Landmark Writings in Western Mathematics 1640–1940* (pp. 366–376). Amsterdam - Boston - Heidelberg - London - New York - Oxford - Paris - San Diego - San Francisco - Singapore - Sydney - Tokyo: Elsevier.

Gray, J. 2008. *Plato's Ghost. The modernist transformation of mathematics*. Princeton: Princeton University Press.

Gray, J. 2010. *Worlds out of nothing. A course in the history of geometry in the 19th century*. London: Springer.

Guicciardini, N. 2015. Editing Newton in Geneva and Rome: The Annotated Edition of the *Principia* by Calandrini, Le Seur and Jacquier. *Annals of Science, 72*(3), 337–380.

Hachette, J.N.P. 1809. *Programmes D'Un Cours De Physique: Ou Précis De Leçons Sur les principaux Phénomènes de la Nature, et sur quelque applications des Mathématiques à la Physique*. Paris: Bernard.

Hachette, J.N.P. 1813. *Traité des surfaces du second degré*. Paris: Klostermann.

Hachette, J.N.P. 1817. Eléments de géométrie a trois dimensions. Paris: Courcier.

Hachette, J.P.N. 1822. *Traité de géométrie descriptive: comprenant les applications de cette géométrie aux ombres, à la perspective et à la stéréotomie*. Paris: Corby-Guillaume.

Halpern, P. 2020. *Synchronicity: The Epic Quest to Understand the Quantum Nature of Cause and Effect*. New York: Basic Books.

Hanna, G. – Sidoli, N. 2006a. Visualization and proof: A brief survey. In A. Simpson (Ed.), *Retirement as Process and Concept*, pp. 101-109. Prague: Karlova Univerzita v Praze.

Hanna, G. – Sidoli, N. 2006b. Visualization and proof: A brief survey of philosophical perspectives. *Zentralblatt für Didaktik der Mathematik*, 39: 73-78.

Haubrichs dos Santos, C. 2015. Étienne Bobillier (1798–1840): percours mathématique, enseignant et professionnel. PhD dissertation, Simon Fraser University.

Hecht, N. 2011. Visualizing the Rotation of a Rigid Body. *The International Conference on Computational & Experimental Engineering and Sciences, 16*(1), 25–26.

Hesse, O. 1863. Jakob Steiner. *Journal für die reine und angewandte Mathematik* 62: 99–100.

Hilbert, D. 1899. *Grundlagen der Geometrie*. Leipzig: Teubner.

Hogendijk, J.P. 1991. Desargues' 'Brouillon project' and the 'Conics' of Apollonius. *Centaurus* 34 (1): 1-43.

Ivins, W.M. 1943. A note on Gerard Desargues, *Scripta Mathematica*, 9: 33-48.

Ivins, W.M. 1947. A note on Desargues' theorem, *Scripta Mathematica*, 13: 203-210.

Ivory, J. 1809. On the Attractions of Homogeneous Ellipsoids. *Philosophical Transactions of the Royal Society of London, 99*, 345–372.

Jonquiéres, E. 1856. *Mélangés de géométrie pure comprenant diverses applications des théories exposées dans le Traité de Géométrie Supérieure de M. Chasles au mouvement infiniment petit d'un corps, aux sections coniques, aux courbes du troisième ordre, etc, et la traduction du Traité de Maclaurin sur les courbes du troisième ordre*. Paris: Mallet-Bachelier.

Jullien, P.M. 1855. *Problèmes de Mécanique rationnelle disposés pour servir d'applications aux principes ensignés dans les cours*, two volumes. Paris: Mallet-Bachelier.

Jung, C.G. 1957–1979. The Collected Works of C.G. Jung, 20 volumes. Princeton: Princeton University Press. London: Routledge & Kegan, Paul. Abbreviated as CW

Jung, C.G. 1911/1912. *Symbols of Transformation*. CW 5.

Jung, C. G. 1918. The Role of the Unconscious. CW 10, pp. 3–28.

Jung, C. G. 1921. *Psychological Types: the psychology of individuation*. (Trans. H. Godwin Baynes). London: Kegan, Paul, Trench, Trubner & C.

Jung, C. G. 1928. On Psychic Energy. CW 8, pp. 3–66.

Jung, C. G. 1934. Archetypes of the collective unconscious. CW 9/1, pp. 3–41.

Jung, C. G. 1936. Concerning the archetypes, with special reference to the anima concept. CW 9, part 1, pp. 54–72

Jung, C. G. 1939. Conscious, Unconscious, and Individuation. CW 9, part 1, pp. 275–289.

Jung, C. G. 1951. *Aion. Researches into the Phenomenology of the Self*. CW 9, part 2.

Kepler, J. 1937–2012. Gesammelte Werke, Van Dyck W., Caspar M, et al. (eds.) Revised April 2013. 22 Vols. München: Deutsche Forschungsgemeineschaft und Beyerische Akademie der Wissenschaften. Beck'sche Verlagsbuchhandlung. Abbreviated as KGW.

Kepler, J. 1596. *Prodromus dissertationum cosmographicarum continens mysterium cosmographicum de admirabili proportione orbium coelestium*. In KGW, I:2-80. Second edition KGW, VIII: 7-128.

Kepler, J. 1604. *Ad Vitellionem paralipomena quibus astronomiae pars optica traditur*. In KGW, II.

Kepler, J. 1619. *Harmonices mundi libri V*. In KGW, VI.

Kepler, J. 1981. *Mysterium Cosmographicum (The Secret of the Universe)*. Translation by A. M. Duncan. Introduction and Commentary by E. J. Aiton. Preface by I. Bernard Cohen. New York: Abaris Book.

Kepler, J. 1984. *Le secret du monde*. Introduction, Traduction et Notes de A. Segonds. Paris: Les Belles Lettres.

Kepler, J. 1997. *The Harmony of the World*. English translation and introduction by E.J. Aiton, A, M. Duncan, J.V. Field. Philadelphia: American Philosophical Society.

Klein, F. 1871, 1873. Ueber di sogennante Nicht-Euklidische Geometrie. *Mathematische Annalen*. First part *4*, 576-625. Second part *6*, 112–145

Kline, M. 1972. *Mathematical thought from ancient to modern times,* 3 volumes. Oxford: Oxford University Press.

Koetsier, T. 2007. Euler and kinematics. In R. E. Bradley & C. E. Sandifer (Eds.), *Leonhard Euler: Life, Work and Legacy* (pp. 167–194). Amsterdam: Elsevier.

Koppelman, E. 1971. "Chasles, Michel." In C. Gillispie (Ed.), *Dictionary of scientific biography*, pp. 212-215. New-York: Charles Scribner's Sons. Freely available at: https://www.encyclopedia.com/science/dictionaries-thesauruses-pictures-and-press-releases/chasles-michel.

Kosmann-Schwarzbach, Y. (Ed.). 2013. *Simeon-Denis Poisson. Les Mathématiques au Service de la Science*. Paris: Édition de l'École polytechnique.

Kötter, E. 1887. *Grundzüge Einer Rein Geometrischen Theorie der Algebraischen Ebenen Curven.* Berlin. Königliche Akademie der Wissenscaften.

Kötter, E. 1901. Die Entwicklung der synthetischen Geometrie von Monge bis auf Staudt (1847). *Jahresbericht der Deutschen mathematiker Vereinigung* 5(2): 1–484.

Lagrange, J. L. 1867–1897. *Œuvres de Lagrange*, in 14 volumes. Paris: Gauthier-Villars.

Lagrange, J.L. 1763. *Recherches sur la libration de la Lune*. In *Œuvres de Lagrange*, 6 : 5–61.

Lagrange, J.L. 1773, 1775. Sur l'attraction des sphéroïdes elliptiques. In *Œuvres de Lagrange*, 3: 619-651.

Lagrange , J.L. 1775, 1777. Additions au Mémoire sur l'attraction des sphéroïdes elliptiques imprimé dans le Volume pour l'Année 1773. In *Œuvres de Lagrange*, 3: 651–658.

Lagrange, J.L. 1780. *Théorie de la libration de la Lune*. In *Œuvres de Lagrange*, 5 : 5–124.

Lagrange, J.L. 1788. *Méchanique analitique*. Paris: La Veuve Desaint.

Lagrange, J. L. 1797, 1813. *Théorie des fonctions analytiques. Œuvres de Lagrange, 9.* (1813 second edition).

Lagrange, J.L. 1811. *Mécanique analytique*. Paris: Courcier.

Lagrange, J.L. 1997. *Analytical Mechanics*. Translated from the *Méchanique analytique, nouvelle edition* of 1811. Translated and edited by A. Boissonnade and V.N. Vagliante. Dordrecht: Springer.

Lamarle, M.E. 1859. *Théorie géométrique des centres et axes instantanés de rotation*. Bruxelles: Hayes.

Lamé, G. 1837. Mémoire sur les surfaces isothermes dans les corps solides homogènes en équilibre de température. *Journal de Mathématiques pures et appliquées*, 2: 147-183.

Lampe, E. 1900. Zur Biographie von Jacob Steiner. *Bibliotheca mathematica* 1: 129–141.

Lange, J. 1899. *Jacob Steiners Lebensjahre in Berlin 1821–1863 nach seinen Personalakten dargestellt*. Berlin: Gaertner.

Lange, L. 1885a. Über das Beharrungsgesetz. *Berichte über Verhandlungen der Königlich Sächsischen Gesellschaft der Wissenschaften. Mathematisch-physikalische Klasse*: 333–351.

Lange, L. 1885b. Über die wissenschaftliche Fassung des Galileischen Beharrungsgesetzes. *Philosophische Studien*, 2: 266-297.

Lange, L. 1885c. Nochmals über das Beharrungsgesetz. *Philosophische Studien*, 2: 539-545.

Lange, L. 1886. *Die geschichtliche Entwicklung des Bewegungsbegriffs und ihr voraussichtliches Endergebnis. Ein Beitrag zur historischen Kritik der mechanischen Prinzipien*. Leipzig: W. Engelmann.

Lange, L. 1902. Das Inertialsystem vor dem Forum der Naturforschung. *Philosophische Studien*, 20: 1-71.

Langins, J. 1990. The École Polytechnique and the French Revolution: Merit, militarization and mathematics. *LLULL*, 13: 91-105.

Langton, S. G. 2007. Euler on Rigid Bodies. In R. E. Bradley & C. E. Sandifer (Eds.), *Leonhard Euler: Life, Work and Legacy* (pp. 195–211). Amsterdam: Elsevier.

Laplace, P.S. 1772, 1776. Recherches sur le Calcul intégral et sur le Système du Monde plus Additions. *Mémoires de Mathématique & de Physique de l'Académie Royale des Sciences*: 267-376; Additions: 533-554.

Laplace, P.S. 1782, 1785. Théorie des attractions des sphéroïdes et de la Figure des Planètes. *Mémoires de Mathématique & de Physique de l'Académie Royale des Sciences*: 113-196.

Laplace, P.S. 1783, 1786. Mémoire sur la figure de la Terre. *Mémoires de Mathématique & de Physique de l'Académie Royale des Sciences*: 17-46.

Laplace, P.S. 1784. *Théorie du Mouvement et de la Figure Elliptique des Planètes*. Paris: Ph.-D. Pierres.

Laplace, P.S. 1787, 1789. Mémoire sur la Théorie de l'Anneau de Saturne. *Mémoires de Mathématique & de Physique de l'Académie Royale des Sciences*: 249-267.

Laplace, P. S. 1798–1825. *Traité de Mécanique Céleste*. Paris: Imprimerie de Crapelet. (Five tomes, I, 1798, II, 1799; III, 1802; IV, 1805; V, 1825).

Laurikainen, K.V. 1988. *Beyond the Atom: The Philosophical Thought of Wolfgang Pauli*. Berlin-Heidelberg: Springer.

Lawrence, S. 2011. Developable Surfaces: Their History and Application. *Nexus Network Journal*, 13: 701-714.

Legendre, A.M. 1782, 1785. Recherches sur l'attraction des sphéroïdes homogènes. *Mémoires de Mathématiques et de Physique présentés par divers Savants Étrangers à l'Académie Royale des Sciences de Paris*, 10: 411-435.

Legendre, A.M. 1788. Mémoire sur les intégrales doubles. *Mémoires de l'Académie Royale des Sciences de Paris*: 454-486.

Le Goff, J.-P. 2005. De l'irruption ou l'invention de l'infini actuel, de l'espace actuellement infini et de l'involution comme invariant numérique, dans l'œuvres de Desargues. In M. Franciosi (Ed.). *Prospettiva e geometria della spazio*, pp. 177-270. Sarzana: Agorà.

Lennes, J.B. 1930. Projective geometry from 1822–1918. PhD dissertation, The University of Montana.

Levy, A. 1830. Mémoire sur quelques propriétés des systèmes de forces. *Correspondance mathématique et physique*, VI: 261-271.

Lhuilier, S. 1828. Recherches polyèdrométriques. *Bibliotéque universelle de Geneve*, 37: 249-264.

Lindorff, D. 2004. *Pauli and Jung: The Meeting of Two Great Minds*. Wheaton: Quest Books.

Liouville, J. 1842. Note a l'occasion du Mémoire de M. Chasles. *Additions a la Connaissance des Temps pour l'an 1845 (1842)*: 34-36.

Lorenat, J. 2015a. Die Freude an der Gestalt: Methods, figures, and practices in early nineteenth century geometry. PhD dissertation under the direction of P. Nabonnand and T. Roque, Université de Lorraine.

Lorenat, J. 2015b. Polemics in Public: Poncelet, Gergonne, Plücker, and the Duality Controversy. *Science in Context*, 28, 4: 545-585.

Lorenat, J. 2015c. Figures real, imagined, and missing in Poncelet, Plücker and Gergonne. *Historia Mathematica*, 42: 155-192.

Lorenat, J. 2016. Synthetic and analytic geometries in the publications of Jakob Steiner and Julius Plücker (1827–1829). *Archive for History of Exact Sciences*, 70, 4: 413-462.

Loria, G. 1896. *Il Passato e il presente delle principali teorie geometriche.* Seconda edizione accresciuta e interamente rifatta. Torino: Clausen.

Lüders, G. 1954. On the Equivalence of Invariance under Time-Reversal and under Particle-Antiparticle Conjugation for Relativistic Field Theories. *Kongelige Danske Videnskabernes Selskab Matematisk-Fysiske Meddelelser, 28*(5), 1–17.

Lüroth, J. 1875. Das Imaginäre in der Geometrie und das Rechnen mit Würfen. Darstellung und Erweiterung der v. Staudt'schen Theorie. *Mathematische Annalen*, 8: 145-214.

Lüroth, J. 1877. Das Imaginäre in der Geometrie und das Rechnen mit Würfen. (Zweite Abhandlung). *Mathematische Annalen*, 11: 84-110.

Macbeth, D. 2010. Diagrammatic reasoning in Euclid's *Elements*. In B. van Kerkhove, J. P. van Bendegem, & J. de Vuyst (Eds.), *Philosophical Perspectives on Mathematical Practice 12* (pp. 235–267). College Publications.

Mach, E. 1919. *The science of mechanics. A critical and historical account of its development.* Chicago-London: The Open Court Publishing Co. Original German text 1883.

Maclaurin, C. 1741. De Causa Physica Fluxus et Refluxus Maris. *Académie des Sciences: Recueil des pièces qui ont remporté les Prix de l'Académie Rorale des Sciencs, ec MDDCXL, Sur le Flux et le Reflux de la Mer*: 193-224. Paris: G. Martin, J.B. Coignard, & les Frères Guerin.

Maclaurin, C. 1742. *A Treatise of fluxions*. Edinburgh: T.W. and T. Ruddimans.

Magri, F. 2013. La dynamique des corps solides de d'Alembert à Poisson. In Y. Kosmann-Schwarzbach (Ed.), *Poisson: La Mathématique au service de la science*: 175-206. Palaiseau : Les éditions de l'Ecole polytechnique.

Manders, K. 2008. The Euclidean diagram. In P. Mancosu (Ed.), *The philosophy of mathematical practice* pp. 112–183. Oxford: Oxford University Press.

Mannheim, A. 1858. Construction des centres de courbure des lignes décrites dans le mouvement d'une figure plane qui glisse sur son plan. *Journal de l'Ecole Polytechnique*, 21: 179-190.

Martin G.E. 1998. *Geometric constructions*. New York: Springer.

Mascheroni, L. 1797. *La geometria del compasso*. Pavia. Presso gli eredi di Pietro Galeazzi.

Metivier, M. – Costabel, P. – Dugac, P. 1981. *Simeon-Denis Poisson et la science de son temps*. Paris: Édition de l'École polytechnique.

Michel, N. 2020a. *Of Words & Numbers. The writing of Generality in the Emergence of Enumerative Geometry* (1852-1893). Thèse de doctorat d'histoire et de philosophie des mathématiques. Dirigée par Karine Chemla et par Ivahn Smadja. Présentée et soutenue publiquement le 24 Septembre 2020. HAL Id: tel-03296439. https://theses.hal.science/tel-03296439.

Michel, N.- Smadja, I. 2021. The Ancients and the Moderns: Chasles on Euclid's Lost Porisms and the Pursuit of Geometry. Http://caphi.univ-nantes.fr/IMG/pdf/porisms_sic_2021-2.pdf.

Miller, A.I. 2010. *Jung, Pauli and the pursuit of a scientific obsession*. New York: Norton & Company.

Miller, N. 2007. *Euclid and his twentieth century rivals: Diagrams in the logic of Euclidean geometry*. Stanford, CA: Center for the Study of Language and Information.

Milne, J.J. 1911. *An elementary treatise on cross-ratio geometry, with historical notes*. Cambridge: Cambridge University Press.

Möbius, A.F. 1827. *Der barycentrische Calcül: ein neues Hilfsmittel zur analytischen Behandlung der Geometrie*. Leipzig: Johann Ambrosius Barth.

Möbius, A.F. 1829. Beweis eines neuen, von Herrn Chasles in der Statik entdeckten Satzes, nebst einigen Zusätzen. *Journal für die reine und angewandte Mathematik*, 4: 179-184.

Möbius, A.F. 1837. *Lehrbuch der Statik* (two parts). Leipzig. Göschen.

Monge, G. 1784–1785. Sur l'expression analytique de la génération des surfaces courbes. *Mémoires de l'Académie Royale des Sciences de Turin*, Première Partie: 19-33.

Monge, G. 1798. *Géométrie descriptive*. Paris: Baudouin.

Monge, G. 1807. *Application de l'analyse à la géométrie*. Paris: Perronneau.

Mozzi, G. 1763. *Discorso matematico sopra il rotolamento momentaneo dei corpi*. Napoli: Donato Campo.

Mumma, J. 2008. Ensuring Generality in Euclid's Diagrammatic Arguments. In: G. Stapelton, J. Howse and J. Lee (Eds.), *Diagrammatic Representation and Inference*. New York: Springer.

Mumma, J. 2012. Constructive geometrical reasoning and diagrams. *Synthese*, 186, 1: 103-119.

Nabonnand, P. 2006. Contributions à l'histoire de la géométrie projective au 19e siècle. Document présenté pour l'HDR. 2006.

Nabonnand, P. 2011a. L'argument de la généralité chez Carnot, Poncelet et Chasles. In *Justifier en mathématiques*, eds. D. Flament and P. Nabonnand, 17–44. Paris: Éditions de la Maison des sciences de l'homme. Consulted free edition with number of pages from 1 to 29, Hal-00637385.

Nabonnand, P. 2011b. Une Géométrie sans figure? In *La Figure et la Lettre*, eds. E. Barbin and P. Lombard, 99-119. Nancy: Presses Universitaires de Nancy - Editions de l'Université de Lorraine. Consulted free edition with number of pages from 1 to 23, Hal-01082518.

Nabonnand, P. 2015. L'étude des propriétés projectives des figures par Poncelet: une modernité explicitement ancrée dans la tradition. https://hal.archives-ouvertes.fr/hal-01266538/document, pp. 1-25.

Nagel, E. 1939. The Formation of Modern Conception of Formal Logic in the Development of Geometry. *Osiris, 7*, 142–223.

Needham, T. 1997. *Visual Complex Analysis*. Oxford: Clarendon Press.

Needham, T. 2021. *Visual Differential Geometry and Forms: A Mathematical Drama in Five Acts*. Princeton: Princeton University Press.

Netz, R. 1998. Greek mathematical diagrams: Their use and their meaning. *For the Learning of Mathematics,* 18(3): 33–39.

Netz, R. 1999. *The Shaping of Deduction in Greek Mathematics: A Study in Cognitive History*. Cambridge: Cambridge University Press.

Neumann, C. 1870. *Über die Principien der Galilei-Newtonschen Theorien*. Leipzig: Teubner.

Neville Greaves, G. 2013. Poisson's ratio over two centuries: challenging hypotheses. *Notes and Records: the Royal Society journal of the history of science*, 67: 37-58.

Newton, I. 1726. *Philosophiae Naturalis Principia Mathematica*. Editio tertia aucta et emendata. London: Innys.

Newton, I. 1729. *The Mathematical Principles of Natural Philosophy*. Translated into English by Andrew Motte. London: Benjamin Motte.

Newton, I. 1739–1742, 1822. *Philosophiae naturalis principia mathematica, auctore Isaaco Newtono, Eq. Aurato. Perpetuis commentariis illustrate, communi studio pp.* In *Thomae le Seur et Francisci Jacquier ex Gallicana Minimorum Familia, matheseos professsorum. Editio nova, summa cura recensita*. Glasgow: J. Duncan.

Obenrauch, F.J. 1897. *Geschichte der darstellenden und projektiven Geometrie mit besonderer Berücksichtigung ihrer Begründung in Frankreich und Deutschland und ihrer wissenschaftlichen Pflege in Österreich*. Brünn: Winiker.

Ostermann, A. – Wanner G. 2012. *Geometry by its history*. Berlin-Heidelberg: Springer.

Panza, M. 2012. The twofold role of diagrams in Euclid's plane geometry. *Synthese*, 186, 1: 55-102.

Pappus. 1877. *Pappi Alexandrini collectionis quae supersunt e libri manus scriptis edidit latina interpretatione et commentariis instruxit Fridericus Hultsch*, Volumen II. Berlin: Weidmann.

Pappus d'Alexandrie. 1933. *La collection mathematique. Ouevre traduite pour la première fois du Grec en Français avec une introduction e des notes par Paul Ver Eecke*. Paris-Bruges: Desclée de Brouwer.

Pappus of Alexandria. 1986. *Book 7 of the* Collection. *Part 1: Introduction, Text, and Translation*. Edited with Translation and Commentary by Alexander Jones. New York: Springer.

Pauli, W. 1948. Modern Examples of Background Physics. In W. Pauli & C. G. Jung (Eds.), *Atom and Archetype. The Pauli-Jung Letters, 1932-1948* (pp. 179–196). Princeton: Princeton University Press. 2014.

Pauli, W. 1950. The Philosophical Significance of the Idea of Complementarity. In W. Pauli (Ed.), *Writings on Physics and Philosophy* (pp. 35–42). Berlin-Heidelberg: Springer. 1994.

Pauli, W. 1952. The Influence of Archetypal Ideas on the Scientific Theories of Kepler. In W. Pauli (Ed.), *Writings on Physics and Philosophy* (pp. 219–279). Berlin-Heidelberg: Springer. 1994.

Pauli, W. 1954. Ideas of the Unconscious from the Standpoint of Natural Science and Epistemology. In W. Pauli (Ed.), *Writings on Physics and Philosophy* (pp. 149–164). Berlin-Heidelberg: Springer. 1994.

Pauli, W. 1955. Exclusion principle, Lorentz group and reflexion of space-time and charge. In W. Pauli, L. Rosenfeld, & V. Weisskopf (Eds.), *Niels Bohr and the Development of Physics* (pp. 30–51). London: Pergamon.

Pauli, W., & Jung, C. G. 2014. *Atom and Archetype. The Pauli-Jung Letters, 1932-1948*. Princeton: Princeton University Press.

Pecot, J-B. 1993. Le problème de l'ellipsoïde et l'analyse harmonique: la controverse entre Legendre et Laplace. *Cahiers du séminaire d'histoire des mathématiques* 2, 3: 113-157

Pedoe, D. 1975. Notes on the history of geometrical ideas II. The principle of duality. *Mathematics Magazine* 48(5): 274–277.

Pisano, R. – Bussotti, P. 2012. Galileo and Kepler: On Theoremata Circa Centrum Gravitatis Solidorum and Mysterium Cosmographicum. *History Research*, 2, 2: 110-145.

Pisano, R. – Bussotti, P. 2016. A Newtonian tale details on notes and proofs in Geneva edition of Newton's *Principia*. *BSHM Bulletin Journal of the British Society for the History of Mathematics*, 31, 3: 160-178.

Pisano, R., & Bussotti, P. 2017. The Fiction of Infinitesimals in Newton's Works. On the Metaphorical Use of Infinitesimals in Newton. *Isonomia – Epistemologica, 9*. Special issue *Reasoning, Metaphors and Science* (edited by F. Marcacci, M.G. Rossi), pp. 141–160.

Pisano, R., & Bussotti, P. 2022. Conceptual Frameworks on Mathematics Applied to Physics in the Newton *Principia* Geneva Edition (1822). *Foundations of Science, 27*, 1127-1182

Plücker, J., 1826a. Géométrie analytique. Recherche graphique du cercle osculateur, pour les lignes du second ordre. *Annales de mathématiques pures et appliquées*, XVII : 69-72.

Plücker, J., 1826b. Géométrie de la règle. Théorèmes et problèmes sur les contacts des sections coniques. *Annales de mathématiques pures et appliquées*, XVII : 37-59.
Plücker, J. 1828–1831. *Analytisch-geometrische Entwicklungen*. Essen: Baedeker.
Plücker, J. 1835. *System der analytischen Geometrie*. Berlin: Duncker und Humblot.
Plücker, J. 1839. *Theorie der algebraischen Curven, gegründet auf eine neue Behandlungsweise der analytischen Geometrie*. Bonn: Adolph Marcus.
Plücker, J.-Schoenflies, A. 1904. Über des wissenschaftlichen Nachlass Julius Plückers. *Mathematische Annalen*, 58: 385-403.
Poincaré, H. 1905. *Science and hypothesis*. New York: Walter Scott.
Poincaré, H. 1908. *La valeur de la science*. Paris: Flammarion.
Poincaré, H. 1917. La science et l'hypothèse. Paris: Flammarion.
Poincaré, H. 1920. *Science et Méthode*. Paris: Flammarion
Poincaré, H. 1958. *The value of science*. New York: Dover.
Poincaré, H. 2003. *Science and Method*. New York: Dover.
Poinsot, L. 1803. *Élémens de Statique*. Paris: Calixte-Volland.
Poinsot, L. 1806a. Mémoire sur la composition des moments et des aires. *Journal de l'École Polytechnique, VI, 13*, 182–205.
Poinsot, L. 1806b. Théorie générale de l'équilibre et du mouvement ses systèmes. *Journal de l'École Polytechnique, VI, 13*, 206–241.
Poinsot, L. 1838. Note de M. Poinsot sur les Remarques qu'on trouve au commencement du *Compte rendu* de la séance précédente. *Comptes rendus hebdomadaires des séances de l'Académie des Sciences*, VI: 869-872.
Poinsot, L. 1851. Théorie nouvelle de la rotations des corps. *Journal de mathématiques pures et appliquées*, 16: 9-129 and 289-336.
Poinsot, L. 1853. *Théorie des cônes circulaires roulants*. Paris: Bachelier.
Poinsot, L. 1861. *Éléments de Statique* (Dixième edn). Paris: Mallet-Bachelier.
Poisson, D. 1811. *Traité de Méchanique*. Paris: Courcier.
Poisson, D. 1812–1813. Mémoire sur la distribution de l'électricité à la surface des corps conducteurs. *Mémoires de la classe des Sciences mathématiques et physiques de l'Institut Impérial de France*, 12: First part pp. 1-92, second part 164-274.
Poisson, D. 1826. Mémoire Sur l'Attraction des Sphéroïdes. *Cannaissance de Tems, ou de Mouvemens Célestes, a l'Usage des Astronomes et des Navigateurs, pour l'An 1829*: 329-380.
Poisson, D. 1827. Mémoire sur la théorie du magnétisme en mouvement. *Mémoires de l'Académie des Sciences*, 6 (1823), 1827: 441-570.
Poisson, D. 1835a. Mémoire sur l'attraction d'un ellipsoïde homogène. *Mémoires de l'Académie Royale des Sciences de l'Institut de France, XIII*, 497-545
Poisson, D. 1835b. *Théorie mathématique de la chaleur*. Paris: Bachelier.
Poncelet, J.V. 1817. Philosophie mathématique. Réflexions sur l'usage de l'analise algébrique dans la géométrie; suivies de la solution de quelques problèmes dépendant da la géométrie de la règle. *Annales de mathématiques pures et appliquées*, 8: 141-145.
Poncelet, J.V. 1818. Sur la loi des signes de position en géometrie, la loi et le principe de continuité, in Poncelet, *Applications d'analyse et de géométrie qui ont servi de principal fondement au traité des propriétés projectives des figures*, t. 2, pp. 167-195. Paris: Gauthier-Villars.
Poncelet, J.V. 1820. Essai sur les propriétés projectives des sections coniques. Présenté à Académie des Sciences de Paris. In Poncelet. 1864. *Applications d'analyse et de géométrie qui ont servi du principal fondement au* Traité des propriétés projectives des figures, 2: 365-454. Paris: Gauthier-Villars.
Poncelet, J.V. 1822. *Traité des propriétés projectives des figures*. Paris: Bachelier.
Poncelet, J. V. 1826–1827. Philosophie mathématique. Analyse d'un mémoire présenté à l'Académie royale des Sciences (Extrait d'un lettre de l'Auteur au Rédacteur des *Annales*). *Annales de mathématiques pures et appliquées, 17*, 265–272.

Poncelet, J.V. 1827. Note sur divers articles du bulletin des sciences de 1826 et de 1827, relatif à la théorie des polaires réciproques, à la dualité des propriétés de situation de l'étendue, etc. *Annales de mathématiques pures et appliquées*, 18: 125-142.

Poncelet, J.V. 1828a. Mémoire sur les centres de moyennes harmoniques [. . .]. *Journal für die reine und angewandte Mathematik* 3: 213–272.

Poncelet, J.V. 1828b. Sur la dualité de situation et sur la théorie des polaires réciproques, 2e article en réponse aux observations de M. Gergonne. *Bulletin des sciences mathématiques, astronomiques, physiques et chimiques*, 9: 292-302.

Poncelet, J.V. 1829a. Mémoire sur la théorie général des polaires réciproques [. . .]. *Journal für die reine und angewandte Mathematik* 4: 1–71.

Poncelet, J.V. 1829b. Réponse de M. Poncelet aux réclamations de M. Plucker, insérées pag. 330, no. 262 du cahier de nov. 1828 du *Bulletin des sciences. Bulletin des sciences mathématiques, astronomiques, physiques et chimiques*, 11 : 330-333.

Poncelet, J. V. 1832. Notes sur quelques principes généraux de transformation des relations métriques des figures, sur la transformation, et spécialement des relations métriques projectives orthogonalement, et d'autre qui le soient coniquement. In J. V. Poncelet, *Traité des propriétés projectives des figures. Tome second* (pp. 332–345). Paris: Gauthier-Villars. 1866.

Poncelet, J.V. 1857. Sur la transformation des propriétés métriques des figures au moyen de la théorie des polaires réciproques. *Comptes rendus hebdomadaires des séances de l'Académie des sciences*, 45: 553–554.

Pontécoulant, P.G. 1829–1834. *Théorie analytique du Système du Monde*. Three volumes. Paris: Bachelier.

Price, B. 1856. *A Treatise on Infinitesimal Calculus, Vol. III. Statics, and Dynamics of a Material Point*. Oxford: Oxford University Press.

Prony, G. Riche baron de. 1799. *Mécanique philosophique, ou analyse raisonnée des diverses parties de la science de l'équilibre et du mouvement*. Paris: Imprimerie de la République.

Quetelet, A. 1825. Sur l'emploi des projections stéréographiques en géométrie par M. G. Dandelin, professuer Ext. a l'Université de Liége. *Correspondance mathématique et physique, I, V,* 256–264, 316–322.

Quetelet, A. 1826. Sur quelques propriétés nouvelles des caustiques secondaires, déduites des projections stéréographiques. *Correspondance mathématique et physique* II: 81–86.

Quetelet, A. 1829. Sur les lignes aplanétiques – Sur les lignes colorées que produit la polarisation dans le plaques de cristal [. . .]; réponse du rédacteur a M. Chasles. *Correspondance mathématique et physique, publiée par A. Quetelet* V: 190–196.

Quetelet, A. 1830. 135. Mémoire de géométrie pure sur les systèmes de forces et les systèmes d'aires planes; et sur les polygones, les polyèdres et les centres des moyennes distances; par M. Chasles. *Bulletin des sciences mathématiques, physiques et chimiques*, XIII: 246-249.

Ramsey, A. S. 1940. *An Introduction to the Theory of Newtonian Attraction*. Cambridge: Cambridge University Press.

Riccardi, P. 1881. *Commemorazione di Michele Chasles. Rendiconto dell'Accademia delle Scienza dell'Istituto di Bologna. Anno Accademico 1880-1881*: 37-50.

Richards, J.L. 2003. The Geometrical Tradition: Mathematics, Space and Reason in the Nineteenth Century. In M.J. Nye, *The Cambridge History of Science, Vol. 5. The Modern Physical and Mathematical Sciences*, pp. 449-467. Cambridge: Cambridge University Press.

Richards, J.L. 2006. Historical Mathematics in the French Eighteenth Century. *Isis*, 97, 4: 700-713.

Robinson, A. 1966. *Non-standard Analysis*. Amsterdam: North-Holland.

Rodrigues, O. 1815. De l'attraction des sphéroïdes. *Correspondance sur l'École Royale Polytechnique*, III: 361-385.

Rodrigues O. 1840. Des lois géométriques qui régissent les déplacements d'un système solide dans l'espace, et de la variation des coordonnées provenant de ces déplacements considérés indépendamment des causes qui peuvent les produire. *Journal de mathématiques pures et appliquées*, 1, 5: 380-440.

Rowe, D. 1997. In search of Steiner's Ghosts: Imaginary elements in nineteenth century geometry. In D. Flament (Ed.), *Le nombre, une hydre àn visages; entre nombres complexes et vecteurs* (pp. 193–208). Paris: Éditions de la Maison des sciences de l'homme.
Saito, K. 2012. Traditions of the diagram, tradition of the text: A case study. *Synthese*, 186, 1: 7-20.
Saito, K., & Sidoli, N. 2012. Diagrams and arguments in ancient Greek mathematics: Lessons drawn from comparison of manuscript diagrams with those in modern critical editions. In K. Chemla (Ed.), *The History of Mathematical Proofs in Ancient Traditions* (pp. 135–162). Cambridge: Cambridge University Press.
Salmon, G. 1852. *A treatise on the higher plane curves*. Dublin: Hodges and Smith.
Schell, W. 1870. *Theorie der Bewegung und Kräfte. Ein Lehrbuch der theoretischen Mechanik, mit besendere Rücksicht auf die Bedürfnisse technischer Hochschulen*. Leipzig: Teubner.
Schneider, M. 1983. "Girard Desargues. The architectural and perspective geometry: A study in the rationalisation of figure". Doctoral Dissertation, Virginia Polytechnic Institute and State University.
Schönflies, A. 1907–1910. Projektive Geometrie. In *Encyclopädie der mathematischen Wissenschaften mit Einschluss ihre Anwendung*. Dritter Band. Ester Teil, erste Häfte, pp. 389-480. Leipzig: Teubner.
Schubring, G. 2005. Conflicts Between Generalization, Rigor, and Intuition Number Concepts Underlying the Development of Analysis in 17th-19th Century France and Germany. New York: Springer.
Servois, F.J. 1810–1811. Solution du premier des deux problèmes proposés à la page 259 de ce volume, et du problème proposé à la page 126 du même volume, *Annales de mathématiques pures et appliquées* 3: 293–302.
Simon, M. 1906. *Über di Entwicklung der Elementar Geometrie im XIX. Jahrhundert*. Leipzig: Teubner.
Smith, H.F. 1935. The influence of French Revolution on Mathematics. *Proceedings of the Iowa Academy of Science*, 42: 139-146.
Sonego, S. 2016. *Classical Mechanics*. Lecture notes of the course of Rational Mechanics held at the Department of Mathematics, Informatics, Physics, University of Udine.
Sonego, S. 2020. *General Relativity*. Lecture notes of the course of Mathematical Physics held at the Department of Mathematics, Informatics, Physics, University of Udine.
Southall, J.P.C. 1922. Aplanatic (or Cartesian) optical surfaces. *Journal of the Franklin Institute*, 193, 5, pp. 609-626.
Staudt (von), G.K.C. 1847. *Geometrie der Lage*. Nürnberg: Bauer und Raspe.
Staudt (von), G.K.C. 1856–1860. *Beiträge zur Geometrie der Lage*. Nürnberg: Korn'schen.
Staudt (von), G.K.C. 1889. *Geometria di posizione*. Torino: Fratelli Bocca.
Steichen, M. 1855. *Un supplément a la géométrie destiné a servir d'introduction a l'étude de la Mécanique*. Bruxelles : Delevingne et Callewaert.
Steiner, J. 1832. *Systematische Entwickelung der Abhängigkeit geometrischer Gestalten von einander [...]*. Berlin: G. Fincke.
Steiner, J. 1833. *Die geometrischen Konstructionen, ausgeführt mittelst der geraden Linie und eines festen Kreises, als Lehrgegenstand auf höheren Unterrichts-Anstalten und zur praktischen Benutzung*. Berlin: Ferdinand Dümmler.
Steiner, J. 1834. Démonstration géométrique d'un théorème relatif à l'attraction d'une couche ellipsoïdique sur un point extérieur. *Journal für die reine und angewandte Mathematik*, XIII: 141-143.
Steiner, J., and Gergonne J.D. 1826–1827. Géométrie pure. Théorie générale des contacts et des intersections des cercles. *Annales de mathématiques pures et appliquées* 17: 285–315.
Stephenson, B. 1987. *Kepler's Physical Astronomy*. New York - Berlin - Heidelberg - London - Paris - Tokyo: Springer.
Stephenson, B. 1994. *The Music of the Heavens. Kepler's Harmonic Astronomy*. Princeton: Princeton University Press.

Straume, E. 2014. A Survey of the Development of Geometry up to 1870. *ArXiv*. https://arxiv.org/abs/1409.1140

Sturm, C. 1824–1825. Recherches analitiques sur les polygones rectilignes plans ou gauches, renfermant la solution de plusieurs questions proposées dans le présent recueil. *Annales de mathématiques pures et appliquées*, 15: 309-344.

Sturm, C. 1842. Note Sur un Mémoire de M. Chasles. *Journal de Mathématiques Pures et Appliquées*, VII: 345-355.

Tagliagambe, S – Malinconico, A. 2011. *Pauli e Jung. Un confronto su materia e psiche*. Milano: Raffaele Cortina.

Tagliagambe, S – Malinconico, A. 2018. *Tempo e sincronicità. Tessere il tempo*. Milano: Mimesis

Targ, S.M. 1979. Entry “Solidification Principle”. *The Great Soviet Encyclopedia*. Translation freely available at https://encyclopedia2.thefreedictionary.com/Solidification+Principle.

Taton, R., 1951a. *L'Oeuvre scientifique de Gaspard Monge*. Paris: Presses Universitaires de France.

Taton, R. 1951b. *L'œuvre mathématique de G. Desargues*. Paris: Presses Universitaires.

Teodorescu, P. P. 2007. *Mechanical Systems, Classical Models: Vol. 1: Particle Mechanics*. Berlin-Heidelberg-New York: Springer.

Thieme, H. 1879. Die Definition der geometrischen Gebilde durch Construction ihrer Polarsysteme. *Zeitschrift für Mathematik und Physik*, XXIV: 221-229

Todhunter, I. 1873. *History of the Theories of Attraction and the Figure of the Earth, from the Time of Newton to that of Laplace*, 2 Vols. London: Macmillan.

Tournés, D. 2012. Diagrams in the Theory of Differential Equations (Eighteenth to Nineteenth Centuries). *Synthese, 186*, 257–288.

Transon, A. 1845. Méthode géométrique pour les rayons de courbure d'une certaine classe de courbes. *Journal de Mathématiques Pures et Appliquées*, 10 :148 - 155

Varignon, P. 1725. *Nouvelle Mechanique ou Statique dont les projet fut donne en MDCLXXXVII*. Deux Tomes. Paris: Claude Jombert.

Veblen, O., & Young, J. W. 1908. A Set of Assumptions for Projective Geometry. *American Journal of Mathematics, 30*(4), 347–380.

Vecchioni, D. 2012. I signori della truffa. Tricase (LE): YOUCAPRINT.

Veronese, G. 1891. *Fondamenti di geometria a più dimensioni e a più specie di unità rettilinee*. Padova: Tipografia del Seminario.

Viola, T. 1946. Per la storia del teorema di Desargues sui triangoli omologici, *Atti della Accademia Nazionale dei Lincei. Rendiconti Classe Scienze Fisiche, Matematiche e Naturali* (8) 1: 570-575.

Vita, V. 1974. Il teorema sul quadrangolo completo nel *Brouillon project* di Desargues. *Archimede* 26: 47-55.

Voelke, J.-D. 2008. Le théorème fondamental de la géométrie projective: evolution de sa preuve entre 1847 et 1900. *Archive for History of Exact Sciences* 62: 243–296.

Voelke, J-D. 2010. Le développement historique du concept d'espace projectif. In L. Biosemat-Martagon (Ed.), *Élements d'une Biographie de L'Espace Projectif* (pp. 207–286).

Westfall, R.S. 1971. *Force in Newton's Physics. The Science of Dynamics in the Seventeenth Century*. London-New York: McDonald-Elsevier.

Williams T.R.-Fyfe K.R. 2010. Rodrigues' spatial kinematics. *Mechanism and Machine Theory*, 45,1: 15-22.

Zacharias, M. 1941. Desargues' Bedeutung für die projective Geometrie. *Deutsche Mathematik*, 5: 446-457.

Mentioned Works by Chasles

Chasles, M. 1813. Théorème de Géométrie. De la génération du paraboloïde hyperboliqué et de l'hyperboloïde à une nappe, assujetties à passer par un quadrilatère gauche *Correspondance sur l'École Royale Politechnique* 2(5): 446–447.

Chasles, M. 1814a. Proposition de Géométrie. *Correspondance sur l'École Royale Politechnique* 3 (1): 6.

Chasles, M. 1814b. Propositions relatives aux Courbes et aux Surfaces du second degré. *Correspondance sur l'École Royale Politechnique* 3(1): 11–17.

Chasles, M. 1816a. Propriétés des diamètres de l'ellipsoïd. *Correspondance sur l'École Royale Politechnique* 3(3): 302–328.

Chasles, M. 1816b. Démonstration des théorèmes sur les surfaces du second degré, énoncés par M. Monge, Correspondance sur l'Ecole Polytechnique, tom. II, p. 319. *Correspondance sur l'École Royale Politechnique* 3(3): 328–342.

Chasles, M. 1817. Exemples de la discussion des équations numériques du second degré à trois variables, d'après la méthode indiquée dans ce Traité des Surfaces du second degré (pag. 241–249, art. 149–154), in J.N.P. Hachette, *Éléments de géométrie a trois dimensions*. Partie algébrique, pp. 259–274, Paris, 1817.

Chasles, M. 1827–1828a. Géométrie pure. Théorèmes sur les sections coniques confocales. *Annales de mathématiques pures et appliquées* 18: 269–276.

Chasles, M. 1827–1828b. Géométrie de situation. Mémoire sur les propriétés des systèmes de sections coniques, situées dans un même plan. *Annales de mathématiques pures et appliquées* 18: 277–301.

Chasles, M. 1827–1828c. Géométrie pure. Mémoire sur les projections stéréographiques, et sur les coniques homothétiques. *Annales de mathématiques pures et appliquées* 18: 305–320.

Chasles, M. 1828a. Extrait d'une lettre de M. Chasles sur les surfaces du second degré. *Correspondance mathématique et physique, publiée par A. Quetelet* IV: 294–295.

Chasles, M. 1828b. Sur une propriété générale des coniques dont un cas particulier relatif à la parabole, a été démontré dans la Correspondance, tom. IV, p. 155. *Correspondance mathématique et physique, publiée par A. Quetelet, IV*, 363–371.

Chasles, M. 1828–1829a. Géométrie de situation. Additions et corrections au mémoire sur les propriétés des systèmes de coniques, inséré à la pag. 277 du précédent volume. *Annales de mathématiques pures et appliquées* 19: 26–32.

Chasles, M. 1828–1829b. Géométrie de situation. Démonstration de quelques propriétés du triangle, de l'angle trièdre et du tétraèdre, considérés par rapport aux lignes et surfaces du second ordre. *Annales de mathématiques pures et appliquées* 19: 65–85.

Chasles, M. 1828–1829c. Géométrie de situation. Recherches sur les projections stéréographiques, et sur diverses propriétés générales des surfaces du second ordre. *Annales de mathématiques pures et appliquées* 19: 157–175.

Chasles, M. 1829a. Propriétés générales des coniques. *Correspondance mathématique et physique, publiée par A. Quetelet* V: 6–22.

Chasles, M. 1829b. Démonstration géométrique des propriétés de la courbe d'intersection d'une sphère et d'un cône de révolution dont le sommet est un point de la sphère. *Correspondance mathématique et physique, publiée par A. Quetelet* V: 44–48.

Chasles, M. 1829c. Sur les lignes dirimantes à deux foyers conjugués. *Correspondance mathématique et physique, publiée par A. Quetelet* V: 116–120.

Chasles, M. 1829d. Mémoire sur les propriétés des diamètres conjugués des hyperboloïdes. *Correspondance mathématique et physique, publiée par A. Quetelet* V: 129–157.

Chasles, M. 1829e. Sur les surfaces du second degré. *Correspondance mathématique et physique, publiée par A. Quetelet* V: 173–174.

Chasles, M. 1829f. Sur les lignes aplanétiques. *Correspondance mathématique et physique, publiée par A. Quetelet* V: 188–190.

Chasles, M. 1829g. Premier Mémoire sur la transformation des relations métriques des figures. *Correspondance mathématique et physique, publiée par A. Quetelet* V: 281–324.

Chasles, M. 1829h. *Recherches de géométrie pure sur les lignes et les surfaces du second degré.* Bruxelles: M. Hayez, Imprimeur de l'Académie Royale.

Chasles, M. 1829i. Sur les courbes du troisième et du quatrième degré (Extrait d'une lettre de M. Chasles au rédacteur). *Correspondance mathématique et physique, publiée par A. Quetelet*, V. The letters are two: first letter 231–233; second letter 234–236.

Chasles, M. 1830a. Second Mémoire sur la transformation parabolique des relations métriques des figures. *Correspondance mathématique et physique, publiée par A. Quetelet, VI*, 1–24.

Chasles, M. 1830b. Lettre de M. Chasles au rédacteur, au sujet d'un Mémoire de M. Plucker, inséré dans le journal de M. Crelle plus Note supplémentaire. *Correspondance mathématique et physique, publiée par A. Quetelet* VI: 81–87.

Chasles, M. 1830c. Mémoire de géométrie pure, sur les systèmes de forces et les systèmes d'aires planes; et sur les polygones, les polyèdres, et les centres des moyennes distances. *Correspondance mathématique et physique, publiée par A. Quetelet* VI: 92–120.

Chasles, M. 1830d. Sur la génération des focales, lettre de M. Chasles. *Correspondance mathématique et physique, publiée par A. Quetelet* VI: 207–208.

Chasles, M. 1830e. Théorèmes généraux sur les diamètres des surfaces du second degré. *Correspondance mathématique et physique, publiée par A. Quetelet* VI: 255–258.

Chasles, M. 1830f. Théorèmes sur les surfaces du second degré. *Correspondance mathématique et physique, publiée par A. Quetelet* VI: 272–273.

Chasles, M. 1830g. Note sur une construction graphique nouvelle des tangentes et des rayons de courbature des courbes géométriques. *Bulletin des sciences mathématiques, physiques et chimiques* XIII: 390–393.

Chasles, M. 1830h. *Mémoire de géométrie pure sur les propriétés générales des cônes du second degré.* M. Hayez, Imprimeur de l'Académie Royale.

Chasles, M. 1830i. Note sur les Propriétés générales du système de deux corps semblables entr'eux, et placés d'une manière quelconque dans l'espace; et sur le déplacement fini, ou infiniment petit, d'un corps solide libre. *Bulletin des sciences mathématiques, astronomiques, physiques et chimiques, XIV*, 321–326.

Chasles, M. 1831. *Mémoire de géométrie pure sur les propriétés générales des coniques sphériques.* M. Hayez, Imprimeur de l'Académie de Bruxelles.

Chasles, M. 1832a. Extrait d'une lettre de M. Chasles a M. Quetelet sur la description de la spirale d'Archimède. *Correspondance mathématique et physique de l'observatoire de Bruxelles publiée par A. Quetelet* VII: 41–43.

Chasles, M. 1832b. Extrait d'une lettre de M. Chasles au rédacteur sur le propriétés des coniques qui ont un foyer commun. *Correspondance mathématique et physique de l'observatoire de Bruxelles publiée par A. Quetelet* VII: 295–297.

Chasles, M. 1832c. Note sur les propriétés générales du système de deux corps semblables entre eux [. . .]. *Correspondance mathématique et physique de l'observatoire de Bruxelles publiée par A. Quetelet* VII: 352–357.

Chasles, M. 1835a. Propositions de géométrie et solution d'une question proposée à la fin de la 2e livraison du tome VII de la Corresp. Math. *Correspondance mathématique et physique de l'observatoire de Bruxelles publiée par A. Quetelet, VIII*, 56–58.

Chasles, M. 1835b. Sur l'hyperboloïde à une nappe, et sur le paraboloide hyperbolique. *Correspondance mathématique et physique de l'observatoire de Bruxelles publiée par A. Quetelet* VIII: 128–134.

Chasles, M. 1836a. Sur les surfaces du second degré qui n'ont pas de foyer. *Journal des mathématiques pures et appliquées* 1: 187–190.

Chasles, M. 1836b. Géométrie. Analogie entre des propositions plane et de Géométrie à trois dimensions – Géométrie de la sphère – Hyperboloïde à une nappe. *Journal des mathématiques pures et appliquées* 1: 324–334.

Chasles, M. 1837a. *Aperçu historique sur l'origine et le développement des méthodes en géométrie particulièrement de celles qui se rapportent a la géométrie moderne*. Consulted edition. Paris: Gauthier-Villars, 1875.

Chasles, M. 1837b. Note sur les équations indéterminées du second degré – Formules d'Euler pour la resolution de l'équation $Cx^2 \mp A = y^2$ – Leur identité avec celles des algébristes indiens et arabes – Démonstration géométrique de ces formules. *Journal des mathématiques pures et appliquées* 2: 37–55.

Chasles, M. 1837c. Note sur un cas particulier de la construction des tangentes aux projections des courbes, pour lequel les méthodes générales sont en défaut [...]. *Journal des mathématiques pures et appliquées* 2: 293–311.

Chasles, M. 1837d. Mémoire sur diverses manières de généralizer les propriétés des diamètres conjugués dans les sections coniques [...]. *Journal des mathématiques pures et appliquées* 2: 388–405.

Chasles, M. 1837e. Mémoire sur l'attraction des ellipsoïdes. *Journal de l'École Polytechnique*, 15 : 244-265.

Chasles, M. 1837f. Mémoire sur l'attraction d'une couche ellipsoïdale infiniment mince et les rapports qui ont lieu entre cette attraction et les lois de la chaleur en mouvement dans un corps en équilibre de température. *Journal de l'École Polytechnique*, 15: 266-316.

Chasles, M. 1837, 1846. *Mémoire sur l'attraction des ellipsoïdes. Solution synthétique pour le cas général d'un ellipsoïde hétérogène et d'un point extérieur*. Paris: Imprimerie Royale.

Chasles, M. 1838. Nouvelle solution du problème de l'attraction d'un ellipsoïde hétérogène sur un point extérieur. *Comptes Rendus des Séances de l'Académie des Sciences*, 6: 902-915.

Chasles, M. 1839. Énoncé de deux théorèmes généraux sur l'attraction des corps et la théorie de la chaleur. Comptes Rendus hebdomadaires des séances de l'Académie des Sciences, 8: 209-211.

Chasles, M. 1840. Nouvelle solution du problème de l'attraction d'un ellipsoïde hétérogène sur un point extérieur. *Journal des mathématiques pures et appliquées* 5: 465-488.

Chasles, M. 1841. *Two geometrical memoirs on the properties of cones of the second degree and of the spherical conics*. Translated from the French, with notes and additions, and an appendix on the application of analysis to spherical geometry, by the Rev. Charles Graves, A,M., M.R.I.A. Fellow and Tutor of Trinity College, Dublin. Dublin: The University Press.

Chasles, M. 1842. Théorèmes généraux sur l'attraction des corps. *Additions a la Connaissance des Temps pour l'an 1845 (1842)*: 18-33

Chasles, M. 1843. Propriétés géométriques relatives au mouvement infiniment petit d'un corps solide libre dans l'espace. *Compte rendu des séances de l'Académie des Sciences*: 1420-1432.

Chasles, M. 1846. Sur les lignes géodésiques et les lignes de courbure des surfaces du second degré. *Journal de mathématiques pures et appliquées* 11: 5–20.

Chasles M. 1847. Théorèmes généraux sur les systèmes de forces et leurs moments. *Journal de mathématiques pures et appliquées*, 1, 12: 213-224.

Chasles, M. 1852. *Traité de Géométrie supérieure*. Paris: Bachelier.

Chasles, M. 1855. Principe de correspondance entre deux objets variables, qui peut être d'un grande usage in Géométrie. *Comptes-Rendus de l'Académie des Sciences*, 41, 1097-1107.

Chasles, M. 1860a. *Les Trois Livres de Porismes d'Euclide*. Paris: Mallet-Bachelier.

Chasles, M. 1860b. Résumé d'un théorie des coniques sphériques homofocales et des surfaces du second ordre homofocales. *Journal de mathématiques pures et appliquées*, 2, 5: 425-454.

Chasles, M. 1860–1861. Propriétés relatives au déplacement fini quelconque, dans l'espace, d'une figure de forme invariable. In five parts. *Comptes rendus hebdomadaires des séances de l'Académie des Sciences*. First part, 51: 855-863. Second part, 51: 905-914. Third part, 52: 77-85. Fourth part: 189-197. Fifth part: 487-501.

Chasles, M. 1864a. Construction des coniques qui satisfont à cinq conditions. Nombres des solutions dans chaque question. *Compes-Rendus de l'Académie des Sciences*, 58: 297-308.

Chasles, M. 1864b. Considérations sur la Méthode générale exposée dans la séance du 15 février.- Différences entre cette méthode et la méthode analytique. Procédés généraux de démonstrations [27] juin. *Compes-Rendus de l'Académie des Sciences*, 58: 1167-1175.
Chasles, M. 1865. *Traité des sections coniques*. Paris: Gauthier-Villars.
Chasles, M. 1870. *Rapport sur les progrès de la géométrie en France*. Paris: Imprimerie Nationale.

Index[1]

[1] Since the name Michel Chasles appears in almost every page, it is not significant for mapping the text. Thence, it is omitted.

P. Bussotti, *Chasles and the Projective Geometry*,
https://doi.org/10.1007/978-3-031-54266-4

Printed in the USA
CPSIA information can be obtained
at www.ICGtesting.com
CBHW071517040824
12687CB00003B/105

9 783031 542657